MICROECONOMICS

SECOND CANADIAN EDITION

David C. Colander
Middlebury College

Peter S. Sephton
Queen's University

Charlene Richter
British Columbia Institute of Technology

Toronto Montréal Burr Ridge, IL Dubuque, IA Madison, WI New York
San Francisco St. Louis Bangkok Bogotá Caracas Kuala Lumpur Lisbon
London Madrid Mexico City Milan New Delhi Santiago Seoul Singapore
Sydney Taipei

McGraw-Hill
Ryerson Limited

A Subsidiary of The ***McGraw-Hill*** *Companies*

Dedicated to the memory of Frank Knight and Thorstein Veblen, both of whose economics have significantly influenced the contents of this book.

Microeconomics, Second Canadian Edition

ISBN: 0-07-090165-1

1 2 3 4 5 6 7 8 9 10 TCP 0 9 8 7 6 5 4 3

Printed and bound in Canada

Cover Art:
- Alexander Calder (1898–1976)
- *Little Ball with Counterweight* ca. 1931
- Painted sheet metal, wire and wood
- 63¾ x 12½ x 12½ in. (161.9 x 31.8 x 31.8 cm.)
- Collection of Whitney Museum of American Art, New York
- Promised 50th Anniversary Gift of Mr. and Mrs. Leonard J. Howich
- © 1999 Estate of Alexander Calder/Artists Rights Society (ARS), New York
- Photography by: Jerry L. Thompson

Interior Artist: Alexander Calder (1898–1976)
- *Yellow Whale*, 1958
- Painted sheet metal, wire and paint, 26 x 25 inches (66 cm x 114.3 cm)
- Collection of Peter and Beverly Lipman
- © 1999 Estate of Alexander Calder/Artists Rights Society (ARS), New York
- Photography by: Jerry L. Thompson

Vice President and Editorial Director: Pat Ferrier
Senior Sponsoring Editor: Lynn Fisher
Developmental Editor: Maria Chu
Economics Editor: Ron Doleman
Marketing Manager: Kelly Smyth
Supervising Editor: Anne Macdonald
Copy Editor: Susan James
Production Coordinator: Jennifer Wilkie
Composition: First Folio Resource Group, Inc.
Cover Design: Greg Devitt
Printer: Transcontinental Printing Group

National Library of Canada Cataloguing in Publication Data

Colander, David C.
Microeconomics/David C. Colander, Peter Sephton, Charlene Richter.—2nd Canadian ed.

Includes index.
ISBN 0-07-090165-1

1. Microeconomics. I. Sephton, Peter S. II. Richter, Charlene, 1956- III. Title.

HB172.C64 2002 338.5 C2002-902249-5

ABOUT THE AUTHORS

David Colander is the Christian A. Johnson Distinguished Professor of Economics at Middlebury College. He has authored, co-authored, or edited 35 books and over 100 articles on a wide range of economic topics.

He earned his B.A. at Columbia College and his M.Phil and Ph.D. at Columbia University. He also studied at the University of Birmingham in England and at Wilhelmsburg Gymnasium in Germany. Professor Colander has taught at Columbia College, Vassar College, and the University of Miami, as well as having been a consultant to Time-Life Films, a consultant to Congress, a Brookings Policy Fellow, and Visiting Scholar at Nuffield College, Oxford.

He belongs to a variety of professional associations and has served on the board of directors and as vice president and president of both the History of Economic Thought Society and the Eastern Economics Association. He has also served on the editorial boards of the *Journal of Economic Perspectives*, *The Journal of Economic Education*, *The Journal of Economic Methodology*, *The Journal of the History of Economic Thought*, and *The Eastern Economics Journal*.

He is married to a pediatrician, Patrice, who has a private practice in Middlebury, Vermont. In their spare time, the Colanders designed and built their oak post-and-beam house on a ridge overlooking the Green Mountains to the east and the Adirondacks to the west. The house is located on the site of a former drive-in movie theatre. (They replaced the speaker poles with fruit trees and used the I-beams from the screen as support for the second story of the carriage house and the garage. Dave's office and library are in the former projection room.)

Peter Sephton is a Professor in the School of Business at Queen's University and the Director of the Queen's National Executive MBA program. He has authored or co-authored several books as well as over three dozen articles on a wide range of topics.

He earned his B.A. at McMaster University and his M.A. and Ph.D. from Queen's University. Professor Sephton has taught at the University of Regina, the University of New Brunswick, and St. Thomas University. He has also been a visiting scholar at the International Monetary Fund as well as the Federal Reserve Bank of St. Louis. He has acted as a consultant to provincial governments and environmental engineering firms. In addition, he has provided advice in legal cases that involved estimating damages for personal injury and fatal accident claims. He has also held positions at the Bank of Canada, the Ontario Ministry of Treasury and Economics, and the Federal Business Development Bank.

Charlene Richter currently teaches in the School of Business at the British Columbia Institute of Technology in Burnaby; teaching various introductory and advanced economics courses to (very bright) students in the Financial Management and Advanced Studies in Business programs. Previously, Charlene worked as an economist with Bell Canada where she analyzed industrial strategy and competition policy. She was also a member of a collaborative research team at the Centre for Trade Policy and Law in Ottawa that quantified the effects of North American free trade on Canadian manufacturing industries, and assessed the significance of international technology transfers between Canadian and U.S. high tech firms. She received B.A. and M.A. degrees from Simon Fraser University, and is a provincially certified teacher holding a British Columbia Professional Teaching Certificate. She has written several economics course manuals and designed and delivered distance education courses. Charlene is the co-author of several academic papers, and a recipient of an award for excellence in teaching. Besides her interest in Canadian industrial policy and international trade, Charlene has many other fascinations: photography, art, architecture, wine, fitness and the stock market. However, Charlene has discovered that her recreational activities are—most unfortunately—subject to irascible and unsympathetic budget constraints.

BRIEF CONTENTS

INTRODUCTION: THINKING LIKE AN ECONOMIST

MICROECONOMICS

CONTENTS

3 THE CANADIAN ECONOMY IN A GLOBAL SETTING 55

4 SUPPLY AND DEMAND 82

MICROECONOMICS

I MICROECONOMICS: THE BASICS

12 MONOPOLY 255

13 MONOPOLISTIC COMPETITION, OLIGOPOLY, AND STRATEGIC PRICING 277

14 GLOBALIZATION, TECHNOLOGY, AND REAL-WORLD COMPETITION 299

IV APPLYING ECONOMIC REASONING TO POLICY

15 GOVERNMENT POLICY AND MARKET FAILURES 319

16 COMPETITION POLICY 340

PREFACE

"I actually enjoyed reading this book. As I read the chapter I felt like the professor was in the room." We received e-mails like these as we began work on this edition. They capture what we believe to be the most distinctive feature of our book—students enjoy reading it.

Our first edition was well received; students liked it, but it was, in large part, a standard book in both tone and structure. Some saw it as a bit idiosyncratic, as it had its own presentation of AS/AD, some unusual metaphors, and too much history and information about institutions.

NEW TO THE SECOND EDITION

We've introduced a number of changes in this edition. One that we're really excited about is the addition of a new co-author, Charlene Richter. Charlene began teaching 20 years ago and is one of the best teachers of economics around. We're excited she's joined us and we know that many of the improvements over the first edition are due directly to Charlene.

The second Canadian edition is the most teachable yet. It is shorter; it reflects many recent changes in the economy and in the profession; many of its more challenging presentations have been simplified; and the production team has done a fantastic job of coordinating a geographically dispersed author team. The following sections review the major changes.

• Shorter

When developing this edition we asked instructors which chapters they assign. The instructors told us they liked many of the chapters but didn't have time to teach everything, so we cut out tangents from chapters so students could focus on the core content. This left a cleaner, shorter, more straightforward presentation of the central ideas of economics. For example, Chapter 3, The Canadian Economy in a Global Setting, combines first-edition Chapters 4 and 5. The combined chapter is much more streamlined and surveys the issues while saving the analysis for the macro core chapters.

Innovation and Globalization

The economy has changed significantly in the last few years and this edition reflects that change. The revisions emphasize the digital revolution, which affects both technology and innovation, and globalization and its effect on growth. Nearly every chapter includes some discussion of one or the other or both. An entire chapter, Chapter 14, Globalization, Technology, and Real-World Competition, focusses on these issues.

Changes in Style and Pedagogy

In response to reviewer feedback, the second edition contains fewer nonstandard terms and presentations. For example, the first edition used metaphors—the invisible handshake and the invisible foot—to describe the social and political forces that influence the economy. Instructors thought they were distracting, so we eliminated them. We still discuss social and political factors—we just don't use the metaphors.

We've also added World Wide Web icons to the margins, which direct students to questions at the end of each chapter that require some on-line research and investigation.

What's Good about the Market?

Is the market a good way to coordinate individuals' activities? Much of this book will be devoted to answering that question. The answer that we, and most Canadian economists, come to is: Yes, it is a reasonable way. True, it has problems; the market can be unfair, mean, and arbitrary, and sometimes it is downright awful. Why then do economists support it? For the same reason that Oliver Wendell Holmes supported democracy—it is a lousy system, but, based on experience with alternatives, it is better than all the others we've thought of.

see page 46

We've also reduced the number of margin definitions of terms so that students can focus on the core material, and we've made sure that these terms are clearly and consistently defined from chapter to chapter.

This edition also presents a learning structure that is clean and logical from the ground up. Here are some examples of the changes: Chapter 4, Supply and Demand, now discusses shift factors of demand and supply separately, and Chapter 5 immediately gives students the opportunity to apply those shift factors to the real world, where multiple shift factors may impact the economy simultaneously. Chapter 7, Taxation and Government Intervention, offers a careful, uncomplicated, and standard explanation of the shift factors of demand, and then shows students how to apply the demand and supply model to the real world.

Learning depends on organization, so we've worked hard to make elements dovetail within chapters. **Learning Objectives**, which match the structure of each chapter, serve as a quick chapter introduction and can be used by students as self-quizzes. Judiciously chosen key terms are carefully defined in context, and the **Chapter Summaries** consolidate main points within the Learning Objectives framework. **Questions for Thought and Review** reinforce the Learning Objectives as they test student comprehension.

2

After reading this chapter, you should be able to:

- Define capitalism and explain how it relies on markets to coordinate economic activities.
- Define socialism and explain how in practice it solves the three coordination problems.
- Describe how economic systems have evolved from the eighth century to today.
- Demonstrate opportunity cost with a production possibility curve.
- State the principle of increasing marginal opportunity cost.

Chapter Summary

- Any economic system must solve three central problems:
  - What, and how much, to produce.
  - How to produce it.
  - For whom to produce it.
- Capitalism is based on private property and the market; socialism is based on individuals' goodwill toward others.
- In capitalism, the what, how, and for whom problems are solved by the market.
- In welfare capitalism, the market, the government, and tradition each rule components of the economy.
- The production possibility curve measures the maximum combination of outputs that can be obtained from a given number of inputs. It embodies the opportunity cost concept.
- In general, in order to get more and more of something, we must give up ever-increasing quantities of something

Pedagogy should reinforce content and help students do well on exams. This requires not only a clearly written book, but also a book that gives students an opportunity to try out their new knowledge. In addition to the critical thinking questions that were a hallmark of the previous edition, this edition includes more fundamental questions so that students can be sure they understand the basics.

Portable tutor: A pedagogical aid that uses the text margins to highlight important concepts and to ask questions that reinforce the Learning Objectives. These margin questions are answered at the end of each chapter.

CAPITALISM

Capitalism is an economic system based on private property and the market. It gives private property rights to individuals, and relies on market forces to coordinate economic activity.

Capitalism is *an economic system based on private property and the market in which, in principle, individuals decide how, what, and for whom to produce.* Under capitalism, individuals are encouraged to follow their own self-interest, while market forces of supply and demand are relied on to coordinate those individual pursuits. Distribution of goods is to each individual according to his or her ability, effort, and inherited property.

Reliance on market forces doesn't mean that political, social, and historical forces play no role in coordinating economic decisions. These other forces do influence how the market works. For example, for a market to exist, government must allocate and defend **private property rights**—*the control a private individual or firm has over an asset or a right.* The concept of private ownership must exist and must be accepted by individuals in society. When you say, "This car is mine," it means that it is unlawful for someone else to take it without your permission. If someone takes it without your permission, he or she is subject to punishment through the legal system.

Reliance on the Market

Q-1 John, your study partner, is telling you that the best way to allocate property rights is through the market. How do you respond?

Markets work through a system of rewards and payments. If you do something, you get paid for doing that something; if you take something, you pay for that something. How much you get is determined by how much you give. This relationship seems fair to most people. But there are instances when it doesn't seem fair. Say someone is unable to work. Should that person get nothing? How about Joe down the street, who was given $10 million by his parents? Is it fair that he gets lots of toys, like Corvettes and skiing

Each chapter ends with a set of **Web questions** that direct students to a variety of Web sites from think tanks to government data sites to business-related sites. These new questions fill many roles: They help students see how the concepts in the chapter really do relate to real-world issues; they familiarize students with the mass of information on the Internet; and they give students a chance to apply the concepts they're learning.

Web Questions 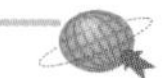

1. The Canada Pension Plan is significant to the evolution of capitalism in Canada. Go to the Human Resources Development Canada home page (www.hrdc-drhc.gc.ca/ips/common/home.shtml) and describe how changes in the plan have moved the Canadian economy toward welfare capitalism. What changes have been made that will alter the nature of CPP? What does this say about the evolution of capitalism today?
2. Starting from the Department of Foreign Affairs and International Trade Web site (www.dfait-maeci.gc.ca/trade/menu-e.asp),
   a. With how many countries does Canada trade?
   b. What are its three largest trading partners?
   c. How does this trade affect the world production possibility curve?
3. Visit the Adam Smith Institute's Web site (www.adamsmith.org/uk) and skim through Book One of *The Wealth of Nations*.
   a. In which chapter does the quotation in the text appear?
   b. What did Smith say limited the division of labour? Why?

In another pedagogical change, we've reorganized the boxed material to fit the theme of "tools, not rules." The boxes in this edition are of three types—Knowing the Tools, Applying the Tools, and Beyond the Tools. Each of the core theory chapters contains a Knowing the Tools box that reviews the chapter's most important concepts, models, and definitions.

KNOWING THE TOOLS

Six Things to Remember When Considering a Demand Curve

- A demand curve had better follow the law of demand: When price rises, quantity demanded falls; and vice versa.
- The horizontal axis—quantity—has a time dimension.
- The quantities are of the same quality.
- The vertical axis—price—assumes all other prices remain the same.
- The curve assumes everything else is held constant.
- Effects of price changes are shown by movements along the demand curve. Effects of anything else on demand (shift factors) are shown by shifts of the entire demand curve.

Applying the Tools boxes provide real-world applications or information related to the chapter.

APPLYING THE TOOLS

A World Economic Geography Quiz

Economic geography isn't much covered in most economics courses because it requires learning enormous numbers of facts, and university and college courses aren't a good place to learn facts. Postsecondary studies are designed to teach you how to interpret and relate facts. Unfortunately, if you don't know facts, much of what you learn now isn't going to do you much good. You'll be relating and interpreting air. The following quiz presents some facts about the world

the characteristics with the country or region.

If you answer 15 or more correctly, you have a reasonably good sense of economic geography. If you don't, we strongly suggest learning more facts. The study guide has other projects, information, and examples. An encyclopedia has even more, and your library has a wealth of information. You could spend the entire term following the economic news carefully, paying attention to

Material that places a concept in a broader or more institutional context appears in Beyond the Tools boxes.

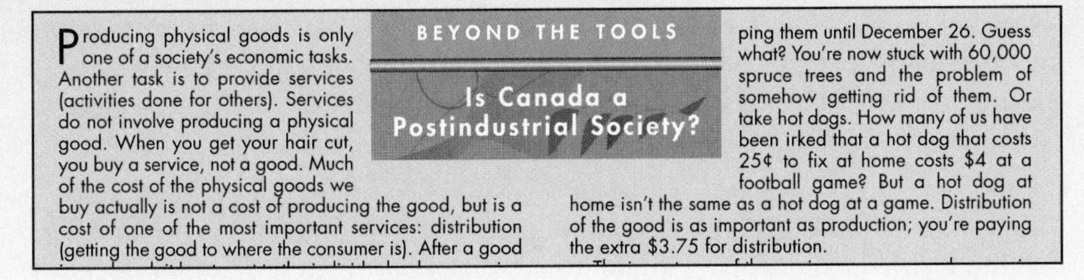

BEYOND THE TOOLS

Is Canada a Postindustrial Society?

Producing physical goods is only one of a society's economic tasks. Another task is to provide services (activities done for others). Services do not involve producing a physical good. When you get your hair cut, you buy a service, not a good. Much of the cost of the physical goods we buy actually is not a cost of producing the good, but is a cost of one of the most important services: distribution (getting the good to where the consumer is). After a good

ping them until December 26. Guess what? You're now stuck with 60,000 spruce trees and the problem of somehow getting rid of them. Or take hot dogs. How many of us have been irked that a hot dog that costs 25¢ to fix at home costs $4 at a football game? But a hot dog at home isn't the same as a hot dog at a game. Distribution of the good is as important as production; you're paying the extra $3.75 for distribution.

These carefully defined boxes help students organize complementary material as they read through the chapters.

We've also added a glossary of colloquial terms so that students whose first language isn't English will broaden and strengthen their understanding of the common usage of English.

Major Changes

The most significant changes involve the reorganization of the structure to introduce policy issues earlier, allowing students to discuss these issues much earlier than before. For example:

- *Early Policy Application.* Chapter 7, Taxation and Government Intervention, which takes up some of the issues of government intervention first introduced in Chapter 5, gives students an opportunity to apply the concepts of consumer and producer surplus learned in Chapter 4 and elasticity learned in Chapter 6. It includes taxation, price floors and ceilings, and quantity restrictions. Efficiency is discussed in relation to all these issues. Politics and rent seeking behaviour are introduced as a fight over who gets what portion of total surplus.
- *Increased Coverage of Technology and Globalization.* Chapter 14, Globalization, Technology, and Real-World Competition, pulls together several issues. The first half of the chapter discusses how goals of real-world firms differ from the pure profit-maximization assumed by the models. The second half discusses how globalization and technological advances have affected real-world competition. The chapter notes how firms have become more specialized, which makes the potential for gain much greater than it used to be, and examines the ways technological advances both affect competition and are affected by it.
- *Enhanced Market Failure Coverage.* Chapter 15, Government Policy and Market Failures, presents a useful policy framework for students. Any time there is a market failure, there is a reason for possible government intervention. This chapter considers three market failures (1) externalities, (2) public goods, and (3) asymmetric information. Each market failure is introduced and its failure considered, using perfect competition as a benchmark. Alternative methods of dealing with each market failure are also discussed. The chapter ends by discussing whether government can address market failures successfully or whether government intervention makes the situation worse.

- *Full Coverage of Canadian Competition Issues.* A new Chapter 16 offers comprehensive coverage of Canadian competition issues. The chapter details Canadian competition law, discusses the economic reasoning behind the legislation, and analyses pertinent cases. The chapter discusses the impact of global competitive forces, and contrasts Canadian competition policy with competition legislation in other countries, notably the United States, and includes an analysis of the Microsoft case.
- *Extended Policy Coverage.* Chapter 17, Microeconomic Policy, Economic Reasoning and Beyond, puts economic reasoning into practice. The first part extends the supply/demand model to a broader cost/benefit framework. The second part considers both the costs and the benefits of using economic reasoning. It notes that markets that are working perfectly (no market failures) might still lead to undesirable outcomes. Three failures of market outcomes are considered: (1) failures due to distributional issues, (2) failures due to human irrationality, and (3) failures due to violations of inalienable rights. The chapter also emphasizes that such failures do not necessarily call for activist policy; policymakers must recognize that government failure exists too. Policy is generally a debate about which failure is worse.

Every chapter has been updated with the latest possible statistics and the most up-to-date policy discussions.

• Design

Besides being different in organization and content, the second Canadian edition also *looks* different. The design is more open and the typeface is more reader-friendly. A lighter colour palette makes the graphs and charts easier to read. The Tools boxes are more integrated so that students are less likely to skip over them.

WHAT WE'VE KEPT

The above discussion may make this seem like a whole new book; it isn't. Microeconomics is still written with the same essential elements that differentiate it from other books. This includes the focus on teaching economic sensibility and the maintenance of the passionate writing style. Finally, institutions and history remain important, so you will still find more historical and institutional issues in this book than in almost any other principles book.

ANCILLARIES

All reviewers agreed that the first edition ancillaries could be improved, so that's what we did. Thus, in this edition we have consolidated and significantly expanded and improved the previous edition's supplements into an accessible, convenient package.

For The Instructor

• The Instructor's Online Learning Centre

The OLC at www.mcgrawhill.ca/college/colander includes a password-protected Web site for Instructors. The site offers downloadable supplements and PageOut, the McGraw-Hill Ryerson course Web site development centre.

• Instructor's CD-ROM

This CD-ROM contains all of the necessary Instructor Supplements, including:

Instructor's Manual

The Instructor's Manual offers eight new features that make class preparation easier than ever. "Chapter Overview" and "What's New" provide a quick review of each chapter. "What's New" will be invaluable when modifying lecture notes to fit the new edition. "Discussion Starters" will help engage students and keep them thinking. "Tips for Teaching Large Sections," offers innovative ideas for teaching very large classes. "Student Stumbling Blocks" provides additional explanations or examples that help clarify difficult concepts. "Ties to the Tools" helps bring those text boxes into the classroom. "Pop Quiz" will help students prepare for exams. The "Case Studies" provide contemporary, real-world economic examples.

Computerized Test Bank

The test banks (micro and macro), have been adapted by Andrew Secord of St. Thomas University. The Test Bank contains over 5,000 questions and each question is categorized by chapter learning objective, level of difficulty (easy, medium, hard), skill being tested (recall, comprehension, application), and type of question (word problem, calculation, graph). They are available in the Diploma electronic test generating system.

PowerPoint Presentations

Adapted by Sonja Novkovic of St. Mary's University, this package includes all text exhibits and key concepts.

• CBC Video Cases

Prepared by Kevin Richter of Douglas College, is a series of video segments drawn from CBC broadcasts. These videos have been chosen to assist students in applying economic concepts to real-world events. A set of instructor notes accompanies each video segment and is available at the Instructor's Online Learning Centre. The video segments will be available in VHS format and through video-streaming from the Online Learning Centre, which is accessible to both instructors and students.

• PageOut

Visit www.mhhe.com/pageout to create a Web page for your course using our resources. PageOut is the McGraw-Hill Ryerson Web site development centre. This Web-page-generation software is free to adopters and is designed to help faculty create an online course, complete with assignments, quizzes, links to relevant Web sites, lecture notes, and more, in a matter of minutes.

In addition, content cartridges are also available for course management systems, such as *WebCT* and *Blackboard*.

For The Student

- **Study Guide [ISBN 007-090197-X]**

Adapted by Oliver Franke of Athabasca University, the Study Guide reviews the main concepts from each chapter and applies those concepts in a variety of ways: short-answer questions, matching terms with definitions, problems and applications, a brain-teaser, multiple-choice questions, and potential essay questions. Since students learn best, not by just knowing the right answer, but by understanding how to get there, each answer comes with an explanation. Timed cumulative pretests help students prepare for exams. Ask for it at your bookstore!

- **Student Online Learning Centre**

Prepared by Oliver Franke of Athabasca University, this electronic learning aid at www.mcgrawhill.ca/college/colander offers a wealth of materials, including CBC Video Cases, Learning Objectives, Online Quizzing (Pre-test and Post-test), Tutorial, Practice Exercises, Web Notes, Sample Exam Questions with answers, Key Terms and Searchable Glossary, PowerPoint slides, and a link to CANSIM II database.

- **GradeSummit [www.gradesummit.com]**

GradeSummit is an Internet-based self-assessment service that offers a variety of ways for students to analyze what they know and don't know. By revealing subject strengths and weaknesses and by providing detailed feedback and direction, GradeSummit enables students to focus their study time on those areas where they are most in need of improvement. GradeSummit provides data about how much students know while they study for an exam —not after they take it. It helps the professor measure an individual student's progress and assess that progress relative to others in their class.

ACKNOWLEDGEMENTS

The entire team at McGraw-Hill Ryerson deserves credit for embracing the idea of this revision and for keeping the project moving in a timely fashion. But Maria Chu, our Developmental Editor, deserves special mention for guiding the manuscript through two revisions with patience and tact. We thank Pat Ferrier, Editorial Director, Lynn Fisher, Senior Sponsoring Editor, and Ron Doleman, Economics Editor, for making this textbook a priority project and for supporting our suggestions for improvement. We would also like to thank Susan James and Susan Broadhurst, Copy Editors, for their attention to detail and accurate copy editing on *Microeconomics* and *Macroeconomics* respectively. We are grateful to Anne Macdonald, Supervising Editor, for transforming the final manuscript into a handsome book and to Kelly Smyth, Marketing Manager, for promoting it with such enthusiasm. And we appreciate the outstanding work of the McGraw-Hill Ryerson production team: Jennifer Wilkie, Production Coordinator, Dianna Little, Art Director, and Greg Devitt, Designer.

This textbook makes use of a wide range of macroeconomic data, mostly from Statistics Canada's CANSIM II database. Statistics Canada information is used with the kind permission of Statistics Canada.

Our sincere thanks go to Study Guide author Oliver Franke of Athabasca University. And to Maurice Tugwell of Acadia University and Audrey Laporte of the University of Toronto, who provided technical reviews of the manuscript.

Last, but not least, we thank the following teachers and colleagues, whose thorough reviews and thoughtful suggestions led to innumerable substantive improvements:

Keir Armstrong
Carleton University

Jeff Davidson
University of Lethbridge

Brian Ferguson
University of Guelph

Oliver Franke
Athabasca University

Raimo Marttala
Malaspina College

Kevin Richter
Douglas College

Gary Riser
Memorial University

Harvey Schwartz
York University

Andrew Secord
St. Thomas University

James Sentance
University of Prince Edward Island

Annie Spears
University of Prince Edward Island

Maurice Tugwell
Acadia University

We would also like to thank our families and friends for allowing us to indulge in long absences while working on the book. Their contributions cannot be quantified, nor can we thank them adequately for their sacrifices. They helped us keep our work in perspective and provided a loving environment in which to work.

In addition, we'd like to thank generations of students for asking good questions. We'd particularly like to thank those students who developed a passionate interest in the economic world around them, and in economics itself.

McGraw-Hill Ryerson
Online Learning Centre

McGraw-Hill Ryerson offers you an online resource that combines the best content with the flexibility and power of the Internet. Organized by chapter, the COLANDER Online Learning Centre (OLC) offers the following features to enhance your learning and understanding of economics:

- Pre-test and Post-test Quizzes
- Web Links
- Student Tutorials
- Practice Exercises
- Sample Exam Questions with Answers
- Interactive Graphing Exercises
- Microsoft® PowerPoint® Powernotes
- CBC Video Cases CBC
- Learning Objectives
- Chapter Summary
- Searchable Key Terms and Glossary
- Links to CANSIM II Database
- EconGraphKit

By connecting to the "real world" through the OLC, you will enjoy a dynamic and rich source of current information that will help you get more from your course and improve your chances for success, both in economics and in the future.

For the Instructor

Downloadable Supplements

All key supplements, including Instructor's Manual and Microsoft® PowerPoint® Presentations, are available, password-protected for instant access!

PageOut

Create your own course Web page for free, quickly and easily. Your professionally designed Web site links directly to OLC material, allows you to post a class syllabus, offers an online gradebook, and much more! Visit www.pageout.net

Online Resources

Primis Online gives you access to our resources in the best medium for your students: printed textbooks or electronic ebooks. There are over 350,000 pages of content available from which you can create customized learning tools from our online database at www.mhhe.com/primis

eServices

McGraw-Hill Ryerson offers a unique services package designed for Canadian faculty. Our mission is to equip providers of higher education with superior tools and resources required for excellence in teaching. For additional information visit http://www.mcgrawhill.ca/highereducation/eservices/

Higher Learning. Forward Thinking.™

ning Centre

For the Student

Interactive Graphing Exercises

Selected chapters include graphs that can be manipulated to solve exercises like those in the textbook. Students can actually see curves shift based on changes to the data.

Online Quizzes

Do you understand the material? You'll know after taking an Online Quiz! Try the Multiple Choice and True/False questions for each chapter. They're auto-graded with feedback and the option to send results directly to faculty. A pre-test and post-test is provided for each chapter.

Microsoft® PowerPoint® Powernotes

View and download presentations created for each text. Great for pre-class preparation and post-class review.

Link to CANSIM II Database Σ-STAT

Provides free-of-charge access to Statistics Canada's Computerized database and information retrieval service through ΣSTAT, Statistics Canada's educational resource that allows you to review socio-economic and demographic data in charts, graphs, and maps.

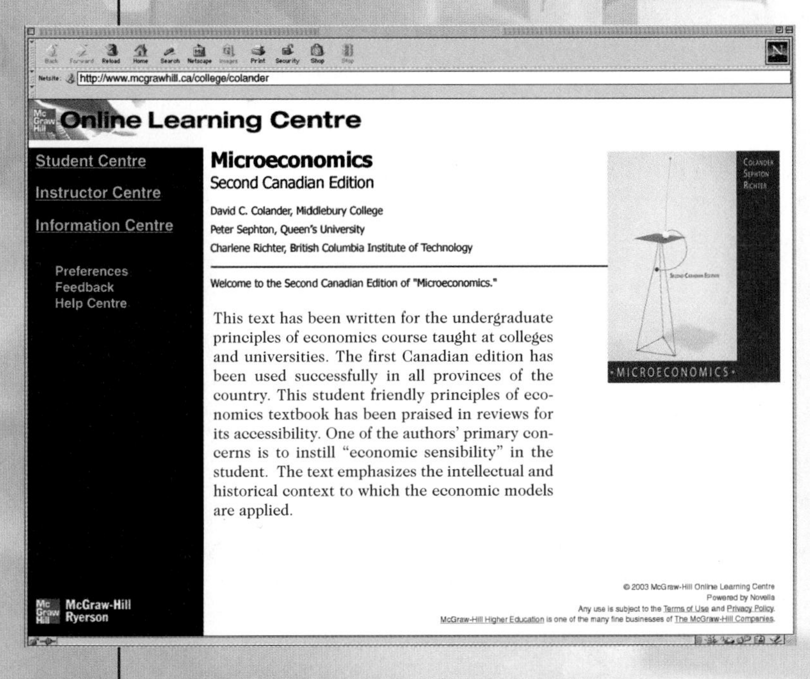

Your Internet companion to the most exciting educational tools on the Web!

The Online Learning Centre can be found at:
www.mcgrawhill.ca/college/colander

INTRODUCTION: THINKING LIKE AN ECONOMIST

I

Part I is an introduction, and an introduction to an introduction seems a little funny. Other sections have introductions, so it seemed a little funny not to have an introduction to Part I, and besides, as you will see, we're a little funny ourselves (which, in turn, has two interpretations; we're sure you'll decide which of the two is appropriate). It will, however, be a very brief introduction, consisting of questions you probably have and some answers to those questions.

SOME QUESTIONS AND ANSWERS

Why study economics?
Because it's neat and interesting and helps provide insight into events that are constantly going on around you.

Why is this book so big?
Because there's a lot of important information in it and because the book is designed so your instructor can pick and choose. You'll likely not be required to read all of it – but once you start it, you'll probably read it all anyhow. (Would you believe?)

Why does this book cost so much?
To answer this question you'll have to read the book.

Will this book make me rich?
No.

Will this book make me happy?
It depends.

This book doesn't seem to be written in a normal textbook style. What gives?
Learning economics is fun, so why write a book that's hard to read? We're passionate when it comes to conveying our love for economics. (OK, we need to get out more). We'll try hard to keep your interest and convince you that thinking like an economist is pretty easy once you get the hang of it.

Will the entire book be like this?
No, the introduction is just trying to rope you in! Much of this book will be hard going. Learning happens to be a difficult task – no pain, no gain. But the authors aren't sadists – we try to make learning as pleasantly painful as possible.

What do the authors' students think of them?
Fair, interesting, and sincerely interested in getting students to learn. (Answer written by some of our students).

So, there you have it. Answers to the questions that you might never have thought of if they hadn't been put in front of you. We hope they give you a sense of the approach we'll use in this book. There are some neat ideas in it. Let's now briefly consider what's in the first five chapters.

A SURVEY OF THE FIRST FIVE CHAPTERS

This first section is really an introduction to the rest of the book. It gives you the background necessary so that the later chapters will make sense. Chapter 1 gives you an overview of the entire field of economics as well as an introduction to our style. Chapter 2 gives you some history of economic systems and shows you how important institutions are. It gives you a sense of how economic forces interact with political and social forces. Chapter 3 introduces you to the institutions of the Canadian economy and discusses how our economy must be viewed in a global setting. Chapters 4 and 5 introduce you to supply and demand, and show you not only the power of those two concepts, but also their limitations.

Now let's get on with the show.

Economics and Economic Reasoning

1

After reading this chapter, you should be able to:

- Define economics and list three coordination problems that an economy must solve.
- Explain how to make decisions by comparing marginal costs and marginal benefits.
- Define opportunity cost and explain its relationship to economic reasoning.
- Explain real-world events in terms of economic forces, social forces, and political forces.
- Differentiate between microeconomics and macroeconomics.
- Distinguish among positive economics, normative economics, and the art of economics.

In my vacations, I visited the poorest quarters of several cities and walked through one street after another, looking at the faces of the poorest people. Next I resolved to make as thorough a study as I could of Political Economy.

Alfred Marshall

When an artist looks at the world, he sees colour. When a musician looks at the world, she hears music. An economist looks at the world and sees a symphony of costs and benefits. The economist's world might not be as colourful or as melodic as the artist's or musician's worlds, but it's more practical. If you want to understand what's going on in the world that's really out there, you need to know economics.

WHAT ECONOMICS IS

Economics is *the study of how human beings coordinate their wants and desires, given the decision-making mechanisms, social customs, and political realities of their society*. One of the key words in the definition of the term "economics" is *coordination*. Coordination can mean many things. In the study of economics, coordination refers to how the three central problems facing any economy are solved. These central problems are:

1. What, and how much, to produce.
2. How to produce it.
3. For whom to produce it.

Three central coordination problems any economy must solve are what to produce, how to produce it, and for whom to produce it.

In answering these questions, economists generally find that individuals want more than is available, given how much they're willing to work. That means that in our economy there is a problem of **scarcity**—*the goods available are too few to satisfy individuals' desires*. Wants are unlimited but resources are limited.

Scarcity has two elements—our wants and our means of fulfilling those wants. These can be interrelated since wants are changeable and partially determined by society. The way we fulfill wants can affect those wants. For example, if you work on Bay Street you will probably want upscale and trendy clothes. If you work in Calgary, you may be quite happy wearing Levi's and cowboy boots.

The degree of scarcity is constantly changing. The quantity of goods, services, and usable resources depends on technology and human action, which underlie production. Individuals' imagination, innovativeness, and willingness to do what needs to be done can greatly increase available goods and resources. Who knows what technologies lie in our future—nanites or micromachines that change atoms into whatever we want could conceivably eliminate scarcity of goods we currently consume. But they would not eliminate scarcity entirely since new wants are constantly developing.

The quantity of goods, services, and usable resources depends on technology and human action.

In all known economies, coordination has involved coercion—limiting people's wants and increasing the amount of work individuals are willing to do to fulfill those wants. The reality is that many people would rather play than help solve society's problems. So the basic economic problem involves inspiring people to do things that other people want them to do, and not to do things that other people don't want them to do. Thus, an alternative definition of economics is that it is the study of how to get people to do things they're not wild about doing (such as studying) and not to do things they are wild about doing (such as eating all the lobster they like), so that the things some people want to do are consistent with the things other people want to do.

To understand an economy you need to learn:

1. *Economic reasoning*.
2. *Economic terminology*.
3. *Economic insights* that economists have about issues, and theories that lead to those insights.
4. Information about *economic institutions*.
5. Information about the *economic policy options* facing society today.

To understand an economy you need to learn:

1. Economic reasoning.
2. Economic terminology.
3. Economic insights.
4. Economic institutions.
5. Economic policy options.

Let's consider each in turn.

A GUIDE TO ECONOMIC REASONING

People trained in economics think in a certain way. They analyze everything critically; they compare the costs and the benefits of every issue and make decisions based on those costs and benefits. For example, say you're trying to decide whether protecting baby seals is a good policy or not. Economists are trained to set their emotions aside and ask: What are the costs of protecting baby seals, and what are the benefits? Thus, they are open to the argument that the benefits of allowing baby seals to be killed might exceed the costs. To think like an economist is to address almost all issues using a cost/benefit approach. Economic reasoning—how to think like an economist, making decisions on the basis of costs and benefits—is the most important lesson you'll learn from this book.

Economic reasoning is making decisions on the basis of costs and benefits.

Economic reasoning, once learned, is infectious. If you're susceptible, being exposed to it will change your life. It will influence your analysis of everything, including issues normally considered outside the scope of economics. For example, you will likely use economic reasoning to decide the possibility of getting a date for Saturday night, and who will pay for dinner. You will likely use it to decide whether to read this book, whether to attend class, whom to marry, and what kind of work to go into after you graduate. This is not to say that economic reasoning will provide all the answers. As you will see throughout this book, real-world questions are inevitably complicated, and economic reasoning simply provides a framework within which to approach a question. In the economic way of thinking, every choice has costs and benefits, and decisions are made by comparing them.

Marginal Costs and Marginal Benefits

The relevant costs and relevant benefits to economic reasoning are the expected *incremental* or additional costs incurred and the expected *incremental* benefits that result from a decision. Economists use the term *marginal* when referring to additional or incremental. Marginal costs and marginal benefits are key concepts.

1.1 
see page 19

A **marginal cost** is *the additional cost to you over and above the costs you have already incurred.* That means eliminating **sunk costs**—*costs that have already been incurred and cannot be recovered*—from the relevant costs when making a decision. Consider, for example, attending class. You've already paid your tuition; it is a sunk cost. So the marginal (or additional) cost of going to class does not include tuition.

Similarly with marginal benefit. A **marginal benefit** is *the additional benefit above what you've already derived.* The marginal benefit of reading this chapter is the *additional* knowledge you get from reading it. If you already knew everything in this chapter before you picked up the book, the marginal benefit of reading it now is zero. The marginal benefit is not zero if by reading the chapter you learn that you are prepared for class; before, you might only have suspected you were prepared.

If the relevant benefits of doing something exceed the relevant costs, do it. If the relevant costs of doing something exceed the relevant benefits, don't do it.

Comparing marginal (additional) costs with marginal (additional) benefits will often tell you how you should adjust your activities to be as well off as possible. Just follow the **economic decision rule:**

If the marginal benefits of doing something exceed the marginal costs, do it.

If the marginal costs of doing something exceed the marginal benefits, don't do it.

As an example, let's consider a discussion we might have with a student who tells us that she is too busy to attend our classes. We respond, "Think about the tuition you've spent for this class—it works out to about $30 a lecture." The student answers that the

Q-1 Say you bought stock A for $10 and stock B for $20. The price of each is currently $15. Assuming taxes are not an issue, which would you sell if you need $15?

Once upon a time, Tanstaafl was made king of all the lands. His first act was to call his economic advisers and tell them to write up all the economic knowledge the society possessed. After years of work, they presented their monumental effort: 25 volumes, each about 400 pages long. But in the interim, King Tanstaafl had become a very busy man, what with running a kingdom of all the lands and everything. Looking at the lengthy volumes, he told his advisers to summarize their findings in one volume.

Despondently, the economists returned to their desks, wondering how they could summarize what they'd been so careful to spell out. After many more years of rewriting, they were finally satisfied with their one-volume effort, and tried to make an appointment to see the king. Unfortunately, affairs of state had become even more pressing than before, and the king couldn't take the time to see them. Instead he sent word to them that he couldn't be bothered with a whole volume, and ordered them, under threat of death (for he had become a tyrant), to reduce the work to one sentence.

The economists returned to their desks, shivering in their sandals and pondering their impossible task. Thinking about their fate if they were not successful, they decided to send out for one last meal. Unfortunately, when they were collecting money to pay for the meal, they discovered they were broke. The disgusted delivery man took the last meal back to the cook, and the economists started down the path to the beheading station. On the way, the delivery man's parting words echoed in their ears. They looked at each other and suddenly they realized the truth. "We're saved!" they screamed. "That's it! That's economic knowledge in one sentence!" They wrote the sentence down and presented it to the king, who thereafter fully understood all economic problems. (He also gave them a good meal.) The sentence?

There **A**in't **N**o **S**uch **T**hing **A**s **A** **F**ree **L**unch—**TANSTAAFL**

book she reads for class is a book that we wrote, and that we wrote it so clearly she fully understands everything. She goes on:

> I've already paid the tuition and whether I go to class or not, I can't get any of the tuition back, so the tuition is a sunk cost and doesn't enter into my decision. The marginal cost to me is what I could be doing with the hour instead of spending it in class. I value my time at $75 an hour [people who understand everything value their time highly], and even though I've heard that your lectures are super, I estimate that the marginal benefit of your class is only $50. The marginal cost, $75, exceeds the marginal benefit, $50, so I don't attend class.

We would congratulate her on her diplomacy and her economic reasoning, but tell her that we give a quiz every week, that students who miss a quiz fail the quiz, that those who fail all the quizzes fail the course, and that those who fail the course do not graduate. In short, she is underestimating the marginal benefits of attending our classes. Correctly estimated, the marginal benefits of attending class exceed the marginal costs. So she should attend classes.

Economics and Passion

Recognizing that everything has a cost is reasonable, but it's a reasonableness that many people don't like. It takes some of the passion out of life. It leads you to consider possibilities like these:

- Saving some people's lives with liver transplants might not be worth the additional cost. The money might be better spent on nutritional programs that would save 20 lives for every 2 lives you might save with transplants.

- Maybe we shouldn't try to eliminate all pollution, because the additional cost of doing so may be too high. To eliminate all pollution might be to forgo too much of some other worthwhile activity.
- Buying a stock that went up 20 percent wasn't necessarily the greatest investment if in doing so you had to forgo some other investment that would have paid you a 30 percent return.
- It might make sense for the automobile industry to save $12 per car by not installing a safety device, even though without the safety device some people will be killed.

Economic reasoning is based on the premise that everything has a cost.

Q-2 Can you think of a reason why a cost/benefit approach to a problem might be inappropriate? Can you give an example?

You get the idea. This kind of reasonableness is often criticized for being cold-hearted. But, not surprisingly, economists disagree; they argue that their reasoning leads to a better society for the majority of people.

Opportunity Cost

Putting economists' cost/benefit rules into practice isn't easy. To do so, you have to be able to choose and measure the costs and benefits correctly. Economists have devised the concept of **opportunity cost** to help you do that. The opportunity cost of undertaking an activity is *the benefit forgone by undertaking that activity*. The benefit forgone is the benefit that you might have gained from choosing the next-best alternative. To obtain the benefit of something, you must give up (forgo) something else—namely, the next-best alternative. All activities that have a next-best alternative have an opportunity cost.

Opportunity cost is the basis of cost/benefit economic reasoning; it is the benefit forgone, or the cost, of the next-best alternative to the activity you've chosen. In economic reasoning, that cost is less than the benefit of what you've chosen.

Let's consider some examples. The opportunity cost of going out once with Natalia (or Nathaniel), the most beautiful woman (attractive man) in the world, might well be losing your solid steady, Margo (Mike). The opportunity cost of cleaning up the environment might be a reduction in the money available to assist low-income individuals. The opportunity cost of having a child might be two boats, three cars, and a two-week vacation each year for five years.

Opportunity costs have always made choice difficult, as we see in the early-19th-century engraving, "One or the Other."

Examples are endless, but let's consider two that are particularly relevant to you: your choice of courses and your decision about how much to study. Let's say you're a full-time student and at the beginning of the term you had to choose four or five courses to take. Taking one precluded taking some other, and the opportunity cost of taking an economics course may well have been not taking a course on theatre. Similarly with studying: you have a limited amount of time to spend studying economics, studying some other subject, sleeping, or partying. The more time you spend on one activity, the less time you have for another. That's opportunity cost.

Notice how neatly the opportunity cost concept takes into account costs and benefits of all other options, and converts these alternative benefits into costs of the decision you're now making.

The relevance of opportunity cost isn't limited to your individual decisions. Opportunity costs are also relevant to government's decisions, which affect everyone in society. A common example is the guns-versus-butter debate. The resources that a society has are limited; therefore, its decision to use those resources to have more guns (more weapons) means that it must have less butter (fewer consumer goods). Thus, when society decides to spent $5 billion more on an improved health care system, the opportunity cost of that decision is $5 billion not spent on helping the homeless, paying off some of the national debt, or providing for national defense.

Q-3 John, your math study partner, has just said that the opportunity cost of studying one math chapter is about 1/40 the price you paid for your math book, since the chapter is about 1/40 of the book. Is he right? Why or why not?

The opportunity cost concept has endless implications. It can even be turned upon itself. For instance, it takes time to think about alternatives; that means that there's a cost to being reasonable, so it's only reasonable to be somewhat unreasonable. If you followed that argument, you've caught the economic bug. If you didn't, don't worry. Just remember the opportunity cost concept for now; we'll infect you with economic thinking in the rest of the book.

Economic and Market Forces

The opportunity cost concept applies to all aspects of life and is fundamental to understanding how society reacts to scarcity. When goods are scarce, those goods must be rationed. That is, a mechanism must be chosen to determine who gets what. Society must deal with the scarcity, thinking about and deciding how to allocate the scarce good.

Q-4 Ali, your study partner, states that charging a user fee for health care is immoral—that health care should be freely available to all individuals in society. How would you respond?

Let's consider some specific real-world rationing mechanisms. Dormitory rooms are often rationed by lottery, and permission to register in popular classes is often rationed by a first-come, first-registered rule. Food in Canada, however, is generally rationed by price. If price did not ration food, there wouldn't be enough food to go around. All scarce goods or rights must be rationed in some fashion. These rationing mechanisms are examples of **economic forces,** *the necessary reactions to scarcity.*

One of the important choices that a society must make is whether to allow these economic forces to operate freely and openly or to try to rein them in. A **market force** is *an economic force that is given relatively free rein by society to work through the market.* Market forces ration by changing prices. When there's a shortage, the price goes up. When there's a surplus, the price goes down. Much of this book will be devoted to analyzing how the market works like an invisible hand, guiding economic forces to coordinate individual actions and allocate scarce resources. The **invisible hand** is *the price mechanism, the rise and fall of prices that guides our actions in a market.*

When an economic force operates through the market, it becomes a market force.

Economic reality is controlled by three forces:

1. Economic forces (the invisible hand);
2. Social and cultural forces; and
3. Political and legal forces.

Societies can't choose whether or not to allow economic forces to operate—economic forces are always operating. However, societies may choose whether to allow market forces to predominate. Other forces play a major role in deciding whether to let market forces operate. Economic reality is determined by a contest among these forces.

Let's consider an example in which social forces prevent an economic force from becoming a market force: the problem of getting a date for Saturday night. If a school (or a society) has significantly more people of one sex than the other (let's say more men than women), some men may well find themselves without a date—that is, men will be in excess supply—and will have to find something else to do, say study or go to a movie by themselves. An "excess supply" person could solve the problem by paying someone to go out with him or her, but that would probably change the nature of the date in unacceptable ways. It would be revolting to the person who offered payment and to the person who was offered payment. That unacceptability is an example of the complex social and cultural norms that guide and limit our activities. People don't try to buy dates because social forces prevent them from doing so.

Social and cultural forces can play a significant role in the economy.

Now let's consider another example in which political and legal influences stop economic forces from becoming market forces. Say you decide that you can make some money delivering mail in your neighborhood. You try to establish a small business, but suddenly you are confronted with the law. Canada Post has a legal exclusive right to deliver regular mail, so you'll be prohibited from delivering regular mail in competition with the post office. Economic forces—the desire to make money—led you to want to enter the business, but in this case political forces quash the invisible hand.

Q-5 Your study partner, Joan, states that market forces are always operative. Is she right? Why or why not?

Often political and social forces work together against the invisible hand. For example, in Canada there aren't enough babies to satisfy all the couples who desire them. Babies born to particular sets of parents are rationed—by luck. Consider a group of parents, all of whom want babies. Those who can, have a baby; those who can't have one, but want one, try to adopt. Adoption agencies ration the available babies. Who gets a

Economic forces are always operative; society may allow market forces to operate.

All too often, students study economics out of context. They're presented with sterile analysis and boring facts to memorize, and are never shown how economics fits into the larger scheme of things. That's bad: it makes economics seem boring—but economics is not boring. Every so often throughout this book, sometimes in the appendixes and sometimes in boxes, we'll step back and put the analysis in perspective, giving you an idea of where the analysis sprang from and its historical context. In educational jargon, this is called *enrichment.*

We begin here with economics itself.

First, its history: In the 1500s there were few universities. Those that existed taught theology, Latin, Greek, philosophy, history, and mathematics. No economics. Then came the *Enlightenment* (about 1700), in which reasoning replaced God as the explanation of why things were the way they were. Pre-Enlightenment thinkers would answer the question, "Why am I poor?" with "Because God wills it." Enlightenment scholars looked for a different explanation. "Because of the nature of land ownership" is one answer they found.

Such reasoned explanations required more knowledge of the way things were, and the amount of information expanded so rapidly that it had to be divided or categorized for an individual to have any hope of knowing a subject. Soon philosophy was subdivided into science and philosophy. In the 1700s, the sciences were split into natural sciences and social sciences. The amount of knowledge kept increasing, and in the late 1800s and early 1900s social science itself split into subdivisions: economics, political science, history, geography, sociology, anthropology, and psychology. Many of the insights about how the economic system worked were codified in Adam Smith's *The Wealth of Nations,* written in 1776. Notice that this was written before economics as a subdiscipline developed, and Adam Smith could also be classified as an anthropologist, a sociologist, a political scientist, and a social philosopher.

Throughout the 18th and 19th centuries, economists such as Adam Smith, Thomas Malthus, John Stuart Mill, David Ricardo, and Karl Marx were more than economists; they were social philosophers who covered all aspects of social science. These writers were subsequently called *classical economists.* Alfred Marshall continued in that classical tradition and his book, *Principles of Economics,* published in the late 1800s, was written with the other social sciences much in evidence. But Marshall also changed the questions economists ask; he focussed on those questions that could be asked in a graphical supply/demand framework.

This book falls solidly in the Marshallian tradition. It sees economics as a way of thinking—as an engine of analysis used to understand real-world phenomena.

Marshallian economics is an art, not a science. It is primarily about policy, not theory. It sees institutions as well as political and social dimensions of reality as important, and it shows you how economics ties in to those dimensions.

baby depends on criteria set by the adoption agency and on the desires of the birth mother, who can often specify the socioeconomic background (and many other characteristics) of the family in which she wants her baby to grow up. That's the economic force in action; it gives more power to the supplier of something that's in short supply.

1.2
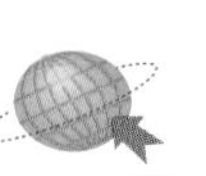
see page 19

If our society allowed individuals to buy and sell babies, that economic force would be translated into a market force. The invisible hand would see to it that the quantity of babies supplied would equal the quantity of babies demanded at some price. The market, not the adoption agencies, would do the rationing.[1]

[1]Even though it's against the law, some babies are nonetheless "sold" on a semilegal market, also called a grey market. In the United States, at the turn of the century, the "market price" for a healthy baby was about U.S. $30,000. If it were legal to sell babies (and if people didn't find it morally repugnant to have babies in order to sell them), the price would be much lower, because there would be a larger supply of babies. (It was not against the law in the U.S. to sell human eggs in the late 1990s, and one human egg was sold for U.S. $50,000. The average price was much lower; it varied with donor characteristics such as academic performance and athletic accomplishments.)

Most people, including us, find the idea of selling babies repugnant. But why? It's the strength of social forces backed up and strengthened by political forces.

What is and isn't allowable differs from one society to another. For example, in Russia, until recently, private businesses were against the law, so not many people started their own businesses. In the United States, until the 1970s, it was against the law to hold gold except in jewelry and for certain limited uses such as dental supplies, so most people refrained from holding gold. Ultimately a country's laws and social norms determine whether the invisible hand will be allowed to work.

Social and political forces are active in all parts of your life. Political forces influence many of your everyday actions. You don't practice medicine without a license; you don't sell body parts or certain addictive drugs. These actions are against the law. But many people do sell alcohol; that's not against the law if you have a permit. Social forces also influence us. You don't make profitable loans to your friends (you don't charge your friends interest); you don't charge your children for their food (parents are supposed to feed their children); many sports and media stars don't sell their autographs (some do, but many consider the practice tacky); you don't lower the wage you'll accept in order to take a job away from someone else. The list is long. You cannot understand economics without understanding the limitations that political and social forces place on economic actions.

In summary, what happens in a society can be seen as the reaction to, and interaction of, these three forces: economic forces, political and legal forces, and social and historical forces. Economics has a role to play in sociology, history, and politics, just as sociology, history, and politics have roles to play in economics.

What happens in society can be seen as a reaction to, and interaction of, economic forces, political forces, social forces, and historical forces.

Economics is about the real world. Throughout this book we'll use the forces just described to talk about real-world events and the interrelationships of economics, history, sociology, and politics.

ECONOMIC TERMINOLOGY

Economic terminology needs little discussion. It simply needs learning. As terms come up, you'll begin to recognize them. Soon you'll begin to understand them, and finally you'll begin to feel comfortable using them. In this book we're trying to describe how economics works in the real world, so we introduce you to many of the terms that occur in business and in discussions of the economy. Whenever possible we'll integrate the introduction of new terms into the discussion so that learning them will seem painless. In fact, we've already introduced you to a number of economic terms: *opportunity cost*, *the invisible hand*, *market forces*, *economic forces*, to name just a few. By the end of the book we'll have introduced you to many more.

ECONOMIC INSIGHTS

Economists have thought about the economy for a long time, so it's not surprising that they've developed some insights into the way it works.

These insights are often based on generalizations, called theories, about the workings of an abstract economy. Theories tie together economists' terminology and knowledge about how an economy operates. Theories are inevitably too abstract to apply in specific cases, and thus a theory is often embodied in an **economic model**—*a framework that places the generalized insights of the theory in a more specific contextual setting*—or in an **economic principle**—*a commonly held economic insight stated as a law or general assump-*

Theories, models, and principles must be combined with a knowledge of real-world economic institutions to arrive at specific policy recommendations.

tion. Then these theories, models, and principles are empirically tested (as best one can) to ensure that they correspond to reality. While these models and principles are less general than theories, they are still usually too general to apply in specific cases. Theories, models, and principles must be combined with a knowledge of real-world economic institutions to arrive at specific policy recommendations. An example? Early in 2002, many stores in Moncton, N.B. were open on Sundays, yet shops in Amherst, N.S., a short drive away, were not allowed to open on Sundays. Our theories and models need to take these institutional features into account if they are going to be of use in answering economic questions.

To see the importance of principles, think back to when you learned to add. You didn't memorize the sum of 147 and 138; instead you learned a principle of addition. The principle says that when adding 147 and 138, you first add 7 + 8, which you memorized was 15. You write down the 5 and carry the 1, which you add to 4 + 3 to get 8. Then add 1 + 1 = 2. So the answer is 285. When you know just one principle, you know how to add millions of combinations of numbers.

The Invisible Hand Theory

In the same way, knowing a theory gives you insight into a wide variety of economic phenomena, even though you don't know the particulars of each phenomenon. For example, much of economic theory deals with the *pricing mechanism* and how the market operates to coordinate *individuals' decisions*. Economists have come to the following insights:

> When the quantity supplied is greater than the quantity demanded, price has a tendency to fall.
>
> When the quantity demanded is greater than the quantity supplied, price has a tendency to rise.

Q-6 There has been a superb growing season and the quantity of tomatoes supplied exceeds the quantity demanded. What is likely to happen to the price of tomatoes?

Using these generalized insights, economists have developed a theory of markets that leads to the further insight that, under certain conditions, markets are efficient. That is, the market will coordinate individuals' decisions, allocating scarce resources to their best possible use. **Efficiency** means *achieving a goal as cheaply as possible*. Economists call this insight the **invisible hand theory**—*a market economy, through the price mechanism, will allocate resources efficiently*.

Theories, and the models used to represent them, are enormously efficient methods of conveying information, but they're also necessarily abstract. They rely on simplifying assumptions, and *if you don't know the assumptions, you don't know the theory*. The result of forgetting assumptions could be similar to what happens if you forget that you're supposed to add numbers in columns. Forgetting that, yet remembering all the steps, can lead to a wildly incorrect answer. For example,

$$\begin{array}{r} 147 \\ +\ \ 138 \\ \hline 1{,}608 \end{array} \text{ is wrong.}$$

Knowing the assumptions of theories and models allows you to progress beyond gut reaction and better understand the strengths and weaknesses of various economic systems. Let's consider a central economic assumption: the assumption that individuals behave rationally—that what they choose reflects what makes them happiest, given the constraints they face, such as level of income. If that assumption doesn't hold, the invisible hand theory doesn't hold.

APPLYING THE TOOLS

Winston Churchill and Lady Astor

There are many stories about Nancy Astor, the first woman elected to Britain's Parliament. A vivacious, fearless American woman, she married into the English aristocracy and, during the 1930s and 1940s, became a bright light on the English social and political scenes, which were already quite bright.

One story told about Lady Astor is that she and Winston Churchill, the unorthodox genius who had a long and distinguished political career and who was Britain's prime minister during World War II, were sitting in a pub having a theoretical discussion about morality. Churchill suggested that as a thought experiment Lady Astor ponder the question: If a man were to promise her a huge amount of money—say a million pounds—for the privilege, would she sleep with him? Lady Astor did ponder the question for a while and finally answered, yes, she would, if the money were guaranteed. Churchill then asked her if she would sleep with him for five pounds. Her response was sharp: "Of course not. What do you think I am—a prostitute?" This time Churchill won the battle of wits by answering, "We have already established that fact; we are now simply negotiating about price."

One moral that economists might draw from this story is that economic incentives, if high enough, can have a powerful influence on behaviour. An equally important moral of the story is that noneconomic incentives can also be very strong. Why do most people feel it's wrong to sell sex for money, even if they would be willing to do so if the price were high enough? Keeping this second moral in mind will significantly increase your economic understanding of real-world events.

Economic Theory and Stories

Economic theory, and the models in which that theory is presented, often developed as a shorthand way of telling a story. These stories are important; they make the theory come alive and convey the insights that give economic theory its power. In this book we present plenty of theories and models, but they're accompanied by stories that provide the context that makes them relevant.

Theory is a shorthand way of telling a story.

At times, because there are many new terms, discussing models and theories takes up much of the presentation time and becomes a bit oppressive. That's the nature of the beast. As Albert Einstein said, "Theories should be as simple as possible, but not more so." When a theory or a model becomes oppressive, pause and think about the underlying story that the theory is meant to convey. That story should make sense and be concrete. If you can't translate the theory into a story, you don't understand the theory.

Often economic theories are presented in mathematical—graphical or algebraic—models. In this book we keep models to a minimum, but we cannot avoid them completely. Graphical models are used so much by economists that they must be included. So part of the course will consist of translating the verbal discussions of the economy into graphical models. To prepare you for these graphical models we have written a brief introduction to the basics of graphical presentation in Appendix A of this chapter.

Microeconomics and Macroeconomics

Economic theory is divided into two parts: microeconomic theory and macroeconomic theory. Microeconomic theory considers economic reasoning from the viewpoint of individuals and firms and builds up from there to an analysis of the whole economy. We define **microeconomics** as *the study of individual choice, and how that choice is influenced by economic forces*. Microeconomics studies such things as the pricing policies of firms, households' decisions of what to buy, and how markets allocate resources among alternative ends. Our discussion of opportunity cost was based on microeconomic theory. The invisible hand theory comes from microeconomics.

Microeconomics is the study of how individual choice is influenced by economic forces.

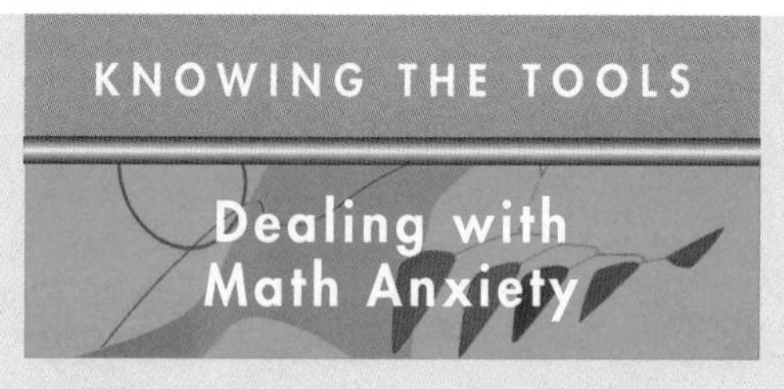

Knowing our own students, we can see the red flags rising, the legs tensing up, the fear flooding over many of you. Here it comes—the math and the graphs.

We wish we could change things by saying to you, "Don't worry—mathematics and graphical analysis are easy." But we can't. That doesn't mean math and graphical analysis aren't wonderful tools that convey ideas neatly and efficiently. They are. But we've had enough teaching experience to know that somewhere back in elementary school some teacher blew it and put about 40 percent of you off mathematics for life. A tool that scares you to death is not useful; it can be a hindrance, not a help, to learning. Nothing your current teacher or we now can say, write, or do is going to completely reassure you, but we'll do our best to relieve your anxiety.

Try to follow the numerical and graphical examples carefully, because they not only cement the knowledge into your minds, they also present in a rigorous manner the ideas we're discussing. The ideas conveyed in the numerical and graphical examples will be explained in words—and the graphical analysis (the type of mathematical explanation most used in introductory economics) generally will simply be a more precise presentation of the accompanying discussion in words. In most economics courses the exams pose the questions in graphical terms, so there's no getting around the need to understand the ideas graphically. And it is easier than you think. (Appendix A at the end of this chapter discusses the basics of graphical analysis.)

Macroeconomics is the study of inflation, unemployment, business cycles, and growth; it focusses on aggregate relationships.

As one builds up from microeconomic analysis to an analysis of the entire economy, everything gets rather complicated. Many economists try to uncomplicate matters by taking a different approach—a macroeconomic approach—first looking at the aggregate, or whole, and then breaking it down into components. We define **macroeconomics** as *the study of the economy as a whole, which includes inflation, unemployment, business cycles, and growth.* Macroeconomics focusses on aggregate relationships, such as how household consumption is related to income and how government policies can affect growth. A micro approach would analyze a person by looking first at each individual cell and then building up. A macro approach would start with the person and then go on to his or her components—arms, legs, fingernails, feelings, and so on. Put simply, microeconomics analyzes from the parts to the whole; macroeconomics analyzes from the whole to the parts.

Microeconomics and macroeconomics are very much interrelated. Clearly, what happens in the economy as a whole is based on individual decisions, but individual decisions are made within an economy and can be understood only within that context. For example, whether a firm decides to expand production capacity will depend on what the owners expect will happen to the demand for their products. Those expectations are determined by macroeconomic conditions. Likewise, decisions by the federal government to change social assistance programs in the mid-1990s had to be made based on how those changes would affect the decisions of millions of individuals. Because microeconomics focusses on the individual and macroeconomics focusses on the whole economy, traditionally microeconomics and macroeconomics are taught separately, even though they are interrelated.

Q.7 Classify the following topics as macroeconomic or microeconomic:

1. The impact of a tax increase on aggregate output.
2. The relationship between two competing firms' pricing behaviour.
3. A farmer's decision to plant soy beans or wheat.
4. The effect of trade on economic growth.

ECONOMIC INSTITUTIONS

To know whether you can apply economic theory to reality, you must know about economic institutions. Corporations, governments, and cultural norms are all economic institutions. Many economic institutions have social, political, and religious dimensions. For example, your job often influences your social standing. In addition, many social institutions, such as the family, have economic functions. If any institution significantly affects economic decisions, we include it as an economic institution because you must understand that institution if you are to understand how the economy functions.

Economic institutions differ significantly among countries. For example, in Germany banks are allowed to own companies; in the United States they cannot. This contributes to a difference in the flow of resources into investment in Germany as compared to the flow in the United States. Or alternatively, in Japan, competition laws (laws under which companies can combine or coordinate their activities) are loose; in Canada they are more restrictive. This causes differences in the nature of competition in the two countries.

Economic institutions sometimes seem to operate in ways quite different than economic theory predicts. For example, economic theory says that prices are determined by supply and demand. However, businesses say that prices are set by rules of thumb—often by what are called cost-plus-markup rules. That is, you determine what your costs are, multiply by 1.4 or 1.5, and the result is the price you set. Economic theory says that supply and demand determine who's hired; experience suggests that hiring is often done on the basis of who you know, not by economic forces.

These apparent contradictions have two complementary explanations. First, economic theory abstracts from many issues. These issues may account for the differences. Second, there's no contradiction; economic principles often affect decisions from behind the scenes. For instance, supply and demand pressures determine what the price markup over cost will be. In all cases, however, to apply economic theory to reality—to gain the full value of economic insights—you've got to have a sense of economic institutions.

To apply economic theory to reality, you've got to have a sense of economic institutions.

ECONOMIC POLICY OPTIONS

Economic policies are *actions (or inaction) taken by government to influence economic actions*. The final goal of the course is to present the economic policy options facing our society today. For example, should the government restrict mergers between firms? Should it run a budget deficit? Should it do something about the international trade surplus? Should it decrease taxes?

We saved this discussion for last because there's no sense talking about policy options unless you know some economic terminology, some economic theory, and something about economic institutions. Once you know something about them, you're in a position to consider the policy options available for dealing with the economic problems our society faces.

Policies operate within institutions, but policies can also influence the institutions within which they operate. Let's consider an example: employment insurance and seasonal workers. In 1956 Canada introduced legislation to protect workers from short periods of unemployment (The Canadian Unemployment Assistance Act). That program evolved to the point where nearly 80 percent of the claimants were people who had previously received benefits. The initial aim of the program was to provide *temporary assistance* to the unemployed, but it had become a means of stabilizing incomes for those working in seasonal industries (fishing, lumber, etc). In the 1990s the federal government changed the program (now called employment insurance) and it tightened up the conditions under which repeat use was allowed. This is not to say that we shouldn't have programs to protect seasonal workers against unemployment; it is only to say that we must build into our policies their effect on institutions. When employment and unemployment insurance programs motivate people towards repeat use, the government might want to consider whether separate programs and premiums should apply to industries dominated by repeat users. Policies affect institutions; with a potential for feedback from institutions to policies.

To carry out economic policy effectively one must understand how institutions might change as a result of the economic policy.

Q-8 True or false? Economists should focus their policy analysis on institutional changes because such policies offer the largest gains.

Some policies are designed to change institutions directly. While these policies are much more difficult to implement than policies that don't, they also offer the largest potential for gain. Let's consider an example. In the 1990s, a number of countries decided to replace socialist institutions with market economies. The result: output in those countries fell enormously as the old institutions fell apart. Eventually, these countries hope, once the new market institutions are predominant, output will bounce back and further gains will be made. The temporary hardship these countries are experiencing shows the enormous difficulty of implementing policies involving major institutional changes.

Objective Policy Analysis

Q-9 John, your study partner, is a free market advocate. He argues that the invisible hand theory tells us that the government should not interfere with the economy. Do you agree? Why or why not?

1.3 

see page 19

Good economic policy analysis is objective; that is, it keeps the analyst's value judgments separate from the analysis. Objective analysis does not say, "This is the way things should be," reflecting a goal established by the analyst. That would be subjective analysis because it would reflect the analyst's view of how things should be. Instead, objective analysis says, "This is the way the economy works, and if society (or the individual or firm for whom you're doing the analysis) wants to achieve a particular goal, this is how it might go about doing so." Objective analysis keeps, or at least tries to keep, subjective views—value judgments—separate.

Positive economics is the study of what is, and how the economy works.

Normative economics is the study of what the goals of the economy should be.

To make clear the distinction between objective and subjective analysis, economists have divided economics into three categories: *positive economics*, *normative economics*, and the *art of economics*. **Positive economics** is *the study of what is, and how the economy works*. It asks such questions as: How does the market for pork bellies work? How do price restrictions affect market forces? These questions fall under the heading of economic theory. **Normative economics** is *the study of what the goals of the economy should be*. Normative economics asks such questions as: What should the distribution of income be? What should tax policy be designed to achieve? In discussing such questions, economists must carefully delineate whose goals they are discussing. One cannot simply assume that one's own goals for society are society's goals.

The *art of economics* is the application of the knowledge learned in positive economics to the achievement of the goals determined in normative economics.

The **art of economics** is *the application of the knowledge learned in positive economics to the achievement of the goals one has determined in normative economics*. It looks at such questions as: To achieve a certain distribution of income, how would you go about it, given the way the economy works?[2] Most policy discussions fall under the art of economics.

Q-10 State whether the following five statements belong in positive economics, normative economics, or the art of economics.

1. We should support the market because it is efficient.
2. Given certain conditions, the market achieves efficient results.
3. Based on past experience and our understanding of markets, if one wants a reasonably efficient result, markets should probably be relied on.
4. The distribution of income should be left to markets.
5. Markets allocate income according to contributions of factors of production.

In each of these three branches of economics, economists separate their own value judgments from their objective analysis as much as possible. The qualifier "as much as possible" is important, since some value judgments inevitably sneak in. We are products of our environment, and the questions we ask, the framework we use, and the way we interpret empirical evidence all embody value judgments and reflect our backgrounds.

Maintaining objectivity is easiest in positive economics, where one is working with abstract models to understand how the economy works. Maintaining objectivity is harder in normative economics. You must always be objective about whose normative values you are using. It's easy to assume that all of society shares your values, but that assumption is often wrong.

It's hardest to maintain objectivity in the art of economics because it embodies the problems of both positive and normative economics. Because noneconomic forces affect policy, to practice the art of economics we must make judgments about how these

[2] This three-part distinction was made back in 1896 by a famous economist, John Neville Keynes, father of John Maynard Keynes, the economist who developed macroeconomics. This distinction was instilled into modern economics by Milton Friedman and Richard Lipsey in the 1950s. They, however, downplayed the art of economics, which J. N. Keynes had seen as central to understanding the economist's role in policy making.

noneconomic forces work. These judgments are likely to embody our own value judgments. So we must be exceedingly careful to be as objective as possible in practicing the art of economics.

Policy and Social and Political Forces

When you think about the policy options facing society, you'll quickly discover that the choice of policy options depends on much more than economic theory. You must take into account historical precedent plus social, cultural, and political forces. In an economics course, we don't have time to analyze these forces in as much depth as we'd like. That's one reason there are separate history, political science, sociology, and anthropology courses.

We don't pretend that these forces play an insignificant role in policy decisions, but specialization is necessary. In economics, we focus the analysis on the invisible hand, and much of economic theory is devoted to how the economy would operate if the invisible hand were the only force operating. But as soon as we apply theory to reality and policy, we must take into account political and social forces as well.

An example will make our point more concrete. Most economists agree that holding down or eliminating tariffs (taxes on imports) and quotas (numerical limitations on imports) makes good economic sense. They strongly advise governments to follow a policy of free trade. Do governments follow free trade policies? Almost invariably they do not. Politics leads society in a different direction. If you're advising a policy maker, you need to point out that these other forces must be taken into account, and how other forces should (if they should) and can (if they can) be integrated with your recommendations.

CONCLUSION

There's tons more that could be said by way of introducing you to economics, but an introduction must remain an introduction. As it is, this chapter should have:

1. Introduced you to economic reasoning.
2. Surveyed what we're going to cover in this book.

This introduction was our opening line. We hope it also conveyed the importance and relevance that belong to economics. If it did, it has served its intended purpose. Economics is tough, but tough can be fun.

Chapter Summary

- The three coordination problems any economy must solve are what to produce, how to produce it, and for whom to produce it. In solving these problems economies have found that there is a problem of scarcity.
- Economic reasoning structures all questions in a cost/benefit frame: If the marginal benefits of doing something exceed the marginal costs, do it. If the marginal costs exceed the marginal benefits, don't do it.
- Sunk costs are not relevant to the economic decision rule.
- The opportunity cost of undertaking an activity is the benefit you might have gained from choosing the next-best alternative.
- "There ain't no such thing as a free lunch" (TANSTAAFL) embodies the opportunity cost concept.
- Economic forces, the forces of scarcity, are always working. Market forces, which ration by changing prices, are not always allowed to work.

- Economic reality is controlled and directed by three types of forces: economic forces, political forces, and social forces.
- Under certain conditions the market, through its price mechanism, will allocate scarce resources efficiently.
- Economics can be divided into microeconomics and macroeconomics. Microeconomics is the study of individual choice and how that choice is influenced by economic forces. Macroeconomics is the study of the economy as a whole, which includes inflation, unemployment, business cycles, and growth.
- Economics can be subdivided into positive economics, normative economics, and the art of economics. Positive economics is the study of what is, normative economics is the study of what should be, and the art of economics relates positive to normative economics.

Key Terms

art of economics (16)
economic decision rule (6)
economic forces (9)
economic model (11)
economic policies (15)
economic principle (11)
economics (5)
efficiency (12)
invisible hand (9)
invisible hand theory (12)
macroeconomics (14)
marginal benefit (6)
marginal cost (6)
market force (9)
microeconomics (13)
normative economics (16)
opportunity cost (8)
positive economics (16)
scarcity (5)
sunk costs (6)

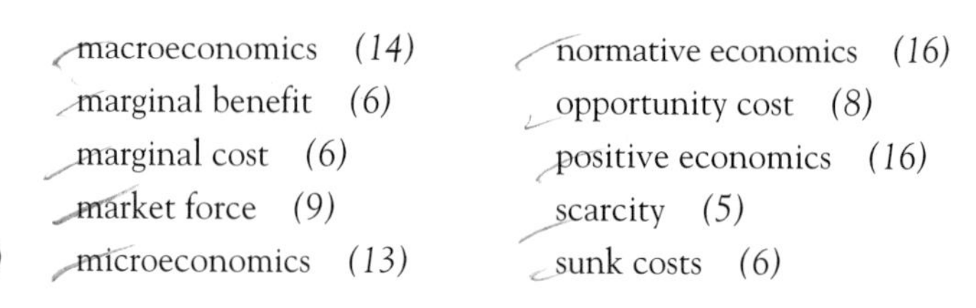

Questions for Thought and Review

1. Why did we focus the definition of economics on coordination rather than on scarcity?
2. List two recent choices you made and explain why you made those choices in terms of marginal benefits and marginal costs.
3. At times we all regret decisions. Does this necessarily mean we did not use the economic decision rule when making the decision?
4. What is the opportunity cost of buying a $20,000 car?
5. Name three ways a limited number of dormitory rooms could be rationed. How would economic forces determine individual behaviour in each? How would social or legal forces determine whether those economic forces become market forces?
6. Give two examples of social forces and explain how they keep economic forces from becoming market forces.
7. Give two examples of political or legal forces and explain how they might interact with the invisible hand.
8. What is an economic model? What besides a model do economists need to make policy recommendations?
9. Does economic theory prove that the free market system is best? Why?
10. List two microeconomic and two macroeconomic problems.
11. Name an economic institution and explain how it either embodies economic principles or affects economic decision making.
12. Is a good economist always objective? Why?

Problems and Exercises

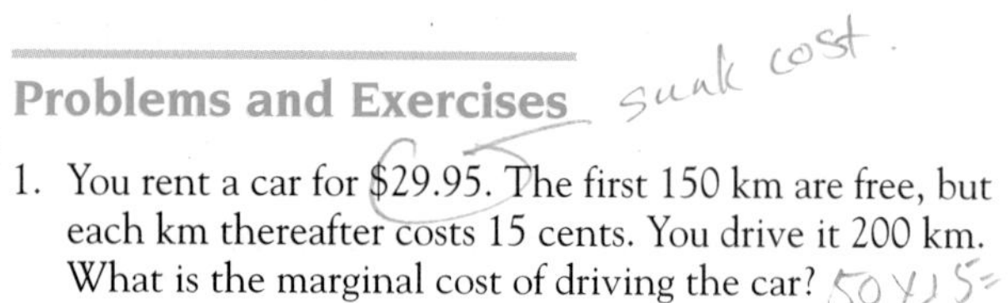

1. You rent a car for $29.95. The first 150 km are free, but each km thereafter costs 15 cents. You drive it 200 km. What is the marginal cost of driving the car?
2. Calculate, using the best estimates you can:
   a. Your opportunity cost of attending school.
   b. Your opportunity cost of taking this course.
   c. Your opportunity cost of attending yesterday's lecture in this course.
3. Individuals have two kidneys but most of us need only one. People who have lost both kidneys through accident or disease must be hooked up to a dialysis machine, which cleanses waste from their bodies. Say a person who has two good kidneys offers to sell one of them to someone whose kidney function has been totally destroyed. The seller asks $30,000 for the kidney, and the person who has lost both kidneys accepts the offer. Who benefits from the deal? Who is hurt? Should a society allow such market transactions? Why?
4. For some years, China has had a one-child-per-family policy. For cultural reasons, there are now many more male than female children born in China. How is this likely to affect who pays the cost of dates in China in 15 or 20 years? Explain your response.

5. In some provinces shopping on Sundays is legal while in others it is not. Can economics be used to explain this fact? Why?
6. Go to two stores: a supermarket and a convenience store.
   a. Write down the cost of a litre of milk in each.
   b. The prices are most likely different. Using the terminology used in this chapter, explain why that is the case and why anyone would buy milk in the store with the higher price.
   c. Do the same exercise with shirts or dresses in Wal-Mart (or its equivalent) and The Bay (or its equivalent).
7. In the mid-1990s, the German comedian Harald Schmidt attempted to mimic the successful David Letterman show in Germany. The format didn't entertain its German audience. What economic policy lesson can be learned from Harald Schmidt's failure?
8. State whether the following statements belong in positive economics, normative economics, or the art of economics.
   a. In a market, when quantity supplied exceeds quantity demanded, price tends to fall.
   b. When determining tax rates, the government should take into account the income needs of individuals.
   c. What society feels is fair is determined largely by cultural norms.
   d. When deciding which rationing mechanism is best (lottery, price, first-come/first-served), one must take into account the goals of society.
   e. Suppose Ontario rations water to farmers at subsidized prices. If Ontario allows the trading of water rights, it will allow economic forces to be a market force.
9. Adam Smith, who wrote *The Wealth of Nations* and is seen as the father of modern economics, also wrote *The Theory of Moral Sentiments*, in which he argued that society would be better off if people weren't so selfish and were more considerate of others. How does this view fit with the discussion of economic reasoning presented in the chapter?

Web Questions

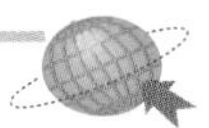

1. Find an employment Web page (an example is www.monster.ca) and search for available jobs using "economist" as a keyword. List five jobs that economists have and write a one-sentence description of each.
2. Use an online periodical (an example is www.canadianbusiness.com) to find two examples of political or legal forces at work. Do those forces keep economic forces from becoming market forces?
3. Using an Internet mapping page (an example is ca.maps.yahoo.com), create a map of your neighborhood and answer the following questions:
   a. How is the map like a model?
   b. What are the limitations of the map?
   c. Could you use this map to determine change in elevation in your neighbourhood? Distance from one place to another? Traffic speed? What do your answers suggest about what to consider when using a map or a model?

Answers to Margin Questions

The numbers in parentheses refer to the page number of each margin question.

1. Since the price of both stocks is now $15, it doesn't matter which one you sell (assuming no differential capital gains taxation). The price you bought them for doesn't matter; it's a sunk cost. Marginal analysis refers to the future gain, so what you expect to happen to future prices of the stocks—not past prices—should determine which stock you decide to sell. (6)
2. A cost/benefit analysis requires that you put a value on a good, and placing a value on a good can be seen as demeaning it. Consider love. Try telling an acquaintance that you'd like to buy his or her spiritual love, and see what response you get. (8)
3. John is wrong. The opportunity cost of reading the chapter is primarily the time you spend reading it. Reading the book prevents you from doing other things. Assuming that you already paid for the book, the original price is no longer part of the opportunity cost; it is a sunk cost. Bygones are bygones. (8)
4. Whenever there is scarcity, the scarce good must be rationed by some means. Free health care has an opportunity cost in other resources. So if health care is not rationed, to get the resources to supply that care, other goods would have to be more tightly rationed than they currently are. It is likely that the opportunity cost of supplying free health care would be larger than most societies would be willing to pay. Hence, some procedures deemed "non-essential" involve a fee not covered by Canada's medicare system. (9)

5. Joan is wrong. Economic forces are always operative; market forces are not. *(9)*
6. According to the invisible hand theory, the price of tomatoes will likely fall. *(12)*
7. (1) Macroeconomics; (2) Microeconomics; (3) Microeconomics; (4) Macroeconomics. *(14)*
8. False. While such changes have the largest gain, they may also have the largest cost. The policies economists should focus on are those that offer the largest net gain—benefits minus costs—to society. *(16)*
9. He is wrong. The invisible hand theory is a positive theory and does not tell us anything about policy. To do so would be to violate Hume's dictum that a "should" cannot be derived from an "is." This is not to say that government should or should not interfere; whether government should interfere is a very difficult question. *(16)*
10. (1) Normative; (2) Positive; (3) Art; (4) Normative; (5) Positive. *(16)*

The Language of Graphs

A picture is worth 1,000 words. Economists, being efficient, like to present ideas in **graphs,** *pictures of points in a coordinate system in which points denote relationships between two or more variables*. But a graph is worth 1,000 words only if the person looking at the graph knows the graphical language. Graphs are usually written on graph paper.

We have enormous sympathy for students who don't understand graphs. A number of our students get thrown for a loop by graphs. They understand the idea, but graphs confuse them. This appendix is for them, and for those of you like them. It's a primer on graphs.

TWO WAYS TO USE GRAPHS

In this book we use graphs in two ways:

1. To present an economic model or theory visually; to show how two variables interrelate.
2. To present real-world data visually. To do this, we use primarily bar charts, line charts, and pie charts.

Actually, these two ways of using graphs are related. They are both ways of presenting visually the *relationship* between two things.

Graphs are built around a number line, or axis, like the one in Figure A1-1(a). The numbers are generally placed in order, equal distances from one another. That number line allows us to represent a number at an appropriate point on the line. For example, point *A* represents the number 4.

The number line in Figure A1-1(a) is drawn horizontally, but it doesn't have to be; it can also be drawn vertically, as in Figure A1-1(b).

Figure A1-1 (a, b, and c) **HORIZONTAL AND VERTICAL NUMBER LINES AND A COORDINATE SYSTEM**

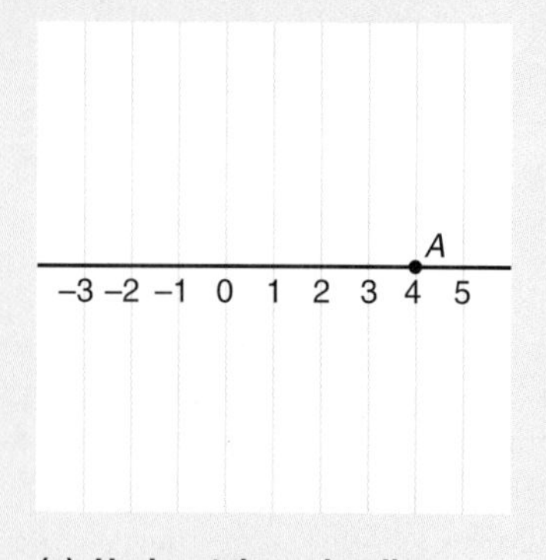

(a) Horizontal number line

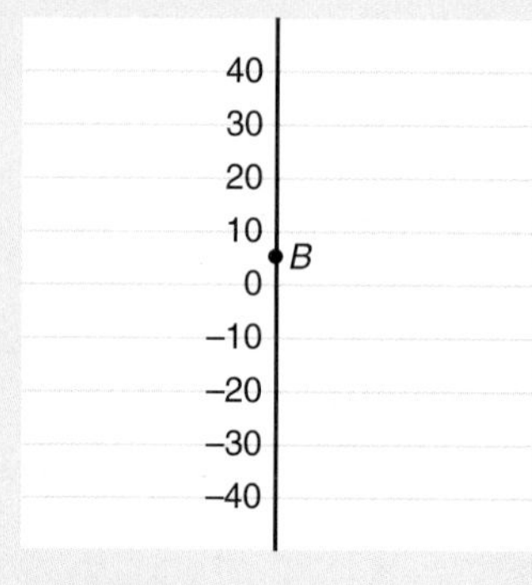

(b) Vertical number line

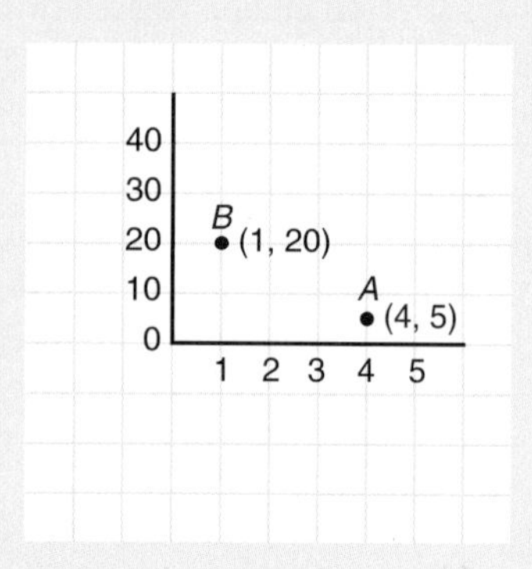

(c) Coordinate system

How we divide our axes, or number lines, into intervals, is up to us. In Figure A1-1(a), we called each interval 1; in Figure A1-1(b), we called each interval 10. Point A appears after 4 intervals of 1 (starting at 0 and reading from left to right), so it represents 4. In Figure A1-1(b), where each interval represents 10, to represent 5, we place point *B* halfway in the interval between 0 and 10.

So far, so good. Graphs combine a vertical and a horizontal number line, as in Figure A1-1(c). When the horizontal and vertical number lines are put together, they're called *axes*. (Each line is an axis. *Axes* is the plural of *axis*.) We now have a **coordinate system**—*a two-dimensional space in which one point represents two numbers*. For example, point A in Figure A1-1(c) represents the numbers (4, 5): 4 on the horizontal number line and 5 on the vertical number line. Point *B* represents the numbers (1, 20). (By convention, the horizontal numbers are always written first.)

Being able to represent two numbers with one point is useful because it allows the relationships between two numbers to be presented visually instead of having to be expressed verbally, which is often cumbersome. For example, say the cost of producing 6 units of something is $4 per unit and the cost of producing 10 units is $3 per unit. By putting both these points on a graph, we can visually see that producing 10 units costs less per unit than producing 6 units.

Another way to use graphs to present real-world data visually is to use the horizontal line to represent time. Say that we let each horizontal interval equal a year. and each vertical interval equal $100 in income. By graphing your income each year, you can obtain a visual representation of how your income has changed over time.

Using Graphs in Economic Modelling

We use graphs throughout the book as we present economic models, or simplifications of reality. A few terms are often used in describing these graphs, and we'll now go over them. Consider Figure A1-2(a), on the next page, which lists the number of pens bought per day (column 2) at various prices (column 1).

We can present the table's information in a graph by combining the pairs of numbers in the two columns of the table and representing, or plotting, them on two axes. We do that in Figure A1-2(b).

By convention, when graphing a relationship between price and quantity, economists place price on the vertical axis and quantity on the horizontal axis.

We can now connect the points, producing a line like the one in Figure A1-2(c). With this line, we interpolate the numbers between the points (which makes for a nice visual presentation). That is, we make the **interpolation assumption**—*the assumption that the relationship between variables is the same between points as it is at the points*. The interpolation assumption allows us to think of a line as a collection of points and therefore to connect the points into a line.

Even though the line in Figure A1-2(c) is straight, economists call any such line drawn on a graph a *curve*. Because it's straight, the curve in A1-2(c) is called a **linear curve**—*a curve that is drawn as a straight line*. Notice that this curve starts high on the left-hand side and goes down to the right. Economists say that any curve that looks like that is *downward-sloping*. They also say that a downward-sloping curve represents an **inverse relationship**—*a relationship between two variables in which whenever one goes up, the other goes down*. In this example, the line demonstrates an inverse relationship between price and quantity—that is, when the price of pens goes up, the quantity bought goes down.

Figure A1-2(d) presents a **nonlinear curve**—*a curve that is drawn as a curved line*. This curve, which really is curved, starts low on the left-hand side and goes up to the right. Economists say any curve that goes up to the right is *upward-sloping*. An upward-sloping curve represents a **direct relationship**—*a relationship in which when one variable goes up, the other goes up too*. The direct relationship we're talking about here is the one between the two variables (what's measured on the horizontal and vertical lines). *Downward-sloping* and *upward-sloping* are terms you need to memorize if you want to keep graphically in your mind the image of the relationships they represent.

Slope

One can, of course, be far more explicit about how much the curve is sloping upward or downward by defining it in terms of **slope**—*the change in the value on the vertical axis divided by the change in the value on the horizontal axis*. Sometimes the slope is presented as "rise over run":

$$\text{Slope} = \frac{\text{Rise}}{\text{Run}} = \frac{\text{Change in value on vertical axis}}{\text{Change in value on horizontal axis}}$$

Slopes of Linear Curves

In Figure A1-3 (on page 23), we present five linear curves and measure their slope. Let's go through an example to show how we can measure slope. To do so, we must pick two points. Let's use points A (6, 8) and *B* (7, 4) on curve *a*. Looking at these points, we see that as we move from 6 to 7 on the horizontal axis, we move from 8 to 4 on the

Figure A1-2 (a, b, c, and d) **A TABLE AND GRAPHS SHOWING THE RELATIONSHIPS BETWEEN PRICE AND QUANTITY**

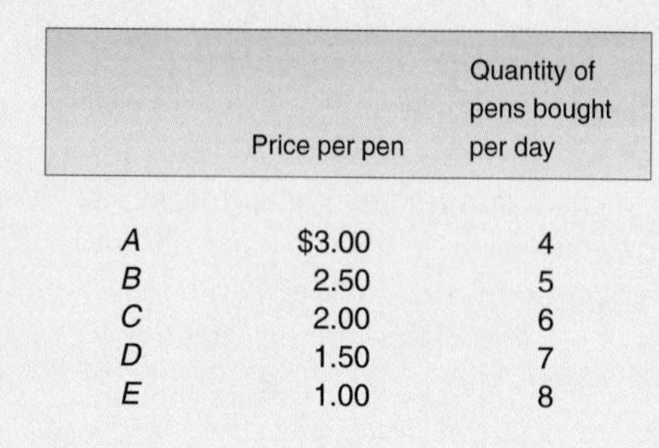

	Price per pen	Quantity of pens bought per day
A	$3.00	4
B	2.50	5
C	2.00	6
D	1.50	7
E	1.00	8

(a) Price quantity table

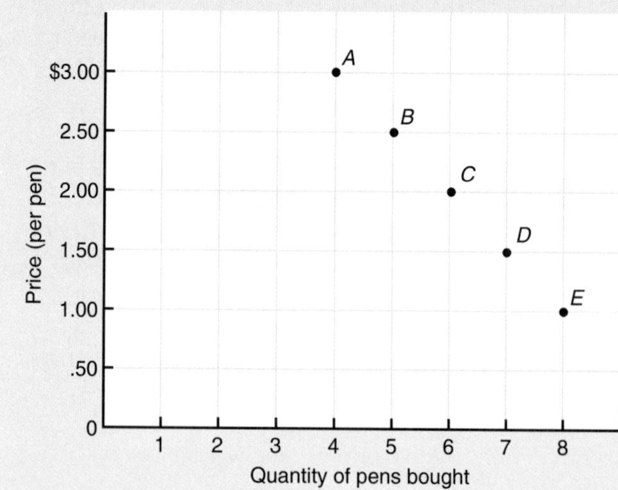

(b) From a table to a graph (1)

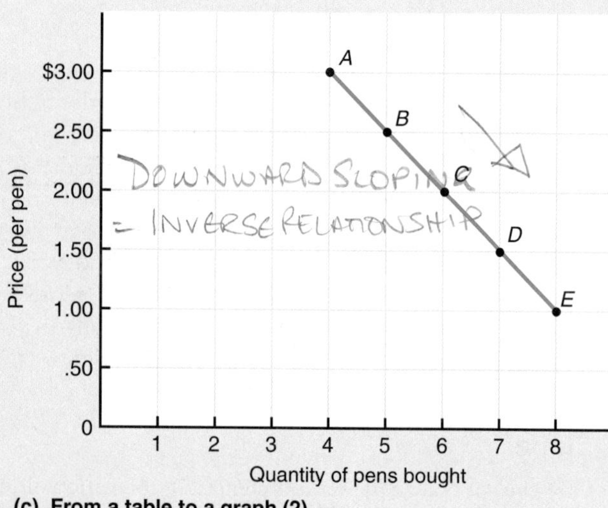

(c) From a table to a graph (2)

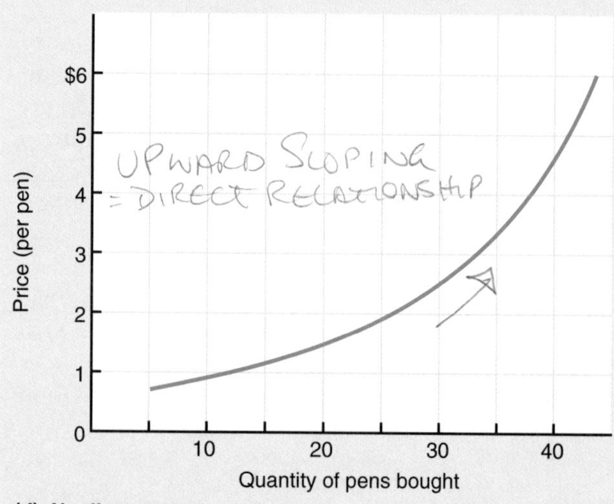

(d) Nonlinear curve

vertical axis. So when the number on the vertical axis falls by 4, the number on the horizontal axis increases by 1. That means the slope is −4 divided by 1, or −4.

Notice that the inverse relationships represented by the two downward-sloping curves, *a* and *b*, have negative slopes, and that the direct relationships represented by the two upward-sloping curves, *c* and *d*, have positive slopes. Notice also that the flatter the curve, the smaller the numerical value of the slope; and the more vertical, or steeper, the curve, the larger the numerical value of the slope. There are two extreme cases:

1. When the curve is horizontal (flat), the slope is zero.
2. When the curve is vertical (straight up and down), the slope is infinite (larger than large).

Knowing the term *slope* and how it's measured lets us describe verbally the pictures we see visually. For example, if we say a curve has a slope of zero, you should picture in your mind a flat line; if we say "a curve with a slope of minus one," you should picture a falling line that makes a 45° angle with the horizontal and vertical axes. (It's the hypotenuse of an isosceles right triangle with the axes as the other two sides.)

Figure A1-3 **SLOPES OF CURVES**

The slope of a curve is determined by rise over run. The slope of curve *a* is shown in the graph. The rest are shown below:

	Rise	÷	Run	=	Slope
b	−1		+2		−.5
c	1		1		1
d	4		1		4
e	1		1		1

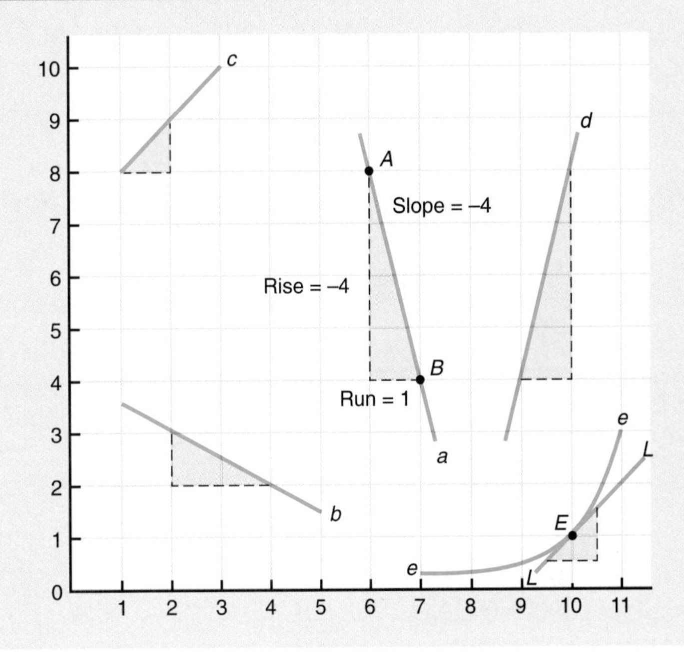

Slopes of Nonlinear Curves

The preceding examples were of *linear (straight) curves.* With *nonlinear curves*—the ones that really do curve—the slope of the curve is constantly changing. As a result, we must talk about the slope of the curve at a particular point, rather than the slope of the whole curve. How can a point have a slope? Well, it can't really, but it can almost, and if that's good enough for mathematicians, it's good enough for us.

Defining the slope of a nonlinear curve is a bit more difficult. The slope at a given point on a nonlinear curve is determined by the slope of a linear (or straight) line that's tangent to that curve. (A line that's tangent to a curve is a line that just touches the curve, and touches it only at one point in the immediate vicinity of the given point.) In Figure A1-3, the line *LL* is tangent to the curve *ee* at point *E*. The slope of that line, and hence the slope of the curve at the one point where the line touches the curve, is +1.

KNOWING THE TOOLS

Inverse and Direct Relationships

Inverse relationship:
When X goes up, Y goes down.
When X goes down, Y goes up.

Direct relationship:
When X goes up, Y goes up.
When X goes down, Y goes down.

X Y X Y

Maximum and Minimum Points

Two points on a nonlinear curve deserve special mention. These points are the ones for which the slope of the curve is zero. We demonstrate those in Figure A1-4(a) and (b). At point *A*, we're at the top of the curve so it's at a maximum point; at point *B*, we're at the bottom of the curve so it's at a minimum point. These maximum and minimum points are referred to often by economists, and it's important to realize that the value of the slope of the curve at each of these points is zero.

There are, of course, many other types of curves, and much more can be said about the curves we've talked about. We won't do so because, for purposes of this course, we won't need to get into those refinements. We've presented as much on graphing as you need to know for this book.

Figure A1-4 (a and b) A MAXIMUM AND A MINIMUM POINT

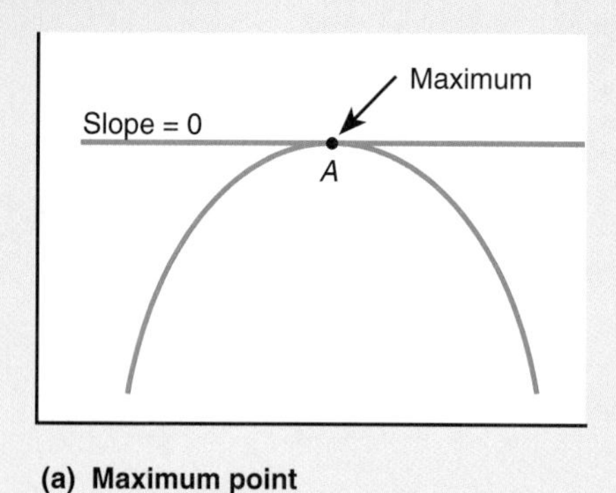

(a) Maximum point

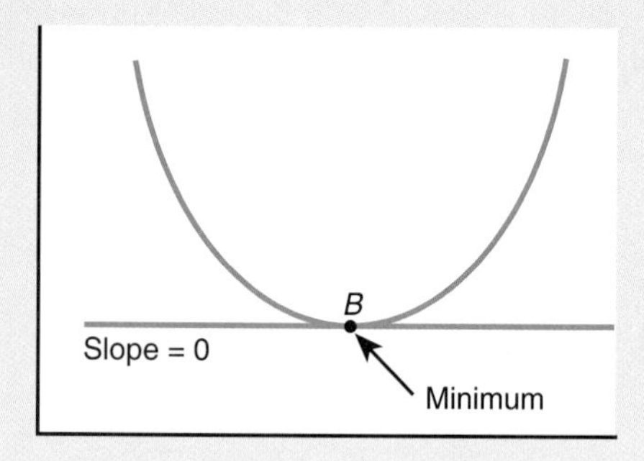

(b) Minimum point

PRESENTING REAL-WORLD DATA IN GRAPHS

The previous discussion covered terms that economists use in presenting models that focus on hypothetical relationships. Economists also use graphs in presenting actual economic data. Say, for example, that you want to show how exports have changed over time. Then you would place years on the horizontal axis (by convention) and exports on the vertical axis, as in Figure A1-5(a) and (b). Having done so, you have a couple of choices: you can draw a **line graph**—*a graph where the data are connected by a continuous line*; or you can make a **bar graph**—*a graph where the area under each point is filled in to look like a bar*. Figure A1-5(a) shows a line graph and Figure A1-5(b) shows a bar graph.

Another type of graph is a **pie chart**—*a circle divided into "pie pieces," where the undivided pie represents the total amount and the pie pieces reflect the percentage of the whole pie that the various components make up*. This type of graph is useful in visually presenting how a total amount is divided. Figure A1-5(c) shows a pie chart, which happens to represent the division of grades on a test we gave. Notice that 5 percent of the students got As.

There are other types of graphs, but they're all variations on line and bar graphs and pie charts. Once you understand these three basic types of graphs, you shouldn't have any trouble understanding the other types.

Figure A1-5 (a, b, and c) PRESENTING INFORMATION VISUALLY

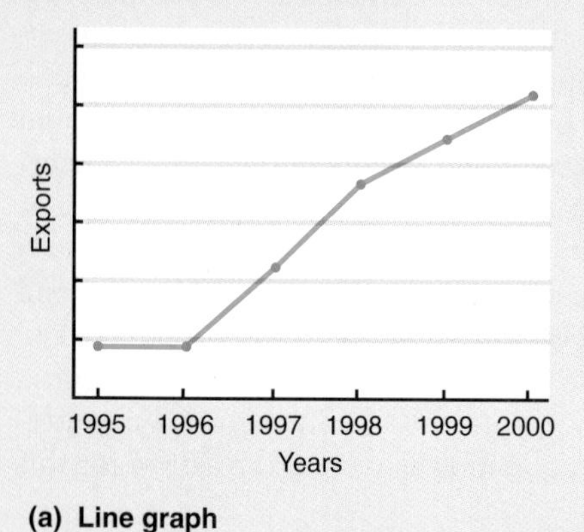

(a) Line graph

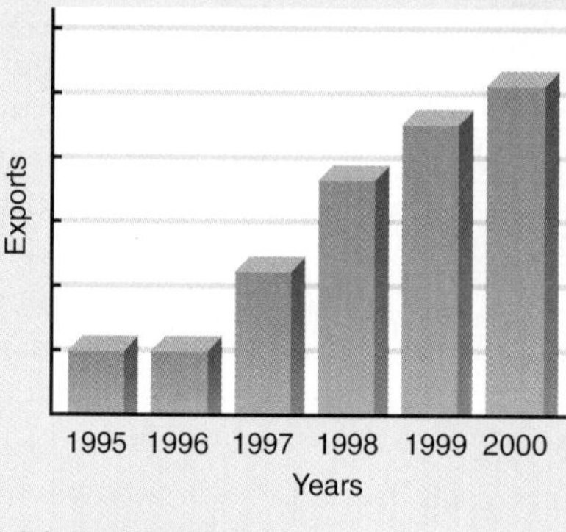

(b) Bar graph

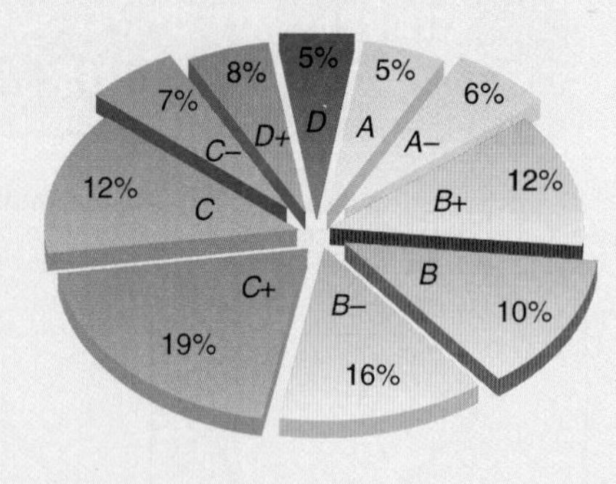

(c) Pie chart

Figure A1-6 (a and b) **THE IMPORTANCE OF SCALES**

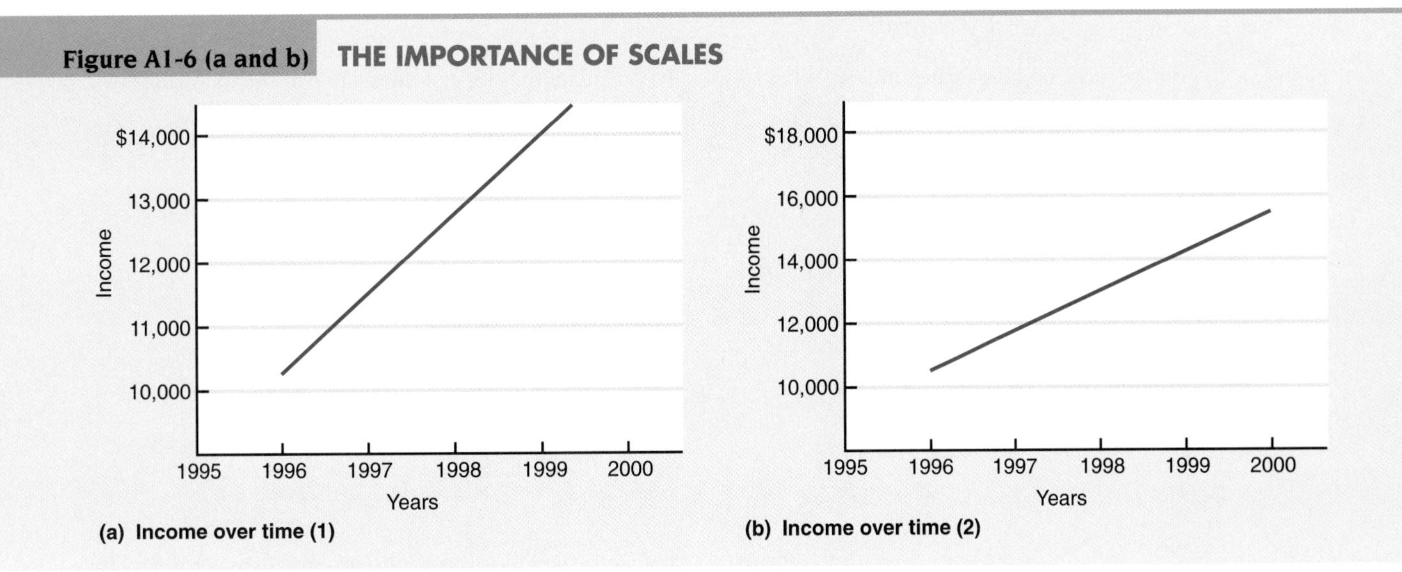

(a) Income over time (1)

(b) Income over time (2)

INTERPRETING GRAPHS ABOUT THE REAL WORLD

Understanding graphs is important because, if you don't, you can easily misinterpret the meaning of graphs. For example, consider the two graphs in Figure A1-6 (a) and (b). Which graph demonstrates the larger rise in income? If you said (a), you're wrong. The intervals in the vertical axes differ, and if you look carefully you'll see that the curves in both graphs represent the same combination of points. So when considering graphs, always make sure you understand the markings on the axes. Only then can you interpret the graph.

Let's now review what we've covered.

- A graph is a picture of points on a coordinate system in which the points denote relationships between numbers.
- A downward-sloping line represents an inverse relationship or a negative slope.
- An upward-sloping line represents a direct relationship or a positive slope.
- Slope is measured by rise over run, or a change of y (the number measured on the vertical axis) over a change in x (the number measured on the horizontal axis).
- The slope of a point on a nonlinear curve is measured by the rise over run of a line tangent to that point.
- At the maximum and minimum points of a nonlinear curve, the value of the slope is zero.
- In reading graphs, one must be careful to understand what's being measured on the vertical and horizontal axes.

Key Terms

bar graph *(24)*
coordinate system *(21)*
direct relationship *(21)*
graph *(20)*
interpolation assumption *(21)*
inverse relationship *(21)*
line graph *(24)*
linear curve *(21)*
nonlinear curve *(21)*
pie chart *(24)*
slope *(21)*

Questions for Thought and Review

1. Create a coordinate space on graph paper and label the following points:
   a. (0, 5)
   b. (−5, −5)
   c. (2, − 3)
   d. (−1, 1)
2. Graph the following costs per unit, and answer the questions that follow.

Horizontal Axis: Output	Vertical Axis: Cost per Unit
1	$30
2	20
3	12
4	6
5	2
6	6
7	12
8	20
9	30

   a. Is the relationship between cost per unit and output linear or nonlinear? Why?
   b. In what range in output is the relationship inverse? In what range in output is the relationship direct?
   c. In what range in output is the slope negative? In what range in output is the slope positive?
   d. What is the slope between 1 and 2 units?
3. Within a coordinate space, draw a line with:
   a. Zero slope.
   b. Infinite slope.
   c. Positive slope.
   d. Negative slope.
4. Calculate the slope of lines *a* to *e* in the following coordinate system.

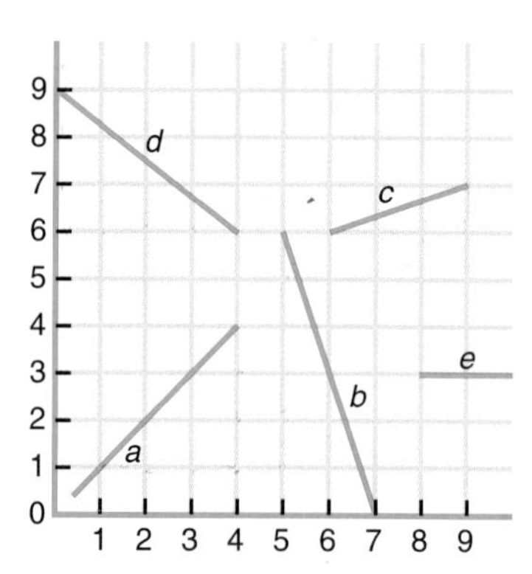

5. Given the following nonlinear curve, answer the following questions:

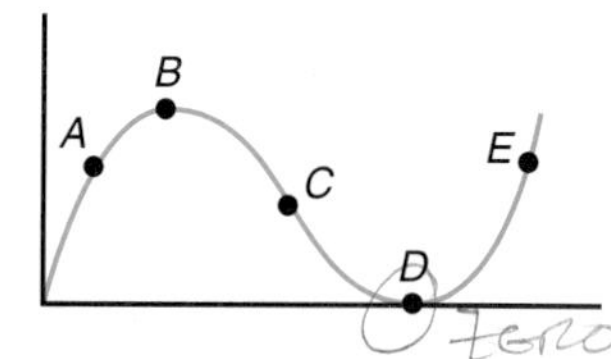

   a. At what point(s) is the slope negative?
   b. At what point(s) is the slope positive?
   c. At what point(s) is the slope zero?
   d. What point is the maximum? What point is the minimum?
6. State what type of graph or chart you would use to show the following real-world data:
   a. Interest rates from 1929 to 2000.
   b. Median income levels of various ethnic groups in Canada.
   c. Total federal expenditures by selected categories.
   d. Total costs of producing 100–800 shoes.

The Economic Organization of Society

2

After reading this chapter, you should be able to:

- Define capitalism and explain how it relies on markets to coordinate economic activities.
- Define socialism and explain how in practice it solves the three coordination problems.
- Describe how economic systems have evolved from the eighth century to today.
- Demonstrate opportunity cost with a production possibility curve.
- State the principle of increasing marginal opportunity cost.
- Explain why economic reasoning must be considered in context that requires a knowledge of history and institutions.
- Relate the concept of comparative advantage to the production possibility curve.

In capitalism man exploits man;
in socialism it's the other way round.

Abba Lerner

Every economy must solve three main coordination problems:

1. What, and how much, to produce.
2. How to produce it.
3. For whom to produce it.

Societies face a universal problem when trying to solve the three problems: Usually what individuals want to do isn't consistent with what "society" wants them to do. Society would often like people to consider what's good for society when making their individual decisions, and to agree that what society wants for them is what they want, too. For example, say society has garbage, and society determines that your neighbourhood is the best place to set up a garbage dump. Even if you agree a garbage dump is needed, you probably won't want it in your neighbourhood. This attitude of **NIMBY** (*Not In My Back Yard*)—*a mindset of approving a project but not wanting it to be nearby*—is found throughout our society.

Individual goals and social goals also conflict when decisions are being made about how much to produce and consume. Individuals generally like to consume much more than they like to produce. So a society must provide incentives for its members to produce more and consume less to alleviate that scarcity. A sure sign that an economic system isn't working is when people perceive that there are important things that need to be done, but are sitting around doing nothing because the system doesn't provide them with the incentive to do them.

The coordination problems faced by society are immense.

How hard is it to make the three decisions we've listed? Imagine for a moment the problem of living in a family: the fights, arguments, and questions that come up. "Do I have to do the dishes?" "Why can't I have piano lessons?" "Bobby got a new sweater. How come I didn't?" "Mom likes you best." Now multiply the size of the family by millions. The same fights, the same arguments, the same questions—only for society the problems are millions of times more complicated than for one family.

How are these complicated coordination problems solved? The two main economic systems the world has used in the past 50 years—capitalism and socialism—answer this question differently.

CAPITALISM

Capitalism is an economic system based on private property and the market. It gives private property rights to individuals, and relies on market forces to coordinate economic activity.

Capitalism is *an economic system based on private property and the market in which, in principle, individuals decide how, what, and for whom to produce*. Under capitalism, individuals are encouraged to follow their own self-interest, while market forces of supply and demand are relied on to coordinate those individual pursuits. Distribution of goods is to each individual according to his or her ability, effort, and inherited property.

Reliance on market forces doesn't mean that political, social, and historical forces play no role in coordinating economic decisions. These other forces do influence how the market works. For example, for a market to exist, government must allocate and defend **private property rights**—*the control a private individual or firm has over an asset or a right*. The concept of private ownership must exist and must be accepted by individuals in society. When you say, "This car is mine," it means that it is unlawful for someone else to take it without your permission. If someone takes it without your permission, he or she is subject to punishment through the legal system.

Reliance on the Market

Q-1 John, your study partner, is telling you that the best way to allocate property rights is through the market. How do you respond?

Markets work through a system of rewards and payments. If you do something, you get paid for doing that something; if you take something, you pay for that something. How much you get is determined by how much you give. This relationship seems fair to most people. But there are instances when it doesn't seem fair. Say someone is unable to work. Should that person get nothing? How about Joe down the street, who was given $10 million by his parents? Is it fair that he gets lots of toys, like Corvettes and skiing trips to Whistler, and doesn't have to work, while the rest of us have to work 40 hours a week and maybe go to school at night?

We'll put those questions about fairness off at this point—they are very difficult questions. For now, all we want to present is the underlying concept of fairness that capitalism embodies: "Them that works, gets; them that don't, starves."[1] In capitalism, individuals are encouraged to follow their own self-interest.

In capitalist economies, individuals are free to do whatever they want as long as it's legal. The market is relied on to see that what people want to get, and want to do, is consistent with what's available. Price is the mechanism through which people's desires are coordinated and goods are rationed. If there's not enough of something to go around, its price goes up; if more of something needs to get done, the price given to individuals willing to do it goes up. If something isn't wanted or doesn't need to be done, its price goes down. Under capitalism, fluctuations in prices play a central role in coordinating individuals' wants.

Under capitalism, fluctuations in prices play a central role in coordinating individuals' wants.

What's Good about the Market?

see page 46

Is the market a good way to coordinate individuals' activities? Much of this book will be devoted to answering that question. The answer that we, and most Canadian economists, come to is: Yes, it is a reasonable way. True, it has problems; the market can be unfair, mean, and arbitrary, and sometimes it is downright awful. Why then do economists support it? For the same reason that Oliver Wendell Holmes supported democracy—it is a lousy system, but, based on experience with alternatives, it is better than all the others we've thought of.

The primary debate among economists is not about using markets; it is about how markets should be structured, and whether they should be modified and adjusted by government regulation. Those are much harder questions, and on these questions, opinions differ enormously.

The primary debate among economists is not about using markets but about how markets are structured.

SOCIALISM

The view that markets are a reasonable way to organize society has not always been shared by all economists. Throughout history strong arguments have been made against markets. These arguments are both philosophical and practical. The philosophical argument against the market is that it brings out the worst in people—it glorifies greed. It encourages people to beat out others rather than to be cooperative. As an alternative some economists have supported **socialism**—which is, in theory, *an economic system based on individuals' goodwill toward others, not on their own self-interest, and in which, in principle, society decides what, how, and for whom to produce.*

Socialism in Theory

You can best understand the idea behind theoretical socialism by thinking about how decisions are made in a family. In most families, benevolent parents decide who gets

[1]How come the authors get to use rotten grammar but scream when they see rotten grammar in your papers? Well, that's fairness for you. Actually, we should say a bit more about writing style. All writers are expected to know correct grammar; if they don't, they don't deserve to be called writers. Once you know grammar, you can individualize your writing style, breaking the rules of grammar where the meter and flow of the writing require it. Right now, you're still proving that you know grammar, so in papers handed in to your instructor, you shouldn't break the rules of grammar until you've proved to the instructor that you know them. We've done lots of books, so our editors give us a bit more leeway than your instructors will give you.

Q-2 Are there any activities in a family that you believe should be allocated by a market? What characteristics do those activities have?

what, based on the needs of each member of the family. When Sabin gets a new coat and his sister Sally doesn't, it's because Sabin needs a coat while Sally already has two coats that fit her and are in good condition. Victor may be slow as molasses, but from his family he still gets as much as his superefficient brother Jerry gets. In fact, Victor may get more than Jerry because he needs extra help.

Markets have little role in most families. In our families, when food is placed on the table we don't bid on what we want, with the highest bidder getting the food. In our families, every person can eat all he or she wants, although if one child eats more than a fair share, that child gets a lecture from us on the importance of sharing. "Be thoughtful; be considerate. Think of others first," are lessons that many families try to teach.

Socialism is, in theory, an economic system that tries to organize society in the same way as most families are organized—all people contribute what they can, and all get what they need.

In theory, socialism is an economic system that tries to organize society in the same way as these families are organized, trying to see that individuals get what they need. Socialism tries to take other people's needs into account and adjust people's own wants in accordance with what's available. In socialist economies, individuals are urged to look out for the other person; if individuals' inherent goodness won't make them consider the general good, government will make them. In contrast, a capitalist economy expects people to be selfish; it relies on markets and competition to direct that selfishness to the general good.[2]

Socialism in Practice

Q-3 Which would be more likely to attempt to foster individualism: socialism or capitalism?

Soviet-style socialism is an economic system that uses administrative control or central planning to solve the coordination problems: what, how, and for whom.

Q-4 What is the difference between socialism in theory and socialism in practice?

Few economists argue directly in favour of greed. Most accept that it would be great if everyone wanted to be good to others. However, they point out that in practice, economic systems based upon people's goodwill have tended to break down. This is certainly true of the major countries that tried socialism starting in the 1900s. In practice, socialist governments had to take a strong role in guiding the economy. Socialism became an economic system based on government ownership of the means of production, with economic activity governed by central planning. What we are describing as "socialism in practice" is often called **Soviet-style socialism**—*an economic system that uses administrative control or central planning to solve the coordination problems: what, how, and for whom*—because it was the system used by the Soviet Union. Under that Soviet-style socialist economic system, government planning boards set society's goals and then directed individuals and firms as to how to achieve those goals.

THE PRODUCTION POSSIBILITY CURVE AND ECONOMIC REASONING

The choices that a society must make are often presented in terms of a production possibility curve. The production possibility curve is related to the concept of opportunity cost that you were introduced to in Chapter 1. It is a tool that shows the trade-offs among choices we make. It can be used nicely to discuss choices societies must make about economic systems. Applying economic reasoning outside of any historical and institutional context, however, has difficulties and we'll discuss these difficulties too.

[2]As you probably surmised, the above distinction is too sharp. Even capitalist societies want people to be selfless, but not too selfless. Children in capitalist societies are generally taught to be selfless at least in dealing with friends and family. The difficulties parents and societies face is finding a midpoint between the two positions: selfless but not too selfless; selfish but not too selfish.

APPLYING THE TOOLS

Tradition and Today's Economy

In a tradition-based society, the social and cultural forces give a society inertia (a tendency to resist change) that predominates over economic and political forces.

"Why did you do it that way?"

"Because that's the way we've always done it."

Tradition-based societies had markets, but they were peripheral, not central, to economic life. In feudal times what was produced, how it was produced, and for whom it was produced were primarily decided by tradition.

In today's Canadian economy, the market plays the central role in economic decisions. But that doesn't mean that tradition is dead. As we said in Chapter 1, tradition still plays a significant role in today's society, and, in many aspects of society, tradition still overwhelms the invisible hand. Consider the following:

1. The persistent view that women should be homemakers rather than factory workers, consumers rather than producers.
2. The raised eyebrows when a man is introduced as a nurse, secretary, homemaker, or member of any other profession conventionally identified as women's work.
3. Society's unwillingness to permit the sale of individuals or body organs.
4. Parents' willingness to care for their children without financial compensation.

Each of these tendencies reflects tradition's influence in Western society. Some are so deep-rooted that we see them as self-evident. Some of tradition's effects we like; others we don't—but we often take them for granted. Economic forces may work against these traditions, but the fact that they're still around indicates the continued strength of tradition in our market economy.

The Production Possibility Table

As we discussed in Chapter 1, the concept of opportunity cost—every decision has a cost in forgone opportunities—lies at the centre of economic reasoning. Opportunity cost can be seen numerically with a **production possibility table**—*a table that lists the maximum combination of outputs that can be obtained from a given number of inputs*. An **output** is simply *a result of an activity*, and an **input** is *what you put into a production process to achieve an output*. For example, your grade in a course is an output and your study time is an input.

A production possibility table lists the maximum combination of outputs that can be obtained from a given number of inputs.

Let's present the study-time/grades example numerically. Say you have exactly 20 hours a week to devote to two courses: economics and history. (So, maybe we're a bit optimistic.) Grades are given numerically and you know that the following relationships exist: if you study 20 hours in economics, you'll get a grade of 100; 18 hours, 94; and so forth.[3]

Let's say that the best you can do in history is a 98 with 20 hours of study a week; 19 hours of study guarantees a 96, and so on. The production possibility table in Figure 2-1(a) shows the highest combination of grades you can get with various allocations of the 20 hours available for studying the two subjects. One possibility is getting 100 in economics and 58 in history. Another is getting 70 in economics and 78 in history.

Notice that the opportunity cost of studying one subject rather than the other is embodied in the production possibility table. The information in the table comes from experience: we are assuming that you've discovered that if you transfer an hour of study

Markets can be as sophisticated as the Toronto Stock Exchange or as informal as a yard sale.

[3]Throughout the book we'll be presenting numerical examples to help you understand the concepts. The numbers we choose are often arbitrary. After all, you have to choose something. As an exercise, you might choose different numbers than we did, numbers that apply to your own life, and work out the argument using those numbers.

BEYOND THE TOOLS

The Rise of Markets in Perspective

Back in the Middle Ages, markets developed spontaneously. "You have something I want; I have something you want. Let's trade" is a basic human attitude we see in all aspects of life. Even children quickly get into trading: chocolate ice cream for vanilla, two Pokémon cards for a ride on a motor scooter. Markets institutionalize such trading by providing a place where people know they can go to trade. New markets are continually being formed. Today there are markets for hockey cards, pork bellies (which become bacon and pork chops), rare coins, and so on. The Internet, with sites like eBay, is expanding markets enormously, allowing ordinary people to trade with people thousands of miles away.

Throughout history, societies have tried to prevent some markets from operating because they feel those markets are ethically wrong or have undesirable side effects. Societies have the power to prevent markets, to make some kinds of markets illegal. In Canada, the addictive drug market, the baby market, and the sex market, to name a few, are illegal. In socialist countries, markets in a much wider range of goods (such as clothes, cars, and soft drinks) and activities (such as private business for individual profit) have been illegal.

But, even if a society prevents the market from operating, society cannot escape the invisible hand. If there's excess supply, there will be downward pressure on prices; if there's excess demand, there will be upward pressure on prices. To maintain an equilibrium in which the quantity supplied does not equal the quantity demanded, a society needs a strong force to prevent the invisible hand from working. In the Middle Ages, that strong force was religion. The Church told people that if they got too far into the market mentality—if they followed their self-interest—they'd go to Hell.

Until recently, in socialist society the state provided the preventive force. The educational system in socialist countries emphasized a communal set of values. They taught students that a member of socialist society does not try to take advantage of other human beings but, rather, lives by the philosophy "From each according to his ability; to each according to his need."

For whatever reason—whether it be that true socialism wasn't really tried, or that people's self-interest is too strong—the "from each according to his ability; to each according to his need" approach didn't work in socialist countries. They have switched (some say succumbed) to greater reliance on the market.

from economics to history, you'll lose 3 points on your grade in economics and gain 2 points in history. Thus, the opportunity cost of a 2-point rise in your history grade is a 3-point decrease in your economics grade.

The Production Possibility Curve

The production possibility curve is a curve measuring the maximum combination of outputs that can be obtained from a given number of inputs.

The information in the production possibility table can also be presented graphically in a diagram called a production possibility curve. **A production possibility curve** is *a curve measuring the maximum combination of outputs that can be obtained from a given number of inputs*. It is a graphical presentation of the opportunity cost concept.

A production possibility curve is created from a production possibility table by mapping the table in a two-dimensional graph. We've taken the information from the table in Figure 2-1(a) and mapped it into Figure 2-1(b). The history grade is mapped, or plotted, on the horizontal axis; the economics grade is on the vertical axis.

As you can see from the bottom row of Figure 2-1(a), if you study economics for all 20 hours and study history for 0 hours, you'll get grades of 100 in economics and 58 in history. Point A in Figure 2-1(b) represents that choice. If you study history for all 20 hours and study economics for 0 hours, you'll get a 98 in history and a 40 in economics. Point *E* represents that choice. Points *B*, *C*, and *D* represent three possible choices between these two extremes.

Notice that the production possibility curve slopes downward from left to right. That means that there is an inverse relationship (a trade-off) between grades in

Figure 2-1 (a and b) **A PRODUCTION POSSIBILITY TABLE AND CURVE FOR GRADES IN ECONOMICS AND HISTORY**

The production possibility table (a) shows the highest combination of grades you can get with only 20 hours available for studying economics and history.

The information in the production possibility table in (a) can be plotted on a graph, as is done in (b). The grade received in economics is on the vertical axis, and the grade received in history is on the horizontal axis.

Hours of study in history	Grade in history	Hours of study in economics	Grade in economics
20	98	0	40
19	96	1	43
18	94	2	46
17	92	3	49
16	90	4	52
15	88	5	55
14	86	6	58
13	84	7	61
12	82	8	64
11	80	9	67
10	78	10	70
9	76	11	73
8	74	12	76
7	72	13	79
6	70	14	82
5	68	15	85
4	66	16	88
3	64	17	91
2	62	18	94
1	60	19	97
0	58	20	100

(a) Production possibility table

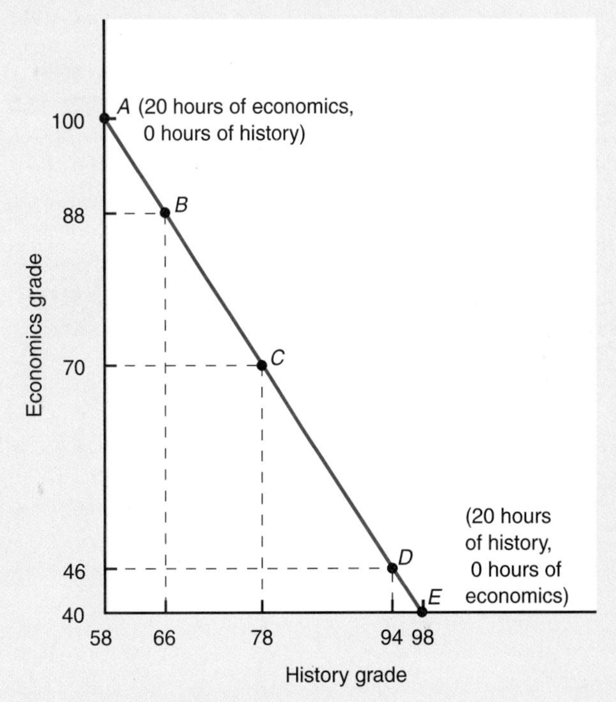

(b) Production possibility curve

economics and grades in history. The better the grade in economics, the worse the grade in history, and vice versa. That downward slope represents the opportunity cost concept—you get more of one benefit only if you get less of another benefit.

The production possibility curve not only demonstrates the opportunity cost concept, it also measures the opportunity cost. For example, in Figure 2-1(b), say you want to raise your grade in history from a 94 to a 98 (move from point *D* to point *E*). The opportunity cost of that 4-point increase would be a 6-point decrease in your economics grade, from 46 to 40.

To summarize, the production possibility curve demonstrates that:

1. There is a limit to what you can achieve, given the existing institutions, resources, and technology.
2. Every choice you make has an opportunity cost. You can get more of something only by giving up something else.

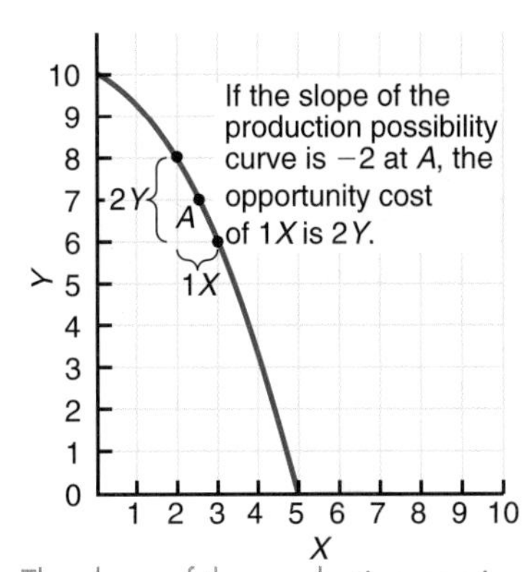

The slope of the production possibility curve tells you the opportunity cost of good *X* in terms of good *Y*. You have to give up 2*Y* to get 1*X* when you're around point *A*.

Q-5 If a production possibility curve is a straight line, is opportunity cost constant?

Increasing Marginal Opportunity Cost We chose an unchanging trade-off in the study-time/grade example because it made the initial presentation of the production possibility curve easier. Since, by assumption, you could always trade two points on your history grade for three points on your economics grade, the production possibility curve was a straight line and opportunity cost was constant. But is that the way we'd expect

Figure 2-2 (a and b) **A PRODUCTION POSSIBILITY TABLE AND CURVE**

The table in (**a**) contains information on the trade-off between the production of burgers and DVDs. This information has been plotted on the graph in (**b**). Notice in (**b**) that as we move along the production possibility curve from *A* to *F*, trading DVDs for burgers, we get fewer and fewer burgers for each pound of DVDs given up. That is, the opportunity cost of choosing burgers over DVDs increases as we increase the production of burgers. This concept is called the principle of increasing marginal opportunity cost. The phenomenon occurs because some resources are better suited for the production of DVDs than for the production of burgers, and we use the better ones first.

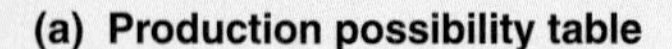

% of resources devoted to production of burgers	Number of burgers	% of resources devoted to production of DVDs	DVDs	Row
0	0	100	15	*A*
20	4	80	14	*B*
40	7	60	12	*C*
60	9	40	9	*D*
80	11	20	5	*E*
100	12	0	0	*F*

(a) Production possibility table

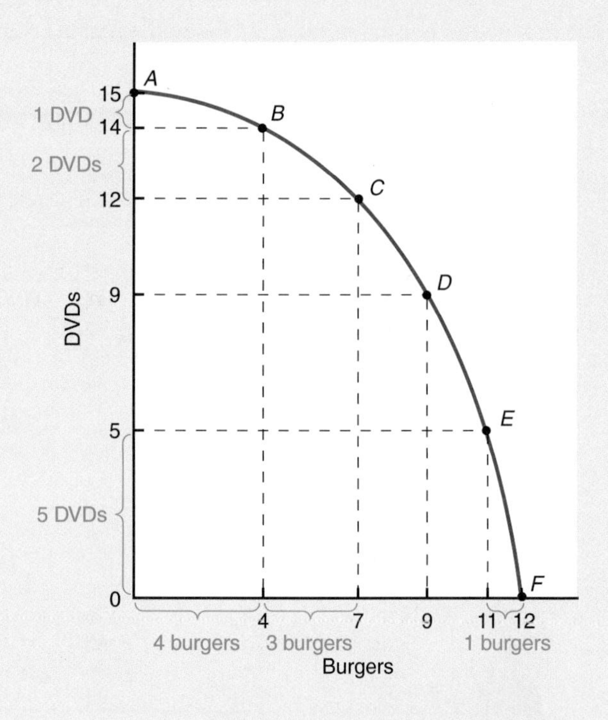

(b) Production possibility curve

reality to be? Probably not. The production possibility curve is generally bowed outward, as in Figure 2-2(b).

Why? Because some resources are better suited for the production of certain kinds of goods than they are for the production of other kinds of goods. To make the answer more concrete, let's talk specifically about society's choice between burgers and DVDs. The graph in Figure 2-2(b) is derived from the table in Figure 2-2(a).

Let's see what the shape of the curve means in terms of numbers. Let's start with society producing only DVDs (point *A*). Giving up a DVD initially gains us a lot of burgers (4), moving us to point *B*. The next 2 DVDs we give up gains us slightly fewer burgers (point *C*). If we continue to trade DVDs for burgers, we find that at point *D* we gain very few burgers from giving up a DVD. The opportunity cost of choosing burgers over DVDs increases as we increase the production of burgers.

Q-6 If no resource had a comparative advantage in the production of any good, what would the shape of the production possibility curve be? Why?

The reason the opportunity cost of burgers increases as we consume more burgers is that some resources are relatively better suited to producing burgers, while others are relatively better suited to producing DVDs. Put in economists' terminology, some resources have a **comparative advantage** over other resources—*the ability to be better suited to the production of one good than to the production of another good*. In this example, some resources have a comparative advantage over other resources in the production of DVDs, while other resources have a comparative advantage in the production of burgers.

When making small amounts of burgers and large amounts of DVDs, in the production of those burgers we use the resources whose comparative advantage is in the

KNOWING THE TOOLS

Production Possibility Curves

Definition	Shape	Shifts	Points In, Out, and On
The production possibility curve is a curve that measures the maximum combination of outputs that can be obtained with a given number of inputs	Most are outward bowed because of increasing marginal opportunity cost; if opportunity cost doesn't change, the production possibility curve is a straight line	Increases in inputs or increases in the productivity of inputs shift the production possibility curve out; decreases have the opposite effect; the production possibility curve shifts along the axis whose input is changing	Points inside the production possibility curve are points of inefficiency; points on the production possibility curve are points of efficiency; points outside the production possibility curve are not obtainable

production of burgers. All other resources are devoted to producing DVDs. Because the resources used in producing burgers aren't good at producing DVDs, we're not giving up many DVDs to get those burgers. As we produce more and more of a good, we must use resources whose comparative advantage is in the production of the other good—in this case, more suitable for producing DVDs than for producing burgers. As we remove resources from the production of DVDs to get the same additional amount of burgers, we must give up increasing numbers of DVDs. An alternative way of saying this is that the opportunity cost of producing burgers becomes greater as the production of burgers increases. As we continue to increase the production of burgers, the opportunity cost of more burgers becomes very high because we're using resources to produce burgers that have a strong comparative advantage for producing DVDs.

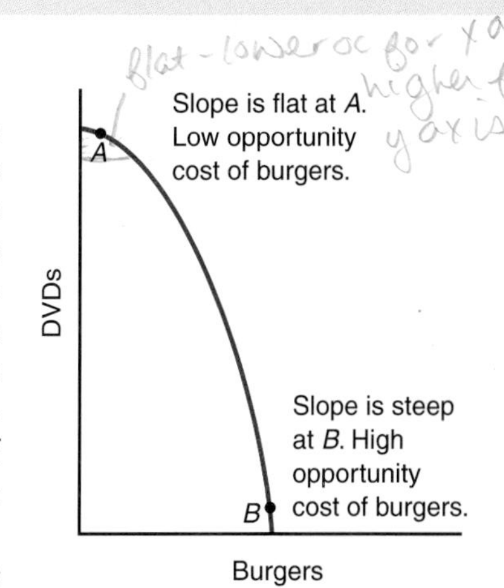

Remember: when the slope is flat, there's a low opportunity cost of burgers (a high opportunity cost of DVDs). When the slope is steep, there's a high opportunity cost of burgers (a low opportunity cost of DVDs).

For many of the choices society must make, opportunity costs tend to increase as we choose more and more of an item. The reason is that resources are not easily adaptable from the production of one good to the production of another. Such a phenomenon about choice is so common, in fact, that it has acquired a name: the **principle of increasing marginal opportunity cost.** That principle states:

> *In order to get more of something, one must give up ever-increasing quantities of something else.*

In other words, initially the opportunity costs of an activity are low, but they increase the more we concentrate on that activity. Sometimes this law is called the flowerpot law because, if it didn't hold, all the world's food could be grown in a flowerpot. But it can't be. As we add more seeds to a fixed amount of soil, there won't be enough nutrients or room for the roots, so output per seed decreases.

The principle of increasing marginal opportunity cost states that opportunity costs increase the more you concentrate on the activity. In order to get more of something, one must give up ever-increasing quantities of something else.

Efficiency We would like, if possible, to get as much output as possible from a given amount of inputs or resources. That's **productive efficiency**—*achieving as much output as possible from a given amount of inputs or resources*. We would like to be efficient. The production possibility curve helps us see what is meant by productive efficiency. Consider point A in Figure 2-3(a), which is inside the production possibility curve. If we are producing at point A, we are using all our resources to produce 6 burgers and 4 DVDs.

Figure 2-3 (a, b, and c) EFFICIENCY, INEFFICIENCY, AND TECHNOLOGICAL CHANGE

The production possibility curve helps us see what is meant by efficiency. At point A, in (**a**), all inputs are used to make 4 DVDs and 6 burgers. This is inefficient since there is a way to obtain more of one without giving up any of the other, that is, to obtain 6 DVDs and 6 burgers (point C) or 8 burgers and 4 DVDs (point *B*). All points inside the production possibility curve are inefficient. With fixed inputs and given technology, we cannot go beyond the production possibility curve. For example, point *D* is unattainable.

A technological change that improves production techniques will shift the production possibility curve outward, as shown in both (**b**) and (**c**). How the curve shifts outward depends on how technology improves. For example, if we become more efficient in the production of both burgers and DVDs, the curve will shift out as in (**b**). If we become more efficient in producing DVDs, but not in producing burgers, then the curve will shift as in (**c**).

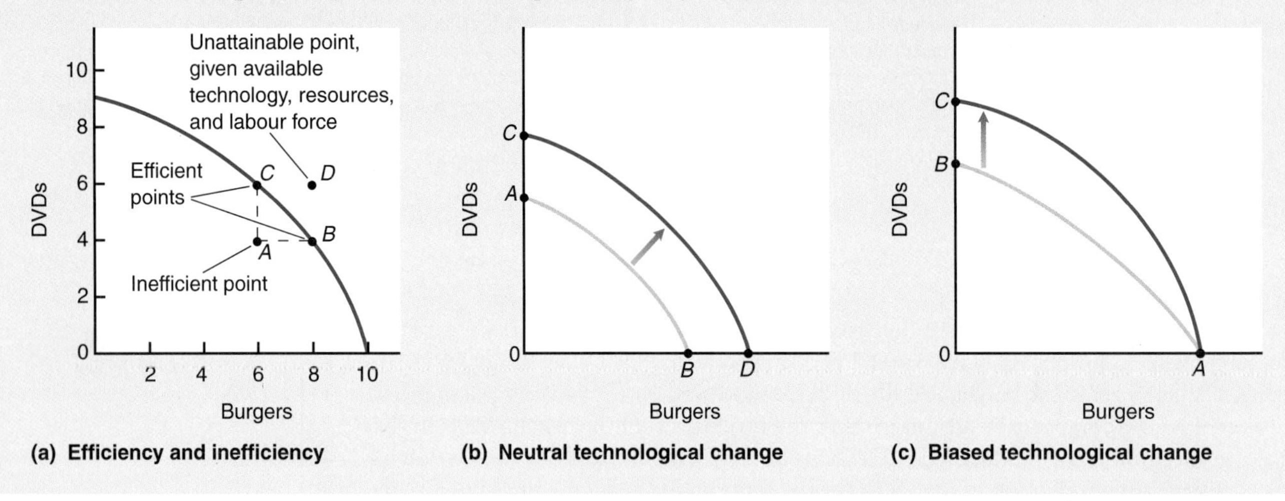

(a) Efficiency and inefficiency **(b) Neutral technological change** **(c) Biased technological change**

Efficiency involves achieving a goal as cheaply as possible. Efficiency has meaning only in relation to a specified goal.

Point A represents **inefficiency**—*getting less output from inputs which, if devoted to some other activity, would produce more output.* That's because with the same inputs we could be getting either 8 burgers and 4 DVDs (point *B*) or 6 DVDs and 6 burgers (point C). As long as we prefer more to less, both points *B* and C represent **efficiency**—*achieving a goal using as few inputs as possible*. We always want to move our production out to a point on the production possibility curve.

Why not move out farther, to point *D*? If we could, we would, but by definition the production possibility curve represents the most output we can get from a certain combination of inputs. So point *D* is unattainable, given our resources and technology.

When technology improves, when more resources are discovered, or when the economic institutions get better at fulfilling our wants, we can get more output with the same inputs. What this means is that when technology or an economic institution improves, the entire production possibility curve shifts outward from *AB* to *CD* in Figure 2-3(b). How the production possibility curve shifts outward depends on how the technology improves. For example, say we become more efficient in producing DVDs, but not more efficient in producing burgers. Then the production possibility curve shifts outward to *AC* in Figure 2-3(c).

Policies that costlessly shift the production possibility curve outward are the most desirable policies because they don't require us to decrease our consumption of one good to get more of another. Alas, such policies are the most infrequent. Improving technology and institutions and discovering more resources are not costless; generally there's an opportunity cost of doing so that must be taken into account.

Figure 2-4 EXAMPLES OF SHIFTS IN PRODUCTION POSSIBILITY CURVES

Each of these curves reflects a different type of shift. Your assignment is to match these shifts with the situations given in the text on pages 37 and 38.

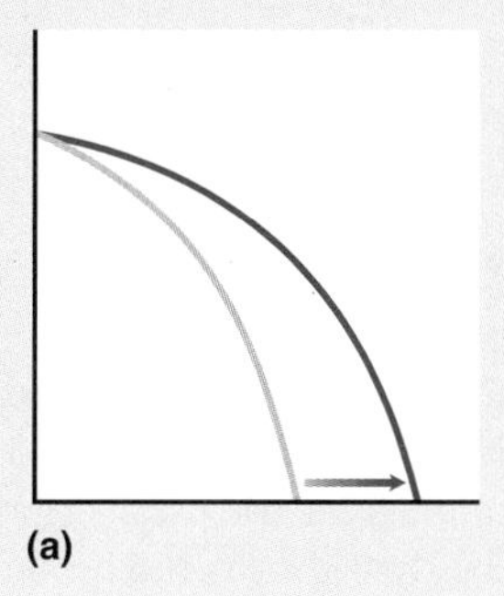
(a)

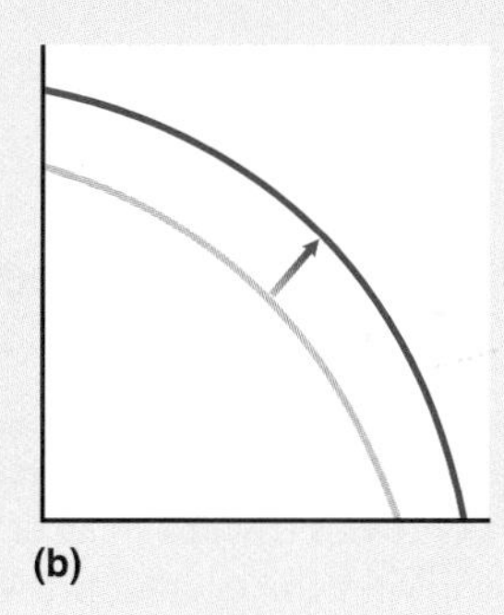
(b)

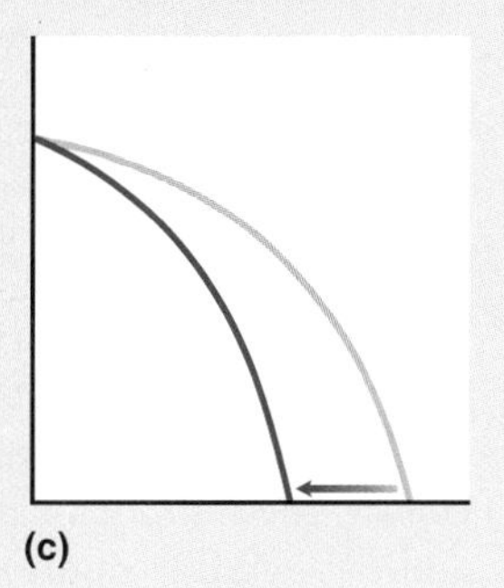
(c)

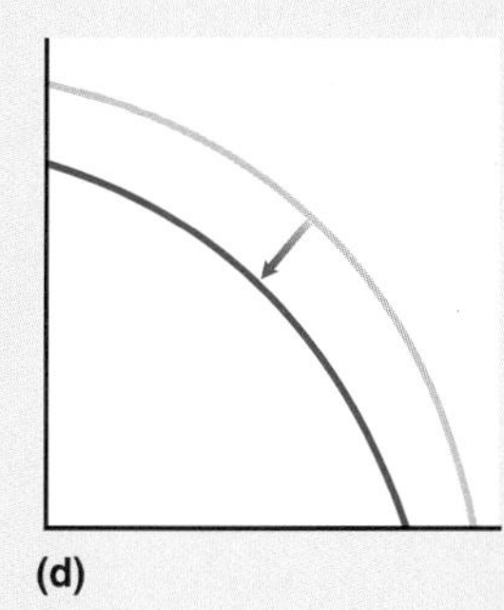
(d)

Distribution and Productive Efficiency In discussing the production possibility curve, we avoided questions of distribution: Who gets what? But such questions cannot be ignored in real-world situations. Specifically, if the method of production is tied to a particular income distribution and choosing one method will help some people but hurt others, we can't say that one method of production is efficient and the other inefficient, even if one method produces more total output than the other. As we stated above, the term *efficiency* involves achieving a goal as cheaply as possible. The term has meaning only in regard to a specified goal. Say, for example, that we have a society of ascetics who believe that consumption above some minimum is immoral. For such a society, producing more for less (productive efficiency) would not be efficient since consumption is not its goal. Or say that we have a society that cares that what is produced is fairly distributed. An increase in output that goes to only one person and not to anyone else would not necessarily be efficient.

Q-7 Your firm is establishing a trucking business in Saudi Arabia. The firm has noticed that women are generally paid much less than men in Saudi Arabia, and the firm suggests that hiring women would be more efficient than hiring men. What should you respond?

In our society, however, most people prefer more to less, and many policies have relatively small distributional consequences. On the basis of the assumption that more is better than less, economists use their own kind of shorthand for such policies and talk about efficiency as identical to productive efficiency—increasing total output. But it's important to remember the assumption under which that shorthand is used: that the distributional effects that accompany the policy aren't undesirable and that we, as a society, prefer more output.

Some Examples of Shifts in the Production Possibility Curve

To see whether you understand the production possibility curve, let us now consider some situations that can be shown with the production possibility curve. In Figure 2-4 we demonstrate four situations with production possibility curves. Below, we list five situations. To test your understanding of the curve, match each situation to one of the curves in Figure 2-4 (a graph can match more than one situation).

1. A new genetic material is found that doubles the speed at which agricultural goods grow.
2. Nanites (micromachines) are perfected that lower the cost of manufactured goods.
3. A meteor hits the earth and destroys half the world's natural resources.

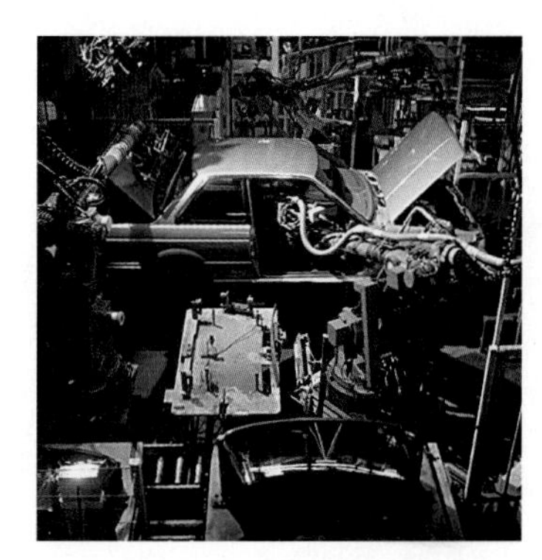

Expanding technology and capacities to produce shifts out the production possibilities curve.

Q-8 When a natural disaster hits the prairies, where most of Canada's wheat is produced, what happens to the Canadian production possibility curve for wheat and butter?

4. A world trade war erupts and trade restrictions increase enormously.
5. Soviet-style socialist countries give up on the socialist system and switch to a market system.

The correct answers are: 1–a; 2–a; 3–d; 4–d; 5–d.

If you got them all right, you are well on your way to understanding the production possibility curve.

The Production Possibility Curve and Economic Systems

Some of you may have rightly wondered about one of the answers in the above examples, specifically, the last one. (If you wondered about the others, a review is in order for you.) The appropriate wondering is the following: According to what we have said previously, the shift by socialist countries toward markets should shift the production possibility curve out, not in; so wouldn't the correct answer be that it shifts the production possibility curve out because it introduces markets that allow trade?

PPC + CHOICES

The answer to that question is: Yes, it should *eventually* shift the production possibility curve out. But in the short and medium run (i.e., within 5 to 10 years) the change will shift it in. The explanation of why this is so brings us back to our discussion of economic systems and allows us to tie together that discussion with the discussion of opportunity costs as represented by the production possibility curve.

The production possibility curve presents choices in a timeless fashion, but most choices are dependent upon previous choices.

The production possibility curve presents choices in a timeless fashion and therefore makes opportunity costs clear-cut; there are two choices, one with a higher cost and one with a lower cost. The reality is that most choices are dependent on other choices; they are made sequentially with a time dimension. With sequential choices you cannot simply reverse your decision. Once you have started on a path, to take another path you have to return to the beginning. Thus, following one path often lowers the costs of options along that path, but it raises the costs of options along another path.

Q-9 Draw a decision tree that shows the choices you have faced when pursuing your undergraduate studies. Which choices were most costly?

Such sequential decisions can best be seen within the framework of a **decision tree**—*a visual description of sequential choices*. A decision tree is shown in Figure 2-5.

Once you make the initial decision to go on path A, the costs of path B options become higher; they include the costs of reversing your path and starting over. The decision trees of life have thousands of branches; each decision you make rules out other paths, or at least increases your costs highly. (Remember that day you decided to blow off your homework? That decision may have changed your future life.)

Another way of putting this same point is that all *decisions are made in context:* What makes sense in one context may not make sense in another. For example, say you're answering the question "Would society be better off if students were taught literature or if

Figure 2-5 A DECISION TREE

Decisions are often made sequentially. Decisions made at low levels of the decision tree preclude, or at least significantly increase the costs of, other decisions.

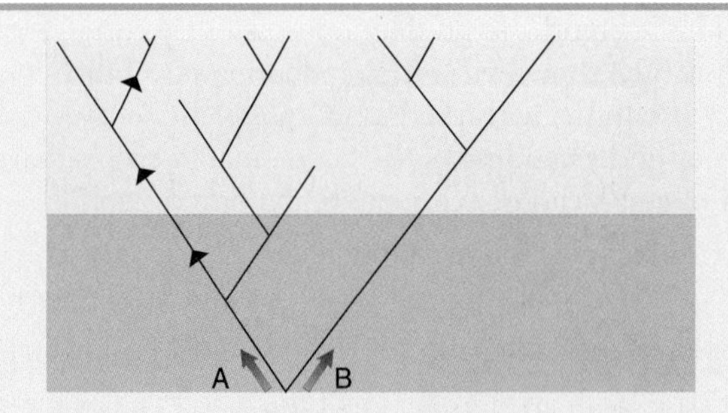

they were taught agriculture?" The answer depends on the institutional context. In a developing country whose goal is large increases in material output, teaching agriculture may make sense. In a developed country, where growth in material output is less important, teaching literature may make sense.

Recognizing the contextual nature of decisions is important when interpreting the production possibility curve. Because decisions are contextual, what the production possibility curve for a particular decision looks like depends on the existing institutions, and the analysis can be applied only in institutional and historical context. The production possibility curve is not a purely technical phenomenon. The curve is an engine of analysis to make contextual choices, not a definitive tool to decide what one should do in all cases.

Because decisions are contextual, what the production possibility curve for a particular decision looks like depends on the existing institutions, and the analysis can be applied only in an institutional and historical context.

From the above discussion, it should be clear where economic systems fit into this discussion. The choice of economic systems is not the type of decision that the production possibility curve is designed to address. Usually, the curve takes the system as given. The production possibility curve is most useful when analyzing questions that involve policy decisions given existing systems and institutions—slight movements in the production of one good or another.

The Production Possibility Curve and Tough Choices

The production possibility curve represents the tough choices society must make. Look at questions such as: Should we save the spotted owl or should we allow logging in the Western forests? Should we expand the government health care system or should we strengthen our national defense system? Should we emphasize policies that allow more consumption now or should we emphasize policies that allow more consumption in the future? Such choices involve difficult trade-offs that can be pictured by the production possibility curve.

The production possibility curve represents the tough choices society must make.

Not everyone recognizes these choices. For example, politicians often talk as if the production possibility curve were nonexistent. They promise voters the world, telling them, "If you elect me, you can have more of everything." When they say that, they obscure the hard choices and increase their probability of getting elected.

Economists do the opposite. They promise little except that life is tough, and they continually point out that seemingly free lunches often involve significant hidden costs. Alas, political candidates who exhibit such reasonableness seldom get elected. Economists' reasonableness has earned economics the nickname *the dismal science*.

Economists continually point out that seemingly free lunches often involve significant hidden costs.

COMPARATIVE ADVANTAGE, SPECIALIZATION, AND TRADE

The same comparative advantage argument we used to explain the shape of the production possibility curve can be used to show how trade (and hence how the markets that facilitate trade) make society better off. Let's start with an international example. Say Canada can produce widgets at a cost of $4 apiece and wadgets at $4 apiece, while Korea can produce widgets at a cost of 300 won (won is the Korean currency) apiece and wadgets at a cost of 100 won apiece. In Canada, the opportunity cost of one widget is one wadget. (Since each costs $4, Canada must reduce its production of wadgets by one to produce another widget.) In Korea, the opportunity cost of a widget is three wadgets, since it costs three times as much to produce a widget as it does to produce a wadget. Because Canada's opportunity cost of producing widgets is lower than Korea's, Canada is said to have a comparative advantage in producing widgets. Similarly, Korea

A country has a comparative advantage in a good if it can produce that good at a lower *opportunity* cost than another country can. (Remember, it is a lower opportunity cost, not necessarily a lower absolute cost.)

is said to have comparative advantage in producing wadgets because its opportunity cost of wadgets is one-third of a widget, while Canada's opportunity cost of wadgets is one widget.

If one country has a comparative advantage in one good, the other country must necessarily have a comparative advantage in the other good. Notice how comparative advantage hinges on opportunity cost, not total cost. Even if one country can produce all goods cheaper than another country, trade between them is still possible since the opportunity costs of various goods differ.

Q-10 Show, using production possibility curves, that Steve and Sarah would be better off specializing in their baking activities, and then trading, rather than baking only for themselves, given the following production possibility tables.

(a) Steve's Production per Day		(b) Sarah's Production per Day	
Loaves of Bread	Dozens of Cookies	Loaves of Bread	Dozens of Cookies
4	0	4	0
3	2	3	1
2	4	2	2
1	6	1	3
0	8	0	4

The same reasoning can be applied to trade among individuals. To keep the analysis simple, let's consider two individuals. Sunder is a whiz at creative writing; he can turn out four creative writing papers a day or one economics paper a day, or any proportional combination thereof. Ti is an economics whiz; she can turn out four economics papers a day or one creative writing paper a day, or any proportional combination thereof. Figure 2-6(a) shows how many papers each can write in a day.

Sunder and Ti's teacher has given each of them a weekly assignment to do four economics and four creative writing papers. Ti spends four days writing four creative writing papers and one day writing four economics papers. Sunder does the reverse. This will leave no time for partying.

The following week their teacher allows Ti and Sunder to collaborate on their assignment. The team must turn in eight creative writing papers and eight economics papers. Ti and Sunder will receive a collective grade. The question facing Ti and Sunder is how to divide up the work. If they work as they did the week before, it will take each of them all week.

Fortunately Ti is taking an economics course and has just learned about comparative advantage. She points out to Sunder that they can do much better if they avail themselves of their comparative advantages. Since Ti has a comparative advantage in economics and Sunder has a comparative advantage in creative writing, it pays for them to specialize in their respective strengths. Let's say they do that. Their new production possibility table per day is shown in Figure 2-6(b). They both specialize, and in one day the team can turn out four economics papers and four creative writing papers. Now it takes them only two days to do their assignment, leaving them three days to party (or to study for other courses, or to work at a part-time job to pay for their education).

The benefits of collaboration and specialization can also be shown with production possibility curves. Production possibility curves show all the possible combinations of dividing up each student's time and possible combinations when dividing up the work.

2.2 see page 46

Sunder and Ti's respective paper-writing production possibility curves per day are shown in Figure 2-6(c). The blue line represents Sunder's production possibility curve; the red line represents Ti's production possibility curve. The curve connecting points A, B, and C shows the production possibilities if Ti and Sunder collaborate. You should be able to explain the individual production possibility curves. The combined curve is a little more complicated. This curve tells you what they can write when working together. If they both spent the day writing economics papers, they would end up with 5 economics papers and no creative writing papers—point A. If instead, they both wrote creative writing papers, they would have 5 creative writing papers and no economics papers—point C. At these endpoints one person is not taking advantage of his or her comparative advantage. If, instead, they take advantage of their own special skills, Sunder would write 4 creative writing papers and Ti would write 4 economics papers in one day—point B. Connecting points A, B, and C shows all combinations of papers both can write together in one day.

Notice that the combined production possibility curve is shifted out significantly. When individuals collaborate, using their comparative advantages, their production

Figure 2-6 (a, b, and c) THE GAINS FROM TRADE

Trade makes people better off. In this figure we compare individuals' combined production possibilities under two alternative assumptions. The table in (a) shows the number of economics and creative writing papers each student can produce in a day. The table in (b) shows their maximum possibilities when the students collaborate. The graph in (c) plots the production possibility curves associated with the tables in (a) and (b). Notice what trade does to the production possibility frontier. The combined curve is bowed out, reflecting the principle of increasing marginal opportunity costs.

(a) Sunder and Ti's individual possibilities

	Economics papers per day	Creative writing papers per day
Sunder	1	4
Ti	4	1

(b) Sunder and Ti's joint possibilities

	Economics papers per day	Creative writing papers per day
A	5	0
B	4	4
C	0	5

(c) Production possibility curves (numbers of papers per day)

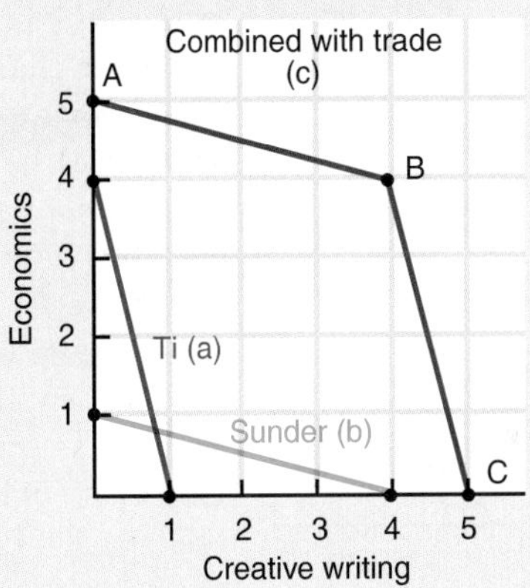

possibilities expand. By collaborating and specializing, each can reach beyond his or her individual production possibility curve. This shifting out of the production possibility curve comes from using comparative advantage and is a geometric representation of the gains to trade.

When individuals trade using their comparative advantages, their combined production possibility curve shifts outward.

The Division of Labour

see page 46

The above examples give a visual sense of the power of markets to make people better off. Markets allow people to trade—to utilize their comparative advantages and specialize in what they do best—and thereby to improve society's combined production possibility curve. Adam Smith, the founder of modern economics, saw this aspect of markets and trade as what differentiated human beings from animals. He wrote:

> This division of labour, from which so many advantages are derived, is not originally the effect of any human wisdom, which foresees and intends that general opulence to which it gives occasion. It is the necessary, though very slow and gradual consequence, of a certain propensity in human nature which has in view no such extensive utility; the propensity to truck, barter, and exchange one thing for another. . . . [This propensity] is common to all men, and to be found in no other race of animals, which seem to know neither this nor any other species of contracts. . . . Nobody ever saw a dog make a fair and deliberate exchange of one bone for another with another dog. Nobody ever saw one animal by its gestures and natural cries signify to another, this is mine, that yours; I am willing to give this for that.

The argument that the division of labour and trade makes individuals better off also holds for countries. Trade shifts out countries' production possibility curves, making them better off. There are, of course, exceptions to this proposition. (Remember all models are dependent on the assumptions of the model.) These exceptions are consid-

Markets allow specialization and the division of labour. They allow individuals to develop their comparative advantages, thereby increasing the production possibilities of society.

ered later in microeconomics courses when trade is looked at in more depth. But the general argument that trade and markets make people better off because they allow individuals and economies to assert their comparative advantages carries through and is a primary reason why economists generally support markets.

Markets, Specialization, and Growth

We can see the effect of markets on our well-being empirically by considering the growth of economies. As you can see from Figure 2-7, for 1700 years the world economy grew very slowly. Then, at the end of the 18th century, the world economy started to grow, and it has grown at an increasing rate since then.

What changed? The introduction of markets and democracy. There's something about markets that leads to economic growth. Markets allow specialization and encourage trade. The bowing out of the production possibilities from trade is part of the story, but a minor part. As individuals compete and specialize they learn by doing, becoming even better at what they do. Markets also foster competition, which pushes individuals to find better ways of doing things. They devise new technologies that further the growth process.

The new millennium is offering new ways for individuals to specialize and compete. More and more businesses are trading on the Internet. For example, schools, such as the University of Phoenix and Athabasca University are providing online competition for traditional schools. Similarly, online bookstores and drugstores are proliferating. As Internet technology becomes built into our economy, we can expect more specialization, more division of labour, and the economic growth that follows.

CONCLUSION

As we will emphasize throughout this book, the tools you will be learning here are powerful but simple, and that simplicity means that much care must be used in their application—to see that the tools fit the situation for which they are used. Businesses tell us that they have much more of a problem with new employees who think they understand things, but actually don't, than they do with new employees who recognize how little they know—who are willing to learn the institutional ropes—and recognize that the time for changing institutions is only when they understand the structural role that the institutions play.

Figure 2-7 GROWTH IN THE PAST TWO MILLENNIA

Source: Angus Maddison, *Monitoring the World Economy*, OECD, 1995; Angus Maddison, "Poor Until 1820," *The Wall Street Journal*, January 11, 1999.

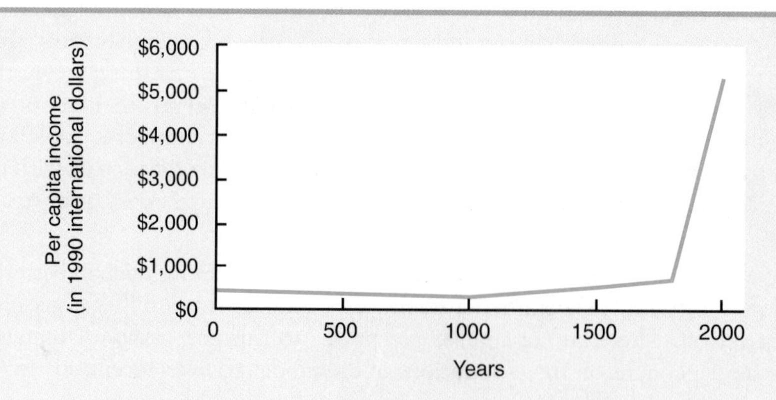

What this means is that economic reasoning is extraordinarily strong, and it really helps you if you apply it in the appropriate time and context. It remains strong when inappropriately applied; but it leads you to the wrong conclusion. The reality is that history and institutions are fundamentally important for making good economic decisions. Economic reasoning alone does not tell anyone, "This is right and this is wrong policy." What economic reasoning does is to provide a framework of analysis that focusses the decision on precisely what are the opportunity costs, and what are the alternative ways of measuring those costs. Fitting that reasoning together with your gut intuitive sense of what is right and wrong is what makes a good economist.

Chapter Summary

- Any economic system must solve three central problems:
  What, and how much, to produce.
  How to produce it.
  For whom to produce it.
- Capitalism is based on private property and the market; socialism is based on individuals' goodwill toward others.
- In capitalism, the what, how, and for whom problems are solved by the market.
- In Soviet-style socialism, the what, how, and for whom problems were solved by government planning boards.
- Political, social, and economic forces are active in both capitalism and socialism.
- Economic systems are in a constant state of evolution.
- Markets use the price mechanism to coordinate economic activity.
- In feudalism, tradition rules; in mercantilism, the government rules; in capitalism, the market rules.
- In welfare capitalism, the market, the government, and tradition each rule components of the economy.
- The production possibility curve measures the maximum combination of outputs that can be obtained from a given number of inputs. It embodies the opportunity cost concept.
- In general, in order to get more and more of something, we must give up ever-increasing quantities of something else. This is the principle of increasing marginal opportunity cost.
- Production possibility curves must be interpreted within the contextual nature of decisions.
- Trade allows people to use their comparative advantage and shift out society's production possibility curve.
- The rise of markets has coincided with significant increases in output. Specialization, trade, and competition have all contributed to the increase.

Key Terms

capitalism *(28)*
comparative advantage *(34)*
decision tree *(38)*
efficiency *(36)*
inefficiency *(36)*
input *(31)*
NIMBY *(28)*
output *(31)*
principle of increasing marginal opportunity cost *(35)*
private property rights *(28)*
production possibility curve *(32)*
production possibility table *(31)*
productive efficiency *(35)*
socialism *(29)*
Soviet-style socialism *(30)*

Questions for Thought and Review

1. What three problems must any economic system solve?
2. How does capitalism solve these three problems?
3. How does Soviet-style socialism solve these three problems?
4. Is capitalism or socialism the better economic system? Why?
5. What arguments can you give for supporting a socialist organization of a family and a capitalist organization of the economy?

6. Design a grade production possibility table and curve that embody the principle of increasing marginal opportunity cost.
7. What would the production possibility curve look like if there were decreasing marginal opportunity costs? Explain. What is an example of decreasing marginal opportunity costs?
8. Show how a production possibility curve would shift if a society became more productive in its output of widgets but less productive in its output of wadgets.
9. How does the theory of comparative advantage relate to production possibility curves?
10. When all people use economic reasoning, inefficiency is impossible, because if the benefit of reducing that inefficiency were greater than the cost, the efficiency would be eliminated. Thus, if people use economic reasoning, it's impossible to be on the interior of a production possibility curve. Is this statement true or false? Why?
11. Why, in the near term, would a switch of socialist countries to a market system cause their production possibility curves to shift in? What institutional and societal characteristics are necessary to carry through that switch successfully?
12. If trade shifts the production possibility curve out, why do some politicians oppose actions to promote trade?

Problems and Exercises

1. Poland, Bulgaria, and Hungary (all former socialist countries) were in the process of changing to a market economy in the 1990s.
   a. Go to the library and find the latest information about their transitions.
   b. Explain what has happened in the markets, political structures, and social customs of those countries.
2. Economists Edward Lazear and Robert Michael have calculated that the average family spends two and a half times as much on each adult as they do on each child.
   a. Does this mean that children are deprived and that the distribution is unfair?
   b. Do you think these percentages change with family income? If so, how?
   c. Do you think that the allocation would be different in a family in a Soviet-style socialist country than in a capitalist country? Why?
3. One of the specific problems Soviet-style socialist economies had was keeping up with capitalist countries technologically.
   a. Can you think of any reason inherent in a centrally planned economy that would make innovation difficult?
   b. Can you think of any reason inherent in a capitalist country that would foster innovation?
   c. Joseph Schumpeter, a famous Harvard economist of the 1930s, predicted that as firms in capitalist societies grew in size they would innovate less. Can you suggest what his argument might have been?
   d. Schumpeter's prediction did not come true. Modern capitalist economies have had enormous innovations. Can you provide explanations as to why?
4. A country has the following production possibility table:

Resources Devoted to Clothing	Output of Clothing	Resources Devoted to Food	Output of Food
100%	20	0%	0
80	16	20	5
60	12	40	9
40	8	60	12
20	4	80	14
0	0	100	15

   a. Draw the country's production possibility curve.
   b. What's happening to marginal opportunity costs as output of food increases?
   c. Say the country gets better at the production of food. What will happen to the production possibility curve?
   d. Say the country gets equally better at producing both food and clothing. What will happen to the production possibility curve?
5. Suppose a country has the following production possibility table:

Resources Devoted to Clothing	Output of Clothing	Resources Devoted to Food	Output of Food
100%	15	0%	0
80	14	20	4
60	12	40	8
40	9	60	12
20	5	80	16
0	0	100	20

   a. Draw the country's production possibility curve.
   b. What is the combined production possibility curve for this country and the country in question 4 if the

countries do not trade and they devote equal proportionate resources to produce each good?

c. What would be the combined production possibility curve of the two countries if they took advantage of their comparative advantages?

6. Assume Canada can produce Toyotas at the cost of $8,000 per car and Chevrolets at $6,000 per car. In Japan, Toyotas can be produced at 1,000,000 yen and Chevrolets at 500,000 yen.
   a. In terms of Chevrolets, what is the opportunity cost of producing Toyotas in each country?
   b. Who has the comparative advantage in producing Chevrolets?
   c. Assume Canadians purchase 50,000 Chevrolets and 30,000 Toyotas each year. The Japanese purchase far fewer of each. Using productive efficiency as the guide, who should most likely produce Chevrolets and who should produce Toyotas, assuming Chevrolets are going to be produced in one country and Toyotas in the other?

7. Lawns produce no crops but occupy more land in Canada than any single crop, such as corn. This means that Canada is operating inefficiently and hence is at a point inside the production possibility curve. Right? If not, what does it mean?

8. Groucho Marx is reported to have said, "The secret of success is honest and fair dealing. If you can fake those, you've got it made." What would likely happen to society's production possibility curve if everyone could fake honesty? Why? (Hint: Remember that society's production possibility curve reflects more than just technical relationships.)

9. Suppose Wombatland has the following production possibilities table:

Label	Resources Devoted to Thingamabobs	Output of Thingamabobs	Resources Devoted to Whatsits	Output of Whatsits
A	0%	0	100	10
B	20	6	80	8
C	40	11	60	6
D	60	15	40	4
E	80	18	20	2
F	100	20	0	0

   a. Draw the production possibilities curve with the quantity of Whatsits on the vertical axis, and mark the points A through F using the labels in the table.
   b. Is the opportunity cost of a Thingamabob constant? How can you tell without making any calculations?
   c. Calculate the opportunity cost of a Thingamabob between points A and B, B and C, C and D, D and E, and E and F.

10. Refer to the data in the table in Problem 9 above. Suppose there is a technological innovation that increases the production of both Thingamabobs and Whatsits by 100 percent.
   a. Draw the new production possibilities curve.
   b. Has technological change affected the opportunity cost of Thingamabobs?
   c. Explain why. Use calculations between points A and B, B and C, and so on to support your answer.

11. The Kingdom of Kent has the following production possibilities table:

Label	Resources Devoted to Dohickies	Output of Dohickies	Resources Devoted to Thingies	Output of Thingies
A	0%	0	100%	980
B	25	70	75	840
C	50	140	50	630
D	75	210	25	350
E	100	280	0	0

   a. Draw the production possibilities curve with the quantity of Dohickies on the horizontal axis and label the points A through E.
   b. Does the opportunity cost of a Dohickie change as more and more Dohickies are produced? If so, how, and why?
   c. Calculate the opportunity cost of a Dohickie between points A and B; B and C; C and D; and D and E. Are you finding what you expected given your answer to part b?
   d. Without performing the calculations, how can you determine the opportunity cost of a Thingie between points E and D; D and C; C and B; and B and A? Should opportunity costs be rising between points E and A? Why?
   e. Now calculate the opportunity cost of a Thingie between the points considered in part d. Is your answer what you expected? Explain.

Web Questions

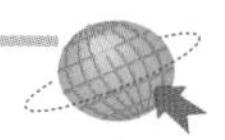

1. The Canada Pension Plan is significant to the evolution of capitalism in Canada. Go to the Human Resources Development Canada home page (www.hrdc-drhc.gc.ca/ips/common/home.shtml) and describe how changes in the plan have moved the Canadian economy toward welfare capitalism. What changes have been made that will alter the nature of CPP? What does this say about the evolution of capitalism today?
2. Starting from the Department of Foreign Affairs and International Trade Web site (www.dfait-maeci.gc.ca/trade/menu-e.asp),
   a. With how many countries does Canada trade?
   b. What are its three largest trading partners?
   c. How does this trade affect the world production possibility curve?
3. Visit the Adam Smith Institute's Web site (www.adamsmith.org/uk) and skim through Book One of *The Wealth of Nations*.
   a. In which chapter does the quotation in the text appear?
   b. What did Smith say limited the division of labour? Why?

Answers to Margin Questions

1. He is wrong. Property rights are required for a market to operate. Once property rights are allocated, the market will allocate goods, but the market cannot distribute the property rights that are required for the market to operate. *(28)*
2. Most families allocate basic needs through control and command. The parents do (or try to do) the controlling and commanding. Generally they are well-intentioned, trying to meet their perception of their children's needs. However, some family activities that are not basic needs might be allocated through the market. For example, if one child wants a go-cart and is willing to do extra work at home in order to get it, go-carts might be allocated through the market, with the child earning chits that can be used for such nonessentials. *(30)*
3. Capitalism places much more emphasis on fostering individualism. Socialism tries to develop a system in which the individual's needs are placed second to society's needs. *(30)*
4. In theory, socialism is an economic system based upon individuals' goodwill. In practice, socialism follows the Soviet model and involves central planning and government ownership of the primary means of production. *(30)*
5. The slope of the production possibilities curve tells us the opportunity cost of one good in terms of another. A straight line has a constant slope, so opportunity costs would be fixed (constant). *(33)*
6. If no resource had a comparative advantage, the production possibility curve would be a straight line connecting the points of maximum production of each product, as in the graph below.

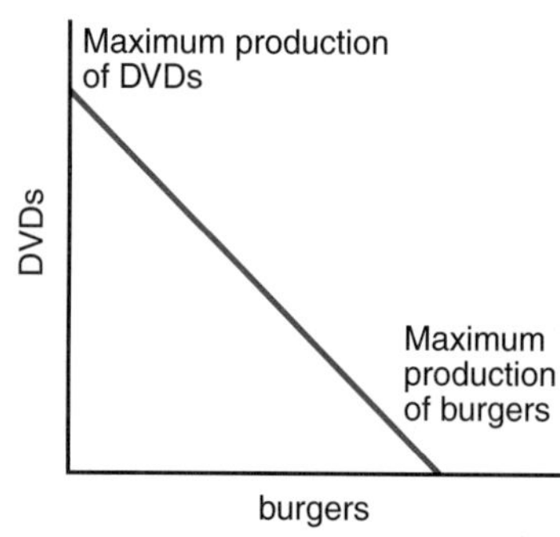

   At all points along this curve, the opportunity cost of producing burgers and DVDs is equal. *(34)*
7. We remind them of the importance of cultural forces. In Saudi Arabia women are not allowed to drive. *(37)*
8. The production possibility curve shifts along the wheat axis as in the graph below. *(38)*

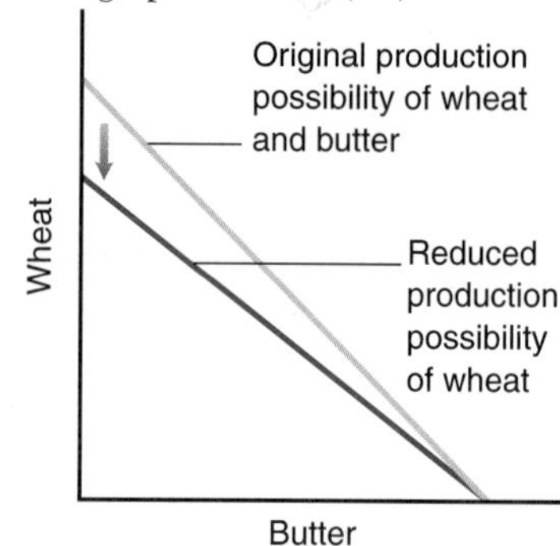

9. A decision tree showing your choices as an undergraduate might look like this: *(38)*

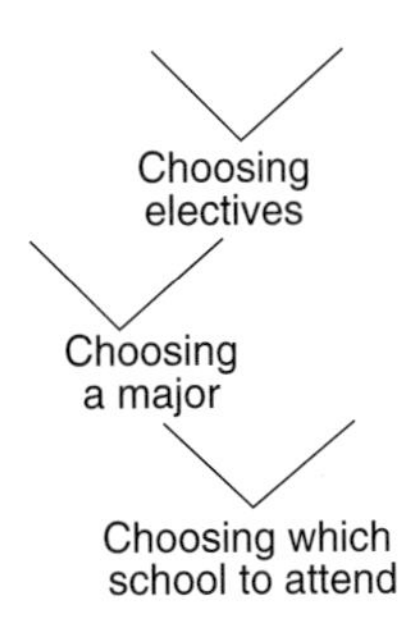

(a) Combined, No Trade		(b) Combined, Trade	
Loaves of Bread	Dozens of Cookies	Loaves of Bread	Dozens of Cookies
8	0	8	0
6	3	6	4
4	6	4	8
2	9	2	10
0	12	0	12

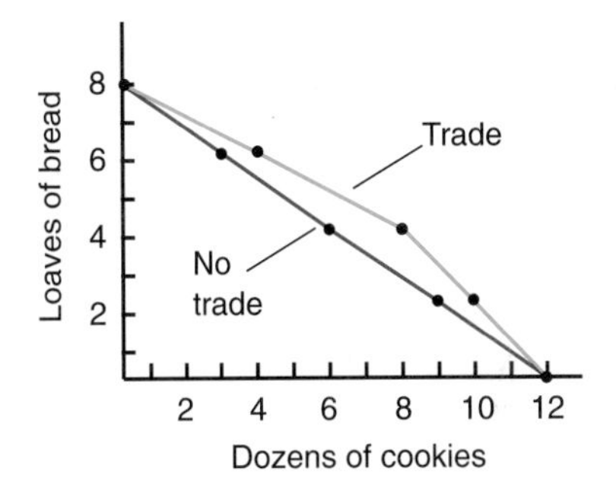

10. If Steve and Sarah do not trade, and divide their time identically, their combined production possibility table is shown in (**a**). If they do collaborate, their production possibility table is shown in (**b**). The corresponding production possibility curves are drawn beneath the tables. The production possibility curve with trade is further to the right than the one without trade. Steve and Sarah can each end up with more bread and cookies if they trade than if they work on their own. *(40)*

The History of Economic Systems

Remember the distinction between market and economic forces: Economic forces have always existed—they operate in all aspects of our lives; but market forces have not always existed. Markets are social creations societies use to coordinate individuals' actions. Markets developed, sometimes spontaneously, sometimes by design, because they offered a better life for at least some—and usually a large majority of—individuals in a society.

To understand why markets developed, it is helpful to look briefly at the history of the economic systems from which our own system descended.

EVOLVING ECONOMIC SYSTEMS

Capitalism and socialism have not existed forever. Capitalism came into widespread existence in the mid-1700s; socialism came into existence in the early 1900s. Before capitalism and socialism, other forms of economic systems existed, including **feudalism**—*an economic system in which traditions rule*. In feudalism if your parents were serfs (small farmers who lived on a manor), you would be a serf. Feudalism dominated the Western world from about the 8th century to the 15th century.

Throughout the feudalistic period merchants and artisans (small manufacturers who produced goods by hand) grew in importance and wealth and, eventually, their increased importance led to a change in the economic system from feudalism to **mercantilism**—*an economic system in which government determines the what, how, and for whom decisions by doling out the rights to undertake certain economic activities*. Mercantilism remained the dominant economic system until the 1700s, when the **Industrial Revolution**—*a time when technology and machines rapidly modernized industrial production and mass-produced goods replaced handmade goods*—led to a decrease in the power of small producers, an increase in the power of capitalists, and eventually to a revolution instituting capitalism as the dominant economic system.

Some economists prefer to call the system that evolved from mercantilism a *market economic system* rather than capitalism. Their justification for doing so is that the key element of the new system is not the power of the capitalists but the central reliance on markets to coordinate economic activities. They argue that the market system is not fundamentally changed if the power shifts from a small group of capitalists to corporations, or even to nonprofit organizations, as long as the market remains central. We agree with their argument, but we prefer the term *capitalism* because of its widespread usage, not because we believe that capitalist control is the central element of our modern market economy.

We mention feudalism and mercantilism because aspects of both continue in economies today. For example, governments in Japan and Germany play significant roles in directing their economies. Their economic systems are sometimes referred to as *neomercantilist economies*.

The Need for Coordination in an Economic System

As economic systems evolved from mercantilism to a market economy, many people asked: Who will coordinate economic activities if the government does not? It was in answering this question that modern economics developed. To answer that question, a British moral philosopher named Adam Smith developed, in his famous book *The Wealth of Nations* (1776), the concept of the invisible hand, and used it to explain how markets could coordinate the economy without the active involvement of government. Smith wrote:

> Man has almost constant occasion for the help of his brethren, and it is in vain for him to expect it from their benevolence only. He will be more likely to prevail, if he can interest their self-love in his favour, and show them that it is for their own advantage to do for him what he requires of them. Whoever offers to another a bargain of any kind proposes to do this. Give me that which I want, and you shall have that which you want, is the meaning of every such offer; and it is in this manner that we obtain from one another the far greater part of those good offices which we stand in need of. It is not from the benevolence of the butcher, the brewer, or the baker, that we expect our dinner, but from their regard to their own interest. We address ourselves, not to their humanity but to their self-love, and never talk to them of our own necessities but of their advantages.

Smith argued that the market's invisible hand would guide suppliers' actions toward the general good. No government coordination was necessary.

Evolutionary Changes within Systems

Revolutionary shifts that give rise to new economic systems are not the only way economic systems change. Systems also evolve. For example, both capitalism and socialism have changed over the years, evolving with changes in social customs, political forces, and the strength of markets. A brief look at their evolution will give us a good sense of the struggles between them that dominate our lives. In the 1930s, during the Great Depression, capitalist countries integrated a number of what might be called socialist institutions into their existing institutions. Distribution of goods was no longer, even in theory, only according to ability; need also played a role. Governments began to play a larger role in the economy, taking control over some of the *how*, *what*, and *for whom* decisions. For example, most capitalist nations established welfare and social security systems, providing an economic safety net for people whose incomes could not fill their needs. Capitalism became what is sometimes called **welfare capitalism**—*an economic system in which the market is allowed to operate but in which government plays dual roles in determining distribution and making the what, how, and for whom decisions*. In some countries the governments played such an important role that they could reasonably be called market-modified socialist economies. Some countries that allow markets today, such as Sweden, call their economic system a socialist, rather than a capitalist, system.

In the 1980s, in socialist countries the reverse process took place: socialism integrated capitalist institutions into its existing institutions. By the 1990s, most of the former socialist countries had given up Soviet-style socialism and had begun to let the market determine the what, how, and for whom decisions. Some government activities in former socialist economies were privatized—sold to the private sector. Similarly in the 1990s, in capitalist countries governments became even more market-oriented and tried to pull back their involvement in the market in favour of private enterprise. For example, governments in Western Europe and the United States tried to eliminate some aspects of their welfare systems. In the 1980s and 1990s economic systems moved more toward the capitalist end of the spectrum.

A Blurring of the Distinction between Capitalism and Socialism

The result of these recent changes has been a blending of economic systems and a blurring of the distinctions between capitalism and socialism. China still ostensibly has a socialist system, but a recent trip to China convinced one of us that it is a very capitalist type of socialism. If the trend toward use of market mechanisms in socialist countries continues, in the 21st century only one general type of economic structure may exist. It won't be pure socialism and it won't be pure capitalism. It will be a blend of the two.

FEUDAL SOCIETY: RULE OF TRADITION

Let's go back in time to the year 1000 when Europe had no nation-states as we now know them. (Ideally, we would have gone back farther and explained other economic systems, but, given the limited space, we had to draw the line somewhere—an example of trade-off.) The predominant economic system at that time was feudalism. There was no coordinated central government, no unified system of law, no national patriotism, no national defense, although a strong religious institution simply called the Church fulfilled some of these roles. There were few towns; most individuals lived in walled manors or "estates." These manors "belonged to" the "lord of the manor." (Occasionally the "lord" was a lady, but not often.) We say "belonged to" rather than "were owned by" because most of the empires or federations at that time were not formal nation-states that could organize, administer, and regulate ownership. No documents or deeds gave ownership of the land to an individual. Instead, tradition ruled, and in normal times nobody questioned the lord's right to the land. The land "belonged to" the lord because the land "belonged to" him—that's the way it was.

Without a central nation-state, the manor served many functions a nation-state would have served had it existed. The lord provided protection, often within a walled area surrounding the manor house or, if the manor was large enough, a castle. He provided administration and decided disputes. He also decided *what* would be done, *how* it would be done, and *who* would get what, but these decisions were limited. In the same way that the land belonged to the lord because that's the way it always had been, what people did and how they did it were determined by what they always had done. Tradition ruled the manor more than the lord did.

Problems of a Tradition-Based Society

Feudalism developed about the 8th and 9th centuries and lasted until about the 15th century, though in isolated countries such as Russia it continued well into the 19th century, and in all European countries its influence lingered for hundreds of years (as late as about 140 years ago in some parts of Germany). Such a long-lived system must have done some things right, and feudalism did: it solved the what, how, and for whom problems in an acceptable way.

But a tradition-based society has problems. In a traditional society, because someone's father was a baker, the son must also be a baker, and because a woman was a homemaker, she wouldn't be allowed to be anything but a homemaker. But what if Joe Blacksmith, Jr., the son of Joe Blacksmith, Sr., is a lousy blacksmith and longs to knead dough, while Joe Baker, Jr., would make a superb blacksmith but hates kneading pastry? Tough. Tradition dictated who did what. In fact, tradition probably arranged things so that we will never know whether Joe Blacksmith, Jr., would have made a superb baker.

As long as a society doesn't change too much, tradition operates reasonably well, although not especially efficiently, in holding the society together. However, when a society must undergo change, tradition does not work. Change means that the things that were done before no longer need to be done, while new things do need to get done. But if no one has traditionally done these new things, then they don't get done. If the change is important but a society can't figure out some way for the new things to get done, the society falls apart. That's what happened to feudal society. It didn't change when change was required.

The life of individuals living on the land, called *serfs*, was difficult, and feudalism was designed to benefit the lord. Some individuals in feudal society just couldn't take life on the manor, and they set off on their own. Because there was no organized police force, they were unlikely to be caught and forced to return to the manor. Going hungry, being killed, or both, however, were frequent fates of an escaped serf. One place to which serfs could safely escape, though, was a town or city—the remains of what in Roman times had been thriving and active cities. These cities, which had been decimated by plagues, plundering bands, and starvation in the preceding centuries, nevertheless remained an escape hatch for runaway serfs because they relied far less on tradition than did manors. City dwellers had to live by their wits; many became merchants who lived predominantly by trading. They were middlemen; they would buy from one group and sell to another.

Trading in towns was an alternative to the traditional feudal order because trading allowed people to have an income independent of the traditional social structure. Markets broke down tradition. Initially merchants traded using barter (exchange of one kind of good for another): silk and spices from the Orient for wheat, flour, and artisan products in Europe. But soon a generalized purchasing power (money) developed as a medium of exchange. Money greatly expanded the possibilities of trading because its use meant that goods no longer needed to be bartered. They could be sold for money, which could then be spent to buy other goods.

In the beginning, land was not traded, but soon the feudal lord who just had to have a silk robe but had no money was saying, "Why not? I'll sell you a small piece of land so I can buy a shipment of silk." Once land became tradable, the traditional base of the feudal society was undermined. Tradition that can be bought and sold is no longer tradition—it's just another commodity.

FROM FEUDALISM TO MERCANTILISM

Toward the end of the Middle Ages (mid-15th century), markets went from being a sideshow, a fair that spiced up people's lives, to being the main event. Over time, some traders and merchants started to amass fortunes that dwarfed those of the feudal lords. Rich traders settled down; existing towns and cities expanded and new towns were formed. As towns grew and as fortunes shifted from feudal lords to merchants, power in society shifted to the towns. And with that shift came a change in society's political and economic structure.

As these traders became stronger politically and economically, they threw their support behind a king (the strongest lord) in the hope that the king would expand their ability to trade. In doing so, they made the king even stronger. Eventually, the king became so powerful that his will prevailed over the will of the other lords and even over the will of the Church. As the king consolidated his power, nation-states as we know them today evolved. *The government became an active influence on economic decision making.*

As markets grew, feudalism evolved into mercantilism. The evolution of feudal systems into mercantilism occurred in somewhat this way: As cities and their markets grew in size and power relative to the feudal manors and the traditional economy, a whole new variety of possible economic activities developed. It was only natural that individuals began to look to a king to establish a new tradition that would determine who would do what. Individuals in particular occupations organized into groups called *guilds*, which were similar to strong labour unions today. These guilds, many of which had financed and supported the king, now expected the king and his government to protect their interests.

As new economic activities, such as trading companies, developed, individuals involved in these activities similarly depended on the king for the right to trade and for help in financing and organizing their activities. For example, in 1492, when Christopher Columbus had the wild idea that by sailing west he could get to the East Indies and trade for their riches, he went to Spain's Queen Isabella and King Ferdinand for financial support.

Since many traders had played and continued to play important roles in financing, establishing, and supporting the king, the king was usually happy to protect their interests. The government doled out the rights to undertake a variety of economic activities. This was the era in which the Hudson's Bay Company received its charter. By the late 1400s, Western Europe had evolved from a feudal to a mercantilist economy.

The mercantilist period was marked by the increased role of government, which could be classified in two ways: by the way it encouraged growth, and by the way it limited growth. Government legitimized and financed a variety of activities, thus encouraging growth. But government also limited economic activity in order to protect the monopolies of those it favoured, thus limiting growth. So mercantilism allowed the market to operate, but it kept the market under its control. The market was not allowed to respond freely to the laws of supply and demand.

FROM MERCANTILISM TO CAPITALISM

Mercantilism provided the source for major growth in Western Europe, but mercantilism also unleashed new tensions within society. Like feudalism, mercantilism limited entry into economic activities. It used a different form of limitation—politics rather than social and cultural tradition—but individuals who were excluded still felt unfairly treated.

The most significant source of tension was the different roles played by craft guilds and owners of new businesses, who were called industrialists or **capitalists**—*businesspeople who have acquired large amounts of money and use it to invest in businesses*. Craft guild members were artists in their own crafts: pottery, shoemaking, and the like. New business owners destroyed the art of production by devising machines to replace hand production. Machines produced goods cheaper and faster than craftsmen.[1] The result was an increase in supply and a downward pressure on the price, which was set by the government. Craftsmen didn't want to be replaced by machines. They argued that machine-manufactured goods didn't have the same quality as hand-crafted goods, and that the new machines would disrupt the economic and social life of the community.

Industrialists were the outsiders with a vested interest in changing the existing system. They wanted the freedom to conduct business as they saw fit. Because of the enormous cost advantage of manufactured goods over hand-crafted goods, a few industrialists overcame government opposition and succeeded within the mercantilist system. They earned their fortunes and became an independent political power.

Once again the economic power base shifted, and two groups competed with each other for power—this time, the guilds and the industrialists. The government had to decide whether the support the industrialists (who wanted government to loosen its power over the country's economic affairs) or the craftsmen and guilds (who argued for strong government limitations and for maintaining traditional values of workmanship). This struggle raged in the 1700s and 1800s. But during this time, governments themselves were changing. This was the Age of Revolutions, and the kings' powers were being limited by democratic reform movements—revolutions supported and financed in large part by the industrialists.

The Need for Coordination in an Economy

Craftsmen argued that coordination of the economy was necessary, and the government had to be involved. If government wasn't going to coordinate economic activity, who would? To answer that question, a British moral philosopher named Adam Smith developed the concept of the invisible hand, in his famous book *The Wealth of Nations* (1776), and used it to explain how markets could coordinate the economy without the active involvement of government.

As stated earlier in the chapter, Smith argued that the market's invisible hand would guide suppliers' actions toward the general good. No government coordination was necessary.

With the help of economists such as Adam Smith, the industrialists' view won out. Government pulled back from its role in guiding the economy and adopted a **laissez-faire** policy—*the economic policy of leaving coordination of individuals' wants to be controlled by the market*. (*Laissez-faire*, a French term, means "Let events take their course; leave things alone.")

The Industrial Revolution

The invisible hand worked; capitalism thrived. Beginning about 1750 and continuing through the late 1800s, machine production increased enormously, almost totally replacing hand production. This phenomenon has been given a name: the Industrial Revolution. The economy grew faster than ever before. Society was forever transformed. New inventions changed all aspects of life. James Watt's steam engine (1769) made manufacturing and travel easier. Eli Whitney's cotton gin (1793) changed the way cotton was processed. James Kay's flying shuttle (1733),[2] James Hargreaves' spinning jenny (1765), and Richard Arkwright's power loom (1769), combined with the steam engine, changed the way cloth was processed and the clothes people wore.

The need to mine vast amounts of coal to provide power to run the machines changed the economic and physical landscapes. The repeating rifle changed the nature of warfare. Modern economic institutions replaced guilds. Stock markets, insurance companies, and corporations all became important. Trading was no longer financed by government; it was privately financed—although government policies, such as colonial policies giving certain companies (such as the famous East India Company) monopoly trading rights with a country's colonies, helped in that trading. The Industrial Revolution, democracy, and capitalism all arose in the middle and late 1700s. By the 1800s, they were part of the institutional landscape of Western society. Capitalism had arrived.

[1]Throughout this section we use *men* to emphasize that these societies were strongly male-dominated. There were almost no businesswomen. In fact, a woman had to turn over her property to a man upon her marriage, and the marriage contract was written as if she were owned by her husband!

[2]The invention of the flying shuttle frustrated the textile industry because it enabled workers to weave so much cloth that the spinners of thread from which the cloth was woven couldn't keep up. This challenge to the textile industry was met by offering a prize to anyone who could invent something to increase the thread spinners' productivity. The prize was won when the spinning jenny was invented.

FROM CAPITALISM TO ~~SOCIALISM~~ *Welfare Capitalism*

Capitalism was marked by significant economic growth in the Western world. But it was also marked by human abuses—18-hour workdays, low wages, children as young as five years old slaving long hours in dirty, dangerous factories and mines—to produce enormous wealth for an elite few. Such conditions and inequalities led to criticism of the capitalist or market economic system.

Marx's Analysis

The best-known critic of this system was Karl Marx, a German philosopher, economist, and sociologist who wrote in the 1800s and who developed an analysis of the dynamics of change in economic systems. Marx argued that economic systems are in a constant state of change, and that capitalism would not last. Workers would revolt, and capitalism would be replaced by a socialist economic system.

Marx saw an economy marked by tensions between economic classes. He saw capitalism as an economic system controlled by the capitalist class (businessmen). His class analysis was that capitalist society is divided into capitalist and worker classes. He said constant tension between these economic classes causes changes in the system. The capitalist class made large profits by exploiting the **proletariat** class—*the working class*—and extracting what he called *surplus value* from workers who, according to Marx's labour theory of value, produced all the value inherent in goods. Surplus value was the additional profit, rent, or interest that, according to Marx's normative views, capitalists added to the price of goods. What economic analysis sees as recognizing a need that society has and fulfilling it, Marx saw as exploitation.

Marx argued that this exploitation would increase as production facilities became larger and larger and as competition among capitalists decreased. At some point, he believed, exploitation would lead to a revolt by the proletariat, who would overthrow their capitalist exploiters.

By the late 1800s, some of what Marx predicted had occurred, although not in the way that he thought it would. Production moved from small to large factories. Corporations developed, and classes became more distinct from one another. Workers were significantly differentiated from owners. Small firms merged and were organized into monopolies and trusts (large combinations of firms). The trusts developed ways to prevent competition among themselves and ways to limit entry of new competitors into the market. Marx was right in his predictions about these developments, but he was wrong in his prediction about society's response to them.

The Revolution that Did Not Occur

Western society's response to the problems of capitalism was not a revolt by the workers. Instead, governments stepped in to stop the worst abuses of capitalism. The hard edges of capitalism were softened.

Evolution, not revolution, was capitalism's destiny. The democratic state did not act, as Marx argued it would, as a mere representative of the capitalist class. Competing pressure groups developed; workers gained political power that offset the economic power of businesses.

In the late 1930s and the 1940s, workers dominated the political agenda. During this time, capitalist economies developed an economic safety net that included government-funded programs, such as public welfare and unemployment insurance, and established an extensive set of regulations affecting all aspects of the economy. Today, depressions are met with direct government policy. Competition laws, regulatory agencies, and social programs of government softened the hard edges of capitalism. Laws were passed prohibiting child labour, mandating a certain minimum wage, and limiting the hours of work. Capitalism became what is sometimes called welfare capitalism.

Due to these developments, federal government spending now accounts for about a fifth of all spending in Canada, and for more than half in some European countries. Were an economist from the late 1800s to return from the grave, he'd probably say socialism, not capitalism, exists in Western societies. Most modern-day economists wouldn't go that far, but they would agree that our economy today is better described as a welfare capitalist economy than as a capitalist, or even a market, economy. Because of these changes, the North American and Western European economies are a far cry from the competitive "capitalist" economy that Karl Marx criticized. Markets operate, but they are constrained by the government: they are regulated.

The concept of *capitalism* was developed to denote a market system controlled by one group in society, the capitalists. Looking at Western societies today, we see that domination by one group no longer characterizes Western economies. Although in theory capitalists control corporations through their ownership of shares of stock, in practice corporations are controlled in large part by managers.

There remains an elite group who control business, but *capitalist* is not a good term to describe them. Managers, not capitalists, exercise primary control over business, and even their control is limited by laws or the fear of laws being passed by governments.

Governments in turn are influenced by a variety of pressure groups. Sometimes one group has more influence; at other times, another. Government policies similarly fluctuate. Sometimes they are pro-worker, sometimes pro-industrialist, sometimes pro-government, and sometimes pro-society.

FROM FEUDALISM TO SOCIALISM

You probably noticed that we crossed out *Socialism* in the previous section's heading and replaced it with *Welfare Capitalism*. That's because capitalism did not evolve to socialism as Karl Marx predicted it would. Instead, Marx's socialist ideas took root in feudalist Russia, a society that the Industrial Revolution had in large part bypassed. Arriving at a different place and a different time than Marx predicted it would, you shouldn't be surprised to read that socialism arrived in a different way than Marx predicted. The proletariat did not revolt to establish socialism. Instead, the First World War, which the Russians were losing, crippled Russia's feudal economy and government. A small group of socialists overthrew the czar (Russia's king) and took over the government in 1917. They quickly pulled Russia out of the war, and then set out to organize a socialist society and economy.

Russian socialists tried to adhere to Marx's ideas, but they found that Marx had concentrated on how capitalist economies operate, not on how a socialist economy should be run. Thus, Russian socialists faced a huge task with little guidance. Their most immediate problem was how to increase production so that the economy could emerge from feudalism into the modern industrial world. In Marx's analysis, capitalism was a necessary stage in the evolution toward the ideal state for a very practical reason. The capitalists exploit the workers, but in doing so capitalists extract the necessary surplus—an amount of production in excess of what is consumed. That surplus had to be extracted in order to provide the factories and machinery upon which a socialist economic system would be built. But since capitalism did not exist in Russia, a true socialist state could not be established immediately. Instead, the socialists created **state socialism**—*an economic system in which government sees to it that people work for the common good until they can be relied upon to do that on their own.*

Socialists saw state socialism as a transition stage to pure socialism. This transition stage still exploited the workers; when Joseph Stalin took power in Russia in the late 1920s, he took the peasants' and small farmers' land and turned it into collective farms. The government then paid farmers low prices for their produce. When farmers balked at the low prices, millions of them were killed.

Simultaneously, Stalin created central planning agencies that told individuals what to produce and how to produce it, and decided for whom things would be produced. During this period, *socialism* became synonymous with *central economic planning*, and Soviet-style socialism became the model of socialism in practice.

Also during this time, Russia took control of a number of neighbouring states and established the Union of Soviet Socialist Republics (USSR), the formal name of the Soviet Union. The Soviet Union also installed Soviet-dominated governments in a number of Eastern European countries. In 1949 most of China, under the rule of Mao Zedong, adopted Soviet-style socialist principles.

Since the late 1980s, the Soviet socialist economic and political structure has fallen apart. The Soviet Union as a political state broke up, and its former republics became autonomous. Eastern European countries were released from Soviet control. Now they faced a new problem: transition from socialism to a market economy. Why did the Soviet socialist economy fall apart? Because workers lacked incentives to work; production was inefficient; consumer goods were either unavailable or of poor quality; and high Soviet officials were exploiting their positions, keeping the best jobs for themselves and moving themselves up in the waiting lists for consumer goods. In short, the parents of the socialist family (the Communist party) were not acting benevolently; they were taking many of the benefits for themselves.

Recent political and economic upheavals in Eastern Europe and the former Soviet Union suggest that the kind of socialism these societies tried did not work. However, that failure does not mean that socialist goals are bad; nor does it mean that no type of socialism can ever work. To overthrow socialist-dominated governments it is not necessary to accept capitalism, and many citizens of these countries are looking for an alternative to both systems. Most, however, want to establish market economies.

FROM SOCIALISM TO . . . ?

The upheavals in the former Soviet Union and Eastern Europe have left China as the only major power using a socialist economic system. But even in China there have been changes, and the Chinese economy is socialist in name only. Almost uncontrolled markets exist in numerous sectors of the economy. These changes have led some socialists to modify their view that state socialism is the path from capitalism to true socialism, and instead to joke: "Socialism is the longest path from capitalism to capitalism."

Key Terms

capitalists *(50)*
laissez-faire *(51)*
proletariat *(52)*
state socialism *(53)*
feudalism *(47)*
Industrial Revolution *(47)*
mercantilism *(47)*
welfare capitalism *(48)*

The Canadian Economy in a Global Setting

3

After reading this chapter, you should be able to:

- Describe how businesses, households, and government interact in a market economy.
- Summarize briefly the advantages and disadvantages of various types of businesses.
- Explain why, even though households have the ultimate power, much of the economic decision making is done by business and government.
- List two general roles of government.
- List the primary areas with which Canada trades.
- State two ways international trade differs from intranational trade.
- List three important global trade organizations.
- List four important international policy organizations.

Letting a hundred flowers blossom and a hundred schools of thought contend is the policy for promoting progress in the arts and sciences and a flourishing socialist culture in our land.

Mao Zedong

The Canadian economic machine generates enormous economic activity and provides a high standard of living (compared to most other countries) for almost all its inhabitants. It also provides economic security for its citizens. Starvation is far from most

Ultimately the Canadian economy's strength is its people and its other resources.

people's minds. Ultimately, what underlies the Canadian economy's strength is its people and its other resources. Canada has vast central plains that are extraordinarily fertile, as are areas in its East and West. It is the world's largest producer, and largest exporter, of zinc. It has excellent ports and almost a million kilometres of highways.

The positive attributes of the Canadian economy don't mean that Canada has no problems. Critics point out that crime and drugs are omnipresent, economic resources such as oil and minerals are declining, the environment is deteriorating, the distribution of income is skewed toward the rich, regional disparities persist even though governments have spent billions to fight them, and enormous economic effort goes into economic gamesmanship (real estate deals, stock market deals, deals about deals) that seems simply to reshuffle existing wealth, not create new wealth.

THE CANADIAN ECONOMY

Q-1 Into what three groups are market economies generally broken up?

Income earned by households goes to purchasing goods and services produced by business.

Figure 3-1 diagrams a market economy such as that of Canada. Notice that it's divided up into three groups: businesses, households, and government. Households supply labour and other factors of production to businesses and are paid by businesses for doing so. The market where this interaction takes place is called a *factor market*. Businesses produce goods and services and sell them to households and government. The market where this interaction takes place is called the *goods market*.

Notice also the arrows going out to and coming in from both business and households. Those arrows represent the connection of an economy to the world economy. It consists of interrelated flows of goods (exports and imports) and money (capital flows). Finally, consider the arrows connecting government with households and business.

Figure 3-1 DIAGRAMMATIC REPRESENTATION OF A MARKET ECONOMY

This circular flow diagram of the economy is a good way to organize your thinking about the aggregate economy. As you can see, the three sectors—households, government, and business—interact in a variety of ways.

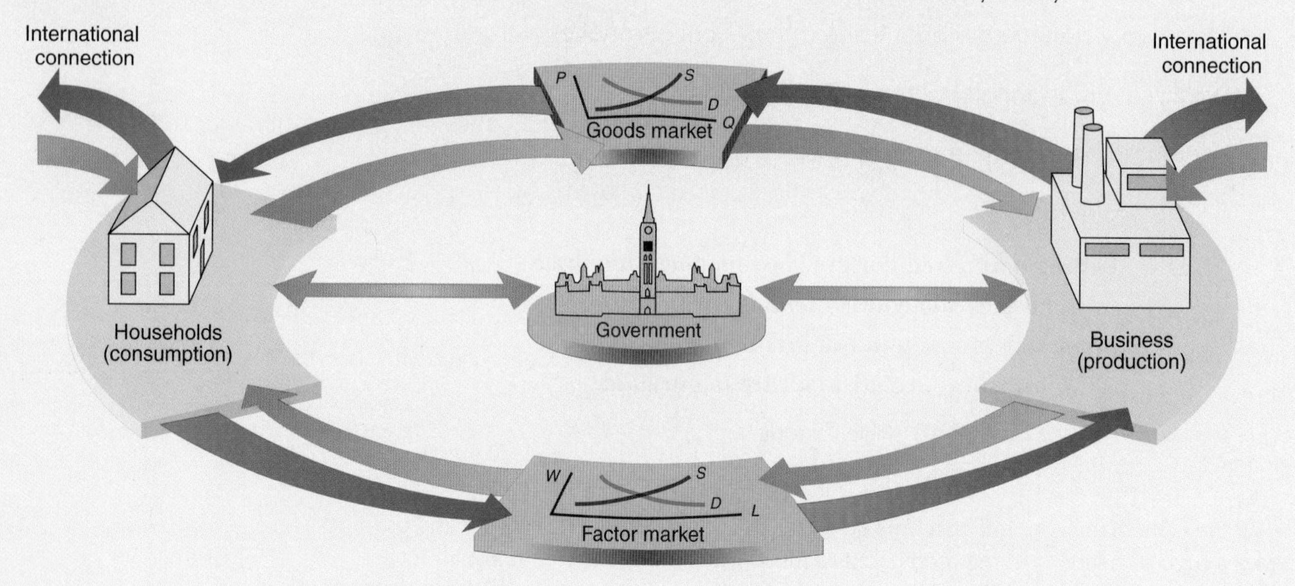

Government taxes business and households. It buys goods and services from business and buys labour services from households. Then, with some of its tax revenue, it provides services (e.g., roads, education) to both business and households and gives some of its tax revenue directly back to individuals. In doing so, it redistributes income. But government also serves a second function. It oversees the interaction of business and households in the goods and factor markets. Government, of course, is not independent. Canada is a democracy, so households vote to determine who will govern.

Now let's look briefly at the individual components.

BUSINESS

Business is responsible for about 80 percent of Canadian production. (Government is responsible for the other 20 percent.) In fact, any time a household decides to produce something, it becomes a business. **Business** is simply the name given to *private producing units in our society*.

see page 77

Businesses in Canada decide *what* to produce, *how* much to produce, and *for whom* to produce it. They make these central economic decisions on the basis of their own feelings, which are influenced by market incentives. Anyone who wants to can start a business, provided he or she can come up with the required cash and meet the necessary regulatory requirements. Each year thousands of businesses are started.

Businesses in Canada decide *what* to produce, *how much* to produce, and *for whom* to produce it.

Don't think of business as something other than people. Businesses are ultimately made up of a group of people organized together to accomplish some end. Although corporations account for a large part of all sales, in terms of numbers of businesses, most are one- or two-person operations. Home-based businesses are easy to start. All you have to do is say you're in business, and you are. However, some businesses require licenses, permits, and approvals from various government agencies. That's one reason why **entrepreneurship** (*the ability to organize and get something done*) is an important part of business.

Consumer Sovereignty and Business

To say that businesses decide what to produce isn't to say that **consumer sovereignty** (*the consumer's wishes rule what's produced*) doesn't reign in Canada. Businesses decide what to produce based on what they believe will sell. A key question a person in Canada should ask about starting a business is: Can I make a profit from it? **Profit** is *what's left over from total revenues after all the appropriate costs have been subtracted*. Businesses that guess correctly what the consumer wants generally make a profit. Businesses that guess wrong generally operate at a loss.

Although businesses decide what to produce, they are guided by consumer sovereignty.

People are free to start businesses for whatever purposes they want. No one asks them: "What's the social value of your term paper assistance business, your Twinkies business, your pornography business, or your textbook publishing business?" Yet the Canadian economic system is designed to channel individuals' desire to make a profit into the general good of society. That's the invisible hand at work. As long as the business doesn't violate a law and does conform to regulations, people in Canada are free to start whatever business they want, if they can get the money to finance it.

Q-2 In Canada the invisible hand ensures that only socially valuable businesses are started. True or false? Why?

Research at Statistics Canada indicates that from 1984 to 1994, at least half of new businesses in Canada went out of business after less than one year, and only one-fifth of them survived a decade. The average business life during this period was about six years. That's the invisible hand — it may be brutal, but it works.[1]

[1]http://www.statcan.ca/english/IPS/Data/61-526-XIE.htm

BEYOND THE TOOLS

Is Canada a Postindustrial Society?

Producing physical goods is only one of a society's economic tasks. Another task is to provide services (activities done for others). Services do not involve producing a physical good. When you get your hair cut, you buy a service, not a good. Much of the cost of the physical goods we buy actually is not a cost of producing the good, but is a cost of one of the most important services: distribution (getting the good to where the consumer is). After a good is produced, it has to get to the individuals who are going to consume it at the time they need it. If the distribution system gets botched up, it's as if the good had never been produced.

Let's consider a couple of examples. Take Christmas trees. Say you're sitting on 60,000 cut spruce trees in New Brunswick, but an ice storm prevents you from shipping them until December 26. Guess what? You're now stuck with 60,000 spruce trees and the problem of somehow getting rid of them. Or take hot dogs. How many of us have been irked that a hot dog that costs 25¢ to fix at home costs $4 at a football game? But a hot dog at home isn't the same as a hot dog at a game. Distribution of the good is as important as production; you're paying the extra $3.75 for distribution.

The importance of the service economy can be seen in modern technology companies. They provide information and methods of handling information, not physical goods. Operating systems, such as Linux and Windows, can be supplied over the Internet; no physical production is necessary. So, yes, it is fair to say that our economy is a postindustrial economy.

Forms of Business

The three primary forms of business are sole proprietorships, partnerships, and corporations. **Sole proprietorships**—*businesses that have only one owner*—are the easiest to start and have the fewest bureaucratic hassles. **Partnerships**—*businesses with two or more owners*—create possibilities for sharing the burden, but they also create unlimited liability for each of the partners. **Corporations**—*businesses that are treated as a person and are legally owned by their stockholders, who are not liable for the actions of the corporate "person"*—are the largest form of business when measured in terms of receipts.

The advantages and disadvantages of each are summarized in the following table:

Advantages and Disadvantages of Various Forms of For-Profit Businesses

	Sole Proprietor	Partnership	Corporation
Advantages	1. Minimum bureaucratic hassle 2. Direct control by owner	1. Ability to share work and risks 2. Relatively easy to form	1. No personal liability 2. Increasing ability to get funds 3. Ability to shed personal income and gain added expenses
Disadvantages	1. Limited ability to get funds 2. Unlimited personal liability	1. Unlimited personal liability (even for partner's blunder) 2. Limited ability to get funds	1. Legal hassle to organize 2. Possible double taxation of income 3. Monitoring problems

Since corporations are the most complex, let's consider them more carefully. When a corporation is formed, it issues **stock** (*certificates of ownership in a company*), which is sold or given to individuals. Proceeds of the sale of that stock make up what is called the *equity capital* of a company. Ownership of stock entitles you to vote in the election of a corporation's directors.

Corporations were developed as institutions to make it easier for company owners to be separated from company management. A corporation provides the owners with **limited liability**—*the stockholder's liability is limited to the amount that stockholder has invested in the company*. With the other two forms of business, owners can lose everything they possess, even if they have only a small amount invested in the company, but in a corporation the owners can lose only what they have invested in that corporation. If you've invested $100, you can lose only $100. In the other kinds of business, even if you've invested only $100, you could lose everything; the business's losses must be covered by the individual owners. Corporations' limited liability makes it easier for them to attract investment capital. Corporations pay taxes, but they also offer their individual owners ways of legally avoiding taxes.

Q.3 Many businesses in Canada end with "Ltd.," while in the United States many end in "Inc." Why?

A corporation's stock can be distributed among a few people or among millions of stockholders. Stocks can be bought and sold either in an independent transaction between two people (an *over-the-counter* trade) or through a broker and a *stock exchange*.

In corporations, ownership is separated from control of the firm. Most stockholders have little input into the decisions a corporation makes. Instead, corporations are often controlled by their managers, who often run them for their own benefit as well as for the owners'. The reason is that owners' control of management is limited.

A large percentage of most corporations' stock is not even controlled by the owners; instead, it is controlled by financial institutions such as mutual funds (financial institutions that invest individuals' money for them) and by pension funds (financial institutions that hold people's money for them until it is to be paid out to them upon their retirement). Thus, ownership of corporations is another step removed from individuals.

Most corporations are controlled by managers, with little effective stockholder control.

Why is the question of who controls a firm important? Because economic theory assumes the goal of business owners is to maximize profits, which would be true of corporations if stockholders made the decisions. Managers don't have the same incentives to maximize profits that owners do. There's pressure on managers to maximize profits, but that pressure can often be weak or ineffective. An example of how firms deal with this problem involves stock options. Many companies give their managers stock options—rights to buy stock at a low price—to encourage them to worry about the price of their company's stock. But these stock options dilute the value of company ownership and decrease profits per share.

Q.4 It is obvious that all for-profit businesses in Canada will maximize profit. True or false? Why?

Finance and Business

Much of what you hear in the news about business concerns financial assets—assets that acquire value from an obligation of someone else to pay. Stocks are one example of a financial asset; bonds are another. Financial assets are traded in markets such as the Toronto Stock Exchange. Trading in financial markets can make people rich (or poor) quickly. Stocks and bonds can also provide a means through which corporations can finance expansions and new investments.

Recently there has been much in the news about one part of these financial markets—initial public offerings (IPOs), in which a company first offers some of its stock to the general public, with the owners keeping a large portion for themselves. This is an example of how financial markets work to fund companies, allowing the economy to grow.

see page 77

Stocks are usually traded on a *stock exchange*—a formal market in which stocks are bought and sold.

In order to buy or sell a stock, you contact a stockbroker (or simply contact the company through the Web—it's cheaper that way) and say you want to buy or sell whatever stock you've decided on—say Ford Motor Company. The commission you're charged for having the broker sell you the stock (or sell it for you) varies. It usually starts at some minimum between $10 and $30, and then is so much per share.

There are a number of stock exchanges. The largest and most familiar one in Canada is the Toronto Stock Exchange.

To judge how stocks as a whole are doing, a number of indexes have been developed. These include the TSX, and in the U.S., Standard and Poor's (S&P 500), the Wilshire Index, the Russell 2000, and the Dow Jones Industrial Average. The TSX replaced the TSE300 index in May 2002. It's the one you are most likely to hear about in the news.

When a share of a corporation's existing stock is sold on the stock exchange, corporations get no money from that sale. The sale is simply a transfer of ownership from one individual (or organization) to another. The only time a corporation gets money from the sale of stock is when it first issues the shares.

HOW TO READ THE STOCK TABLES

Get all your quotes throughout the day at www.nationalpost.com

Lines in boldface indicate that the stock closed at least 5% higher or lower than the previous board lot closing price. Stocks must close at a minimum $1 and trade at least 500 shares to qualify.

Underlined stocks have traded 500% or more above their 60-day average daily volume.

1. Up/down arrows indicate a new 52-week high or low in the day's trading
2. 52-week high/low: Highest and lowest inter-day price reached in the previous 52 weeks
3. Stock names have been abbreviated
4. Ticker: Basic trading symbol for primary issues (usually common)
5. Dividend: Indicated annual rate. See footnotes
6. Yield %: Annual dividend rate or amount paid in past 12 months as a percentage of closing price in past 12 months
7. P/E: Price earnings ratio, closing price divided by earnings per share in past 12 months. Figures reported in US$ converted to C$
8. Volume: Number of shares traded in 00s; **z** - odd lot; **e** - exact no. of shares
9. High: Highest inter-day trading price
10. Low: Lowest inter-day trading price

	1	2	2	3	4	5	6	7	8	9	10	11	12
		52W high	52W low	Stock	Ticker	Div	Yield %	P/E	Vol 00s	High	Low	Close	Net chg
	↑x	29.25	21.15	MaxStockna	MAX	f0.50	1.96	7.4	210501	27.25	27.05	27.20	+0.10
	n	**39.25**	**31.15**	**MaxStockna..**	**MAX**	**f1.00**	**2.83**	**10.3**	**210501**	**37.25**	**37.05**	**37.20**	**+0.10**
	↓s	49.25	41.15	MaxStockna	MAX	f1.50	3.30	13.2	210501	47.25	47.05	47.20	+0.[illegible]

11. Close: Closing price
12. Net change: Change between board lot closing price and previous board lot closing price

If a Canadian listed stock doesn't trade, its last bid and ask price can be found in the bid/ask table

Footnotes
***** - traded in $US **x** - stock is trading ex-dividend **n** - stock is newly listed on exchange in past year **s** - stock has split in past year **c** - stock has consolidated in past year **a** - spinoff company distributed as shares **‡** - shares carry unusual voting rights **†** - denotes tier I stocks on CDNX (tier III stocks on CDNX have a **Y** as first letter of ticker)

Dividend footnotes
r - dividend in arrears **u** - US$ **p** - paid in the past 12 months including extras **y** - dividend paid in stock, cash equivalent **f** - floating rate, annualized **v** - variable rate, annualized based on last payment

Data supplied by CGI
905-479-STAR and FP DataGroup
416-350-6500
Historical Nasdaq supplied by DataStream

How to read the options, index options, futures prices and futures options tables

P/C - Option put or call. Futures prices open interest reflects previous trading day. **CBOT** - Chicago Board of Trade, **CDNX** - Canadian Venture Exchange, **CME** - Chicago Mercantile Exchange, **COMEX** - New York Commodity Exchange, **FINEX** - Financial Instruments Exchange, **IMM** - International Money Market, **CSCE** - Coffee, Sugar, Cocoa Exchange, **KBOT** - Kansas City Board of Trade, **MPLS** - Minneapolis Grain Exchange, **NYCE** - New York Cotton Exchange, **NYME** - New York Mercantile Exchange, **NYFE** - New York Futures Exchange, **r** - option not traded, **s** - no option offered, **TSE** - Toronto Stock Exchange, **TFE** - Toronto Futures Exchange, **WPG** - Winnipeg Commodity Exchange.

Source: *The National Post*, used with permission.

In the late 1990s, the stock prices of many of the ".com" (read: dot com) companies, which are firms in some way related to the Internet, exploded. The stocks sold for enormous values—values that many economists felt could not last. (See Appendix A at the end of this chapter for a discussion of how economists value financial assets.) They saw what is called a financial bubble—a situation in which financial assets' valuations are based on what almost everyone believes are insupportable expectations—and were expecting the bubble to burst, or at least to shrink down to a supportable size. That, in fact, is what happened late in 2000 and during 2001. Some firms that had been trading for $150 a share fell to $7. Most market-watchers think the bubble has burst, although late in 2001 some investment gurus predicted stock prices would continue to fall well into 2002.

E-Commerce and the Digital Economy

To say that many of the Internet companies were overvalued is not to say that e-commerce—buying and selling over the Internet—will not make an enormous difference to the economy, and to the nature of business. Almost all economists believe it will. E-commerce adds heavy competition to the economy, and will place pressure on existing firms to compete with low prices. If they don't, they will lose their customers. More and more individuals are buying cars, books, prescription drugs, and even groceries on the Web, and the range of what is bought will expand. One reason is that the Internet removes the importance of geography and location for firms. In doing so it reduces the value of firms whose comparative advantage depends on geography, while increasing the value of firms whose comparative advantage does not depend on geography.

Some goods are more conducive to being sold on the Internet than others. For example, there are certain goods customers want to see before they buy them. Until virtual reality is perfected, that is impossible with the Internet. The practice of going to see goods at local stores but then buying them on the Internet will likely lead to some changes in the nature of businesses. Consider cars; because the Internet allows car buyers to compare prices and get a better deal than the local dealer generally gives, more and more people are buying cars over the Internet. Before they do so, however, they generally visit the local car dealership to see and test-drive the car. As Internet car buying expands, such practices are unlikely to continue. The future may well include car showrooms that charge you to test-drive, or even look at, a car (with the charge being rebated by the selling company if you actually buy the car).

E-commerce has grown enormously in the early 2000s, and a terminology developed that is captured in the table below. Initially, B2B (business selling to business) activity grew, but B2C (business selling to consumers) was also growing. Auction sites C2C (consumers selling to consumers) and bidding sites for consumers (C2B) were also beginning to grow.

B2B	B2C
C2B	C2C

HOUSEHOLDS

The second classification we'll consider in this overview of Canadian economic institutions is households. **Households** *(groups of individuals living together and making joint decisions)* are the most powerful economic institution. They ultimately influence government and business, the other two economic institutions. Households' votes in the political arena determine government policy; their decisions about supplying labour and capital determine what businesses will have available to work with; and their spending decisions or expenditures (the "votes" they cast with their dollars) determine what business will be able to sell.

In the economy, households vote with their dollars.

The Power of Households

While the ultimate power does in principle lie with the people and households, much of that power has been assigned to representatives. As we discussed above, corporations are only partially responsive to owners of their stocks, and much of that ownership is once-removed from individuals. Ownership of 1,000 shares in a company with a total of 2 million shares isn't going to get you any influence over the corporation's activities. As a stockholder, you simply accept what the corporation does.

A major decision that corporations make independently of their stockholders concerns what to produce. Ultimately households decide whether to buy what business produces, but business spends a lot of money telling us what services we want, what products make us "with it," what books we want to read, and the like. Most economists believe that consumer sovereignty reigns—that we are not fooled or controlled by advertising. Still, it is an open question in some economists' minds whether we, the people, control business or the business representatives control people.

Consumer sovereignty reigns, but it works indirectly by influencing businesses.

Because of this assignment of power to other institutions, in many spheres of the economy households are not active producers of output but merely passive recipients of income, primarily in their role as suppliers of labour.

Suppliers of Labour

The largest source of household income is wages and salaries (the income households get from labour). Households supply the labour with which businesses produce and government governs. The total Canadian labour force is about 16 million people, about 7 percent (1.1 million) of whom were unemployed in 2001. The average Canadian work week is about 39 hours. The average pay in Canada was nearly $700 per week, which translates to about $18.00 per hour. Of course, that average has enormous variability and depends on the occupation and region of the country where one is employed. For example, lawyers often earn $100,000 per year; physicians earn about $150,000 per year; and CEOs of large corporations may make $2 million per year or more. A beginning McDonald's employee generally makes about $12,000 per year.

The table below shows current and predicted labour market conditions (in 2004) by skill level and skill type. The skill level refers to the minimum level of education and training required to work in those occupations, while the skill type refers to a broad industry category. Primary industry and sales and services are both expected to decline, whereas labour market conditions in almost every other market are expected to remain at their current levels or improve.

Skill Types	Skill Levels											
	Managerial		Professional		Technical, paraprofessional & skilled		Intermediate		Labouring & elemental		All	
	Current	2004	Current	2004	Current	2004	Current	2004	Current	2004	Current	2004
Business, Finance & Administration	Good	Good	Good	Good	Good	Good	Fair	Fair	--	--	Good	Good
Natural & Applied Sciences	Good	Good	Good	Good	Good	Good	--	--	--	--	Good	Good
Health	Good	Good	Good	Good	Fair	Fair	Fair	Fair	--	--	Good	Good
Social Science, Education, Government Service & Religion	Good	Good	Fair	Fair	Fair	Fair	--	--	--	--	Fair	Fair
Art, Culture, Recreation & Sport	Good	Good	Fair	Fair	Fair	Fair	--	--	--	--	Fair	Fair
Sales & Services	Good	Good	--	--	Fair	Good	Fair	Limited	Limited	Limited	Fair	Limited
Trades, Transport & Equipment Operators	Fair	Good	--	--	Fair	Fair	Fair	Fair	Limited	Limited	Fair	Fair
Primary Industry	Good	Fair	--	--	Fair	Fair	Limited	Limited	Limited	Limited	Fair	Fair
Processing, Manufacturing & Utilities	Good	Good	--	--	Good	Good	Fair	Fair	Limited	Limited	Fair	Fair
All	Good	Good	Good	Good	Fair	Fair	Fair	Fair	Limited	Limited	Fair	Fair

Source: Human Resources Development Canada, Job Futures 2000 World of Work, http://jobfutures.ca/doc/jf/lmo/part1/en/table1.shtml

GOVERNMENT

The third major Canadian economic institution we'll consider is government. Government plays two general roles in the economy. It's both a referee (setting the rules that determine relations between business and households) and an actor (collecting money in taxes and spending that money on its own projects, such as health care and education). Let's first consider government's role as an actor.

Government as an Actor

Canada has a federal government system, which means we have various levels of government (federal, provincial, and local), each with its own powers. Together they consume about 20 percent of the country's total output and employ about 800,000 individuals. The various levels of government also have a number of programs that redistribute income through taxation or through an array of social welfare and assistance programs designed to help specific groups.

Provincial and Local Government Provincial and local governments employ over 450,000 people and spend over $250 billion a year. As you can see in Figure 3-2(a), provincial and local governments get much of their income from taxes: property taxes, sales taxes, and provincial income taxes. They spend their tax revenues on social services, health care, and education (education through high school is available free in public schools), and roads, as Figure 3-2(b) shows.

Federal Government Probably the best way to get an initial feel for the federal government and its size is to look at the various categories of its tax revenues and expenditures in Figure 3-3(a). Notice income taxes make up about 63 percent of the federal government's revenue, while sales taxes make up about 20 percent. That's more than 80 percent of the federal government's revenues, most of which shows up as a deduction from your paycheque. In Figure 3-3(b), notice that the federal government's two largest categories of spending are social services and debt charges, with transfer payments well behind.

see page 77

Q.5 The largest percentage of federal expenditure is in what general category?

What are transfer payments? You'll notice there are two categories of revenues in Figure 3-2(a) listed as general purpose transfers and special purpose transfers. **Transfer payments** are payments by governments to individuals that are not in return for goods and services, with the federal government making transfers to the provinces, and the provinces making transfers to the municipalities.

Figure 3-2 (a and b) INCOME AND EXPENDITURES OF PROVINCIAL AND LOCAL GOVERNMENTS, 2001

The charts give you a sense of the importance of provincial and local governments—where they get (a), and where they spend (b), their revenues.

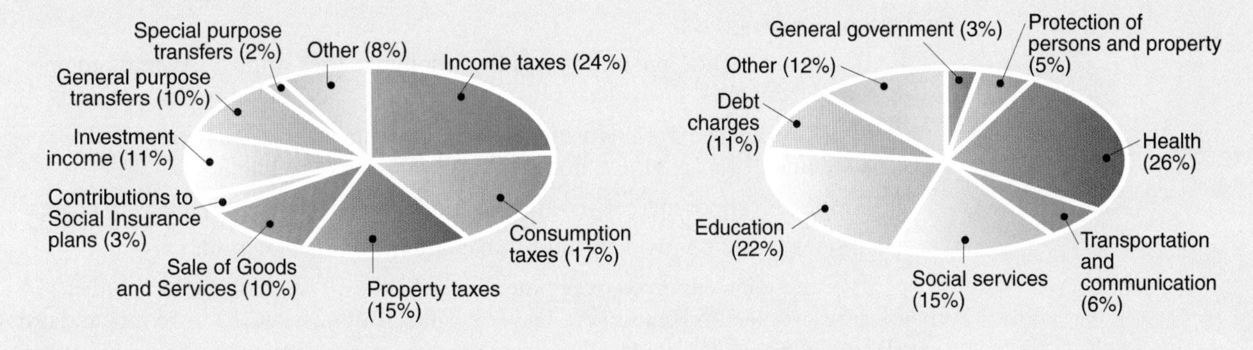

Source: Statistics Canada, Cansim II, Table 385-001

Figure 3-3 (a and b) INCOME AND EXPENDITURES OF THE FEDERAL GOVERNMENT, 2001

The pie charts show the sources and uses of federal government revenue. It is important to note that, when the government runs a deficit, expenditures exceed income and the difference is made up by borrowing, so the size of the income and expenditure pies may not be equal.

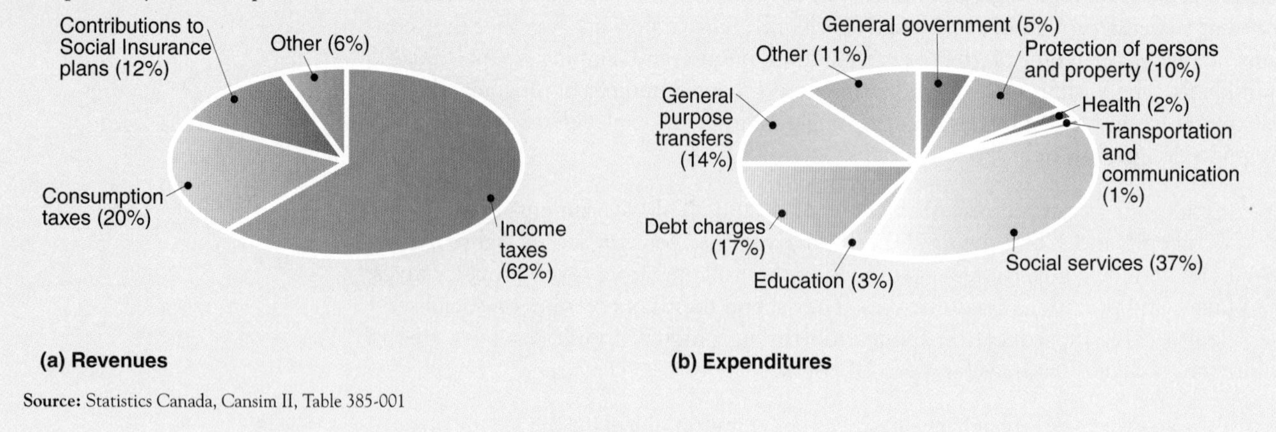

Source: Statistics Canada, Cansim II, Table 385-001

General purpose transfers are equalization payments that are meant to reduce disparities between the "have" and the "have not" provinces. They are supposed to allow all provinces to provide similar services at comparable levels of taxation. Special purpose transfers are primarily aimed at funding social spending on health care, welfare, and post-secondary education. The special purpose transfers are known as the Canada Health and Social Transfer.

Government as a Referee

Even if government spending made up only a small proportion of total expenditures, government would still be central to the study of economics. The reason is that, in a market economy, government sets the rules of interaction between households and businesses, and acts as a referee, changing the rules when it sees fit. Government decides whether economic forces will be allowed to operate freely.

Some examples of Canadian laws regulating the interaction between households and businesses today are:

1. Businesses are not free to hire and fire whomever they want. They must comply with labour laws. Even closing a plant requires notice for many kinds of firms.
2. Many working conditions are subject to government regulation: safety rules, wage rules, overtime rules, hours-of-work rules, and the like.
3. Businesses cannot meet with other businesses to agree on prices they will charge.
4. In some businesses workers must join a union to work at certain jobs.

Most of these laws evolved over time. Up until the 1930s, household members, in their roles as workers and consumers, had few rights. Businesses were free to hire and fire at will and, if they chose, to deceive and take advantage of consumers.

Over time, new laws to curb business abuses have been passed, and government agencies have been formed to enforce these laws. Many people think the pendulum has swung too far the other way. They believe businesses are saddled with too many regulatory burdens.

Figure 3-4 THE CANADIAN ECONOMY IN A GLOBAL SETTING

One big question that we'll address throughout this book is: What referee role should the government play in an economy? For example, should government use its taxing powers to redistribute income from the rich to the poor? Should it allow mergers between companies? Should it regulate air traffic? Should it regulate prices? Should it attempt to stabilize fluctuations of aggregate income?

THE GLOBAL SETTING

We now turn to a discussion of how international issues affect the Canadian economy. Nearly every decision facing Canada or a firm operating in Canada involves international issues. The Canadian economy is integrated with the world economy, and we cannot reasonably discuss Canadian economic issues without discussing the role that international considerations play in these issues.

International issues must be taken into account in just about any economic decision a country or a firm faces.

Probably the best way to put the Canadian economy into perspective is to look at a map of the world, such as the one in Figure 3-4. Notice the physical size of Canada. Compared to the world, it's relatively small. Alternatively, consider population; the Canadian population is about 31 million; the world population is 6 billion, so the Canadian population is about 0.5 percent of the world population. The importance of Canada increases somewhat when we consider that Canadian production accounts for nearly 2 percent of world output, but it is important for Canadians to remember that there is a whole world out there, and Canada is simply a part of that world.

Global Corporations

Global corporations are corporations with substantial operations on both the production and sales sides in more than one country.

The importance of the world to the Canadian economy can also be seen by considering the clothes on your back. Most likely they were made abroad. Similarly with the cars you drive. It's likely that half of you drive a car that was made abroad. Of course, it's often difficult to tell. Just because a car has a Japanese or German name doesn't mean that it was produced abroad. Some Japanese and German companies have manufacturing plants in Canada. Others, such as Chrysler and Daimler Benz (they make Mercedes) have merged. When goods are produced by **global corporations** (*corporations with substantial operations in both production and sales in more than one country*), corporate names don't always tell much about where a good is produced. As global corporations' importance has grown, most manufacturing decisions have come to be made in reference to the international market, not the Canadian domestic market.

Global corporations offer enormous benefits for countries. They create jobs; they bring new ideas and new technologies to a country, and they provide competition for domestic companies, keeping them on their toes. But global corporations also pose a number of problems for governments. One is their implication for domestic and international policy. A domestic corporation exists within a country and can be dealt with using policy measures within that country. A global corporation exists within many countries and there is no global government to regulate or control it. If it doesn't like the policies in one country—say taxes are too high or regulations too tight—it can shift its operations to other countries.

At times it seems that global corporations are governments unto themselves, especially in relation to poorer countries. In terms of sales, a number of global corporations are larger than the economies of middle-size countries. For example, General Motors has a total annual revenue of U.S. $177 billion; Ecuador has a total annual income of U.S. $60 billion. This comparison is not quite accurate, since sales do not necessarily reflect power; but when a company's decisions can significantly affect what happens in a country's economy, that company has significant economic power.

When global corporations have such power, it is not surprising that they can sometimes dominate a country. The corporation can use its expertise and experience to direct a small country to do its bidding rather than the other way around.

Before you condemn globals, remember: Globals don't have it so easy either. Customs and laws differ among countries. Trying to meet the differing laws and ambiguous limits of acceptable action in various countries is often impossible. For example, in many countries bribery is an acceptable part of doing business. If you want to get permission to sell a good, you must pay the appropriate officials *baksheesh* (as it's called in Egypt) or *la mordita* (as it's called in Mexico). In Canada, such payments are called bribes and are illegal. Given these differing laws, the only way a Canadian company can do business in some foreign countries is to break Canadian laws.

Moreover, global corporations often work to maintain close ties among countries and to reduce international tension. If part of your company is in an Eastern European country and part is in Canada, you want the two countries to be friends. So beware of making judgments about whether global corporations are good or bad. They're both simultaneously.

International Trade

The volume and value of world trade have grown substantially over the last century but there have been significant fluctuations around this trend. Sometimes trade has grown rapidly; at other times it has grown slowly or has even fallen off.

In part, fluctuations in world trade result from fluctuations in world output. When output rises, international trade rises; when output falls, international trade falls.

APPLYING THE TOOLS

A World Economic Geography Quiz

Economic geography isn't much covered in most economics courses because it requires learning enormous numbers of facts, and university and college courses aren't a good place to learn facts. Postsecondary studies are designed to teach you how to interpret and relate facts. Unfortunately, if you don't know facts, much of what you learn now isn't going to do you much good. You'll be relating and interpreting air. The following quiz presents some facts about the world economy. Below we list characteristics of 20 countries or regions in random order. Beneath the characteristics, in alphabetical order, we list 20 countries or regions. Associate the characteristics with the country or region.

If you answer 15 or more correctly, you have a reasonably good sense of economic geography. If you don't, we strongly suggest learning more facts. The study guide has other projects, information, and examples. An encyclopedia has even more, and your library has a wealth of information. You could spend the entire term following the economic news carefully, paying attention to where various commodities are produced, and picturing in your mind a map whenever you hear about an economic event.

1. Former British colony, now small independent island country famous for producing rum.
2. Large sandy country contains world's largest known oil reserves.
3. Very large country with few people; produces 25 percent of the world's wool.
4. Temperate country ideal for producing wheat, soybeans, fruits, vegetables, wine, and meat.
5. Small tropical country produces abundant coffee and bananas.
6. Has world's largest population and world's largest hydropower potential.
7. Second-largest country in Europe; famous for wine and romance.
8. Former Belgian colony with vast copper mines.
9. European country; exports luxury clothing, footwear, and automobiles.
10. Country that has depleted many of its natural resources but that has the highest level of GDP of any country in the world.
11. Long, narrow country of four main islands; most thickly populated country in the world; exports majority of the world's electronics products.
12. Recently politically reunified country; one important product is steel.
13. Second-largest country in the world; leading zinc exporter.
14. European country for centuries politically repressed; now becoming industrialized; chemicals are one of its leading exports.
15. 96 percent of its people live on 4 percent of the land; much of the world's finest cotton comes from here.
16. Politically and radically troubled African nation has world's largest concentration of gold.
17. Huge, heavily populated country eats most of what it raises but is a major tea exporter.
18. Country that is a top producer of oil and gold; has recently undergone major political and economic changes.
19. Has only about 50 people per square mile but lots of trees; timber and fish exporter.
20. Sliver of a country on Europe's Atlantic coast; by far the world's largest exporter of cork.

A. Argentina
B. Australia
C. Barbados
D. Canada
E. China
F. Costa Rica
G. Egypt
H. France
I. Germany
J. India
K. Italy
L. Japan
M. Portugal
N. Russia
O. Saudi Arabia
P. South Africa
Q. Spain
R. Sweden
S. United States
T. Democratic Republic of the Congo

Answers: 1–C, 2–O, 3–B, 4–A, 5–F, 6–E, 7–H, 8–T, 9–K, 10–S, 11–L, 12–I, 13–D, 14–Q, 15–G, 16–P, 17–J, 18–N, 19–R, 20–M.

Fluctuations in world trade are also explained in part by trade restrictions that countries have imposed from time to time. For example, decreases in world income during the Depression caused a large decrease in trade, but that decrease was exacerbated by a worldwide increase in trade restrictions during the 1930s.

Differences in the Importance of Trade The importance of international trade to countries' economies differs widely, as we can see in the table below, which presents the importance of the shares of **exports**—*the value of goods sold abroad*—and **imports**—*the value of goods purchased abroad*—for various countries.

Canada's major trading partner is the United States.

	GDP*	Export Ratio	Import Ratio
United States	$10 486	11%	14%
Canada	703	43	40
Netherlands	382	66	59
Germany	1 858	29	27
United Kingdom	1 417	26	27
Italy	1 097	28	22
France	1 314	26	23
Japan	4 177	9	8

*Numbers in billions of U.S. dollars.
Source: *The World Development Report,* The World Bank, 2001 (http://www.worldbank.org) and individual country Web pages.

Q-6 Among the countries listed in the table, which has the lowest exports and imports as a percentage of GDP?

Among the countries listed, the Netherlands has the highest amount of exports compared to GDP; the United States and Japan have the lowest.

The Netherlands' imports are also the highest as a percentage of GDP. Japan has the lowest. The relationship between a country's imports and its exports is no coincidence. For most countries, imports and exports roughly equal one another, though in any particular year that equality can be rough indeed. For Canada in recent years, exports have generally significantly exceeded imports. But that situation can't continue forever, as we'll discuss.

Total trade figures provide us with only part of the international trade picture. We must also look at what types of goods are traded and with whom that trade is conducted.

What Canada Trades and with Whom The majority of Canadian exports and imports involve significant amounts of manufactured goods. This isn't unusual, since much of international trade is in manufactured goods.

Figure 3-5 shows the regions with which Canada trades. Exports to the United States and the European Union made up the largest percentage of total Canadian exports to individual countries in 2001. Countries from which Canada imports major quantities include the United States and the regions of the European Union and the Pacific Rim. Thus, the countries we export to are also the countries we import from.

One reason a Canadian economics course must consider international issues early on is the Canadian **balance of trade**—the difference between the value of goods and services Canada exports and the value of goods and services it imports. When *imports exceed exports* the country is running a balance of **trade deficit**; when it *exports more than it imports*, it runs a balance of **trade surplus**. A large share of Canadian production is destined for foreign markets so the balance of trade is an important measure in Canada.

We have to be careful about terminology when we look at Canada's international position because the balance of trade contains two components. The first is the

Figure 3-5 (a and b) CANADIAN EXPORTS AND IMPORTS BY REGION, 2000

Major regions that trade with Canada include the United States, the European Union, and the Pacific Rim.

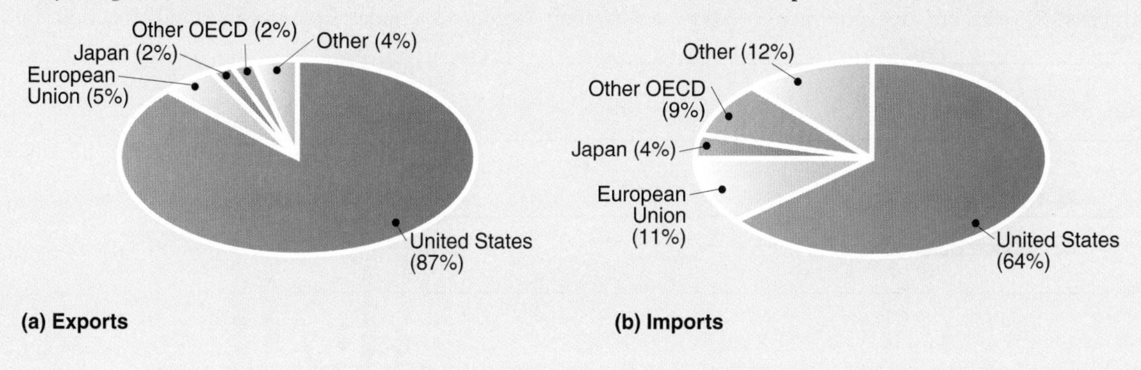

merchandise trade balance—it *measures the difference between exports and imports of goods.* The second component is the **services balance**, which *measures trade in services.* Canada has been a net exporter of goods since the early 1970s but we've been a net importer of services since the 1950s. It may sound funny, but when your cousin Fred goes to Florida on vacation, he's importing the services the Florida vacation provides. The Canadian goods and services trade balance has cycled up and down as the balance of trade and the services balance have changed over time. Figure 3-6 (on the next page) shows how this looks for the last 30 years or so.

Debtor and Creditor Nations Running a trade surplus isn't necessarily a bad thing, nor is running a trade deficit. In fact, if you were a country, running a trade deficit would be rather nice. It would probably mean that you are consuming (importing) more than you're producing (exporting). How can you do that? By living off past savings, getting support from your parents or spouse, or borrowing.

Countries have the same options. They can live off foreign aid, past savings, or loans. The **current account balance** *measures trade in goods and services and includes the interest payments we make to foreigners for the use of their savings.* From 1986 until 1998, Canada ran a current account deficit (with the exception of a small surplus in 1996), which meant it was borrowing from abroad and selling off assets—financial assets such as stocks and bonds, or real assets such as real estate and corporations. Since the assets of Canada total many billions of dollars, we could run trade deficits for many years. The problem with doing so is that it requires that we continually borrow from abroad, and the interest payments we make to foreigners for the use of their funds don't contribute to Canadian incomes.

In 1999 and 2000, Canada had a current account surplus, which means we were exporting more than we were importing. As we go to press, figures for 2001 suggest we also ran a surplus, but since the figures are subject to sometimes substantial revisions, they are preliminary. We don't need to borrow as much, or look for friendly support. If we look south, we'll see that the United States is the world's largest debtor. It has a huge trade deficit — it has borrowed more from abroad than it has lent abroad.

Q-7 Will a debtor nation necessarily be running a trade deficit?

Figure 3-6 **THE CANADIAN BALANCE OF TRADE**

The balance of trade is composed of the merchandise trade balance and the services balance. The Canadian services balance has been negative for the past 30 years, but we export more goods than we import, leading to a trade surplus for most of this period.

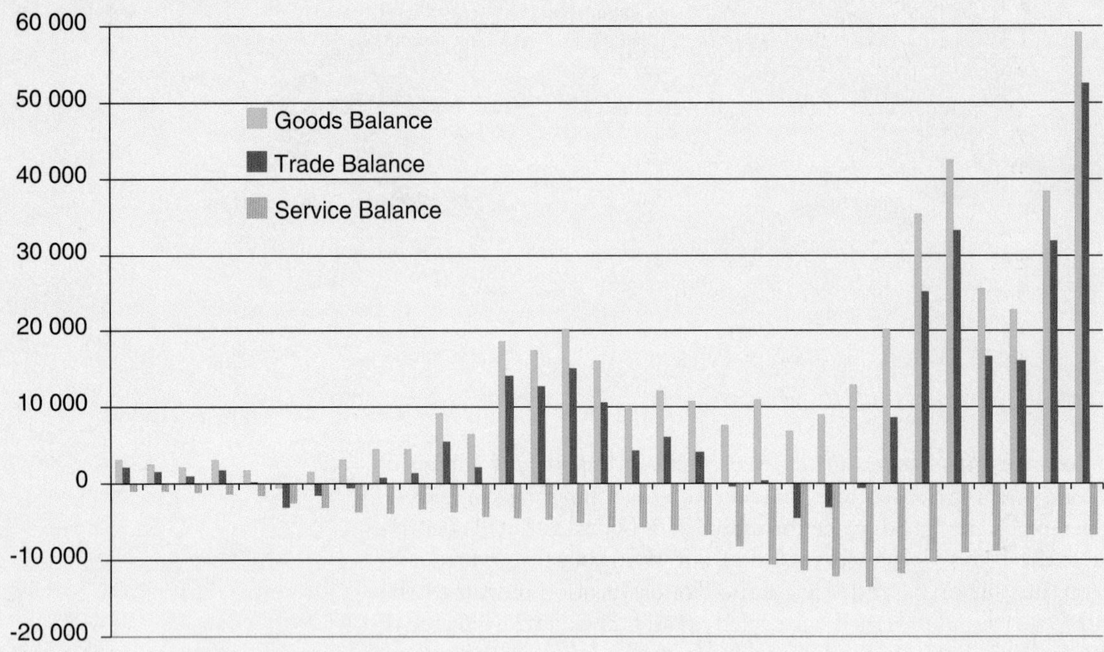

Source: Statistics Canada, Cansim II Table 380–0027.

BEYOND THE TOOLS

International Issues in Perspective

Let's consider an example from history. In the Middle Ages, Greek ideas and philosophy were lost to Europe when hordes of barbarians swept over the continent. These ideas and that philosophy were only rediscovered in the Renaissance as a by-product of trade between the Italian merchant cities and the Middle East. (The Greek ideas that had spread to the Middle East were protected from European upheavals.) *Renaissance* means rebirth; a rebirth in Europe of Greek learning. Many of our traditions and sensibilities are based on those of the Renaissance, and that Renaissance was caused, or at least significantly influenced, by international trade. Had there been no trade, our entire philosophy of life might have been different.

Fernand Braudel, a French historian, has provided wonderful examples of the broader implications for trade. For instance, he argued that the effects of international trade, specifically Sir Walter Raleigh's introduction of the potato into England from South America in 1588, had more long-term consequences than the celebrated 1588 battle between the English navy and Spanish Armada.

Another example, which Braudel did not live to see, is the major change in socialist countries in the 1990s. Through the 1960s China, the Soviet Union, and the Eastern European countries were relatively closed societies—behind the Iron Curtain. That changed in the 1970s and 1980s as these socialist countries opened up trade with the West as a way to speed up their own economic development. That trade, and the resulting increased contact with the West, gave the people of those countries a better sense of the material goods to be had in the West. That trade also spread Western ideas of the proper organization of government and the economy to these societies. A strong argument can be made that along with trade came the seeds of discontent that changed those societies and their economies forever.

In economics courses we do not focus on these broader cultural issues but instead focus on relatively technical issues such as the reasons for trade and the implications of tariffs. But keep in the back of your mind these broader implications as you go through the various components of international economics. They add a dimension to the story that otherwise might be forgotten.

KNOWING THE TOOLS

Our International Competitors

The world economy is often divided into three main areas or trading blocs: The Americas, Europe and Africa, and East Asia. These trading blocs are shown in the map below.

The table below gives you a sense of the similarities and differences in the economies of the United States, Japan, and Europe.

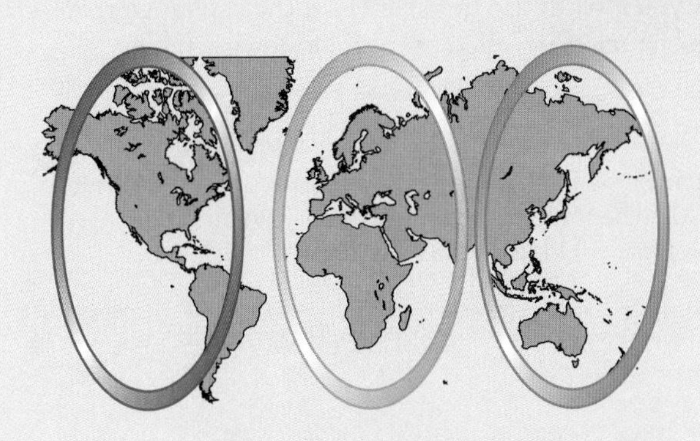

The three dominant economies in these trading blocs are the Unites States, Germany, and Japan. Each area has a major currency. In the Americas, it is the U.S. dollar; in Europe it is the euro, a new currency recently created by the European Union; and in East Asia it is the Japanese yen.

	Canada	United States	Japan	Europe
Area (square km)	9,976,140	9,629,091	377,835	2,371,321
Population (million)	31.6	278	126.8	292
GDP (billion US$)	774.7	9,963	3,150	€6,430
GDP per capita	$24,500	$35,800	$24,800	€22,000
Natural resources	Timber, fish, metal ore, natural gas	Coal, copper, lead, and others	Very few	Coal, iron ore, natural gas, fish, and others
Exports/GDP	35%	8%	14%	33%
Imports/GDP	31%	12%	11%	28%
Currency value ($C per foreign currency unit, January 25, 2002)	1.00	1.61	0.01196	1.39

Source: CIA World Factbook (www.cia.gov/cia/publication/factbook), European Central Bank. All values are in U.S. dollars unless otherwise noted.

HOW INTERNATIONAL TRADE DIFFERS FROM DOMESTIC TRADE

*Inter*national trade differs from *intra*national (domestic) trade in two ways. First, international trade involves potential barriers to the flow of imports and exports. Before they can be sold in Canada, international goods must be sent through customs; that is, when they enter Canada they are inspected by Canadian officials and usually charged fees, known as *customs*. A company in Toronto can produce output to sell in any province without worrying that its right to sell will be limited; a producer outside the Canadian boundary cannot. At any time, a foreign producer's right to sell in Canada can be limited by government-imposed **quotas** (*limitations on how much of a good can be shipped into a country*), **tariffs** (*taxes on imports*), and **nontariff barriers** (*indirect regulatory restrictions on imports and exports*).

Two ways in which *inter*national trade differs from *intra*national (domestic) trade are:

1. International trade involves potential barriers to trade; and
2. International trade involves multiple currencies.

Q-8 What is the difference between a quota and a tariff?

The last category, indirect regulatory restrictions on imports and exports, may be unfamiliar to you, so let's consider an example. U.S. building codes require that plywood have fewer than, say, three flaws per sheet. Canadian building codes require that plywood have fewer than, say, five flaws per sheet. The different building codes are a nontariff barrier, making trade in building materials between the United States and Canada difficult.

The second way international trade differs from domestic or intranational trade is countries' use of different currencies. When people in one country sell something to people in another, they must find a way to exchange currencies as well as goods. **Foreign exchange markets** (*markets where one currency can be exchanged for another*) have developed to provide this service.

A foreign exchange market is a market in which one currency can be exchanged for another.

How many dollars will a Canadian have to pay to get a given amount of the currency of another country? That depends on the supply of and demand for that currency. To find out what you'd have to pay, you look in the newspaper for the **exchange rate**—*the rate at which one currency is traded for another*—such as in the following table:

A Foreign Exchange Rate Table

3.4 
see page 77

From the exchange rate table, you learn how much a dollar is worth in other countries. For example, on this day, Feb. 28, 2002, the table tells you how many dollars other currencies can buy. For example, one Euro could buy 1.3896 dollars.

FOREIGN EXCHANGE BY COUNTRY

Supplied BMO Nesbitt Burns Capital Markets – indicative noon rates

Currency	in C$	in US$	Daily % chg	Currency	in C$	in US$	Daily % chg
Antigua, Gr. EC $	0.6010	0.3745	nil	Kuwait (Dinar)	5.2118	3.2478	nil
Argentina (Peso)	0.74464	0.46404	-0.70	Lebanon (Pound)	0.001060	0.000661	-0.02
Austria (Euro)	1.3896	0.8660	0.19	Luxemb. Euro	1.38960	0.86596	0.19
Bahamas (Dollar)	1.6047	1.0000	nil	Malaysia (Ringgit)	0.4223	0.2632	-0.01
Bahrain (Dinar)	4.2565	2.6525	nil	Malta (Lira)	3.4817	2.1697	0.07
Barbados (Dollar)	0.8064	0.5025	nil	Netherlands (Euro)	1.3896	0.5460	0.19
Belgium (Euro)	1.3896	0.8660	0.19	Neth. Ant. Guilder	0.9015	0.5618	nil
Bermuda (Dollar)	1.6047	1.0000	nil	New Zealand $	8.6767	0.4217	0.52
Brazil (Real)	0.6802	0.4239	0.25	Norway (Krone)	0.1801	0.1122	0.07
Bulgaria (Lev)	0.713835	0.4448	0.29	Pakistan (Rupee)	0.02677	0.01668	-0.08
Chile (Peso)	0.002388	0.001488	0.31	Panama (Balloa)	1.6047	1.0000	nil
Colombia (Peso)	0.000697	0.000434	0.22	Philippines (Peso)	0.03137	0.01955	0.23
Costa Rica Colon	0.004624	0.002881	-0.04	Poland (Zloty)	0.3798	0.2367	-0.73
Cuba (Peso)	0.0764	0.0476	nil	Portugal (Euro)	1.3896	0.8660	0.19
Cyprus (Pound)	2.4145	1.5047	0.32	Peru (New Sol)	0.46265	0.28831	0.07
Czech (Koruna)	0.0440	0.0274	0.56	Romania (Leu)	0.000049	0.000031	-0.23
Denmark (Krone)	0.1870	0.1165	0.19	Russia (Ruble)	0.051836	0.032303	nil
Dominican Rep Peso	0.0973	0.0606	nil	Saudi Arabia Riyal	0.4279	0.2667	nil
Egypt (Pound)	0.3473	0.2165	-0.22	Slovakia (Koruna)	0.0334	0.0208	0.49
Finland (Euro)	1.3896	0.8660	0.19	Slovenia (Tolar)	0.0062	0.0039	0.23
Greece (Euro)	1.3896	0.8660	0.19	S. Africa (Rand)	0.1415	0.0882	0.45
Guyana (Dollar)	0.00889	0.00554	nil	S. Korea (Won)	0.001221	0.000761	0.27
Hong Kong (Dollar)	0.2058	0.1282	nil	Spain (Euro)	1.3896	0.8660	0.19
Hungary (Forint)	0.00567	0.00354	0.09	Sri Lanka (Rupee)	0.01717	0.01070	nil
India (Rupee)	0.03298	0.02055	nil	Sweden (Krona)	0.1533	0.0956	0.14
Indonesia Rupiah	0.000158	0.000099	0.39	Taiwan (Dollar)	0.04584	0.0286	nil
Ireland (Euro)	1.3896	0.8660	0.19	Thailand (Baht)	0.03674	0.0229	0.23
Israel (N Shekel)	0.3444	0.2146	-1.16	Trinidad & Tob: $	0.2626	0.1637	nil
Italy (Euro)	1.3896	0.8660	0.19	Turkey (Lira)	0.0000012	0.0000007	1.08
Jamaica (Dollar)	0.03385	0.02110	-0.11	Venezuela Bolivar	0.001581	0.00099	3.94
Jordan (Dinar)	2.2665	1.4124	nil	Spec Draw Right SDR	1.9924	1.2416	0.06

Source: Reprinted by permission of *The National Post*, March 1, 2002.

By looking at an exchange rate table, you can determine how much various goods will likely cost in different countries.

If you want shekels, you'll have to pay about 34¢ apiece. If you want punt, one rand will cost you $0.1415. (If you're wondering what shekels and rand are, look at the table above.)

Unless you collect currencies, the reason you want the currency of another country is that you want to buy something that country produces or an existing asset of that country. Say you want to buy a Hyundai car that costs 13,476,000 South Korean won. Looking at the table, you see that the exchange rate is $1 for 819 won. Dividing 819 into 13,476,000 won tells you that you need $16,454 to buy the car. So before you can buy the Hyundai, somebody must go to a foreign exchange market with $16,454 and exchange those dollars for 13.476 million won. Only then can the car be bought in Canada. Most final buyers don't do this; the importer does it for them. But whenever a foreign good is bought, someone must trade currencies.

Q-9 You are going to Chile and plan to exchange $100. Based on the foreign exchange rate table in the text, how many Chilean pesos will you receive?

INSTITUTIONS SUPPORTING FREE TRADE

As we stated in Chapter 2, economists generally like markets and favour trade being as free as possible. They argue that trade allows specialization and the division of labour. When each country follows its comparative advantage, production is more efficient, and the production possibility curve shifts out. These views mean that most economists, liberal and conservative alike, generally oppose international trade restrictions.

Free Trade Organizations

Despite political pressures to restrict trade, governments have generally tried to follow economists' advice and have entered into a variety of international agreements and organizations. The most important is the **World Trade Organization (WTO),** which is *an organization committed to getting countries to agree not to impose new tariffs or other trade restrictions except under certain limited conditions*. The WTO is the successor to the **General Agreement on Tariffs and Trade (GATT),** *an agreement among many subscribing countries on certain conditions of international trade*, to which you will still occasionally see references, even though the WTO has taken its place. One of the differences between the WTO and GATT is that the WTO includes some enforcement mechanisms.

Important international economic organizations include the WTO, GATT, the EU, and NAFTA.

The push for free trade has a geographic dimension, which includes **free trade associations**—*groups of countries that have reduced or eliminated trade barriers among themselves*. For example, the **European Union** is *a free trade association of 15 western European countries*, 12 of which have adopted a single currency called the euro. Other groups have loose trading relationships because of cultural or historical reasons. These loose trading relationships are sometimes called trading zones. For example, many European countries maintain close trading ties with many of their former colonies in Africa, where they fit into a number of overlapping trading zones. European companies tend to see that area as their turf. Similarly, the United States, Canada, and Mexico have created the **North American Free Trade Agreement (NAFTA),** *a U.S.–Canada–Mexico free trade zone that is phasing in reductions in tariffs*. Canada has entered into a free trade agreement with Chile, and continues to build close ties in South America, making the Western hemisphere another trading zone. Another example of a trading zone is that of Japan and its economic ties with other Far East countries; Japanese companies often see that area as their commercial domain.

These trading zones overlap, sometimes on many levels. For instance, Australia and England, Portugal and Brazil, and the United States and Saudi Arabia are tied together for historical or political reasons, and those ties lead to increased trade between them that seems to deviate from the above trading zones. Similarly, as companies become more and more global, it is harder and harder to associate companies with particular countries. Let us give an example: Do you know who the largest exporters of cars from the United States are? The answer is: Japanese automobile companies!

Thus, there is no hard-and-fast specification of trading zones, and knowing history and politics is important to understanding many of the relationships.

International Economic Policy Organizations

Just as international trade differs from domestic trade, so does international economic policy differ from domestic economic policy. When economists talk about Canadian economic policy, they generally refer to what the Canadian federal government can do to achieve certain goals. In theory, at least, the Canadian federal government has both the power and the legal right of compulsion to make Canadian citizens do what it says. It can tax, it can redistribute income, it can regulate, and it can enforce property rights.

There is no international counterpart to a nation's federal government. Any meeting of a group of countries to discuss trade policies is voluntary. No international body has powers of compulsion. Hence, international problems must be dealt with through negotiation, consensus, bullying, and concessions.

3.5 see page 77

To discourage bullying and to encourage negotiation and consensus, governments have developed a variety of international institutions to promote negotiations and coordinate economic relations among countries. These include the United Nations (UN), the World Bank, the World Court, and the International Monetary Fund (IMF). These organizations have a variety of goals. For example, the **World Bank** is *a multinational, international financial institution that works with developing countries to secure low-interest loans*, channelling such loans to them to foster economic growth. The **International Monetary Fund (IMF)** is *a multinational, international financial institution concerned primarily with monetary issues*. It deals with international financial arrangements. When developing countries encountered financial problems in the 1980s and had large international debts that they could not pay, the IMF helped work out repayment plans.

In addition to these formal institutions, there are informal meetings of various countries. These include the **Group of Five,** which *meets to promote negotiations and coordinate economic relations among countries*. The Five are Japan, Germany, Britain, France, and the United States. The **Group of Eight** also *meets to promote negotiations and coordinate economic relations among countries*. The Eight are the five countries just named plus Canada, Italy, and Russia.

Since governmental membership in international organizations is voluntary, their power is limited.

Since governmental membership in international organizations is voluntary, their power is limited. When Canada doesn't like a World Court ruling, it simply states that it isn't going to follow the ruling. When the United States is unhappy with what the United Nations is doing, it withholds some of its dues. Other countries do the same from time to time. Other member countries complain, but can do little to force compliance. It doesn't work that way domestically. If you decide you don't like Canadian policy and refuse to pay your taxes, you'll wind up in jail.

Q-10 If Canada chooses not to follow a World Court decision, what are the consequences?

What keeps nations somewhat in line when it comes to international rules is a moral tradition: Countries want to (or at least want to look as if they want to) do what's "right." Countries will sometimes follow international rules to keep international opinion favourable to them. But perceived national self-interest often overrides international scruples.

CONCLUSION

This has been a whirlwind tour of the Canadian economy and its global setting. The Canadian economy in the 21st century is a global economy with links through both its trade sector and its financial sector. To understand it, you must understand its components—business, households, and government—and their interrelationship.

The economy is undergoing significant changes because of technological change. E-commerce is growing exponentially and is making markets more global. On the

Internet the location of a trade doesn't matter. Countries, however, pose barriers to trade, and there will likely be much conflict as the push for free trade comes up against national boundaries.

Chapter Summary

- A diagram of the Canadian market economy shows the connections among businesses, households, and government. It also shows the Canadian economic connection to other countries.
- In Canada, businesses make the *what, how much,* and *for whom* decisions.
- Although businesses decide what to produce, they succeed or fail depending on their ability to meet consumers' desires. That's consumer sovereignty.
- The three main forms of business are corporations, sole proprietorships, and partnerships. Each has its advantages and disadvantages.
- Government plays two general roles in the economy: (1) as a referee, and (2) as an actor.
- Although households are the most powerful economic institution, they have assigned much of their power to government and business. Economics focusses on households' role as the supplier of labour.
- To understand the Canadian economy, one must understand its role in the world economy.
- Global corporations are corporations with significant operations in more than one country. They are increasing in importance.
- The areas with which Canada trades include the United States, the Pacific Rim countries, and the European Union.
- Canadian exports of goods and services have significantly exceeded imports for many years.
- International trade differs from domestic trade because (1) there are potential barriers to trade and (2) countries use different currencies.
- The rate at which one currency trades for another is that currency's exchange rate.
- Governments of many countries have formed free trade associations that agree to reduce barriers to trade among members. Three well-known free trade organizations are the WTO, NAFTA, and the European Union.
- International policy coordination must be achieved through consensus among nations.

Key Terms

balance of trade *(68)*
business *(57)*
consumer sovereignty *(57)*
corporations *(58)*
current account balance *(69)*
entrepreneurship *(57)*
European Union *(73)*
exchange rate *(72)*
exports *(68)*
foreign exchange markets *(72)*
free trade associations *(73)*
General Agreement on Tariffs and Trade (GATT) *(73)*
global corporations *(66)*
Group of Five *(74)*
Group of Eight *(74)*
households *(61)*
imports *(68)*
International Monetary Fund (IMF) *(74)*
limited liability *(59)*
merchandise trade balance *(69)*
nontariff barriers *(71)*
North American Free Trade Agreement (NAFTA) *(73)*
partnerships *(58)*
profit *(57)*
quotas *(71)*
services balance *(69)*
sole proprietorships *(58)*
stock *(58)*
tariffs *(71)*
trade deficit *(68)*
trade surplus *(68)*
World Bank *(74)*
World Trade Organization (WTO) *(73)*

Questions for Thought and Review

1. Why does an economy's strength ultimately reside in its people?
2. A market system is often said to be based on consumer sovereignty—the consumer determines what's to be produced. Yet business decides what's to be produced. Can these two views be reconciled? How? If not, why?
3. Why is entrepreneurship a central part of any business?

4. You're starting a software company in which you plan to sell software to your fellow students. What form of business organization would you choose? Why?
5. What are the two largest categories of federal government expenditures?
6. A good measure of a country's importance to the world economy is its area and population. True or false? Why?
7. What are the two ways in which international trade differs from domestic trade?
8. If one Canadian dollar will buy .67 Swiss francs, how many Canadian dollars will one Swiss franc buy?
9. The U.S. economy is falling apart because the United States is the biggest debtor nation in the world. Discuss.
10. Why do most economists oppose trade restrictions?
11. What is the relationship between GATT and the WTO?
12. Look up a recent foreign currency exchange rate table from *The National Post*.
    a. How many Egyptian pounds will you receive for $100?
    b. Say you want to buy a Volvo directly from Sweden. The foreign car dealer quotes a price of 235,794 Swedish krona. How many dollars will you have to exchange to purchase the Volvo?

Problems and Exercises

1. Go to a store in your community.
    a. Ask what limitations the owners faced in starting their business.
    b. Were these limitations necessary?
    c. Should there have been more or fewer limitations?
    d. Under what heading of reasons for government intervention would you put each of the limitations?
    e. Ask what kinds of taxes the business pays and what benefits it believes it gets for those taxes.
    f. Is it satisfied with the existing situation? Why? What would it change?
2. You've been appointed to a county counterterrorist squad. Your assignment is to work up a set of plans to stop a group of 10 terrorists the government believes are going to disrupt the economy as much as possible with explosives.
    a. List their five most likely targets in your county, city, or town.
    b. What counterterrorist action would you take?
    c. How would you advise the economy to adjust to a successful attack on each of the targets?
3. Tom Rollins heads a new venture called Teaching Co. He has taped lectures at the top universities, packaged the lectures on audio- and videocassettes, and sells them for $90 and $150 per eight-hour series.
    a. Discuss whether such an idea could be expanded to include college courses that one could take at home.
    b. What are the technical, social, and economic issues involved?
    c. If it is technically possible and cost-effective, will the new venture be a success?
4. This is a library research question.
    a. What are the primary exports of Brazil, Honduras, Italy, Pakistan, and Nigeria?
    b. Which countries produce most of the world's tin, rubber, potatoes, wheat, marble, and refrigerators?
5. This is an entrepreneurial research question. You'd be amazed what information is out there if you use a bit of initiative.
    a. Does the largest company in your relevant geographic area (town, city, whatever) have an export division? Why or why not?
    b. If you were an adviser to the company, would you suggest expanding or contracting its export division? Why or why not?
    c. Go to a store and look at 10 products at random. How many were made in Canada? Give a probable explanation of why they were produced where they were.
6. Exchange rates can be found in a variety of sources including the Internet.
    a. Using the exchange rate table on page 72, determine the Canadian dollar equivalent of the:
        (1) Euro
        (2) Zloty
        (3) Rand
        (4) Forint
    b. Determine the most recent dollar equivalent of those same currencies. (Use the Web or a recent newspaper.)
    c. Using the information in *a* or *b*, calculate the number of dollars you could get from one unit of each of the above currencies.

Web Questions

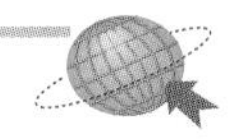

1. Go to the Nortel home page (www.nortelnetworks.com) and answer the following questions:
   a. Is Nortel a sole proprietorship, partnership, or corporation? For what reasons do you suspect it has chosen that form of business?
   b. Is Nortel a global corporation? Explain your answer.
   c. Are the shares of Nortel publicly traded?
2. Visit the Toronto Stock Exchange Web site (www.tse.com) and answer the following questions:
   a. How many IPOs were there last year on the TSE?
   b. How many firms' shares are included in the TSE 300?
   c. Find the area of the Web site that lists job openings. What kinds of qualifications are required for a position at the TSE?
3. Visit the Federal Department of Finance Web site (www.fin.gc.ca) and find the section on transfer payments to the provinces. Answer the following questions:
   a. Which provinces have most recently been recipients of equalization payments?
   b. How are equalization payments calculated?
   c. Have Canadian Health and Social Transfers (CHST) risen recently? Why?
   d. What changes do the provinces want made to the CHST?
4. Visit the Bank of Canada's Web site (www.bankofcanada.ca) and answer the following questions:
   a. What is the value of the Canadian dollar against the Euro?
   b. What is the exchange rate between the Canadian and the U.S. dollars?
   c. Using your answers to *a* and *b*, what should be the exchange rate between the U.S. dollar and the Euro? Is the actual rate (check another Web site or a local newspaper) close to this figure? Can you explain why?
5. Go to the World Bank's home page (www.worldbank.org) to answer the following questions:
   a. What is the World Bank?
   b. What goals has the World Bank set for the new millennium?
   c. How does the World Bank finance its activities?
   d. Which two regions of the world get the largest loans from the World Bank?

Answers to Margin Questions

1. Market economies are generally broken up into businesses, households, and government. *(56)*
2. False. In Canada individuals are free to start any type of business they want, provided it doesn't violate the law. The invisible hand sees to it that only those businesses that customers want earn a profit. The others lose money and eventually go out of business, so in that sense only businesses that customers want stay in business. *(57)*
3. In Canada "Ltd." denotes limited liability—in the United States "Inc." is short for incorporated. Although it is also possible to create a limited liability corporation in the U.S. the two forms of business differ in the way their activities are taxed. Check out www.4inc.com/choices.htm for more. *(59)*
4. While profits are important to business, because of internal monitoring problems it is not clear that managers maximize profit. They may waste profit potential in high-priced benefits for themselves and in inefficiency generally. The market, however, provides a limit on inefficiency, and firms that exceed that limit and have losses go out of business. *(59)*
5. The largest percentage of federal expenditure is for social services. *(63)*
6. Japan has the lowest exports and imports as a percentage of GDP, followed closely by the United States. *(68)*
7. A debtor nation will not necessarily be running a trade deficit. *Debt* refers to accumulated past deficits. If a country had accumulated large deficits in the past, it could run a surplus now but still be a debtor nation. *(69)*
8. A quota is a quantitative limitation on trade. A tariff is a type of tax on imports. *(71)*
9. You will receive 41,876 pesos. *(73)*
10. The World Court has no enforcement mechanism. Thus, when a country refuses to follow the Court's decisions, the country cannot be directly punished except through indirect international pressures. *(74)*

Valuing Stocks and Bonds

A financial asset's worth comes from the stream of income it will pay in the future. With financial assets like bonds, that stream of income can be calculated rather precisely. With stocks, where the stream of income is a percentage of the firm's profits, which fluctuate significantly, the stream of future income is uncertain and valuations depends significantly on expectations.

Let's start by considering some generally held beliefs among economists and financial experts. The first is that an average share of stock in a company in a mature industry sells for somewhere between 15 and 20 times its normal profits. The second is that bond prices rise as market interest rates fall, and fall as market interest rates rise. The first step in understanding where the beliefs come from is to recognize that $1 today is not equal to $1 next year. Why? Because if you have $1 today you can invest it and earn interest (say 10 percent per year), and next year you will have $1.10, not $1. So if the annual interest rate is 10 percent, $1.10 next year is worth $1 today; alternatively, $1 next year is worth roughly 91 cents today. A dollar two years in the future is worth even less today, and dollars 30 years in the future are worth very little today.

Present value is *a method of translating a flow of future income or savings into its current worth*. For example, say a smooth-talking, high-pressure salesperson is wining and dining you. "Isn't that amazing?" the salesman says. "My company will pay $10 a year not only to you, but also to your great-great-great-grandchildren, and more, for 500 years—thousands of dollars in all. And I will sell this annuity—this promise to pay money at periodic intervals in the future—to you for a payment to me now of only $800, but you must act fast. After tonight the price will rise to $2,000."

Do you buy it? Our rhetoric suggests that the answer should be no—but can you explain why? And what price *would* you be willing to pay?

To decide how much an annuity is worth, you need some way of valuing that $10 per year. *You can't simply add up the $10 five hundred times*. Doing so is wrong. Instead you must *discount* all future dollars by the interest rate in the economy. Discounting is required because a dollar in the future is not worth a dollar now.

If you have $1 now, you can take that dollar, put it in the bank, and in a year you will have that dollar plus interest. If the interest rate you can get from the bank is 5 percent, that dollar will grow to $1.05 a year from now. That means also that if the interest rate in the economy is 5 percent, if you have 95 cents now, in a year it will be worth $.9975 (5% × $.95 = $.0475). Reversing the reasoning, $1 one year in the future is worth 95 cents today. So the present value of $1 one year in the future at a 5 percent interest rate is 95 cents.

A dollar *two* years from now is worth even less today. Carry out that same reasoning and you'll find that if the interest rate is 5 percent, $1 two years from now is worth approximately 90 cents today. Why? Because you could take 90 cents now, put it in the bank at 5 percent interest, and in two years have $1.

THE PRESENT VALUE FORMULA

Carrying out such reasoning for every case would be a real pain. But luckily, there's a formula and a table that can be used to determine the present value (PV) of future income. The formula is:

$$PV = A_1/(1 + i) + A_2/(1 + i)^2 + A_3/(1 + i)^3 + \dots + A_n/(1 + i)^n$$

where

A_n = the amount of money received n periods in the future

i = the interest rate in the economy (assumed constant)

Solving this formula for any time period longer than one or two years is complicated. To deal with it, people either use a business calculator or a present value table like that in Table A3-1.

Table A3-1(a) gives the present value of a single dollar at some time in the future at various interest rates. Notice a couple of things about the chart. First, the further into the future one goes, the lower the present value. Second, the higher the interest rate, the lower the present value. At a 12 percent interest rate, $1 fifty years from now has a present value of essentially zero.

Table A3-1(b) is an annuity table; it tells us how much a constant stream of income for a specific number of years is worth. Notice that as the interest rate rises, the value of an annuity falls. At an 18 percent interest rate, $1 per year for 50 years has a present value of \$5.55. To get the value of amounts other than \$1, one simply multiplies the entry in the table by the amount. For example, \$10 per year for 50 years at 18 percent interest is 10 × \$5.55, or \$55.50.

As you can see, the interest rate in the economy is a key to present value. *You must know the interest rate to know the value of money over time*. The higher the current (and assumed constant) interest rate, the more a given amount of money in the present will be worth in the future. Or, alternatively, the higher the current interest rate, the less a given amount of money in the future will be worth in the present.

SOME RULES OF THUMB FOR DETERMINING PRESENT VALUE

Sometimes you don't have a present value table or a business calculator handy. For those times, there are a few rules of thumb and simplified formulas for which you don't need either a present value table or a calculator. Let's consider two of them: the infinite annuity rule and the rule of 72.

The Annuity Rule

To find the present value of an annuity that will pay \$1 for an infinite number of years in the future when the interest rate is 5 percent, we simply divide \$1 by 5 percent (.05). Doing so gives us \$20. So at 5 percent, \$1 a year paid to you forever has a present value of \$20. The **annuity rule** is that *the present value of any annuity is the annual income it yields divided by the interest rate*. Our general annuity rule for any annuity is expressed as:

$$PV = X/i$$

That is, the present value of an infinite flow of income, X, is that income divided by the interest rate, i.

Most of the time, people don't offer to sell you annuities for the infinite future. A typical annuity runs for 30, 40, or 50 years. However, the annuity rule is still useful. As you can see from the present value table, in 30 years at a 9 percent interest rate, the present value of \$1 isn't much (it's 8 cents), so we can use this infinite flow formula as an approximation of long-lasting, but less than infinite, flows of future income. We simply subtract a little bit from what we get with our formula. The longer the time period, the less we subtract. For example, say you are wondering what \$200 a year for 40 years is worth when the interest rate is 8 percent. Dividing \$200 by .08 gives \$2,500, so we know the annuity must be worth a bit less than \$2,500. (It's actually worth \$2,411.)

The annuity rule allows us to answer the question posed at the beginning of this section: How much is \$10 a year for 500 years worth right now? The answer is that it depends on the interest rate you could earn on a specified amount of money now. If the interest rate is 10 percent, the maximum you should be willing to pay for that 500-year \$10 annuity is \$100:

$$\$10/.10 = \$100$$

If the interest rate is 5 percent, the most you should pay is \$200 (\$10/.05 = \$200). So now you know why you

TABLE A3-1 (a and b) Sample Present Value and Annuity Tables

	Interest Rate						
Year	3%	4%	6%	9%	12%	15%	18%
1	\$0.97	\$0.96	\$0.94	\$0.92	\$0.89	\$0.87	\$0.85
2	0.94	0.92	0.89	0.84	0.80	0.76	0.72
3	0.92	0.89	0.84	0.77	0.71	0.66	0.61
4	0.89	0.85	0.79	0.71	0.64	0.57	0.52
5	0.86	0.82	0.75	0.65	0.57	0.50	0.44
6	0.84	0.79	0.70	0.60	0.51	0.43	0.37
7	0.81	0.76	0.67	0.55	0.45	0.38	0.31
8	0.79	0.73	0.63	0.50	0.40	0.33	0.27
9	0.77	0.70	0.59	0.46	0.36	0.28	0.23
10	0.74	0.68	0.56	0.42	0.32	0.25	0.19
15	0.64	0.56	0.42	0.27	0.18	0.12	0.08
20	0.55	0.46	0.31	0.18	0.10	0.06	0.04
30	0.41	0.31	0.17	0.08	0.03	0.02	0.01
40	0.31	0.21	0.10	0.03	0.01	0.00	0.00
50	0.23	0.14	0.05	0.01	0.00	0.00	0.00

(a) Present value table (value now of \$1 to be received *x* years in the future)
The present value table converts a future amount into a present amount.

	Interest Rate						
Number of years	3%	4%	6%	9%	12%	15%	18%
1	\$0.97	\$0.96	\$0.94	\$0.92	\$0.89	\$0.87	\$0.85
2	1.91	1.89	1.83	1.76	1.69	1.63	1.57
3	2.83	2.78	2.67	2.53	2.40	2.28	2.17
4	3.72	3.63	3.47	3.24	3.04	2.85	2.69
5	4.58	4.45	4.21	3.89	3.60	3.35	3.13
6	5.42	5.24	4.92	4.49	4.11	3.78	3.50
7	6.23	6.00	5.58	5.03	4.56	4.16	3.81
8	7.02	6.73	6.21	5.53	4.97	4.49	4.08
9	7.79	7.44	6.80	6.00	5.33	4.77	4.30
10	8.53	8.11	7.36	6.42	5.65	5.02	4.49
15	11.94	11.12	9.71	8.06	6.81	5.85	5.09
20	14.88	13.59	11.47	9.13	7.47	6.26	5.35
30	19.60	17.29	13.76	10.27	8.06	6.57	5.52
40	23.11	19.79	15.05	10.76	8.24	6.64	5.55
50	25.73	21.48	15.76	10.96	8.30	6.66	5.55

(b) Annuity table (value now of \$1 per year to be received for *x* years)
The annuity table converts a known stream of income into a present amount.

The failure to understand the concept of present value often shows up in the popular press. Here are three examples.

Headline: **COURT SETTLEMENT IS $40,000,000**

Inside story: The money will be paid out over a 40-year period.

Actual value: $11,925,000 (8 percent interest rate).

Headline: **DISABLED WIDOW WINS $25 MILLION LOTTERY**

Inside story: The money will be paid over 20 years.

Actual value: $13,254,499 (8 percent interest rate).

Headline: **BOND ISSUE TO COST CITY TAXPAYERS $68 MILLION**

Inside story: The $68 million is the total of interest and principal payments. The interest is paid yearly; the principal won't be paid back to the bond purchasers until 30 years from now.

Actual cost: $20,000,000 (8 percent interest rate).

Such stories are common. Be on the lookout for them as you read the newspaper or watch the evening news.

should have said no to that supersalesman who offered it to you for $800.

The Rule of 72

A second rule of thumb for determining present values of shorter time periods is the **rule of 72,** which states:

The number of years it takes for a certain amount to double in value is equal to 72 divided by the rate of interest.

Say, for example, that the interest rate is 4 percent. How long will it take for your $100 to become $200? Dividing 72 by 4 gives 18, so the answer is 18 years. Conversely, the present value of $200 at a 4 percent interest rate 18 years in the future is about $100. (Actually it's $102.67.)

Alternatively, say that you will receive $1,000 in 10 years. Is it worth paying $500 for that amount now if the interest rate is 9 percent? Using the rule of 72, we know that at a 9 percent interest rate it will take about eight years for $500 to double:

$$72/9 = 8$$

so the future value of $500 in 10 years is more than $1,000. It's probably about $1,200. (Actually it's $1,184.) So if the interest rate in the economy is 9 percent, it's not worth paying $500 now in order to get that $1,000 in 10 years. By investing that same $500 today at 9 percent, you can have $1,184 in 10 years.

THE IMPORTANCE OF PRESENT VALUE

Many business decisions require such present value calculations. In almost any business, you'll be looking at flows of income in the future and comparing them to present costs or to other flows of money in the future.

Generally, however, when most people calculate present value they don't use any of the formulas. They pull out a handy business calculator, press in the numbers to calculate the present value, and watch while the calculator graphically displays the results.

Let's now use our knowledge of present value to explain the two observations at the beginning of the appendix. Since all financial assets can be broken down into promises to pay certain amounts at certain times in the future, we can determine their value with the present value formula. If the asset is a bond, it consists of a stream of income payments over a number of years and the repayment of the face value of the bond. Each year's interest payment and the eventual repayment of the face value must be calculated separately, and then the results must be added together. If the financial asset is a share of stock, the valuation is a bit less clear since a stock does not guarantee the payment of anything definite—just a share of the profits. No profits, no payment. So, with stocks, expectations of profits are of central importance. Let's consider an example: Say a share of stock is earning $1 per share per year and is expected to continue to earn that long into the future. Using the annuity rule and an interest rate of 6.5 percent, the present value of that future stream of expected earnings is about 1/.065, or a bit more than $15. Assuming profits are expected to grow slightly, that would mean that the stock should sell for somewhere around $20, or 20 times its profit per share, which is the explanation to the first observation in the appendix.

To see the answer to the second, say the interest rate rises to 10 percent. Then the value of the stock or bond that is earning a fixed amount—in this case $1 per share—

will go down to $10. Interest rate up, value of stock or bond down. This is the explanation of the second observation.

There is nothing immutable in the above reasoning. For example, if promises to pay aren't trustworthy, you don't put the amount that's promised into your calculation; you put in the amount you actually expect to receive. That's why when a company or a country looks as if it's going to default on loans or stop paying dividends, the value of its bonds and stock will fall considerably. For example, in the late 1980s many people thought Brazil would default on its bonds. That expectation caused the price of Brazilian bonds to fall to about 30 percent on the dollar. Then in the 1990s, when people believed total default was less likely, the value rose.

Of course, the expectations could go in the opposite direction. Say that the interest rate is 10 percent, and that you expect a company's profit, which is now $1 per share, to grow by 10 percent per year. In that case, since expected profit growth is as high as the interest rate, the current value of the stock is infinite. It is such expectations of future profit growth that have fueled the Internet stock craze and have caused the valuation of firms with no current profits (indeed, many are experiencing significant losses) at multiples of sales of 300 or more. Financial valuations based on such optimistic expectations are the reason most economists considered the stock market in Internet stocks to be overvalued in the late 1990s.

ASSET PRICES, INTEREST RATES, AND THE ECONOMY

This appendix isn't meant to cover the intricacies of valuation over time. That's done in a finance course. The point is to help you understand the relationship between interest rates and asset prices. Central to valuing stocks or bonds is the present value formula. From that we know that increases in interest rates (because they make future flows of income coming from an asset worth less now) make financial asset prices fall, and that decreases in interest rates (because they make the future flow of income coming from an asset worth more now) make financial asset prices rise.

Key Terms

annuity rule *(79)*

present value *(78)*

rule of 72 *(80)*

Questions for Thought and Review

1. How much is $50 to be received 50 years from now worth if the interest rate is 5 percent? (Use Figure A3-1).
2. How much is $50 to be received 50 years from now worth if the interest rate is 10 percent? (Use Figure A3-1).
3. Your employer offers you a choice of two bonus packages: $1,400 today or $2,000 five years from now. Assuming a 5 percent rate of interest, which is the better value? Assuming an interest rate of 10 percent, which is the better value?
4. Suppose the price of a one-year 10 percent coupon bond with a $100 face value is $98.
   a. Are market interest rates likely to be above or below 10 percent? Explain.
   b. What is the bond's yield or return?
   c. If market interest rates fell, what would happen to the price of the bond?
5. Explain in words why the present value of $100 to be received in 10 years would decline as the interest rate rises.
6. A 6 percent bond will pay you $1,060 one year from now. The interest rate in the economy is 10 percent. How much is that bond worth now?
7. You are to receive $100 a year for the next 40 years. How much is it worth now if the current interest rate in the economy is 6 percent? (Use annuity table.)
8. You are to receive $200, thirty years from now. About how much is it worth now? (The interest rate is 3 percent.)
9. A salesperson calls you up and offers you $200 a year for life. If the interest rate is 7 percent, how much should you be willing to pay for that annuity?
10. The same salesperson offers you a lump sum of $20,000 in 10 years. How much should you be willing to pay? (The interest rate is still 7 percent.)
11. What is the present value of a cash flow of $100 per year forever (a perpetuity), assuming:

    The interest rate is 10 percent.

    The interest rate is 5 percent.

    The interest rate is 20 percent.

    a. Working with those same three interest rates, what are the future values of $100 today in one year? How about in two years?
    b. Working with those same three interest rates, how long will it take you to double your money?

Supply and Demand

4

After reading this chapter, you should be able to:

- State the law of demand.
- Explain the importance of substitution to the laws of supply and demand.
- Distinguish a shift in demand from a movement along the demand curve.
- Draw a demand curve from a demand table.
- State the law of supply.
- Distinguish a shift in supply from a movement along the supply curve.
- Draw a supply curve from a supply table.
- Explain how the law of demand and the law of supply interact to bring about equilibrium.
- Show how equilibrium maximizes consumer and producer surplus.

***Teach a parrot the terms* supply *and* demand**
and you've got an economist.

Thomas Carlyle

Supply and demand. Supply and demand. Roll the phrase around in your mouth, savour it like a good wine. *Supply* and *demand* are the most-used words in economics. And for good reason. They provide a good off-the-cuff answer for any economic question. Try it. Why are bacon and oranges so expensive this winter? *Supply and demand*.

Why are interest rates falling? *Supply and demand.*

Why can't I find decent wool socks anymore? *Supply and demand.*

The importance of the interplay of supply and demand makes it only natural that, early in any economics course, you must learn about supply and demand. Let's start with demand.

DEMAND

People want lots of things; they "demand" much less than they want because demand means a willingness and capacity to pay. Unless you are willing and able to pay for it, you may *want* it, but you don't *demand* it. For example, we want to own fancy cars. But, we must admit, we're not willing to do what's necessary to own one. If we really wanted one, we'd mortgage everything we own, increase our income by doubling the number of hours we work, not buy anything else, and get that car. But we don't do any of those things, so at the going price, $360,000, we do not demand a Maserati. Sure, we'd buy one if it cost $10,000, but from our actions it's clear that, at $360,000, we don't demand it. This points to an important aspect of demand: The quantity you demand at a low price differs from the quantity you demand at a high price. Specifically, the quantity you demand varies inversely—in the opposite direction—with price.

Prices are the tool by which the market coordinates individuals' desires and limits how much people are willing to buy—how much they demand. When goods become scarce, the market reduces the quantity of those scarce goods people demand; as their prices go up, people buy fewer goods. As goods become abundant, their prices go down, and people want more of them. The invisible hand—the price mechanism—sees to it that what people demand (do what's necessary to get) matches what's available. In doing so, the invisible hand coordinates individuals' demands.

Prices are the tool by which the market coordinates individual desires.

The Law of Demand

The ideas expressed above are the foundation of the **law of demand:**

Quantity demanded rises as price falls, other things constant.

Or alternatively:

Quantity demanded falls as price rises, other things constant.

This law is fundamental to the invisible hand's ability to coordinate individuals' desires: as prices change, people change how much of a particular good they're willing to buy.

The law of demand states that the quantity of a good demanded is inversely related to the good's price. When price goes up, quantity demanded goes down. When price goes down, quantity demanded goes up.

What accounts for the law of demand? Individuals' tendency to substitute other goods for goods whose price has gone up. If the price of CDs rises from $15 to $20 but the price of cassette tapes stays at $9.99, you're more likely to buy that new Christina Aguilera recording on cassette than on CD.

To see that the law of demand makes intuitive sense, just think of something you'd really like but can't afford. If the price is cut in half, you—and other consumers—will become more likely to buy it. Quantity demanded goes up as price goes down.

Just to be sure you've got it, let's consider a real world example: scalpers and the demand for hockey tickets. Standing outside a sold-out game between Montreal and Pittsburgh in Montreal, we saw scalpers trying to sell tickets for $100 a seat. There were few takers — that is, there was little demand at that price. The sellers saw that they had set too high a price and they started calling out lower prices. As the price dropped to $60, then $50, quantity demanded increased; when the price dropped to $35, quantity demanded soared. That's the law of demand in action.

Figure 4-1 A SAMPLE DEMAND CURVE

The law of demand states that the quantity demanded of a good is inversely related to the price of that good, other things constant. As the price of a good goes up, the quantity demanded goes down, so the demand curve is downward sloping.

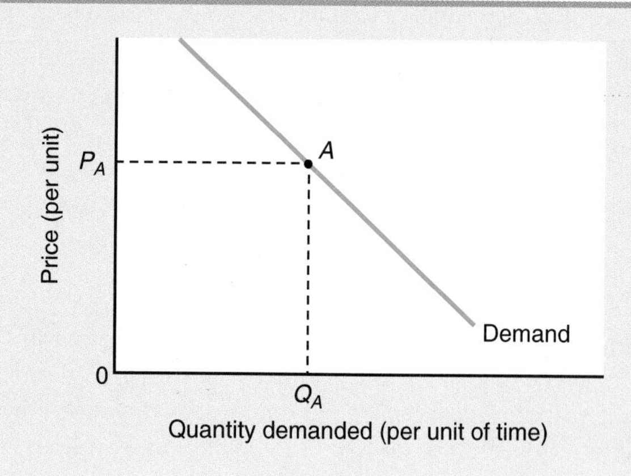

The Demand Curve

Q-1 Why does the demand curve slope downward?

A **demand curve** is *the graphic representation of the relationship between price and quantity demanded*. Figure 4-1 shows a demand curve.

As you can see, in graphical terms, the law of demand states that as the price goes up, the quantity demanded goes down, other things constant. An alternative way of saying the same thing is that price and quantity demanded are inversely related, so the demand curve slopes downward to the right.

"Other things constant" places a limitation on the application of the law of demand.

inversely: one quantity increases while the other goes down.

Notice that in stating the law of demand, we put in the qualification "other things constant." That's three extra words, and unless they were important we wouldn't have put them in. But what does "other things constant" mean? Say that over a period of two years, the price of cars rises as the number of cars purchased likewise rises. That seems to violate the law of demand, since the number of cars purchased should have fallen in response to the rise in price. Looking at the data more closely, however, we see that a third factor has also changed: individuals' income has increased. As income increases, people buy more cars, increasing the demand for cars.

The increase in price works as the law of demand states—it decreases the number of cars bought. But in this case, income doesn't remain constant; it increases. That rise in income increases the demand for cars. That increase in demand outweighs the decrease in quantity demanded that results from a rise in price, so ultimately more cars are sold. If you want to study the effect of price alone—which is what the law of demand refers to—you must make adjustments to hold income constant when you make your study. That's why the qualifying phrase "other things constant" is an important part of the law of demand.

The other things that are held constant include individuals' tastes, prices of other goods, and even the weather. Those other factors must remain constant if you're to make a valid study of the effect of an increase in the price of a good on the quantity demanded. In practice, it's impossible to keep all other things constant, so you have to be careful when you say that when price goes up, quantity demanded goes down. Quantity demanded is likely to go down, but it's always possible that something besides price has changed.

Shifts in Demand versus Movements along a Demand Curve

To distinguish between the effects of changes in a good's price and the effects of other factors on how much of a good is demanded, economists have developed the following precise terminology—terminology that inevitably shows up on exams. The first distinction to make is between demand and quantity demanded.

- **Demand** refers to *a schedule of quantities of a good that will be bought per unit of time at various prices, other things constant.*
- **Quantity demanded** refers to *a specific amount that will be demanded per unit of time at a specific price, other things constant.*

In graphical terms, the term *demand* refers to the entire demand curve. Demand tells how much of a good will be bought *at various prices*. *Quantity demanded* refers to a point on a demand curve, such as point A in Figure 4-1. This terminology allows us to distinguish between *changes in quantity demanded* and *shifts in demand*. A change in the quantity demanded refers to the effect of a price change on the quantity demanded. It refers to a **movement along a demand curve**—*the graphical representation of the effect of a change in price on the quantity demanded*. A **shift in demand** refers to *the effect of anything other than price on demand*.

Q-2 In the 1980s and early 1990s, as animal rights activists made wearing fur coats déclassé, the __________ decreased. Should the missing words be "demand for furs" or "quantity of furs demanded"?

Shift Factors of Demand

Shift factors of demand are factors that cause shifts in the demand curve. A change in anything besides a good's price causes a shift of the entire demand curve.

Important shift factors of demand include:

1. Society's income.
2. The prices of other goods.
3. Tastes.
4. Expectations.
5. Population.

Income From our example above of "the other things constant" qualification, we saw that a rise in income increases the demand for goods. For most goods this is true. As individuals' income rises, they can afford more of the goods they want.

Price of Other Goods Because people make their buying decisions based on the price of related goods, demand will be affected by the prices of other goods. Suppose the price of jeans rose from $25 to $35, but the price of khakis remained at $25. Next time you need pants, you're apt to try khakis instead of jeans. They are substitutes. When two goods are substitutes, if the price of one of the goods falls while the other price remains unchanged, there will be an increase in the quantity demanded of the good whose price fell, and a reduction in the demand for the good whose price remained fixed.

Tastes An old saying goes: "There's no accounting for taste." Of course, many advertisers believe otherwise. Changes in taste can affect the demand for a good without a change in price. As you become older, you may find that your taste for rock concerts has changed to a taste for an evening at the opera or local philharmonic.

Q-3 Explain the effect of each of the following on the demand for new computers:

1. The price of computers falls by 30 percent.
2. Total income in the economy rises.

Expectations Expectations will also affect demand. Expectations can cover a lot. If you expect your income to rise in the future, you're bound to start spending some of it today. If you expect the price of computers to fall soon, you may put off buying one until later.

These aren't the only shift factors. In fact anything—except the price of the good itself—that affects demand (and many things do) is a shift factor. While economists agree these shift factors are important, they believe that no shift factor influences how much is demanded as consistently as does price of the specific item. That's what makes economists focus first on price as they try to understand the world. That's why economists make the law of demand central to their analysis.

Figure 4-2 SHIFT IN DEMAND VERSUS A CHANGE IN QUANTITY DEMANDED

A rise in a good's price results in a reduction in quantity demanded and is shown by a movement up along a demand curve from point A to point B in (**a**). A change in any other factor besides price that affects demand leads to a shift in the entire demand curve as shown in (**b**).

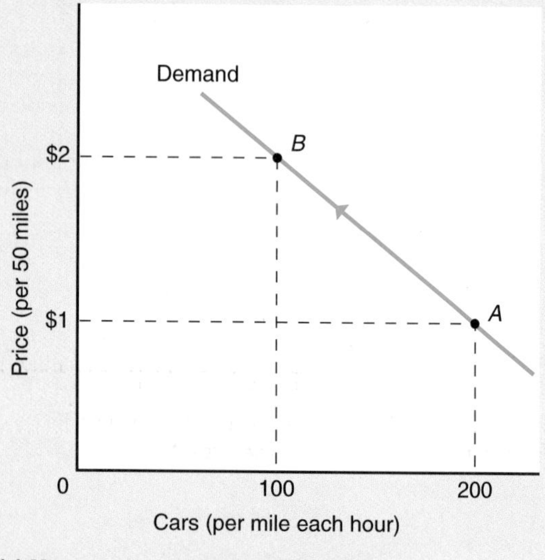

(a) Movement along a demand curve

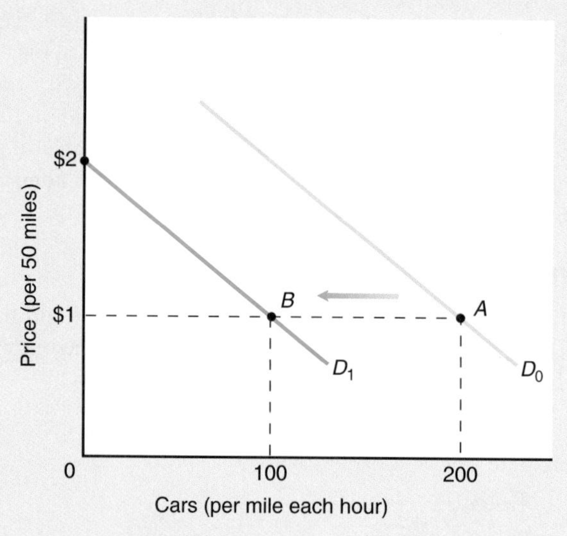

(b) Shift in demand

Population Finally, population will also affect demand. If there is an increase in population, there will be a higher quantity demanded at every price. If population falls, as it did in Newfoundland's outports in the mid-1900s, demand falls. It's that simple.

Change in price causes a movement along a demand curve; a change in a shift factor causes a shift in demand.

To make sure you understand the difference between a movement along a demand curve and a shift in demand, let's consider an example. Singapore has one of the highest numbers of cars per mile of road. This means that congestion is considerable. Singapore has adopted two policies to reduce road use: It increased the fee charged to use roads, and it provided an expanded public transportation system. Both policies reduced congestion. Figure 4-2(a) shows that increasing the toll charged to use roads from $1 to $2 per 50 miles of road reduces quantity demanded from 200 to 100 cars per mile every hour (a movement along the demand curve). Figure 4-2(b) shows that providing alternative methods of transportation such as buses and subways will shift the demand curve for roads. Demand for road use shifts to the left so that at the $1 fee, demand drops from 200 to 100 cars per mile every hour (a shift in the demand curve).

A Review

Let's test your understanding by having you specify what happens to your demand curve for videocassettes in the following examples: First, let's say you buy a DVD player. Next, let's say that the price of videocassettes falls; and finally, say that you won $1 million in a lottery. What happens to the demand for videocassettes in each case? If you answered: It shifts in; it remains unchanged; and it shifts out—you've got it.

Figure 4-3 (a and b) FROM A DEMAND TABLE TO A DEMAND CURVE

The demand table in (**a**) is translated into a demand curve in (**b**). Each combination of price and quantity in the table corresponds to a point on the curve. For example, point A on the graph represents row A in the table: Marie demands 9 videocassette rentals at a price of 50 cents. A demand curve is constructed by plotting all points from the demand table and connecting the points by a line.

	Price per cassette	Cassette rentals demanded per week
A	$0.50	9
B	1.00	8
C	2.00	6
D	3.00	4
E	4.00	2

(a) A demand table

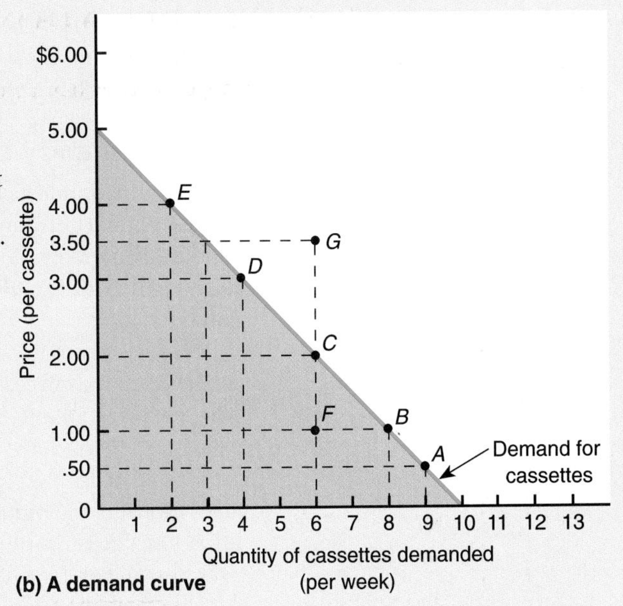

(b) A demand curve

The Demand Table

As we emphasized in Chapter 1, introductory economics depends heavily on graphs and graphical analysis—translating ideas into graphs and back into words. So let's graph the demand curve.

Figure 4-3(a), a demand table, describes Marie's demand for renting videocassettes. For example, at a price of $2, Marie will rent (buy the use of) 6 cassettes per week and at a price of 50 cents she will rent 9.

There are four points about the relationship between the number of videos Marie rents and the price of renting them that are worth mentioning. First, the relationship follows the law of demand: as the rental price rises, quantity demanded decreases. Second, quantity demanded has a specific *time dimension* to it. In this example demand refers to the number of cassette rentals per week. Without the time dimension, the table wouldn't provide us with any useful information. Nine cassette rentals per year is quite a different concept from 9 cassette rentals per week. Third, Marie's cassette rentals are interchangeable—the 9th cassette rental doesn't significantly differ from the 1st, 3rd, or any other cassette rental. The fourth point is already familiar to you: The schedule assumes that everything else is held constant.

From a Demand Table to a Demand Curve

Figure 4-3(b) translates the demand table in Figure 4-3(a) into a graph. Point A (quantity = 9, price = $.50) is graphed first at the (9, $.50) coordinates. Next we plot points B, C, D, and E in the same manner and connect the resulting dots with a solid line. The result is the demand curve, which graphically conveys the same information that's in the demand table. Notice that the demand curve is downward sloping (from left to right), indicating that the law of demand holds in the example.

The demand curve represents the *maximum price* that an individual will pay for various quantities of a good; the individual will happily pay less. For example, say someone

The demand curve represents the maximum price that an individual will pay.

offers Marie 6 cassette rentals at a price of $1 each (point *F* of Figure 4-3(b)). Will she accept? Sure; she'll pay any price within the shaded area to the left of the demand curve. But if someone offers her 6 rentals at $3.50 each (point G), she won't accept. At a rental price of $3.50 apiece, she's willing to buy only 3 cassette rentals.

Individual and Market Demand Curves

Normally, economists talk about market demand curves rather than individual demand curves. A **market demand curve** is *the horizontal sum of all individual demand curves*. Market demand curves are what most firms are interested in. Firms don't care whether individual A or individual B buys their goods; they only care that *someone* buys their goods.

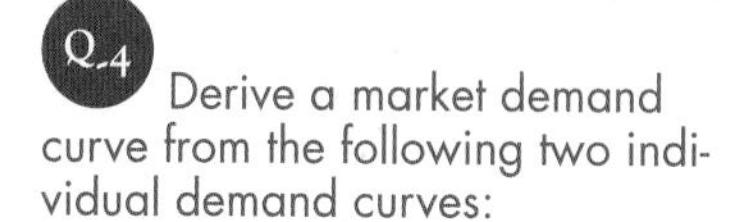

Q-4 Derive a market demand curve from the following two individual demand curves:

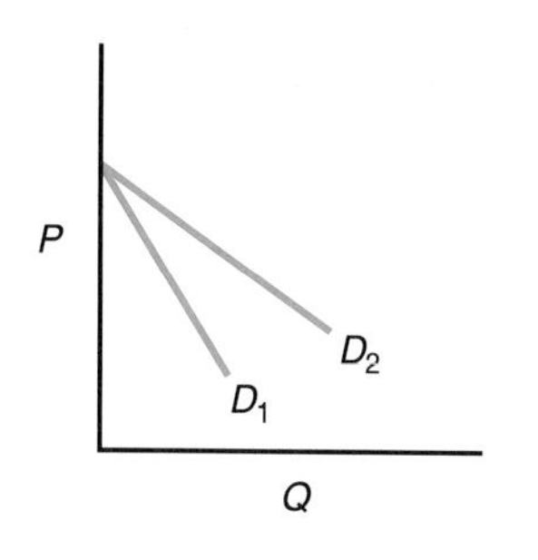

It's a good graphical exercise to add individual demand curves together to create a market demand curve. We do that in Figure 4-4. In it we assume that the market consists of three buyers, Marie, Pierre, and Cathy, whose demand tables are given in Figure 4-4(a). Marie and Pierre have demand tables similar to the demand tables discussed previously. At a price of $3 each, Marie rents 4 cassettes; at a price of $2, she rents 6. Cathy is an all-or-nothing individual. She rents 1 cassette as long as the price is equal to or below $1; otherwise she rents nothing. If you plot Cathy's demand curve, it's a vertical line. However, the law of demand still holds: as price increases, quantity demanded decreases.

The quantity demanded by each consumer is listed in columns 2, 3, and 4 of Figure 4-4(a). Column 5 shows total market demand; each entry is the horizontal sum of the entries in columns 2, 3, and 4. For example, at a price of $3 apiece (row *F*), Marie

Figure 4-4 (a and b) FROM INDIVIDUAL DEMANDS TO A MARKET DEMAND CURVE

The table (**a**) shows the demand schedules for Marie, Pierre, and Cathy. Together they make up the market for videocassette rentals. Their total quantity demanded (market demand) for videocassette rentals at each price is given in column 5. As you can see in (**b**), Marie's, Pierre's, and Cathy's demand curves can be added together to get the total market demand curve. For example, at a price of $2, Cathy demands 0, Pierre demands 3, and Marie demands 6, for a market demand of 9 (point *D*).

	(1)	(2)	(3)	(4)	(5)
	Price (per cassette)	Marie's demand	Pierre 's demand	Cathy's demand	Market demand
A	$0.50	9	6	1	16
B	1.00	8	5	1	14
C	1.50	7	4	0	11
D	2.00	6	3	0	9
E	2.50	5	2	0	7
F	3.00	4	1	0	5
G	3.50	3	0	0	3
H	4.00	2	0	0	2

(a) A demand table

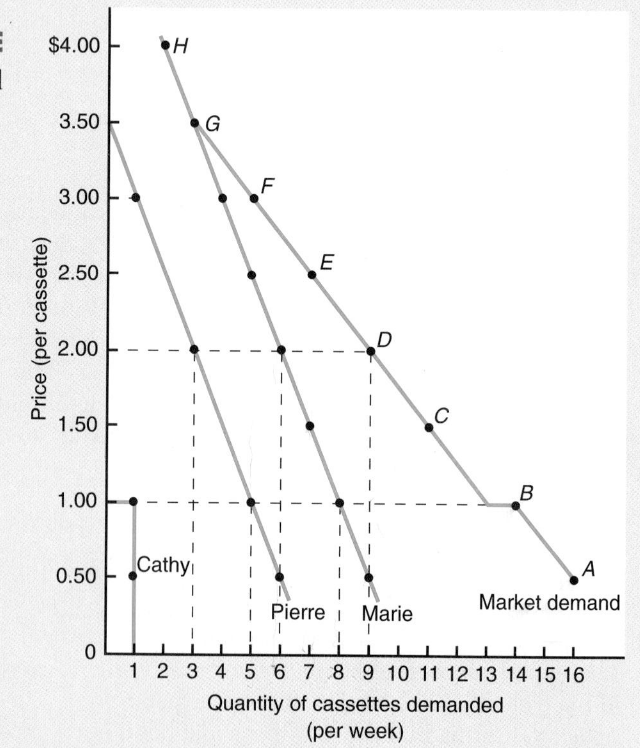

(b) Adding demand curves

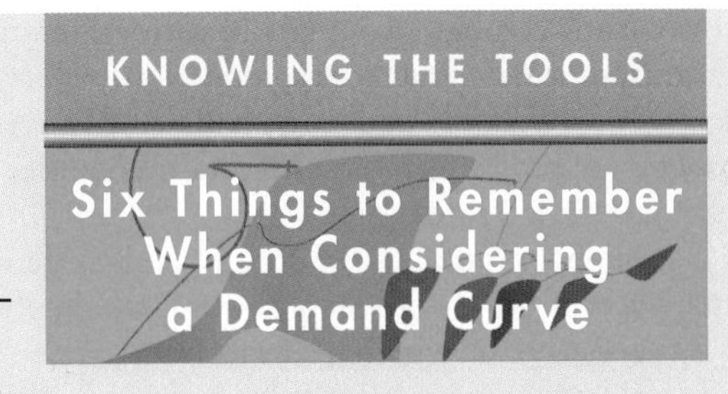

- A demand curve had better follow the law of demand: When price rises, quantity demanded falls; and vice versa.
- The horizontal axis—quantity—has a time dimension.
- The quantities are of the same quality.
- The vertical axis—price—assumes all other prices remain the same.
- The curve assumes everything else is held constant.
- Effects of price changes are shown by movements along the demand curve. Effects of anything else on demand (shift factors) are shown by shifts of the entire demand curve.

demands 4 cassette rentals, Pierre demands 1, and Cathy demands 0, for a total market demand of 5 cassette rentals.

Figure 4-4(b) shows three demand curves: one each for Marie, Pierre, and Cathy. The market, or total, demand curve is the horizontal sum of the individual demand curves. To see that this is the case, notice that if we take the quantity demanded at $1 by Marie (8), Pierre (5), and Cathy (1), they sum to 14, which is point *B* (14, $1) on the market demand curve. We can do that for each price. Alternatively, we can simply add the individual quantities demanded, given in the demand tables, prior to graphing (which we do in column 5 of Figure 4-4(a)), and graph that total in relation to price. Not surprisingly, we get the same total market demand curve.

In practice, of course, firms don't measure individual demand curves, so they don't sum them up in this fashion. Instead, they estimate total demand. Still, summing up individual demand curves is a useful exercise because it shows you how the market demand curve is the sum (the horizontal sum, graphically speaking) of the individual demand curves, and it gives you a good sense of where market demand curves come from. It also shows you that, even if individuals don't respond to small changes in price, the market demand curve can still be smooth and downward sloping. That's because, for the market, the law of demand is based on two phenomena:

1. At lower prices, existing demanders buy more.
2. At lower prices, new demanders (some all-or-nothing demanders like Cathy) enter the market.

For the market, the law of demand is based on two phenomena:

1. At lower prices, existing demanders buy more.
2. At lower prices, new demanders enter the market.

SUPPLY

In one sense, supply is the mirror image of demand. Individuals control the factors of production—inputs, or resources, necessary to produce goods. Individuals' supply of these factors to the market mirrors other individuals' demand for those factors. For example, say you decide you want to rest rather than weed your garden. You hire someone to do the weeding; you demand labour. Someone else decides she would prefer more income instead of more rest; she supplies labour to you. You trade money for labour; she trades labour for money. Her supply is the mirror image of your demand.

For a large number of goods and services, however, the supply process is more complicated than demand. For many goods there's an intermediate step in supply: individuals supply factors of production to firms.

Let's consider a simple example. Say you're a taco technician. You supply your labour to the factor market. The taco company demands your labour (hires you). The taco company combines your labour with other inputs like meat, cheese, beans, and tables, and produces many tacos (production), which it supplies to customers in the goods market. For produced goods, supply depends not only on individuals' decisions to supply factors of production but also on firms' ability to produce—to transform those factors of production into usable goods.

Supply of produced goods involves a much more complicated process than demand and is divided into analysis of factors of production and the transformation of those factors into goods.

The supply process of produced goods is generally complicated. Often there are many layers of firms—production firms, wholesale firms, distribution firms, and retailing firms—each of which passes on in-process goods to the next layer of firms. Real-world production and supply of produced goods is a multistage process.

The supply of nonproduced goods is more direct. Individuals supply their labour in the form of services directly to the goods market. For example, an independent contractor may repair your washing machine. That contractor supplies his labour directly to you.

Thus, the analysis of the supply of produced goods has two parts: an analysis of the supply of factors of production to households and to firms, and an analysis of one process by which firms transform those factors of production into usable goods and services.

The Law of Supply

In talking about supply, the same convention exists that we used for demand. Supply refers to the various quantities offered for sale at various prices. Quantity supplied refers to a specific quantity offered for sale at a specific price.

There's a law of supply that corresponds to the law of demand. The **law of supply** states:

Quantity supplied rises as price rises, other things constant.

Or alternatively:

Quantity supplied falls as price falls, other things constant.

Price regulates quantity supplied just as it regulates quantity demanded. Like the law of demand, the law of supply is fundamental to the invisible hand's (the market's) ability to coordinate individuals' actions.

What accounts for the law of supply? When the price of a good rises, individuals and firms can rearrange their activities in order to supply more of that good to the market. The law of supply is based on a firm's ability to substitute production of one good for another, or vice versa. If the price of corn rises and the price of wheat has not changed, farmers will grow less wheat and more corn, other things constant.

With firms, there's a second explanation of the law of supply. Assuming firms' costs are constant, a higher price means higher profits (the difference between a firm's revenues and its costs). The expectation of those higher profits leads it to increased output as price rises, which is what the law of supply states.

As crude oil prices rise, the incentive to produce more oil rises.

The Supply Curve

A **supply curve** is *the graphical representation of the relationship between price and quantity supplied.* A supply curve is shown graphically in Figure 4-5.

Notice how the supply curve slopes upward to the right. That upward slope captures the law of supply. It tells us that the quantity supplied varies *directly*—in the same direction—with the price.

As with the law of demand, the law of supply assumes other things are held constant. Thus, if the price of wheat rises and quantity supplied falls, you'll look for something else that changed—for example, a drought might have caused a drop in supply. Your explanation would go as follows: Had there been no drought, the quantity supplied would have increased in response to the rise in price, but because there was a drought, the supply decreased, which caused prices to rise.

Figure 4-5 A SAMPLE SUPPLY CURVE

The supply curve demonstrates graphically the law of supply, which states that the quantity supplied of a good is directly related to that good's price, other things constant. As the price of a good goes up, the quantity supplied also goes up, so the supply curve is upward sloping.

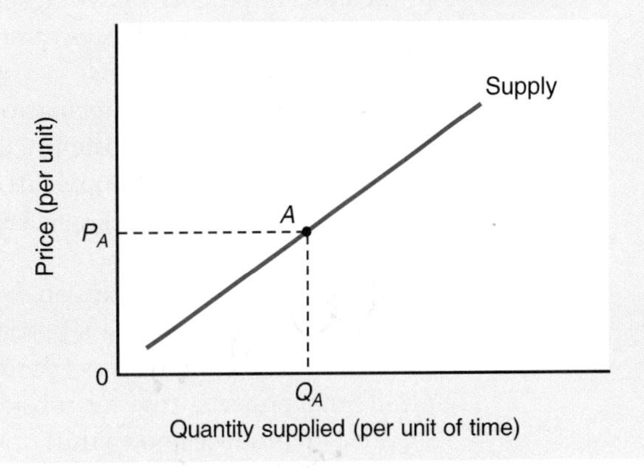

As with the law of demand, the law of supply represents economists' off-the-cuff response to the question "What happens to quantity supplied if price rises?" If the law seems to be violated, economists search for some other variable that has changed. As was the case with demand, these other variables that might change are called shift factors.

Shifts in Supply versus Movements along a Supply Curve

The same distinctions in terms made for demand apply to supply.

> **Supply** refers to *a schedule of quantities a seller is willing to sell per unit of time at various prices, other things constant.*
>
> **Quantity supplied** refers to *a specific amount that will be supplied at a specific price.*

In graphical terms, supply refers to the entire supply curve because a supply curve tells us how much will be offered for sale at various prices. "Quantity supplied" refers to a point on a supply curve, such as point A in Figure 4-5.

The second distinction that is important to make is between the effects of a change in a good's price and the effects of shift factors on how much of a good is supplied. Changes in price cause changes in quantity supplied; such changes are represented by a **movement along a supply curve**—*the graphic representation of the effect of a change in a good's price on the quantity supplied.* If the amount supplied is affected by anything other than that good's price, that is, by a shift factor of supply, there will be a **shift in supply**—*the graphic representation of the effect of a change in a factor other than price on supply.*

Q.5 In the 1980s and 1990s, as animal activists caused a decrease in the demand for fur coats, the prices of furs fell. This made __________ decline. Should the missing words be "the supply" or "the quantity supplied"?

Shift Factors of Supply

Other factors besides a good's price that affect how much will be supplied include the price of inputs used in production, technology, expectations, and taxes and subsidies. Let's see how.

Price of Inputs Firms produce to earn a profit. Since their profit is tied to costs, it's no surprise that costs will affect how much a firm is willing to supply. If costs rise, profits will decline, and a firm has less incentive to supply. Supply falls when the price of inputs rises. If costs rise substantially, a firm might even shut down.

Technology Advances in technology change the production process, reducing the number of inputs needed to produce a given supply of goods. Thus, a technological advance that reduces the number of workers will reduce costs of production. A reduction in the costs of production increases profits and leads suppliers to increase production. Advances in technology increase supply.

Expectations Supplier expectations are an important factor in the production decision. If a supplier expects the price of her good to rise at some time in the future, she may store some of today's supply in order to sell it later and reap higher profits, decreasing supply now and increasing it later.

Taxes and Subsidies Taxes on supplies increase the cost of production by requiring a firm to pay the government a portion of the income from products or services sold. Because taxes increase the cost of production, profit declines and suppliers will reduce supply. The opposite is true for subsidies. Subsidies are payments by the government to suppliers to produce goods; thus, they reduce the cost of production. Subsidies increase supply. Taxes on suppliers reduce supply.

These aren't the only shift factors. As was the case with demand, a shift factor of supply is anything that affects supply, other than its price.

Q-6 Explain the effect of each of the following on the supply of romance novels:

1. The price of paper rises by 20 percent.
2. Government increases the sales tax on all books by 5 percentage points.

Shift in Supply versus a Movement along a Supply Curve

The same "movement along" and "shift of" distinction that we developed for demand exists for supply. To make that distinction clear, let's consider an example: the supply of oil. In 1990 and 1991, world oil prices in U.S. dollars rose from $15 to $36 a barrel when oil production in the Persian Gulf was disrupted by the Iraqi invasion of Kuwait. Oil producers, seeing that they could sell their oil at a higher price, increased oil production. As the price of oil rose, domestic producers increased the quantity of oil supplied. The change in domestic quantity supplied in response to the rise in world oil prices is illustrated in Figure 4-6(a) as a movement up along the domestic supply curve

Figure 4-6 SHIFT IN SUPPLY VERSUS CHANGE IN QUANTITY SUPPLIED

A change in quantity supplied results from a change in price and is shown by a movement along a supply curve like the movement from point A to point B in (**a**). A shift in supply—a shift in the entire supply curve—brought about by a change in a nonprice factor is shown in (**b**).

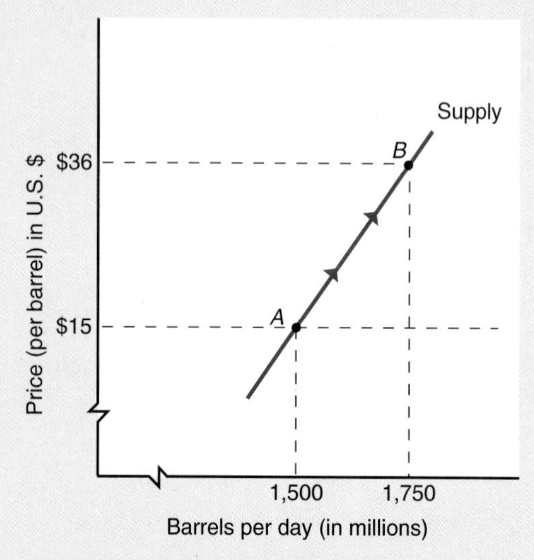

(a) Movement along a supply curve

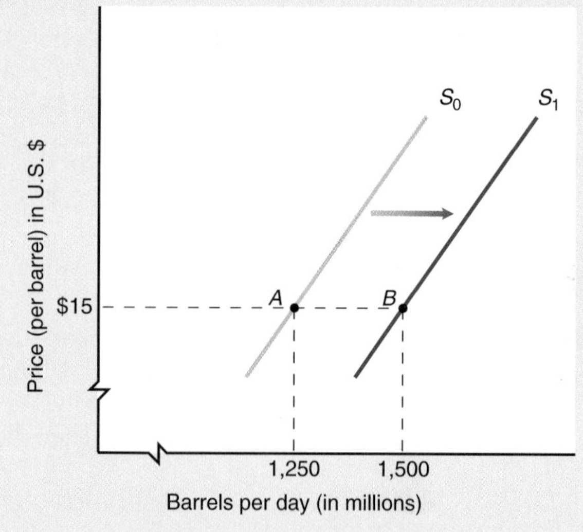

(b) Shift in supply

from point A to point B. At $15 a barrel, producers supplied 1,500 million barrels of oil a day, and at $36 a barrel they supplied 1,750 million barrels per day.

Earlier, in the 1980s, technological advances in horizontal drilling more than doubled the amount of oil that could be extracted from some oil fields. Technological innovations such as this reduced the cost of supplying oil and shifted the supply of oil to the right as shown in Figure 4-6(b). Before the innovation, suppliers were willing to provide 1,250 million barrels of oil per day at U.S. $15 a barrel. After the innovation, suppliers were willing to supply 1,500 million barrels of oil per day at U.S. $15 a barrel.

A Review

To be sure you understand shifts in supply, explain what is likely to happen to your supply curve for labour in the following cases: (1) You suddenly decide that you absolutely need a new car. (2) You suddenly won a million dollars in the lottery. And finally, (3) the wage you could earn doubled. If you came up with the answers: shift out, shift in, and no change—you've got it down. If not, it's time for a review.

Do we see such shifts in the supply curve often? Yes. A good example is computers. For the past 30 years, technological changes have continually shifted the supply curve for computers out.

The Supply Table

Remember Figure 4-4(a)'s demand table for cassette rentals. In Figure 4-7(a), columns 2 (Ann), 3 (Barry), and 4 (Charlie), we follow the same reasoning to construct a supply

Figure 4-7 (a and b) FROM INDIVIDUAL SUPPLIES TO A MARKET SUPPLY

As with market demand, market supply is determined by adding all quantities supplied at a given price. Three suppliers—Ann, Barry, and Charlie—make up the market of videocassette suppliers. The total market supply is the sum of their individual supplies at each price, shown in column 5 of **(a)**.

Each of the individual supply curves and the market supply curve have been plotted in **(b)**. Notice how the market supply curve is the horizontal sum of the individual supply curves.

Quantities supplied	(1) Price (per cassette)	(2) Ann's supply	(3) Barry's supply	(4) Charlie's supply	(5) Market supply
A	$0.00	0	0	0	0
B	0.50	1	0	0	1
C	1.00	2	1	0	3
D	1.50	3	2	0	5
E	2.00	4	3	0	7
F	2.50	5	4	0	9
G	3.00	6	5	0	11
H	3.50	7	5	2	14
I	4.00	8	5	2	15

(a) A supply table

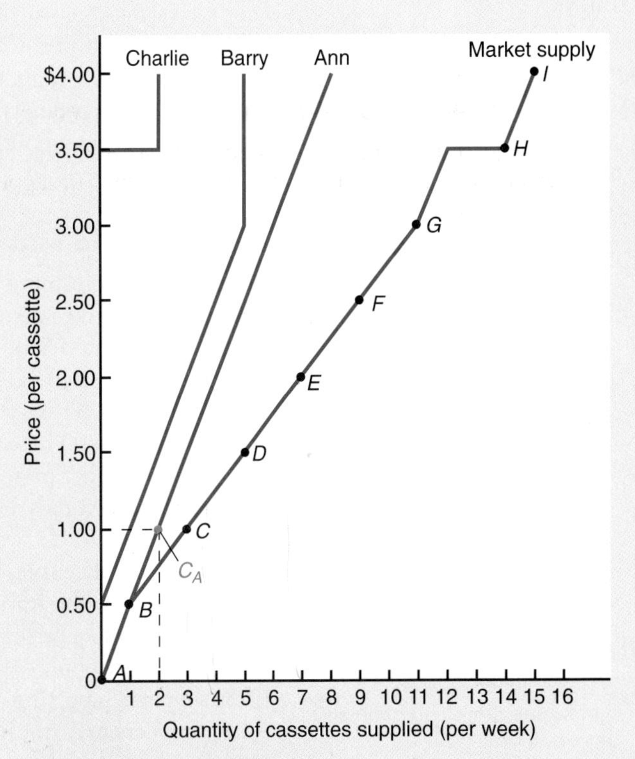

(b) Adding supply curves

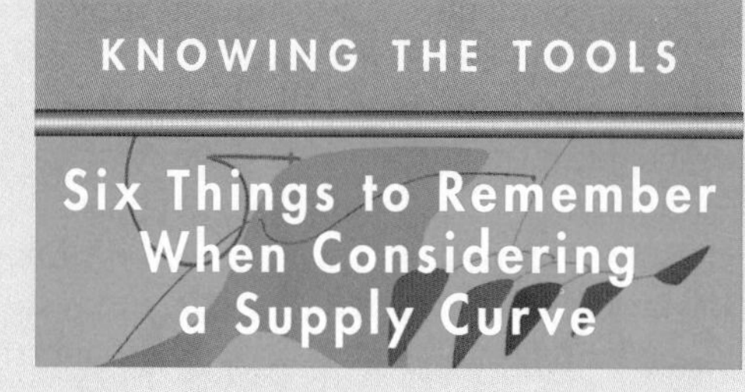

- A supply curve follows the law of supply. When price rises, quantity supplied increases, and vice versa.
- The horizontal axis—quantity—has a time dimension.
- The quantities are of the same quality.
- The vertical axis—price—assumes all other prices remain constant.
- The curve assumes everything else is constant.
- Effects of price changes are shown by movements along the supply curve. Effects of nonprice determinants of supply are shown by shifts of the entire supply curve.

table for three hypothetical cassette suppliers. Each supplier follows the law of supply: When price rises, each supplies more, or at least as much as each did at a lower price.

Q-7 Derive the market supply curve from the following two individual supply curves.

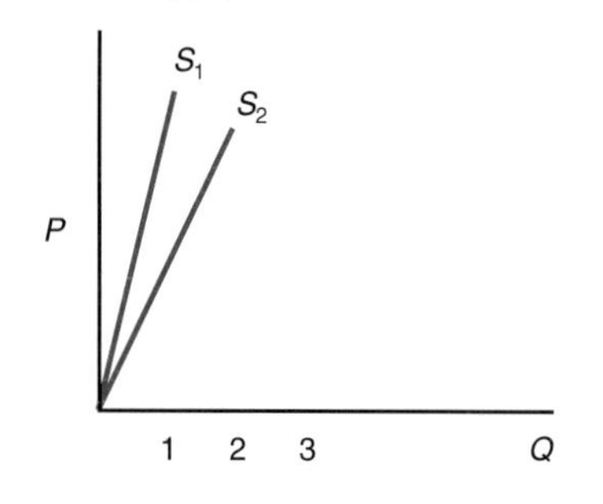

From a Supply Table to a Supply Curve

Figure 4-7(b) takes the information in Figure 4-7(a)'s supply table and translates it into a graph of each supplier's supply curve. For instance, point C_A on Ann's supply curve corresponds to the information in columns 1 and 2, row C. Point C_A is at a price of $1 per cassette and a quantity of 2 cassettes per week. Notice that Ann's supply curve is upward sloping, meaning that price is positively related to quantity. Charlie's and Barry's supply curves are similarly derived.

The supply curve represents the set of *minimum* prices an individual seller will accept for various quantities of a good. The market's invisible hand stops suppliers from charging more than the market price. If suppliers could escape the market's invisible hand and charge a higher price, they would gladly do so. Unfortunately for them, and fortunately for consumers, a higher price encourages other suppliers to begin selling cassettes. Competing suppliers' entry into the market sets a limit on the price any supplier can charge.

Individual and Market Supply Curves

The market supply curve is derived from individual supply curves in precisely the same way that the market demand curve was. To emphasize the symmetry, we've made the three suppliers quite similar to the three demanders. Ann (column 2) will supply 2 at $1; if price goes up to $2, she increases her supply to 4. Barry (column 3) begins supplying at $1, and at $3 supplies 5, the most he'll supply regardless of how high price rises. Charlie (column 4) has only two units to supply. At a price of $3.50 he'll supply that quantity, but higher prices won't get him to supply any more.

The **market supply curve** is *the horizontal sum of all individual supply curves*. In Figure 4-7(a) (column 5), we add together Ann's, Barry's, and Charlie's supply to arrive at the market supply curve, which is graphed in Figure 4-7(b). Notice that each point on it corresponds to the information in columns 1 and 5 for each row. For example, point *H* corresponds to a price of $3.50 and a quantity of 14.

The law of supply is based on two phenomena:

1. At higher prices, existing suppliers supply more.
2. At higher prices, new suppliers enter the market.

The market supply curve's upward slope is determined by two different sources: by existing suppliers supplying more and by new suppliers entering the market. Sometimes existing suppliers may not be willing to increase their quantity supplied in response to an increase in prices, but a rise in price often brings brand-new suppliers into the market. For example, a rise in teachers' salaries will have little effect on the amount of teaching current teachers do, but it will increase the number of people choosing to be teachers.

THE MARRIAGE OF SUPPLY AND DEMAND

Thomas Carlyle, the English historian who dubbed economics "the dismal science," also wrote this chapter's introductory tidbit. "Teach a parrot the words *supply* and *demand* and you've got an economist." In earlier chapters, we tried to convince you that economics is *not* dismal. In the rest of this chapter, we hope to convince you that, while supply and demand are important to economics, parrots don't make good economists. If students think that when they've learned the terms *supply* and *demand* they've learned economics, they're mistaken. Those terms are just labels for the ideas behind supply and demand, and it's the ideas that are important. What matters about supply and demand isn't the labels but how the concepts interact. For instance, what happens if a freeze kills the blossoms on the orange trees? The quantity of oranges supplied isn't expected to equal the quantity demanded. It's in understanding the interaction of supply and demand that economics becomes interesting and relevant.

During the 1990s, an overproduction of wheat led to excess supply and downward pressure on glodal wheat prices.

Excess Supply

When you have a market in which neither suppliers nor consumers can collude and in which prices are free to adjust, economists have a good answer for the question: What happens if quantity supplied doesn't equal quantity demanded? If there is **excess supply** (a surplus), *quantity supplied is greater than quantity demanded,* and some suppliers won't be able to sell all their goods. Each supplier will think: "Gee, if I offer to sell it for a bit less, I'll be the lucky one who sells my goods; someone else will be stuck with not selling their goods." But because all suppliers with excess goods will be thinking the same thing, the price in the market will fall. As that happens, consumers will increase their quantity demanded. So the movement toward equilibrium created initially by excess supply will be on both the supply and demand sides.

Excess Demand

The reverse is also true. Say that instead of excess supply, there's **excess demand** (a shortage)—*quantity demanded is greater than quantity supplied*. There are more consumers who want the good than there are suppliers selling the good. Let's consider what's likely to go through demanders' minds. They'll likely call long-lost friends who just happen to be sellers of that good and tell them it's good to talk to them and, by the way, don't they want to sell that . . . ? Suppliers will be rather pleased that so many of their old friends have remembered them, but they'll also likely see the connection between excess demand and their friends' thoughtfulness. To stop their phones from ringing all the time, they'll likely raise their price. The reverse is true for excess supply. It's amazing how friendly suppliers become to potential consumers when there's excess supply.

Q-8 Explain what a sudden popularity of "Economics Professor" brand casual wear would likely do to prices of that brand.

Price Adjusts

This tendency for prices to rise when the quantity demanded exceeds the quantity supplied and for prices to fall when the quantity supplied exceeds the quantity demanded is a central element to understanding supply and demand. So remember:

When quantity demanded is greater than quantity supplied, prices tend to rise.
When quantity supplied is greater than quantity demanded, prices tend to fall.

Two other things to note about supply and demand are (1) the greater the difference between quantity supplied and quantity demanded, the more pressure there is for prices to rise or fall, and (2) when quantity demanded equals quantity supplied, the market is in equilibrium.

see page 102

People's tendencies to change prices exist as long as there's some difference between quantity supplied and quantity demanded. But the change in price brings the laws of supply and demand into play. As price falls, quantity supplied decreases as some suppliers leave the business (the law of supply). And as some people who originally weren't really interested in buying the good think, "Well, at this low price, maybe I do want to buy," quantity demanded increases (the law of demand). Similarly, when price rises, quantity supplied will increase (the law of supply) and quantity demanded will decrease (the law of demand).

Whenever quantity supplied and quantity demanded are unequal, price tends to change. If, however, quantity supplied and quantity demanded are equal, price will stay the same because no one will have an incentive to change.

The Graphical Marriage of Supply and Demand

Figure 4-8 shows supply and demand curves for cassette rentals and demonstrates the force of the invisible hand. Let's consider what will happen to the price of cassettes in three cases:

1. When the price is $3.50 each;
2. When the price is $1.50 each; and
3. When the price is $2.50 each.

Q-9 In a flood, it is ironic that usable water supplies tend to decline because the pumps and water lines are damaged. What will a flood likely do to the prices of bottled water?

1. When price is $3.50, quantity supplied is 7 and quantity demanded is only 3. Excess supply is 4. Individual consumers can get all they want, but most suppliers can't sell all they wish; they'll be stuck with cassettes that they'd like to rent. Suppliers will tend to offer their goods at a lower price and demanders, who see plenty of suppliers out there, will bargain harder for an even lower price. Both these forces will push the price as indicated by the A arrows in Figure 4-8.

Now let's start from the other side.

2. Say price is $1.50. The situation is now reversed. Quantity supplied is 3 and quantity demanded is 7. Excess demand is 4. Now it's consumers who can't get

Figure 4-8 **THE MARRIAGE OF SUPPLY AND DEMAND**

Combining Ann's supply from Figure 4-7 and Marie's demand from Figure 4-4, let's see the force of the invisible hand. When there is excess demand there is upward pressure on price. When there is excess supply there is downward pressure on price. Understanding these pressures is essential to understanding how to apply economics to reality.

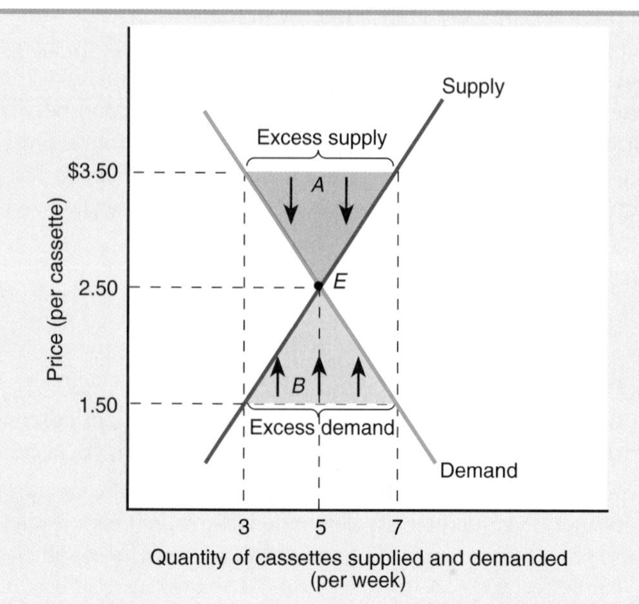

what they want and suppliers who are in the strong bargaining position. The pressures will be on price to rise in the direction of the *B* arrows in Figure 4-8.

3. At $2.50, price is at its equilibrium: quantity supplied equals quantity demanded. Suppliers offer to sell 5 and consumers want to buy 5, so there's no pressure on price to rise or fall. Price will tend to remain where it is (point *E* in Figure 4-8). Notice that the equilibrium price is where the supply and demand curves intersect.

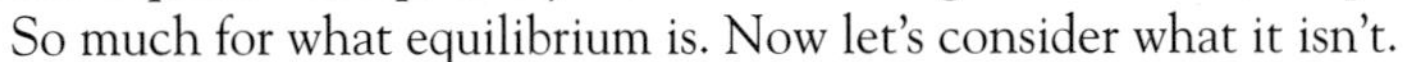

EQUILIBRIUM

The concept of equilibrium appears often throughout this text. You need to understand what equilibrium is and what it isn't.

What Equilibrium Is

see page 102

The concept itself comes from physics—classical mechanics. **Equilibrium** is *a concept in which opposing dynamic forces cancel each other out*. For example, a hot-air balloon is in equilibrium when the upward force exerted by the hot air in the balloon equals the downward pressure exerted on the balloon by gravity. In supply and demand analysis, equilibrium means that the upward pressure on price is exactly offset by the downward pressure on price. **Equilibrium price** is *the price toward which the invisible hand drives the market*. **Equilibrium quantity** is *the amount bought and sold at the equilibrium price*.

So much for what equilibrium is. Now let's consider what it isn't.

What Equilibrium Isn't

First, equilibrium isn't a state of the world. It's a characteristic of the model—the framework you use to look at the world. The same situation could be seen as an equilibrium in one framework and as a disequilibrium in another. Say you're describing a car that's speeding along at 100 kilometres an hour. That car is changing position relative to objects on the ground. Its movement could be, and generally is, described as if it were in disequilibrium. However, if you consider this car relative to another car going 100 kilometres an hour, the cars could be modelled as being in equilibrium because their positions relative to each other aren't changing.

Equilibrium is not inherently good or bad.

Second, equilibrium isn't inherently good or bad. It's simply a state in which dynamic pressures offset each other. Some equilibria are awful. Say two countries are engaged in a nuclear war against each other and both sides are blown away. An equilibrium will have been reached, but there's nothing good about it.

Desirable Characteristics of Supply/Demand Equilibrium

While there is nothing necessarily good about equilibrium, the supply/demand equilibrium has certain desirable characteristics that are very important when applying economic analysis. To see those desirable characteristics, let's consider what the demand curve and supply curve are telling us. Each of these curves tells us how much individuals would be willing to pay (in the case of demand) or accept (in the case of supply) for a good. Thus, in Figure 4-9(a) (on the next page) a consumer at quantity 2 would be willing to pay $8 for a good, and the supplier would be willing to sell it for $2.

If the consumer pays less than what he's willing to pay, he walks away better off. Thus, the distance between the demand curve and the price he pays is a net gain for the consumer. Economists call this net benefit **consumer surplus**—*the value the consumer*

Figure 4-9 (a and b) CONSUMER AND PRODUCER SURPLUS

Market equilibrium price and quantity maximizes the combination of consumer surplus (shown in blue) and producer surplus (shown in red) as demonstrated in **(a)**. When price deviates from its equilibrium as in **(b)**, combined consumer and producer surplus falls. The grey shaded region shows the loss of total surplus when price is $1 higher than equilibrium price.

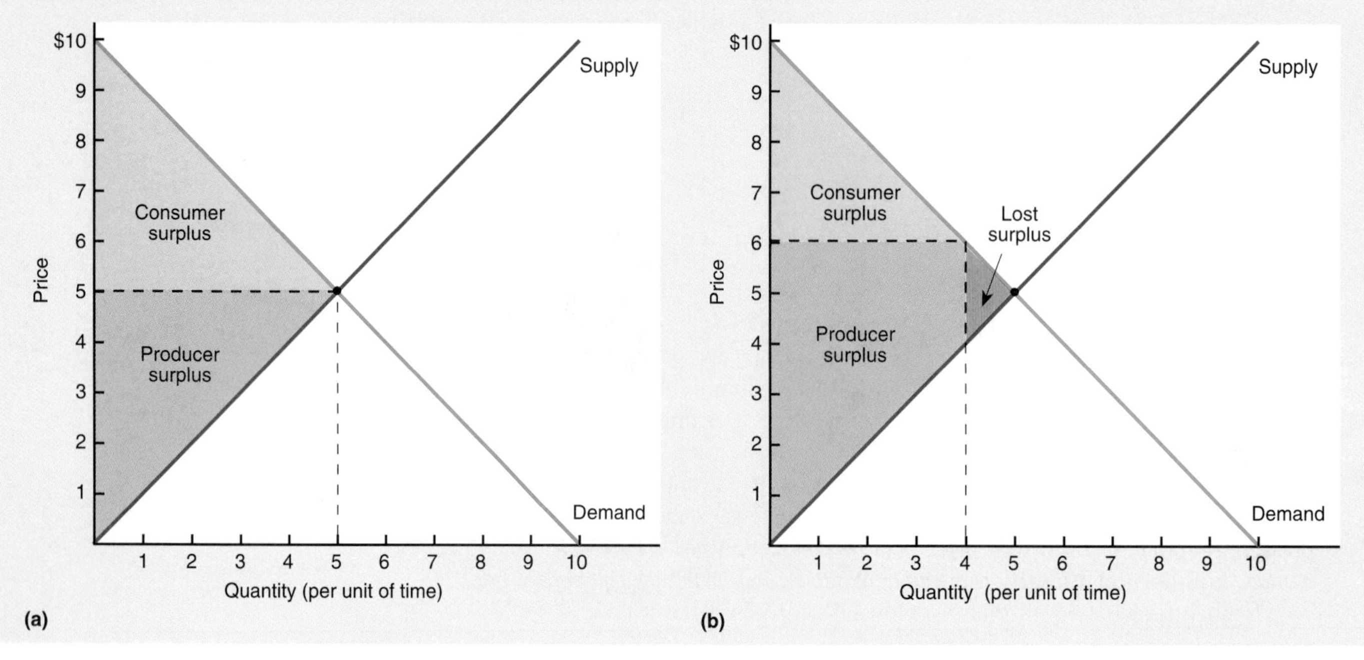

gets from buying a product, less its price. It is represented by the area underneath the demand curve and above the price that an individual pays. Thus, with the price at equilibrium ($5), consumer surplus is represented by the blue area.

Similarly, if a producer receives more than the price she would be willing to sell it for, she too receives a net benefit. Economists call this gain **producer surplus**—*the price the producer sells a product for, less the cost of producing it*. We use this cost information to construct a supply curve, so producer surplus is represented by the area above the supply curve but below the price the producer receives. Thus, with the price at equilibrium ($5), producer surplus is represented by the red area.

Q-10 If price moves from disequilibrium to equilibrium, what happens to the combination of producer and consumer surplus in the market?

What's good about equilibrium is that it makes the combination of consumer and producer surpluses as large as it can be. To see this, say that for some reason the equilibrium price is held at $6. Consumers will demand only 4 units of the good, and some suppliers are not able to sell all the goods they would like. The combined producer and consumer surplus will decrease, as shown in Figure 4-9(b). The grey triangle represents lost consumer and producer surplus. In general, a deviation of price from equilibrium lowers the combination of producer and consumer surplus. This is one of the reasons economists support markets and why we teach the supply/demand model. It gives us a visual sense of what is good about markets: By allowing trade, markets maximize the combination of consumer and producer surplus. How the government might intervene in a market to alter the distribution of consumer surplus and producer surplus takes us into the realm of normative economics.

APPLYING THE TOOLS

The Supply and Demand for Children

In Chapter 1, we distinguished between an economic force and a market force. Economic forces are operative in all aspects of our lives; market forces are economic forces that are allowed to be expressed through a market. Our examples in this chapter are of market forces—of goods sold in a market—but supply and demand can also be used to analyze situations in which economic, but not market, forces operate. An economist who is adept at this is Gary Becker of the University of Chicago. He has applied supply and demand analysis to a wide range of issues, even the supply and demand for children.

Becker doesn't argue that children should be bought and sold. But he does argue that economic considerations play a large role in people's decisions on how many children to have. In farming communities, children can be productive early in life; by age six or seven, they can work on a farm. In an advanced industrial community, children provide pleasure but generally don't contribute productively to family income. Even getting them to help around the house can be difficult.

Becker argues that since the price of having children is lower for a farming society than for an industrial society, farming societies will have more children per family. Quantity of children demanded will be larger. And that's what we find. Developing countries that rely primarily on farming often have three, four, or more children per family. Industrial societies average fewer than two children per family.

To fix the ideas of consumer and producer surplus in your mind, let's consider a couple of real-world examples. Think about the water you drink. What does it cost? Almost nothing. Given that water is readily available, it has a low price. But since you'd die from thirst if you had no water, you are getting an enormous amount of consumer surplus from that water. Next, consider a ballet dancer who loves the ballet so much he'd dance for free. But he finds that people are willing to pay to see him and that he can receive $400 a performance. He is receiving producer surplus.

CONCLUSION

Throughout the book we'll be presenting examples of supply and demand. So we'll end this chapter here because its intended purposes have been served. What were those intended purposes? First, we exposed you to enough economic terminology and economic thinking to allow you to proceed to our more complicated examples. Second, we have set your mind to work putting the events around you into a supply/demand framework. Doing that will give you new insights into the events that shape all our lives. Once you incorporate the supply/demand framework into your way of looking at the world, you will have made an important step toward thinking like an economist.

see page 102

Chapter Summary

- The law of demand states that quantity demanded rises as price falls, other things constant.
- The law of supply states that quantity supplied rises as price rises, other things constant.
- Factors that affect supply and demand other than price are called shift factors. Shift factors of demand include income, prices of other goods, tastes, population, and expectations. Shift factors of supply include the price of inputs, technology, expectations, and taxes and subsidies.
- A change in quantity demanded (supplied) is a movement along the demand (supply) curve. A change in demand (supply) is a shift of the entire demand (supply) curve.
- The laws of supply and demand hold true because individuals can substitute.

- A market demand (supply) curve is the horizontal sum of all individual demand (supply) curves.
- When quantity demanded is greater than quantity supplied, prices tend to rise. When quantity supplied is greater than quantity demanded, prices tend to fall.
- When quantity supplied equals quantity demanded, prices have no tendency to change. This is equilibrium.
- Equilibrium maximizes the combination of consumer surplus and producer surplus. Consumer surplus is the net benefit a consumer gets from purchasing a good, while producer surplus is the net benefit a producer gets from selling a good.

Key Terms

consumer surplus (97)
demand (85)
demand curve (84)
equilibrium (97)
equilibrium price (97)
equilibrium quantity (97)
excess demand (95)
excess supply (95)
law of demand (83)
law of supply (90)
market demand curve (88)
market supply curve (94)
movement along a demand curve (85)
movement along a supply curve (91)
producer surplus (98)
quantity demanded (85)
quantity supplied (91)
shift in demand (85)
shift in supply (91)
supply (91)
supply curve (90)

Questions for Thought and Review

1. State the law of demand. Why is price inversely related to quantity demanded?
2. State the law of supply. Why is price directly related to quantity supplied?
3. List four shift factors of demand and explain how each affects demand.
4. Distinguish the effect of a shift factor of demand on the demand curve from the effect of a change in price on the demand curve.
5. Draw a market demand curve from the following demand table.

P	*Q*
37	20
47	15
57	10
67	5

6. Draw a demand curve from the following demand table.

P	D_1	D_2	D_3
37	20	4	8
47	15	2	7
57	10	0	6
67	5	0	5

7. Danielle has just stated that normally, as price rises, supply will increase. Her teacher grimaces. Why?
8. List four shift factors of supply and explain how each affects supply.
9. Draw a market supply curve from the following supply table.

P	S_1	S_2	S_3
37	0	4	14
47	0	8	16
57	10	12	18
67	10	16	20

10. It has just been reported that eating meat is bad for your health. Using supply and demand curves, demonstrate the report's likely effect on the price and quantity of steak sold in the market.
11. Explain why the combination of consumer and producer surplus is not maximized if there is either excess demand or supply.
12. Use economic reasoning to explain why nearly every purchase you make provides you with consumer surplus.

Problems and Exercises

1. You're given the following individual demand tables for comic books.

Price	Jean	Liz	Connie
$ 2	4	36	24
4	4	32	20
6	0	28	16
8	0	24	12
10	0	20	8
12	0	16	4
14	0	12	0
16	0	8	0

   a. Determine the market demand table.
   b. Graph the individual and market demand curves.
   c. If the current market price is $4, what's the total market demand? What happens to total market demand if price rises to $8?
   d. Say that an advertising campaign increases demand by 50 percent. Illustrate graphically what will happen to the individual and market demand curves.

2. Draw hypothetical supply and demand curves for tea. Show how the equilibrium price and quantity will be affected by each of the following occurrences:
   a. Bad weather wreaks havoc with the tea crop.
   b. A medical report implying tea is bad for your health is published.
   c. A technological innovation lowers the cost of producing tea.
   d. Consumers' income falls.

3. This is a question concerning what economists call the *identification problem*. Say you go out and find figures on the quantity bought of various products. You will find something like the following:

Product	Year	Quantity	Average Price
VCRs	1998	100,000	$210
	1999	110,000	220
	2000	125,000	225
	2001	140,000	215
	2002	135,000	215
	2003	160,000	220

   Plot these figures on a graph.
   a. Have you plotted a supply curve, a demand curve, or what?
   b. If we assume that the market for VCRs is competitive, what information must you know to determine whether these are points on a supply curve or on a demand curve?
   c. Say you know that the market is one in which suppliers set the price and allow the quantity to vary. Could you then say anything more about the curves you have plotted?
   d. What information about shift factors would you expect to find to make these points reflect the law of demand?

4. You're a commodity trader and you've just heard a report that the winter wheat harvest will be 2.09 billion bushels, a 44 percent jump, rather than an expected 35 percent jump to 1.96 billion bushels.
   a. What would you expect would happen to wheat prices?
   b. Demonstrate graphically the effect you suggested in *a*.

5. In Canada, gasoline costs consumers about $0.80 per litre. In Italy it costs consumers about $2 per litre. What effect does this price differential likely have on:
   a. The size of cars in Canada and in Italy?
   b. The use of public transportation in Canada and in Italy?
   c. The fuel efficiency of cars in Canada and in Italy? What would be the effect of raising the price of gasoline in Canada to $2 per litre?

6. Use the graph below to answer the following questions:

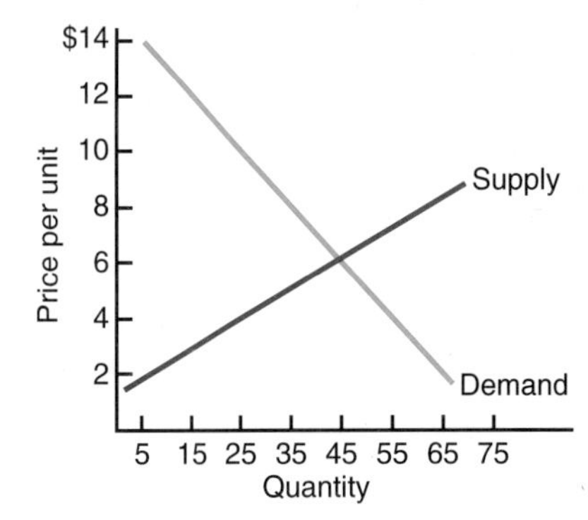

   a. What is equilibrium price and quantity?
   b. What is producer surplus when the market is in equilibrium?
   c. What is consumer surplus when the market is in equilibrium?
   d. If price were held at $10 a unit, what is consumer and producer surplus?

7. The following graph shows the market for apples.

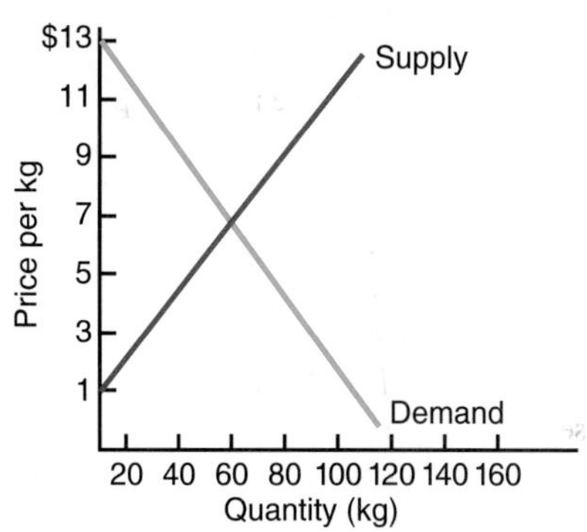

a. What are equilibrium price and quantity?
b. Determine producer surplus and consumer surplus at the equilibrium price.
c. If the government places a price floor at $9, what is the quantity traded in the market?
d. What are producer and consumer surplus at this controlled price?
e. What is the lost surplus resulting from the price floor?

Web Questions

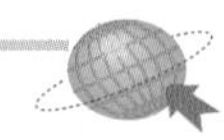

1. Go to the World Bank's Health, Nutrition and Population home page (devdata.worldbank.org/hnpstats/) and find data about Canada's population in 2000 and projections for 2010, 2020, and 2035. What do you expect to happen to the proportion of the population over 65? Report your findings. Other things constant, what do you expect will happen in the next 50 years to the relative demand and supply for each of the following, being careful to distinguish between shifts of and a movement along a curve:
   a. Nursing homes.
   b. Prescription medication.
   c. Baby high chairs.
   d. Postsecondary education.
2. Go to Natural Resources Canada's Energy Policy Branch home page (www.nrcan.gc.ca/es/epb/eng/enghome.htm)and answer the following questions:
   a. List the factors that are expected to affect demand and supply for energy in the near term. How will each factor affect demand? Supply?
   b. What is the Energy Policy Branch's forecast for world oil prices? Show graphically how the factors listed in your answer to (a) are consistent with the Energy Policy Branch's forecast. Label all shifts in demand and supply.
   c. Describe and explain the Energy Policy Branch's forecast for the price of gasoline, heating oil and natural gas. Be sure to mention the factors that are affecting the forecast.
3. Go to the Canadian Taxpayers Federation home page (www.taxpayer.com) and look up sales tax rates for the 10 provinces.
   a. Which province(s) have no sales tax? Which province(s) have the highest sales tax?
   b. Show graphically the effect of sales tax on supply, demand, equilibrium quantity, and equilibrium price.
   c. Name two neighbouring provinces that have significantly different sales tax rates. How does that affect the supply or demand for goods in those provinces?

Answers to Margin Questions

1. The demand curve slopes downward because price and quantity demanded are inversely related. As the price of a good rises, people switch to purchasing other goods whose prices have not risen as much. *(84)*
2. *Demand for furs*. The other possibility, *quantity demanded*, is used to refer to movements along (not shifts of) the demand curve. *(85)*
3. (1) The decline in price will increase the quantity of computers demanded (movement down along the demand curve); (2) With more income, demand for computers will rise (shift of the demand curve to the right). *(85)*
4. When adding two demand curves, you sum them horizontally, as in the accompanying diagram. *(88)*

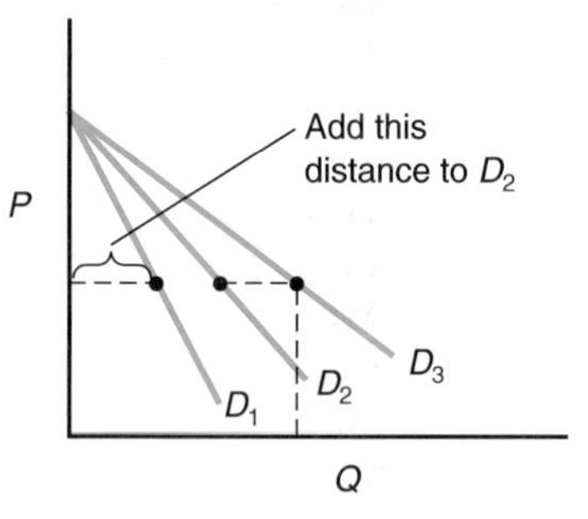

5. "The quantity supplied" declined because there was a movement along the supply curve. The supply curve itself remained unchanged. *(91)*
6. (1) The supply of romance novels declines since paper is an input to production (supply shifts to the left); (2) the supply of romance novels declines since the tax increases the cost to the producer (supply shifts to the left). *(92)*
7. When adding two supply curves, sum horizontally the two individual supply curves, as in the diagram below. *(94)*

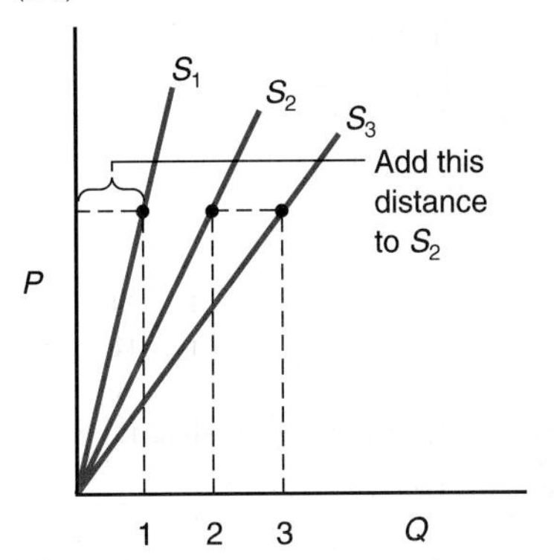

8. Customers will flock to stores demanding that funky "Economics Professor" look, creating excess demand. This excess demand will soon catch the attention of suppliers, and prices will be pushed upward. *(95)*
9. As substitutes—tap water—decrease, demand for bottled water increases enormously, and there will be upward pressure on prices. Social and political forces will, however, likely work in the opposite direction—against "profiteering" in people's misery. *(96)*
10. The combination of consumer and producer surplus will increase since there will be no lost surplus at the equilibrium price. *(98)*

Using Supply and Demand

5

After reading this chapter, you should be able to:

- Show the effect of a shift in demand and supply on equilibrium price and quantity.
- Explain real-world events using supply and demand.
- Demonstrate the effect of a price ceiling and a price floor on a market.
- Explain the effect of taxes, tariffs, and quotas on equilibrium price and quantity.
- State the limitations of demand and supply analysis.
- State six roles of government.

It is by invisible hands that we are bent and tortured worst.

Nietzsche

In the last chapter we introduced you to the concepts of supply and demand. In this chapter we will (1) show you the power of supply and demand, (2) show you how the invisible hand interacts with social and political forces to change the outcome of supply and demand analysis; and (3) discuss how one must adjust supply and demand analysis with other issues kept at the back of one's mind.

THE POWER OF SUPPLY AND DEMAND

To ensure that you understand the supply and demand graphs throughout the book, and can apply them, let's go through an example. Figure 5-1(a) deals with an increase in demand. Figure 5-1(b) deals with a decrease in supply.

Figure 5-1 (a and b) SHIFTS IN SUPPLY AND DEMAND

When there is an increase in demand (the demand curve shifts outward), there is upward pressure on the price, as shown in (**a**). If demand increases from D_0 to D_1, the quantity of cassette rentals that was demanded at a price of $2.25, 8, increases to 10, but the quantity supplied remains at 8. This excess demand tends to cause prices to rise. Eventually, a new equilibrium is reached at the price of $2.50, where the quantity supplied and the quantity demanded is 9 (point *B*).

If supply of cassette rentals decreases, then the entire supply curve shifts inward to the left, as shown in (**b**), from S_0 to S_1. At the price of $2.25, the quantity supplied has now decreased to 6 cassettes, but the quantity demanded has remained at 8 cassettes. The excess demand tends to force the price upward. Eventually, an equilibrium is reached at the price of $2.50 and quantity 7 (point C).

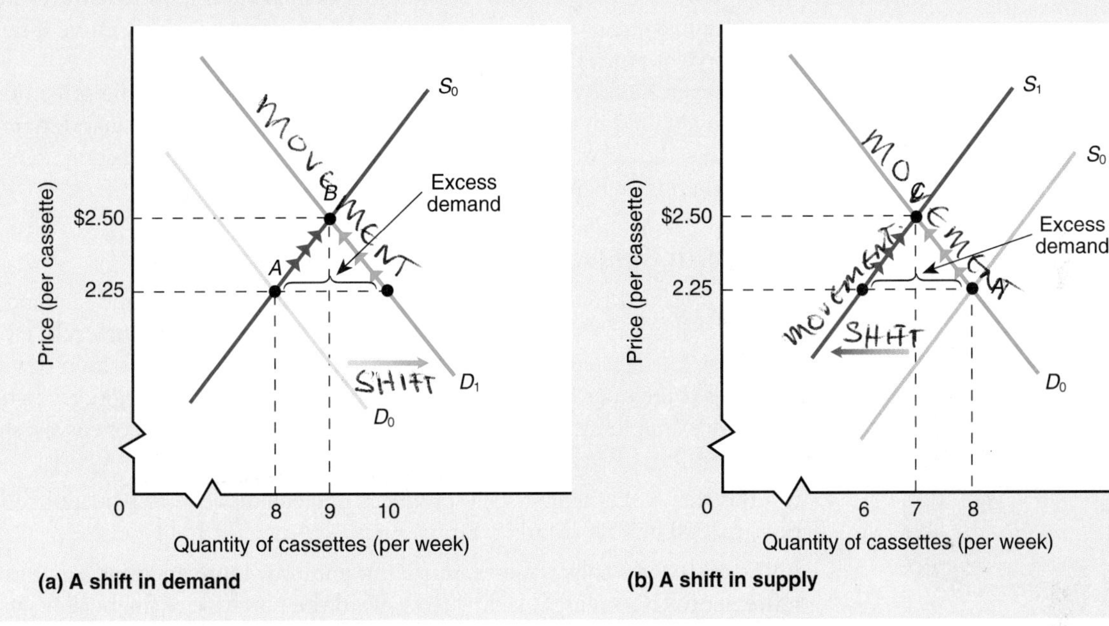

Let's consider again the supply and demand for videocassette rentals from Chapter 4. In Figure 5-1(a), the supply is S_0 and initial demand is D_0. They meet at an equilibrium price of $2.25 per cassette and an equilibrium quantity of 8 cassettes per week (point A). Now say that the demand for cassette rentals increases from D_0 to D_1. At a price of $2.25, the quantity of cassette rentals supplied will be 8 and the quantity demanded will be 10; excess demand of 2 exists.

The excess demand pushes prices upward in the direction of the small arrows, decreasing the quantity demanded and increasing the quantity supplied. As it does so, movement takes place along both the supply curve and the demand curve.

The upward push on price decreases the gap between the quantity supplied and the quantity demanded. As the gap decreases, the upward pressure decreases, but as long as that gap exists at all, price will be pushed upward until the new equilibrium price ($2.50) and new quantity (9) are reached (point *B*). At point *B*, quantity supplied equals quantity demanded. So the market is in equilibrium. Notice that the adjustment is twofold: The higher price brings about equilibrium by both increasing the quantity supplied (from 8 to 9) and decreasing the quantity demanded (from 10 to 9).

Figure 5-1(b) begins with the same situation that we started with in Figure 5-1(a); the initial equilibrium quantity and price are 8 cassettes per week and $2.25 per cassette (point A). In this example, however, instead of demand increasing, let's assume supply decreases—say because some suppliers change what they like to do, and decide they will

Q-1 Demonstrate graphically the effect of a heavy frost in Nova Scotia on the equilibrium quantity and price of apples.

Q-2 Say a hormone has been discovered that increases cows' milk production by 20 percent. Demonstrate graphically what effect this discovery would have on the price and quantity of milk sold in a market.

no longer supply cassettes. That means that the entire supply curve shifts inward to the left (from S_0 to S_1). At the initial equilibrium price of $2.25, the quantity demanded is greater than the quantity supplied. Two more cassettes are demanded than are supplied. (Excess demand = 2.)

Q-3 Demonstrate graphically the likely effect of an increase in the price of gas on the equilibrium quantity and price of compact cars.

This excess demand exerts upward pressure on price. Price is pushed in the direction of the small arrows. As the price rises, the upward pressure on price is reduced but will still exist until the new equilibrium price, $2.50, and new quantity, 7, are reached. At $2.50, the quantity supplied equals the quantity demanded. The adjustment has involved a movement along the demand curve and the new supply curve. As price rises, quantity supplied is adjusted upward and quantity demanded is adjusted downward until quantity supplied equals quantity demanded where the new supply curve intersects the demand curve at point C, an equilibrium of 7 and $2.50.

Here is an exercise for you to try. Demonstrate graphically how the price of computers could have fallen dramatically in the past 10 years, even as demand increased. (Hint: Supply has shifted even more, so even at lower prices, far more computers have been supplied than were being supplied 10 years ago.)

Six Real-World Examples

If this orange orchard was damaged, supply would be reduced, thereby putting upward pressure on orange prices.

Now that we've been through a generic example of shifts in supply and demand, let's consider some real-world examples. Below are six events. After reading each, try your hand at explaining what happened, using supply and demand curves. To help you in the process, Figure 5-2 provides some diagrams. *Before* reading our explanation, try to match the shifts to the examples. In each, be careful to explain which curve, or curves, shifted and how those shifts affected equilibrium price and quantity.

1. Brazil is the world's largest sugar producer. Inclement weather reduced production in 2000 by 15%. Market: Sugar.
2. In the mid-1990s baby boomers started to put away more and more savings for retirement. This saving was directed toward the purchase of financial assets, driving up the price of stocks. Market: Financial assets.
3. The majority of golfers in Korea prefer to use the newest American-made golf clubs. The Korean government, in an effort to protect domestic golf club producers, imposed a 20 percent luxury tax on imported American clubs. Market: American-made golf clubs in Korea.
4. Rice is crucial to Indonesia's nutritional needs and its rituals. In 1997, drought, pestilence, and a financial crash led to disruptions in the availability of rice. Its price rose so high that in 1998 more than a quarter of all Indonesians could not buy enough market-priced rice to meet their daily needs. Government programs to deliver subsidized rice were insufficient to bring the price of rice back to affordable levels. Market: Rice in Indonesia.
5. In late summer 1998, U.S. farmers were hard pressed to find enough seasonal farmhands. Why? El Niño's weather patterns compressed the harvest season. Grape, apple, and peach growers, who usually harvested at different times, were competing for the same workers. In addition, stronger efforts by authorities had reduced the flow of illegal workers to the United States. Market: Farm labourers.
6. Every Christmas a new toy becomes the craze. In 1997 it was Tickle Me Elmo and in 1998 it was Furby. Before Christmas Day, these toys were hard to find and sold for as much as 10 times their retail price on what is called the black market. Here we use the Furby as the example. Toymaker Tiger, along with retailers, worked up initial interest in Furby in late November, advertising the

Figure 5-2 (*a–f*)

In this exhibit, six shifts of supply and demand are shown. Your task is to match them with the events listed in the chapter.

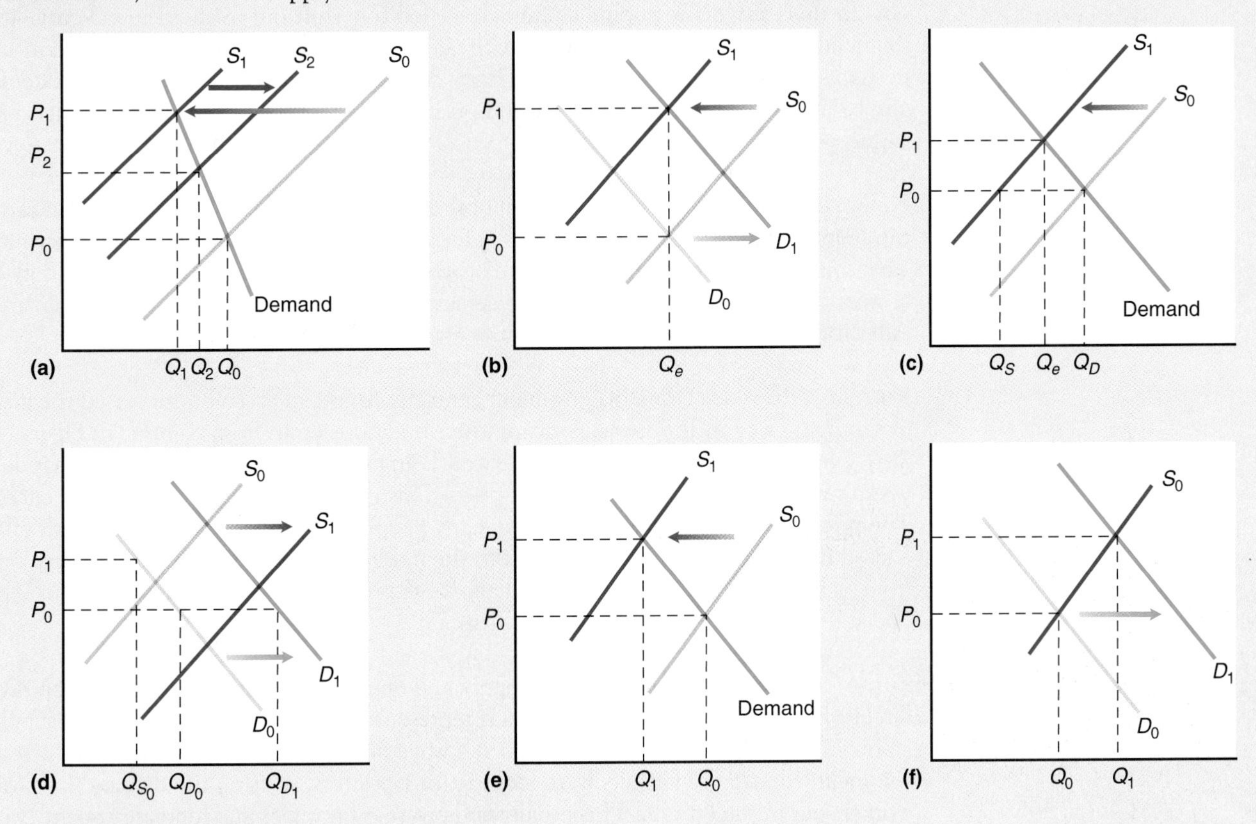

Answers: 1:c; 2:f; 3:e; 4:a; 5:b; 6:d.

limited supply. As early as 2:00 A.M., lines formed at the stores carrying Furbies. Some shoppers (including "toy scouts") were able to buy Furbies then resell them the same afternoon for as much as $300 apiece. Even with the shortage, retailers kept the price at its preset advertised price and producers continued to limit distribution. Newspapers carried stories about the lines and black market prices, intensifying demand for Furbies, which became even harder to come by. Days before Christmas, the supplier increased shipments of Furbies to meet the increased demand. Customers felt "lucky" when they were able to find Furbies with so few days left before Christmas, and for only $30 instead of $300 on the black market. Market: Furbies in 1998.

Sugar Shock The weather is invariably uncooperative. Nearly every year, some market is hit with a crop-damaging freeze, too little precipitation, or even too much rain. This is a shift factor of supply because it raises the cost of supplying sugar. The bad weather in 2000 shifted the supply curve for Brazilian sugar in, as shown in Figure 5-2(c). At the original price, quantity demanded exceeded quantity supplied and the invisible hand of the market pressured the price to rise until quantity demanded equalled quantity supplied.

Financial Assets and the Baby Boomers The postwar population swell we call the baby boom resulted in increased demand for all sorts of products as the boomers graduated, then bought houses, and now are demanding more health care and financial assets. In this case, demographic changes have led to a shift out in the demand curve for financial assets, resulting in a rise in stock market prices and an increase in the quantity of stocks and mutual funds supplied. This is depicted in Figure 5-2(f). This figure could also be used to describe the huge rise in housing prices in the 1980s as baby boomers began to purchase houses.

Excise Taxes In Chapter 4's discussion of shift factors, we explained that taxes levied on the supplier will reduce supply. The 20 percent luxury tax will shift the supply curve in. That some golfers use their old clubs and others look elsewhere to buy clubs is substitution at work, and a movement up along the demand curve. Figure 5-2(e) shows this scenario. After the tax, price rises to P_1 and quantity of clubs sold declines to Q_1.

Rice in Indonesia Drought, pestilence, and the financial crash all increased the cost of supplying rice in Indonesia, shifting the supply of rice in from S_0 to S_1 in Figure 5-2(a). Since rice is so important to the well-being of Indonesians, quantity demanded doesn't change much with changes in price. This is shown by the steep demand curve. The price rose to levels unaffordable to many people. In response, the government purchased imported rice and distributed it to the market. This shifted the supply curve out from S_1 to S_2. Since the price was still above its previous level, we know that this second shift in supply is smaller than the first.

Farm Labourers In this case both supply and demand shift, but this time in opposite directions. The previous year's demand is represented in Figure 5-2(b) by D_0 and supply is shown by S_0. Q_e labourers were hired at a wage of P_0. The compressed harvesting season meant that more farmers were looking for labourers, shifting the demand for farm workers out from D_0 to D_1. This put upward pressure on wages and increased quantity of labour supplied. Simultaneously, however, the supply of farm workers shifted in from S_0 to S_1 as the authorities increased border patrols. This put further upward pressure on wages and reduced the quantity of labour demanded. Wages are clearly bid up, in this case to P_1. The effect on the number of labourers hired, however, depends on the relative size of the demand and supply shifts. As we have drawn it, the quantity of labourers hired returns to the quantity of the previous year, Q_e. If the supply shift were greater than the shift in demand, the number of labourers would have declined. If it were smaller, the number of labourers would have risen.

Christmas Toys In this example, both supply and demand shift in the same direction. The initial market is shown by D_0 and S_0 in Figure 5-2(d). The price of $30 (shown by P_0) was below the equilibrium price and a shortage of $Q_{D_0} - Q_{S_0}$ existed. The black market price of $300 (shown by P_1) is shown by the amount that consumers are willing to pay for the quantity supplied, Q_{S_0}. As the craze for the toy intensified following the free newspaper publicity of the lines and black market prices, demand shifted out to D_1. Price was kept at $30 and the shortage became even greater, $Q_{D_1} - Q_{S_0}$. When Tiger made more Furbies available, supply shifted to S_1, eliminating most, but not all, of the shortage. At least one Walmart employee was injured in the mad rush to obtain a Furby.

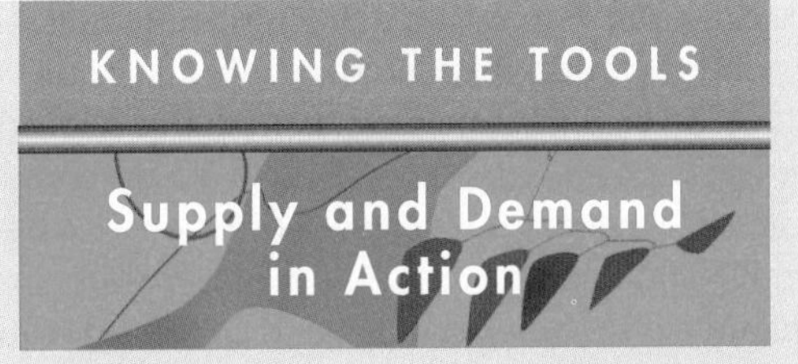

Sorting out the effects of the shifts of supply or demand or both can be confusing. Here are some helpful hints to keep things straight:

- Draw the initial demand and supply curves and label them. The equilibrium price and quantity is where these curves intersect. Label them.
- If only price has changed, no curves will shift and a shortage or surplus will result.
- If a nonprice factor affects demand, determine the direction demand has shifted and add the new demand curve. Do the same for supply.
- Equilibrium price and quantity is where the new demand and supply curves intersect. Label them.
- Compare the initial equilibrium price and quantity to the new equilibrium price and quantity.

See if you can describe what happened in the three graphs below.

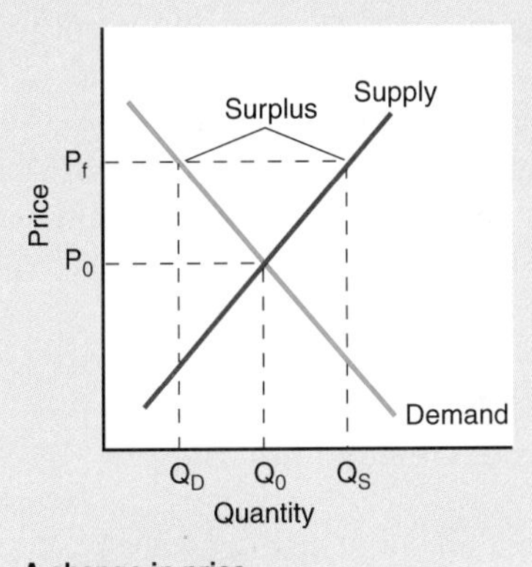

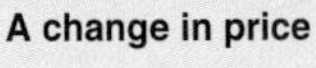

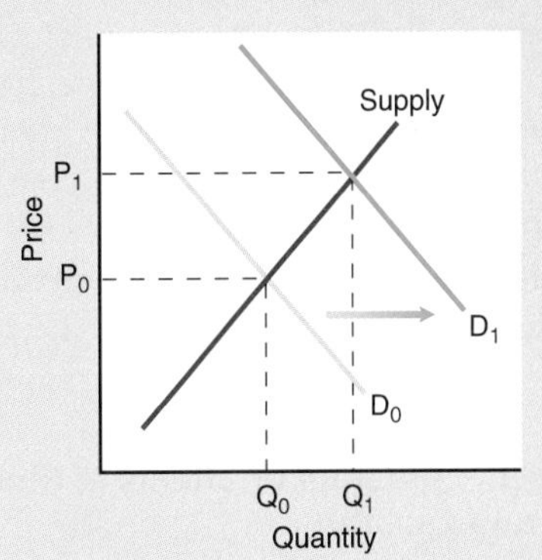

A shift in demand

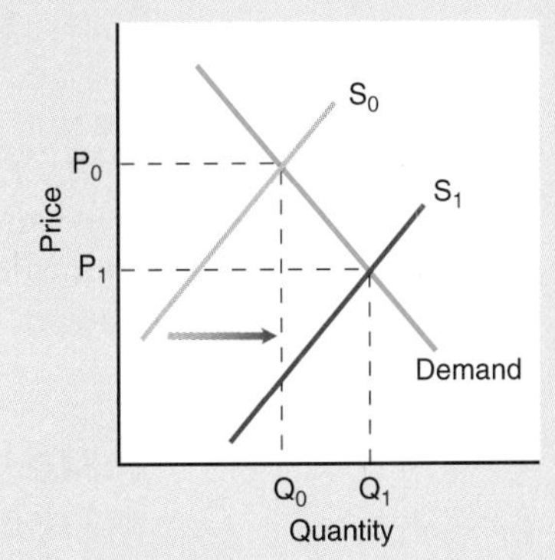

A shift in supply

A Review

As you can see, supply and demand analysis can get quite complicated. That is why you must separate shifts in demand and supply from movements along the supply and demand curves. Remember: Anything that affects demand and supply other than price of the good under consideration will shift the curves. Changes in the price of the good under consideration result in movements along the curves. Another thing to recognize is that when both curves are shifting you can get a change in price but little change in quantity, or a change in quantity but little change in price.

Anything other than price that affects demand or supply will shift the curves.

To test your understanding, we'll now give you six generic results from the interaction of supply and demand. Your job is to decide what shifts produced those results. This exercise is a variation of the first. It goes over the same issues, but this time without the graphs. On the left-hand side of the table below, we list combinations of movements of observed prices and quantities, labeling them 1–6. On the right we give six shifts in supply and demand, labeling them *a–f*.

If you don't confuse your "shifts of" with your "movements along," supply and demand provide good off-the-cuff answers for many economic questions.

Price and Quantity Changes

1.	P↑	Q↑
2.	P↑	Q↓
3.	P↑	Q?
4.	P↓	Q?
5.	P?	Q↑
6.	P↓	Q↓

Shifts in Supply and Demand

a. Supply shifts in. No change in demand.
b. Demand shifts out. Supply shifts in.
c. Demand shifts in. No change in supply.
d. Demand shifts out. Supply shifts out.
e. Demand shifts out. No change in supply.
f. Demand shifts in. Supply shifts out.

You are to match the shifts with the price and quantity movements that best fit each described shift, using each shift and movement only once. Our recommendation to you

Q-4 If both demand and supply shift in, what happens to price and quantity?

is to draw the graphs that are described in *a–f*, decide what happens to price and quantity, and then find the match in 1–6.

Now that you've worked them, let us give you the answers we came up with. They are: 1:*e*; 2:*a*; 3:*b*; 4:*f*; 5:*d*; 6:*c*. How did we come up with the answers? We did what we suggested you do—took each of the scenarios on the right and predicted what happens to price and quantity. For case *a*, supply shifts in and there is a movement up along the demand curve. Since the demand curve is downward sloping, the price rises and quantity declines. This matches number *2* on the left. For case *b*, demand shifts out. Along the original supply curve, price and quantity would rise. But supply shifts in, leading to even higher prices, but lower quantity. What happens to quantity is unclear, so the match must be number *3*. For case *c*, demand shifts in. There is movement down along the supply curve with lower price and lower quantity. This matches number 6. For case *d*, demand shifts out and supply shifts out. As demand shifts out, we move along the supply curve to the right and price and quantity rise. But supply shifts out too, and we move out along the new demand curve. Price declines, erasing some or all of the previous rise, and the quantity rises even more. This matches number 5.

We'll leave it up to you to confirm our answers to *e* and *f*. Notice that when supply and demand both shift, the change in either price or quantity is uncertain—it depends on the direction and the relative size of the shifts. As a summary, we present a diagrammatic of the combinations in Table 5-1.

TABLE 5-1 Diagram of Effects of Shifts of Demand and Supply on Price and Quantity

This table provides a summary of the effects of shifts in supply and demand on price and quantity. Notice that when both curves shift, the effect on either price or quantity depends on the relative size of the shifts.

	No change in supply.	Supply shifts out.	Supply shifts in.
No change in demand.	No change.	P↓ Q↑ Price declines and quantity rises.	P↑ Q↓ Price rises. Quantity declines.
Demand shifts out.	P↑ Q↑ Price rises. Quantity rises.	P? Q↑ Quantity rises. Price could be higher or lower depending upon relative size of shifts.	P↑ Q? Price rises. Quantity could rise or fall depending upon relative size of shifts.
Demand shifts in.	P↓ Q↓ Price declines. Quantity declines.	P↓ Q? Price declines. Quantity could rise or fall depending upon relative size of shifts.	P? Q↓ Quantity declines. Price rises or falls depending upon relative size of shifts.

GOVERNMENT INTERVENTIONS: PRICE CEILINGS AND PRICE FLOORS

People don't always like the market-determined price. When prices fall, sellers look to government for ways to hold prices up; when prices rise, buyers look to government for ways to hold prices down. Let's now consider the effect of such actions. Let's start with an example of the price being held down.

Price Ceilings

When government wants to hold prices down, it imposes a **price ceiling**—*a government-imposed limit on how high a price can be charged*. Rent control is an example of a price ceiling. (For the price ceiling to be effective, it must be below the equilibrium price, and throughout this discussion we shall assume that it is.)

Specifically, let's consider rent control in Paris in 1948. **Rent control** is *a price ceiling on rents set by government*. During the First World War, to stabilize housing prices and help out those fighting for France, rents were frozen. Upon the return of veterans, the freeze was held in the interest of society. In 1926, rent control was reviewed but by that time, lifting the controls would have resulted in huge increases in rents. Rents were allowed to rise only slightly. Again, during the Second World War, rents were frozen. Right after the end of the Second World War rent was capped at $2.50 a month. Without rent control, rent would have been $17 a month.

see page 126

This was a good situation for those occupying apartments, but it had drawbacks. For instance, there was an enormous shortage of apartments. The situation is shown in Figure 5-3.

What were the results of the rent control besides the shortage of apartments? More than 80 percent of Parisians had no private bathrooms and 20 percent had no running water. Since rental properties weren't profitable, no new buildings were being constructed and existing buildings weren't kept in repair. From 1914 (before the First World War) to 1948, the housing stock increased by only 10 percent. Many couldn't find housing in Paris. Couples lived with their in-laws. Existing apartments had to be rationed in some way. To get into a rent-controlled apartment, individuals paid bribes of up to

Figure 5-3 RENT CONTROL IN PARIS

A price ceiling imposed on housing rent in Paris during the First World War created a shortage of housing when the war ended and veterans returned home. The shortage would have been eliminated if rents had been allowed to rise to $17 per month.

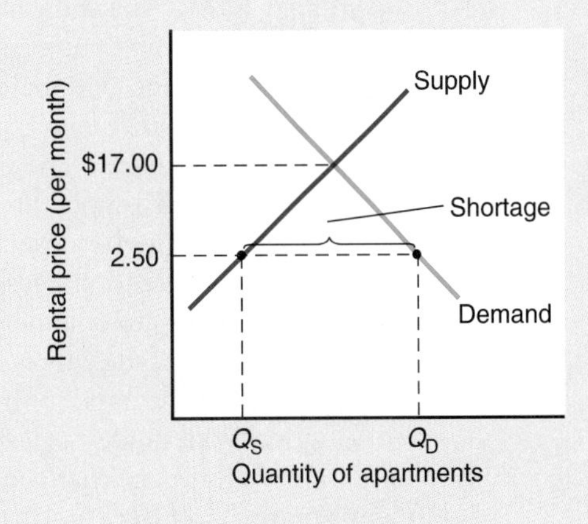

$1,500 per room or watched the obituaries and then simply moved in their furniture before anyone else did. Eventually the situation got so bad that rent controls were lifted.

The system of rent controls is not only of historical interest. Below we list some phenomena that existed recently.

Q-5 What is the effect of the price ceiling, P_c, shown in the graph below on price and quantity?

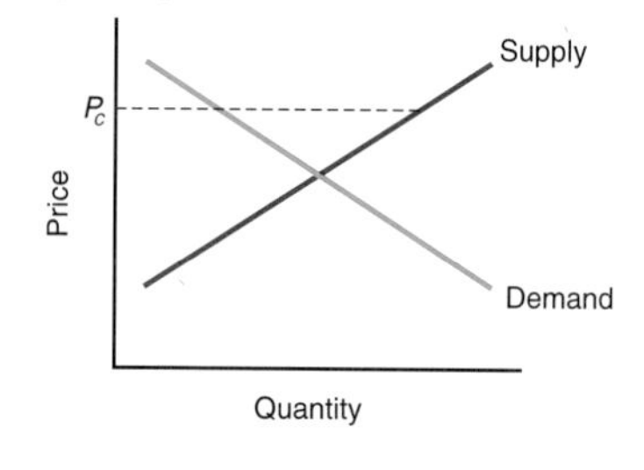

1. A couple pays $350 a month for a two-bedroom downtown apartment with a solarium and two terraces, while another individual pays $1,200 a month for a studio apartment shared with two roommates.
2. The vacancy rate for apartments in Montreal was 1.5 percent in 2000. Anything under 5 percent is considered a housing emergency.
3. Would-be tenants make payments, called key money, to current tenants or landlords to get apartments.

Your assignment is to explain how these phenomena might have come about, and to demonstrate, with supply and demand, the situation that likely caused them.

Now that you have done your assignment (you have, haven't you?), let us give you our answers so that you can check them with your answers.

The situation is identical with that presented above in Figure 5-3. Take the first item. The couple lives in a rent-controlled apartment while the individual with roommates does not. If rent control were eliminated, rent on the downtown apartment would rise and rent on the studio would most likely decline. Item 2: The housing emergency is a result of rent control. Below-market rent results in excess demand and little vacancy. Item 3: New residents must search for a long time to find apartments to rent, and many discover that illegal payments to landlords are the only way to obtain a rent-controlled apartment. Key money is a side payment for a rent-controlled apartment. Because of the limited supply of apartments, individuals are willing to pay far more than the controlled price. Landlords can use other methods of rationing the limited supply of apartments—instituting first-come, first-served policies, and, in practice, selecting tenants based on gender, race, or other personal characteristics, even though such discriminatory selection is illegal.

With price ceilings, existing goods are no longer rationed entirely by price. Other methods of rationing existing goods arise called nonprice rationing.

Before we move away from our discussion of rent controls, there's a dynamic issue we need to explicitly consider. In the long run an increase in rents should increase the quantity of apartments supplied as building owners convert commercial properties to residential use and erect new buildings, but since it takes time to construct a new apartment building or convert existing structures to apartment use, in the short run the supply of apartments is relatively fixed. This suggests we should use two supply curves to examine how rent controls differ in their impacts in the short-run and the long-run. You can see this in Figure 5–4. The short run supply curve is vertical to illustrate that the supply of apartments is fixed at a point in time, while the long run supply curve slopes upward to show that higher rents will increase the quantity of units supplied in the long run.

Suppose the market is in equilibrium at point A at a price of $750 per month for a one bedroom apartment. The introduction of rent controls at the initial price creates excess demand in the short run since the quantity demanded at the controlled price of $500 per month is higher than the fixed quantity supplied. Over time, landlords will have little incentive to maintain existing properties and may decide to convert their apartment buildings to commercial use structures that are not subject to rent controls. Over the long term, the supply of apartments will fall to S_S' in Figure 5–4, leading to a permanent excess demand for housing.

What does our model suggest will happen if rent controls are relaxed? In the short-term, rents will rise to what people are willing to pay, given the existing stock of apartments: point B in Figure 5–4. Over time, as landlords adjust to the new higher rents by building new units, the shortages will be eliminated. This is what is happening today in

Figure 5-4 RENT CONTROLS OVER TIME

In the short run the supply of rental units is relatively fixed and given by S_S. The long-run supply curve demonstrates that landowners will increase the quantity of apartments supplied if rents rise, over time. The initial equilibrium is at A.

Rent controls will set the price below the equilibrium, leading to excess demand in the short run. The shortages will grow over time as landlords decide to convert existing apartments to commercial use or refuse to maintain their current units. The short-run supply curve will shift left to S_S'. If rent controls are removed, in the short run rents will rise to what the market will bear at point B. Landlords respond by increasing supply over time, shifting the short-run supply curve back to its initial position.

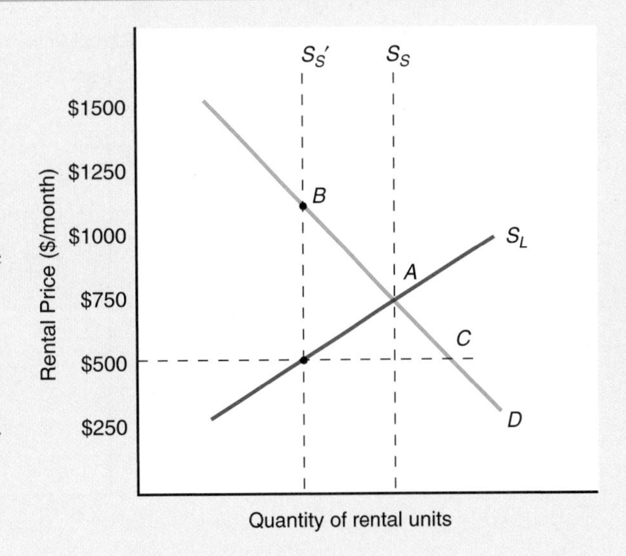

Ontario. Rent controls were introduced there in 1975. In 1998 the Ontario government passed the *Ontario Tenants Act*, which, among other things, relaxed rent controls. In the first three months rents rose by over 7 percent; in 1999 they rose by over 5 percent; and they just keep rising. The *Act* provides a guideline for rent increases for 2001 of 2.9 percent and in 2002 of 3.9 percent. Actual increases depend on how long you've been in your apartment. For new tenants, landlords can charge "what the market will bear." If you rent an apartment in Ontario, has your rent risen by the amount suggested by the guideline this year? Check out the Ontario Rental Housing Tribunal Web site for more on some Ontario evidence (http://www.orht.gov.on.ca/home.html).

If rent controls create a wedge between quantity demanded and quantity supplied, why would a government ever introduce them? Well, consider Toronto's bid for the 2008 Olympics. The demand for housing was expected to swell in anticipation of the games, and remain high until shortly after the closing ceremony. Figure 5–5 demonstrates that there would be a temporary increase in the demand for rental units with demand shifting to D′. Given the existing supply of rental units and the fact that landlords weren't expected to construct a large number of new buildings (since after the games, demand for rental units was expected to fall back to its initial level), rents would rise from R_0 to R_1, and then fall back to R_0 after the games. Landlords would earn a temporary windfall of the shaded area. The authorities might have viewed this as an undesirable distribution of income and a negative consequence of holding the games in Toronto. People already living in Toronto would face the higher rents, as would those temporarily trying to find accommodations in the area. To stop this from happening, the authorities could put in place a system of rent controls. If the price ceiling were set at the initial price, landlords wouldn't gain, existing renters wouldn't lose, and the controls would create a temporary shortage of housing. Those wishing to rent units in the area might have to commute from other areas (like Buffalo) or seek alternative accommodations (such as living at campsites in their RV).

Figure 5-5 WHEN RENT CONTROLS WORK

A temporary increase in demand transfers the shaded area from tenants to landlords. The government can stop this by placing rent controls at the initial price R_0, creating a temporary shortage of AB.

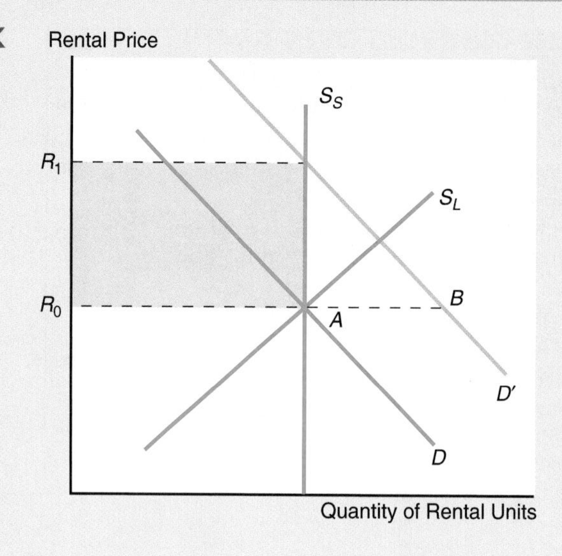

Q-6 What is the effect of the price floor, P_f, shown in the graph below on price and quantity?

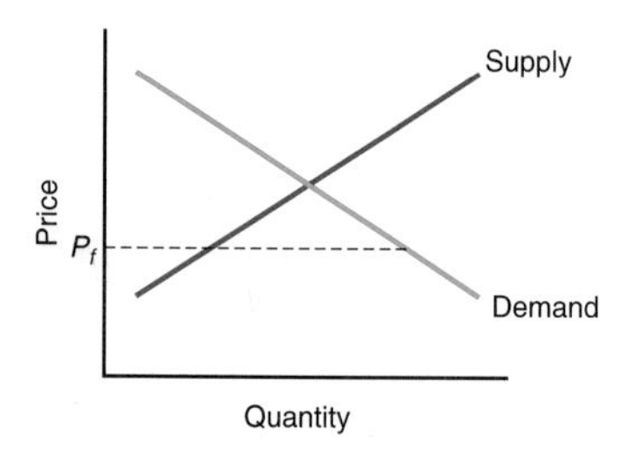

Price Floors

Sometimes political forces favour suppliers, sometimes consumers. So let us now go briefly through a case when the government is trying to favour suppliers by attempting to prevent the price from falling below a certain level. **Price floors**—*government-imposed limits on how low a price can be charged*—do just this. An example of a price floor is the minimum wage. Both individual provinces and the federal government impose **minimum wage laws**—*laws specifying the lowest wage a firm can legally pay an employee*. These price floors were initially set to provide a level of income that would allow a worker to cover their basic necessities. British Columbia was the first province to enact minimum wage legislation (in 1925), with the federal minimum wage set at the provincial level. In late 2001 these wages ranged from $5.50 per hour in Newfoundland to $8.00 per hour in British Columbia. (In 1965 the B.C. minimum wage was $1 per hour; with inflation, that's a much smaller increase than it looks). At an average of 2000 hours worked a year, many believe these wage are too low to provide for the basic necessities of life. The minimum wage affects thousands of workers who are mostly unskilled. The market-determined equilibrium wage for skilled workers is generally above the minimum wage.

The effect of a minimum wage on the unskilled labour market is shown in Figure 5-6. The government-set minimum wage is above equilibrium, as shown by W_{min}. At the market-determined equilibrium wage W_e, the quantity of labour supplied and demanded equals Q_e. At the higher minimum wage, the quantity of labour supplied rises to Q_1 and the quantity of labour demanded declines to Q_2. There is an excess supply of workers (a shortage of jobs) represented by the difference $Q_2 - Q_1$. This represents people who are looking for work but cannot find it.

The minimum wage helps some people and hurts others.

Who wins and who loses from a minimum wage? The minimum wage improves the wages of the Q_2 workers who are able to find work. Without the minimum wage, they would have earned W_e per hour. The minimum wage hurts those, however, who cannot find work at the minimum wage but who are willing to work, and would have been hired, at the market-determined wage. These workers are represented by the distance

Figure 5-6 A MINIMUM WAGE

A minimum wage, W_{min}, set above equilibrium wage, W_e, helps those who are able to find work, shown by Q_2, but hurts those who would have been employed at the equilibrium wage but can no longer find employment, shown by $Q_e - Q_2$. A minimum wage also hurts producers who have higher costs of production and consumers who may face higher product prices.

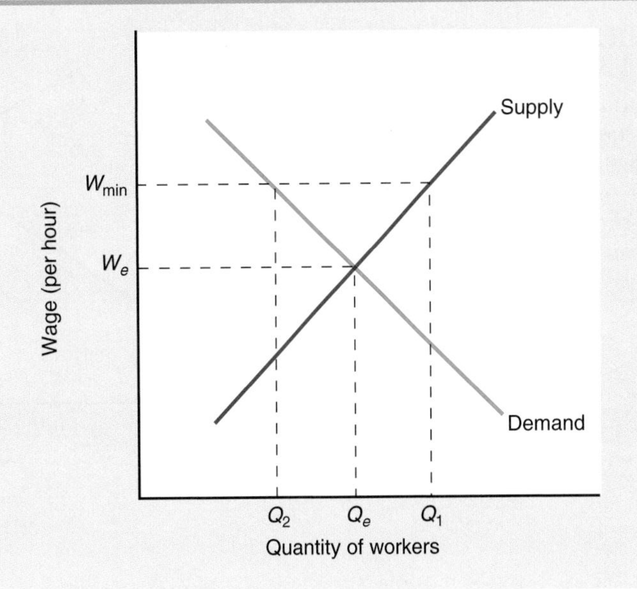

$Q_e - Q_2$ in Figure 5-6. The minimum wage also hurts firms who now must pay their workers more, increasing the cost of production. The minimum wage also hurts consumers to the extent that firms are able to pass that increase in production cost on in the form of higher product prices.

Late in 2001 the government of British Columbia introduced new legislation to create two different minimum wages in the province. It raised the general minimum wage to $8.00 an hour and introduced a "first-job rate" of $6.00 an hour for the first 500 hours. After the first 500 hours, workers would then be entitled to the general minimum wage of $8.00 an hour. This was an attempt to encourage employers to hire and train new workers, recognizing that it may take several months to fully train a new employee. Those people already earning the minimum wage would continue to earn $8.00 an hour; only new hires would face the $6.00 wage floor.

see page 126

Some commentators suggested that the introduction of a $6.00 first-job rate would stimulate employment, particularly of youth, while others argued that after 500 hours, many young workers will receive their layoff notice, only to be replaced by new entrants into the labour force. Other provinces are watching closely to see whether they should follow suit.

All economists agree that the above analysis is logical and correct. But they disagree about whether governments should have minimum wage laws. One reason is that the empirical effects of minimum wage laws are difficult to determine, since "other things" are never remaining constant. A second reason is that some real-world labour markets are not sufficiently competitive to fit the supply/demand model. The third reason is that the minimum wage affects the economy in ways that some economists see as desirable and others see as undesirable. We point this out to remind you that the supply/demand framework is a tool to be used to analyze issues. It does not provide final answers about policy. That's where the art of economics—blending positive economic analysis with normative issues—comes into play. (In microeconomics, economists explore the policy issues of interventions in markets much more carefully.)

Figure 5-7 THE EFFECT OF AN EXCISE TAX

An excise tax on suppliers shifts the entire supply curve up by the amount of the tax. Since at a price equal to the original price plus the tax there is excess supply, the price of the good rises by less than the tax.

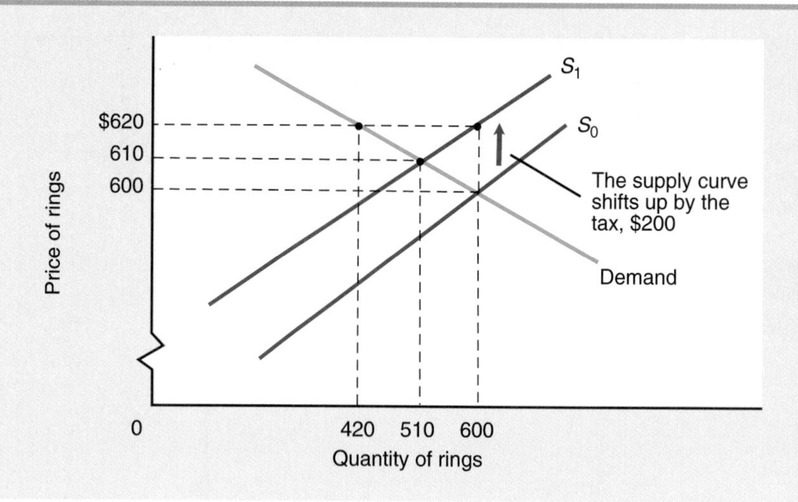

GOVERNMENT INTERVENTIONS: TAXES, TARIFFS, AND QUOTAS

Let's now consider an example of the government entering into a market and modifying the results of supply/demand analysis in the form of a tax. An **excise tax** is *a tax that is levied on a specific good*. The luxury tax on expensive cars that the United States imposed in 1991 is an example. A **tariff** is *an excise tax on an imported good*. What effect will excise taxes and tariffs have on the price and quantity in a market?

A tax on suppliers shifts the supply curve up by the amount of the tax.

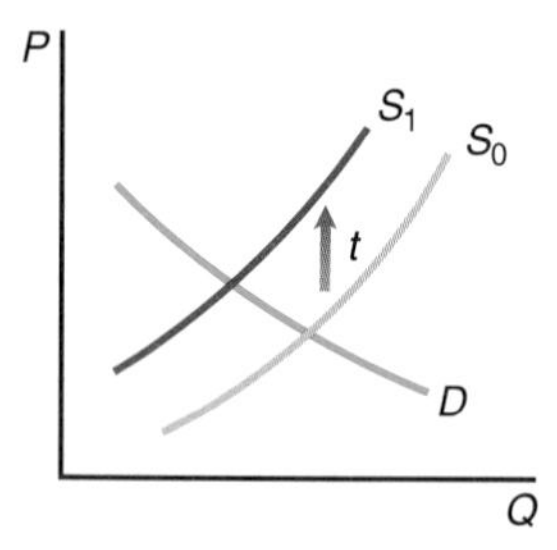

Q-7 Your study partner, Umar, has just stated that a tax on demanders of $2 per unit will raise the equilibrium price from $4 to $6. How do you respond?

Excise Taxes and Tariffs

To lend some sense of reality, let's take the example of the only luxury tax in Canada still in existence—that on jewellery manufactured in Canada. This tax was paid by the supplier. Say the price of a ring before the luxury tax was $600, and 600 rings were sold at that price. Now the government places a tax of $20 on such rings. What will the new price of the ring be, and how many will be sold?

If you were about to answer "The new price will be $620," be careful. Ask yourself whether we would have given you that question if the answer were that easy. By looking at supply and demand curves in Figure 5-7 you can see why $620 is the wrong answer.

To supply 600 rings, suppliers must be fully compensated for the tax. So the tax of $20 on the supplier shifts the supply curve up from S_0 to S_1. However, at $620, consumers are not willing to purchase 600 rings. They are willing to purchase only 420 rings. Quantity supplied exceeds quantity demanded at $620. Suppliers lower their prices until quantity supplied equals quantity demanded at $610, the new equilibrium price. Consumers increase the quantity of rings they are willing to purchase to 510, still less than the original 600 at $600. Why? At the higher price of $610 some people choose not to buy rings and others purchase their rings manufactured outside Canada.

Notice that at the new equilibrium the new price is $610, not $620. The reason is that at the higher price, the quantity of rings people demand is less. This is a movement up along a demand curve to the left. Excise taxes reduce the quantity of goods demanded. That's why jewellers remain up in arms that the tax has not been repealed and why the revenue generated from the tax was less than expected. Instead of collecting $20 × 600 ($12,000), revenue collected was only $10 × 510 ($5100).

Figure 5-8 (a and b) **THE RELATIONSHIP BETWEEN A QUOTA AND A TARIFF**

Figure **(a)** shows the effect of a quota of 5 million cars on the price of Japanese cars sold. Price rises from $20,000 to $27,000. Figure **(b)** demonstrates that a tax of $11,000 on each Japanese car sold in Canada has the same effect as the quota shown in **(a)**. The difference between a quota and a tariff lies in who gets the revenue. With a quota, the firm selling the good gets the revenue as additional profits. With a tariff, the government gets the revenue shown by the shaded box in **(b)**.

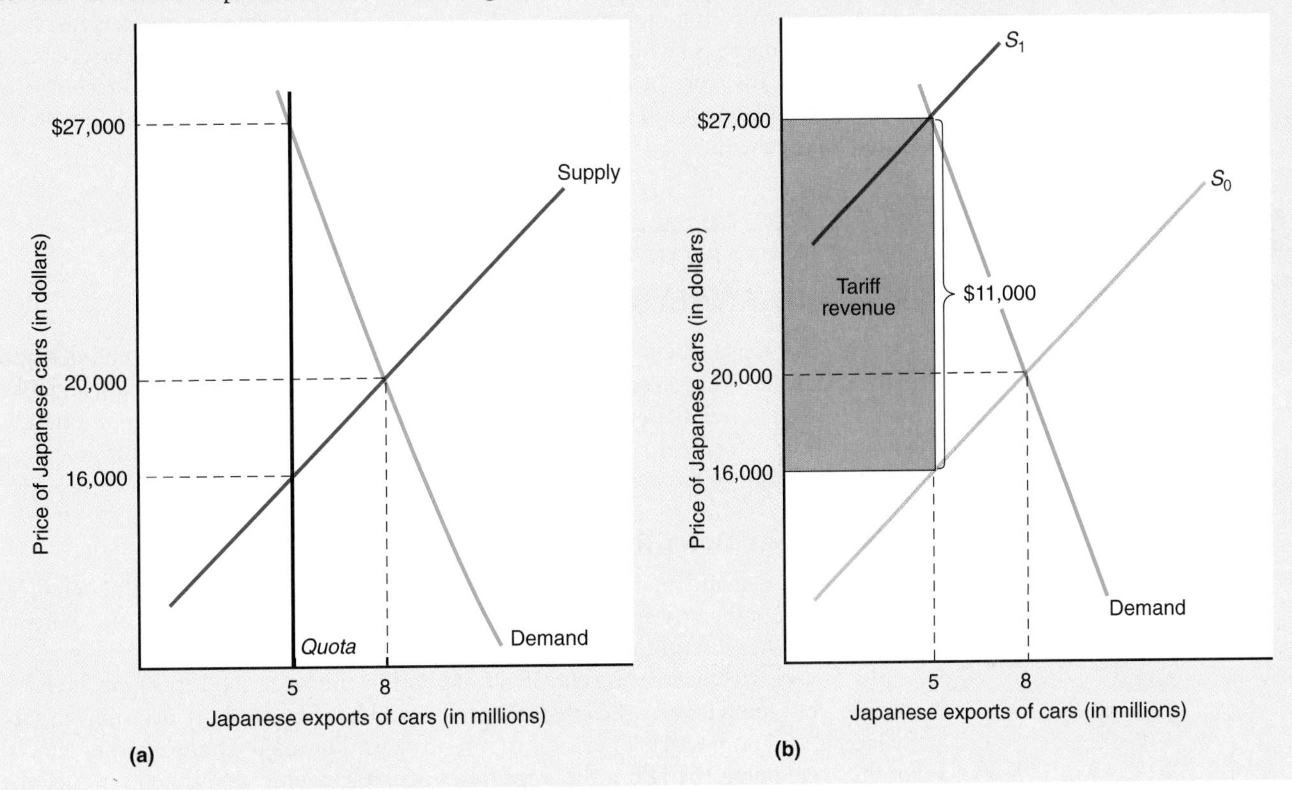

Quotas

The next example we want to consider is the imposition of a quantity control—a legal restriction on the quantity that can be bought or sold. The most common type of quantity control is the international **quota**—*a quantitative restriction on the amount that one country can export to another*. Suppose Canada wanted to restrict imports of Japanese cars. We show the effect in Figure 5-8(a).

The market price of a Japanese car is $20,000. At that price, Canadian consumers demand, and Japanese car makers are willing to supply, 8 million cars. But, when Canada places an import quota of, say, 5 million cars into Canada, Canadian consumers are willing to pay $27,000 a car even though Japanese firms are willing to sell them for $16,000 apiece. Of course, sellers will accept what consumers are willing to pay, and the price of Japanese cars rises to $27,000. Notice what the effect of the quota is. It raises the price of Japanese cars in Canada.

The Relationship between a Quota and a Tariff

Above we considered the effect of both quotas and taxes, and we noted that a tariff is a type of excise tax. Both devices increase price and reduce quantity. Let's now compare the effects of the quota above with the results of a tariff that would have achieved the same reduction in output.

A quota of Q_1 has the same effect as a tax t.

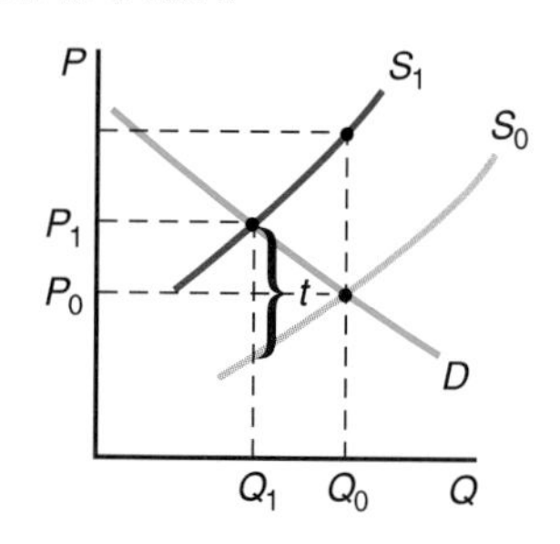

We show the effects of such a tariff in Figure 5-8(b). There you can see that a tariff of $11,000 shifts the supply curve in sufficiently to end up with the same sale of 5 million Japanese cars obtained with the quota. Consumers still pay $27,000 for Japanese cars.

Is there a difference between the two cases? The answer is yes. In the quota case, the higher price brings in more revenue to the company. In the tariff case, the revenue goes to the government. So which of the two do you think companies favour? The quota, of course—it means more profits as long as your company is one of those to receive the rights to fill those quotas. In fact, once quotas are instituted, Japanese firms compete intensely to get them.

Q-8 Why do firms prefer quotas to tariffs?

THE LIMITATIONS OF SUPPLY AND DEMAND ANALYSIS

We hope the above discussion has convinced you of the power of supply and demand analysis. Now let's discuss some of its limitations. Supply and demand are tools, and, like most tools, they help us enormously when used appropriately. Used inappropriately, however, they can reduce our understanding. Throughout the book we'll introduce you to the limitations of the tools, but let us discuss an important one here.

Other Things Don't Remain Constant

In supply and demand analysis other things are assumed equal. If other things change, then one cannot directly apply supply/demand analysis. Sometimes supply and demand are interconnected, making it impossible to hold other things constant. Let's take an example. Say we are considering the effect of a fall in the wage rate. In supply/demand analysis, you would look at the effect that fall would have on workers' decisions to supply labour, and on business's decision to hire workers. But might there be other effects? For instance, might the fall in the wage lower people's income, and thereby reduce the demand for firms' goods? And might that reduction affect firms' demand for workers? If these effects do occur, and are important enough to affect the result—those effects have to be added to the analysis in order for you to have a complete analysis.

Deciding whether such issues are important enough to affect the result requires a knowledge of the structure of the economy. All actions have a multitude of ripple or feedback effects—they create waves, like those that spread out from a stone thrown into a pool. These waves bounce back, and then hit other waves, which create new waves, which . . . . The art of applying supply/demand analysis is determining which of these multitude of ripple effects affect the analysis sufficiently that they must be added back in to the analysis.

Q-9 When determining the effect of a shift factor on price and quantity, in which of the following markets could you likely assume that other things will remain constant?

1. Market for eggs.
2. Labour market.
3. World oil market.
4. Market for luxury boats.

There is no single answer to the question of which ripples must be included, and much debate among economists involves which ripple effects to include. But there are some general rules. Supply/demand analysis, used without adjustment, is most appropriate for questions where the goods are a small percentage of the entire economy. That is when the other-things-constant assumption will most likely hold. As soon as one starts analyzing goods that are a large percentage of the entire economy, the other-things-constant assumption is likely not to hold true. The reason is found in the **fallacy of composition**—*the false assumption that what is true for a part will also be true for the whole.*

The Fallacy of Composition

Let's consider the fallacy of composition more carefully. When you are analyzing one individual's actions, small effects of an individual's actions can reasonably be assumed to be too small to change the results of supply/demand analysis. However, when you are analyzing the whole, or a significant portion of the whole, these small effects add up to a big effect. To see this, think of one supplier lowering the price of his or her good. People will substitute that good for other goods, and the quantity of the good demanded will increase. But what if all suppliers lower their prices? Since all prices have gone down, why should consumers switch? The substitution story can't be used in the aggregate. There are many such examples.

The fallacy of composition is the false assumption that what is true for a part will also be true for the whole.

An understanding of the fallacy of composition is of central relevance to macroeconomics. In the aggregate, whenever firms produce (whenever they supply), they create income (demand for their goods). So in macro, when supply changes, demand changes. This interdependence is one of the primary reasons we have a separate macroeconomics. In macroeconomics, the other-things-constant assumption central to microeconomic supply/demand analysis cannot hold. In macroeconomics, output creates income and income is spent on output. When the output side increases, so does the income side.

It is to account for these interdependencies that we have a separate macro analysis and micro analysis. In macro we use curves whose underlying foundations are much more complicated than the supply and demand curves we use in micro.

Macroeconomic analysis involves curves similar to, but very different from, microeconomic demand and supply curves.

One final comment: The fact that there may be an interdependence between supply and demand does not mean that you can't use supply and demand analysis; it simply means that you must modify its results with the interdependency that, if you've done the analysis correctly, you've kept in the back of your head. Thus using supply and demand analysis is generally a step in any good economic analysis, but you must remember that it may be only a step.

The fallacy of composition gives you some insight into why economists separate micro from macro. One of the important side effects of decisions that must be considered in macro, but not in micro, is the side effect of spending decisions. Your spending decision is someone else's income, and someone else's spending decision creates your income. The circular flow diagram presented in Chapter 3 (Figure 3-1) demonstrated that interconnection.

Q-10 Why is the fallacy of composition relevant for macroeconomic issues?

In our economy people can spend what they want; there is no direct mechanism in the economy that coordinates our spending decisions. Thus, the composite of all spending decisions may be either lots of spending or little spending. As that spending changes, output changes, and the economy experiences a business cycle.

THE ROLES OF GOVERNMENT

It is important to recognize both the strengths and weaknesses of supply and demand analysis because much of this book will involve a consideration of government policy issues within a supply/demand framework. To set the foundation for that discussion we will conclude this chapter by considering the general roles of a government within a market economy. This will establish the framework for consideration of public policy in later chapters.

Government's roles in a market economy include (1) providing a stable set of institutions and rules, (2) promoting effective and workable competition, (3) correcting for externalities, (4) ensuring economic stability and growth, (5) providing public goods, and (6) adjusting for undesired market results.

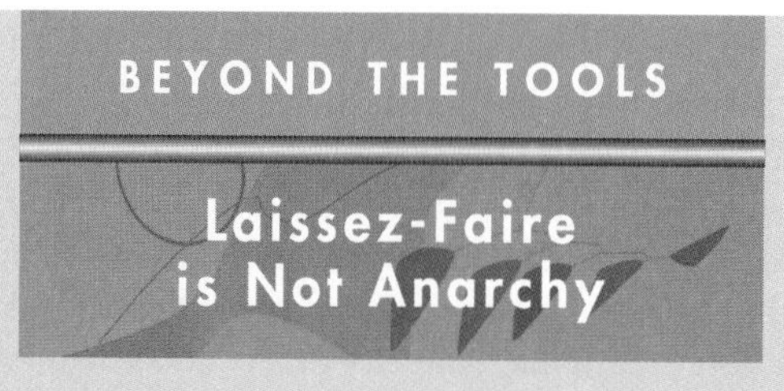

Most reasons for government intervention discussed in this chapter are debatable.

There is, however, one governmental role that even the strongest laissez-faire advocates generally accept. That role is for government to set up an appropriate institutional and legal structure within which markets can operate.

The reason there's little debate about this role is that all economists recognize that markets do not operate when there is anarchy. They require institutional structures that determine the rules of ownership, what types of trade are allowable, how contracts will be enforced, and what productive institutions are most desirable.

Before anyone conducts business, he or she needs to know the rules of the game and must have a reasonable expectation that those rules will not be changed. The operation of the modern economy requires that contractual arrangements be made among individuals. These contractual arrangements must be enforced if the economy is to operate effectively.

Economists differ significantly as to what the rules for such a system should be and whether any rules that already exist should be modified. Even if the rules are currently perceived as unfair, it can be argued that they should be kept in place. Individuals have already made decisions based on those rules, and it's unfair to them to change the rules in the middle of the game.

Stability of rules is a benefit to society. When the rules are perceived as unfair and changing them is also perceived as unfair, the government must find a balance between these two degrees of unfairness. Government often finds itself in that difficult position. Thus, while there's little debate about government's role in providing some institutional framework, there's heated debate about which framework is most appropriate.

Roles of government in a market economy are:

1. providing a stable set of institutions and rules;
2. promoting effective and workable competition;
3. correcting for externalities;
4. ensuring economic stability and growth;
5. providing public goods; and
6. adjusting for undesired market results.

Provide a Stable Set of Institutions and Rules

A basic role of government is to provide a stable institutional framework that includes the set of laws specifying what can and cannot be done as well as a mechanism to enforce those laws. Before people conduct business, they need to know the rules of the game and have a reasonable belief about what those rules will be in the future. These rules can initially develop spontaneously, but as society becomes more complex, the rules must be codified; enforcement mechanisms must be established. The modern market economy requires enforceable complex contractual arrangements among individuals. Where governments don't provide a stable institutional framework, as often happens in developing and transitional countries, economic activity is difficult; usually such economies are stagnant. Russia in the late 1990s is an example. As various groups fought for political control, the Russian economy stagnated.

Promote Effective and Workable Competition

In a market economy the forces of monopoly and competition are always in conflict, and the government must decide what role it is to play in protecting or promoting competition. Historically, consumer sentiment runs against monopoly power. **Monopoly power** is *the ability of individuals or firms currently in business to prevent other individuals or firms from entering the same kind of business*. Monopoly power gives existing firms and individuals the ability to raise their prices. Similarly, individuals' or firms' ability to enter freely into business activities is generally seen as good. Government's job is to promote policies that prevent excess monopoly power from limiting competition. It needs competition policies that work.

What makes this a difficult function for government is that most individuals and firms believe that competition is far better for the other guy than it is for themselves, that their own monopolies are necessary monopolies, and that competition facing them is unfair competition. For example, most farmers support competition, but these same farmers also support government farm subsidies (payments by government to producers based on production levels) some marketing boards (which set price floors), and import restrictions. Likewise, most firms support competition, but these same firms also support tariffs, which protect them from foreign competition. Most professionals, such as architects and engineers, support competition, but they also support professional licensing, which limits the number of competitors who can enter their field. Now, as you will see in reading the newspapers, there are always arguments for limiting entry into fields. The job of the government is to determine whether these arguments are strong enough to overcome the negative effects those limitations have on competition.

This isn't an easy task. Take the recent merger of Canadian Airlines and Air Canada. It gave Air Canada a monopoly on some domestic routes. Was this fair to the other air carriers? The government believed so, partly because the Canadian market is relatively small and geographically distinct from many others. Air Canada needed the benefits accruing to a monopolist to remain a viable business. Other air carriers responded by searching for niche markets—leisure travel, short-notice travel—to compete against Air Canada. The events of September 11, 2001 notwithstanding, Air Canada reacted by creating its own no-frills carrier, Air Canada Tango. Whether Tango will be capable of competing against Westjet and other carriers remains to be seen, but this example highlights the complexity of designing competition policies that work.

Correct for Externalities

When there are externalities, there is a potential role for government.

When two people freely enter into a trade agreement, they both believe that they will benefit from the trade. But unless they're required to do so, traders are unlikely to take into account any effect that an action may have on a third party. Economists call *the effect of a decision on a third party not taken into account by the decision maker* an **externality.** An externality can be positive (in which case society as a whole benefits from the trade between the two parties) or negative (in which case society as a whole is harmed by the trade between the two parties).

An example of a positive externality is education. When people educate themselves, all of society benefits, since better-educated people usually make better citizens and are better equipped to figure out new approaches to solving problems—approaches that benefit society as a whole. An example of a negative externality is pollution. Air conditioners emit a small amount of chlorofluorocarbons into the earth's atmosphere and contribute to the destruction of the ozone layer. Since the ozone layer protects all living things by filtering out some of the sun's harmful ultraviolet light rays, having a thinner layer of ozone can contribute to cancer and other harmful or fatal conditions. Neither the firms that produce the air conditioners nor the consumers who buy them take those effects into account. This means that the destruction of the ozone layer is an externality—the result of an effect that is not taken into account by market participants.

When there are externalities, there is a potential role for government to adjust the market result. If one's goal is to benefit society as much as possible, actions with positive externalities should be encouraged and actions with negative externalities should be restricted. Governments can step in and change the rules so that the actors must take into account the effect of their actions on society as a whole. We emphasize that the role is a potential one for two reasons. The first is that government often has difficulty dealing with externalities in such a way that society gains. For example, even if the government totally banned products that emit chlorofluorocarbons, other countries might not do the same and the ozone layer would continue to be destroyed. The second reason is that government is an institution that reflects, and is often guided by, politics and vested interests. It's not clear that, given the political realities, government intervention to correct externalities would improve the situation. In later chapters we'll have a lot more to say about government's role in correcting for externalities.

Environmental damage is a negative externality.

Ensure Economic Stability and Growth

In addition to providing general stability, government has the potential role of providing economic stability. If it's possible, most people would agree that government should prevent large fluctuations in the level of economic activity, maintain a relatively constant price level, and provide an economic environment conducive to economic growth. These aims are generally considered macroeconomic goals. They're justified as appropriate aims for government to pursue because they involve **macroeconomic externalities** *(externalities that affect the levels of unemployment, inflation, or growth in the economy as a whole)*.

A macroeconomic externality is the effect of an individual decision that affects the levels of unemployment, inflation, or growth in an economy as a whole, but is not taken into account by the individual decision maker.

Here's how a macro externality could occur. When individuals decide how much to spend, they don't take into account the effects of their decision on others; thus, there may be too much or too little spending. Too little spending often leads to unemployment. But in making their spending decisions, people don't take into account the fact that spending less might create unemployment. So their spending decisions can involve a macro externality. Similarly, when people raise their price and don't consider the effect on inflation, they too might be creating a macro externality.

Provide for Public Goods

A **public good** is *a good that if supplied to one person must be supplied to all and whose consumption by one individual does not prevent its consumption by another individual*. In contrast, a **private good** is *a good that, when consumed by one individual, cannot be consumed by another individual*. An example of a private good is an apple; once you eat that apple, no one else can consume it. National defense is generally considered a public good.

There are very few pure public goods, but many goods have public good aspects to them, and in general economists use the term *public good* to describe goods that are most efficiently provided collectively rather than privately. Parks, playgrounds, roads, and national defense are examples. Let's consider national defense more closely. For technological reasons national defense must protect all individuals in an area; a missile system cannot protect some houses in an area without protecting others nearby.

Everyone agrees that national defense is needed, but not everyone takes part in it. If someone else defends the country, you're defended for free; you can be a **free rider**—*a person who participates in something for free because others have paid for it*. Because self-interested people would like to enjoy the benefits of national defense while letting someone else pay for it, everyone has an incentive to be a free rider. But if everyone tries to be a free rider, there won't be any national defense. In such cases government can step in and require that everyone pay part of the cost of national defense, reducing the free rider problem.

Adjust for Undesired Market Results

A controversial role for government is to adjust the results of the market when those market results are seen as socially undesirable. An example is income distribution. Many people believe the government should see to it that income is "fairly" distributed. Determining what's fair is a difficult philosophical question. Let's consider two of the many manifestations of the fairness problem. Should the government use a **progressive tax** *(a tax whose rates increase as a person's income increases)* to redistribute money from the rich to the poor? (A progressive income tax schedule might tax individuals at a rate of 15 percent for income up to $20,000; at 25 percent for income between $20,000 and $40,000; and at 35 percent for every dollar earned over $40,000.) Or should government impose a **regressive tax** *(a tax whose rates decrease as income rises)* to redistribute money from the poor to the rich? Or should government impose a flat or **proportional tax** *(a tax whose rates are constant at all income levels, no matter what a taxpayer's total* *annual income is*)? Such a tax might be, say, 25 percent of every dollar of income. Canada has chosen a progressive income tax, while contributions for employment insurance and the Canada Pension Plan are a proportional tax up to a specified earned income. Economists can tell government the effects of various types of taxes and forms of taxation, but we can't tell government what's fair. That is for the people, through the government, to decide.

Another example of this role involves having government decide what's best for people, independently of their desires. The market allows individuals to decide. But what if people don't know what's best for themselves? Or what if they do know but don't act on that knowledge? For example, people might know that addictive drugs are bad for them, but because of peer pressure, or because they just don't care, they may take drugs anyway. Government action prohibiting such activities through laws or high taxes may then be warranted. *Goods or activities that government believes are bad for people even though they choose to use the goods or engage in the activities* are called **demerit goods or activities.** Illegal drugs are demerit goods and using addictive drugs is a demerit activity.

With merit and demerit goods, individuals are assumed not to be doing what is in their self-interest.

Alternatively, there are some activities that government believes are good for people, even if people may not choose to engage in them. For example, government may believe that going to the opera or contributing to charity is a good activity. But in Canada only a small percentage of people go to the opera, and not everyone in Canada contributes to charity. Similarly, government may believe that whole-wheat bread is more nutritious than white bread. But many consumers prefer white bread. Goods like whole-wheat bread and activities like contributing to charity are known as **merit goods or activities**—*goods and activities that government believes are good for you even though you may not choose to engage in the activities or consume the goods*. Government sometimes provides support for them through subsidies or tax benefits.

Market Failures and Government Failures

The reasons for government intervention are often summed up in the phrase *market failure*. **Market failures** are *situations in which the market does not lead to a desired result*. In the real world, market failures are pervasive—the market is always failing in one way or another. But the fact that there are market failures does not mean that government intervention will improve the situation. There are also **government failures**—*situations in which the government intervenes and makes things worse*. Government failures are pervasive in the government—the government is always failing in one way or another. So real-world policy makers usually end up choosing which failure—market failure or government failure—will be least problematic.

CONCLUSION

We will conclude the chapter here. We will talk much more about these roles of government in later chapters. For now, we simply want you to understand the general policy framework. When you combine that general policy framework with the supply/demand framework that we presented in Chapter 4 and the first part of this chapter, you have a good foundation for understanding the economic way of thinking about policy issues.

Chapter Summary

- When the demand curve shifts to the right (left), equilibrium price rises (declines) and equilibrium quantity rises (falls).
- When the supply curve shifts to the right (left), equilibrium price declines (rises) and equilibrium quantity rises (falls).
- By minding your *P*s and *Q*s—the shifts of and movements along curves—you can describe almost all events in terms of supply and demand.
- A price ceiling is a government-imposed limit on how high a price can be charged. Price ceilings below market price create shortages.
- A price floor is a government-imposed limit on how low a price can be charged. Price floors above market price create surpluses.
- Taxes and tariffs paid by suppliers shift the supply curve up by the amount of the tax or tariff. They raise the equilibrium price (inclusive of tax) and decrease the equilibrium quantity.
- Quotas restrict the quantity of goods that one country can export to another. They increase equilibrium price and reduce equilibrium quantity. The effect of a quota on equilibrium price and quantity is the same as a tariff. The difference between the two is who gets the additional revenue.
- In macro, small side effects that can be assumed away in micro are multiplied enormously. Thus, they can significantly change the results and cannot be ignored. To ignore them is to fall into the fallacy of composition.
- Six roles of government are (1) to provide a stable set of institutions and rules, (2) to promote effective and workable competition, (3) to correct for externalities, (4) to ensure economic stability and growth, (5) to provide public goods, and (6) to adjust for undesired market results.

Key Terms

demerit goods or activities *(123)*
excise tax *(116)*
externality *(121)*
fallacy of composition *(118)*
free rider *(122)*
government failure *(123)*
macroeconomic externality *(122)*
market failure *(123)*
merit goods or activities *(123)*
minimum wage laws *(114)*
monopoly power *(120)*
price ceiling *(111)*
price floor *(114)*
private good *(122)*
progressive tax *(122)*
proportional tax *(122)*
public good *(122)*
quota *(117)*
regressive tax *(122)*
rent control *(111)*
tariff *(116)*

Questions for Thought and Review

1. Say that price and quantity both fell. What would you say was the most likely cause?
2. Say that price fell and quantity remained constant. What would you say was the most likely cause?
3. Demonstrate graphically the effect of a price ceiling.
4. Demonstrate graphically why rent controls might increase the total payment that new renters pay for an apartment.
5. Demonstrate graphically the effect of a price floor.
6. Graphically show the effects of a minimum wage on the number of unemployed.
7. Oftentimes, to be considered for a job, you have to know someone in the firm. What does this observation tell you about the wage paid for that job?
8. In most developing countries, there are long lines of taxis at airports, and these taxis often wait two or three hours. What does this tell you about the price in that market? Demonstrate with supply and demand analysis.
9. Supply/demand analysis states that equilibrium occurs where quantity supplied equals quantity demanded, but in Canadian agricultural markets quantity supplied almost always exceeds quantity demanded. How can this be?
10. Demonstrate graphically the effect of a tax of $4 per unit on equilibrium price and quantity.
11. Using a graph like the one you drew for question 10 above, show graphically a quota that leads to the same price and quantity.
12. You've set up the rules for a game and started the game, but now realize that the rules are unfair. Should you change the rules?

13. Say the government establishes rights to pollute so that without a pollution permit you aren't allowed to emit pollutants into the air, water, or soil. Firms are allowed to buy and sell these rights. In what way will this correct for an externality?
14. What are six roles of government?

Problems and Exercises

1. The Canadian government has supported the price of sugar produced by Canadian sugar producers by placing a tariff on sugar imported into Canada. The tariff is effective because Canada consumes more sugar than it produces.
   a. Using supply/demand analysis, demonstrate how the tariff increases the price of domestic sugar.
   b. What other import policy could the government implement to have the same effect as the tariff?
   c. If Canada were to eliminate the tariff, how would this affect the Canadian sugar market?
2. "Scalping" is the name given to the buying of tickets at a low price and reselling them at a high price. The following information about a Grey Cup game appeared in your local newspaper. At the beginning of the season:
   a. Tickets sell for $27 and are sold out in preseason.
   b. Halfway through the season, both front-runners have maintained unbeaten records. Resale price of tickets rises to $200.
   c. One week before the game, both conference finalists have remained unbeaten and are ranked 1 and 2. Ticket price rises to $600.
   d. Three days before the game, price falls to $400.

   Demonstrate, using supply/demand analysis and words, what might have happened to cause these fluctuations in price.
3. In some localities "scalping" is against the law, although enforcement of these laws is spotty (difficult).
   a. Using supply/demand analysis and words, demonstrate what a weakly enforced antiscalping law would likely do to the price of tickets.
   b. Using supply/demand analysis and words, demonstrate what a strongly enforced antiscalping law would likely do to the price of tickets.
4. Apartments in large Canadian cities like Toronto are often hard to find. One of the major reasons is that there are, or were, rent controls.
   a. Demonstrate graphically how rent controls could make apartments hard to find.
   b. Often one can get an apartment if one makes a side payment to the current tenant. Can you explain why?
   c. What would be the likely effect of eliminating rent controls?
   d. What is the political appeal of rent controls?
5. Until recently, angora goat wool (mohair) has been designated as a strategic commodity (it used to be utilized in some military clothing). Because of that, in 1992 for every dollar's worth of mohair sold to manufacturers, ranchers received $3.60.
   a. Demonstrate graphically the effect of the elimination of this designation and subsidy.
   b. Explain why the program was likely kept in existence for so long.
   c. Say that a politician has suggested that the government should pass a law that requires all consumers to pay a price for angora goat wool high enough so that the sellers of that wool would receive $3.60 more than the market price. Demonstrate the effect of the law graphically. Would consumers support it? How about suppliers?
6. The technology is now developing so that road use can be priced by computer. A computer in the surface of the road picks up a signal from your car and automatically charges you for the use of the road.
   a. How could this technological change contribute to ending bottlenecks and rush hour congestion?
   b. What are some of the problems that might develop with such a system?
   c. How would your transportation habits likely change if you had to pay to use roads?
7. Suppose your province established a licensing requirement for all beauticians last year, and your neighbour continued to operate her salon out of her basement, without obtaining a license. After a particularly bad visit, one of her clients reports her to the authorities, and she is fined and forced to close her home business.
   a. Why would the province have created the licensing requirement in the first place?
   b. What options might you propose to change the system?
   c. What will be the political difficulties of implementing those options?

Web Questions 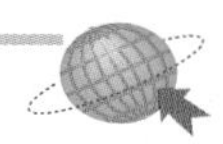

1. Many provinces have recently removed rent controls on apartments. Economic reasoning suggests this should raise the quantity of apartments supplied as landlords are able to charge higher prices. The Canadian Centre for Policy Alternatives argues this is fine in theory, but it won't work "in the real world." Visit the Manitoba home page (www.policyalternatives.ca/manitoba) and find the Fast Facts article "Lifting Rent Controls Would Not Renew Inner City." After reading the article, answer the following questions.
   a. What does the traditional demand and supply model predict will happen to the price of renting an apartment if controls are removed?
   b. Why does the CCPA think this won't work?
   c. What do they think the government should do?
2. Go to the Fraser Institute's home page (www.fraserinstitute.ca) and search for the article "The Economics of Minimum Wage Laws" by Marc Law. Using that article, answer the following questions:
   a. What happened to the proportion of young and old workers earing the minimum wage between 1988 and 1995? Within the standard supply/demand framework, how does this affect unemployment resulting from the minimum wage?
   b. Who is affected by the minimum wage?
   c. What effect does the article say that increasing the minimum wage will have on the distribution of income? What evidence does the author cite?

Answers to Margin Questions 

1. A heavy frost in Nova Scotia will decrease the supply of apples, increasing the price and decreasing the quantity demanded, as in the accompanying graph. *(105)*

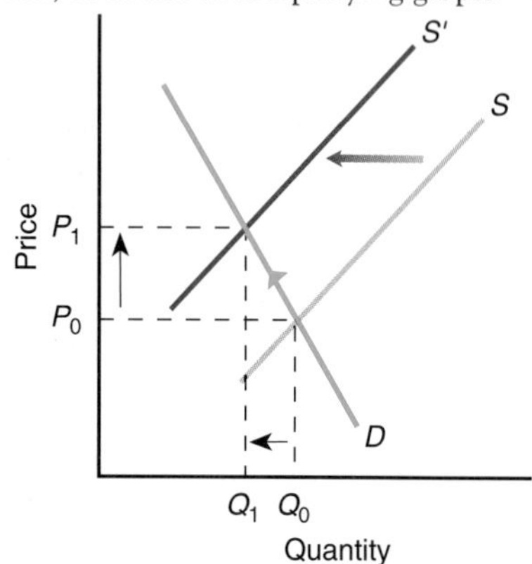

2. A discovery of a hormone that will increase cows' milk production by 20 percent will increase the supply of milk, pushing the price down and increasing the quantity demanded, as in the accompanying graph. *(105)*

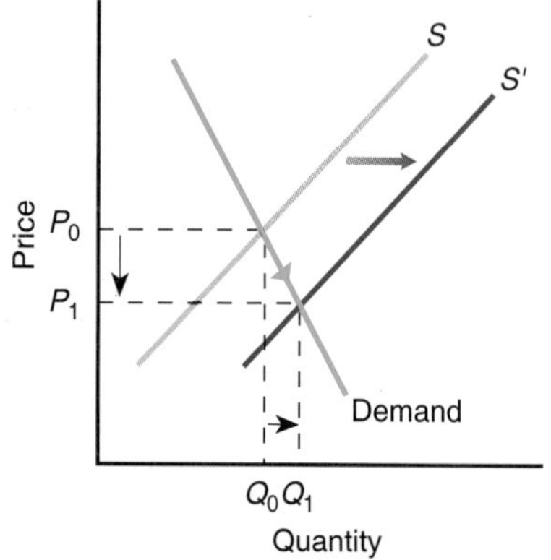

3. An increase in the price of gas will likely increase the demand for compact cars, increasing their price and increasing the quantity supplied, as in the accompanying graph. *(106)*

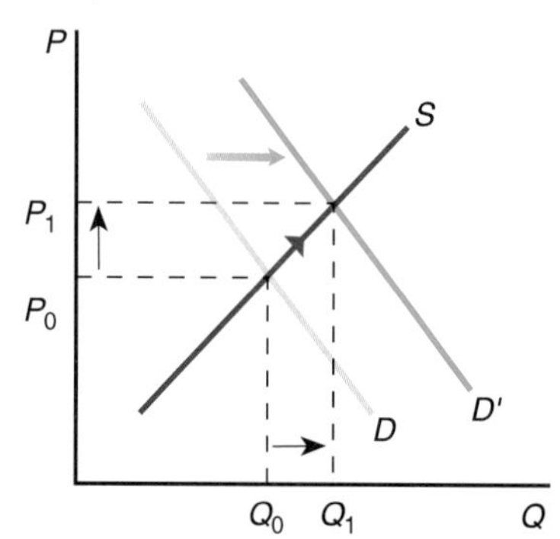

4. Quantity decreases but it is unclear what happens to price. *(110)*
5. Since the price ceiling is above the equilibrium price, it will have no effect on the market-determined equilibrium price and quantity. *(112)*
6. Since the price floor is below the equilibrium price, it will have no effect on the market-determined equilibrium price and quantity. *(114)*
7. We state that the tax will most likely raise the price by less than $2 since the tax will cause the quantity demand to decrease. This will decrease quantity supplied, and hence decrease the price the suppliers receive. In the diagram below, Q falls from Q_0 to Q_1 and the price the supplier receives falls from $4 to $3, making the final price $5, not $6. *(116)*

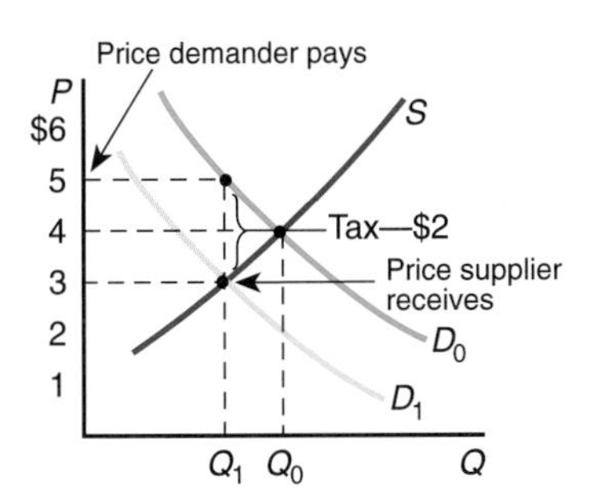

8. Firms prefer quotas to tariffs because with quotas, they receive additional profit; with tariffs, the government receives tax revenues. *(118)*
9. Other things are most likely to remain constant in the egg and luxury boat markets because each is a small percentage of the whole economy. Factors that affect the world oil market and the labour market will have ripple effects that must be taken into account in any analysis. *(118)*
10. The fallacy of composition is relevant for macroeconomic issues because it reminds us that, in the aggregate, small effects that are immaterial for micro issues can add up and be material. *(119)*

Algebraic Representation of Supply, Demand, and Equilibrium

In Chapters 4 and 5, we discussed demand, supply, and the determination of equilibrium price and quantity in words and graphs. These concepts can also be presented in equations. In this appendix we do so, using straight-line supply and demand curves.

THE LAWS OF SUPPLY AND DEMAND IN EQUATIONS

Since the law of supply states that quantity supplied is positively related to price, the slope of an equation specifying a supply curve is positive. (The quantity intercept term is generally less than zero since suppliers are generally unwilling to supply a good at a price less than zero.) An example of a supply equation is:

$$Q_S = -5 + 2P$$

where Q_S is units supplied and P is the price of each unit in dollars per unit. The law of demand states that as price rises, quantity demanded declines. Price and quantity are negatively related, so a demand curve has a negative slope. An example of a demand equation is:

$$Q_D = 10 - P$$

where Q_D is units demanded and P is the price of each unit in dollars per unit.

Determination of Equilibrium

The equilibrium price and quantity can be determined in three steps using these two equations. To find the equilibrium price and quantity for these particular demand and supply curves, you must find the quantity and price that solve both equations simultaneously.

Step 1: Set the quantity demanded equal to quantity supplied:

$$Q_S = Q_D \rightarrow -5 + 2P = 10 - P$$

Step 2: Solve for the price by rearranging terms. Doing so gives:

$$3P = 15$$

$$P = \$5$$

Thus, equilibrium price is $5.

Step 3: To find equilibrium quantity, you can substitute $5 for P in either the demand or supply equation. Let's do it for supply: $Q_S = -5 + (2 \times 5) = 5$ units. we'll leave it to you to confirm that the quantity you obtain by substituting $P = \$5$ in the demand equation is also 5 units.

The answer could also be found graphically. The supply and demand curves specified by these equations are depicted in Figure A5-1. As you can see, demand and supply intersect; quantity demanded equals quantity supplied at a quantity of 5 units and a price of $5.

Figure A5-1 SUPPLY AND DEMAND EQUILIBRIUM

The algebra in this appendix leads to the same results as the geometry in the chapter. Equilibrium occurs where quantity supplied equals quantity demanded.

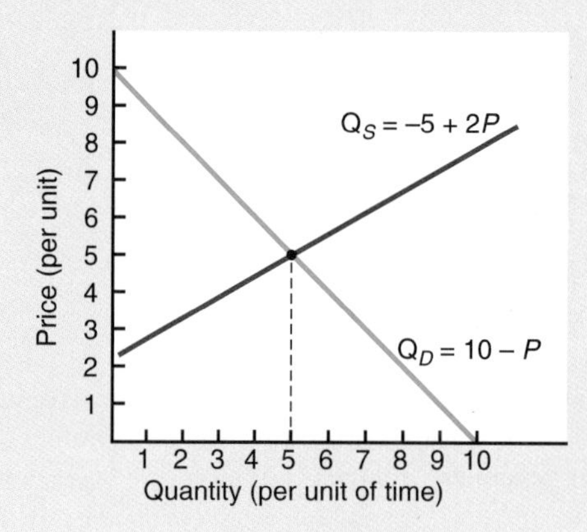

Movements along a Demand and Supply Curve

The demand and supply curves above represent schedules of quantities demanded and supplied at various prices. Movements along each can be represented by selecting various prices and solving for quantity demanded and supplied. Let's create a supply and demand table using the above equations—supply: $Q_S = -5 + 2P$; demand: $Q_D = 10 - P$.

P	$Q_S = -5 + 2P$	$Q_D = 10 - P$
$ 0	−5	10
1	−3	9
2	−1	8
3	1	7
4	3	6
5	5	5
6	7	4
7	9	3
8	11	2
9	13	1
10	15	0

As you move down the rows, you are moving up along the supply schedule, as shown by increasing quantity supplied, and moving down along the demand schedule, as shown by decreasing quantity demanded. Just to confirm your equilibrium quantity and price calculations, notice that at a price of $5, quantity demanded equals quantity supplied.

Shifts of a Demand and Supply Schedule

What would happen if suppliers' expectations changed so that they would be willing to sell more goods at every price? This shift factor of supply would shift the entire supply curve out to the right. Let's say that at every price, quantity supplied increases by three. Mathematically the new equation would be $Q_S = -2 + 2P$. The quantity intercept increases by 3. What would you expect to happen to equilibrium price and quantity? Let's solve the equations mathematically first.

Step 1: To determine equilibrium price, set the new quantity supplied equal to quantity demanded:

$$10 - P = -2 + 2P$$

Step 2: Solve for the equilibrium price:

$$12 = 3P$$

$$P = \$4$$

Step 3: To determine equilibrium quantity, substitute P in either the demand or supply equation:

$$Q_D = 10 - (1 \times 4) = 6 \text{ units}$$

$$Q_S = -2 + (2 \times 4) = 6 \text{ units}$$

Equilibrium price declined to $4 and equilibrium quantity rose to 6, just as you would expect with a rightward shift in a supply curve.

Now let's suppose that demand shifts out to the right. Here we would expect both equilibrium price and equilibrium quantity to rise. We begin with our original supply and demand curves—supply: $Q_S = -5 + 2P$; demand: $Q_D = 10 - P$. Let's say at every price, the quantity demanded rises by 3. The new equation for demand would be $Q_D = 13 - P$. You may want to solve this equation for various prices to confirm that at every price, quantity demanded rises by 3. Let's solve the equations for equilibrium price and quantity.

Step 1: Set the quantities equal to one another:

$$13 - P = -5 + 2P$$

Step 2: Solve for equilibrium price:

$18 = 3P$

$P = \$6$

Step 3: Substitute P in either the demand or supply equation:

$Q_D = 13 - (1 \times 6) = 7$ units

$Q_S = -5 + (2 \times 6) = 7$ units

Equilibrium price rose to \$6 and equilibrium quantity rose to 7 units, just as you would expect with a rightward shift in a demand curve.

Just to make sure you've got it, we will do two more examples. First, suppose the demand and supply equations for wheat per year in Canada can be specified as follows (notice that the slope is negative for the demand curve and positive for the supply curve):

$Q_D = 500 - 2P$

$Q_S = -100 + 4P$

P is the price in dollars per thousand bushels and Q is the quantity of wheat in thousands of bushels. Remember that the units must always be stated. What is the equilibrium price and quantity?

Step 1: Set the quantities equal to one another:

$500 - 2P = -100 + 4P$

Step 2: Solve for equilibrium price:

$600 = 6P$

$P = \$100$

Step 3: Substitute P in either the demand or supply equation:

$Q_D = 500 - (2 \times 100) = 300$

$Q_S = -100 + (4 \times 100) = 300$

Equilibrium quantity is 300,000 (300 thousand) bushels.

As our final example, take a look at Marie's demand curve depicted in Figure 4-4(b) in Chapter 4. Can you write an equation that represents the demand curve in that figure? It is $Q_D = 10 - 2P$. At a price of zero, the quantity of cassette rentals Marie demands is 10, and for every increase in price of \$1, the quantity she demands falls by 2. Now look at Ann's supply curve shown in Figure 4-7(b) in Chapter 4. Ann's supply curve mathematically is $Q_S = 2P$. At a zero price, the quantity Ann supplies is zero, and for every \$1 increase in price, the quantity she supplies rises by 2. What is the equilibrium price and quantity?

Step 1: Set the quantities equal to one another:

$10 - 2P = 2P$

Step 2: Solve for equilibrium price:

$4P = 10$

$P = \$2.5$

Step 3: Substitute P in either the demand or supply equation:

$Q_D = 10 - (2 \times 2.5) = 5$, or

$Q_S = 2 \times 2.5 = 5$ cassettes per week

Ann is willing to supply 5 cassettes per week at \$2.50 per rental and Marie demands 5 cassettes at \$2.50 per cassette rental. Remember that in Figure 4-8 in Chapter 4, we showed you graphically the equilibrium quantity and price of Marie's demand curve and Ann's supply curve. We'll leave it up to you to check that the graphic solution in Figure 4-8 is the same as the mathematical solution we came up with here.

PRICE CEILINGS AND PRICE FLOORS

Let's now consider a price ceiling and price floor. We start with the supply and demand curves:

$Q_S = -5 + 2P$

$Q_D = 10 - P$

This gave us the solution:

$P = 5$

$Q = 5$

Now, say that a price ceiling of \$4 is imposed. Would you expect a shortage or a surplus? If you said "shortage" you're doing well. If not, review the chapter before continuing with this appendix. To find out how much the shortage is we must find out how much will be supplied and how much will be demanded at the price ceiling. Substituting \$4 for price in both lets us see that $Q_S = 3$ units and $Q_D = 6$ units. There will be a shortage of 3 units. Next, let's consider a price floor of \$6. To determine the surplus we follow the same exercise. Substituting \$6 into the two equations gives a quantity supplied of 7 units and a quantity demanded of 4 units, so there is a surplus of 3 units.

TAXES AND SUBSIDIES

Next, let's consider the effect of a tax of \$1 placed on the supplier. That tax would decrease the price received by suppliers by \$1. In other words:

$Q_S = -5 + 2(P - 1)$

Multiplying the terms in parentheses by 2 and collecting terms results in

$Q_S = -7 + 2P$

This supply equation has the same slope as in the previous case, but a new intercept term—just what you'd expect. To determine the new equilibrium price and quantity, follow steps 1 to 3 discussed earlier. Setting this new equation equal to demand and solving for price gives

$$P = 5\tfrac{2}{3}$$

Substituting this price into the demand and supply equations tells us equilibrium quantity:

$$Q_S = Q_D = 4\tfrac{1}{3} \text{ units}$$

Of that price, the supplier must pay $1 in tax, so the price the supplier receives net of tax is $4⅔.

Next, let's say that the tax was put on the demander rather than on the supplier. In that case, the tax increases the price for demanders by $1 and the demand equation becomes

$$Q_D = 10 - (P + 1), \text{ or}$$

$$Q_D = 9 - P$$

Again solving for equilibrium price and quantity requires setting the demand and supply equations equal to one another and solving for price. We leave the steps to you. The result is:

$$P = 4\tfrac{2}{3}$$

This is the price the supplier receives. The price demanders pay is $5⅔. The equilibrium quantity will be 4⅓ units.

These are the same results we got in the previous cases, showing that, given the assumptions, it doesn't matter who actually pays the tax: The effect on equilibrium price and quantity is identical no matter who pays it.

QUOTAS

Finally, let's consider the effect of a quota of 4⅓ placed on the market. Since a quota limits the quantity supplied, as long as the quota is less than the market equilibrium quantity the supply equation becomes:

$$Q_S = 4\tfrac{1}{3}$$

where Q_S is the actual amount supplied. The price that the market will arrive at for this quantity is determined by the demand curve. To find that price substitute the quantity 4⅓ into the demand equation ($Q_D = 10 - P$):

$$4\tfrac{1}{3} = 10 - P$$

and solve for P:

$$P = 5\tfrac{2}{3}$$

Since consumers are willing to pay 5⅔, this is what suppliers will receive. The price that suppliers would have been willing to accept for a quantity of 4⅓ is 4⅔. This can be found by substituting the amount of the quota in the supply equation:

$$4\tfrac{1}{3} = -5 + 2P$$

and solving for P:

$$2P = 9\tfrac{1}{3}$$

$$P = 4\tfrac{2}{3}$$

Notice that this result is very similar to the tax. For demanders it is identical; they pay $5⅔ and receive 4⅓ units. For suppliers, however, the situation is much preferable; instead of receiving a price of $4⅔, the amount they received with the tax, they receive $5⅔. With a quota, suppliers receive the "implicit tax revenue" that results from the higher price.

Questions for Thought and Review

1, 2, 4, 5, 9

1. Suppose the demand and supply for milk is described by the following equations: $Q_D = 600 - 100P$; $Q_S = -150 + 150P$, where P is price in dollars, Q_D is quantity demanded in millions of litres per year, and Q_S is quantity supplied in millions of litres per year.
   a. Create demand and supply tables corresponding to these equations.
   b. Graph supply and demand and determine equilibrium price and quantity.
   c. Confirm your answer to (b) by solving the equations mathematically.
2. Suppose a growth hormone is introduced that allows dairy farmers to offer 125 million more litres of milk per year at each price.
   a. Construct new demand and supply curves reflecting this change. Describe with words what happened to the supply curve and to the demand curve.
   b. Graph the new curves and determine equilibrium price and quantity.
   c. Determine equilibrium price and quantity by solving the equations mathematically.
   d. Suppose the government set the price of milk at $3 a litre. Demonstrate the effect of this regulation on the

market for milk. What is quantity demanded? What is quantity supplied?

3. Write demand and supply equations that represent demand, D_0, and supply, S_0, in Figure 5-1(a) in the chapter.
   a. Solve for equilibrium price and quantity mathematically.
   b. Rewrite the demand equation to reflect the shift in demand to D_1. What happens to equilibrium price and quantity as shown in Figure 5-1(a) in the chapter? Confirm by solving the equations for equilibrium price and quantity.
4. a. How is a shift in demand reflected in a demand equation?
   b. How is a shift in supply reflected in a supply equation?
   c. How is a movement along a demand (supply) curve reflected in a demand (supply) equation?
5. Suppose the demand and supply for milk is described by the following equations: $Q_D = 600 - 100P$; $Q_S = -150 + 150P$, where P is the price in dollars; Q_D is quantity demanded in millions of litres per year; and Q_S is quantity supplied in millions of litres per year.
   a. Solve for equilibrium price and quantity of milk.
   b. Would a government-set price of $4 create a surplus or a shortage of milk? How much? Is $4 a price ceiling or a price floor?
6. Suppose the government imposes a $1 per litre of milk tax on dairy farmers. Using the demand and supply equations from question 1:
   a. What is the effect of the tax on the supply equation? The demand equation?
   b. What is the new equilibrium price and quantity?
   c. How much do dairy farmers receive per litre of milk after the tax? How much do demanders pay?
7. Repeat question 6(a) to 6(c) assuming the tax is placed on the buyers of milk. Does it matter who pays the tax?
8. Repeat question 6(a) to 6(c) assuming the government pays a subsidy of $1 per litre of milk to farmers.
9. Suppose the demand for cassettes is represented by $Q_D = 16 - 4P$, and the supply of cassettes is represented by $Q_S = 4P - 1$. Determine if each of the following is a price floor, price ceiling, or neither. In each case, determine the shortage or surplus.
   a. $P = \$3$.
   b. $P = \$1.50$.
   c. $P = \$2.25$.
   d. $P = \$2.50$.

MICROECONOMICS

II

> In my vacations, I visited the poorest quarters of several cities and walked through one street after another, looking at the faces of the poorest people. Next I resolved to make as thorough a study as I could of Political Economy.

You may remember having already seen this quotation from Alfred Marshall. It began the first chapter. We chose this beginning for two reasons. First, it gives what we believe to be the best reason to study economics. Second, the quotation is from one of the economic giants of all times. His *Principles of Economics* was the economists' bible in the late 1800s and early 1900s. How important was Marshall? It was Marshall who first used the supply and demand curves as an engine of analysis.

We repeat this quotation here in the introduction to the microeconomics section because for Marshall economics was microeconomics, and it is his vision of economics that underlies this book's approach to microeconomics. For Marshall, economics was an art that was meant to be applied—used to explain why things were the way they were, and what we could do about them. He had little use for esoteric theory that didn't lead to a direct application to a real-world problem. Reflecting on the state of economics in 1906, Marshall wrote to a friend:

> I had a growing feeling in the later years of my work at the subject that a good mathematical theorem dealing with economic hypotheses was very unlikely to be good economics: and I went more and more on the rules—(1) Use mathematics as a shorthand language, rather than as an engine of inquiry. (2) Keep to them until you have done. (3) Translate into English. (4) Then illustrate by examples that are important in real life. (5) Burn the mathematics. (6) If you can't succeed in (4), burn (3). This last I did often. (From a letter from Marshall to A. L. Bowley, reprinted in A. C. Pigou, *Memorials of Alfred Marshall*, p. 427.)

Marshall didn't feel this way about mathematical economics because he couldn't do mathematics. He was trained as a formal mathematician, and he was a good one. But, for him, mathematics wasn't economics, and the real world was too messy to have applied to it much of the fancy mathematical economic work that some of his fellow economists were doing. Marshall recognized the influence of market, political, and social forces and believed that all three had to be taken into account in applying economic reasoning to reality.

Since 1906, when Marshall wrote this letter, the economics profession has moved away from its Marshallian roots. The profession has found other heroes who have created a mathematical foundation for economics that's both impressive and stultifying. Mathematical economics that has only the slightest connection to the real world has overwhelmed much of the real-world economics that Marshall followed. That's sad.

Not to worry. You won't see this kind of mathematical economics in these microeconomic chapters. The chapters follow the Marshallian methodology and present the minimum of formal theory necessary to apply the concepts of economics to the real world, and then they do just that: start talking about real-world issues.

Section I, Microeconomics: The Basics (Chapters 6 and 7), and Section II, Foundations of Supply and Demand (Chapters 8–10), present the background theory necessary to understand the economic way of thinking. These sections introduce you to the foundations of economic reasoning and to some central terms and ideas of microeconomics. But even in these theoretical chapters the focus is on intuition, policy, and on putting the economic approach to problems into perspective, rather than on presenting technique for the sake of technique. Section III, Market Structure and Policy (Chapters 11–14), introduces you to various market structures, providing you with a way of approaching real-world markets.

Section IV, Applying Economic Reasoning to Policy (Chapters 15–17), provides an overview of the implications of microeconomics for policy. Section V, Policy Issues in Depth (Chapters 18 and 19), looks more carefully at two important issues—agriculture policy and trade policy.

Section VI, Factor Markets (Chapters 20 and 21), looks at a particular set of markets—factor markets. These markets play a central role in determining the distribution of income. These chapters won't tell you how to get rich (you'll have to wait for the sequel for that), but they will give you new insights into how labour markets work.

Describing Demand and Supply: Elasticities

6

After reading this chapter, you should be able to:

- Use the terms *price elasticity of demand* and *price elasticity of supply* to describe the responsiveness of quantity demanded and quantity supplied to changes in price.
- Calculate elasticity graphically and numerically.
- Explain the importance of substitution in determining the price elasticity of demand and the price elasticity of supply.
- Relate price elasticity of demand to total revenue.
- Interpret the four main elasticity calculations, price elasticity of demand, income elasticity of demand, cross-price elasticity of demand, and price elasticity of supply.
- State how other elasticity concepts are useful in describing the effect of shift factors on demand.
- Explain how the concept of *elasticity* makes supply and demand analysis more useful.

The master economist must understand symbols and speak in words. He must contemplate the particular in terms of the general, and touch abstract and concrete in the same flight of thought.

J. M. Keynes

Chapters 4 and 5 explained the concepts of supply and demand. This chapter introduces you to the concept of elasticity, which describes the magnitude of consumer's or firms' response to a change in some significant variable affecting them. For example, we may be interested in knowing how consumers respond to changes in price (price elasticity of demand), changes in income (income elasticity of demand), or changes in the price of a related good (cross-price elasticity of demand).

PRICE ELASTICITY OF DEMAND

The most commonly used elasticity concept is price elasticity of demand. **Price elasticity of demand** measures the responsiveness of consumers to a change in the price of the product. The elasticity coefficient compares the relative magnitudes of the price and quantity changes by calculating *the percentage change in quantity demanded divided by the percentage change in price:*

$$\varepsilon^D = \frac{\text{Percentage change in quantity demanded}}{\text{Percentage change in price}}$$

Economists use the Greek symbol Epsilon, ε, to represent price elasticities. The superscript "D" indicates that it is a price elasticity of demand.

Let's consider a numerical example. The **Law of Demand** states that *quantity demanded is inversely related to price.* That is, when the product price rises, quantity demanded falls, and when product price falls, quantity demanded rises. Say the price of a good rises by 10 percent and, in response, quantity demanded falls by 20 percent. The price elasticity of demand is 2 (−20 percent/10 percent). Because quantity demanded is inversely related to price, the calculation for the price elasticity of demand always comes out negative. So we can use our understanding of the Law of Demand, and report price elasticity of demand as a positive number. Mathematically, we take the absolute value of the elasticity coefficient: $\varepsilon^D = |-2| = 2$. Thus, a *larger* number for price elasticity of demand means quantity demanded is *more responsive* to price.

Price elasticity of demand measures consumers' response to a price change, and is calculated as the percentage change in quantity demanded divided by the percentage change in price.

Classifying Demand as Elastic or Inelastic

Economists usually describe demand by the terms *elastic* and *inelastic*. Formally, demand is **elastic** if *the percentage change in quantity is greater than the percentage change in price* ($\varepsilon^D > 1$). Alternately, demand is **inelastic** if *the percentage change in quantity is less than the percentage change in price* ($\varepsilon^D < 1$). In the example, we calculated a price elasticity of demand equal to 2. You can think of this as a one percent change in price leading to a two percent change in quantity demanded. Since the quantity response is greater than the price change (in percentages), demand is said to be elastic (responsive).

Q.1 If price elasticity of demand equals 4.21, what would we call demand: elastic or inelastic?

Demand Elastic
$\Delta Q > \Delta P$
Demand Inelastic
$\Delta Q < \Delta P$

Elasticity Is Independent of Units

Before continuing, notice that price elasticity of demand measures the *percentage*, not the *absolute*, change in variables. Using percentages allows us to have a measure of responsiveness that is independent of units and thus makes comparisons of responsiveness of different goods easier. Compare a $10 increase in the price of a $2,000 computer with a $10 increase in the price of a shirt, from $20 to $30. Such a comparison of absolute numbers is not very helpful. A $10 increase in the price of a computer is not likely to have a very noticeable effect; however, a $10 increase in the price of shirts is significant, and is likely to have a significant effect on the quantity of shirts bought.

Percentages allow us to have a measure of responsiveness that is independent of units, and thus make comparisons of responsiveness of different goods easier.

CALCULATING ELASTICITIES

The Mid-point Formula

Consider the following example:

	Price	Quantity Demanded
Situation A	\$20	9,000
Situation B	\$26	7,000

Suppose the price rises from \$20 to \$26, and in response, the quantity demanded falls from 9,000 to 7,000. This is illustrated in Figure 6–1(a). Is the consumer's behaviour responsive (elastic demand) or unresponsive (inelastic demand) to the change in price?

Calculate: $\varepsilon^D = \dfrac{\text{percentage change in quantity demanded}}{\text{percentage change in price}}$

To find the percentage change in quantity demanded, compute:

$$\frac{Q_B - Q_A}{\text{average } Q} = \frac{7{,}000 - 9{,}000}{(7{,}000 + 9{,}000) \div 2}$$

and to find the percentage change in price, compute:

$$\frac{P_B - P_A}{\text{average } P} = \frac{26 - 20}{(26 + 20) \div 2}$$

Figure 6-1(a, and b) GRAPHS OF ELASTICITY OF DEMAND

In **(a)** we are calculating the elasticity of the arc between A and B. We essentially find the midpoint and use that midpoint in our calculations of percentage changes. This gives us a percentage change in price of 26 percent and a percentage change in quantity of 33 percent, which gives us an elasticity of 1.3. In **(b)**, the calculations are left for you to do.

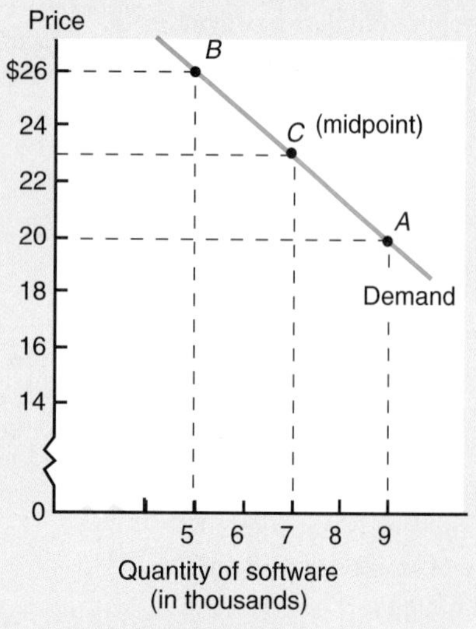

(a) Elasticity of demand

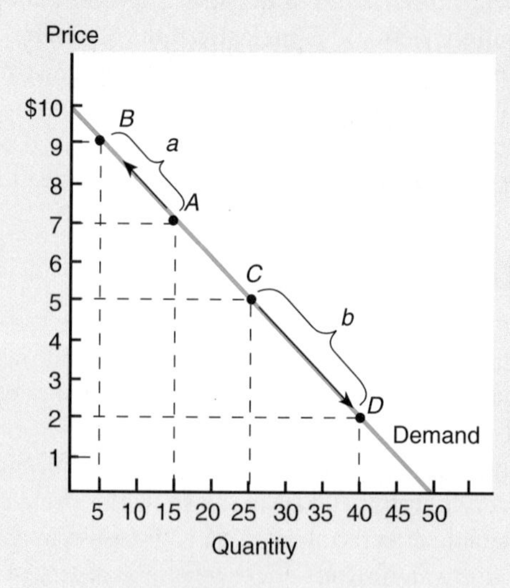

(b) Some examples

(Answers to (b): $a = 4$; $b = 0.54$)

Therefore, to calculate the price elasticity of demand, compute:

$$E^D = \frac{\dfrac{Q_B - Q_A}{\text{average } Q}}{\dfrac{P_B - P_A}{\text{average } P}} = \frac{\dfrac{7{,}000 - 9{,}000}{8000}}{\dfrac{26-20}{23}} - \frac{\dfrac{-2{,}000}{8{,}000}}{\dfrac{6}{23}} - \frac{-.96}{.26} - |-2.19| \text{ (inelastic)}$$

Notice the negative sign in the calculation. Try this: Reverse Situations A and B, recalculate the elasticity, and see where the negative sign appears.

The End-point Problem

Why is elasticity calculated using average quantity and average price? To see why, compare the following two cases.

Economists use the mid-point formula to avoid the end-point problem.

Case 1: Price Rises As we move from Situation A to Situation B, the price of the product rises from \$20 to \$26 and quantity purchased falls from 9,000 to 7,000. To determine the magnitude of the response by consumers, calculate the price elasticity of demand. Using the "original" (Situation A) price and quantity as the base to calculate the percentage changes, price has risen by 30%:

$$\frac{P_B - P_A}{P_B} = \frac{26 - 20}{26} = -.3$$

and quantity has fallen by 22%:

$$\frac{Q_B - Q_A}{Q_A} = \frac{7{,}000 - 9{,}000}{9{,}000} = -.22.$$

Using these numbers, demand is unresponsive to the price increase:

$$\varepsilon^D = \frac{\text{percentage change in quantity demanded}}{\text{percentage change in price}} = |-.74| = .74 \text{ (inelastic)}$$

Case 2: Price Falls Consider the opposite case. Suppose price was \$26 and 7,000 units were being bought (Situation B). What happens when price falls to \$20? Quantity demanded rises to 9,000 (Situation A). Calculate the price elasticity of demand as we move from Situation B to Situation A. Using the "original" (now Situation B) price and quantity as the base to calculate the percentage changes, price has fallen by 23%:

$$\frac{P_B - P_A}{P_B} = \frac{26 - 20}{26} = -.23$$

while quantity has risen by 29%:

$$\frac{Q_A - Q_B}{Q_B} = \frac{9{,}000 - 7{,}000}{7{,}000} = .29$$

In the case of a price decrease, demand is responsive:

$$\varepsilon^D = \frac{\text{percentage change in quantity demanded}}{\text{percentage change in price}} = |-1.24| = 1.24 \text{ (elastic)}$$

In this example, it seems to matter whether the price of the product is rising or falling. This happens because we are calculating what happens over a range of price and quantity values, and the greater the difference between the two prices (or the two quantities), the larger the discrepancy between the "price-rising" and the "price-falling" elasticities. To adjust for this problem, we use the mid-point formula introduced earlier.

Q-2 What is the approximate elasticity between points *A* and *B* on the graph below?

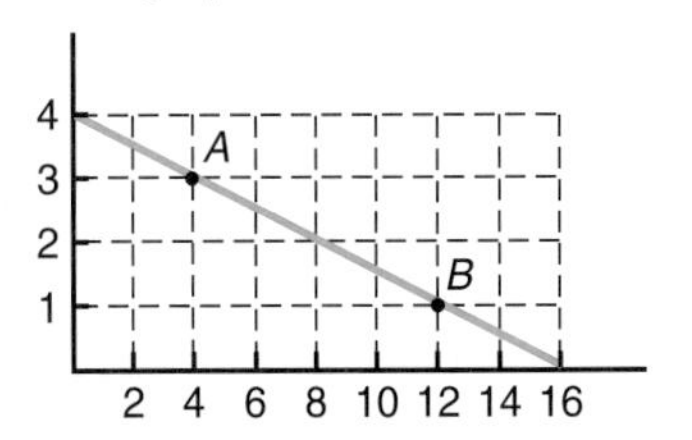

KNOWING THE TOOLS

Calculating Elasticity at a Point

The text has explained how to calculate elasticity of a range. Let's now turn to a method of calculating the elasticity at a specific point, rather than over a range. The approach we will use is to create a line segment around the point and use the mid-point elasticity formula to calculate elasticity. Let's go through an example.

Say you want to determine the elasticity at point *A* in Figure (a). First create a line segment, with Point A as the segment's midpoint. The segment can be of any length. In Figure (a) the segment begins 4 units on the quantity axis before the quantity at *A* and extends 4 units beyond the quantity at *A*. Thus, the quantity arc extends from 20 to 28. Next, determine which price arc matches the chosen quantities. In the example, the price that corresponds to the lower quantity is $3, and the price that corresponds to the higher quantity is $5.

We now have the line segment, so we are ready to use the mid-point elasticity formula to calculate the elasticity at point *A*: Percentage change in quantity = [(28 − 20)/24] × 100 = 33 percent. Percentage change in price = [(5 − 3)/4] × 100 = 50 percent. Elasticity at point *A* is 33/50 = 0.66.

To see that you've got the calculation down, in Figure (b) there are two more points. Your assignment is to calculate these two elasticities yourself. The answers are upside down below the diagram.

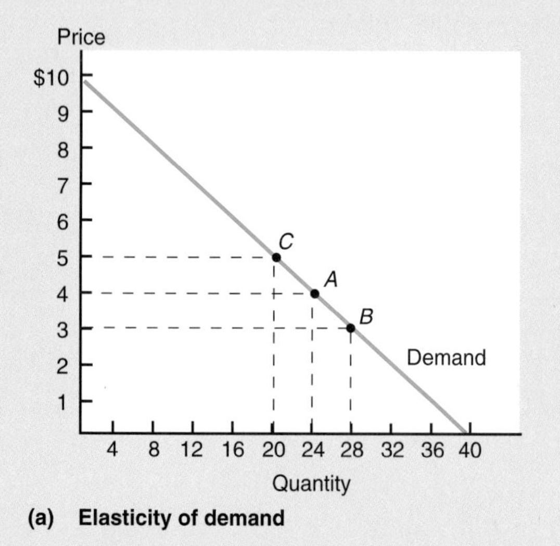

(a) Elasticity of demand

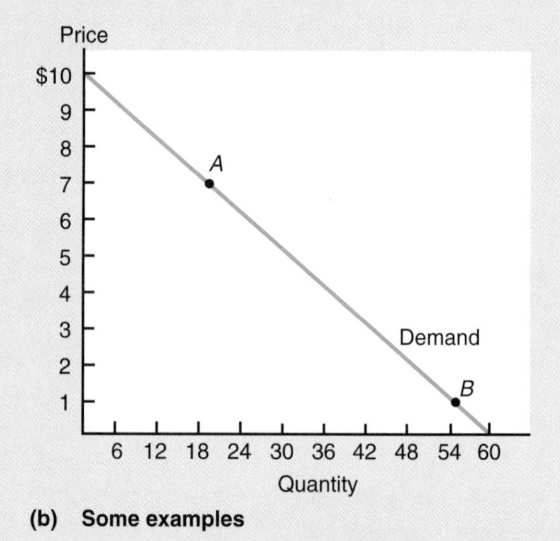

(b) Some examples

(Answers to (b): *A* = 2.33; *B* = .11

The mid-point formula uses the average price (and average quantity) rather than either of the end-points, which means that it doesn't matter whether price rises or falls, the calculation will always yield the same value for elasticity. All elasticities are calculated in this way.

Learning the mechanics of calculating elasticities takes practice, so Figure 6-1(b) gives you additional examples, for practice. Using the mid-point formula, calculate the price elasticity of demand:

a. From A to B

b. From C to D

The answers can be found upside down at the bottom of the figure.

ELASTICITY AND DEMAND CURVES

There are two important points about elasticity and demand curves. First, elasticity is related to (but is not the same as) slope. Second, elasticity changes along a straight-line demand curve.

Elasticity Is Not the Same as Slope

Generally the relationship between elasticity and slope is the following: The steeper the curve, the less elastic is demand. At the limit, a demand curve could be vertical (most steep), as shown in Figure 6-2(a), or horizontal (least steep), shown in Figure 6-2(b).

Q-3 What do the percentage changes in quantity demanded and in price need to be to obtain $\varepsilon^D = 0$?

The vertical demand curve shown in Figure 6-2(a) demonstrates how a change in price leads to no change in quantity demanded. Economists describe this curve as **perfectly inelastic**—*quantity does not respond at all to changes in price* ($\varepsilon^D = 0$). Curves that are vertical are perfectly inelastic. The demand curve shown in Figure 6-2(b), in contrast, is horizontal. A change in price from above or below P_0 results in an infinitely large increase in quantity demanded. This curve is **perfectly elastic**, reflecting the fact that *quantity responds enormously to changes in price* ($\varepsilon^D = \infty$). Horizontal curves are perfectly elastic. In general, steeper (more vertical) curves at a given point are more *in*elastic than less steep (more horizontal) curves at a given point.

Consider Figure 6-3 on the next page. For the price interval \$3 to \$4 for demand curve D_1, the price elasticity of demand is:

$$\varepsilon^D = \frac{\text{percentage change in quantity demanded}}{\text{percentage change in price}} = \frac{\dfrac{Q_C - Q_A}{\text{average } Q}}{\dfrac{P_C - P_A}{\text{average } P}}$$

$$= \frac{\dfrac{35-40}{37.5}}{\dfrac{4-3}{3.50}} = \left|\frac{-.133}{.286}\right| = .47 \text{ (inelastic)}$$

Figure 6-2 (a and b) ELASTICITIES AND DEMAND CURVES

In (a) and (b), two special elasticity cases are shown. A perfectly inelastic curve is vertical; a perfectly elastic curve is horizontal.

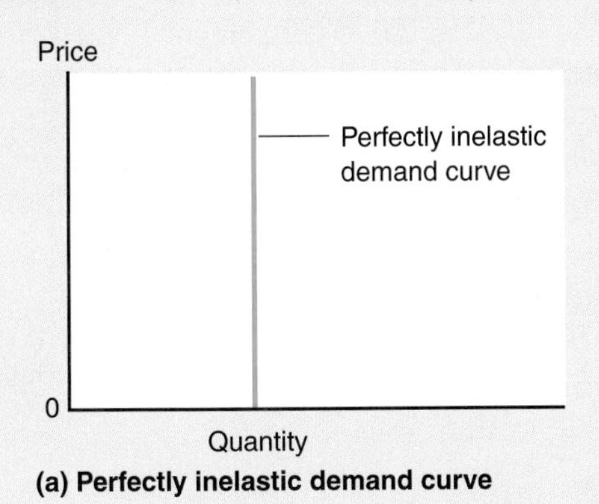

(a) Perfectly inelastic demand curve

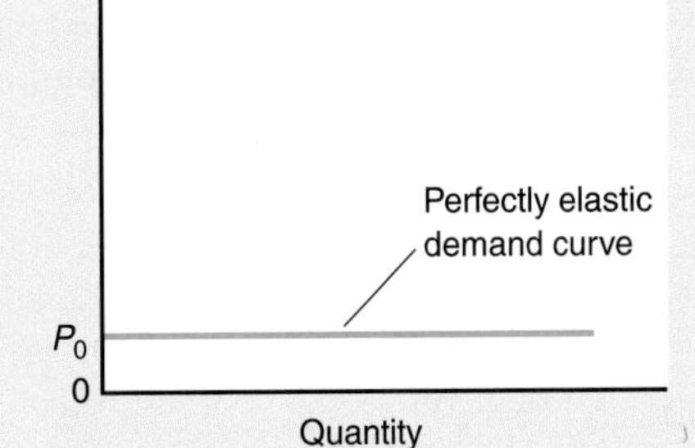

(b) Perfectly elastic demand curve

Figure 6-3 CALCULATING ELASTICITIES

From a given point, a steeper demand curve will be less elastic. Over the $3 to $4 price interval, the price elasticity of demand for D_1 is .47 (inelastic), while for D_2, the price elasticity of demand is 4.2 (elastic).

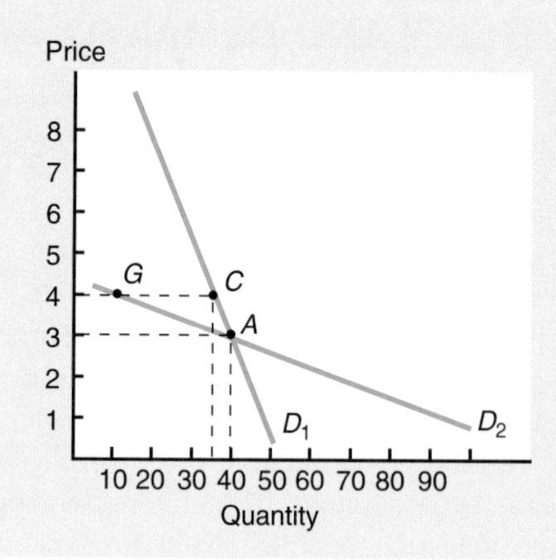

For the same price interval, the price elasticity of demand for demand curve D_2 is:

$$\varepsilon^D = \frac{\frac{Q_G - Q_A}{\text{average } Q}}{\frac{P_G - P_A}{\text{average } P}} = \frac{\frac{10-40}{25}}{\frac{4-3}{3.50}} = \left|\frac{-1.2}{.286}\right| = 4.2 \text{ (elastic)}$$

We see that in the $3 to $4 price range, demand curve D_1 is inelastic. When price rose by 28.6%, quantity demanded fell by 5 units, or 13.3%, which indicates that consumers are unresponsive to the price change. For demand curve D_2, the 28.6% rise in price resulted in a decrease in quantity demanded of 30 units, or 120%, which is highly responsive.

Elasticity is not the same as slope. Elasticity changes along a straight-line curve, but slope does not.

Elasticity Changes along Straight-Line Curves

Figure 6-4 shows how elasticity changes along a demand curve. At the price intercept of the demand curve, demand is perfectly elastic ($\varepsilon^D = \infty$); elasticity becomes smaller as price declines until it becomes perfectly inelastic ($\varepsilon^D = 0$) at the quantity intercept. At one point along the demand curve, between an elasticity of infinity and zero, $\varepsilon^D = 1$. Demand is **unit elastic**—*the percentage change in quantity equals the percentage change in price*. Demand is unit elastic at a price of $5. (To confirm this, calculate elasticity of demand between $4 and $6. The percentage change in price is $(2/5) \times 100 = 40$ percent, and the percentage change in quantity is $(2/5) \times 100 = 40$ percent.) The point at which demand is unit elastic divides the demand curve into two sections—an elastic portion ($\varepsilon^D > 1$) above the point at which demand is unit elastic and an inelastic portion ($\varepsilon^D < 1$) below the point at which demand is unit elastic.

Interpreting Elasticities

From most to least elastic, the five elastic terms are: perfectly elastic ($\varepsilon^D = \infty$); elastic ($\varepsilon^D > 1$); unit elastic ($\varepsilon^D = 1$); inelastic ($\varepsilon^D < 1$); and perfectly inelastic ($\varepsilon^D = 0$).

We know that as price rises, consumers buy less (Law of Demand). Price elasticity of demand tells us whether consumers reduce their purchases by a lot (elastic demand) or by a little (inelastic demand). See Table 6-1 for more interpretations of price elasticity of demand.

Figure 6-4 **ELASTICITY ALONG A STRAIGHT-LINE DEMAND CURVE**

Elasticity varies from infinity at the vertical axis intercept to zero at the horizontal axis intercept.

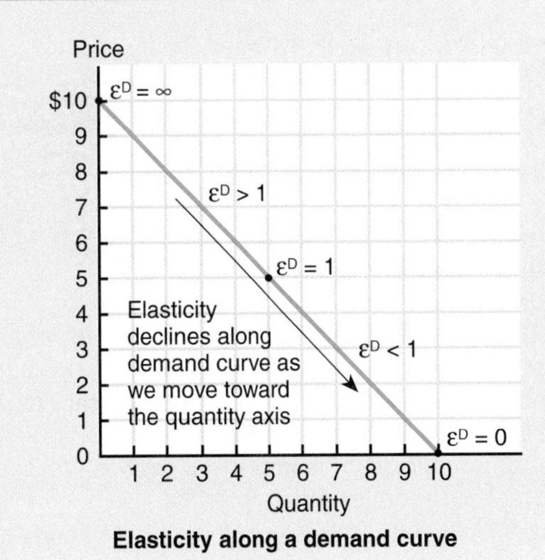

Elasticity along a demand curve

TABLE 6-1 Interpreting Price Elasticity of Demand

ε^D	Description of Demand	Interpretation
$\varepsilon^D = \infty$	Perfectly Elastic	Quantity responds enormously to changes in price.
$\varepsilon^D > 1$	Elastic	Consumers are responsive to price changes. The percentage change in quantity demanded exceeds the percentage change in price.
$\varepsilon^D = 1$	Unit Elastic	The percentage change in quantity demanded equals the percentage change in price.
$\varepsilon^D < 1$	Inelastic	Consumers are unresponsive to changes in price. The percentage change in quantity demanded is less than the percentage change in price.
$\varepsilon^D = 0$	Perfectly Inelastic	Consumers are completely unresponsive to price. Quantity demanded does not respond at all to changes in price.

SUBSTITUTION AND PRICE ELASTICITY OF DEMAND

How responsive quantity demanded will be to changes in price can be summed up in one word: substitution. As a general rule, the more substitutes a good has, the more elastic is its demand.

6.1 *see page 158*

The reasoning is as follows: If a good has many or close substitutes, a rise in the price of that good means that consumers can easily shift consumption to those substitute goods. If a satisfactory substitute is available, a rise in a good's price will have a large effect on the

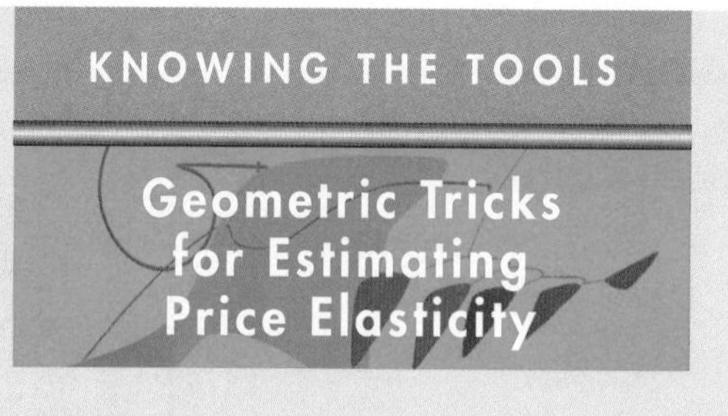

There is a trick that is useful in determining whether a point on a straight-line demand curve is elastic or inelastic: (1) Determine where the demand curve intersects the price and quantity axes. (2) Find the point midway between the origin and the point at which the demand line meets the quantity axis, and draw a vertical line back up to the demand curve. The point where it intersects the demand curve will have an elasticity of 1 (unit elastic); all points to the left of that line will be elastic, and all points to the right of it will be inelastic.

This point of unit elasticity is important, as we will see in the discussion of monopolies in Chapter 12.

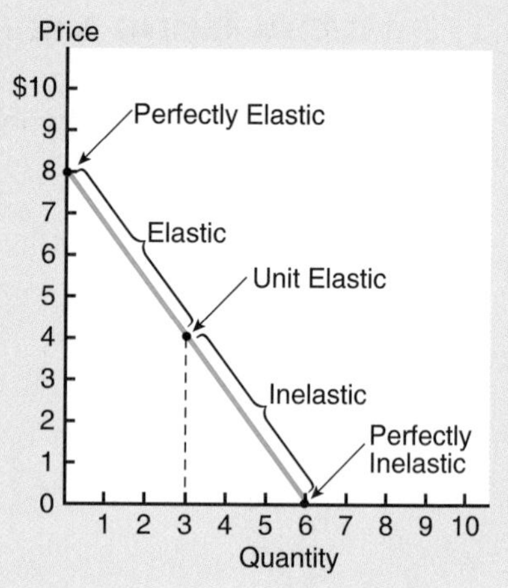

Q-4 Name a good which has many substitutes. Name a good which has few, but close, substitutes.

The more substitutes, the more elastic the demand.

More substitutes become available in the long run.

Q-5 What are four important factors affecting the number of substitutes a good has?

quantity demanded. For example, orange juice is probably a satisfactory substitute for apple juice. If most people agree, the demand for apple juice would be very elastic.

The number of substitutes a good has is affected by many factors. Four of the most important are:

1. **Time to Adjust.** There are more substitutes in the long run than in the short run. That's because the long run provides more alternatives. For example, let's consider the World War II period, when the price of rubber went up considerably. In the short run, there were few substitutes; the demand for rubber was inelastic. In the long run, however, the rise in the price of rubber stimulated research for alternatives. Many alternatives were found. Today automobile tires, which were all made of rubber at the time World War II broke out, are almost entirely made from synthetic materials. In the long run, demand was very elastic. So, the larger the time interval considered, or the longer the run, the more elastic is the demand for the good.
2. **Luxuries versus Necessities.** The less a good is a necessity, the more elastic is its demand. Because by definition *one cannot do without necessities*, they tend to have fewer substitutes than luxuries. Insulin for a diabetic is a necessity; the demand is highly inelastic. Chocolate Ecstasy cake, however, is a luxury. A variety of other luxuries can be substituted for it (for example, cheesecake or a ball game).
3. **Narrow or Broad Definition.** As the definition of a good becomes more specific, demand becomes more elastic. If the good we're talking about is broadly defined (say, transportation), there are not many substitutes and demand will be inelastic. If you want to get from A to B, you need transportation. If the definition of the good is narrowed—say, to "travel by bus"—there are more substitutes. Instead of taking a bus you can walk, ride your bicycle, or drive your car. In that case, demand is more elastic.
4. **Budget Proportion.** You would be more responsive to changes in the price of goods which represent a large proportion of your budget than for goods which represent a small proportion of your budget. It's not worth spending a lot of time figuring out whether there's a good substitute for an item that costs very

little relative to your total expenditures. If you wanted to buy five pencils, it wouldn't really matter whether the price was 5¢ or 10¢ each. Pencils are not a big item in your budget, so demand is inelastic. It is, however, worth spending lots of time looking for substitutes for goods that take a large portion of your income. Most people shop around for the lowest price on expensive items like televisions, fridges, and cars. The demand for such goods tends to be more elastic.

Pencils are not a big-budget item.

Empirical Estimates of Elasticities

Table 6-2 presents empirical estimates of price elasticity of demand. Notice that, as expected, different estimates are provided for the short- and long-run elasticities of each good. Also notice that the elasticity estimates are for the entire country; estimates for a specific regions would likely be higher (more elastic).

Taking an example from Table 6-2, notice that the long-run demand for movies is elastic. If movie theatres raise their prices, it's relatively easy for individuals simply to stay home and watch television. Movies have close substitutes, so we would expect the demand to be relatively elastic.

In the long run, demand generally becomes more elastic.

As a second example, in the short run the demand for electricity is highly inelastic. Either people have electrical appliances or they don't. In the long run, however, it becomes elastic since people can shift to gas for cooking and oil for heating, and can buy more energy-efficient appliances. As an exercise, you might see if you can explain why each of the other goods listed in the table has the elasticity of demand reported.

PRICE ELASTICITY OF DEMAND AND TOTAL REVENUE

Knowing elasticity of demand is useful to firms because from it they can tell what happens to total revenue when they raise or lower their prices. The total revenue a supplier receives is the price he or she charges times the quantity he or she sells. (Total revenue

Total Revenue: the total amount of money a firm receives from selling its product. TR = P × Q.

TABLE 6-2 Short-Run and Long-Run Elasticities of Demand

	Price Elasticity	
Product	Short-Run	Long-Run
Tobacco products	0.46	1.89
Electricity (for household consumption)	0.13	1.89
Health services	0.20	0.92
Toys (nondurable)	0.30	1.02
Movies/motion pictures	0.87	3.67
Foreign travel	0.14	1.77
Beer	.56	1.39
Wine	.68	.84
University tuition	.52	—
Rail transit	.62	1.59

Sources: Hendrik S. Houthakker and Lester D. Taylor, *Consumer Demand in the United States: Analyses and Projections,* 2nd ed. (Cambridge, Mass.: Harvard University Press, 1970); W. S. Comanor and T. A. Wilson, *Advertising and Market Power* (Cambridge, Mass.: Harvard University Press, 1974); Shermon Folland, Allen C. Goodman, and Miron Stano, *The Economics of Health Care* (New York: Macmillan, 1993); Yu Hsing and Hui S. Change, "Testing Increasing Sensitivity of Enrollment at Private Institutions to Tuition and Other Costs," *The American Economist* 41, no. 1 (Spring 1996); Richard Voith, "The Long-Run Elasticity of Demand for Commuter Rail Transportation," *Journal of Urban Economics* 30 (1991).

equals total quantity sold multiplied by price of good.) Elasticity tells sellers what will happen to total revenue if their price changes.

Suppose a firm knew that the demand for its product was 2.7 (elastic). This means that consumers are responsive to price changes, and will reduce their quantities by a large amount when price rises. What happens to a firm's total revenue if it raises the price by 5%? An elasticity of 2.7 means that a 1% rise in price will lead to a 2.7 fall in quantity demanded. We can expect that a 5% price increase will cause a 13.5% ($5 \times 2.7\%$) decrease in quantity sold. How will this affect the firm's total revenue? The firm gains on the 5% increase in price, but loses on the 13.5 decrease in quantity sold, thus total revenue must decrease:

$$P \times Q = TR$$
$$\uparrow P \times \downarrow Q = \downarrow TR$$

The opposite is true for a product whose demand is inelastic. A firm can raise price by 5%, quantity sold falls by less than 5%, and the firm's total revenue increases:

$$P \times Q = TR$$
$$\uparrow P \times \downarrow Q = \uparrow TR$$

Total revenue ($P \times Q$) is represented by the area under the demand curve at the given price and quantity. In Figure 6-5 (a), at point *E* on the demand curve, price is \$8 and the quantity sold is 2. Total revenue is \$8 × 2 units = \$16. This is represented by Areas A + *B*. If the price rises from \$8 to \$9, the quantity demanded falls from 2 to 1 (point *F*). Total revenue is now \$9 × 1 unit = \$9. Total revenue is now represented by Areas A + C. As you can see, the gain (Area C) is much smaller than the loss (Area *B*).

Figure 6-5 (a, b, and c) ELASTICITY AND TOTAL REVENUE

Total revenue is measured by the rectangle produced by extending lines from the demand curve to the price and quantity axes. The change in total revenue resulting from a change in price can be estimated by comparing the sizes of the before and after rectangles. If price is being raised, total revenue increases by rectangle C and decreases by rectangle B. As you can see, the effect of a price rise on total revenue differs significantly at different points on a demand curve; (**a**) shows an elastic point, (**b**) shows an inelastic point, and (**c**) shows a unitary elastic point.

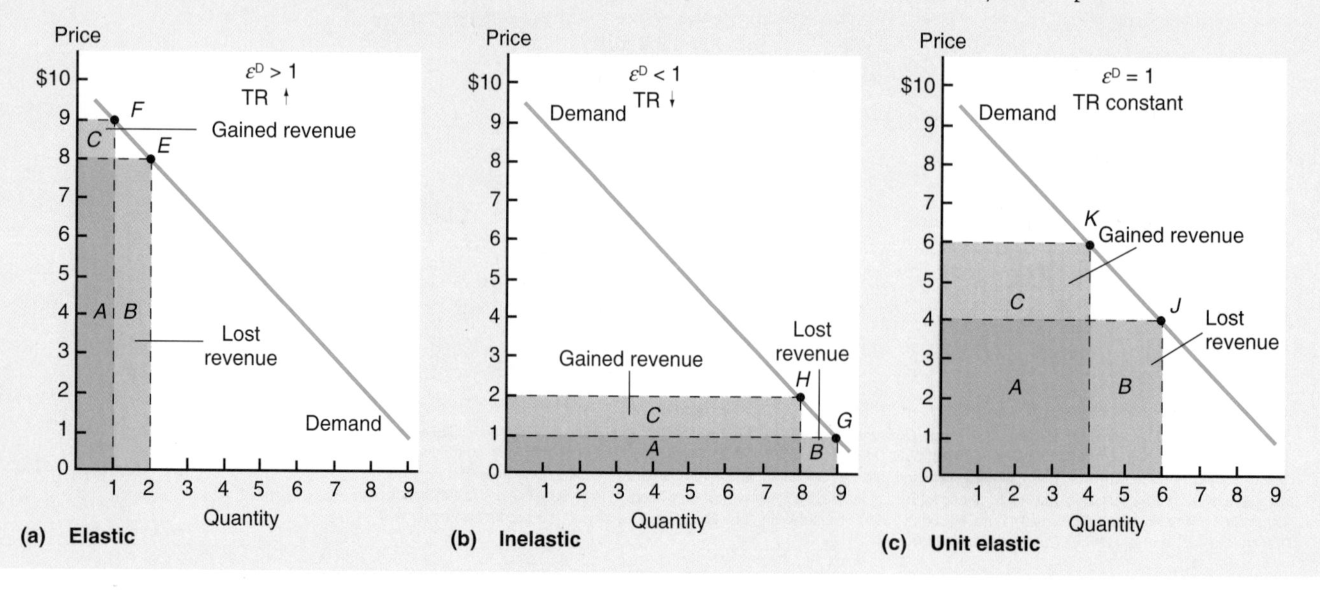

At this point, demand is elastic: consumers' response to the 11.76% increase in price is tremendous (66.7%), so total revenue decreases significantly.

In Figure 6-5(b), the slope of the demand curve is the same as in Figure 6-5(a), but we begin at a different point on the demand curve (point G). If we raise our price from \$1 to \$2, quantity demanded falls from 9 to 8. The gain (Area C) is much greater than the loss (Area *B*). Total revenue increases significantly because the demand curve at point G is highly inelastic.

Finally, in Figure 6-5 (c), at Point *J* on the demand curve, if price rises from \$4 to \$6, quantity demanded will fall from 6 to 4 units. The revenue gained from the higher price (Area C) exactly equals the revenue lost as sales decreased (Area *B*), so there is no change in the firm's total revenue. Revenues gained exactly offset revenues lost if demand is unit elastic.

Summary

If demand is elastic ($\varepsilon^D > 1$), a rise in price lowers total revenue. (Price and total revenue move in opposite directions.)

If demand is unit elastic ($\varepsilon^D = 1$), a rise in price leaves total revenue unchanged.

If demand is inelastic ($\varepsilon^D < 1$), a rise in price increases total revenue. (Price and total revenue move in the same direction.)

Q-6 If demand is inelastic and a firm raises price, what happens to total revenue?

Total Revenue along a Demand Curve

The way in which elasticity changes along a demand curve and its relationship to total revenue can be seen in Figure 6-6 on the next page. When output is zero, total revenue is zero; similarly when price is zero, total revenue is zero. That accounts for the two end points of the total revenue curve in Figure 6-6(b). Let's say we start at a price of zero, where the demand meets the horizontal axis. As we increase price through the inelastic range, even though quantity demanded decreases, total revenue increases. As we continue to raise price, the increases in total revenue become smaller until finally, after output of Q_0, total revenue actually starts decreasing. As we continue to raise price through the elastic range of the demand curve, total revenue continues to decrease at a faster and faster rate until finally, at zero output, total revenue is zero again. As you can see, total revenue is maximized at the quantity Q_0, where the price elasticity of demand is 1 (Unit elastic demand).

With elastic demand, a rise in price decreases total revenue. With inelastic demand, a rise in price increases total revenue.

Elasticity of Individual and Market Demand

In thinking about elasticity of demand, keep in mind the point made in Chapter 4: The market demand curve is the horizontal summation of individual demand curves; some individuals have highly inelastic demands and others have highly elastic demands. A slight rise in the price of a good will cause some people to stop buying the good, but won't affect other people's quantity demanded for the good at all. Market demand elasticity is influenced both by how many people drop out totally (reduce their quantity to zero) and by how much an existing consumer marginally reduces his or her quantity demanded.

If a firm can somehow separate the people with less elastic demand from those with more elastic demand, it can charge more to the individuals with inelastic demands and less to individuals with elastic demands. Economists call this price discrimination. We see firms throughout the economy trying to use price discrimination. Let's consider three examples.

Figure 6-6 (a and b) **HOW TOTAL REVENUE CHANGES**

Total revenue is at a maximum when elasticity equals one, as you can see in (a) and (b). When demand is elastic, total revenue decreases with an increase in price. When demand is inelastic, total revenue increases with an increase in price.

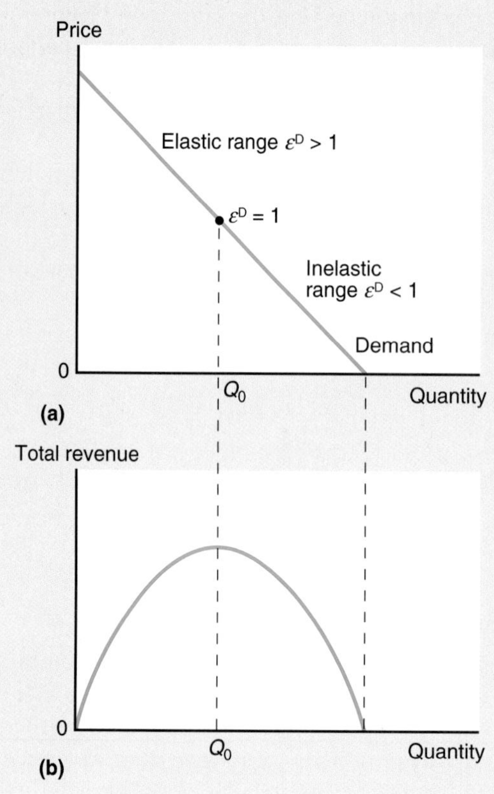

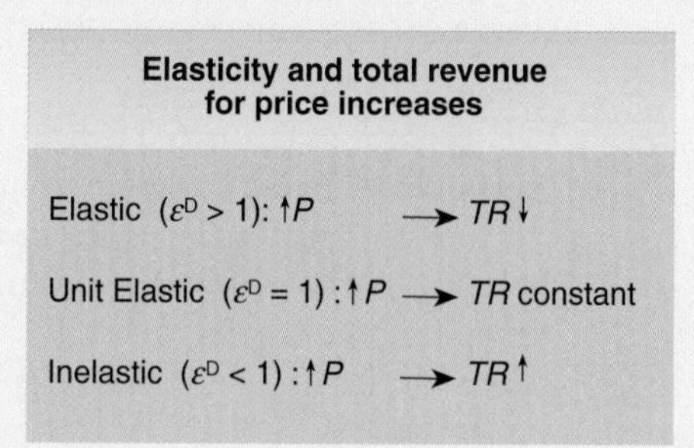

Elasticity and total revenue for price increases	
Elastic ($\varepsilon^D > 1$): ↑P	→ TR↓
Unit Elastic ($\varepsilon^D = 1$): ↑P	→ TR constant
Inelastic ($\varepsilon^D < 1$): ↑P	→ TR↑

Firms have a strong incentive to separate out people with less elastic demand and charge them a higher price.

New cars are considered luxuries.

1. Airlines' Saturday stay-over specials. If you stay over a Saturday night, usually you can get a much lower airline fare than if you don't. The reason is that business travellers have inelastic demands and don't like to stay over Saturday nights, while pleasure travellers have more elastic demands. By requiring individuals to stay over Saturday night, airlines can separate out businesspeople and charge them more.
2. Selling new cars. Most new cars don't sell at the listed price. They sell at a discount. Salespeople are trained to separate out comparison shoppers (who have more elastic demands) from impulse buyers (who have inelastic demands). By not listing the selling price of cars, so that the discount can be worked out in individual negotiations, salespeople can charge more to customers who have inelastic demands.
3. The almost-continual-sale phenomenon. Some items, such as washing machines, go on sale rather often. Why don't suppliers sell them at a low price all the time? Because some buyers whose washing machines break down have inelastic demand. They can't wait, so they'll pay the "unreduced" price. Others have elastic demands; they can wait for the sale. By running sales (even though they're frequent sales), sellers can separate consumers with inelastic demand curves from consumers with elastic demand curves.

APPLYING THE TOOLS

Empirically Measuring Elasticities

In the text, the discussion of determining elasticity concentrates on the technical aspects of the calculation. It assumes we know what point we are at on the supply and demand curve. In the real world, economists don't have the luxury of that knowledge. The data points they use involve interactions of supply and demand, and they must use statistical tools to ensure that they are holding "other things constant." Specifically, to determine points on a demand curve we must vary supply (and nothing else); to determine points on a supply curve we must vary demand (and nothing else).

In practice, holding everything else constant is difficult to do, which means real-world estimates of elasticity are often less than perfect. The tables presented in the text are some economists' estimates, but there are often disputes and technical issues that could lead to different estimates.

Where do firms get the information they need to calculate elasticities? Think of the grocery store where you can get a special buyer's card; you show it to the checkout clerk and you get all the discounts. And the card is free! Those grocery stores are not just being nice. When the clerk scans your purchases in, the store gets information that is forwarded to a central processing unit that can see how people react to different prices. This information is valuable; it allows firms to fine-tune their pricing—raising prices on goods for which the demand is inelastic, and lowering prices on goods for which the demand is elastic.

Alternatively, think of the warranty cards that you send in when you buy a new computer or a new TV. The information goes into the firms' information bases and is used by their economists in future price-setting decisions.

OTHER ELASTICITIES OF DEMAND

There are many other elasticities besides price elasticity of demand. Two other demand elasticities are important in describing consumer behaviour. These elasticity concepts tell you how much the demand curve will shift when there is a change in a shift factor.

Income Elasticity of Demand

The most commonly used elasticity is *income elasticity of demand*. **Income elasticity of demand** tells us how much demand will change with a change in consumers' income, and is calculated as *the percentage change in quantity demanded divided by the percentage change in income*. Economists use the Greek symbol eta (pronounced "Ayta") to represent income elasticity.

Income elasticity of demand shows the responsiveness of demand to changes in income.

Q-7 Why can we **not** take absolute value of the income elasticity of demand?

$$\eta = \frac{\text{Percentage change in quantity demanded}}{\text{Percentage change in income}}$$

Income elasticity of demand tells us how demand responds to changes in income. An increase in income generally increases one's consumption of almost all goods, although the increase may be greater for some goods than for others. **Normal goods**—*goods whose consumption increases with an increase in income*—have income elasticities greater than zero.

Q-8 If a good's consumption increases with an increase in income, what type of good would you call it?

Normal goods are usually divided into two categories, called luxuries and necessities. **Luxuries** are *goods that have an income elasticity greater than 1*—their percentage increase in demand is greater than the percentage increase in income.

For example, consider consumers' responses to different products as incomes rise form $29,000 to $30,500:

Case 1: $\eta > 1$, Income Elastic Good (Luxury). In response to higher incomes, consumers increase their wine consumption from 18 to 25 bottles per year. The income elasticity of demand for wine is

$$\eta = \frac{\text{Percentage change in quantity demanded}}{\text{Percentage change in income}} = \frac{\dfrac{Q_2 - Q_1}{\text{average } Q}}{\dfrac{I_2 - I_1}{\text{average } I}} = \frac{\dfrac{25 - 18}{21.5}}{\dfrac{1500}{29{,}750}}$$

$$= \frac{.33}{.05} = 6.5 \text{ (Income elastic)}$$

So, a 5% increase in income results in an enthusiastic 33% increase in wine consumption.

Case 2: $0 < \eta < 1$, Income Inelastic Good (Necessity). As their incomes rise, consumers' toothpaste purchases rise by .5%. The elasticity calculation indicates that toothpaste consumption does not respond very much to changes in income:

$$\eta = \frac{\text{Percentage change in quantity demanded}}{\text{Percentage change in income}} = \frac{+.5\%}{+5\%} = .1 \text{ (income inelastic)}$$

The consumption of a necessity rises by a smaller proportion than the rise in income.

It is even possible that an increase in income can cause a *decrease* in the consumption of a particular good. These goods have a negative income elasticity of demand. The term applied to such goods is **inferior goods**—*goods whose consumption decreases when income increases*.

Case 3: $\eta < 0$, Inferior Good. As incomes rise, consumers' purchases of generic (store-brand) cereals falls by 3%. In this case, the income elasticity of demand is percentage change in quantity demanded:

$$\eta = \frac{\text{Percentage change in quantity demanded}}{\text{Percentage change in income}} = \frac{-3\%}{+5\%}$$

$$= -.6 \text{ (Inferior)}$$

6.2

see page 158

Table 6-3 presents some income elasticities for different groups of goods, while Table 6-4 shows the relationships described above in Cases 1-3. In the short run, people often save high proportions of their increases in income, so most goods, other than impulse-bought goods, have low income elasticities. To avoid this problem, economists generally focus on long-run income elasticities. Notice which goods are necessities (the ones with long-run income elasticities less than 1). Notice also which goods are luxuries (the ones with elasticities greater than 1).

Finally, notice the one good with a negative income elasticity—food produced and consumed on farms. As mentioned above, such goods are called inferior goods. As income rises, people buy proportionately less of such goods.

Cross-Price Elasticity of Demand

Cross-price elasticity of demand is another frequently used elasticity concept because a rise in the price of a product often affects people's purchases of *other* products. We can measure the impact of such a price increase on the demand for related products by calculating the cross-price elasticity of demand between two products.

Cross-price elasticity of demand measures *how* consumers respond to changes in the price of a related product, and *how strongly* they respond.

TABLE 6-3 Income Elasticities for Selected Goods

Commodity	Income Elasticity Short-Run	Income Elasticity Long-Run
Motion pictures	0.81	3.41
Foreign travel	0.24	3.09
Tobacco products	0.21	0.86
Food produced and consumed on farms	−0.61	—
Furniture	2.60	0.53
Jewellery and watches	1.00	1.64
Beer	—	0.84
Hard liquor	—	2.5
Dental services	—	1.6

Sources: Hendrik S. Houthakker and Lester D. Taylor, *Consumer Demand in the United States: Analyses and Projections,* 2nd ed. (Cambridge, Mass.: Harvard University Press, 1970); E. A. Selvanthan, "Cross-Country Alcohol Consumption: An Application of the Rotterdam Demand System," *Applied Economics* 23 (1991); Shermon Folland, Allen C. Goodman, and Miron Stano, *The Economics of Health Care* (New York: Macmillan, 1993); Yu Hsing and Hui S. Chang, "Testing Increasing Sensitivity of Enrollment at Private Institutions to Tuition and Other Costs," *The American Economist* 41, no. 1 (Spring 1996).

TABLE 6-4 Interpreting Income Elasticity of Demand:

Coefficient	Interpretation	Description
$\eta > 0$	Normal good	$\uparrow I \rightarrow \uparrow Qd$
	Two cases of Normal good:	
	$0 < \eta < 1$	Income inelastic normal good ("necessity")
	$\eta > 1$	Income elastic normal good ("superior" good)
$\eta < 0$	Inferior good	$\uparrow I \rightarrow \downarrow Qd$

Cross-price elasticity of demand shows the responsiveness of demand to changes in prices of other goods.

Cross-price elasticity of demand is computed by *dividing the percentage change in quantity demanded by the percentage change in the price of another good.* Put another way,

$$\varepsilon^{xy} = \frac{\text{Percentage change in quantity demanded}}{\text{Percentage change in price of another good}}$$

The numerator (top) captures *quantity* information about Good X. The denominator (bottom) captures *price* information about Good Y. Notice that this is another price elasticity, so we use ε. Say the price of Toyotas rises. What is likely to happen to the demand for Fords? It is likely to rise, so the cross-price elasticity between the two is positive.

$$\varepsilon^{xy} = \frac{\text{Percentage change in quantity demanded of Fords}}{\text{Percentage change in price of Toyotas}}$$

$$= \frac{\uparrow}{\uparrow} = \text{positive (Substitute goods)}$$

Positive cross-price elasticities of demand mean the goods are **substitutes**—*goods that can be used in place of one another.* When the price of a good goes up, the demand for the substitute goes up. Most goods have substitutes, so most cross-price elasticities are positive. Consider another example: Say the price of hot dogs rises; what is likely to happen to the demand for ketchup? The rise in the price of hot dogs will lower the consumption of both hot dogs and ketchup.

Taste test: Perfect substitutes?

KNOWING THE TOOLS

A Review of the Alternative Elasticity Terms

Income elasticity of demand tells us what happens to the demand for a good when consumer incomes change.

$$\eta = \frac{\text{Percentage change in quantity demanded}}{\text{Percentage change in income}}$$

Normal good: Income elasticity of demand is positive.

Luxury: Income elasticity is greater than 1.

Necessity: Income elasticity is between 0 and 1.

Inferior good: Income elasticity of demand is negative.

Cross-price elasticity of demand tells us what happens to the demand for a good when the price of a related good changes.

$$\varepsilon^{xy} = \frac{\text{Percentage change in quantity demanded}}{\text{Percentage change in price of a related good}}$$

Complement: Cross-price elasticity of demand is negative.

Substitute: Cross-price elasticity of demand is positive.

$$\varepsilon^{xy} = \frac{\text{Percentage change in quantity demanded for ketchup}}{\text{Percentage change in price of hot dogs}}$$

$$= \frac{\downarrow}{\uparrow} = \text{negative (complementary goods)}$$

Substitutes have positive cross-price elasticities; complements have negative cross-price elasticities.

Ketchup and hot dogs are complements. **Complements** are *goods that are used in conjunction with other goods*. A rise in the price of a good will decrease the demand for its complement and a fall in the price of a good will increase the demand for its complement, so the cross-price elasticity of complements is negative.

What does it mean if the cross-price elasticity of demand equals zero? The calculation implies that a change in the price of Good Y (paperclips) has no effect on the demand for Good X (socks):

$$\varepsilon^{xy} = \frac{\text{Percentage change in quantity demanded of Good X}}{\text{Percentage change in price of Good Y}}$$

$$= \frac{0}{\uparrow} = 0 \text{ (unrelated goods)}$$

It must be the case that the two goods are not related.

The relationships in cross-price elasticities of demand are shown in Table 6-5, while some estimates of cross-price elasticities of demand are shown in Table 6-6. You can see that the strongest substitutes are European autos for North American and Asian autos. A 10 percent fall in the price of North American and Asian autos leads to a 6 percent

TABLE 6-5 Interpreting Cross-Price Elasticity of Demand:

Coefficient	Interpretation	Description
$\varepsilon^{xy} > 0$	Substitute Goods	$\downarrow P_Y \rightarrow \downarrow Q_x$
$\varepsilon^{xy} < 0$	Complementary Goods	$\downarrow P_Y \rightarrow \uparrow Q_x$
$\varepsilon^{xy} = 0$	Unrelated Goods	$\downarrow P_Y \rightarrow \Delta Q_x = 0$

Figure 6-7 (a and b) CALCULATING INCOME AND CROSS-PRICE ELASTICITIES OF DEMAND

Shift factors, such as income or price of another good, shift the entire demand curve. To calculate these elasticities, we see how much demand will shift at a constant price and then calculate the relevant elasticities.

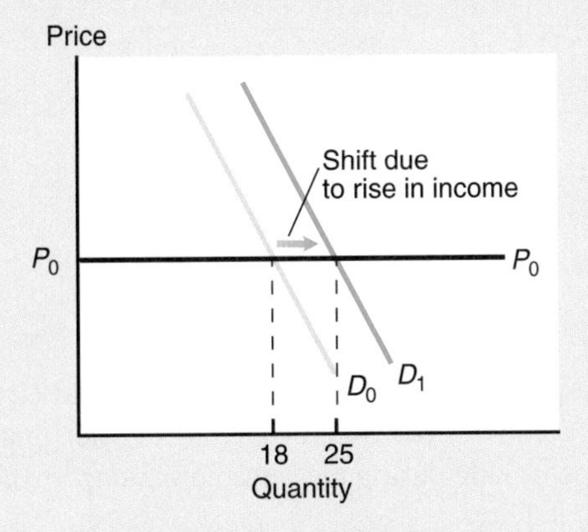

(a) Calculating income elasticity of demand

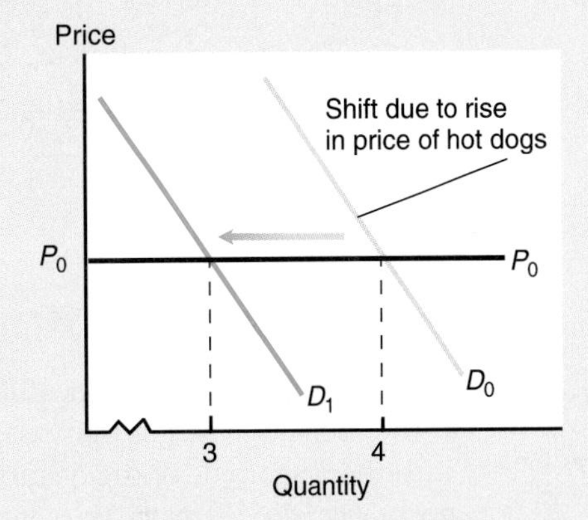

(b) Calculating cross-price elasticity of demand

TABLE 6-6 Cross-Price Elasticities

Commodities	Cross-Price Elasticity
Beef in response to price changes in pork	0.11
Beef in response to price changes in chicken	0.02
North American automobiles in response to price changes in European and Asian automobiles	0.28
European automobiles in response to price changes in North American and Asian automobiles	0.61
Beer in response to price changes in wine	0.23
Hard liquor in response to price changes in beer	−0.11

Sources: J. A. Johnson and E. H. Oksanen, "Socioeconomic Determinants of the Consumption of Alcoholic Beverages," *Applied Economics* (1974); Patrick S. McCarthy, "Market Price and Income Elasticities of New Vehicle Demand," *Review of Economics and Statistics* (August 1996); Kuo S. Huang, "Nutrient Elasticities in a Complete Food Demand System," *American Journal of Agricultural Economics* (February 1996).

fall in the quantity of European autos demanded. Hard liquor and beer are complements. If the price of beer falls by 10 percent, the quantity of hard liquor demanded increases by 1.1 percent.

Graphing Income and Cross-price Elasticities of Demand

Figure 6-7 above demonstrates two examples. In Figure 6-7(a), income has risen by 5 percent, increasing the demand for wine at a constant price, P_0, from 18 to 25. In our earlier calculations, we found that the income elasticity of demand was 6.5, which means the demand for wine is income elastic (Luxury good).

Figure 6-7(b) shows what happens to the demand for ketchup when hot dogs rose from 99¢ to $1.49, causing a decrease in the quantity of hot dogs eaten. Because fewer hot dogs are eaten, the quantity of ketchup demanded has fallen from 4 bottles to 3 bottles per year. The cross-price elasticity of demand is

$$\varepsilon^{xy} = \frac{\text{Percentage change in quantity demanded for ketchup}}{\text{Percentage change in price of hot dogs}}$$

$$= \frac{\dfrac{Q_2^K - Q_1^K}{\text{average } Q^K}}{\dfrac{P_2^{HD} - P_1^{HD}}{\text{average } P^{HD}}} = \frac{\dfrac{3-4}{3.5}}{\dfrac{1.49-.99}{1.24}} = \frac{-.29}{.4} = -.7 \text{ (complements)}$$

PRICE ELASTICITY OF SUPPLY

Price elasticity of supply measures firms' response to a price change, and is calculated as the percentage change in quantity supplied divided by the percentage change in price.

Q-9 If when price rises by 4 percent, quantity supplied rises by 8 percent, what is the price elasticity of supply?

Price elasticity of supply measures the responsiveness of firms to a change in the price of the product. The elasticity coefficient compares the magnitude of the price change with the magnitude of the quantity change by calculating *the percentage change in quantity supplied divided by the percentage change in price:*

$$\varepsilon^S = \frac{\text{Percentage change in quantity supplied}}{\text{Percentage change in price}}$$

Let's consider two examples. Say that when price falls by 5 percent, quantity supplied falls by 2 percent. In this case, the price elasticity of supply is 0.4 (2 percent/5 percent). And, say the price goes up by 10 percent and in response the quantity supplied rises by 15 percent. Price elasticity of supply is 1.5 (15 percent/10 percent).

Classifying Supply as Elastic or Inelastic

The commonsense interpretation of these calculations is the following: An *inelastic* supply means that the quantity supplied doesn't change much with a change in price (our first calculation). Consider a rise in the price of land. The amount of land supplied won't change much, so the supply of land is inelastic. An *elastic* supply means that quantity supplied changes by a larger percentage than the percentage change in price (our second calculation). For example, say the price of pencils doubles. What will happen to the quantity of pencils supplied if their price doubles? Likely it it will more than double, which means that the supply of pencils is elastic. (Hint: How easy or difficult is it to increase production?)

Substitution and Supply

The longer the time period considered, the more elastic the supply.

The same general issues involving substitution are relevant when considering determinants of the elasticity of supply. But when it comes to supply, economists focus on time rather than on other factors because time plays such a central role in determining supply elasticity. The general rule is: The longer the time period considered, the more elastic is the supply curve. The reasoning is the same as with demand; in the long run there are more alternatives, so it is easier (less costly) for suppliers to change and produce other goods.

To emphasize the importance of time, economists distinguish three time periods relevant to supply:

1. In the instantaneous period, quantity supplied is fixed, so supply is perfectly inelastic. This supply is sometimes called the momentary supply.

2. In the short run, some substitution is possible, so the short-run supply curve is somewhat elastic.
3. In the long run, significant substitution is possible; the supply curve becomes very elastic.

Q-10 Is supply generally more elastic in the short run or in the long run?

In determining the elasticity of supply, one must, however, remember an additional factor: Many supplied goods are produced, so we must take into account how easy it is to increase production of those same goods. For example, if the cost per unit of producing a good is constant, its supply is likely highly elastic. But if production is more complicated, it may be difficult for firms to quickly increase the quantity of goods supplied, so supply would likely be less elastic.

Empirical Estimates of Elasticities

There are many fewer empirical measurements of supply than there are of demand elasticities. The reason concerns the structure of markets of produced goods, and the complicated nature of production. Most retail markets have seller-set or posted prices—you go to the store and pay the listed price of toothpaste. You can buy as much as you want at that price, so in a sense the supply of toothpaste (and most retail goods) is perfectly elastic until the store runs out, whereupon the supply becomes perfectly inelastic. But in another sense there is no supply curve, since the selling price is determined by the seller's pricing strategy, not by the market.

We do find empirical measurements of supply elasticities in factor markets, such as the market for labour services. For example, economist David Blau has estimated that the supply of child care workers is elastic—it may be as high as 1.9—which means that a 10 percent rise in the wages paid to child care workers brings about a 19 percent increase in the quantity of child care workers. More generally, economists have estimated that the labour supply elasticity of heads of households is about 0.1, and for secondary workers is about 1.1. A good test of whether you intuitively understand elasticities is whether you can explain why the latter is more elastic.

Other areas in which elasticities of supply are estimated are agricultural and raw materials markets. Estimating supply elasticities here is possible because these goods are often sold in auction markets where price is directly determined by supply and demand, rather than in posted-price markets. In these markets, economists have generally found that the short-run supplies are highly inelastic, and that the long-run supplies are highly elastic.

THE POWER OF SUPPLY AND DEMAND ANALYSIS

Now consider the power of supply and demand analysis when it is combined with the concept of elasticity.

Elasticity and Shifting Supply and Demand

Knowing the elasticity of the supply and demand curves allows us to be more specific about the effects of shifts in supply and demand. Figure 6-8 (a) illustrates the world market for oil. When supply is reduced, since the demand for oil is inelastic, price rises sharply, but the quantity demanded falls by only a small amount, and revenues increase for oil producers.

World demand for oil is inelastic.

Figure 6-8 (a and b) EFFECTS OF SHIFTS IN SUPPLY ON PRICE AND QUANTITY

In both (**a**) and (**b**) the supply shifts from S_0 to S_1. Initial price is P_0 and quantity is Q_0. The new equilibrium is P_1 and Q_1. In (**a**), demand is inelastic and the quantity effects are relatively small. In (**b**), the demand is more elastic and the quantity effects are much larger. In general the effects of shifts in supply on equilibrium quantity and price are determined by the elasticity of demand. When demand is inelastic, price changes are large and quantity changes are small. A useful exercise is to go through the same two cases for demand shifts, showing how the quantity effect is determined by the elasticity of supply.

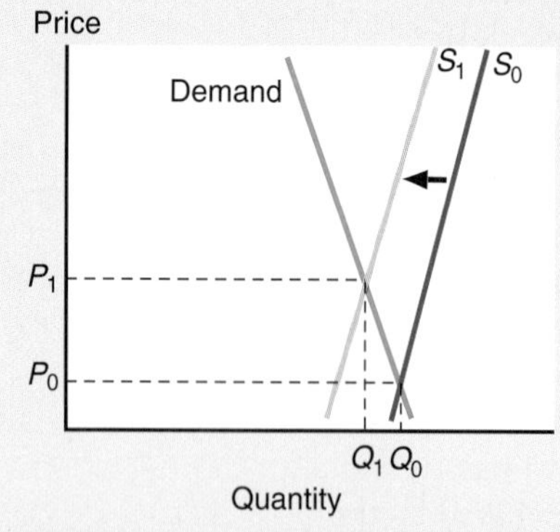

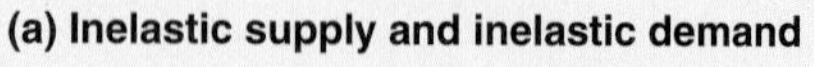

(a) Inelastic supply and inelastic demand

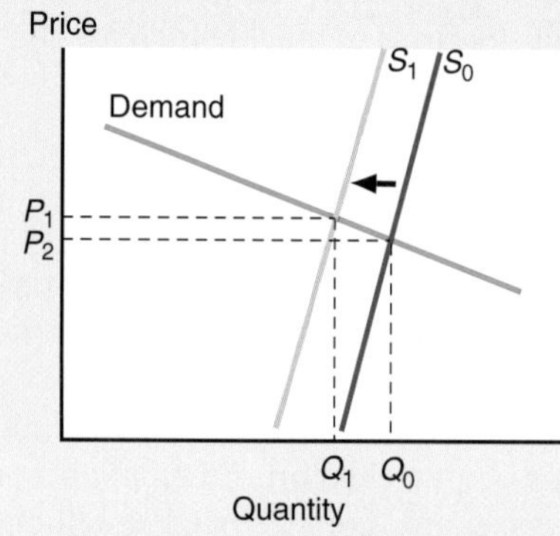

(b) Inelastic supply and elastic demand

Figure 6-8 reviews the effect of shifts in supply with various elasticities.

Figure 6-8(b) demonstrates the relative effects of supply shifts on equilibrium price and quantity under different assumptions about elasticity. As you can see, the more elastic the demand, the greater the effect of a supply shift on quantity, and the smaller the effect on price. With an identical reduction in supply (S_1 shifts left by the same horizontal distance as in Figure 6-8 (a)), people switch to cheaper alternatives, so quantity falls by a large amount and price rises only slightly. Going through a similar exercise for demand shifts with various supply elasticities is also a useful exercise. You will see that the more elastic the supply, the greater the effect of a demand shift on quantity, and the smaller the effect on price.

To be sure that you have understood elasticity, illustrate situations 1, 2, and 3 in a Demand and Supply diagram, and match them with the three descriptions a, b, and c.

Situation:	Result:
1. Demand highly elastic; supply shifts out.	a. Price rises significantly; quantity hardly changes at all.
2. Supply highly inelastic; demand shifts out.	b. Price remains almost constant; quantity increases enormously.
3. Demand is highly inelastic; supply shifts out.	c. Price falls significantly; quantity hardly changes at all.

The answers are 1–b; 2–a; 3–c.

CONCLUSION

We can use demand and supply analysis to explain how various events will impact a market's equilibrium (price and quantity). By introducing the concept of elasticity into the analysis, we can sometimes dramatically improve our ability to explain certain events. For example, in a standard supply and demand diagram, a leftward shift of the supply curve will raise price and reduce quantity. However, we can do a much better job of explaining what has happened several times in the world oil market if we remember that the demand for oil (and the supply of oil) is inelastic. Re-drawing our diagram with steep demand and supply curves to capture this information, we immediately see that the OPEC supply reduction (shift left) has a tremendous effect on price, while quantity changes very little. After the initial effect, however, alternatives emerge, and in the long run, demand (and supply) becomes more elastic. Our analysis of the oil market is greatly improved by including our knowledge of demand and supply elasticities.

Chapter Summary

- Elasticity measures the responsiveness of quantity to a change in some variable that affects quantity, such as price or income. The most common elasticity concept used is price elasticity of demand.
- The Law of Demand tells us that as price falls, consumers buy more. Price elasticity of demand tells us whether they buy a lot more (elastic demand) or only a little more (inelastic demand).
- To calculate the price elasticity of demand coefficient using the mid-point formula:

$$\varepsilon^D = \frac{\text{Percentage change in quantity demanded}}{\text{Percentage change in price}} = \frac{\dfrac{Q_2 - Q_1}{\text{average } Q}}{\dfrac{P_2 - P_1}{\text{average } P}}$$

- Because of the Law of Demand, price elasticity of demand is reported as a positive coefficient (take absolute value).
- Price elasticity of demand is interpreted as: elastic ($\varepsilon^D > 1$), perfectly elastic ($\varepsilon^D = \infty$), inelastic ($\varepsilon^D < 1$), perfectly inelastic ($\varepsilon^D = 0$), and unit elastic ($\varepsilon^D = 1$).
- The more substitutes a good has, the greater its elasticity.
- Factors affecting the number of substitutes in demand are (1) time interval considered, (2) whether the good is a luxury or a necessity, (3) how narrowly the good is defined, and (4) the proportion of one's budget the good represents.
- Elasticity changes along straight-line demand and supply curves. Demand becomes less elastic as we move down along a demand curve.
- When a supplier raises price: if demand is inelastic, total revenue increases; if demand is elastic, total revenue decreases; if demand is unit elastic, total revenue remains constant.
- Income elasticity of demand measures how consumers' purchase patterns change with income, and is calculated as

$$\eta = \frac{\text{Percentage change in quantity demanded}}{\text{Percentage change in income}} = \frac{\dfrac{Q_2 - Q_1}{\text{average } Q}}{\dfrac{I_2 - I_1}{\text{average } I}}$$

- Income elasticity of demand is usually positive because the consumption of most goods rises with income. If $0 < \eta < 1$, the response is income-inelastic and the good is considered a necessity; if $\eta > 1$, the response is income-elastic and the good is a luxury.
- Income elasticity of demand for inferior goods is negative ($\eta < 0$) because the consumption of these goods *falls* with income.

- Cross-price elasticity of demand measures the relationship between two products, and is calculated as

$$\varepsilon^{xy} = \frac{\text{Percentage change in quantity demanded}}{\text{Percentage change in price}} = \frac{\dfrac{Q_2^X - Q_1^X}{\text{average } Q^X}}{\dfrac{P_2^Y - P_1^Y}{\text{average } P^Y}}$$

- A positive cross-price elasticity of demand coefficient means the goods are substitutes; a negative value means the goods are complements. A coefficient value of zero means the goods are not related.
- Price elasticity of supply is calculated as

$$\varepsilon^S = \frac{\text{Percentage change in quantity supplied}}{\text{Percentage change in price}} = \frac{\dfrac{Q_2 - Q_1}{\text{average } Q}}{\dfrac{P_2 - P_1}{\text{average } P}}$$

- The most important factor affecting the number of substitutes in supply is time. As the time interval lengthens, supply becomes more elastic.
- Knowing elasticities allows us to be more precise about the qualitative effects that shifts in demand and supply have on prices and quantities.

Key Terms

coefficient *(149)*
complements *(150)*
cross-price elasticity of demand *(149)*
elastic *(135)*
income elasticity of demand *(147)*
inelastic *(135)*
inferior goods *(148)*
Law of Demand *(135)*
luxuries *(147)*
necessity *(142)*
normal goods *(147)*
perfectly elastic *(139)*
perfectly inelastic *(139)*
price elasticity of demand *(135)*
price elasticity of supply *(152)*
substitutes *(149)*
unit elastic *(140)*

Questions for Thought and Review 1-5, 7, 9, 11

1. Determine the price elasticity of demand if, in response to an increase in price of 10 percent, quantity demanded decreases by 20 percent. Is demand elastic or inelastic?
2. A firm has just increased its price by 5 percent over last year's price, and it found that quantity sold remained the same. The firm comes to you and wants to know its price elasticity of demand. How would you calculate it? What additional information would you search for before you did your calculation?
3. In order to encourage support for the local hockey team, the local pizza shop, Pete's a Pie, offered to lower price by $1 for every goal scored by the team in their next game. Cheered on by (hungry) fans, the team managed to score 6 goals in the game, and pizzas were selling for $2 each. The quantity of pizzas demanded soared from 1 per hour to 100 per hour. What was the price elasticity of demand for pizzas at Pete's?
4. Which of the following pairs of goods would you expect to have a greater price elasticity of demand?
   a. Cars, transportation.
   b. Housing, leisure travel.
   c. Rubber during World War II, rubber during the late 20th century.
5. Rank these products in order from most elastic to least elastic:
   a. apple
   b. MacIntosh apple at Wren's Market.
   c. food
   d. MacIntosh apples
   e. fruit
6. Why would an economist be more hesitant about making an elastic estimate of the effect of an increase in price of 1 percent than an increase in price of 50 percent?
7. Demand for top-ranked MBA programs is generally considered to be highly inelastic. What does this suggest about tuition increases for these programs in the future?
8. Once a book has been written, would an author facing an inelastic demand curve for the book prefer to raise or lower the book's price? Why?
9. Garage sale kits usually suggest that you mark all of your items with prices. Using your knowledge of price elastic-

ity of demand, can you suggest why you might do better by not putting prices on your items?

10. If a firm faces an elastic demand, it should hesitate to raise price. Lowering price, however, *possibly* increases profits (total revenue minus total cost). Why is the word *possibly* important?
11. For each of the following goods, state whether it is a normal good, a luxury, a necessity, or an inferior good. Explain your answers.
    a. Vodka.
    b. Table salt.
    c. Furniture.
    d. Perfume.
    e. Sausage.
    f. Sugar.
12. For each of the following pairs of goods, state whether the cross-price elasticity is likely positive, negative, or zero.
    a. Lettuce, carrots.
    b. Housing, furniture.
    c. Karaoke, fax machine.
    d. Jeans, formal suits.
13. If there were only two goods in the world, can you say whether they would be complements or substitutes? Explain your answer.
14. How is elasticity related to the revenue from a sales tax?

Problems and Exercises

1. A major cereal producer decides to lower price from \$3.60 to \$3 per 700 g box.
    a. If quantity demanded increases by 18 percent, what is the price elasticity of demand?
    b. What if, instead of lowering its price, the cereal producer had increased the size of the box from 700 g to 830 g? What would you expect that the response would have been? Why?
2. In the 1960s coffee came in 1-pound (454 g) cans. Today, most coffee comes in 300 g cans.
    a. Can you think of an explanation why?
    b. Can you think of other products besides coffee whose standard size has shrunk? (Often the standard size is supplemented by a "super-size" alternative.)
3. Using graph paper, and a proper scale on both axes, graph the demand curve for an almost (but not exactly) perfectly elastic demand.
    a. Calculate the slope of your demand curve.
    b. Calculate the percentage change in price and the percentage change in quantity.
    c. Why is the price elasticity at demand such a large number?
4. Economists William Hunter and Mary Rosenbaum wrote an article in which they estimated the demand elasticity for motor fuel to be between 0.4 and 0.85.
    a. If the price rises 10 percent and the initial quantity sold is 10 million gallons, what is the range of estimates of the new quantity demanded?
    b. In carrying out their estimates they came up with different elasticity estimates for rises in price than for falls in price, with an increase in price having a larger elasticity than a decrease in price. What hypothesis might you propose for their findings?
5. Use the price elasticity of demand concept to figure out what a constant-elasticity demand curve would look like.
6. Calculate the elasticity of the designated ranges of supply and demand curves on the following graph.

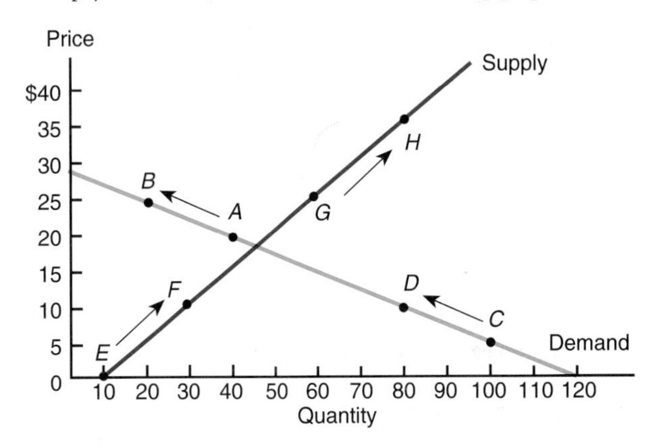

7. Calculate the price elasticities of the designated points on the following graph, using the mid-point formula. (Reread the box "Calculating Elasticity at a Point.")

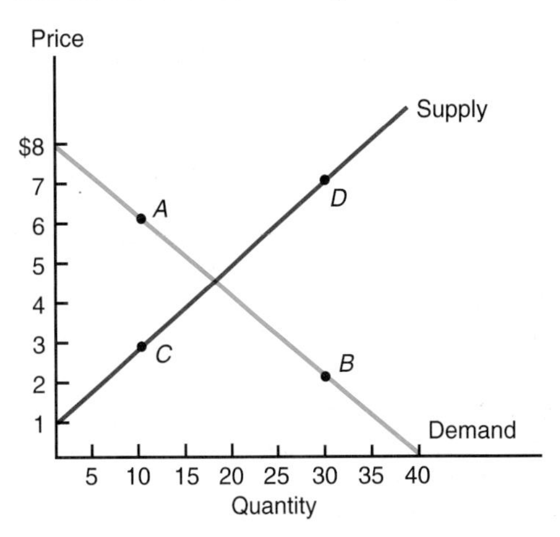

8. Calculate the income elasticities of demand for the following:
   a. Income rises by 20 percent; demand rises by 10 percent.
   b. Income rises from $30,000 to $40,000; demand increases (at a constant price) from 16 to 19.
   c. Income rises by 20%, consumption of potatoes drops from 18 to 12 bags a year.
9. Suppose the price of Good A rises from $9 to $15. In response, the quantity of Good B purchases falls from 25 to 23 units per week.
   a. Calculate and interpret the cross-price elasticity of demand.
   b. In the appropriate diagrams, illustrate what happens to both products.
10. In the 1990s the market price for live worms was $1.17 a dozen. Then a drought hit.
   a. Demonstrate graphically what happened to the price and quantity of worms sold.
   b. If the price rose to $1.75 and the quantity sold fell from 90,000 to 60,000, what would your estimate of the elasticity of demand be?

Web Questions

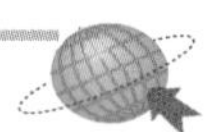

1. Everyone complains about the price of gasoline. Go to www.taxpayer.com and review the information on gasoline taxes. Answer the following questions regarding taxation:
   a. What is the average price of gasoline in your province? What proportion of the price represents tax?
   b. Which province has the highest gas taxes? which has the lowest? Can you explain why this is the case?
   c. What does the tax rate on gas suggest about the price elasticity of demand for gasoline? Would you agree?
   d. Who is most likely to be affected by gas tax increases?
   e. What is the long run price elasticity of demand for gasoline? Explain.
2. Go to www.ebay.ca
   a. What does this site do?
   b. Find an item fitting each of the following categories:
      1. Price elastic
      2. Income elastic
      3. Income inelastic
      4. Inferior
   c. Which category of good is most commonly represented?

Answers to Margin Questions

1. If price elasticity of demand is greater than 1, by definition demand is elastic. *(135)*
2. Using the mid-point formula, the percentage change in quantity is 100 (8/8 × 100) and the percentage change in price is 100 (2/2 × 100). Elasticity therefore, is approximately 1 (100/100). *(137)*
3. Referring to the price elasticity of demand formula, we are looking for

$$\frac{\text{percentage change in quantity demanded}}{\text{percentage change in price}} = 0.$$

   Demand is perfectly inelastic, so an x% change in price will cause no (zero) change in the quantity demanded, so it must be the case that for *any* percentage change in price, the percentage change in quantity demanded is 0. *(139)*
4. Many substitutes: Apple, T-shirt, mystery novel, and Ford Mustang all have *many* substitutes. Few, but close Substitutes: Coke and Pepsi are often thought to be close substitutes; however, any cola or any beverage could be a substitute. One of a few close substitutes for an Economics course at University would be an Economics course at College. *(142)*
5. Four factors affecting the number of substitutes in demand are (1) time interval considered, (2) whether the good is a luxury or a necessity, (3) how specifically the good is defined, and (4) the proportion of one's budget the price of the good is. *(142)*
6. If demand is inelastic, total revenue increases with an increase in price. *(145)*
7. We do not take the absolute value of the income elasticity of demand because knowing the sign is important. A positive value ($\eta > 0$) indicates a normal good, while a negative value ($\eta < 0$) indicates an inferior good. (It was the Law of Demand which allowed us to take the absolute value of the price elasticity of demand.) (147)
8. If consumption increases with an increase in income, the good is a normal good. *(147)*
9. Price elasticity of supply = Percentage change in quantity supplied divided by percentage change in price = 8/4 = 2. *(152)*
10. Supply is generally more elastic in the long run because there are more alternative goods and services for producers to produce. *(153)*

Taxation and Government Intervention

7

After reading this chapter, you should be able to:

- Demonstrate the cost of taxation to consumers and producers.
- Distinguish between the benefit principle and the ability-to-pay principle of taxation.
- Explain why the person who physically pays the tax is not necessarily the person who bears the burden of the tax.
- Demonstrate how an effective price ceiling is the equivalent of a tax on producers and a subsidy to consumers.
- Define rent seeking and show how it is related to elasticity.

Collecting more taxes than is absolutely necessary is legalized robbery.

Calvin Coolidge

Chapter 6 introduced elasticity and demonstrated how supply and demand analysis becomes even more powerful when combined with the concept of elasticity. This chapter uses the demand and supply framework to analyze two policy issues: taxation and government intervention in markets.

TAXATION AND GOVERNMENT

Each year, the Fraser Institute calculates and announces Tax Freedom Day, which identifies the day on which the average working Canadian has worked enough days to pay

Many jurisdictions rely on property taxes for their revenues.

his or her annual total tax bill. Due to tax cuts by both federal and various provincial governments, in 2001, the average Canadian stopped "working for the government" and started "working for himself" on June 28, five days earlier than in 2000, when Tax Freedom Day fell on July 4. So, in the first year of the millennium, Canadians can celebrate Canada Day tax free. All of the money you earn for the rest of the year is yours to spend or save as you wish.

Why is this day important? For the fiscal year 1999–2000, federal government revenues from income taxes were $108.3 billion, representing 62.8% of total revenue collected. Including other forms of tax (for example, sales tax), the total revenue from tax sources was $144.3 billion, accounting for 83.6% of the $172.5 billion in federal revenues collected. Although we recognize that government needs tax revenues to function, and we realize that the market also needs government to function, in most cases there is no direct link between payment of taxes and receipt of benefits, so we Canadians are not entirely convinced that the government spends our money as well as we would spend it. Tax freedom, however, doesn't necessarily translate into personal freedom. For example, public expenditures enhance our freedom to travel on roads and highways, to advance our knowledge in schools and universities and colleges, and to enjoy recreation in public parks and campgrounds. In calculating its date, the Fraser Institute doesn't measure the benefits that Canadians receive from government services, just the cost of providing these services.

Taxes are a necessary pain.

The government needs taxes to function, and the market likewise needs government. When governments do not have a well-functioning tax system, as is the case in some developing and transitional economies such as Russia's, the government is unable to provide the institutional structure markets need to work effectively. The connection between taxes and the roles of government led American jurist Oliver Wendell Holmes, Jr., to state that taxes are the price we pay for civilization.

Tax rates depend on what goods and services government provides. So, having more government-provided goods and services means having higher taxes. The taxes can be low if government plays a minor role in the market—simply providing an institutional framework—or they can be high if government plays a major role such as providing education and health care to all citizens. Figure 7-1 gives you a sense of tax rates in various countries.

Figure 7-1 HIGHEST TAX RATES ON WAGE INCOME (1998)

This figure shows the highest tax rates on wage income for various countries, including Canada. Notice that these figures only measure tax on income. Many U.S. states have low or no income tax (for example, Washington state), but instead collect most of their revenues from property and sales taxes, which are generally considered to be more regressive than income tax.

Source: OECD.

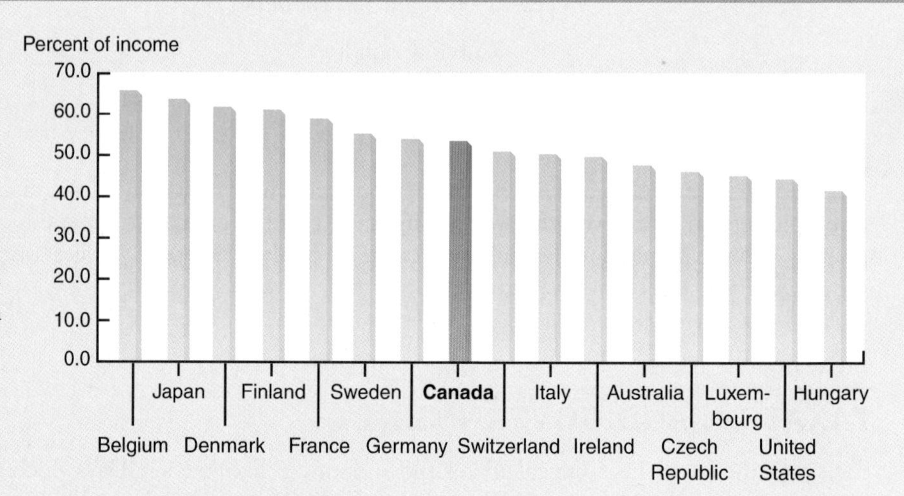

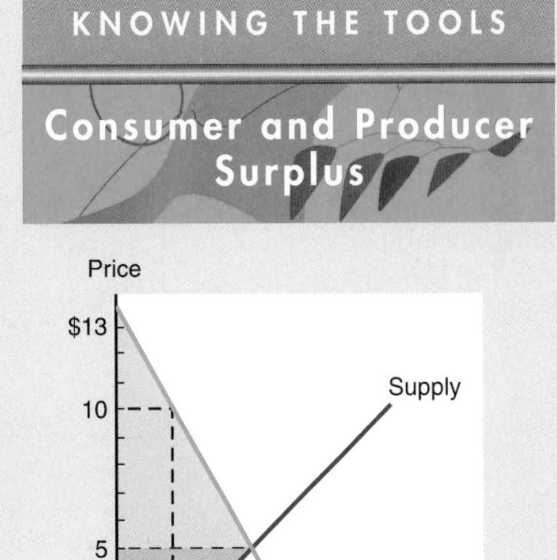

In a market most consumers pay less than they would have been willing to pay, and most producers receive more than they would have been willing to accept. The difference between what consumers would be willing to pay and what they do pay is **consumer surplus**; the difference between what producers receive and what they would be willing to receive is **producer surplus**.

Consumer surplus and producer surplus are shown graphically at the right. Consumers would be willing to pay $10 for the second unit, and suppliers would be willing to sell the second unit for $2.

The distance between the demand curve (what the consumer is willing to pay) and the price the consumer actually pays is consumer surplus. When price is at equilibrium ($5), consumer surplus is represented by the blue shaded area. The distance between the supply curve (what the supplier is willing to sell for) and the price the supplier receives is producer surplus. When price is at equilibrium ($5), producer surplus is represented by the red shaded area. At the market equilibrium price, the combination of consumer and producer surpluses is as large as it can be.

How Much Should Government Tax?

How should a country decide how much to tax? An economist would answer that question by asking another question: What are the costs and the benefits of taxation? The benefits are the gains to society that result from the goods and services government provides when fulfilling its roles in a market economy:

1. Provide a stable set of institutions and rules.
2. Promote effective and workable competition.
3. Correct for externalities.
4. Ensure economic stability and growth.
5. Provide public goods.
6. Adjust for undesired market results.

Public versus private provision: How do we pay for parks?

The Costs of Taxation The costs of taxation to society include the direct cost of the revenue paid to government, the loss of consumer and producer surplus caused by the tax, and administrative costs of collecting the tax. Figure 7-2 provides the basic framework for understanding the costs of taxation. For a given good, a per unit tax t paid by the supplier increases the price at which suppliers are willing to sell that good. The supplier now needs enough revenues to cover the cost of production plus the additional "cost" of the tax, so the supply curve shifts up by the amount of the tax, t, from S_0 to S_1. The equilibrium price of the good rises to P_1, and the quantity sold declines to Q_1.

A tax paid by the supplier shifts the supply curve up by the amount of the tax.

Before the tax, consumers pay P_0 and producers keep P_0, and consumer surplus is represented by areas $A + B + C$ and producer surplus is represented by areas $D + E + F$. With the tax t, equilibrium price rises to P_1 and equilibrium quantity falls to Q_1. Consumers now pay P_1, but producers only keep $P_1 - t$. Tax revenue paid equals the tax

Q-1 Why does equilibrium market price not rise by the full amount of the tax?

Figure 7-2 THE COSTS OF TAXATION

A per unit tax t paid by the supplier shifts the supply curve up from S_0 to S_1. Equilibrium price rises from P_0 to P_1, and equilibrium quantity falls from Q_0 to Q_1. Before the tax, consumer surplus is represented by areas A, B, and C, and after the tax by area A. Before the tax, producer surplus is represented by areas D, E, and F, and after the tax, by area F. Government collects tax shown by areas B and D. The tax imposes a deadweight loss, represented by the welfare loss triangle of areas C and E.

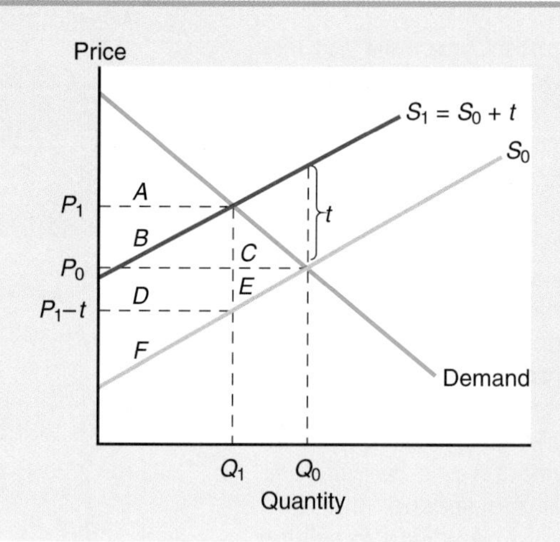

t times equilibrium quantity Q_1, or areas $B + D$, and consumer surplus equals area A only.

Q-2 Demonstrate the welfare loss of a tax when the supply is highly elastic and the demand is highly inelastic.

The total cost to consumers and producers, however, is more than the amount of the tax revenue. Consumers pay area B in tax revenue and also lose area C in consumer surplus. Producers pay area D in tax revenue and lose area E in producer surplus. The triangular area $C + E$ represents an additional cost of taxation in excess of the revenue paid to government. It is lost consumer and producer surplus that is not gained by government: It arises because the tax distorts price incentives. Producers, after accounting for the tax, receive price $P_0 + t$ for their product, and therefore reduce the quantity they supply. Buyers, on the other hand, see the market price rise to P_1 with the tax, and according to the Law of Demand, reduce their quantity demanded. This creates a deadweight loss equal to Areas $C + E$: Suppliers who previously supplied product to market (at price P_0) no longer do so (and these resources are reallocated to another use), and buyers who previously bought product (at price P_0) no longer do so. This lost opportunity to trade results in a deadweight loss which no one can claim. *The loss of consumer and producer surplus from a tax is known as* **deadweight loss.** Deadweight loss is shown graphically by the **welfare loss triangle**—*a geometric representation of the welfare cost in terms of misallocated resources caused by a deviation from a supply/demand equilibrium*. Keep in mind that with the tax, quantity sold declines. The loss of welfare therefore represents a loss for those consumers and producers who would have traded without the tax but do not trade with the tax.

Consumer surplus: the difference between what consumers were willing to pay and what they actually had to pay. It is the area under the Demand curve, above the equilibrium price.

Producer surplus: the difference between the price producers received and the price at which they were willing to supply the product. It is the area above the supply curve, below the equilibrium price.

There are additional cost of taxation. Resources must be allocated by government to collect the tax and by individuals to comply with it. Firms and individuals either spend hours filling out income tax forms or pay others to do so. Firms hire accountants and lawyers to take full advantage of any tax allowances. These types of administration costs can equal as much as 5 percent of the total tax revenue paid to government. Like the tax itself, these costs increase the price at which producers are willing to sell their goods, reducing quantity sold and further increasing welfare loss.

The Benefits of Taxation The benefits of taxes consist of the goods and services that government provides. Some government-provided goods are part of the basic institutional structure of a market economy and must be supplied if the market is to function effectively. The basic legal system is an example. Other goods are provided by government because they have the qualities of a public good; fire and police services are examples.

Still other goods are provided by government for reasons of equity or because they create positive externalities. Private markets may provide some of these goods, but perhaps not distribute them equitably or in sufficient quantities. Education is a good example; health care is another.

Measuring the benefits of government-supplied goods is difficult because they are often supplied at a zero price. In a properly-functioning market, the fact that someone will pay $5 for a good means that the value of the good to that person must be at least $5. All that can be said about publicly-supplied goods and services is that consumers value them at a price of at least $0. Because of this difficulty, along with the problem of deciding whether certain goods should be provided by government at all, the task of choosing which goods and services to provide and how to finance them is always surrounded by debate.

Two Principles of Taxation

In making decisions about taxes, government follows two general principles: the *benefit principle*, and the *ability-to-pay principle*.

The Benefit Principle The **benefit principle** of taxation follows the same principle as the market: *The individuals who receive the benefit of a good or service should pay the cost (opportunity cost) of the resources used to produce the good. In a market, that cost is represented by the price; for a public good, the cost is represented by the amount of the tax.* An example of a tax based on the benefit principle is the gasoline tax used to finance road construction. Since the users of gasoline are using the roads, there is a connection between use and payment; however, unlike a toll, the tax is not direct—people do not pay each time they use the road.

Another example is airport use taxes, which provide funding for the building of airports. Along with many airports in Canada and worldwide, the Vancouver Airport Authority charges passengers[1] departing Vancouver International Airport (YVR) a fee. The Airport Improvement Fee, as it is called, was imposed in 1992 in order to help fund construction of new facilities to serve growing air traffic volumes. YVR charges $5 for British Columbia and Yukon destinations, $10 for North American departures (including Hawaii and Mexico), and $15 for destinations outside of North America. Calgary, Edmonton, and Montreal also charge airport departure fees; many U.S. airports impose a passenger facility charge (PFC) of about CA$4.20 per airport (up to a maximum of four airports); and most major airports in Asia, Australia and New Zealand also levy a fee, in the range of CA$20 to CA$30.

The difference between paying for goods through taxes and paying for them in a market is that, with government-supplied goods, consumers do not directly pay for the goods and consequently may not reveal their preference for the good. The connection between benefit and cost in the decision to buy a publicly provided good is much less visible than it is with a privately supplied good, and this can create problems.

With government-supplied goods individuals indirectly reveal their preference for the good.

The Ability-to-Pay Principle The **ability-to-pay principle** of taxation ignores the connection between use of the good and payment for it. The ability-to-pay principle taxes people according to their ability to pay for the collective set of goods and services society desires. The Canadian income tax system employs the principle of ability-to-

Q-3 If the tax on the first $20,000 is zero and thereafter the tax is 20 percent, what average tax rate would a person earning $30,000 pay?

[1] Passengers on connecting flights and children under two years of age are exempt.

7.1 
see page 175

pay because it is a **progressive tax**, meaning that *the more income a person earns, the greater the amount of tax that is paid.* Tax collection appeals to society's standards of justice. In proposing progressive taxes, Canadians have implicitly embraced the notion that equity of distribution can be improved by redistributing incomes from one group of individuals (high income earners) to another (low income earners). In 2001, in the annual Ekos "Rethinking Government" poll, Canadians ranked health care, education, child poverty, and national debt reduction as more important than tax cuts. This parallels the economic finding that the market system does a relatively poor job of distributing incomes.

In an ability-to-pay tax regime, the amount of tax paid is independent of whether a particular individual would actually use any of the publicly-supplied goods and services. However, although an individual may not travel by bus to work every day — and thus perhaps be unwilling to support bus service through his or her taxes — the individual will benefit indirectly from bus service as likely there are fewer cars on the road during rush hour, making his or her own commute easier. *The individuals who are most able to bear the burden of the tax should pay the tax.* Generally this principle is interpreted as supporting a progressive income tax—one whose rate increases with increases in ability to pay.

The reasoning behind the ability-to-pay principle is that similar tax rates represent a smaller sacrifice for those with higher incomes compared to those with lower incomes. The wealthy should pay more because they can. Since the ability-to-pay principle of taxation makes no attempt to relate the benefit to the tax, the income tax revenue is used to finance a wide variety of government activities such as health care, education, welfare, and defense.

The two principles of taxation, the benefit principle and the ability-to-pay principle, are often in conflict with each other.

Difficulty of Applying the Principles of Taxation The two principles of taxation discussed above are the ones that economists and policymakers use most often to assess tax structures. Unfortunately, they are not easy to apply and often conflict. For example, benefits of the goods that government supplies are often difficult to assign, making the benefit principle ambiguous.

When taxes are based on the ability-to-pay principle, many high-income individuals ask why they should pay disproportionately more for goods such as income-support programs that are benefiting others, not them. The conflict between these two principles, combined with people's general desire to have "the other person" pay the taxes while they receive the benefits, often leads to significant debates about what and how to tax. Thus, we see debates about whether education should be financed by a *property tax*—a tax on houses and land—or by a *sales tax*—a tax on goods and services sold to consumers. Property and sales taxes are generally considered to be **regressive taxes** since they *impose a proportionally greater cost on lower income people*. Should income be taxed, or should only specific goods, like gasoline, be taxed? Should we use a corporate income tax or the general income tax?

The elasticity concept helps provide insight into the above debates. Remember that the more broadly the good is defined, the more inelastic is the demand. Thus, the demand for all goods is much more inelastic than the demand for a particular good, since individuals cannot switch their consumption out of all goods but can switch from one specific good to another. This means that if government wants to have as little effect on individual actions as possible—or, in the language of consumer and producer surplus, if government wants to minimize the welfare loss—then it should tax goods with inelastic supplies or demands. Broad-based taxes such as income and sales taxes accomplish this. Most countries use a broad-based income tax, value-added tax, or general sales tax

If the government wants to minimize the deadweight loss it should tax goods with inelastic supplies and demands.

The issue of taxation has been debated for centuries. Over a hundred years ago, an early political economist, John Stuart Mill (1806–1873), analyzed the relationships between justice, social utility, and taxation. In his famous essay on political economy, *Utilitarianism*, published in 1861, Mill outlined the basic standards of justice that guide society's implementation of its tax system. The scope of issues included in his analysis is much broader than we usually see in economic analysis today. Here is an excerpt:

> One opinion is that payment to the state should be in numerical proportion to pecuniary means. Others think that justice dictates what they term graduated taxation — taking a higher percentage from those who have more to spare. In point of natural justice a strong case might be made for disregarding means altogether, and taking the same absolute sum (whenever it could be got) from everyone; as the subscribers to a mess or to a club all pay the same sum for the same privileges, whether they can all equally afford it or not. Since the protection (it might be said) of law and government is afforded to and is equally required by all, there is no injustice in making all buy it at the same price. It is reckoned justice, not injustice, that a dealer should charge to all customers the same price for the same article, not a price varying according to their means of payment. This doctrine, as applied to taxation, finds no advocates because it conflicts strongly with man's feeling of humanity and of social expediency; but the principle of justice which it invokes is as true and as binding as those which can be appealed to against it.

Taxation has always been a complex issue, and how we Canadians resolve the questions of what and how to tax is related to the same sense of community and justice that also guides most of our other collective decisions.

as their primary source of tax revenue because it produces the smallest distortions in the markets, and therefore creates the fewest incentives for people to adjust their behaviour.

Who Bears the Burden of a Tax?

Generally we don't like to pay taxes, and there are usually political debates about whom government should tax. For example, should Employment Insurance premiums be paid by workers or by the company that hires them? Or does it matter? The supply and demand framework gives an unexpected answer to this question.

Tax Burden Depends on Relative Elasticity An **excise tax** is *a tax levied on a specific good*. An excise tax can be levied on the consumer or on the seller; this means that either the buyer or the seller will be responsible for "paying" the tax.

The person who *physically pays* the tax, however, is not necessarily the person who *bears the burden* of the tax. The **tax incidence** (*who bears the burden of the tax*) depends on who can most easily change their behaviour in response to the tax. This is the individual who has the greater elasticity.

Figure 7-3(a) (on the next page) demonstrates the incidence of a per unit tax. A $10,000 per unit tax levied on the supplier means that the supplier would need to charge an additional $10,000 per unit in order to cover the cost of production plus the (new) tax, so the supply curve shifts up from S_0 to S_1. That reduces quantity supplied and quantity demanded by 90—from 600 to 510. Is the firm able to place the entire burden of the tax on the consumer? The interaction of supply and demand yields a new market equilibrium quantity of 510 units traded, and a new market price of $65,000. Suppliers are able to shift $5,000 of the total $10,000 per unit tax onto consumers, leaving the suppliers the burden of the remaining $5,000.

Since the demand and supply elasticities at the market equilibrium are identical, the tax burden will be shared equally. Specifically, suppliers sold and consumers purchased 90 fewer boats, a 15 percent reduction. The suppliers' price fell by about 8 percent while the consumers' price rose by about 8 percent, meaning the elasticity of both supply and demand was approximately 1.9.[2] With equal elasticities, the tax burden will be divided equally.

Q-4 Show the tax burden in the case where the demand is elastic and the supply is inelastic.

In reality the tax burden is rarely shared equally because elasticities are rarely equal. The burden of the tax follows this general rule: *The relatively more inelastic is supply or demand, the larger the portion of the tax burden one will bear.* If demand were more inelastic, sellers would have been able to sell the boats at a higher price and could have passed more of the tax along to the buyers.

Tax burden is distributed by relative elasticities.

Figure 7-3(b) shows what the divisions would have been had the demand curve been highly inelastic. In this case the price would rise more, the supplier would pay a lower proportion of the tax (the red area), and the consumer would pay a much larger proportion (the blue area). The general rule about elasticities and the tax burden is this: If demand is more inelastic than supply, then consumers will pay a higher percentage of the tax; if supply is more inelastic than demand, then suppliers will pay a higher share. This rule makes sense—*elasticity is a measure of how easy it is for the supplier or the consumer to change behaviour and substitute another good.* Table 7-1 shows who bears the burden of tax in Figure 7-3 (a) and (b).

Q-5 Illustrate the tax incidence for these products:
a. cigarettes
b. perfume
c. food

This can lead to some unexpected consequences of taxation. For example, a luxury tax on boats might be implemented as a way to tax the wealthy. However, their demand is relatively elastic. They can purchase other luxury items or purchase boats from foreign

[2]There will be slight variations in the measured elasticities depending on how they are calculated. The precise equality holds only for point elasticities, and we are using mid-point elasticities (see Chapter 6).

Figure 7-3 (a, b, and c) WHO BEARS THE BURDEN OF A TAX

In the general case, tax incidence is determined by the relative elasticities of supply and demand. In (a) and (c), demand and supply elasticities are identical. The blue shaded area shows the burden on the consumer; the red shaded area shows the burden on the supplier. This split occurs regardless of who actually pays the tax, as can be seen by noticing that the burden of the tax is equal in (a) where the supplier pays the tax and in (c) where the consumer pays the tax. In (b) you can see how consumers with an inelastic demand bear a greater burden of the tax.

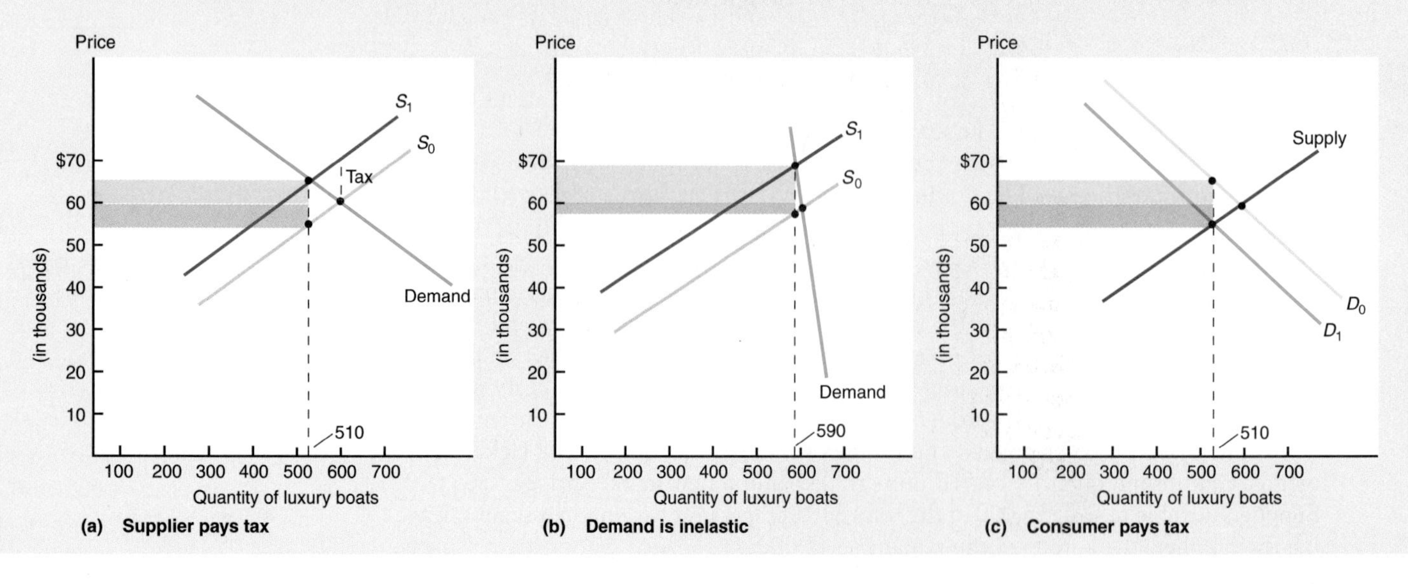

firms. Domestic boat manufacturers' supply is inelastic. As a result, if they tried to pass on the cost increase to consumers, their sales would plummet. They would have to lower their price by almost as much as the tax, which means that they would bear most of the burden of the tax.

High substitution in the demand for luxury goods.

TABLE 7-1 Tax Incidence

Although the total tax paid is identical, the burden of the tax depends on the relative elasticities of the demand and supply curves.

	BEFORE TAX: PRICE	AFTER TAX: PRICE	TOTAL TAX PAID
FIGURE 7-3(a)			
Buyer	$60,000 (paid)	$65,000 (paid)	$5,000 (by buyer)
Seller	$60,000 (received)	$55,000 (received)	$5,000 (by seller)
			$10,000
FIGURE 7-3(b)			
Buyer	$60,000 (paid)	$69,000 (paid)	$9,000 (by buyer)
Seller	$60,000 (received)	$59,000 (received)	$1,000 (by seller)
			$10,000

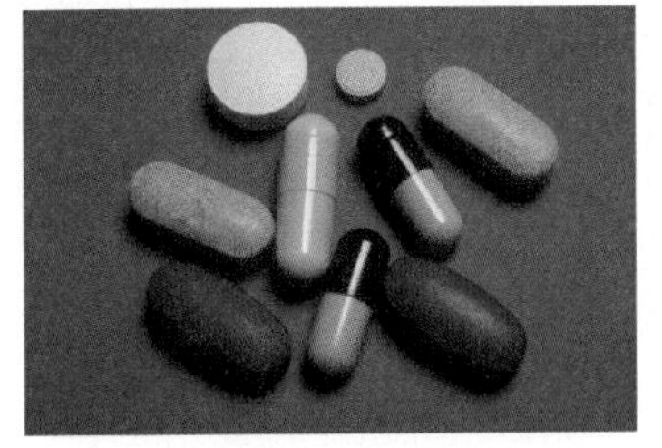

Demand for life-saving medicines is inelastic. Should we tax these?

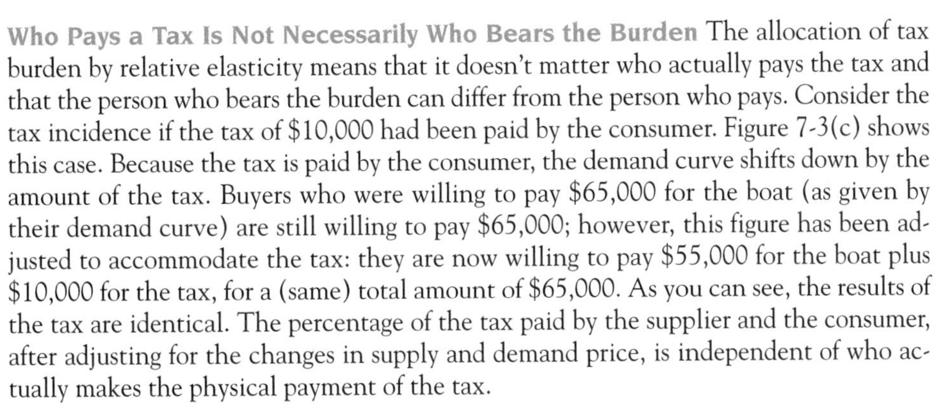

Who Pays a Tax Is Not Necessarily Who Bears the Burden The allocation of tax burden by relative elasticity means that it doesn't matter who actually pays the tax and that the person who bears the burden can differ from the person who pays. Consider the tax incidence if the tax of $10,000 had been paid by the consumer. Figure 7-3(c) shows this case. Because the tax is paid by the consumer, the demand curve shifts down by the amount of the tax. Buyers who were willing to pay $65,000 for the boat (as given by their demand curve) are still willing to pay $65,000; however, this figure has been adjusted to accommodate the tax: they are now willing to pay $55,000 for the boat plus $10,000 for the tax, for a (same) total amount of $65,000. As you can see, the results of the tax are identical. The percentage of the tax paid by the supplier and the consumer, after adjusting for the changes in supply and demand price, is independent of who actually makes the physical payment of the tax.

The burden is independent of who physically pays the tax.

Tax Incidence and Current Policy Debates

Now let's consider two policy questions in relation to what we have learned about tax incidence.

Employment Insurance Premiums In 2001, an employee's Employment Insurance premiums were 2.25% of his or her insurable earnings, up to a maximum of $877.50 per year. Employers paid 1.4 times the amount paid by the employee; however, the fact that the larger share of the tax was paid by employers does not mean that the burden of the tax lay with the employer. On average, labour supply tends to be less elastic than labour demand. This means that the burden falls primarily on the employees, even though employees see only their own statutory portion of the EI premium on their pay stub.

Q-6 If Employment Insurance premiums were paid only by employees, what would likely happen to workers' pretax pay?

Suppose the government were to place the entire tax on the employer and eliminate the tax on the employee. What will be the effect? Our tax incidence analysis tells us that, ultimately, it will have no effect. Wages paid to employees will fall to compensate employers for the cost of the tax. This example shows that who is assessed the tax can be quite different than who actually bears the burden, or incidence, of the tax. The truth is that, no matter who is "responsible" for the tax, the burden will be borne by

APPLYING THE TOOLS

What Goods Should Be Taxed?

What goods should be taxed depends on the goal of government. If the goal is to fund a program with as little loss as possible in consumer and producer surplus, then the government should tax a good whose supply or demand is inelastic. If the goal is to change behaviour, taxes will be most effective if demand or supply is elastic. As a quick review, use the following table:

Goal of Government	Most Effective When
Raise revenue, limit deadweight loss	Demand or supply is inelastic
Change behaviour	Demand or supply is elastic

Distributional issues must also be considered when determining what goods are to be taxed. In general, the group with the relatively more inelastic supply or demand will bear a greater portion of the tax. The following table reviews these conclusions:

Elasticity	Who Bears the Burden?
Demand inelastic and supply elastic	Consumers
Supply inelastic and demand elastic	Producers
Both supply and demand elastic	Shared; but the group whose supply or demand is more inelastic bears more

those with the most inelastic supply or demand, because they have no way of substituting out of paying the tax.

Figure 7–4 illustrates the result. Firm's demand curve for labour shifts down by the amount of the tax, creating a "wedge" between the wage the employer sees and the wage the employee sees. Employers see the market equilibrium wage rises to W_F and the quantity of labour employed falls to L_2. Now the firm pays W_F with the following breakdown: W_F - W_L represents the payment for the EI premium, and workers take home W_L. With an inelastic labour supply, the majority of the EI "tax" is "paid" by the employees, even though the employer (firm) is responsible for paying it.

Figure 7-4 BURDEN OF THE EMPLOYMENT INSURANCE PREMIUM

When the employer is responsible for Employment Insurance premiums, their demand for labour will shift from D_0 to D_1. The employer "pays" W_F in total, with $W_F - W_L$ representing the amount of the EI premium, and W_L representing the worker's actual wage. The total EI premium paid equals $(W_F - W_L)$ times L_2, and is borne largely by the employees.

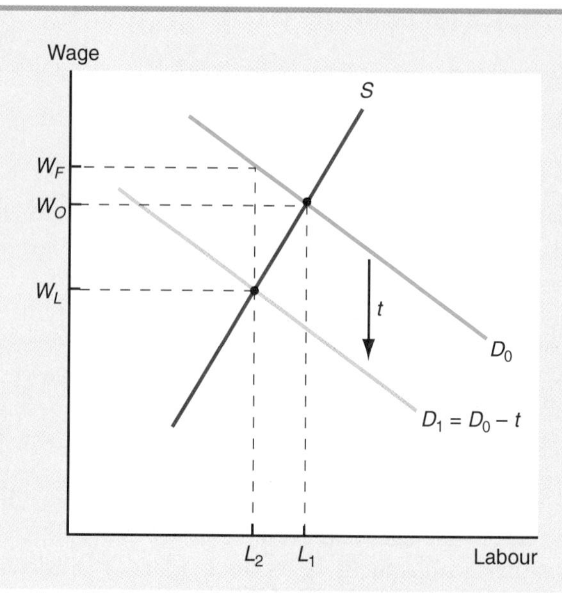

Although economically it will not make a difference who pays the tax, politically it may be a popular proposal because individuals generally look at statutory assessment, not incidence. Politics often focusses on surface appearance; what is good politics is not always good economics.

What makes sense politically does not always make sense economically.

Sales Taxes Our second policy question concerns sales taxes paid by retailers on the basis of their sales revenue. Since sales taxes are broadly defined, most consumers have little ability to substitute. Demand is inelastic and consumers bear the greater burden of the tax. Although stores could simply incorporate the tax into the price of their goods, most stores add the tax onto the bill after the initial sale is calculated, to make you aware of the tax. Again, it doesn't matter whether the tax is assessed on the store or on the customer.

Most Canadians spend almost 100% of their disposable incomes in Canada, and consequently have very little ability to substitute out of goods and services which are taxed. Many wealthy Canadians, however, spend part of the year outside Canada in regions where there is little or no tax on consumption. They are able to substitute away from some consumption taxes. However, with the growth of the Internet, many Canadians now are able to access tax-free jurisdictions (cyberspace), and thus substitute away from Canadian consumption taxes also.

GOVERNMENT INTERVENTION

Taxes are not the only way government affects our lives. For example, government establishes laws that dictate what we can do, what prices we can charge for goods, and what working conditions are and are not acceptable. The concept of elasticity can help us evaluate these interventions. Using the producer and consumer surplus framework, such interventions can be seen as a combination of tax and subsidy that does not show up on government books.

Government Intervention as Implicit Taxation

To see how government intervention in the market can be seen as a combination tax and subsidy, consider two types of price controls: price ceilings and price floors.

Price Ceilings and Floors A **price ceiling** is *a government-set maximum price which the market price cannot exceed. Generally, the price ceiling is set below the market equilibrium price*. It is in essence an implicit tax on producers and an implicit subsidy to consumers. Consider the effect of a price ceiling on producer and consumer surplus, shown in Figure 7-5(a) on the next page.

Q-7 Demonstrate the impact of an effective price ceiling on producer and consumer surplus when both supply and demand are highly inelastic.

If the price were at the market equilibrium price, P_0, the total surplus would be the combination of the areas A through F (consumer surplus = A + B + C, and producer surplus = D + E + F). But with an effective price ceiling P_1, the quantity supplied falls from Q_0 to Q_1. The combined producer and consumer surplus is reduced by triangles C and E. The loss of surplus represents those individuals who would like to make trades—the individuals represented by the demand and supply curves between Q_1 and Q_0—but cannot do so because of the price ceiling.

This loss of consumer and producer surplus is identical to the welfare loss from taxation. That is not a coincidence. The price ceiling is a combination implicit tax on suppliers, shown by area *D*, and implicit subsidy to consumers of that same area. It is as if government places a tax on suppliers when they buy the good, and then gives that tax

Figure 7-5 (a and b) EFFECT OF PRICE CONTROLS ON CONSUMER AND PRODUCER SURPLUS

Price floors and price ceilings create deadweight loss just as taxes do. In (**a**) we see how a price ceiling, P_1, transfers surplus D from producers to consumers. Price ceilings are equivalent to a tax (worth D) on producers and a subsidy (worth D) to consumers. In (**b**) we see how a price floor, P_2, transfers surplus B from consumers to producers. With the price floor and with the price ceiling, areas C and E represent the deadweight loss.

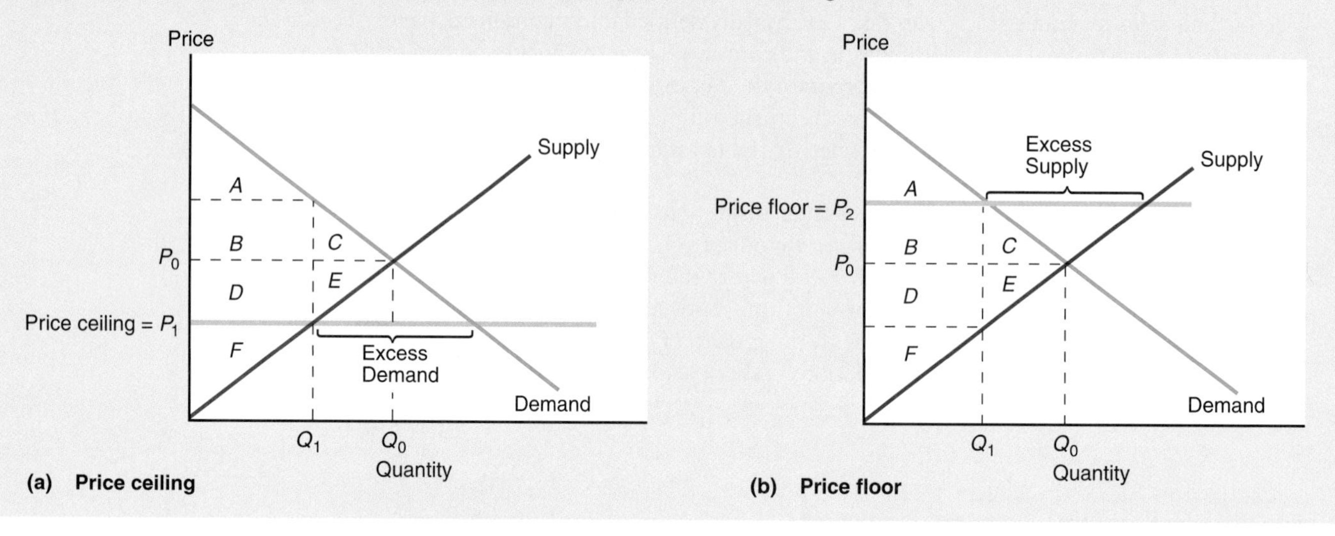

A price ceiling is a combination implicit tax on suppliers and implicit subsidy on consumers.

revenue to consumers when they consume the good. Q_1 is the amount bought under the price ceiling, P_1. Previously, consumers paid P_0 for these units, but since they now only pay P_1 for them, there is a transfer of surplus from sellers (the "tax") to buyers (the "subsidy").

Price floors have the opposite effect on the distribution of consumer and producer surplus. **Price floors**—*government-set minimum prices*—transfer surplus from consumers to producers. Price floors can be seen as a tax on consumers of area B and a subsidy to producers of that same area, as shown in Figure 7-5(b). Price floors also create a deadweight loss, shown by the welfare loss triangle, areas C and E.

Price ceilings create shortages; taxes do not.

The Difference between Taxes and Price Controls While the effects of taxation and controls are similar, there is an important difference: *Price ceilings create shortages; taxes do not*. The reason is that taxes leave people free to choose how much they want to supply and consume as long as they pay the tax. Thus, taxes create a wedge between the price the consumers pay and the price the suppliers receive. That difference is just enough to equate quantity demanded with quantity supplied.

Since with price ceilings the price consumers pay equals the price suppliers receive, the desired quantity demanded will not be equal to the desired quantity supplied since the price is not the market equilibrium price. Only at the market equilibrium price will quantity demanded equal quantity supplied. At all other prices (including ceiling or floor prices), there will be an imbalance between the quantity demanded and the quantity supplied ($Q^D \neq Q^S$). In these cases where the price mechanism does an incomplete job, an alternative method of allocation must be found. In the case of price ceilings, alternative distribution methods such as line-ups (first-come, first-served), lotteries, and ration coupons may be used. In some instances, the shortage causes a Black Market to develop, in which the scarce good is re-sold in a (generally illegal) secondary market at a higher price.

Rent Seeking, Politics, and Elasticities

If price controls reduce total producer and consumer surplus, why do governments institute them? The answer is that *people care more about their own surplus than they do about total surplus*. Price ceilings redistribute surplus from producers to consumers, so if the consumers hold the balance of political power, there will be strong pressures to create price ceilings. Alternatively, if the suppliers hold the political power, there will be strong pressures to create price floors.

The possibility of transferring surplus from one set of individuals to another causes people to devote time and resources to this activity. Individuals have a strong incentive (potential redirection of available surplus to them) to spend resources lobbying government to institute policies that increase their own surplus. Others have an equally strong incentive to spend resources to counteract those lobbying efforts. *Activities designed to transfer surplus from one group to another* are called **rent-seeking activities.** Rent-seeking expends valuable resources in unproductive activities. **Public choice economists**—*economists who integrate an economic analysis of politics with their analysis of the economy*—argue that resources spent on rent seeking are significant, and that much of the transfer of surplus that occurs through government intervention represents an enormous waste of resources. They argue that the taxes and the benefits of government programs offset each other and do not help society significantly, but they do cost resources. These economists point out that much of the redistribution through government is not from upper income to lower income groups, but is from one group of the middle class to another group of the middle class.

Q-8 Would a firm's research and development expenditures be classified as rent seeking?

see page 175

Q-9 How can an increase in productivity harm suppliers?

Inelastic Demand and Incentives to Restrict Supply To understand the rent-seeking process a bit better, let's look more carefully at the incentives that consumers and producers have to lobby government to intervene in the market. We'll begin with suppliers. A classic example of the political pressures to limit supply is found in agricultural markets. Within the past century new machinery, new methods of farming, and hybrid seeds have increased the productivity of farmers tremendously. You might think that because farmers can now produce more at a lower cost, they'd be better off. Consider Figure 7-6. As advances in productivity increase supply, they do increase the quantity sold, but they also result in lower prices. This is the "competitive result" of markets. Price is bid down to its lowest possible level, indicated by the (new) supply curve in Figure 7-6. The previous lowest possible price was given by the old supply curve. Because food is a necessity and has few substitutes, the demand for many agricultural goods is

Figure 7-6 INELASTIC DEMAND AND THE INCENTIVE TO RESTRICT SUPPLY

When demand is inelastic, increases in productivity that shift the supply curve to the right result in lower revenue for suppliers. Although suppliers gain area *B*, they lose the much larger area A. Suppliers have an incentive to restrict supply and raise price when demand is inelastic because, by doing so, they will increase their revenues.

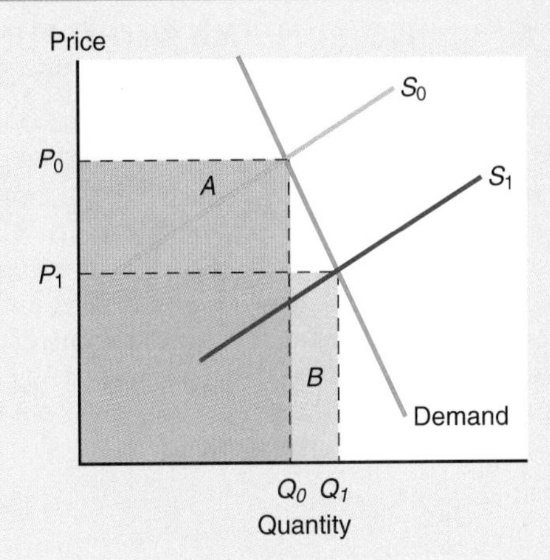

Technology has changed the fortunes of Canadian farmers.

inelastic. Since demand is inelastic, the price declines by a greater proportion than the quantity sold increases, so total revenue declines and the farmers are actually worse off.

Because of the increase in supply, price decreases from P_0 to P_1 and quantity sold increases from Q_0 to Q_1. The farmer's revenue rises by area B but falls by the larger area A. To counteract this trend, farmers have an incentive to encourage government to restrict supply or create a price floor, thereby raising their revenue. The benefits of limiting competition are greatest for suppliers when demand is inelastic because price will rise proportionately more than quantity will fall.

This simple example provides us with an important insight about how markets work and how the politics of government intervention work. Inelastic demand creates an enormous incentive for suppliers to pressure government to limit the quantity supplied or for suppliers to get together to find ways to limit the quantity supplied. The more inelastic demand is, the more suppliers have to gain by restricting supply.

The more elastic supply and demand, the larger the surplus or shortage created by price controls.

Long-Run Problems of Price Controls In the long run, both supply and demand tend to be much more elastic than in the short run. This means that price controls will cause only small shortages or surpluses in the short run, but large ones in the long run. Consider the rent-control example in Figure 7-7. In the short run, both supply and demand are inelastic. Without price controls, if demand shifts from D_0 to D_1, price (rents) will rise significantly, from P_0 to P_1.

However, in the long run, additional apartments will be built and other existing buildings will be converted into apartments. Long run supply (S_1) is much more elastic. Faced with additional competition, landlords will lower their price to P_2. In the long run price will fall, (rise by less than in the short run) and the available number of apartments will increase.

Q-10 Why do price controls tend to create ongoing shortages or surpluses in the long run?

The economics of the situation is clear; the politics is not so clear. In large part, it is the rise in price that brings in new competitors and increases in supply. But if the government imposes price controls, the long-run incentives for competitors (new suppliers) to enter the market will be eliminated. The political problems arise because politics generally responds to short-run pressures. In the short run, the price ceiling will not create significant problems. But in the long run, landlords will convert their rent-controlled apartments to different uses, fewer resources will be spent keeping up existing apartments, and fewer new apartments will be built. In the long run, the shortage becomes even more severe. The government's imposition of a price ceiling prevents the market from achieving a more desirable long-run equilibrium, in which output has expanded and price has fallen from its initially high level.

Figure 7-7 LONG-RUN AND SHORT-RUN EFFECTS OF PRICE CONTROLS

This exhibit shows how an increase in demand from D_0 to D_1 raises equilibrium price from P_0 to P_1. As time progresses, supply becomes more elastic, which is shown as the long run supply curve, S_1. Equilibrium price falls to P_2. If a price ceiling of P_0 had been imposed, the incentive for suppliers to build more apartments would be eliminated. Instead, landlords would convert their apartments to different uses and potential landlords would choose to build fewer apartments, and the shortage resulting from the price control ($Q_3 - Q_0$) would remain.

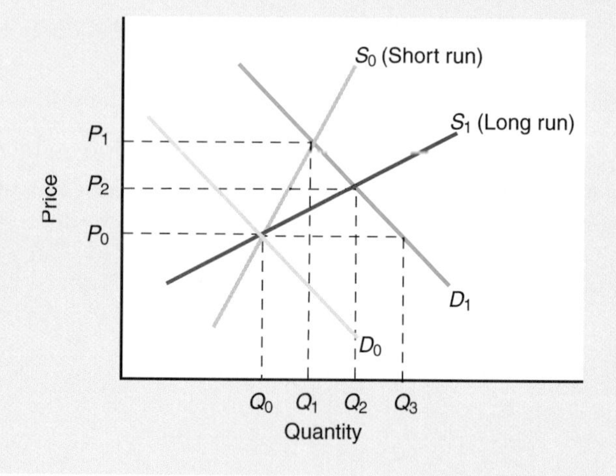

CONCLUSION

Government is a part of our life. Economic theory doesn't say government should or shouldn't play any particular role in the economy or what the taxes should be. Those decisions depend on normative judgments of the society on the relevant costs and benefits. What economic theory does is to identify the costs and benefits. For example, in the case of taxes, economists can show that the cost of taxation in terms of lost surplus is independent of who actually pays the tax.

The central problem of political economy is how to evaluate the role of government in society. Government plays an important role in our society, providing services, distributing income, and providing the infrastructure necessary for markets to operate. This infrastructure would include such items as a system of laws and a police and judicial system to enforce those laws, rules for the economy's financial structure, and physical structures like roads and bridges. Government provides many public goods that enhance the market's productive ability, such as education and health care. Government also intervenes in markets to bring about certain desired outcomes. One example of such intervention is rent control. However, rent controls distort market incentives, and cause us to question the desirability of government intervention in the housing market, and in markets in general. Taxation is another issue. Taxes put a "wedge" between the price paid by consumers and the price received by suppliers, and create a deadweight loss.

This chapter illustrates how elasticity enhances the economic analysis. Although economists tend to suggest that freely functioning markets should be left to operate without interference, it is not always clear that market incentives alone will provide the outcome society would prefer in each case. In many cases, economics presents an accurate, reliable analysis of an issue, but the questions that need answering concern our desires for Canadian society as a whole. These questions lie in the realm of political economy.

Chapter Summary

- Government taxes firms and individuals in order to carry out its roles in a market economy. A government will maximize benefits to society only if it chooses to tax when the marginal benefit of the goods and services provided with the revenue of the tax exceeds the cost of the tax.
- Taxes create a loss of consumer and producer surplus known as deadweight loss, which graphically is represented by the welfare loss triangle.
- The cost of taxation to consumers and producers includes the actual tax paid, the deadweight loss, and the cost of administering the tax.
- Government follows both the benefit principle (individuals who receive the benefit should pay the tax) and the ability-to-pay principle (individuals who are most able to pay should pay the tax) when deciding how to levy taxes.
- Who bears the burden of the tax depends on the relative elasticities of demand and supply. The more inelastic one will bear the larger the burden of the tax.
- Although Employment Insurance premiums are levied on both employers and employees, the supply of labour tends to be more inelastic than the demand for labour, so workers bear the greater burden of these taxes.
- Price ceilings and price floors, like taxes, result in loss of consumer and producer surplus.
- Price ceilings transfer surplus from producers to consumers and therefore are equivalent to a tax on producers and a subsidy to consumers. Price floors have the opposite effect; they are a tax on consumers and a subsidy to producers.
- Rent-seeking activities are designed to transfer surplus from one group to another. Rent-seeking activities are generally considered to be directly unproductive uses of (scarce) resources, which sometimes yield a net result of zero as efforts at rent-seeking exactly offset each other.
- The effects of price controls worsen over time because the incentives for market adjustment have been distorted.

Key Terms

ability-to-pay principle *(163)*
benefit principle *(163)*
consumer surplus *(161)*
deadweight loss *(162)*
excise tax *(165)*
price ceiling *(169)*
price floor *(170)*
producer surplus *(161)*
progressive tax *(164)*
public choice economists *(171)*
regressive tax *(164)*
rent-seeking activities *(171)*
tax incidence *(165)*
welfare loss triangle *(162)*

Questions for Thought and Review

1. Name one local tax that is based on the ability-to-pay principle and one local tax that is based on the benefit principle. State your reason for categorizing the taxes as you did.
2. How is elasticity related to the revenue from a sales tax?
3. If the federal government wanted to tax a good and suppliers were strong lobbyists, but consumers were not, would government prefer supply or demand to be more inelastic? Why?
4. What types of goods would you recommend government tax if it wants the tax to result in minimum welfare loss? Name a few examples.
5. Suppose demand for cigarettes is inelastic and the supply of cigarettes is elastic. Who would bear the larger burden of a tax placed on cigarettes?
6. If the demand for a good is perfectly elastic and the supply is elastic, who will bear the burden of a tax on the good paid by consumers?
7. Which good would an economist normally recommend taxing if government wanted to minimize welfare loss and maximize revenue: a good with an elastic or inelastic supply? Why?
8. Should tenants who rent apartments worry that increases in property taxes will increase their rent? Does your answer change when considering the long run?
9. Can you explain the tax system that led to this building style, which was common in old Eastern European cities?

10. In which case would the shortage resulting from a price ceiling be greater—when supply is inelastic or elastic? Explain your answer.
11. Define rent seeking. Do firms have a greater incentive to engage in rent-seeking behaviour when demand is elastic or when it is inelastic?

Problems and Exercises

1. A political leader comes to you and wonders from whom she will get the most complaints if she institutes a price ceiling when demand is inelastic and supply is elastic.
   a. How do you respond?
   b. Demonstrate why your answer is correct.
2. Suppose the government established a requirement that everyone consume 10 percent more beets than he or she is currently consuming.
   a. Show graphically the welfare loss that would occur.
   b. If someone shows you that the welfare loss is small, say 0.5 percent of the cost of beets, and that eating beets improves people's health, would you support a beet-eating requirement? Why or why not?
3. Demonstrate the welfare loss of:
   a. A restriction on output when supply is perfectly elastic.
   b. A tax *t* placed on suppliers that shifts up a supply curve.
   c. A subsidy *s* given to suppliers that shifts down a supply curve.
   d. A restriction on output when demand is perfectly elastic.
4. Because of the negative incentive effect that taxes have on goods with elastic supply, in the late 1980s Margaret Thatcher (then prime minister of Great Britain) changed the property tax to a poll tax (a tax at a set rate that every individual must pay).
   a. Show why the poll tax is preferable to a property tax in terms of consumer and producer surplus.
   b. What do you think the real-life consequences of the poll tax were?

5. Demonstrate how a price floor is like a tax on consumers and a subsidy to suppliers.
   a. Who gets the revenue in the case of a tax? Label the area that illustrates the tax.
   b. Who gets the revenue in the case of a price floor? Label the transfer of surplus from consumers to suppliers.
   c. Label welfare loss of the tax and the price floor.
6. Suppose government imposed a minimum wage above equilibrium wage.
   a. What do you expect to happen to the resulting shortage of jobs as time progresses? (Assume that inflation and economic growth are both zero.)
   b. What do you expect to happen to the producer surplus transferred to minimum-wage earners as time progresses?
7. New York City has issued 11,797 taxi licenses, called "medallions," and has not changed that number since 1937.
   a. What does that limitation likely do to the price of taxi medallions? Demonstrate graphically.
   b. If the New York City Taxi Commission eliminated a rule that required medallion owners who own only one cab to drive their cabs full time, what would likely happen to the price of the medallion?
   c. If New York City decreased the number of medallions by 1,000 by revoking 1,000 medallions, what would happen to the price of remaining medallions?
   d. What would the political response to such a revocation be?

Web Questions

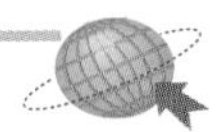

1. Alberta has a plan to eliminate provincial personal income tax within forty years. Go to www.taxpayer.com and find an article by Jean-Francois Wen entitled "Eliminating Albert's Personal Income Tax."
   a. What is the Alberta Heritage Savings Trust Fund? what is its main source of funding? What is the value of the fund?
   b. How do Albertans propose to eliminate provincial personal income tax? What role does the Heritage Fund play?
   c. What are three critical assumptions made in calculating how soon Alberta can eliminate income tax? Briefly describe some of the issues.
   d. What effect would the elimination of provincial personal income tax have in Alberta? In Canada?
2. Henry George, an American economist living in the 1800s, had very specific views on taxation. Read about these by clicking on *taxation* on the Henry George Institute's home page at www.henrygeorge.org and answer the following questions:
   a. What are the four criteria of a good tax according to Classical economists?
   b. In what way do broad-based taxes fulfill those criteria? In what way do they fail to fulfill those criteria?
   c. What good in our economy, according to Henry George, best fits the four criteria?
   d. Demonstrate the welfare loss associated with the taxation of the good Henry George believes should be taxed. Who will bear the burden of the tax?

Answers to Margin Questions

1. According to the law of Demand, as price rises (due to the tax), consumers reduce their quantity demanded. If suppliers were to raise price by the full amount of the tax (to $P_0 + t$), an excess supply (E.S.) would develop, and suppliers would be forced to lower price in order to sell all of their product. Price will fall until quantity supplied equals quantity demanded, which occurs at P_1. *(161)*

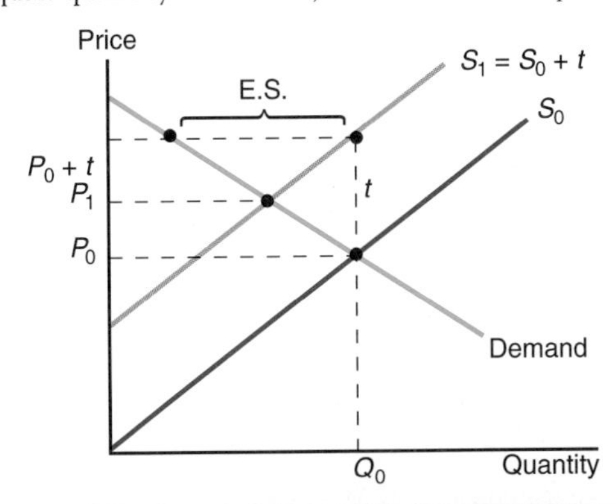

2. Welfare loss when supply is highly elastic and demand is highly inelastic is shown by the shaded triangle in the graph below. The supply curve shifts up by the amount of the tax. Since equilibrium quantity changes very little, from Q_0 to Q_1, welfare loss is very small. *(162)*

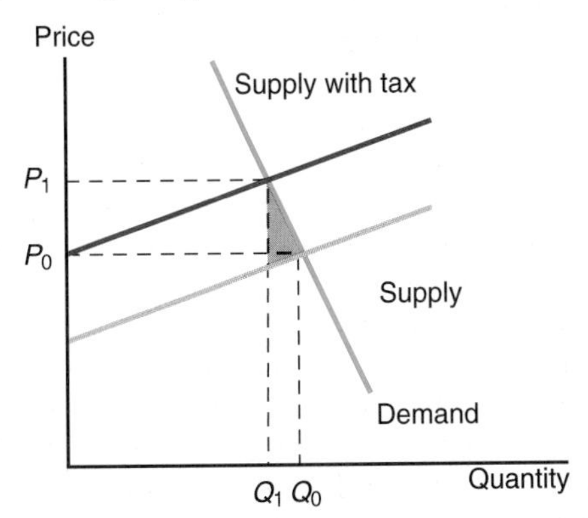

3. The average tax would be total tax divided by total income, or \$2,000 (20% of \$10,000) divided by \$30,000 = 6.7%. *(163)*
4. The tax burden when demand is elastic and supply is inelastic is shown below. The blue shaded region is the tax burden paid by consumers. The red shaded region is the tax burden paid by producers. Producers pay a larger burden of the tax because they are unable to change their production (supply is inelastic). *(166)*

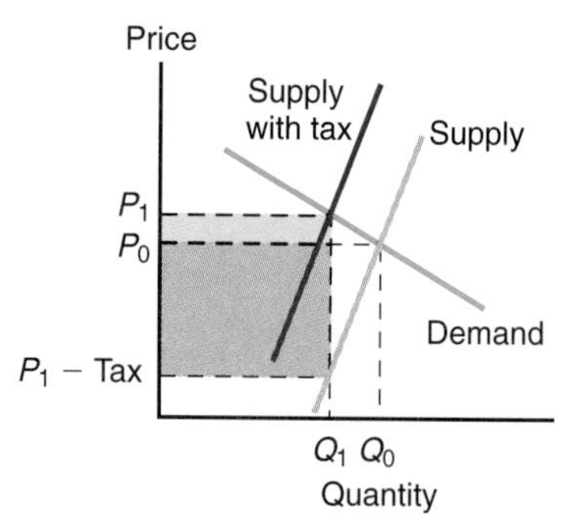

5. The diagram for (a) cigarettes and (c) food will look like figure 7-3(b). The diagram for (b) perfume will look like figure 7-3(a). *(166)*
6. If the entire amount of the tax were levied on employees, their before-tax income would rise because employers would have to compensate their employees for the increased taxes they would have to physically pay. The burden of the taxation does not depend on who pays the tax. It depends on relative elasticities. *(167)*
7. The effect of a price ceiling below equilibrium price when demand and supply are inelastic is shown in the following graph. Quantity demanded exceeds quantity supplied, but because demand and supply are both inelastic, the shortage is not big. Likewise, the welfare loss triangle, shown by the shaded area in the graph, is not large. *(169)*

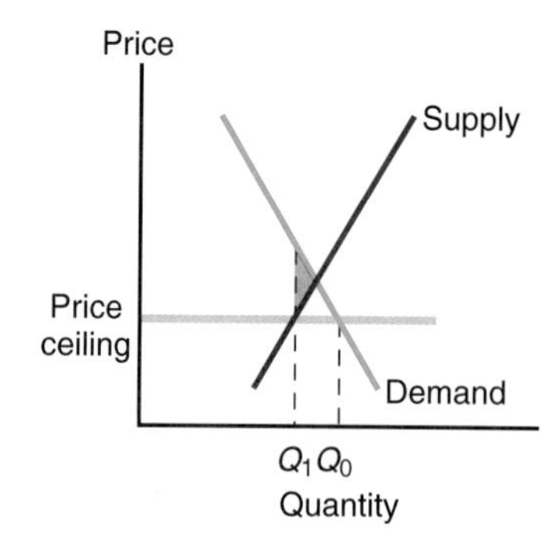

8. No. Research and development expenditures are an effort to increase technology to either lower production costs or discover a new product that can be marketed. If the firm can get a patent on that new product, the firm will have a monopoly and be able to restrict supply, transferring surplus from consumers to themselves, but this is not rent seeking. Rent-seeking activities are designed to transfer surplus from one group to another given current technology. They are unproductive. *(171)*
9. If suppliers were selling a product for which demand is inelastic, increases in productivity would result in a drop in price that would be proportionately greater than the rise in equilibrium quantity. Total revenue would decline for suppliers. *(171)*
10. Price controls tend to create ongoing shortages and surpluses in the long run because they prevent market forces from working. *(172)*

The Logic of Individual Choice: The Foundation of Supply and Demand

8

After reading this chapter, you should be able to:

- Explain why economists can believe there are many explanations of individual choice but nonetheless focus on self-interest.
- Discuss the principle of diminishing marginal utility.
- Summarize the principle of rational choice.
- Explain why, when $MU_x/P_x = MU_y/P_y$, a consumer is maximizing total utility.
- Explain how the principle of rational choice accounts for the Law of Demand.
- Illustrate how economists use indifference curves to represent people's preferences.
- Demonstrate using indifference curves how a person maximizes utility given a limited income.

The theory of economics must begin with a correct theory of consumption.

Stanley Jevons

The analysis of how individuals make choices is central to microeconomics. It is the foundation of economic reasoning and it gives economics much of its power. The first part of this chapter introduces the foundation and relates the analysis to the real world, giving a sense of when the model is useful and when it's not. The second part of the chapter builds a model based on the concept of rational choice, and illustrates how individuals are able to make consumption decisions given a limited income.

UTILITY THEORY AND INDIVIDUAL CHOICE

Different sciences have various explanations for why people behave as they do. For example, Freudian psychology tells us that we do what we do because of an internal conflict between the id, ego, and superego, that while other psychologists tell us it's a search for approval by our peers. Heavy psychological explanations are difficult to apply. Economists use an easier underlying psychological foundation which is consistent with observed behaviour, self-interest. People do what they do because it's in their self-interest.

Economists' analysis of individual choice doesn't deny individual differences.

Economists' analysis of individual choice doesn't deny individual differences. That's obvious in what we buy. On certain items we're cheap; on others we're big spenders. For example, people clip coupons to save 40 cents on cereal but then spend $40 on a haircut. People save 50 cents a pound by buying a low grade of meat but then spend $20 on a bottle of wine, $75 on dinner at a restaurant, or $60 for a concert ticket.

Consuming goods and services provides "utils."

But through it all comes a certain rationality. Much of what people do reflects their rational self-interest. That's why economists start their analysis of individual choice with a relatively simple, but powerful, underlying psychological foundation.

Using that simple concept of self-interest, two things determine what people do: the pleasure people get from doing or consuming something, and the price of doing or consuming that something. Price is the tool the market uses to bring the quantity supplied equal to the quantity demanded. Changes in price provide incentives for people to change what they're doing. Through those incentives the invisible hand guides us all. To understand economics you must understand how price affects our choices. That's why we focus on the effect of price on the quantity demanded. We want to understand how a change in price will affect what we do.

In summary, economists' theory of rational choice is a simple, but powerful, theory that shows how these two things—pleasure and price—are related.

Measuring Pleasure

Economists' analysis of individual choice starts with the proposition that individuals try to get as much pleasure as possible out of life. The goods and services we consume provide value (or satisfaction) to us. Any individual's goal is to maximize the amount of satisfaction or "happiness" they receive through consuming goods and services. By making the right choices in what and how much to consume of various products, consumers maximize their happiness, or utility. **Utility** is *the pleasure or satisfaction or happiness that one receives from consuming a good or service.* Utility serves as the basis for analyzing individual choice.

8.1 

see page 193

Utility refers to the satisfaction one gets from consuming a good or service.

How do we measure utility? If a person "likes" apples and muffins, he or she will get utility from eating them. We might report that consuming an apple yields 2 units of happiness, called "utils." We might also report that consuming a muffin yields 6 units of happiness, or 6 utils. However, we cannot interpret this to mean that the muffin yields three times the happiness, or utility, than the apple does; utility tells us that the muffin provides more satisfaction than the apple, but not how much more satisfaction. The measure of utility is ordinal, which means that we can only rank an individual's preferences. The following utility figures all provide the same ordinal (ranking) information: "muffin is preferred to apple":

A "util" is a unit created by economists to "measure" utility.

Product	Ranking (in utils) Individual 1	Individual 2	Individual 3
apple	2	75,000	115
muffin	6	81,000	15,353

Utility is personal and individual. Utility cannot be compared across individuals, so we also cannot say that Individual 2 likes apples (and muffins too!) more than Individuals 1 or 3. All that can be said is that each person prefers a muffin to an apple (ignoring any price differences in the two goods). An individual's preferences are revealed by his or her choices of goods.

Q-1 One of the assumptions of economists' theory of choice is that utility must be measured. True or false? Explain.

Total Utility and Marginal Utility

Total utility refers to *the total satisfaction one gets from consuming a product.* **Marginal utility** refers to *the satisfaction one gets from consuming one additional unit of a product above and beyond what one has consumed up to that point.* For example, eating a whole bag of tortilla chips might give you 4,700 units of utility. Consuming the first 15 chips may have given you 4,697 units of utility. Consuming the last chip might give you an additional 3 units of utility. The 4,700 is total utility; the 3 is the marginal utility of eating that last chip.

It is important to distinguish between marginal and total utility.

An example of the relationship between total utility and marginal utility is given in Figure 8-1 on the next page. Let's say that the marginal utility of the 1st slice of pizza is 14. At this point, since you've eaten only 1 slice, the total utility is also 14. Let's also say that the marginal utility of the 2nd slice of pizza is 12, which means that the total utility of eating 2 slices of pizza is 26 (14 + 12). Similarly for the 3rd, 4th, and 5th slices of pizza, whose marginal utilities are 10, 8, and 6, respectively. The total utility of your eating those 5 pieces of pizza is the sum of the marginal utilities you get from eating each of the 5 slices. The fifth row of column 2 of Figure 8-1(a) shows that sum.

Negative marginal utility—how many pieces are "too much?"

Notice that marginal utility shows up between the lines. It is the utility associated with changing consumption levels. For example, the marginal utility of changing from 1 to 2 slices of pizza is 12. The relationship between total and marginal utility can also be seen graphically. In Figure 8-1(b) we graph total utility (column 2 of the utility table) on the vertical axis, and the number of slices of pizza (column 1 of the utility table) on the horizontal axis. As you can see, total utility increases up to 7 slices of pizza; after 8 slices it starts decreasing—after 8 pieces of pizza you're so stuffed that you can't eat another slice.

In Figure 8-1(c) we graph marginal utility (column 3 of the utility table) on the vertical axis and slices of pizza (column 1) on the horizontal axis. Notice how marginal utility decreases while total utility increases. When marginal utility is zero (between 7 and 8 slices), total utility stops increasing. Beyond this point, marginal utility is negative and total utility decreases. An additional slice of pizza will actually make you worse off.

Q-2 If the total utility curve is a straight line—that is, does not exhibit diminishing marginal utility—what will the marginal utility curve look like?

Diminishing Marginal Utility

The marginal utility that a person gets from each additional slice of pizza decreases with each slice of pizza eaten. Economists believe that the shapes of these curves accurately describe the pattern of people's enjoyment. They call that pattern the **principle of diminishing marginal utility:**

Figure 8-1 (a, b, and c) MARGINAL AND TOTAL UTILITY

Marginal utility tends to decrease as consumption of a good increases. Notice how the information in the table (a) can be presented graphically in two different ways. The two different ways are, however, related. The downward slope of the marginal utility curve (c) is reflected in the total utility curve bowed downwards in (b). Notice that marginal utility relates to changes in quantity so the marginal utility line is graphed at the halfway point. For example, in (c), between 7 and 8, marginal utility becomes zero.

Number of pizza slices	Total utility	Marginal utility
		14
1	14	
		12
2	26	
		10
3	36	
		8
4	44	
		6
5	50	
		4
6	54	
		2
7	56	
		0
8	56	
		−2
9	54	

(a) Utility table

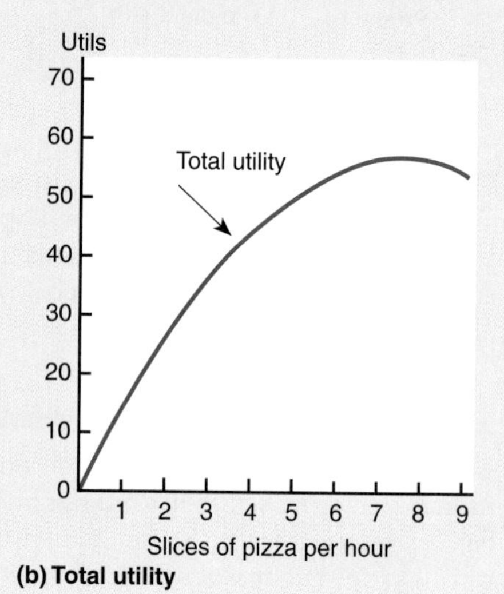

(b) Total utility

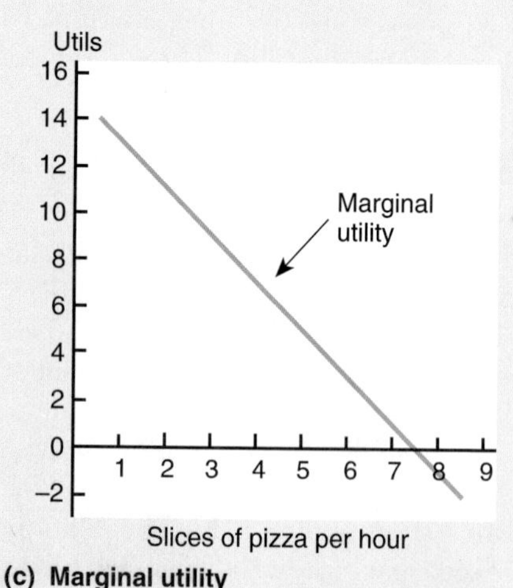

(c) Marginal utility

As you consume more of a good, at some point the marginal utility received from an additional unit of a good begins to decrease with each additional unit consumed.

The principle of diminishing marginal utility states that, at some point, the marginal utility received from an additional unit of a good begins to decrease with each additional unit consumed.

As individuals increase their consumption of a good, at some point, consuming another unit of the product will simply not yield as much additional pleasure as consuming the preceding unit.

Suppose you've ordered a large pizza and bite into the first slice. You are hungry, and the first slice is delicious. But, if you're eating it all by yourself, eventually you'll get less additional enjoyment from eating additional slices. In other words, the marginal utility you get is going to decrease with each additional slice of pizza you consume. That's the principle of diminishing marginal utility.

Notice that the principle of diminishing marginal utility does not say that you don't enjoy consuming more of a good; it simply states that as you consume more of the good, you enjoy the additional units less than you did the initial units. A fourth slice of pizza still tastes good, but it's not as satisfying as the third slice. At some point, however, marginal utility can become negative.

Q.3 Consuming more of a good generally increases its marginal utility. True or false? Explain.

RATIONAL CHOICE AND MARGINAL UTILITY

The analysis of rational choice is the analysis of how individuals choose goods in order to maximize total utility, and how maximizing total utility can be accomplished by considering marginal utility. That analysis begins with the premise that rational individuals want as much satisfaction as they can get from their available income. The term *rational* in economics means, specifically, that people prefer more to less and will make choices that give them as much satisfaction as possible. The problem is that people face an income constraint. They must choose among the alternatives. How do they do that?

Because people face an income constraint, they must choose among alternatives.

Some Choices

Consider the choice between spending another dollar on a slice of pizza that gives you an additional 41 units of utility or spending another dollar on a hero sandwich that gives you an additional 30 units of utility.

8.2
see page 193

Since the slice of pizza and the sandwich both cost $1, and the pizza gives you more units of utility than the sandwich, the pizza is the rational choice. If you spend $1 on the sandwich rather than the pizza, you're losing 11 units of utility and not making yourself as happy as you could be. You're being irrational. Any choice (for the same amount of money) that doesn't give you as much utility as possible is an irrational choice.

But now let's say that the price of the sandwich falls to 50 cents so that you can buy two sandwiches for the same price you previously had to pay for only one. Let's also say that two sandwiches would give you 56 units of utility (not 2 × 30 = 60—remember the principle of diminishing marginal utility). Which would now be the more rational choice? The two sandwiches, because their 56 units of utility are 15 more than you would get from that dollar spent on one slice of pizza.

Another way of thinking about your choice is to recognize that essentially what you're doing is buying units of utility. Obviously you want to get the most for your money, so you choose goods that have the highest units of utility per unit of cost.

Q-4 Which is the rational choice—watching one hour of MTV that gives you 20 units of utility or watching a two-hour movie that gives you 30 units of utility?

The Principle of Rational Choice

The **principle of rational choice** is as follows: *Spend your money on those goods that give you the most marginal utility (MU) per dollar.*

The principle of rational choice tells us to spend our money on those goods that give us the most marginal utility per dollar.

If $\frac{MU_x}{P_x} > \frac{MU_y}{P_y}$, choose to consume an additional unit of good x.

If $\frac{MU_x}{P_x} < \frac{MU_y}{P_y}$, choose to consume an additional unit of good y.

By substituting the marginal utilities and prices of goods into these formulas, you can always decide which good it makes more sense to consume. Consume the one with the highest marginal utility per dollar.

Q-5 True or False? You are maximizing total utility only when the marginal utility of all goods is zero. Explain your answer.

MAXIMIZING UTILITY AND EQUILIBRIUM

When do you stop adjusting your consumption? The principle of rational choice tells you to keep adjusting your spending if the marginal utility per dollar (*MU/P*) of two goods differs. The only time you don't adjust your spending is when there is no clear winner. *When the ratios of the marginal utility to price of the two goods are equal*, you're maximizing utility; this is the **utility-maximizing rule:**

The utility maximizing rule: Consume the quantities of Goods X and Y where:

$$\frac{MU_x}{P_x} = \frac{MU_y}{P_y}$$

If $\frac{MU_x}{P_x} = \frac{MU_y}{P_y}$, *you're maximizing utility.*

When you're maximizing utility, you're in equilibrium. To understand how, by adjusting your spending, you can achieve equilibrium, it's important to remember the principle of diminishing marginal utility. As we consume more of an item, the marginal utility we get from the last unit consumed decreases. Conversely, as we consume *less* of an item, the marginal utility we get from the last unit consumed *increases*. (The principle of diminishing marginal utility also operates in reverse.)

Maximizing utility on a hot day.

An Example of Maximizing Utility

Table 8-1 offers an example of how a person maximizes utility through his or her choice of goods to consume. In this example, we have $10 to spend on ice cream cones and hamburgers. The choice is between ice cream at $1 a cone and hamburgers at $2 apiece. In the table you can see the principle of diminishing marginal utility in action. The marginal utility (*MU*) obtained from either good decreases as more of it is consumed. Notice that marginal utility (*MU*) becomes negative after 5 hamburgers or 6 ice cream cones.

The *MU*/*P* columns tell you the *MU* per dollar spent on each of the items. By following the rule that we choose the good with the higher marginal utility per dollar, we can quickly determine the optimal choice.

Let's start by considering what we'd do with our first $2. Clearly we'd choose two ice cream cones. Doing so would give us 29 + 17 = 46 units of utility, compared to 20 units of utility if we spent the $2 on a hamburger. How about our next $2? Again the choice is clear; the 10 units of utility per dollar from the hamburger are plainly better than the 7 units of utility per dollar we can get from ice cream cones. So we buy 2 ice cream cones and 1 hamburger with our first $4.

TABLE 8-1 Maximizing Utility

This table provides the information needed to make simultaneous decisions. Notice that the marginal utility we get from another unit of the good declines as we consume more of it. To maximize utility, adjust your choices until the marginal utility of all goods is equal.

Hamburgers (*P* = $2)				Ice Cream (*P* = $1)			
Q	*TU*	*MU*	*MU/P*	*Q*	*TU*	*MU*	*MU/P*
0	0			0	0		
		20	10			29	29
1	20			1	29		
		12	6			17	17
2	32			2	46		
		6	3			7	7
3	38			3	53		
		3	1.5			3	3
4	41			4	56		
		0	0			1	1
5	41			5	57		
		−5	−2.5			0	0
6	36			6	57		
		−10	−5			−4	−4
7	26			7	53		

Table 8-2 summarizes our choices as we spend our $10. At equilibrium, the marginal utilities per dollar are the same for both goods and we're maximizing total utility. Our total utility is 94.

Why do these choices make sense? Because they give us the most total utility for the $10 we have to spend. We've followed the utility-maximizing rule: Maximize utility by adjusting your choices until the marginal utilities per dollar are the same. These choices make the marginal utility per dollar between the last hamburger and the last ice cream cone equal. The marginal utility per dollar we get from our last hamburger is:

$$\frac{MU}{P} = \frac{6}{\$2} = 3$$

The marginal utility per dollar we get from our last ice cream cone is:

$$\frac{MU}{P} = \frac{3}{\$1} = 3$$

The marginal utility per dollar of each choice is equal, so we know we can't do any better. For any other choice we would get less total utility, so we could increase our total utility by switching our choices.

Extending the Principle of Rational Choice

Our example involved only two goods, but the reasoning can be extended to the choice among many goods. That general principle of rational choice is to consume more of the good that provides a higher marginal utility per dollar.

When $\frac{MU_x}{P_x} > \frac{MU_z}{P_z}$, consume more of good *x*.

When $\frac{MU_y}{P_y} > \frac{MU_z}{P_z}$, consume more of good *y*.

Stop adjusting your consumption when the marginal utilities per dollar are equal across all goods.

So the general utility-maximizing rule is that you are maximizing utility when the marginal utilities per dollar of all goods consumed are equal.

TABLE 8-2 Maximizing Utility

In order to maximize utility, individuals make successive comparisons of marginal utility relative to price, and choose goods such that they receive the most utility per dollar spent on each good.

Total amount spent	Purchase	MU/P	MU
$1	ice cream cone	29	29
$2	2nd ice cream cone	17	17
$4	hamburger	10	20
$5	3rd ice cream cone	7	7
$7	2nd hamburger	6	12
$9	3rd hamburger	3	6
$10	4th ice cream cone	3	3
			total utility = 94 utils

When $\frac{MU_x}{P_x} = \frac{MU_y}{P_y} = \frac{MU_z}{P_z}$ you are maximizing utility.

When this rule is met, the consumer is in equilibrium; the cost per additional unit of utility is equal for all goods and the consumer is as well off as it is possible to be.

Q-6 If you are initially in equilibrium and the price of one good rises, how would you adjust your consumption to return to equilibrium?

Notice that the rule does not say that the rational consumer should consume a good until its marginal utility reaches zero. The consumer doesn't have enough money to reach this point. Consumers face an income constraint and they must maximize utility subject to their budget.

Opportunity Cost

Opportunity cost is the benefit forgone of the next-best alternative. The opportunity cost of a forgone alternative is essentially the marginal utility per dollar you forgo.

To say

$$\frac{MU_x}{P_x} > \frac{MU_y}{P_y}$$

is to say that the opportunity cost of not consuming good x is greater than the opportunity cost of not consuming good y. So you consume x.

The principle of rational choice states that, to maximize utility, choose goods until the opportunity costs of all alternatives are equal.

When all the marginal utilities per dollar spent are equal, the opportunity cost of all the alternatives are equal. In reality people don't use the utility terminology, and, indeed, a specific measure of utility doesn't exist. But we make choices all the time based on the price of goods relative to the benefit they provide. Instead of utility terminology, people use the "really need" terminology. They say they want to buy a new (or newer) car rather than take a vacation because they *really need* the car. To say you "really need" the car is the equivalent of saying the marginal utility of the car is higher than the opportunity cost of other choices, such as taking a vacation. So the general rule works even if most people don't use the word *utility*. The more you "really, really need" something, the higher its marginal utility.

Q-7 If the opportunity cost of consuming good x is greater than the opportunity cost of consuming good y, which good has the higher marginal utility per dollar?

RATIONAL CHOICE AND THE LAW OF DEMAND

How does the rule for maximizing utility relate to the Law of Demand? The Law of Demand says that the quantity demanded of a good is inversely related to its price. That is, when the price of a good goes up, the quantity we consume of it goes down.

Now consider the Law of Demand in relation to the principle of rational choice. When the price of a good goes up, the marginal utility *per dollar* from that good goes down. Initially we maximize utility by consuming where:

$$\frac{MU_x}{P_x} = \frac{MU_y}{P_y}$$

When the price of good y goes up, then:

$$\frac{MU_x}{P_x} > \frac{MU_y}{P_y}$$

According to the principle of rational choice, if there is diminishing marginal utility and the price of a good goes up, we consume less of that good. Hence, the principle of rational choice leads to the Law of Demand.

Our utility-maximizing rule is no longer satisfied. We now get more utility per dollar from good x, so it would make sense for us to buy more good x and less good y. As we increase the amount of good x, MU_x falls (diminishing marginal utility), and as we decrease the amount of good y, MU_y rises (diminishing marginal utility, again). Once again, we would reach a point where

$$\frac{MU_x}{P_x} = \frac{MU_y}{P_y}.$$

We would be maximizing utility, but would be buying more x and less y than we did before.

So when the price of a good goes up, we would choose to consume less of that good. The principle of rational choice underlies the Law of Demand:

Quantity demanded rises as price falls, other things constant.

Or alternatively:

Quantity demanded falls as price rises, other things constant.

This discussion of marginal utility and rational choice shows the relationship between marginal utility and the price we're willing to pay. When marginal utility is high, as it is with diamonds, the price we're willing to pay is high. When marginal utility is low, as it is with tap water, the price we're willing to pay is low. Since our demand for a good is an expression of our willingness to pay for it, quantity demanded is related to marginal utility.

APPLYING ECONOMISTS' THEORY OF CHOICE TO THE REAL WORLD

Understanding a theory involves more than understanding how a theory works; it also involves understanding the limits the assumptions underlying the theory place on the use of the theory.

The Cost of Decision Making

The principle of rational choice makes reasonably good intuitive sense when we limit our examples to two or three choices. But in reality, people make hundreds of choices every day. It's difficult to believe that we're going to apply the principles of rational choice to all those decisions.

Q-8 Bounded rationality violates the principle of rational choice. True or false?

The decisions we make range from mundane, such as where to buy coffee, to life-defining, such as what subject to study at college or university. Some decisions are difficult to make because of the level of complexity involved, or because we lack necessary information, or because there is some degree of uncertainty involved in the process or the outcome. Each decision requires us to use our limited cognitive ability to receive, process, store, and retrieve information. Because of the many factors involved in each decision we make, we economize on our decision making, and employ a variety of simple rules for some of our decisions.

Following the work of Nobel Prize winner Herbert Simon, a number of economists have come to believe that, to make real-world decisions, most people use *bounded rationality*—rationality adjusted for our limitations. They argue that many of our decisions are made using simple conventions.

Q-9 Using the principle of rational choice, explain why a change in tastes will shift a demand curve.

1. **Price conveys information.** Experience has shown us that higher-priced goods tend to be better than lower-priced goods, so, instead of weighing the opportunity costs of a particular choice, we can use this simple rule to make a quick decision. We rely on price to convey information about quality.
2. **Follow the leader.** We can't be an expert in everything, so sometimes we just do what others are doing. Clothes manufacturers try to exploit this decision rule with their advertising efforts. By convincing you that "everyone" is

wearing a particular style, they induce you to bypass the long process of weighing all the opportunity costs of what to wear. It's easier to flip through a magazine to see the new styles than it is to accurately weigh the costs of each designer's clothing style, fabric choice, colour, and cut.

Habit explains many of our daily activities.

3. Habit. Habit explains a lot of our choices. We usually go for coffee at 10:00 at Sam's. Yesterday (or last week, or six months ago), we did the marginal utility calculation, and chose Sam's. Since nothing is particularly different about today's choice, we can rely on our previous judgment and economize on our decision making by choosing Sam's again.
4. Custom. Ginette chose to study Economics because her parents ran a small business while she was growing up; her cousin became a plumber because his Dad got him into the union. In many Canadian homes, meat and potatoes are served for supper, while in Japan, rice is served. Employing the rule of custom can ease the burden of decision, particularly for repeat decisions, like what to serve for dinner.

Economists take into account changes in tastes as shift factors of demand.

Although these conventions are convenient and commonly used, it is important to remember that utility calculations underlie these decision making shortcuts. If the underlying utility maximization changes significantly, we would expect these simple decision-making rules to change, too.

MAXIMIZING UTILITY USING INDIFFERENCE CURVES

Maximizing Utility

Jaz is a junk food devotee. She lives on two goods: chocolate bars, which cost $1 each, and pop, which sells for 50 cents a can. Jaz is trying to get as much pleasure as possible, given her income, which is $10. Alternatively expressed, Jaz is trying to maximize her utility, given an income constraint.

Graphing the Income Constraint

How can we illustrate her income constraint (the $10 maximum she has to spend)? The graph we'll use will have chocolate bars on the vertical axis and cans of pop on the horizontal axis, as in Figure 8-2. Since a chocolate bar costs $1, if she spends all of her $10 income on chocolate bars she can get 10 bars (point *A* in Figure 8-2). If she spends it all on pop, she can get 20 cans of pop (point *B*). This gives us two points.

But what if she wants some combination of pop and chocolate bars? If she buys only 9 chocolate bars, she has $1 left with which to buy 2 cans of pop, so 9 chocolate bars and 2 cans of pop is also possible. All of the other combinations can be determined in this way. However, because the prices of the two goods are fixed, she faces a constant tradeoff; she can always give up 1 chocolate bar to get 2 cans of pop. The slope of the budget line also must be constant at

The budget line shows the various combinations of goods an individual can afford.

$$\frac{\text{rise}}{\text{run}} = \frac{\text{chocolate bars}}{\text{pop}} = \frac{-1}{2}$$

so we can simply join the two endpoints to obtain Jaz's budget line quickly. This curve illustrates her **income constraint**, or **budget line**, *a curve that shows us the various combinations of goods an individual can buy with a given income.*

Figure 8-2 GRAPHING THE INCOME CONSTRAINT

The budget line represents all the combinations of pop and chocolate bars that it is possible for Jaz to buy with her income of $10. If she spends all of it on chocolate bars, she can buy 10; if she spends all of it on pop, she can buy 20. The budget line has a constant slope because, at $1 for a chocolate bar and 50¢ for a can of pop, she can always trade 2 cans of pop for one chocolate bar.

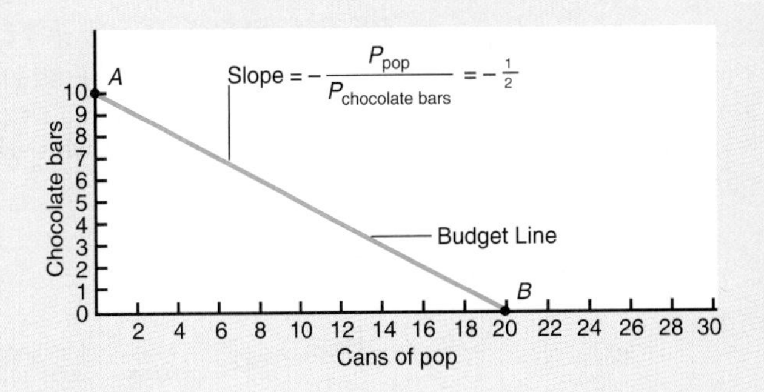

Graphing the Indifference Curve

Jaz is trying to get as much utility as she can from her $10. Suppose Jaz had 14 chocolate bars and 4 cans of pop (point A in Figure 8-3(a) on the next page). If we took away 4 of those chocolate bars (so now she had 10), how many cans of pop would we have to give her so that she would be just as happy as before we took away the 4 chocolate bars?

Since she's got lots of chocolate bars and few cans of pop, her answer is probably "Not too many; say, 1 can of pop." This means that she would be just as happy to have 10 chocolate bars and 5 cans of pop (point *B*) as she would to have 14 chocolate bars and 4 cans of pop (point *A*). That is, she is **indifferent** between having 10 chocolate bars and 5 cans of pop or 14 chocolate bars and 4 cans of pop. Repeating this experiment yields other indifferent combinations, represented by points *C*, *D*, and *E*. She is equally happy to have *any* of the combinations of pop and chocolate bars on the **indifference curve,** *a curve that shows combinations of goods amongst which an individual is indifferent*.

An indifference curve shows the various combinations of goods which yield the same level of utility.

Let's consider the shape of this curve. First, it's downward sloping. That simply says that if you take something away from Jaz, you've got to give her something in return if you want to keep her indifferent between what she had before and what she has now. The absolute value of the slope of an indifference curve is the **marginal rate of substitution**—*the rate at which one good must be added when the other is taken away in order to keep the individual indifferent between the two combinations*.

Second, it's bowed inward. That's because as Jaz gets more and more of one good, she is willing to give up lots of it to get more of the relatively scarce good. This is reflected in the slope of the indifference curve. The marginal rate of substitution is the rate at which she is willing to trade one good for another, and it is given by the slope of her indifference curve:

$$\left| \text{Slope} \right| = \frac{MU_{\text{pop}}}{MU_{\text{chocolate bars}}} = \text{Marginal rate of substitution}$$

Q-10 Explain and illustrate why two indifference curves cannot cross.

Technically, the reasoning for the indifference curve being bowed inward is called the **law of diminishing marginal rate of substitution**—which tells us that *as you get more and more of a good, if some of that good is taken away, then the marginal addition of another good you need to keep you on your indifference curve gets less and less*.

At point A, she has lots of chocolate bars (low marginal utility – remember diminishing marginal utility), but relatively few cans of pop (high marginal utility), so she

Figure 8-3 (a and b) JAZ'S INDIFFERENCE CURVES

In (a), Jaz reports that combinations *A*, *B*, *C*, *D*, and *E* all yield the same amount of utility; Jaz is indifferent as to which bundle of goods she gets. Plotting all of these points creates one indifference curve. In (b), the bundles of goods forming indifference curve U_3 give Jaz higher utility, while the combinations of goods forming indifference curve U_1 provide less utility than bundles *A*, *B*, *C*, *D*, or *E* on U_2.

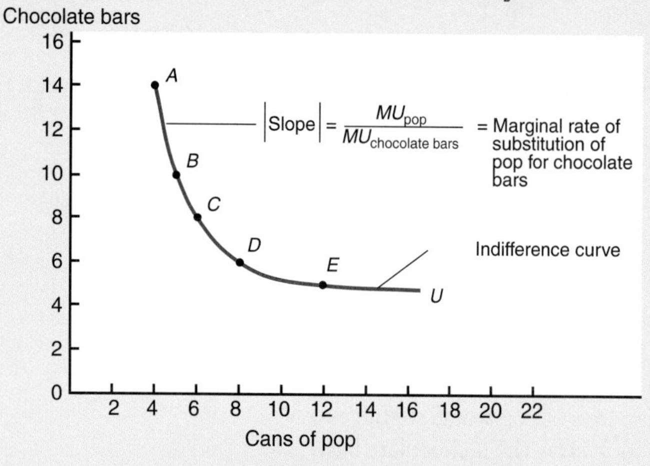

Chocolate bars	Cans of pop	Point
14	4	*A*
10	5	*B*
8	6	*C*
6	8	*D*
5	12	*E*

(a) One indifference curve

(b) A map of indifference curves

would be willing to trade a large amount of chocolate bars to obtain just a few cans of pop. At point E, however, she has a lot of pop (low MU), but few chocolate bars (high MU), so she would be willing to trade lots of pop in order to obtain more chocolate bars. As she gives up pop, the marginal utility of pop will rise, and as she gains chocolate bars, the marginal utility of chocolate bars will fall. To maximize her utility, she wants to find the combination of chocolate bars and pop where

$$\frac{MU_{\text{chocolate bars}}}{P_{\text{chocolate bars}}} = \frac{MU_{\text{pop}}}{P_{\text{pop}}}$$

Jaz also prefers more to less, so if there was a combination which contained 6 chocolate bars and *more than* 7 cans of pop, she would prefer it to point G on indifference curve U_2. She would also prefer a combination which contained 8 cans of pop and more than 7 chocolate bars. She will prefer any combination of chocolate bars and pop which gives her at least as much of each good as she had plus more of one or both. Figure 8-4 illustrates the area containing combinations of chocolate bars and pop which are preferred to point G. Any indifference curve which goes through this area will be preferred to the indifference curve containing point G. By this reasoning, we obtain a complete map of Jaz's indifference curves. Assuming she prefers more to less, her goal is

Figure 8-4 COMBINING INDIFFERENCE CURVES AND THE BUDGET LINE

Constrained by her income, Jaz maximizes her utility by choosing a bundle of chocolate bars and pop from the highest indifference curve she can afford. From any point (such as Point G), all combinations above or to the right (such as points C or K) are preferred. This is the principle of more is preferred to less (MPL). Given her budget constraint, the highest indifference curve she can reach is U_2 by choosing to consume **bundle** *D*, containing 6 chocolate bars and 8 cans of pop.

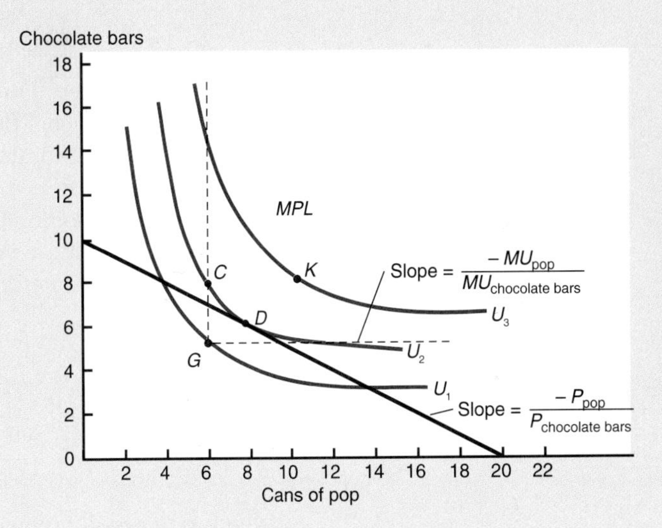

to maximize utility by choosing a point on the highest indifference curve she can reach with her budget.

Combining Indifference Curves and Budget Line

How many chocolate bars and cans of pop will Jaz buy if she has $10, given her preferences, which are described by the indifference curves in Figure 8-3(b).

To answer that question, we must put the income line of Figure 8-2 and the indifference curves of Figure 8-3(b) together, as we do in Figure 8-4.

As we discussed, Jaz's problem is to get to as high an indifference curve as possible, given her income constraint. Let's first ask if she should move to point *K* (8 chocolate bars and 10 cans of soda). That looks like a good point. But you should quickly recognize that she can't get to point *K*; her budget line won't let her. (She doesn't have enough money.) What about point G (7 chocolate bars and 6 cans of soda)? She can afford that combination; it's on her budget line. The problem with point G is the following: **More is preferred to less**. She'd rather be at point C since point C has more chocolate bars and the same amount of soda (8 chocolate bars and 6 cans of soda). But she can't reach point C. Notice, however, that she is indifferent between point C and point *D*, so point *D* (6 chocolate bars and 8 cans of soda), which she *can* reach given her income constraint, is also preferred to point G.

Notice that, at Point G, the marginal rate of substitution (slope of the indifference curve), which represents the rate at which she is willing to trade chocolate bars for pop, exceeds relative prices (slope of the budget line), which is the rate at which she *can* trade chocolate bars for pop:

$$\frac{MU_{pop}}{MU_{chocolate\ bars}} > \frac{P_{pop}}{P_{chocolate\ bars}}$$

Rewriting this, we see that she would get more utility per dollar spent on pop than she would per dollar spent on chocolate bars:

$$\frac{MU_{pop}}{P_{pop}} > \frac{MU_{chocolate\ bars}}{P_{chocolate\ bars}}$$

She could do better than Point G. She could increase her utility by trading (giving up) chocolate bars for pop. By trading at market prices ($P_{pop} / P_{chocolate\ bars}$), she moves along her budget line to Point *D*, where the MU per dollar spent on each good is equal and utility is maximized.

Utility maximization occurs at the point where the (highest) indifference curve is tangent to the budget line.

The same reasoning holds for all other points. At point *D*, the indifference curve and the budget line are **tangent**. The slope of the income line ($-P_{pop}/P_{chocolate\ bars}$) equals the slope of the indifference curve ($-MU_{pop}/MU_{chocolate\ bars}$). Equating those slopes gives

$$\frac{P_{pop}}{P_{chocolate\ bars}} = \frac{MU_{pop}}{MU_{chocolate\ bars}}$$

or,

$$\frac{MU_{chocolate\ bars}}{P_{chocolate\ bars}} = \frac{MU_{pop}}{P_{pop}}$$

which is the equilibrium condition of our principle of rational choice.

CONCLUSION

All consumers make choices that maximize utility, although most of us do not usually think in terms of a formal economic model when we make our decisions. Even in choosing which tie to wear with which shirt, or which church to attend, or whether to study medicine or structural engineering, we perform at least a quick analysis of the relevant costs and benefits of the alternatives before we make a decision. In that analysis, whatever way we measure "utility," we have to assume that the person is trying to choose the "best" alternative when he or she is faced with a decision. So, any decision can be analyzed in terms of utility maximization, as long as we broaden our thinking about how we measure the "utility" of the choice. The "best" choice, then, is the one which yields the highest "utility." For example, in deciding that you want to help your grandmother clean up her yard instead of going bungee jumping with your friends simply means that visiting your grandmother confers greater "benefits" (however you measure them) than jumping off a bridge. Similarly with other choices, whether it is giving to charity, buying a wild and crazy lime green suit, or driving within the speed limit, it is how you measure the benefits of your decision that matters.

Chapter Summary

- Total utility is the satisfaction obtained from consuming a product; marginal utility is the satisfaction obtained from consuming one additional unit of a product.
- The principle of diminishing marginal utility states that after some point, the marginal utility of consuming more of the good will fall.
- The principle of rational choice is:

  If $\frac{MU_x}{P_x} > \frac{MU_y}{P_y}$, choose to consume more of good *x*.

  If $\frac{MU_x}{P_x} < \frac{MU_y}{P_y}$, choose to consume more of good *y*.
- The utility-maximizing rule says:

  If $\frac{MU_x}{P_x} = \frac{MU_y}{P_y}$, you're maximizing utility; you're indifferent between good *x* and good *y*.
- Unless $\frac{MU_x}{P_x} = \frac{MU_y}{P_y}$, an individual can rearrange his or her consumption to increase total utility.

- Opportunity cost is essentially the marginal utility per dollar one forgoes from the consumption of the next-best alternative.
- The law of demand can be derived from the principle of rational choice.
- If you're in equilibrium and the price of a good rises, you'll reduce your consumption of that good to re-establish equilibrium.
- To apply economists' analysis of choice to the real world, we must carefully consider, and adjust for, the underlying assumptions, such as costlessness of decision making and given tastes.
- Individuals are forced to maximize utility subject to an income constraint.
- Indifference curves represent an individual's preferences, and the budget line represents his or her income constraint.
- A consumer maximizes utility by choosing a point on the highest indifference curve he or she can reach (afford).
- Utility maximization occurs at the tangency of the budget line with an indifference curve. At this point,

$$\frac{MU_x}{MU_y} = \frac{P_x}{P_y}$$

(the slope of the indifference curve equals the slope of the budget line), or,

$$\frac{MU_x}{P_x} = \frac{MU_y}{P_y}$$

(the marginal utility per dollar spent on good x equals the marginal utility per dollar spent on good y).

Key Terms

budget line *(186)*
bundle *(189)*
diminishing marginal rate of substitution *(187)*
diminishing marginal utility *(179)*
income constraint *(186)*
indifference curve *(187)*
marginal rate of substitution *(187)*
marginal utility *(179)*
more is preferred to less *(189)*
rational choice *(181)*
tangency *(190)*
total utility *(179)*
util *(178)*
utility *(178)*
utility-maximizing rule *(181)*

Questions for Thought and Review

1. Explain how marginal utility differs from total utility.
2. According to the principle of diminishing marginal utility, how does marginal utility change as more of a good is consumed? As less of a good is consumed?
3. How would the world be different if the principle of diminishing marginal utility seldom held true?
4. It is sometimes said that an economist is a person who knows the price of everything but the value of nothing. Is this statement true or false? Why?
5. Assign a measure of utility to your studying for various courses. Do your study habits follow the principle of rational choice?
6. What key psychological assumptions do economists make in their theory of individual choice?
7. Explain your motivation for four personal decisions you have made in the past year, using economists' model of individual choice.
8. State the law of demand and explain how it relates to the principle of rational choice.
9. There is a small but growing movement known as "voluntary simplicity," which is founded on the belief in a simple life of working less and spending less. Do people who belong to this movement follow the principle of rational choice?
10. Although the percentage of people who say they are "very happy" hasn't changed much in the last five decades, the number of products produced and consumed per person has risen tremendously. How can this be?
11. Early Classical economists found the following "diamond/water" paradox perplexing: "Why is water, which is so useful and necessary, so cheap, when diamonds, which are so useless and unnecessary, so expensive?" Using the utility concept, explain why it is not really a paradox.
12. Give an example of a recent purchase for which you used a simple convention in your decision-making process. Did your decision follow the principle of rational choice? Explain.
13. What would an indifference curve look like if the marginal rate of substitution were zero? If it were constant?
14. What might an indifference curve look like if the law of diminishing marginal utility did not hold?

Problems and Exercises

1. Complete the following table of Kevin's utility from drinking milk and answer the questions below.

Glasses of Milk	Total Utility	Marginal Utility
0	____	
		10
1	____	
		12
2	22	
		____
3	32	
		8
4	____	
		4
5	____	
		____
6	44	
		____
7	42	

   a. At what point does marginal utility begin to fall?
   b. Will Kevin consume the 7th glass of milk? Explain your answer.
   c. True or false? Kevin will be following the utility-maximizing rule by consuming 2 glasses of milk. Explain your answer.

2. The following table gives the price and total units of three goods: A, B, and C.

Good	Price	Total Utility							
		1	2	3	4	5	6	7	8
A	$10	200	380	530	630	680	700	630	430
B	2	20	34	46	56	64	72	78	82
C	6	50	60	70	80	90	100	90	80

START WITH NO UTILITY $0

As closely as possible, determine how much of the three goods you would buy with $20. Explain why you chose what you did.

3. The following table gives the marginal utility of John's consumption of three goods: A, B, and C.

Units of Consumption	*MU* of A	*MU* of B	*MU* of C
1			
	20	25	45
2			
	18	20	30
3			
	16	15	24
4			
	14	10	18
5			
	12	8	15
6			
	10	6	12

   a. Good A costs $2 per unit, good B costs $1, and good C costs $3. How many units of each should a consumer with $12 buy to maximize his or her utility?
   b. How will the answer change if the price of B rises to $2?
   c. How about if the price of C is 50 cents but the other prices are as in *a*?

4. The total utility of your consumption of widgets is 40; it changes by 2 with each change in widgets consumed. The total utility of your consumption of wadgets is also 40 but changes by 3 with each change in wadgets consumed. The price of widgets is $2 and the price of wadgets is $3. How many widgets and wadgets should you consume?

5. Nobel Prize–winning economist George Stigler explains how the famous British economist Phillip Wicksteed decided where to live. His two loves were fresh farm eggs, which were more easily obtained the farther from London he was, and visits from friends, which decreased the farther he moved away from London. Given these two loves, describe the decision rule that you would have expected Wicksteed to follow.

6. You are buying your spouse, significant other, or close friend a ring. You decide to show your reasonableness, and buy a cubic zirconium ring that sells at 1/50 the cost of a mined diamond and that any normal person could not tell from a mined diamond just by looking at it. In fact, the zirconium will have more brilliance and fewer occlusions (imperfections) than a mined diamond.
   a. How will your spouse (significant other, close friend) likely react?
   b. Why?
   c. Is this reaction justified?

7. Suppose Bryan Adams CDs cost $10 apiece and Burton Cummings CDs cost $5 apiece. You have $40 to spend on CDs. The marginal utility that you derive from additional CDs is as follows:

Number of CDs	Bryan Adams	Burton Cummings
0		
	60	30
1		
	40	28
2		
	30	24
3		
	20	20
4		
	10	10
5		

How many of each CD would you buy? Suppose the price of a Burton Cummings CD rises to $10. How many of each CD would you buy? Use this to show how the principle of rational choice leads to the law of demand.

8. Zachary has $5 to spend on two goods: video games and hot dogs. Hot dogs cost $1 apiece while video games cost 50 cents apiece.
   a. Draw a graph of Zachary's income constraint, placing videos on the Y axis.
   b. Suppose the price of hot dogs falls to 50 cents apiece. Draw the new income constraint.
   c. Suppose Zachary now has $8 to spend. Draw the new income constraint using the prices from *b*.
9. Zachary's indifference curves are shown in the following graph. Determine on which indifference curve Zachary will be, given the income constraints in *a*, *b*, and *c* from problem 8.

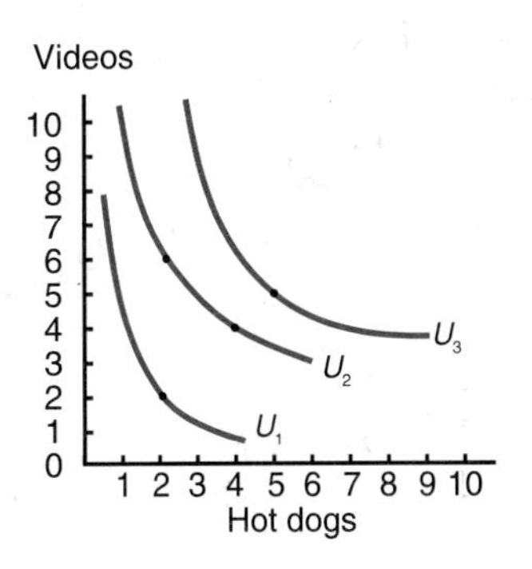

   a. Given a choice, which income constraint would Zachary prefer most? Least?
   b. What is the marginal rate of substitution of hot dogs for videos at each of the combinations chosen with income constraints *a*, *b*, and *c* in problem 8?

Web Questions

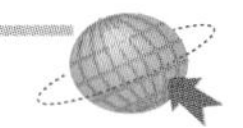

1. Go to www.travelocity.ca
   a. Find a selection of prices of airline fares between two cities for a period of one month ahead of time and staying over a Saturday night. If there are differences in the prices explain why they likely differ.
   b. Now shorten your stay, and do not include a Saturday night stay. What happens to the prices? Explain why.
   c. Now find the price of the same flight you had in *a* only this time booking only three days ahead. What happens to prices? Explain why.
2. Go to www.iwon.com
   a. What does this site do?
   b. Why do they give out a $10,000 daily prize and a $10 million yearly prize for using the site?
   c. What advertisements were shown there?
   d. What does the existence of these advertisements suggest about economists' assumption that tastes are fixed?

Answers to Margin Questions

1. False. Economists' theory of choice does not require them to measure utility. It only requires that the marginal utility of one good be compared to the marginal utility of another. *(179)*
2. If the total utility curve is a straight line, the marginal utility curve will be flat with a slope of zero, since marginal utility would not change with additional units. *(179)*
3. False. The principle of diminishing marginal utility is that as one increases consumption of a good, the good's marginal utility decreases. *(180)*
4. Given a choice between the two, the rational choice is to watch MTV for one hour since it provides the highest marginal utility per hour. *(181)*
5. False. You are maximizing total utility when the marginal utilities per dollar are the same for all goods. This does not have to be where marginal utility is zero. *(181)*
6. If I am currently in equilibrium, then $\frac{MU_x}{P_x} = \frac{MU_y}{P_y} = \frac{MU_z}{P_z}$ for all goods I consume. If the price of one good goes up, I will decrease my consumption of that good and increase the consumption of other goods until the equilibrium is met again where $\frac{MU_x}{P_x} = \frac{MU_y}{P_y} = \frac{MU_z}{P_z}$. *(184)*
7. Good *y* has the higher marginal utility per dollar since the opportunity cost of consuming good *x* is the marginal utility per dollar of consuming good *y*. *(184)*
8. This could be true or false. It depends on how you interpret bounded rationality. If it is interpreted within a costless decision-making environment, it does violate the principle of rational choice since there is no reason to be less than rational. If, however, it is interpreted within a costly decision-making environment, then you

can be making decisions within a range because the marginal cost of increasing the range of choices exceeds the marginal benefit of doing so, and in that case bounded rationality is consistent with the principle of rational choice. Information is not costless. *(185)*

9. If a person is in equilibrium and a change in tastes leads to an increase in the marginal utility for one good, he will increase consumption of that good to reestablish equilibrium. A change in tastes will shift a demand curve because it will cause a change in quantity consumed without a change in the good's price. *(185)*
10. Indifference curves cannot cross each other (intersect) because of the principle of more is preferred to less (MPL). Consider the following example.

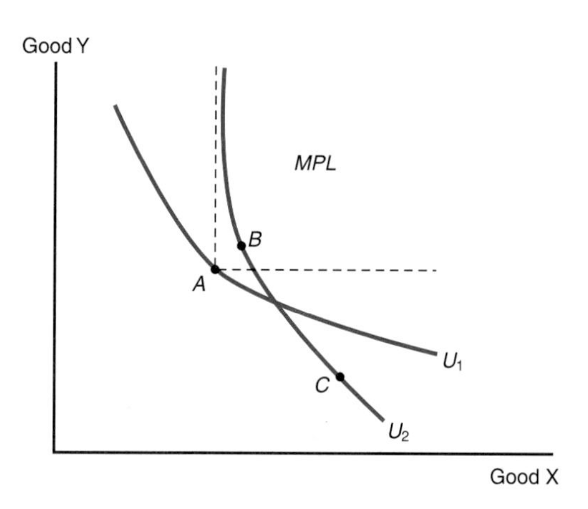

Due to MPL, point *B* is preferred to point A. Since points B and C lie on the same indifference curve, they must yield the same utility. So, if B is preferred to A, then C must be preferred to A also. This is a clear violation of the MPL principle. *(187)*

Describing Consumer Preferences Using Indifference Curves

INCOME-EXPANSION PATH

Consumers maximize utility by choosing a bundle of goods on the highest indifference curve they can achieve given their income. When incomes rise, individuals are able to consume bundles of goods on higher indifference curves. As an individual's income increases, we can trace out his consumption choices. In Figure A8-1(a), increases in income are represented by rightward shifts of the budget line, from B_1, to B_2, to B_3. Notice that the slope does not change because relative prices do not change. As the individual's income rises, utility maximization occurs at the tangency of higher and higher budget lines and indifference curves. Tracing out the utility-maximizing choices yields the Income-Expansion Path (also known as the Income-Consumption Line). Both Good X and Good Y are normal goods: As income rises, more of both goods are consumed. The Income-Expansion Path for an inferior good (Good X) is presented in Figure A8-1 (b).

ENGEL CURVE

Figure A8–2 illustrates the relationship between income and the consumption of *one* good. Goods X_1 and X_2 are normal goods because their consumption rises as income rises. Good X_1 is a luxury good (income-elastic normal good) since the person spends a greater and greater proportion of his income on it, while Good X_2 is a necessity (income-inelastic normal good) since the fraction of income spent on X_2 rises less than proportionately to income. Good X_3 becomes an inferior good, since as income rises beyond a certain point, less X_3 is bought. These curves are known as Engel Curves.

PRICE-EXPANSION PATH

When relative prices change, consumers adjust their purchases, substituting into the good which is relatively cheaper and substituting away from the good which is now relatively more expensive. Figure A8-3 shows how a con-

Figure A8-1 (a and b) INCOME-EXPANSION PATH FOR NORMAL GOOD AND INFERIOR GOOD

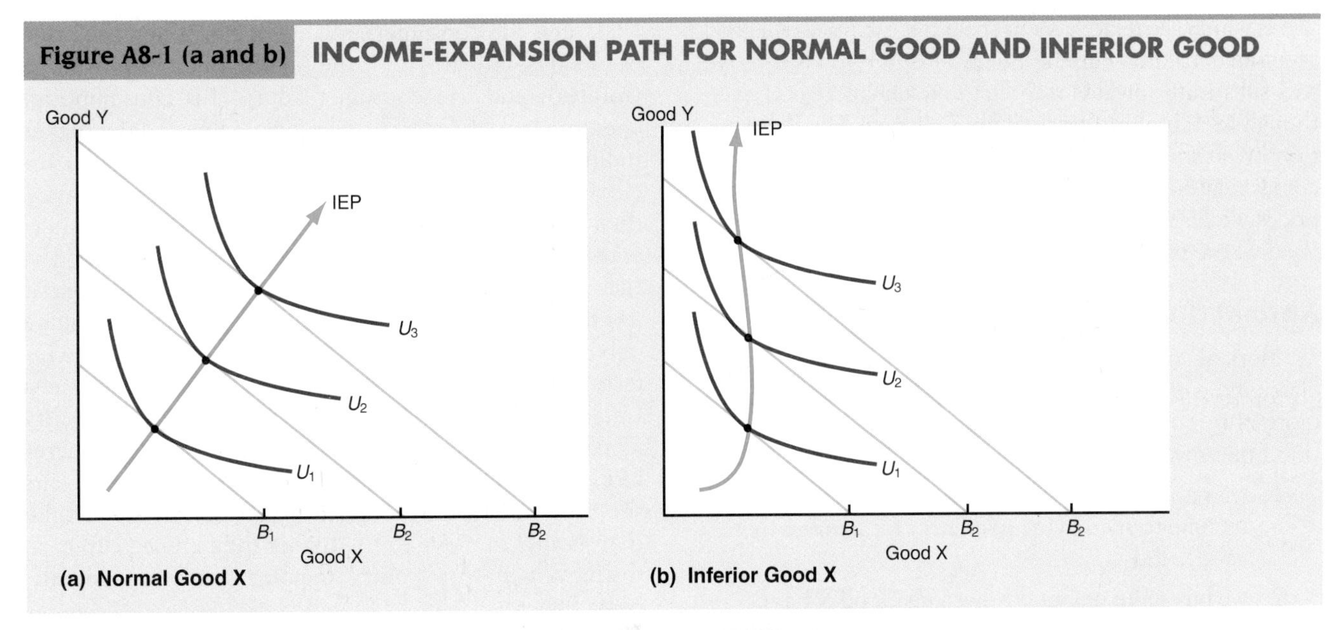

Figure A8-2 ENGEL CURVES

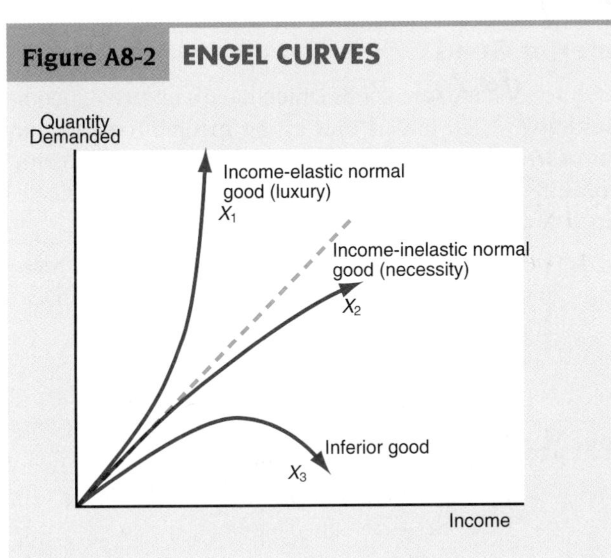

Figure A8-3 PRICE-EXPANSION PATH

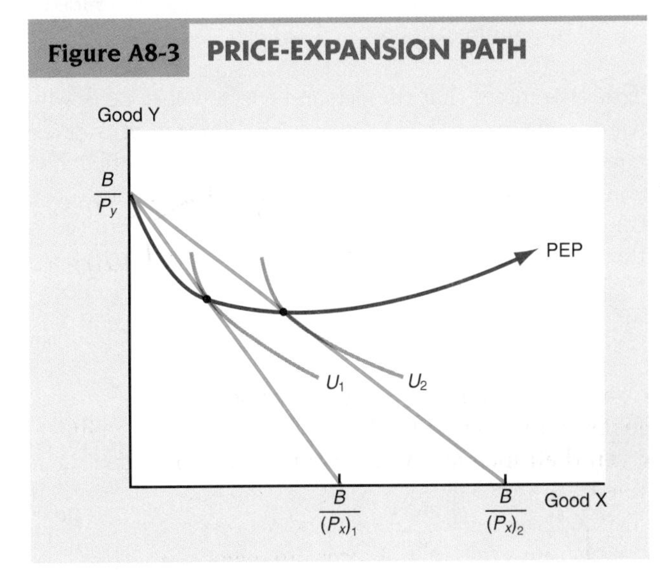

sumer responds to a fall in the price of Good X, buying more X and less Y. Tracing out his utility-maximization choices yields his Price-Expansion Path (also known as the Price-Consumption Line).

INCOME AND SUBSTITUTION EFFECTS

The Law of Demand states that there is an inverse relationship between price and quantity demanded, so we observe that as the price of Good X falls, quantity demanded rises. As the price of Good X falls, however, two effects occur, the Income Effect and the Substitution Effect.

Income Effect. The Income Effect reflects our increased ability to buy goods when prices fall. As the price of a good falls, our purchasing power rises and we can buy more goods, including the good whose price has fallen.

Substitution Effect. The Substitution Effect captures our willingness to switch consumption away from goods whose relative price has risen, and towards goods whose relative price has fallen. Realize that we substitute towards those goods which are relatively cheaper, so if the price of a good remains constant, but the prices of other goods rise, we will substitute towards the relatively cheaper good even though its (absolute) price did not change. The Substitution Effect states that when the price of a good falls, it becomes relatively more attractive than other goods (whose prices we assume do not change), so we buy more of the good whose price has fallen.

Normal Good

Normal goods are goods which have a positive income elasticity, which means that as our income rises, we buy more of it. For a normal good, the Income and Substitution Effects work together: As the price of Good X falls,

1. we buy more of Good X because we have more real income to spend (Income Effect for a normal good), and
2. we buy more of Good X because Good X has become relatively cheaper than other goods (Substitution Effect).

This guarantees that the demand curve will slope downwards. Figure A8-4 (a) isolates the impact of a price change into the two effects for a normal good. Notice that you can trace out both the Income-Expansion Path (goes through points *F* and G) and Price-Expansion Path (goes through points *E* and G).

To start, the consumer is maximizing utility by choosing bundle *E*. The price of Good Y is P_y, the price of X is $(P_x)_1$, and the consumer's income is $B. When the price of Good X falls to $(P_x)_2$, the budget line changes slope (rotates), and the consumer adjusts his consumption, choosing bundle G on the new budget line and the higher indifference curve, U_2. The movement to bundle G includes the Income and Substitution Effects. Let's separate them. The consumer's ability to reach the higher indifference curve results from the Income Effect. Removing the Income Effect places the consumer back on his original indifference curve U_1. Facing the new prices (new slope of budget line), but without the benefit of the Income Effect, what bundle of goods would he choose? At the new (relative) prices, the consumer would maximize utility along his original indifference curve at point *F*. Therefore, the movement from point *E* to point *F* represents the Substitution Effect. When we restore his income, he then is able to move to the higher indifference curve, so the movement from point *F* to point G represents the Income Effect.

Inferior Good

Inferior goods are goods which have a negative income elasticity, which means that as our income rises, we buy *less* of it. For an inferior good, the Income and Substitution Effects work in opposite directions. As the price of Good X falls,

1. we buy less of Good X because we have more real income to spend (Income Effect for an inferior good), but

Figure A8-4 (a and b) **INCOME AND SUBSTITUTION EFFECTS.**

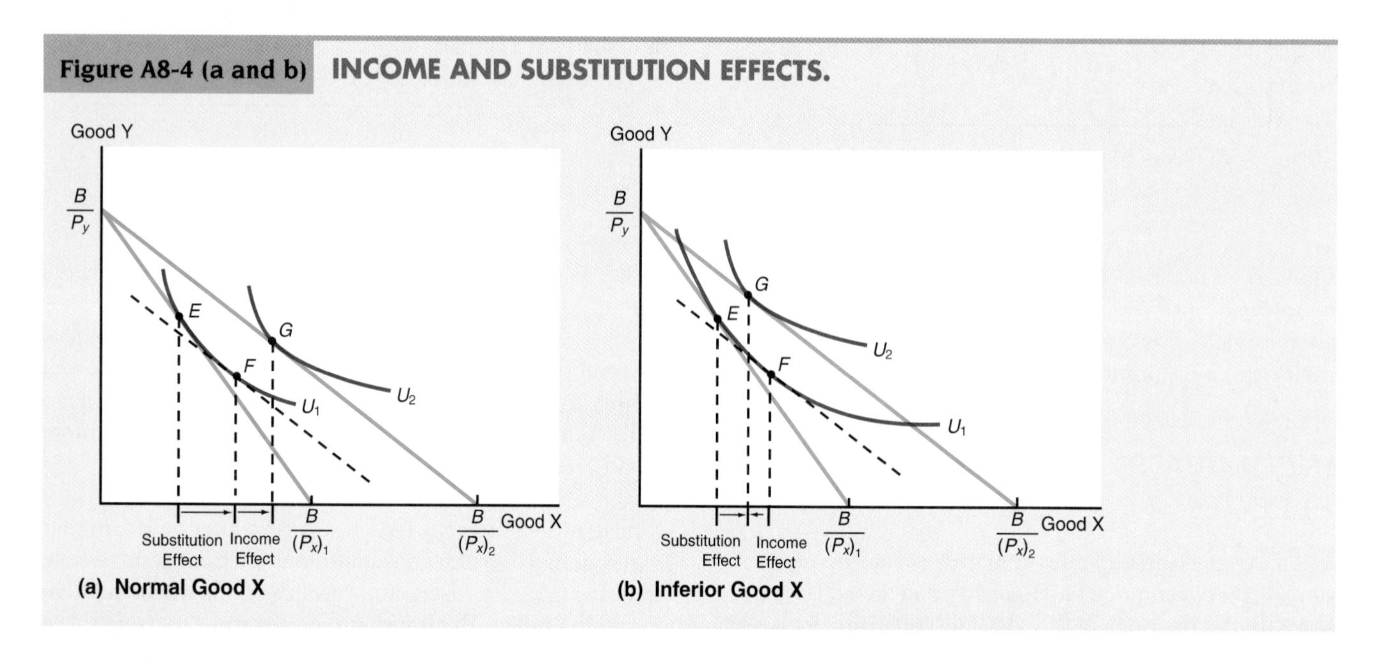

Figure A8-5 (a and b) DERIVING THE DEMAND CURVE FOR GOOD X

Lowering the price of good X from P_1 to P_2 to P_3 increases the consumption (quantity demanded) of Good X from X_1 to X_2 to X_3. Plotting these price-quantity combinations in (b) yields the demand curve for Good X.

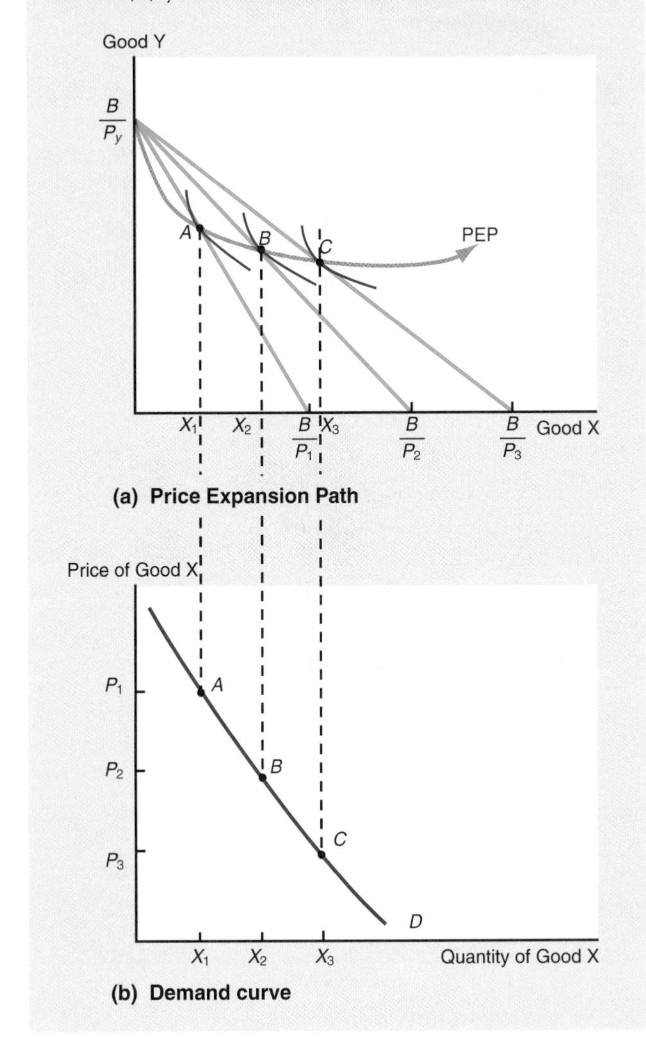

2. we buy more of Good X because Good X has become relatively cheaper than other goods (Substitution Effect).

Usually, the Substitution Effect dominates, so the demand curve slopes downward. Figure A8-4 (b) isolates the impact of a price change into the two effects for an inferior good. Keeping the individual on his original indifference curve U_1 and allowing relative prices to change allows only the Substitution Effect, so he is able to move from point E to point F. Restoring his income allows him to move to the higher indifference curve, U_2. Because Good X is inferior, as income rises, less of it is bought, and in this case, the Income Effect works in the opposite direction to the Substitution Effect. In rare cases, *the Income Effect may even dominate the Substitution Effect, and the demand curve will slope upwards*. This would be the case of a **Giffen good**.

PRICE-EXPANSION PATH AND THE DEMAND CURVE

The demand curve illustrates the relationship between price and quantity demanded. We can derive the demand curve using the information from the Price-Expansion path. Figure A8-5 (a and b) illustrates how this is done. Holding all else constant, let's change the price of Good X. In (a), when the price of Good X is high (P_1), a small amount of X is bought (X_1). As the price of X falls (P_2, then P_3), more X is bought (X_2, then X_3). (Remember that if *only* Good Y were bought, the same amount of it could be purchased since the price of Y has not changed, so the Good Y intercept does not change.) Transferring the price and quantity information to (b), we plot points for P_1 and X_1, for P_2 and X_2, and for P_3 and X_3 to create the demand curve for Good X.

Key Terms

Engel curve *(194)*
Giffen good *(197)*
Income Effect *(195)*
Income-Expansion Path *(194)*
Price-Expansion Path *(194)*
Substitution Effect *(195)*

Production and Cost Analysis I

9

After reading this chapter, you should be able to:

- Differentiate economic profit from accounting profit.
- Distinguish between long-run and short-run production.
- State the law of diminishing marginal productivity.
- Calculate fixed costs, variable costs, marginal costs, total costs, average fixed costs, average variable costs, and average total costs, given the appropriate information.
- Distinguish the various kinds of cost curves and describe the relationships among them.
- Explain why the marginal and average cost curves are U-shaped.
- Explain why the marginal cost curve always goes through the minimum point of an average cost curve.

Production is not the application of tools to materials, but logic to work.

Peter Drucker

The ability of market economies to supply material goods and services to members of their societies makes them the envy of many other societies and is one of the strongest arguments for using the market as a means of organizing society. Somehow markets are able to channel individuals' imagination, creativity, and drive into the production of material goods and services that other people want. Markets do this by giving people incentives to supply goods and services to the market.

Ultimately all supply comes from individuals. Individuals control the factors of production such as land, labour, and capital. Why do individuals supply these factors to the market? Because they want something in return. This means that industry's ability to supply goods is dependent on individuals' willingness to supply the factors of production they control.

Production occurs when factors are transformed into goods and services.

The analysis of supply is more complicated than the analysis of demand. In the supply process, people first offer their factors of production to the market. Then the factors are transformed by firms, such as GM or Corel, into goods that consumers want. **Production** occurs when factors of production (inputs) are transformed into goods and services. It is the name given to that *transformation of factors into goods and services*.

The supply of factors of production (considered in detail in later chapters) is different from the supply of produced goods. This chapter assumes that the prices of factors of production are constant, which simplifies the analysis of the supply of produced goods enormously. There's no problem with doing this as long as we remember that behind any produced good are individuals' factor supplies. Ultimately people, not firms, are responsible for supply.

Two chapters (this chapter and the next) consider production, costs, and supply. This chapter introduces the production process and short-run cost analysis. The next chapter focusses on long-run costs and how cost analysis is used in the real world.

Firms are the basic unit of analysis on the supply side; consumers are the basic unit of analysis on the demand side of the market.

THE ROLE OF THE FIRM

With goods that already exist, like housing and labour, the law of supply is rather intuitive. Their supply to the market depends on people's opportunity costs of keeping them for themselves and of supplying them to the market. But many of the things we buy (such as VCRs, cars, and jackets) don't already exist; they must be produced. The supply of such goods depends on production.

Firms:
1. Organize factors of production,
2. produce goods and services, and/or
3. sell produced goods and services.

A key concept in production is the firm. A **firm** is *an economic institution that transforms factors of production into goods and services*. A firm (1) organizes factors of production, and/or (2) produces goods, and/or (3) sells produced goods to individuals, businesses, or government.

Which combination of activities a firm will undertake depends on the cost of undertaking each activity relative to the cost of subcontracting the work out to another firm. When the firm only organizes production, the firm's sole role is organization. In most firms, more and more of the organizational structure of business is being separated from the production process. As cost structures change because of technological advances such as the Internet, an increasing number of firms can effectively provide organizational activities for production-oriented firms.

More and more of the organizational structure of business is being separated from the production process.

The Firm and the Market

The firm operates within a market, but, simultaneously, it abandons the market, since it replaces the market with command and control. How an economy operates—which activities are organized through markets, and which activities are organized through firms—depends on *transaction costs* (costs of undertaking trades through the market) and the rent or command over resources that organizers can appropriate to themselves by organizing production in a certain way. Ronald Coase won a Nobel Prize in 1991 for groundbreaking work on the nature of the firm and transactions costs.

Firms replace the market with command and control.

In Chapter 3 we discussed the types of firms that exist in real life. They include sole proprietorships, partnerships, corporations, for-profit firms, nonprofit firms, and co-operatives. These various firms are the production organizations that translate factors of production into consumer goods.

APPLYING THE TOOLS

Value Added and the Calculation of Total Production

This book (like all economics textbooks) treats production as if it were a one-stage process—as if a single firm transformed a factor of production into a consumer good. This keeps the analysis manageable. Reality, however, is more complicated, since most goods go through numerous stages of production.

For example, consider the production of desks. One firm transforms raw materials into usable raw materials (iron ore into steel); another firm transforms usable raw materials into more usable inputs (steel into steel rods, bolts, and nuts); another firm transforms those inputs into desks, which it sells wholesale to a general distributor, which then sells them to a retailer, which sells them to consumers. Many goods go through five or six stages of production and distribution. As a result, adding up all the sales of all the firms would overstate how much total production was taking place in the economy.

To figure out how much total production is actually taking place, economists use the concept of *value added.* Value added is the contribution that each stage of production makes to the final value of a good. A firm's value added is determined by subtracting the value (cost) of the inputs bought from other firms from the value of the firm's total output. For example, if a desk assembly firm spends $4,000 of its revenue on component parts and sells its output for $6,000, its value added is $2,000.

When adding up all the stages of production, the value added of all the firms involved must equal exactly 100 percent, and no more, of the total output. When discussing "a firm's" production of a good, think of that firm as a composite firm consisting of all the firms contributing to the production and distribution of that product.

Why is it important to remember that there are various stages of production? Because it emphasizes how complicated producing a good really is. If any one stage is interrupted, the good doesn't get to the consumer. Producing a better mousetrap isn't enough. The firm must also be able to get it out to consumers and let them know that it's a better mousetrap. Many people's dreams of supplying a better product to the market have been squashed by this reality.

Firms Maximize Profit

In analyzing demand, the basic unit of analysis was the individual consumer. The consumer's goal was to maximize utility given an income constraint. On the supply side, the basic unit of analysis is the firm. Its goal is to maximize profit, which is defined as:

Profit = *Total revenue* − *Total cost*

In accounting, **total revenue** *equals total sales times price*; if a firm sells 1,000 pairs of earrings at $5 each, its total revenue is $5,000. For an accountant, total costs are the wages paid to labour, rent paid to owners of capital, interest paid to lenders, and actual payments to other factors of production. If the firm paid $2,000 to employees to make the earrings and $1,000 for the materials, total cost is $3,000.

Explicit costs: a firm's costs that are quantifiable and usually represent an actual payment made by the firm.

In determining what to include in total revenue and total cost, accountants focus on revenues and **explicit costs**. That's because they must have *quantifiable measures*—for example, receipts, that go into a firm's income statement. For this reason, you can think of **accounting profit** *as revenue minus explicit cost*. The accounting profit for the earring firm is $2,000.

Economists measure costs differently and therefore have a different measure of profit. Economists include both explicit and implicit costs. Their measure of profit is revenues minus both explicit and implicit costs.

Accounting focusses on explicit costs; economics focusses on both explicit and implicit costs.

What are implicit costs? **Implicit costs** include *the opportunity costs of the factors of production* provided by the owners of the business. Say that the owner of the earring firm could have earned $1,500 working elsewhere if he did not own the earring firm. The opportunity cost of working in his own business is $1,500. It must be covered, so it is an implicit cost of doing business and is included as a cost. For economists, **total cost** is *ex-*

plicit payments to the factors of production plus the opportunity cost of the factors provided by the owners of the firm. Total cost of the earring firm is $3,000 in explicit cost and $1,500 in implicit cost, or $4,500. Generally implicit costs have no receipts and must be estimated, which is why accountants do not include them.

Economic profit = *Revenue* – (*Explicit and implicit cost*)

So in this case, economic profit is $5,000 – ($3,000 + $1,500) = $500. The difference between accounting profit and economic profit really has to do with measurability. Implicit costs must be estimated, and the estimations can sometimes be inexact. General accounting rules do not permit such inexactness because it might allow firms to misstate their profit, something accounting rules are designed to avoid.

Implicit costs: a firm's expenses that are difficult to quantify because there is no explicit payment made for the use of the resource, measured as opportunity cost.

THE PRODUCTION PROCESS

Supply represents the market's ability to provide the goods and services people want. Underlying supply is production, which firms control.

The Long Run and the Short Run

The production process is generally divided into a *long-run* planning decision, in which a firm chooses the least expensive method of producing from among all possible methods, and a *short-run* adjustment decision, in which a firm adjusts its production within the constraints of previous (long run) decisions. Generally, the firm adjusts its production within the constraints of its current production facility, which was determined by an earlier "long-run" decision.

In a **long-run decision** *a firm chooses among all possible production techniques*. This means that it can choose the size of the plant it wants, the type of machines it wants, and the location it wants. The firm has fewer options in a **short-run decision,** in which *the firm is constrained by past choices in regard to what production decisions it can make*.

A long-run decision is a decision in which the firm can choose among all possible production techniques. A short-run decision is a decision in which the firm is constrained generally by its current facility in regard to what production decisions it can make.

The terms *long run* and *short run* do not necessarily refer to specific periods of time independent of the nature of the production process. They refer to the degree of flexibility the firm has in changing the level of output. In the long run, by definition, the firm can vary the inputs as much as it wants. In the short run, flexibility is limited. The short run, however, is not a specific length of time. The short run refers to the period of time during which one factor of production cannot be changed. Generally, the production facility (the 'plant') is fixed in short run. Depending on the industry, the short run can vary dramatically in length of time. For a corner grocery store, it only takes three or four months to build or expand the facility (to change "plant size"). It would require only a few months to draw up building plans, have them approved by the city, and complete the new structure. The short run is only a short period of time in this case. For an oil refinery to expand its facility, however, would require a much longer time. The short run for an oil refinery can be as long as four or five years. *In the long run all inputs are variable; in the short run some inputs are fixed.*

The short run for an oil refinery may be five years or longer.

Production Tables and Production Functions

How a firm combines factors of production to produce goods and services can be presented in a **production table** (*a table showing the output resulting from various combinations of factors of production or inputs*).

Real-world production tables are complicated. They often involve hundreds of inputs, hundreds of outputs, and millions of possible combinations of inputs and outputs.

Studying these various combinations and determining which is best requires expertise and experience. Business schools devote entire courses to it (operations research and production analysis); engineering schools devote entire specialties to it (industrial engineering).

Studying the problems and answering the questions that surround production make up much of what a firm does: What combination of outputs should it produce? What combination of inputs should it use? What combination of techniques should it use? What new techniques should it explore? To answer these questions, the managers of a firm look at a production table.

The marginal product is the additional output resulting from an additional input, other inputs constant; the average product is the total output divided by the quantity of the input.

Most of the production decisions firms make are short-run decisions involving changes in output at a given production facility. The firm can increase or decrease production by adjusting the amounts of the variable inputs, such as labour or materials, used with the fixed factor, the plant. The production table in Figure 9-1 contains production information for a particular plant, including machinery. Columns 1 and 2 of the table tell us how output of earrings varies as the variable input (the number of workers) changes. For example, you can see that with 3 workers the firm can produce 17 pairs of earrings. Column 3 tells us workers' **marginal product** *(the additional output that will result from an additional worker, other inputs constant)*. Column 4 tells us the workers' **average product** *(output per worker)*.

It is important to distinguish marginal product from average product. Workers' average product is the total output divided by the number of workers. For example, let's consider the case of 5 workers. Total output is 28, so average product is 5.6 (28 divided by 5). To find the marginal product we must ask how much additional output will result if we change the number of workers. For example, if we change from 4 to 5 workers, the additional worker's marginal product will be 5; if we change from 5 to 6, the additional worker's marginal product will be 3. Marginal product occurs as we *change* production; that's why the marginal products are written *between* each level of output.

The information in a production table is often summarized in a production function. A **production function** is *the relationship between the inputs (factors of production) and outputs*. Specifically, the production function tells the maximum amount of output that can be derived from a given number of inputs. Figure 9-1(b) is the production function that displays the information in the production table in Figure 9-1(a). The number of workers is on the horizontal axis and the output of earrings is on the vertical axis.

The Law of Diminishing Marginal Productivity

Q-1 What are the normal shapes of marginal productivity and average productivity curves?

Figure 9-1(c) graphs the workers' average and marginal productivities from the production function in Figure 9-1(b). (Alternatively you can determine those graphs by plotting columns 3 and 4 from the table in Figure 9-1(a).) Notice that both marginal and average productivities are initially increasing, but that eventually they both decrease. Between 7 and 8 workers, the marginal productivity of workers actually becomes negative.

This means that initially this production function exhibits increasing marginal productivity (TP curve increases at an accelerating rate between 0 and 2½ workers), and then it exhibits *diminishing marginal productivity* (TP curve increases at a decelerating rate between 2½ and 7½ workers). Eventually the production function exhibits negative marginal productivity (TP curve falls after 7½ workers). You can confirm that the productivity information in (b) and (c) correspond. Technically the marginal productivity curve is a graph of the slope of the total product curve.

The most important area of these relationships is the area of diminishing marginal productivity and falling average product (between 4 and 7.5 workers). Why? Because that's the most likely area for a firm to operate in. For example, if it's in the first range and marginal productivity is increasing, a firm can increase its existing workers' output

Figure 9-1 (a, b, and c) **A PRODUCTION TABLE AND PRODUCTION FUNCTION**

The production function in (**b**) is a graph of the production table in (**a**). Its shape reflects the underlying production technology. The graph in (**c**) shows the marginal and average product. Notice that when marginal product is increasing, the production function is bowed upward; when marginal product is decreasing, the production function is bowed downward, and when marginal product is zero, the production function is at its highest point. Firms are interested in producing where both average product and marginal product are positive and falling, which starts at 4 workers and ends at 7.5 workers.

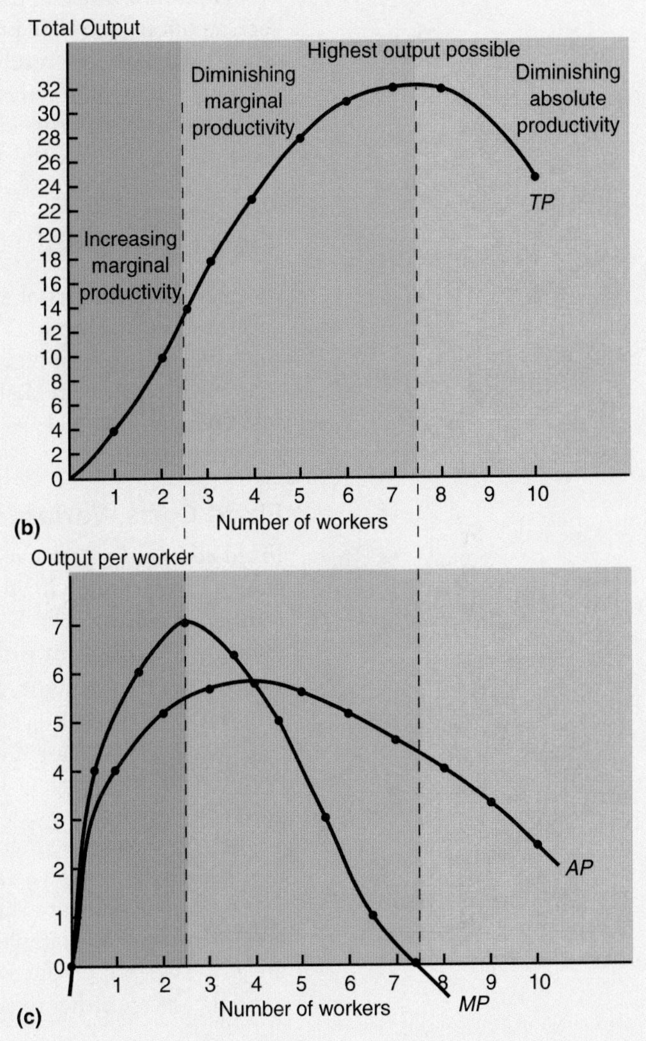

Number of workers	Total output	Marginal product (change in total output)	Average product (total product/ number of workers)	
1	4	4	4	Increasing marginal productivity
2	10	6	5	
3	17	7	5.7	Diminishing marginal productivity
4	23	6	5.8	
5	28	5	5.6	
6	31	3	5.2	
7	32	1	4.6	
8	32	0	4.0	Diminishing absolute productivity
9	30	−2	3.3	
10	25	−5	2.5	

(a)

by hiring more workers; it will have a strong incentive to do so and get out of that range. Similarly, if hiring an additional worker actually cuts total output (as it does when marginal productivity is negative), the firm would be crazy to hire that worker. So it stays out of that range.

Q-2 Firms are likely to operate on what portion of the marginal productivity curve?

This range of the relationship between fixed and variable inputs is so important that economists have formulated a law that describes what happens in production processes when firms reach this range—when more and more of one input is added to a fixed amount of another input. The **law of diminishing marginal productivity** states that *as more and more of a variable input is added to an existing fixed input, eventually the additional output obtained from that additional input is going to fall.*

The law of diminishing marginal productivity states that as more and more of a variable input is added to an existing fixed input, after some point the additional output one gets from the additional input will fall.

The law of diminishing marginal productivity is sometimes called the *flowerpot law* because if it didn't hold true, the world's entire food supply could be grown in one flowerpot. In the absence of diminishing marginal productivity, we could take a flowerpot and keep adding seeds to it, getting more and more food per seed until we had enough to feed the world. In reality, however, a given flowerpot is capable of producing only so much food, no matter how many seeds we add to it. At some point, as we add more and more seeds, each additional seed will produce less additional food than did the seed before it. Eventually the pot may even reach a stage of diminishing absolute productivity, in which the total output, not simply the output per unit of input, decreases as inputs are increased (negative marginal productivity). In comparison, the limit to a firm's production (in the short run) is generally the size of its facility: its "plant size".

THE COSTS OF PRODUCTION

Owners and managers of a firm probably discuss costs far more than anything else. Invariably costs are too high and the firm is trying to figure out ways to lower them. But the concept *costs* is ambiguous; there are many different types of costs and it's important to know what they are. Let's consider some of the most important categories of costs in reference to Table 9-1, which shows costs associated with making between 3 and 32 pairs of earrings.

Fixed Costs, Variable Costs, and Total Costs

9.1 
see page 213

Fixed costs are *costs that are spent and cannot be changed in the period of time under consideration*. There are no fixed costs in the long run since all inputs are variable and hence their costs are variable. Fixed costs are costs which do not vary with the level of production. Suppose your firm rents a space in the local mall. Whether you sell a lot of items or only a few items this month, or even if you lock up and go on vacation for a month, you still must pay your rent.

Fixed costs are shown in column 2 of Table 9-1. Notice that fixed costs remain the same ($50) regardless of the level of production. As you can see, it doesn't matter whether output is 15 or 20 or zero; fixed costs are always $50.

Besides buying the machine, the silversmith must also hire workers. These workers are the earring firm's **variable costs**—*costs that change as output changes*. The earring firm's variable costs are shown in column 3. Notice that as output increases, variable costs increase. For example, when the firm produces 10 pairs of earrings, variable costs are $108; when it produces 16, variable costs rise to $150.

All costs are either fixed or variable, so the *total cost* is the sum of the fixed and variable costs:

$TC = FC + VC$

$$TC = FC + VC$$

The earring firm's total costs are presented in column 4. Each entry in column 4 is the sum of the entries in columns 2 and 3 in the same row. For example, to produce 16 pairs of earrings, fixed costs are $50 and variable costs are $150, so total cost is $200.

Average Total Cost, Average Fixed Cost, and Average Variable Cost

Total cost, fixed cost, and variable cost are important, but much of a firm's concern is over per unit, or average costs. The next distinction we need to make is between total cost and average cost. To arrive at the earring firm's average costs, we simply divide the total cost by the quantity produced. Each of the three costs, total cost, fixed cost, and variable cost, has a corresponding average cost.

Average cost equals total cost divided by quantity.

For example, **average total cost** (often called average cost) equals *total cost divided by the quantity produced*. Thus:

$$ATC = TC/Q$$

Average fixed cost equals *fixed cost divided by quantity produced:*

$$AFC = FC/Q$$

TABLE 9-1 The Cost of Producing Earrings

1	2	3	4	5	6	7	8
Output (*Q*)	Fixed Costs (*FC*)	Variable Costs (*VC*)	Total Costs (*TC*) (*FC* + *VC*)	Marginal Costs (*MC*) (Change in total costs/ Change in output)	Average Fixed Costs (*AFC*) (*FC*/*Q*)	Average Variable Costs (*AVC*) (*VC*/*Q*)	Average Total Costs (*ATC*) (*AFC* + *AVC*)
0	50	0	50		–	–	–
3	50	38	88	$12	$16.67	$12.66	$29.33
4	50	50	100		12.50	12.50	25.00
9	50	100	150	8	5.56	11.11	16.67
10	50	108	158		5.00	10.80	15.80
16	50	150	200	7	3.13	9.38	12.50
17	50	157	207		2.94	9.24	12.18
22	50	200	250	10	2.27	9.09	11.36
23	50	210	260		2.17	9.13	11.30
27	50	255	305	15	1.85	9.44	11.30
28	50	270	320		1.79	9.64	11.43
32	50	350	400		1.56	10.94	12.50

Average variable cost equals *variable cost divided by quantity produced:*

$$AVC = VC/Q$$

Average fixed cost and average variable cost are shown in columns 6 and 7 of Table 9-1. The most important average cost concept, average total cost, is shown in column 8. Average total cost is the sum of average fixed cost and average variable cost:

$$ATC = AFC + AVC$$

The average total cost of producing 16 pairs of earrings is $12.50. It can be calculated by dividing total cost ($200) by output (16).

Marginal Cost

All these costs are important to our earring firm but in deciding how many pairs of earrings to produce, marginal cost, which appears in column 5, is most important.[1] **Marginal cost** equals the change (increase) in total cost from a change (increase) in the level of output by one unit:

$$MC = \Delta TC/\Delta Q$$

Let's find marginal cost by considering what happens if our earring firm increases production by one unit—from 9 to 10. Looking again at Table 9-1, we see that the total cost rises from $150 to $158. In this case the marginal cost of producing the 10th unit is $8. Since it represents the cost of changing output from 9 to 10 units, marginal cost goes in between the two outputs (that is, at the midpoint).

Q-3 If total costs are 400, fixed costs are 0, and output is 10, what are average variable costs?

GRAPHING COST CURVES

We can look at the firm's cost information in two ways, as total costs or as per unit costs. We can illustrate these in a graph, putting quantity on the horizontal axis and a dollar measure of various costs on the vertical axis.

Fixed costs like rent do not vary in the short run; they must be paid, even if output is zero.

[1]Since only selected output levels are shown, not all entries have marginal costs. For a marginal cost to exist, there must be a marginal change, a change by only one unit.

The production process transforms factors of production into the supply of goods and services.

The marginal cost curve goes through the minimum points of the average total cost curve and average variable cost curve; each of these curves is U-shaped. The average fixed cost curve decreases continuously.

Q-4 Draw a graph of both the marginal cost curve and the average cost curve.

Q-5 What determines the distance between the average total cost and the average variable cost?

Maximizing productivity or minimizing cost?

Total Cost Curves

Figure 9-2(a) graphs the total cost, total fixed cost, and total variable costs of all the levels of output given in Table 9-1. The total cost curve is determined by plotting the quantities in column 1 and the corresponding dollar figures in column 4. For example, point *L* corresponds to a quantity of 10 and a total cost of $158. Notice that the total cost curve is upward sloping: Increasing output increases total cost.

The total fixed cost curve is determined by plotting column 1 and column 2 on the graph. The total variable cost curve is determined by plotting column 1 and column 3.

As you can see, the total variable cost curve has the same shape as the total cost curve: Increasing output increases variable cost. This isn't surprising, since the total cost curve is the vertical summation of total fixed cost and total variable cost. For example, at output 10, total fixed cost equals $50 (point M); total variable cost equals $108 (point O); and total cost equals $158 (point *L*).

Per Unit Cost Curves: Average and Marginal Cost Curves

Figure 9-2(b) presents the average fixed cost curve, average total cost curve (or average cost curve, as it's generally called), average variable cost curve, and marginal cost curve associated with the cost figures in Table 9-1. Each point on the four curves represents a combination of two corresponding entries in Table 9-1. Points on the average variable cost curve are determined by plotting the quantities in column 1 and the corresponding dollar figures in column 7. Points on the average fixed cost curve are determined by plotting column 1 and column 6. Points on the average total cost curve are determined by plotting column 1 and the corresponding entries in column 8. Finally, the marginal cost curve is determined by plotting the entries in column 1 and the corresponding entries in column 5. Notice that the marginal cost figures in column 5 correspond to the quantity midpoints in column 1, so MC is plotted in between the quantities. For example, the marginal cost of changing output from 9 to 10 units is $8, so we plot $8 at 9.5 units. As was the case with the total cost curves, all the firm's owner need do is look at this graph to find the various costs associated with different levels of output.

One reason the graphical visualization of cost curves is important is that the graphs of the curves give us a good sense of what happens to costs as we change output.

Downward-Sloping Shape of the Average Fixed Cost Curve

Let's start our consideration with average fixed cost. Average fixed cost is decreasing throughout. The average fixed cost curve looks like a child's slide: It starts out with a steep decline; then it becomes flatter and flatter. What this tells us about production is straightforward: As output increases, the same fixed cost can be spread over a wider range of output, so average fixed cost falls. Average fixed cost initially falls quickly but then falls more and more slowly. As the denominator increases while the numerator stays the same, the denominator's increase has a smaller and smaller effect.

The U Shape of the Average and Marginal Cost Curves

Why do the average and marginal cost curves have the shapes they do? Or expressed another way, how does our analysis of production relate to our analysis of costs? Remember Figure 9-1. Cost analysis is simply another way of considering production analysis. The laws governing costs are the same laws that govern productivity.

In the short run, output can only be raised by increasing the variable input. But as more and more of a variable input is added to a fixed input, the law of diminishing marginal productivity enters in. First, marginal and then average productivities fall. See Figure 9-1(c). The key here is that when marginal productivity falls, marginal cost must

Figure 9-2 (a and b) **TOTAL AND PER UNIT OUTPUT COST CURVES**

Total fixed costs, shown in (**a**), are always constant; they don't change with output. All other total costs increase with output. As output gets high, costs tend to increase faster, that is, the rate of increase has a tendency to increase. In (b), the average fixed cost curve is downward sloping; the average variable cost curve and average total cost curve are U-shaped. The MC curve goes through the minimum points of the *AVC* and *ATC* curves. (The *AFC* curve is often not drawn since *AFC* represents the vertical distance between the *AVC* and *ATC*.)

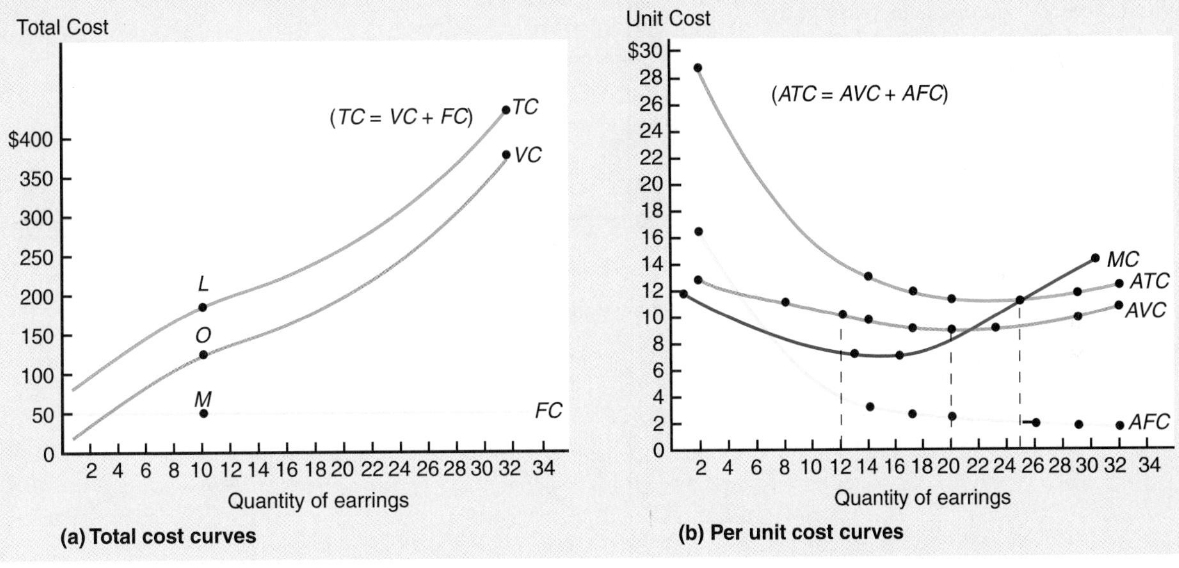

(a) Total cost curves **(b) Per unit cost curves**

rise, and when average productivity of the variable input falls, average variable cost must rise. So saying that productivity falls is equivalent to saying that cost rises.

Eventually the law of diminishing marginal productivity prevails and eventually both the marginal cost curve and the average cost curve must begin to rise. Generally, at low levels of production, marginal and average productivities tend to be increasing, which means that marginal cost and average variable cost are initially falling. If they're falling initially and rising eventually, at some point they must be neither rising nor falling. This results in U-shaped marginal cost and average variable cost curves.

As more and more of a variable input is added to a fixed input, the law of diminishing marginal productivity causes marginal and average productivities to fall. As these fall, marginal and average costs rise.

As you can see in Figure 9-2(b), the average total cost curve has the same general U shape as the average variable cost curve. It has the same U shape because it is the vertical summation of the average fixed cost curve and the average variable cost curve.

Average total cost initially falls faster and then rises more slowly than average variable cost. If we increased output enormously, the average variable cost curve and the average total cost curve would almost meet.

Q-6 If you increase output enormously, what two cost curves would almost meet?

The Relationship between the Marginal Productivity and Marginal Cost Curves

Let's now consider the relationship between marginal product and marginal cost. Figure 9-3(a) contains a marginal cost curve and an average variable cost curve. Initially costs are falling. MC reaches a minimum point at an output of 12, after which the cost of additional units begins to rise. At an output of 21, AVC begins to rise as MC exceeds AVC.

Figure 9-3(b) repeats the average and marginal productivity curves from Figure 9-1(c).

If $MP > AP$, then AP is rising.
If $MP < AP$, then AP is falling.

Figure 9-3 (a and b) **THE RELATIONSHIP BETWEEN PRODUCTIVITY AND COSTS**

The shapes of the cost curves are mirror-image reflections of the shapes of the corresponding productivity curves. When one is increasing, the other is decreasing; when one is at a minimum, the other is at a maximum.

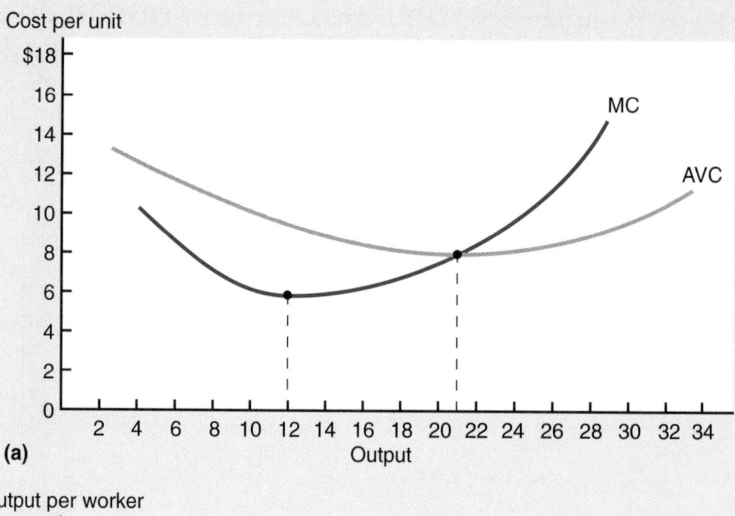

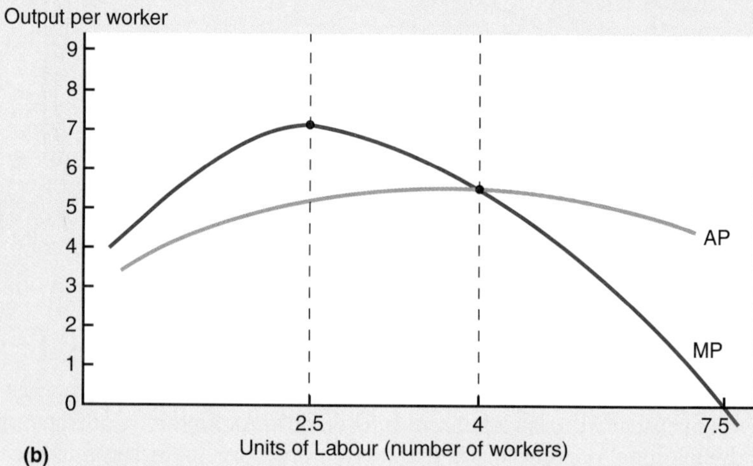

Q-7 When the marginal cost equals the minimum point of the average variable cost, what is true about the average productivity and marginal productivity of workers?

When the productivity curves are falling, the corresponding cost curves are rising.

If you look at Figure 9-3 (a and b) carefully, you'll see that one is simply the mirror image of the other. The minimum point of the average variable cost curve (output = 21) is the same level of output as the maximum point of the average productivity curve. The minimum point of the marginal cost curve (output = 12) occurs when the marginal productivity curve is at its maximum point. The minimum point of the average variable cost curve (output = 21) occurs when the average productivity curve is at its maximum point. So, when the productivity curves are rising, the corresponding cost curves are falling. As productivity rises, costs per unit decrease, and as productivity decreases, costs per unit increase.

The Relationship between the Marginal Cost and Average Cost Curves

9.2 see page 213

Let's consider some of the important relationships between the marginal cost curve on the one hand and the average variable cost and average total cost curves on the other. These relationships are shown graphically for a different production process in Figure 9-4.

Figure 9-4 THE RELATIONSHIP OF MARGINAL COST CURVE TO AVERAGE VARIABLE COST AND AVERAGE TOTAL COST CURVES

The marginal cost curve goes through the minimum point of both the average variable cost curve and the average total cost curve. Thus, there is a small range where average total costs are falling and average variable costs are rising.

If MC < *ATC*, then *ATC* is falling.
If MC = *ATC*, then *ATC* is at its low point.
If MC > *ATC*, then *ATC* is rising.

If MC < *AVC*, then *AVC* is falling.
If MC = *AVC*, then *AVC* is at its low point.
If MC > *AVC*, then *AVC* is rising.

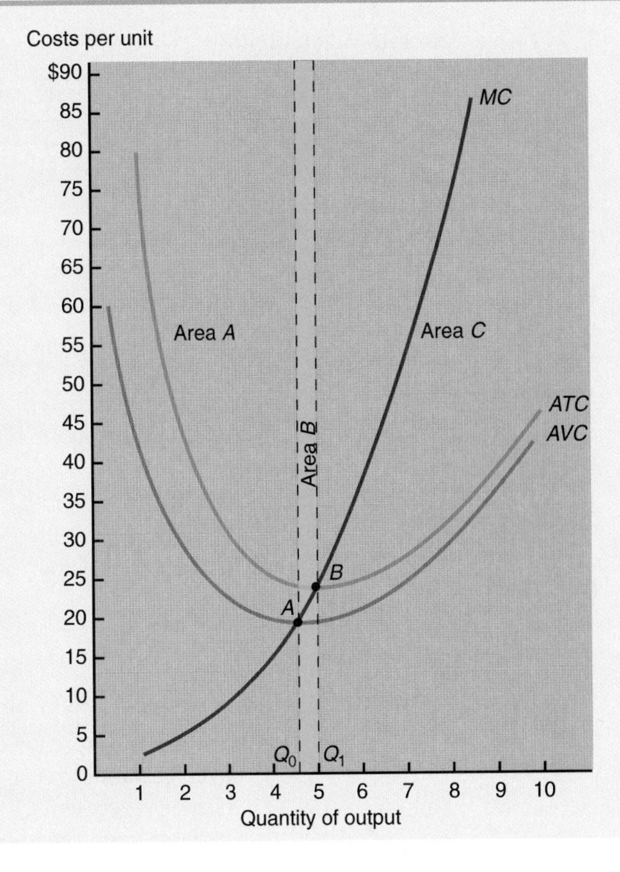

Let's first look at the relationship between marginal cost and average total cost. In areas A and B, at output below 5, even though marginal cost is rising, average total cost is falling. Why? Because in areas A and B the marginal cost curve is below the average total cost curve. At point B, where average total cost is at its lowest, the marginal cost curve intersects the average total cost curve. In area C, above output 5, where average total cost is rising, the marginal cost curve is above the *ATC* curve.

The position of marginal cost relative to average total cost tells us whether average total cost is rising or falling.

If MC < *ATC*, then *ATC* is falling.

If MC = *ATC*, then *ATC* is at its minimum point.

If MC > *ATC*, then *ATC* is rising.

To understand why this is, think of it in terms of your grade point average. If you have a B average and you get a C on the next test (that is, your marginal grade is a C), your grade point average will fall below a B. Your marginal grade is below your average grade, so your average grade is falling. If you get a C+ on the next exam (that is, your marginal grade is a C+), *even though your marginal grade has risen from a C to a C+*, your grade point average will still fall. Why? Because your marginal grade is below your average grade.

This general relationship also holds for marginal cost and average variable cost.

If MC < *AVC*, then *AVC* is falling.

If MC = *AVC*, then *AVC* is at its minimum point.

When marginal cost exceeds average cost, average cost must be rising. When marginal cost is less than average cost, average cost must be falling. This relationship explains why marginal cost curves always intersects either of the average cost curves at the minimums of the average cost curves.

Marginal cost is relevant for production decisions; average cost is relevant for profit calculations.

Q-8 If marginal costs are increasing, what is happening to average total costs?

Q-9 Is there a relationship between productivity and cost? Explain.

We've covered a lot of costs and cost curves quickly, so a review is in order. First, let's list the cost concepts and their definitions.

Total cost: the sum of all costs.

Variable cost: cost of variable inputs. Variable cost does not include fixed cost.

Fixed cost: cost that is already spent and cannot be recovered. (It exists only in the short run.)

Marginal cost: the additional cost resulting from a one-unit increase in output.

Average total cost: total cost divided by total output (*TC/Q*).

Average variable cost: variable cost divided by total output (*VC/Q*).

Average fixed cost: fixed cost divided by total output (*FC/Q*).

Each of these costs is represented by a curve. Some of these curves have specific relationships to the other cost curves.

1. *MC* intersects *AVC* and *ATC* at their minimum points.
2. If $MC > AVC$, then *AVC* is rising. If $MC < AVC$, then *AVC* is falling.
3. If $MC > ATC$, then *ATC* is rising. If $MC < ATC$, then *ATC* is falling.
4. *ATC:* a U-shaped curve higher than the *AVC.*
5. *AVC:* a U-shaped curve, with the minimum point slightly to the left of the ATC's minimum point.
6. *AFC:* a downward-sloping curve that starts high, initially decreases rapidly, and then decreases slowly.

If $MC > AVC$, then *AVC* is rising.

The minimum of the average total cost curve is to the right of the minimum of the average variable cost curve. This relationship is best seen in area *B* of Figure 9-4, when output is between Q_0 and Q_1. In this area the marginal cost curve is above the average variable cost curve, so average variable cost is rising; but the MC curve is below the average total cost curve, so average total cost is falling.

Q-10 Why does the marginal cost curve intersect the average total cost curve at the minimum point?

The intuitive explanation for what is represented in this area is that average total cost includes average variable cost, but it also includes average fixed cost, which is (always) falling.

CONCLUSION

A firm is a legal entity that combines the factors of production to produce the various goods and services which are available in an economy. The goal of a firm is to maximize profit, and it does so by making a series of long run and short run decisions. In the long run, since all options are open to it, the firm may choose any combination of inputs that will maximize its profit. In the short run, however, its options are limited by previous decisions. Short-run decisions generally involve the firm maximizing profit by choosing output, subject to an earlier binding decision, an earlier "long-run" decision regarding production technique and plant size.

So in the long run, the firm can set up any production facility and use any combination of inputs it desires, but in the short run, the firm is limited to decisions involving the particular facility it has already set up. In both the long run and the short run, however, the firm's goal is to maximize profit by producing at an economically efficient point.

Chapter Summary

- Accounting profit is revenue less explicit cost. Economists include explicit and implicit cost in their determination of profit.
- Implicit costs include opportunity cost of time and capital provided by the owners of the firm.
- In the long run a firm can choose among all possible production techniques; in the short run it is constrained in its choices.
- The law of diminishing marginal productivity states that as more and more of a variable input is added to a fixed input, the additional output the firm gets will eventually be decreasing.
- Costs are generally divided into fixed costs, variable costs, and total costs. $TC = FC + VC$;
- MC = change in TC for one-unit change in output
  $MC = \Delta TC/\Delta Q$
- $AFC = FC/Q$
  $AVC = VC/Q$
  $ATC = TC/Q$
  $ATC = AFC + AVC$
- The average variable cost curve and marginal cost curve are mirror images of the average product curve and the marginal product curve, respectively.
- The law of diminishing marginal productivity causes marginal and average costs to rise.
- If $MC < ATC$, then ATC is falling.
  If $MC = ATC$, then ATC is constant.
  If $MC > ATC$, then ATC is rising.
- The marginal cost curve goes through the minimum points of the average variable cost curve and average total cost curve.

Key Terms

accounting profit *(200)*
average fixed cost *(204)*
average product *(202)*
average total cost *(204)*
average variable cost *(205)*
economic profit *(201)*
explicit cost *(200)*
firm *(199)*
fixed costs *(204)*
implicit cost *(200)*
law of diminishing marginal productivity *(203)*
long-run decision *(201)*
marginal cost *(205)*
marginal product *(202)*
production *(199)*
production function *(202)*
production table *(201)*
profit *(200)*
short-run decision *(201)*
total cost *(200)*
total revenue *(200)*
variable costs *(204)*

Questions for Thought and Review

1. What costs and revenues do economists include when calculating profit that accountants don't include? Give an example of each.
2. "There is no long run; there are only short and shorter runs." Evaluate that statement.
3. What is the difference between marginal product and average product?
4. If average product is falling, what is happening to short-run average variable cost?
5. If marginal cost is increasing, what do we know about average cost?
6. If average productivity falls, will marginal cost necessarily rise? How about average cost?
7. Say that neither labour nor machines are fixed but that there is a 50 percent quick-order premium paid to both workers and machines for delivery of them in the short run. Once you buy them, they cannot be returned, however. What do your short-run marginal cost and short-run average total cost curves look like?
8. If machines are variable and labour fixed, how will the general shapes of the short-run average cost curve and marginal cost curve change?
9. If you increase production to an infinitely large level, the average variable cost and the average total cost will merge. Why?
10. Explain whether the following statements are true or false: (a) Supplying labour depends on opportunity costs because labour already exists. (b) Supplying goods that need to be produced does not depend on opportunity costs since they do not already exist.
11. Explain how studying for an exam is subject to the law of diminishing marginal productivity.
12. Labour costs are 17.5 percent of revenue per vehicle for General Motors. In union negotiations in the late

1990s, GM attempted to cut its workforce to increase productivity. Together with the reductions they expected in jobs, GM officials hoped to make its North American operations fully competitive with its U.S. and Japanese rivals on total costs. Why are productivity gains so important to GM?

Problems and Exercises

1. Sarah's cookies are the best in the world. She has been offered a job by Cookie Monster, Inc., to come to work for them at $125,000 per year. Currently, she is producing her own cookies, and she has revenues of $260,000 per year. Her costs are $40,000 for labour, $10,000 for rent, $35,000 for ingredients, and $5,000 for utilities. She has $100,000 of her own money invested in the operation, which, if she leaves, can be sold for $40,000 that she can invest at 10 percent per year.
   a. Calculate her accounting and economic profits.
   b. Advise her as to what she should do.
2. Economan believes in free enterprise. He sets up a firm on extraterrestrial business affairs. The rent of the building is $4,000, the cost of the two secretaries is $40,000, and the cost of electricity and gas comes to $5,000. There's a great demand for his information, and his total revenue amounts to $100,000. By working in the firm, though, Economan forfeits the $50,000 he could earn by working for the Friendly Space Agency and the $4,000 he could have earned as interest had he saved his funds instead of putting them in this business. Is he making a profit or loss by an accountant's definitions of profit and loss? How about by an economist's definition?
3. Calculate, complete the table, and graph the *AFC*, *AVC*, *AC*, and *MC*:

Units	*FC*	*VC*
0	$100	$ 0
1	100	40
2	100	60
3	100	70
4	100	85
5	100	130

4. An economic consultant is presented with the following total product table.

Labour	*TP*
1	5
2	15
3	30
4	36
5	40

   a. Derive a table for average variable costs. The price of labour is $15 per hour.
   b. Show that the graph of the average productivity curve and average variable cost curve are mirror images of each other.
   c. Show the marginal productivity curve for labour inputs between 1 and 5.
   d. Show that the marginal productivity curve and marginal cost curve are mirror images of each other.
5. A firm has fixed costs of $100 and variable costs of the following:

Output	1	2	3	4	5	6	7	8	9
Variable costs	$35	75	110	140	175	215	260	315	390

   a. Graph the *AFC*, *ATC*, *AVC*, and *MC* curves.
   b. Explain the relationship between the *MC* curve and the two average cost curves.
   c. Say fixed costs dropped to $50. Graph the new *AFC*, *ATC*, *AVC*, and *MC* curves.
   d. Which curves shifted in *c*? Why?
6. Say that a firm has fixed costs of $200 and constant average variable costs of $25.
   a. Graph the *AFC*, *ATC*, *AVC*, and *MC* curves.
   b. Explain why the curves have the shapes they do.
   c. What law is not operative for this firm?
   d. Say that instead of remaining a constant $25, average variable costs increase by $5 for each unit, so that the cost of 1 is $25, the cost of 2 is $30, the cost of 3 is $35, and so on. Graph the *AFC*, *ATC*, *AVC*, and *MC* curves associated with these costs.
   e. Explain how costs would have to increase in *d* in order for the curves to have the "normal" shapes of the curves presented in the text.
7. Explain how each of the following will affect the average fixed cost, average variable cost, average total cost, and marginal cost curves faced by a steel manufacturer:
   a. New union agreement increases hourly pay.
   b. Local government imposes an annual lump-sum tax per plant.
   c. Federal government imposes a "stack tax" on emission of air pollutants by steel mills.
   d. New steel-making technology increases productivity of every worker.

Web Questions

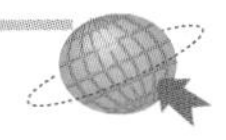

1. Go to www.travelocity.ca and find out how much it would cost to rent a car for a week, to drive from your city to a city in another province.
   a. Fill in the following cost table:

Miles	Total Cost	Marginal Cost	Average Fixed Cost	Average Variable Cost
0	____	____	____	____
500	____	____	____	____
1,000	____	____	____	____
1,500	____	____	____	____
2,500	____	____	____	____

2. Go to the Agriculture, Food, and Rural Development home page at www.agric.gov.ab.ca and click on their Cost of Production Calculator.
   a. Follow the steps to calculate the cost of "spring wheat" in "black" soil.
   b. What is the yield per acre? How much is total revenue?
   c. What are the main expenses involved in the production of the crop? Is the crop profitable?
   d. Choose another "crop" and "soil" combination. Is this a more profitable venture?
   e. Click on the Farm Machinery Cost Calculator. Select from the equipment listed. What is the purchase price? What is the value of the machinery after ten years? What is the annual operating cost? Do the figures surprise you?

Answers to Margin Questions

1. Normally the marginal productivity curve and average productivity curve are both inverted U shapes. *(202)*
2. Firms are likely to operate on the downward-sloping portion of the marginal productivity curve because on the upward-sloping portion, firms could increase workers' output by hiring more workers. A firm will continue to hire more workers at least to the point where diminishing marginal productivity sets in. *(203)*
3. Average variable costs would be 40. *(205)*
4. As you can see in the graph, both these curves are U-shaped and the marginal cost curve goes through the average cost curve at the minimum point of the average cost curve. *(206)*

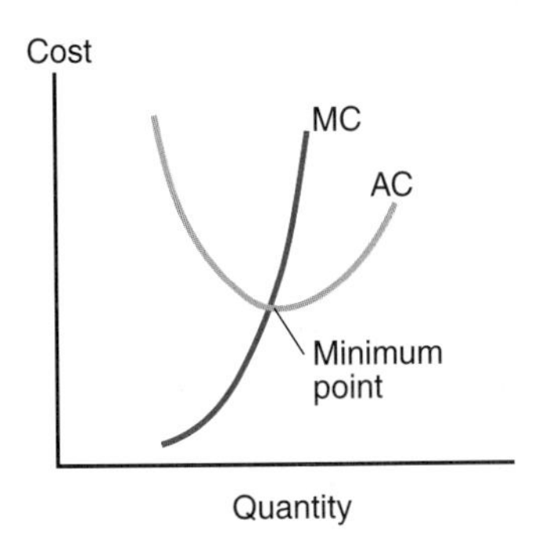

5. The distance between the average total cost and the average variable cost is determined by the average fixed cost at that quantity. If quantity increases, the average fixed cost decreases, so the two curves get closer and closer together. *(206)*
6. As output increases, the average total costs and average variable costs come closer and closer together. *(207)*
7. Since the average productivity and marginal productivity of workers are the mirror images of average costs and marginal costs, and when the marginal costs and average costs intersect the two are equal, it follows that the average productivity and marginal productivity of workers must be equal at that point. *(208)*
8. It is impossible to say what is happening to average total costs on the basis of what is happening to marginal costs. It is the position of marginal costs relative to average total costs that is important. *(209)*
9. Yes. As productivity increases, costs decrease. Specifically, when marginal product is rising, marginal cost is falling, and when average productivity is rising, average cost is falling. *(209)*
10. The marginal cost curve intersects the average total cost curve at the minimum point because once the marginal cost exceeds average total costs, the average total costs must necessarily begin to rise, and vice versa. *(210)*

Production and Cost Analysis II

10

After reading this chapter, you should be able to:

- Distinguish technical efficiency from economic efficiency.
- Explain how economies of scale and diseconomies of scale influence the shape of long-run cost curves.
- State the envelope relationship between short-run cost curves and long-run cost curves.
- Explain the role of the entrepreneur in translating cost of production to supply.
- Discuss some of the problems of using cost analysis in the real world.

Economic efficiency consists of making things that are worth more than they cost.

J. M. Clark

The short run is a time period in which some inputs are fixed. In this chapter we will consider firms' long-run decisions and the determinants of the long-run cost curves, and talk about applying cost analysis to the real world.

MAKING LONG-RUN PRODUCTION DECISIONS

Firms have many more options in the long run than they do in the short run. They can change any input they want. Plant size is not given; neither is the technology available given.

To make their long-run decisions, firms look at the costs of the various inputs and the various production technologies available for combining those inputs, and then decide which combination offers the lowest cost. Even simple production decisions involve complicated questions. These decisions are made on the basis of the expected costs, and expected usefulness, of inputs.

Farming techniques in Japan differ due to the relative scarcity of agricultural land.

TECHNICAL EFFICIENCY AND ECONOMIC EFFICIENCY

When choosing among existing production technologies in the long run, firms are interested in the lowest cost, or most economically efficient, methods of production. They consider all technically efficient methods and compare their costs. The terms *economically efficient* and *technically efficient* differ in meaning. **Technical efficiency** in production means that *the fewest possible inputs are used to produce a given output*.

Many different production processes can be technically efficient. For example, say you know that to produce 100 tons of wheat you can use 10 workers and 1 acre or use 1 worker and 100 acres. Which of these two production techniques is more efficient? Both could be technically efficient since neither involves the use of more of both inputs than the other technique. But that doesn't mean that both are equally economically efficient. That question can't be answered unless you know the relative costs of the two inputs. If an acre of land rents for $1 million and each worker costs $10 a day, our answer likely will be different than if land rents for $40 an acre and each worker costs $100 a day. The **economically efficient** method of production is *the method that produces a given level of output at the lowest possible cost*.

Q-1 True or false? If a process is economically efficient it is also technically efficient. Explain your answer.

In long-run production decisions, firms will look at all available production technologies and choose the technology that, given the available inputs and their prices, is the economically efficient way to produce. The production technique chosen will reflect the prices of the various factors of production. Those prices, in turn, reflect the factors' relative scarcities.

Consider the difference in agricultural practices between Japan and Canada. Because Canada has vast tracts of arable land (80 acres per person), agricultural land is relatively inexpensive in Canada. Just north of Saskatoon, Saskatchewan, for example, an acre of land sells for roughly $1000 / acre for grain farming, and up to $3000 / acre for more intensive agriculture, such as greenhouse operations. In Japan, because of its small geographic size combined with a large population (.71 acres per person), land is very expensive, selling for $15,000 / acre. Because of this difference in the price of inputs, production techniques use land much more intensively in Japan than in Canada. Similarly with China: Labour is more abundant and capital is scarcer, so production techniques in China use capital much more intensively than it is used in Canada. Whereas China would use hundreds of workers, and very little machinery to build a road, Canada would use three or four people along with three machines. Both countries are being economically efficient, but because costs of inputs differ, the economically efficient method of production differs. Thus, the economically efficient method of production is the technically efficient method of production that has the lowest cost. (For a further, graphical analysis of economic efficiency, see Appendix A.)

Q-2 Why does China use more labour-intensive techniques than Canada does?

DETERMINANTS OF THE SHAPE OF THE LONG-RUN COST CURVE

The shape of the long-run cost curve is due to the existence of economies and diseconomies of scale.

In Chapter 9 we saw that the law of diminishing marginal productivity accounted for the shape of the short-run average cost curve. The firm was adding more of a variable input to a fixed input. However, the law of diminishing marginal productivity doesn't

In the late 1980s the normal production run of a U.S. automobile was 200,000 units per year. Why was it so high? Because of indivisible setup costs of the then-current production technology. In order to reduce those indivisible setup costs to an acceptable level, the production level per year had to equal at least 200,000 or the car was considered an economic failure. The Pontiac Fiero, a sporty two-seater, was dropped in 1988 because it didn't sell well enough to sustain that production level.

But what is an indivisible setup cost depends on the structure of production. Japanese companies structured production differently from U.S. companies and had a much lower level of indivisible setup costs. For example, at just about the same time as Pontiac dropped the Fiero, a Japanese company, Mazda, entered the market with the Miata, another sporty two-seater. Because Mazda's assembly line was designed to handle different sizes and shapes of vehicles (which permits economies of scope, discussed later in the chapter), its minimum profitable production level for the Miata was about 30,000, not 200,000. This alternative structure of production made it possible for the Miata to do well in a market that buys a total of about 40,000 two-seater sports coupes annually. In response to this competition, U.S. companies followed suit; they changed production methods and now indivisible setup costs make up a smaller portion of the total costs, thereby lowering their minimum efficient level of production.

Large scale production can justify the use of very expensive, but highly productive, specialized capital.

In the production of steel, the cost of a blast furnace is an indivisible setup cost that requires a minimum level of production to be economically feasible.

apply to the long run, since in the long run all inputs are variable. The most important determinants of what is economically efficient in the long run are economies of scale and diseconomies of scale. Let's consider each of these in turn and see what effect they will have on the shape of the long-run average cost curve.

Economies of Scale (Increasing Returns to Scale)

We say that there are **economies of scale** in production *when long-run average total costs decrease as output increases*. For example, if producing 40,000 VCRs costs the firm $16 million ($400 each), but producing 200,000 VCRs costs the firm $40 million ($200 each), there are significant economies of scale associated with choosing to produce 200,000 rather than 40,000 VCRs. Economies of scale are cost savings (economies) associated with larger scale production.

Despite its large geographic size, Canada is a small market, so at low levels of production, economies of scale are extremely important because many production techniques require a certain minimum level of output to be useful. For example, say you want to produce a pound of steel. You can't just build a mini blast furnace, stick in some coke and iron ore, and come out with a single pound of steel. The smallest technically efficient blast furnaces have a production capacity measured in tons per hour, not pounds per year. The cost of the blast furnace is said to be an **indivisible setup cost**—*the cost of an indivisible input for which a certain minimum amount of production must be undertaken before the input becomes economically feasible to use*.

Economies of scale occur whenever inputs do not need to be increased in proportion to the increase in output. Marketing, distribution, personnel, and accounting functions, and research and development represent areas in a firm where economies of scale are likely to occur. Unit costs fall as these costs are spread over a larger and larger output. A firm can increase its production without having to increase marketing costs, for example, and a larger firm does not need a proportionally larger personnel or accounting department than a smaller firm, so economies of scale are captured by the large firm. As output increases, cost per unit falls. This can also be seen as increased productivity: the firm experiences **increasing returns to scale** (IRTS). Doubling the inputs more than doubles the output. With a larger output, a firm is also able to take advantage of

specialized labour and **specialized capital**. Specialized labour is more productive but is more expensive. A small firm needs people who are flexible and can perform a range of duties. Although specialized labour is more productive (in a limited range of functions), only a firm with sufficiently large output can economically justify the higher expense. It is the same with specialized capital. Although it is more productive, firms with small outputs cannot spread the cost over a large enough output to justify the more expensive, more productive, equipment. Larger firms can also economize on **management costs**. As the firm expands, the number of managers expands less quickly, thereby generating additional cost savings.

Q-3 Why are larger production runs often cheaper per unit than smaller production runs?

Indivisible setup costs are important because they are the source of many real-world economies of scale: As output increases, the costs per unit of output decrease. Cost per unit of a small production run is higher than cost per unit of a large production run.

In the long run, all inputs are variable, so only economies of scale can influence the shape of the long-run cost curve.

Figure 10-1(a) demonstrates a normal long-run production table; Figure 10-1(b) shows the related typical shape of a long-run average cost curve. (Notice that there are no fixed costs. In the long run, all costs are variable.) Economies of scale account for the downward-sloping part. Cost per unit of output is decreasing.

Because of the importance of economies of scale, businesspeople often talk of a minimum efficient level, or scale (MES), of production. What they mean by minimum efficient scale of production is that, given the price at which they expect to be able to sell a good, the indivisible setup costs, for example, are so high that production runs of less than a certain size don't make economic sense. Thus, the **minimum efficient scale of production** is *the amount of production that spreads costs out sufficiently for a firm to undertake production profitably*. At this point, the market has expanded to a size large enough for firms to take advantage of all economies of scale. The minimum efficient level of production is where the average total costs are at a minimum. The implication of economies of scale is that, in some industries, firms must be of a certain (large) size in order to compete successfully.

see page 227

Figure 10-1 (a and b) A TYPICAL LONG-RUN AVERAGE TOTAL COST TABLE AND CURVE

In the long run, average costs initially fall because of economies of scale; then they are constant for a while, and finally they tend to rise due to diseconomies of scale.

Quantity	Total Costs of Labour	Total Costs of Machines	Total Costs = $TC_L + TC_M$	Average Total Costs = TC/Q
11	$381	$254	$635	$58
12	390	260	650	54
13	402	268	670	52
14	420	280	700	50
15	450	300	750	50
16	480	320	800	50
17	510	340	850	50
18	549	366	915	51
19	600	400	1,000	53
20	666	444	1,110	56

(a) Long-run production table

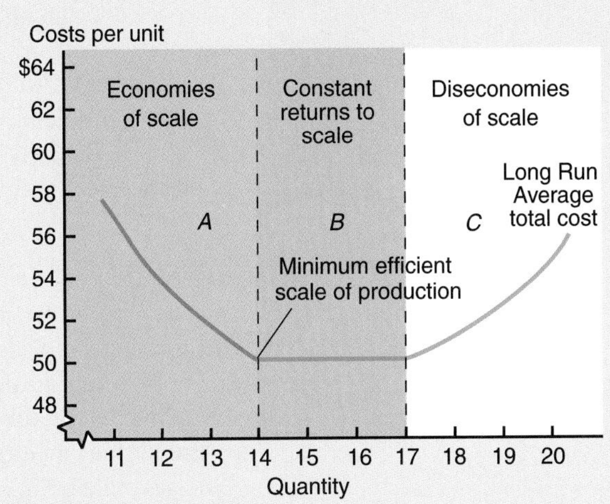

(b) Long-run average total cost curve

Diseconomies of Scale (Decreasing Returns to Scale)

Diminishing marginal productivity refers to the decline in productivity caused by increasing units of a variable input being added to a fixed input. Diseconomies of scale refers to the decreases in productivity which occur when there are equal increases of all inputs (no input is fixed).

Notice that on the right side of Figure 10-1(b) the long-run average cost curve is upward sloping. Average cost is increasing. We say that there are **diseconomies of scale** in production *when long-run average total cost increases as output increases*. For example, if producing 200,000 VCRs costs the firm $40 million ($200 each) and producing 400,000 VCRs costs the firm $100 million ($250 each), there are diseconomies of scale associated with choosing to produce 400,000 rather than 200,000. Diseconomies of scale usually, but not always, start occurring as firms get large.

Diseconomies occur for a number of reasons as a firm increases in size. Coordination of a large firm is more difficult, and as each division needs to know what the rest of the company is doing, information costs and communication costs rise as a firm grows. There may also be some duplication of services; for example, various divisions may each perform record-keeping or personnel duties, or may engage in separate marketing activities. This duplication of functions raises the average cost of the production in larger firms, and the firm is said to be experiencing **decreasing returns to scale (DRTS)**. A *doubling of inputs results in a less-than-doubling of output due to diseconomies setting in*. Other diseconomies of scale include increased monitoring costs and decreased team spirit as a firm gets bigger.

Q-4 If production involved each person working on a computer (for example, designing software), what would the long run average total cost curve look like?

Monitoring costs are *the costs incurred by the firm to ensure that employees do what they're supposed to do*. If you're producing something yourself, the job generally is done the way you want it done; monitoring costs are zero. However, as the scale of production increases, you have to hire people to help you produce. This means that if the job is to be done the way you want it done, you have to monitor (supervise) your employees' performance. The cost of monitoring can increase significantly as output increases; it's a major contributor to diseconomies of scale. Most big firms have several layers of bureaucracy devoted to monitoring employees. The job of middle managers is, to a large extent, monitoring.

As firms become larger, monitoring costs increase and achieving team spirit is more difficult.

People often prefer the team spirit of working at a smaller company. **Team spirit** is *the feelings of friendship and being part of a team that bring out people's best efforts*. When the team spirit or morale is lost, production slows considerably. The larger the firm is, and the less employees in one division feel connected to the rest of the production process, the more difficult it is to maintain team spirit.

Team skills are an important requirement for today's skilled workers.

An important reason diseconomies of scale can come about is that the bigger things get, the more checks and balances are needed to ensure that the right hand and the left hand are coordinated. The larger the organization, the more checks and balances and the more paperwork.

Some large firms manage to solve these problems and thus avoid diseconomies of scale. But problems of monitoring and loss of team spirit often limit the size of firms. They underlie diseconomies of scale in which relatively less output is produced for a given increase in inputs, so that per-unit costs of output increase.

Constant Returns to Scale

Sometimes in a range of output a firm does not experience either economies of scale or diseconomies of scale. In this range there are **constant returns to scale** (CRTS) where *long-run average total costs do not change with an increase in output*. There are no further cost reductions resulting from increasing the scale of production, but costs do not begin to rise in this range either. Constant returns to scale are shown by the flat portion of the average total cost curve in Figure 10-1(b). Constant returns to scale occur when production techniques can be replicated again and again to increase output. Both medium-sized and large firms can compete successfully in an industry characterized by constant returns to scale.

TABLE 10-1 Summary of Returns to Scale.

Returns to Scale	Doubling inputs results in:	Slope of the LRAC
Increasing returns to scale (IRTS; economies of scale)	Output more than doubles	downward
Constant returns to scale (CRTS)	Output exactly doubles	horizontal
Decreasing returns to scale (DRTS; diseconomies of scale)	Output less than doubles	upward

The Importance of Economies and Diseconomies of Scale

Economies and diseconomies of scale, summed up in Table 10-1, play important roles in real-world long-run production decisions. Economies of scale underlie firms' attempts to expand their markets either at home or abroad. If they can make and sell more at lower per-unit costs, they will make more profits. Diseconomies of scale prevent a firm from expanding and can lead corporate raiders to buy the firm and break it up, in the hope that the smaller production units will be more efficient, thus eliminating some of the diseconomies of scale.

Economies and diseconomies of scale play important roles in real-world long-run production decisions.

ENVELOPE RELATIONSHIP

Since in the long run all inputs are variable, while in the short run some inputs are fixed, long-run cost will always be less than or equal to short-run cost at the same level of output. To see this, let's consider a firm that had planned to produce 100 but now adjusts its plan to produce 150. We know that in the long run the firm would produce 150 by choosing the lowest-cost method of production. In the short run the firm faces a constraint: It must use its existing plant to produce the additional output. All expansion must be done by increasing only the variable input. This constraint must increase average cost (or at least not decrease it) compared to what average cost would have been had the firm planned to produce that level to begin with. Additional constraints increase cost. The **envelope relationship** is *the relationship between long-run and short-run average total costs*. It is shown in Figure 10-2, on the next page.

Q-5 Why is the short-run average cost curve a U-shaped curve?

Q-6 Why is the long-run average total cost curve generally considered to be a U-shaped curve?

see page 227

Each plant's short-run average total cost curve touches (is tangent to) the long-run average total cost curve at one, and only one, output level; at all other output levels, short-run average cost exceeds long-run average cost. The long-run average total cost curve is an envelope of short-run average total cost curves.

The intuitive reason why the short-run average total cost curves always lie above the long-run average cost curve is simple. In the short run, you have chosen a plant; that plant is fixed, and its costs for that period are part of your average fixed costs. Changes must be made within the confines of that plant. In the long run you can change everything, choosing the combination of inputs in the most efficient manner. The more options you have to choose from, the lower the costs of production. Put another way: constraints always raise costs. So in the long run, costs must be lower.

The envelope relationship is the relationship explaining that, at the planned output level, short-run average total cost equals long-run average total cost, but at all other levels of output, short-run average total cost is higher than long-run average total cost.

In analyzing the short-run cost curves of the three plants shown in Figure 10-2, notice that the minimum point of the $SRAC_1$ (short run average total cost curve for Plant 1) does not lie on the LRAC (long run average total cost curve). The tangency of $SRAC_1$ and LRAC occurs to the left of the minimum point of $SRAC_1$, at output Q_1. The same is true for Plant 2: the minimum $SRAC_2$ lies above the LRAC since $SRAC_2$ and LRAC are tangent at output Q_2. At output Q_3, both $SRAC_3$ and LRAC are at a minimum. This point represents the **minimum efficient scale (MES)** of production, and is *the least-cost production level of a firm*.

KNOWING THE TOOLS

Distinguishing Diseconomies of Scale from Diminishing Marginal Productivity

As pointed out in the text, the shapes of the short-run average cost curve and the long-run average cost curve are similar. But the reasons underlying those shapes are quite different. It is important to emphasize that difference. In the short run, some inputs are fixed; in the long run, all inputs vary.

Average Total Cost

Why is the short-run average total cost curve U-shaped? What accounts for its shape is what's happening to the marginal productivity of each additional unit of input *keeping all other inputs fixed.* Since costs are based on inputs, how much an input contributes to output directly affects the costs of production. Average total costs in the short run fall initially because of the assumption of increasing marginal productivity: additional inputs are able to produce increasing increments of output. An example of increasing marginal productivity is a 5 percent increase in the quantity of labour, holding capital constant, leading to a 5 percent increase in output, and the next 5 percent increase in labour leading to a 10 percent increase in output.

Eventually, marginal productivity falls and the short-run average cost curve slopes upward. Adding more of *one* factor of production, holding the others constant, contributes less and less to output, causing marginal costs, and eventually average costs, to rise. An example of diminishing marginal productivity is a 5 percent increase in the quantity of labour, holding capital constant, leading to a 10 percent increase in output and the next 5 percent increase, holding capital constant, leading to a 6 percent increase in output. The assumption of initially increasing marginal productivity and eventually diminishing marginal productivity leads to the U shape of the short-run average cost curve.

Long-Run Average Total Cost Curve

Now consider the long-run average total cost curve. Its shape is determined by what's happening to returns to scale. Returns to scale are not about how changes in one input affect output. Instead, they involve changing *all inputs equally.* If there are economies of scale, increasing all factors of production equally, say by 5 percent, leads to a greater increase in output, say by 8 percent.

The assumption economists make in the long run is that initially economies of scale cause the long-run average total cost curve to slope downward. Eventually, however, there are diseconomies of scale. That is, increasing all inputs equally, say by 5 percent, leads to a smaller increase in output, say by 3 percent. Diseconomies of scale cause average total costs to rise and the long-run average total cost curve to slope upward. If there are neither economies of scale nor diseconomies of scale, the average total cost curve is flat because inputs and output both are changing by equal proportions. An example of constant returns to scale is a 5 percent increase in all inputs leading to a 5 percent increase in output. With constant returns to scale, average costs do not change.

The assumption we make about production in the long run is that there are first increasing, then constant, and finally decreasing returns to scale. That assumption about returns to scale accounts for the U shape of the long-run average cost curve.

Figure 10-2 ENVELOPE OF SHORT-RUN AVERAGE TOTAL COST CURVES

The long-run average total cost curve is an envelope of the short-run average total cost curves. Each short-run average total cost curve touches the long-run average total cost curve at only one point. (*SR* stands for short run; *LR* stands for long run.)

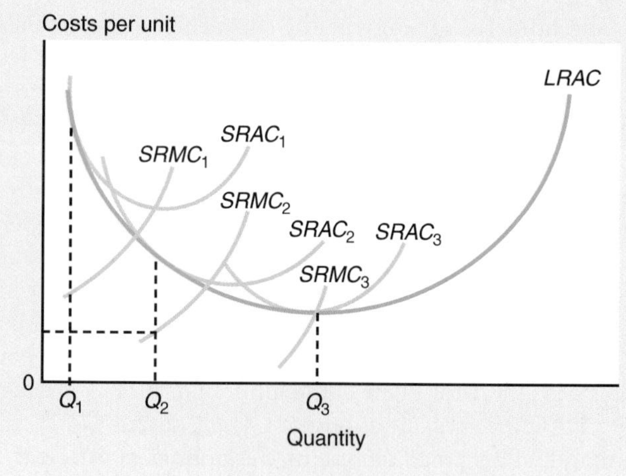

APPLYING THE TOOLS

Why Are Textbooks So Long?

Understanding costs and their structure will help you understand why introductory textbooks are so long—and why their length is to your advantage.

The majority of the costs of a book are fixed costs in relation to the length of the book. The initial costs in terms of length are about 20 percent of the total price of the book. So increasing the length of the book increases costs slightly. But the longer length allows the writer to include more issues that some professors want, and many professors will not consider using the book unless it includes these issues. That means that greater length can allow publishers to sell more books, allowing the fixed costs to be divided over more output. This decrease in fixed cost per unit can lower average total cost more than the increased the length of the book increases average total costs. So if the added length increases the number of users, the additional length can lower the average cost of the book.

ENTREPRENEURIAL ACTIVITY AND THE SUPPLY DECISION

This chapter and the preceding one discussed the technical nature of costs and production. The next chapter will relate costs of production to the supply of goods. As a bridge between the two chapters, let's consider the entrepreneur, who establishes the relationship between costs and the supply decision, and some of the problems of using cost analysis in the real world.

In thinking about the connection between cost and supply, one fundamental insight is that the revenue received for a good must be greater than the planned cost of producing it. Otherwise why would anyone supply it? The difference between the expected price of a good and the expected average total cost of producing it is the supplier's expected economic profit per unit. It's profit that underlies the dynamics of production in a market economy.

The difference between the expected price of a good and the expected average total cost of producing it is the supplier's expected economic profit per unit. The expected revenue per unit must exceed the opportunity cost of supplying the good for a good to be supplied.

Cost curves do not automatically become supply curves. To move from cost to supply, entrepreneurial initiative is needed. An **entrepreneur** is *an individual who sees an opportunity to sell an item at a price higher than the average cost of producing it*. The entrepreneur is the organizer of production and the one who visualizes the demand and convinces the individuals who own the factors of production that they want to produce that good. Businesses work hard at maintaining the entrepreneurial spirit in their employees. The greater the difference between price and average total cost, the greater the entrepreneur's incentive to tackle the organizational problems of production and supply the good.

Q-7 Why is the role of the entrepreneur central to the production process in the economy?

USING COST ANALYSIS IN THE REAL WORLD

In real life, cost analysis is difficult because actual production processes are marked by economies of scope, learning by doing, technological change, unmeasured costs, joint costs, indivisible costs, uncertainty, asymmetries, and multiple planning and adjustment periods with many different short runs.

Economies of Scope

The cost of production of one product often depends on what other products a firm is producing. Economists say that in the production of two goods there are **economies of scope** *when the costs of producing products are interdependent so that it's less costly for a firm to produce one good when it's already producing another*. For example, once a firm has set

Globalization has opened up new opportunities for production and for sales.

up a large marketing department to sell cereal, the department might be able to use its expertise in marketing a different product—say, dog food. A firm that sells gasoline can simultaneously use its gas station attendants to sell pop, milk, magazines, coffee, and donuts, and take advantage of economies of scope.

Q-8 What is the difference between an economy of scope and an economy of scale?

Economies of scope play an important role in firms' decisions of what combination of goods to produce. They look for both economies of scope and economies of scale. When you read about firms' mergers, think about whether the combination of their products will generate economies of scope. Many otherwise unexplainable mergers between seemingly incompatible firms can be explained by economies of scope.

By allowing firms to segment the production process, globalization has made economies of scope even more important to firms in their production decisions. Low-cost labour in other countries has led North American firms to locate their manufacturing processes in those countries and to concentrate domestic activities on other aspects of production. Production is more than simply manufacturing; the costs of product development, marketing, advertising, and distribution are often larger components of the cost of a good than are manufacturing costs. Each of these involves special knowledge and expertise, and companies specializing in the marketing, advertising, and distribution aspects of the production process, and taking advantage of low-cost manufacturing elsewhere are making themselves highly competitive. Often they expand into new areas, taking advantage of economies of scope in distribution and marketing.

Nike is a company which has transformed itself. It is a U.S. marketing and distribution company; it outsources all its production to affiliated companies. Nike is expanding, but not in the production of shoes. It is expanding into leisure clothing, where it hopes economies of scope in its marketing and distribution specialties will bring it success.

Nike is only one of many examples. The large wage differentials in the global economy are causing firms to continually reinvent themselves—to shed aspects of their business where they do not have a comparative advantage, and to add new businesses where their abilities can achieve synergies and economies of scope.

Learning by Doing and Technological Change

Q-9 Does learning by doing cause the cost curve to be downward sloping?

Many firms estimate that worker productivity grows 1 to 2 percent a year because of learning by doing.

The production techniques available to real-world firms are constantly changing because of *learning by doing* and *technological change*. These changes occur over time and cannot be accurately predicted. **Learning by doing** simply means that *as we do something, we learn what works and what doesn't, and over time we become more proficient (better) at it*. Many firms estimate that output per unit of input will increase by 1 or 2 percent a year, even if no changes in inputs or technologies occur, as employees learn by doing.

The concept of learning by doing emphasizes the importance of past efforts in developing cost advantages. Learning by doing applies to even the smallest job in a firm. Every time workers begin a new task, they must learn how best to perform it. The **learning curve** *represents the process of workers becoming better (more productive) at a specific job as they gain experience from doing it repeatedly*.

Production then: The nature of production has changed considerably in the last 80 years. This picture shows a 1933 production line in which people did the work as the goods moved along the line.

External Economies In any industry there are *external forces at work which are capable of reducing costs for all firms belonging to the industry*. These **external economies** are beyond the control of the individual firm, and commonly are the result of increased economic activity in a sector. Consider the high tech firms located in "Silicon Valley" in California, or in "Silicon Valley North" just outside Ottawa. Because of their close geographic proximity, many services can be provided to the firms at lower cost than if the firms were spread out across the country. Cost savings thus become available as all types of firms locate close by in order to service the high tech firms; for example, office supply firms, restaurants, and financial services companies. Commercial realtors specialize in serving the needs of the high tech community, and highly specialized firms spring up to serve specific high tech niches. Outsourcing is very easily accomplished due to the close proximity of the specialty firms. Because of the increased level of economic activity in the region, all firms' per unit costs are lower, and the Long Run Average Cost curve (LRAC) shifts down. Any industry can benefit from these types of economies; external economies is the primary reason firms prefer to locate in industrial parks.

Technological change is *an increase in the range of production techniques that provides new ways of producing goods*. That is, technological change offers an increase in the known range of production. The standard long-run model takes technology as a given. However, technological change can fundamentally alter the nature of production costs, which affects firms' decisions and production.

Technological change can fundamentally alter the nature of production costs.

In some industries technological change is occurring so fast that it overwhelms all other cost issues. Industries where this is occurring are known as high tech industries, and include the telecommunications, biotechnology, and computer industries and the Internet. The computer industry has followed Moore's law, which states that the cost of computing will fall by half every 18 months. Indeed, that has happened since the 1980s. With costs falling so rapidly in the high tech industries, because of learning by doing and technological change, all other cost components are overwhelmed, and, instead of costs increasing as output rises significantly, as might result from diseconomies of scale, costs steadily decrease.

The fall in the cost of computer chips has affected other industries as well. All types of household goods that use computer technology—including telephones, refrigerators, automobiles, TVs, and compact disc players—are undergoing enormous change. For instance, VCRs are now almost extinct; they have been replaced by digital video disc (DVD) players—where all images are transmitted digitally. Computer technology has also fundamentally changed automobiles; they are much more efficient and reliable and their price has fallen because of computer technology. Technological change drives costs down, and can overwhelm diseconomies of scale, causing prices to fall more and more.

Technological change and learning by doing are intricately related.

Technological change occurs in all industries, not only "high-tech" industries.

Many Dimensions

The only dimension of output considered in the standard model is quantity. Many, if not most, decisions that firms make involve more than what quantity to produce. They involve multidimensional questions like "Should we change the quality? Should we change the wrapper? Should we improve our shipping speed? Should we increase our inventory?" Each of these questions relates to a different dimension of the production decision and each has its own marginal costs. Thus, there isn't just one marginal cost; there are dozens of them. Good economic decisions take all relevant margins into account.

Good economic decisions take all relevant margins into account.

The reason that the standard model is important is that each of these questions can be analyzed using the same reasoning used in the standard model.

Unmeasured Opportunity Costs

If asked, "In what area of decision making do businesses most often fail to use economic reasoning?" most economists would say costs. The relevant costs for a firm are not generally the costs you'll find reported in its accounts.

Why the difference? Economists include all opportunity costs, explicit and implicit. Accountants, who have to measure firms' costs and provide actual dollar figures, only include explicit costs—costs that are reasonably precisely measurable.

Consider a new home-based business which produces 1,000 widgets[1] that sell at $3 each for a total revenue of $3,000. To produce these widgets the business had to buy $1,000 worth of widgetgoo, which the owner hand-shapes into widgets. An accountant

Production now: The nature of production has changed considerably in the last 80 years. Here is a modern production line. Robots do much of the work.

[1]What are widgets? They are wonderful little gadgets, but no one knows what they look like or what they are used for.

"Factories run by numbers. Numbers to calculate profit and losses; to analyze the costs of new products; and to chart corporate strategy. But a lot of managers are relying on the wrong numbers.

As they adopt new manufacturing techniques like computer-aided design, just-in-time stock management, and total quality control, many firms are discovering that their existing account systems also need dragging into the 1990s. Unless the bean-counters join the manufacturing revolution, traditional cost accounting will have little place in the factory of the future."

The above quote introduced an article in *The Economist* (March 3, 1990, p. 61) describing a conference on managerial or cost accounting (the application of cost analysis to managerial decisions). Unlike *financial accounting* (which involves keeping track of income, assets, and liabilities), managerial accounting is used to help managers determine the cost of producing products and plan future investment. It's the direct application of microeconomics to production.

In the 1980s and 1990s cost accounting changed enormously. The leaders of this change—such as Robert Kaplan of the Harvard Business School—argue that cost accounting systems based on traditional concepts of fixed and variable costs consistently lead firms to make the wrong decisions. They argue that in today's manufacturing, direct labour costs have fallen substantially—in many industries to only 2 or 3 percent of the total cost—and overhead costs have risen substantially. This change in costs facing firms requires a much more careful division among types of overhead costs, and a recognition that what should and should not be assigned as a cost to a particular product differs with each decision.

These developments in managerial accounting require an even deeper understanding of costs than accountants have previously needed. As one firm's director of manufacturing was quoted in *The Economist* article, "Unless management accountants move fast [to incorporate these new concepts], they will be almost without use to the manufacturing manager."

would say that the total cost of producing 1,000 widgets was $1,000 and that the firm's profit was $2,000. That's because an accountant uses explicit costs that can be measured.

Q-10 As the owner of the firm, Jim pays himself $1,000. All other expenses of the firm add up to $2,000. What would an economist say are the total costs for Jim's firm?

Economic profit is different. An economist, looking at that same example, would point out that the accountant's calculation doesn't take into account the time and effort that the owner put into making the widgets. While a person's time involves no explicit cost in money, it does involve an opportunity cost, the forgone income that the owner could have made by spending that time working in another job. If the business takes 200 hours of the person's time and the person could have earned $12 an hour working for someone else, then the person is forgoing $2,400 in income. Economists include that implicit cost in their concept of cost. When that implicit cost is included, what looks like a $2,000 profit becomes a $400 economic loss.

The Standard Model as a Framework

Despite its limitations, the standard model provides a good framework for cost analysis.

The standard model can be expanded to include these real-world complications. Learning the standard model, however, provides the rudiments of cost analysis, a superb framework for starting to think about real-world cost measurement.

CONCLUSION AND A LOOK AHEAD

Many Canadian industries experience significant economies of scale and scope resulting from a combination of production technology and a small Canadian market. However, in considering Canadian industries, it is important to realize that, while a firm has more options in the short run than in the long run, many factors are beyond the firm's control. Economies of scale mean that there is benefit to being large; firms that achieve their minimum efficient scale (MES) have an absolute cost advantage over all other firms. Innovation and new product development are important dynamic competitive matters for firms; however, the speed of technological change in today's economy is redefining how firms adapt in the short run and in the long run.

Improvements in technology lower costs, and this results in more and cheaper products for consumers. In the short run, the firm's cost curves are seen to shift down to reflect the lower costs, and in the long run, the entire long run average total cost curve (LRAC) shifts down, enabling the firm to gain cost savings in addition to the usual economies of scale. As globalization allows firms to grow, they are able to take advantage of a pair of cost savings: economies of scale, which is seen as a movement down along the LRAC towards the minimum efficient scale, and technological change, which shifts the entire LRAC curve downwards. The combination of economies of scale and technological advances has been a very important element in increasing the availability and lowering the costs (prices) of many consumer products, including electronics, telecommunications services, computers, the Internet, automobiles, small appliances, and more. In particular, through the decade of the 1990s and well into the next, technology has enhanced the quantity and quality and lowered the prices of most of the products in the market. These changes we see in the marketplace are the results of the series of short and long run decisions made by firms competing in today's economy.

Increasing marginal productivity is common.

Chapter Summary

- An economically efficient production process must be technically efficient, but a technically efficient process may not be economically efficient.
- The long-run average total cost curve is U-shaped. Economies of scale initially cause average total cost to decrease; diseconomies eventually cause average total cost to increase.
- Economies of scale and diseconomies of scale occur in the long run as all inputs are increased proportionately.
- For economies of scale (IRTS), doubling all inputs more-than-doubles output.
- For CRTS, doubling all inputs exactly doubles output.
- For diseconomies of scale (DRTS), doubling all inputs less-than-doubles output.
- The marginal cost and short-run average cost curves slope upward because of diminishing marginal productivity. The long-run average cost curve slopes upward because of diseconomies of scale.
- There is an envelope relationship between short-run average cost curves and long-run average cost curves. The short-run average cost curves represent constrained choices, and therefore are always above the long-run average cost curve.
- An entrepreneur is an individual who sees an opportunity to sell an item at a price higher than the average cost of producing it.
- Once we start applying cost analysis to the real world, we must include a variety of other dimensions of costs that the standard quantity-choosing model does not cover.
- Costs in the real world are affected by economies of scope, learning by doing, technological change, the many dimensions to output, and unmeasured opportunity costs.

Key Terms

constant returns to scale *(218)*
CRTS *(218)*
diseconomies of scale *(218)*
DRTS *(218)*
economically efficient *(215)*
economies of scale *(216)*
economies of scope *(221)*
entrepreneur *(221)*
envelope relationship *(219)*
external economies *(223)*
indivisible setup cost *(216)*
IRTS *(216)*
learning by doing *(222)*
learning curve *(222)*
minimum efficient level of production *(217)*
monitoring costs *(218)*
team spirit *(218)*
technical efficiency *(215)*
technological change *(223)*

Questions for Thought and Review

1. Distinguish technical efficiency from economic efficiency.
2. A student has just written on an exam that in the long run fixed cost will make the average total cost curve slope downward. Why will the professor mark it incorrect?
3. What inputs do you use in studying this book? What would the long-run average total cost and marginal cost curves for studying look like? Why?
4. Describe the economies of scale for an industry where both large and small firms complete with each other. Illustrate.
5. When economist Jacob Viner first developed the envelope relationship, he told his draftsman to make sure that all the marginal cost curves went through both (1) the minimum point of the short-run average cost curve and (2) the point where the short-run average total cost curve was tangent to the long-run average total cost curve. The draftsman told him it couldn't be done. Viner told him to do it anyhow. Why was the draftsman right?
6. What is the role of the entrepreneur in translating cost of production into supply?
7. Your average total cost is $40; the price you receive for the good is $12. Should you keep on producing the good? Why?
8. A student has just written on an exam that technological change will mean that the cost curve is downward sloping. Why did the teacher mark it wrong?
9. If you were describing the marginal cost of an additional car driving on a road, what costs would you look at? What is the likely shape of the marginal cost curve?
10. The cost of setting up a steel mill is enormous, costing an estimated $1.5 billion to build. Using this information and the cost concepts from the chapter, explain the following quotation: "To make operations even marginally profitable, big steel makers must run full-out. It's like a car that is more efficient at 100 kilometres an hour than in stop-and-go traffic at 50."

Problems and Exercises

1. Consider the production function for a piano-moving company:

Capital	Labour	Total Product	Marginal Product
1 truck	0 workers		
1 truck	1 worker		
1 truck	2 workers		
1 truck	3 workers		
1 truck	4 workers		

   a. In the table, fill in reasonable figures for total product (pianos moved per day) and for marginal product.
   b. Why is there initially increasing marginal product, and then diminishing marginal product?

2. Ziti wishes to build a new golf course, and faces the following choices:

Golf Pros	Golf Course #1 (small)	Golf Course #2 (medium)	Golf Course #3 (large)
1	19 players	23 players	43 players
2	39	49	116
3	54	73	176
4	62	94	225
5	65	110	254

   a. Assuming that golf pros earn an annual salary of $38,000, calculate the average total cost and marginal cost for each choice of golf course, small, medium, and large.The golf courses are only closed on Christmas Day.
   b. On the same diagram, graph the short run average total costs and marginal costs for each golf course. Draw in the long run average total cost curve.
   c. Describe the economies of scale that the typical golf course would experience.

3. A pair of shoes that wholesale for $28.79 has approximately the following costs:

Manufacturing labour	$2.25
Materials	4.95
Factory overhead, operating expenses, and profit	8.50
Sales costs	4.50
Advertising	2.93
Research and development	2.00
Interest	.33
Net income	3.33
Total	$28.79

   a. Which of these costs would likely be a variable cost?
   b. Which would likely be a fixed cost?
   c. If output were to rise, what would likely happen to average total costs? Why?

4. Draw a long-run average cost curve.
   a. Why does it slope downward initially?
   b. Why does it eventually slope upward?
   c. How would your answer to *a* and *b* differ if you had drawn a short-run average cost curve?
   d. How large is the fixed cost component of the long-run average cost curve?
   e. If there were constant returns to scale everywhere, what would the long-run average cost curve look like?
5. A major issue of contention at many colleges concerns the cost of meals that is rebated when a student does not sign up for the meal plan. The administration usually says that it should rebate only the marginal cost of the food alone, which it calculates at, say, $1.25 per meal. Students say that the marginal cost should include more costs, such as the saved space from fewer students using the facilities and the reduced labour expenses on food preparation. This can raise the marginal cost to $6.00.
   a. Who is correct, the administration or the students?
   b. How might your answer differ if this argument were being conducted in the planning stage, before the dining hall is built?
   c. If you accept the $1.25 figure of a person not eating, how could you justify using a higher figure of about $6.00 for the cost of feeding a guest at the dining hall, as many schools do?

Web Questions

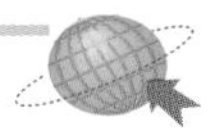

1. Go to the Internet site About.com (www.about.com) and answer the following questions:
   a. What services does About.com provide? What does it charge for those services?
   b. What are the fixed costs associated with running About.com? What is the marginal cost of providing About.com to one additional person?
   c. List the advertising you saw while visiting About.com.
   d. Under what circumstances will an entrepreneur supply a good to the market? What incentives do the owners of About.com have to supply this product on the Internet? (Hint: Look at your answer to *c*).
2. Go to www.about.com and search for an article entitled, "Frequent Flyers and History of Air Wars in Canada." Read the article and answer the following questions. You may also wish to perform a search on the word, "mergers."
   a. Why did Air Canada and Canadian Airlines International merge?
   b. Along what portion of the long-run average cost curve were Air Canada and Canadian Air International operating when they were separate companies? Along what portion do they operate as a merged company?
   c. Why does the airline industry experience economies of scale? Explain.

Answers to Margin Questions

1. True. Since an economically efficient method of production is that method which produces a given level of output at the lowest possible cost, it must also use as few inputs as possible. It is also technically efficient. *(215)*
2. China uses more labour-intensive techniques than Canada because the price of labour is much lower in China than in Canada. Both countries are producing economically efficiently. *(215)*
3. Larger production runs are generally cheaper per unit than smaller production runs because of indivisible setup costs, which do not vary with the size of the run. *(217)*
4. Because the same technical process could be used over and over again at the same cost, the long-run average cost curve would never become upward sloping. *(218)*
5. The short-run average total cost curve initially slopes downward because of large average fixed costs, then begins sloping upward because of diminishing marginal productivity, giving it a U shape. *(219)*
6. The long-run average total cost curve is generally considered to be U-shaped because initially there are economies of scale and, for large amounts of production, there are diseconomies of scale. *(219)*
7. Economic activity does not just happen. Some dynamic, driven individual must instigate production. That dynamic individual is called an entrepreneur. *(221)*
8. Economies of scale are economies that occur because of increases in the amount of a good one is producing. Economies of scope occur when producing different types of goods lowers the cost of each of those goods. *(222)*
9. Learning by doing causes a shift in the cost curve because it is a change in the technical characteristics of production. It does not cause the cost curve to be downward sloping—it causes it to shift downward. *(222)*
10. An economist would say that he doesn't know what total cost is without knowing what Jim could have earned if he had undertaken another activity besides running his business. Just because he paid himself $1,000 doesn't mean that $1,000 is his opportunity cost. *(224)*

Isocost/Isoquant Analysis

In the long run, a firm can vary all of the factors of production. One of the decisions firms face in this long run is which combination of factors of production to use. Economic efficiency involves choosing those factors so that the cost of production is at a minimum.

In analyzing this choice of which combination of factors to use, economists have developed a graphical technique called *isocost/isoquant analysis*, the prefix "iso" meaning "equal." In Figure A10-1, any point on that graph represents a combination of machines and labour that can produce a certain amount of output, say pairs of earrings. For example, point C represents 10 machines and 6 units of labour being used to produce 60 pairs of earrings. Any point in the blue shaded area represents more of one or both factors and any point in the red shaded area represents less of one or both factors.

THE ISOQUANT CURVE

The firm's problem is to figure out how to produce its output—let's say it believes it can sell an output of 60 pairs of earrings. Figure A10-2 shows graphically all the combinations of machines and labour that can produce 60 pairs of earrings. An **isoquant curve** (meaning "equal quantity") is *a curve that represents combinations of factors of production that result in equal amounts of output*. All points on an isoquant curve represent combinations of machines and labour which will produce the same amount of output. So, given a level of output, a firm can identify all of the technically efficient combinations which will produce that output. Suppose a firm can produce 60 pairs of earrings with the following combination of labour and machines:

	Labour	Machines	Pairs of Earrings
A	3	20	60
B	4	15	60
C	6	10	60
D	10	6	60
E	15	4	60
F	20	3	60

This table shows the technical limits of production. It shows that the firm can use, for example, 3 units of labour and 20 machines or 20 units of labour and 3 machines to produce 60 pairs of earrings. The isoquant curve for producing 60 pairs in Figure A10-2 (Points A to F) corresponds to rows A to F in the table.

At point A the firm is producing 60 pairs of earrings using 20 machines and 3 workers. If the firm wants to

Figure A10-1 DRAWING THE ISOCOST/ ISOQUANT GRAPH

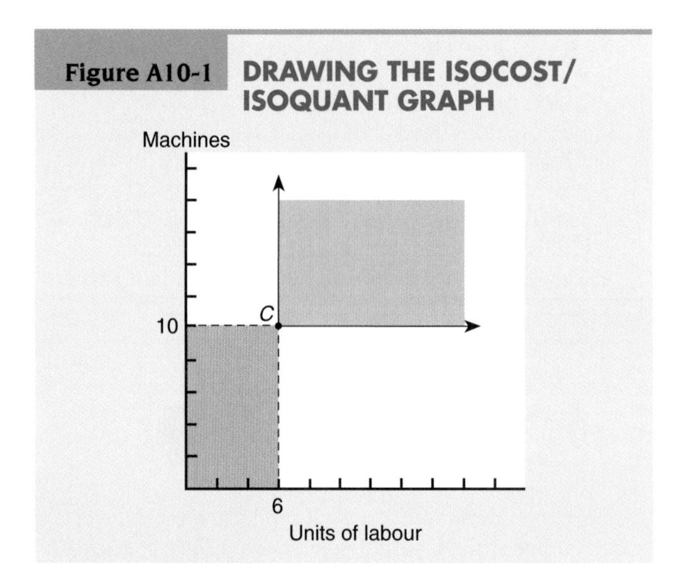

Figure A10-2 ISOQUANT CURVE FOR 60 PAIRS OF EARRINGS

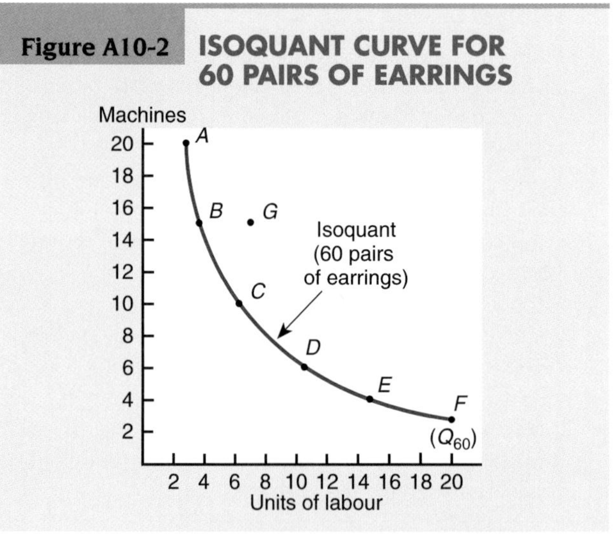

reduce the number of machines by 5, it must increase the number of units of labour by 1 to keep output constant. Doing so moves the firm to point *B*, producing 60 pairs of earrings with 15 machines and 4 workers. Point *B* is a more labour-intensive production method than Point A. At any point on this isoquant curve the firm is being technically efficient—it is using as few resources as possible to produce 60 pairs of earrings. Thus, it would never want to produce 60 at a point like G, because that point uses more inputs. Point G represents a technically inefficient method of production.

Isoquants can be determined for many different levels of output. Doing so will result in an **isoquant map,** *a set of isoquant curves that show technically efficient combinations of inputs that can produce different levels of output*. Such a map for output levels of 40, 60, and 100 is shown in Figure A10-3.

Each curve represents a different level of output. Isoquant Q_{40} is the lower level of output, 40, and the isoquant Q_{100} is the highest level of output. When a firm chooses an output level, it is choosing one of those isoquants. The chosen isoquant represents the technically efficient combinations of resources that can produce the desired output.

The numbers in the production table and the shape of the curve are consistent with the law of diminishing marginal productivity, which means the curve is bowed inward. That is because as the firm increases the use of one factor more and more, it must use fewer and fewer units of the other factor to keep output constant. Thus, the numbers tell us that if a firm wants to keep output constant, as it adds more and more of one factor (and less of the other factor), it has to use relatively more of that factor. This reflects the technical considerations embodied in the law of diminishing marginal productivity.

Figure A10-3 AN ISOQUANT MAP

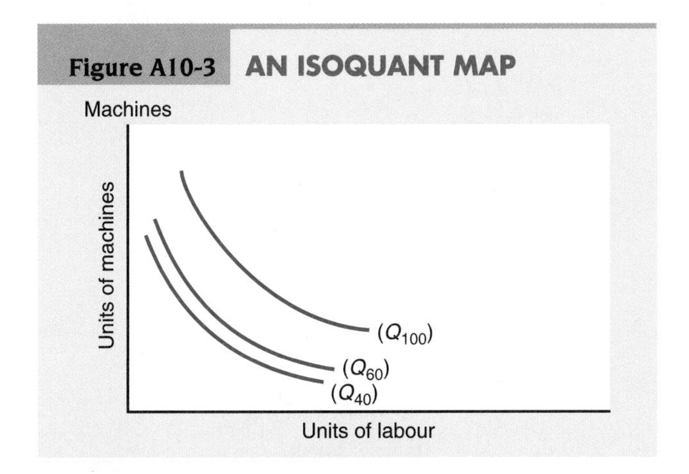

The rate at which one factor must be added to compensate for the loss of another factor, to keep output constant, is called the **marginal rate of substitution.** To say that there is diminishing marginal productivity is to say that there is a diminishing marginal rate of substitution. It is because the table assumes a diminishing marginal rate of substitution that the isoquant curve is bowed inward.

The slope of the isoquant curve is the marginal rate of substitution. To be exact, the absolute value of the slope at a point on the isoquant curve equals the ratio of the marginal productivity of labour to the marginal productivity of machines:

$$|\text{Slope}| = MP_{\text{labour}}/MP_{\text{machines}}$$
$$= \text{Marginal rate of substitution}$$

With this equation, you can see why the isoquant is downward sloping. As the firm moves from point A to point *F* it is using more labour and fewer machines. Because of the law of diminishing marginal productivity, as the firm moves from A to *F*, the marginal productivity of labour decreases and the marginal productivity of machines increases. The slope of the isoquant falls since the marginal rate of substitution is decreasing.

Let's consider a specific example. Say in Figure A10-2 the firm is producing at point *B*. If it cuts its input by 5 machines but also wants to keep output constant, it must increase labour by 2 (move from point *B* to point C). So the marginal rate of substitution of labour for machines between points *B* and C must be 5/2 or 2.5.

THE ISOCOST LINE

So far we have only talked about technical efficiency. To move to economic efficiency, we have to include the costs of production. We do so with the **isocost line** ("equal cost")—*a line that represents alternative combinations of factors of production that have the same costs*. Each point on an isocost line represents a combination of factors of production which cost an equal amount.

To draw the isocost line you must know the cost per unit of each input as well as the amount the firm has chosen to spend on production. Say labour costs $5 a unit and machinery costs $3 a unit and that the firm has chosen to spend $60. Say the firm decides to spend the entire $60 on labour. Since labour costs $5 a unit, if the firm spends all of the $60 on labour it can buy 12 units of labour. This alternative is represented by point A in Figure A10-4.

Alternatively, since machines cost $3 a unit, if the firm chooses to spend all of the $60 on machines it can buy 20 machines (point *B* in Figure A10-4). This gives us two

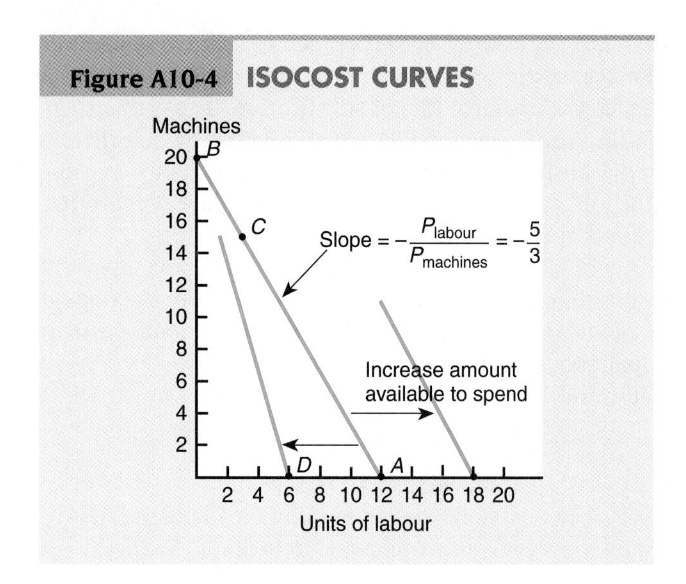

points on the isocost curve. Of course, the assumption of diminishing marginal rates of substitution makes it highly unlikely that firm would want to produce at either of these points. Instead, it would likely use some combination of inputs. Because factor prices are fixed, there is a constant trade-off of $5: $3, so the slope of the isocost line is constant. By connecting the endpoints, we can see the various combinations of inputs that also cost $60.

To see that this is indeed the case, say the firm starts with 20 machines and no labour. If the firm wants to use some combination of labour and machinery, it can give up some machines and use the money it saves by using fewer machines to purchase units of labour. Let's say it gives up 5 machines, leaving it with 15. That means it has $15 to spend on labour, for which it can buy 3 units of labour. That means 15 machines and 3 units of labour is another combination of labour and machines that cost the firm $60. This means that point C is also a point on the isocost line. Thus, the line connecting *A* and *B* is the $60 isocost line.

What would happen to the isocost line if the firm chooses to increase its spending on production to $90? To see the effect we go through the same exercise as before: If it spent it all on labour, it could buy 18 units of labour. If it spent it all on machines, it could buy 30 units of machinery. Connecting these points will give us a curve to the right of and parallel to the original curve. It has the same slope because the relative prices of the factors of production, which determines the slope, have not changed.

What happens to the isocost line if the price of labour rises to $10 a unit? The isocost curve becomes steeper, shifting along the labour axis to point *D* while remaining anchored along the machinery axis until the slope is −10/3. In general, the absolute value of the slope of the isocost curve is the ratio of the price of the factor of production on the *x*-axis to the price of the factor of production on the *y*-axis. That means that as the price of a factor rises, the end point of the isocost curve shifts in on the axis on which that factor is measured.

CHOOSING THE ECONOMICALLY EFFICIENT POINT OF PRODUCTION

Now let's move on to a consideration of the economically efficient combination of resources to produce 60 pairs of earrings with $60. To do that we must put the isoquant cost curve from Figure A10-2 and the isocost curve from Figure A10-4 together. We do so in Figure A10-5.

The problem for the firm is to produce the 60 pairs of earrings as cheaply as possible. Given the level of production it has chosen, it wants to produce using the least-cost combination of the factors of production.

Let's now find the least-cost combination of inputs to produce 60 pairs of earrings. Let's say that, initially, the firm chooses point A on its isoquant curve—that's at 15 machines and 4 workers. That produces 60 pairs of earrings, but has a cost of $45 + $20 = $65. The firm can produce 60 pairs of earrings if it is willing to spend $65.

The firm could choose a more labour-intensive production method, and increase the number of workers to 6 and reduce the number of machines to 10. Doing so still produces 60 pairs of earrings, since C is a point on the isoquant

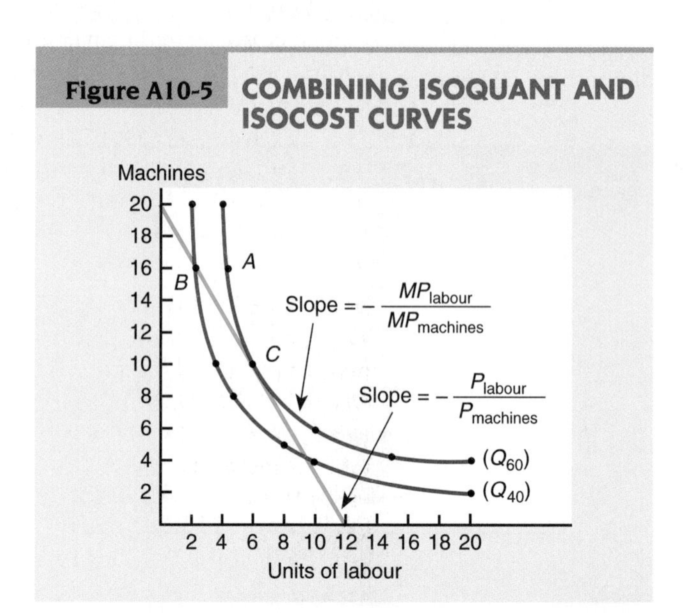

curve, but the strategy reduces the cost from $65 at point A to $60 (10 machines at $3 = $30, and 6 workers at $5 = $30). So the firm is producing 60 pairs of earrings at a cost of $60, operating at the economically efficient point—point C.

Let's talk about the characteristics of point C. Point C is the point where the isoquant curve is tangent to the isocost curve—the point at which the slope of the isoquant curve ($-MP_L/MP_M$) equals the slope of the isocost curve ($-P_L/P_M$). That is, $-MP_L/MP_M = -P_L/P_M$. This can be rewritten as:

$$MP_L/P_L = MP_M/P_M$$

What this equation says is that when the additional output per dollar spent on labour equals the additional output per dollar spent on machines, the firm is operating efficiently. It makes sense. If the additional output per dollar spent on labour exceeded the additional output per dollar spent on machines, the firm would do better by increasing its use of labour and decreasing its use of machines.

Point C represents the combination of labour and machines that will result in the highest output given the isocost curve facing the firm. To put it in technical terms, the firm is operating at an economically efficient point, where marginal rate of substitution equals the ratio of the factor prices. Any point other than C on the isocost curve will cost $60 but produce fewer than 60 pairs of earrings for example, B. Any point other than C on the isoquant curve will produce 60 pairs of earrings but will cost more than $60. Only C is the economically efficient given the factor costs.

Key Terms

isocost line *(229)*
isoquant curve *(228)*
isoquant map *(228)*
marginal rate of substitution *(229)*

Questions for Thought and Review

1. What happens to the marginal rate of substitution as a firm increases the use of one input, keeping output constant? What accounts for this?
2. Draw an isocost curve for a firm that has $100 to spend on producing jeans. Input includes labour and materials. Labour costs $8 and materials cost $4 a unit. How does each of the following affect the isocost curve? Show your answer graphically.
   a. Production budget doubles.
   b. Cost of materials rises to $10 a unit.
   c. Costs of labour and materials each rise by 25 percent.
3. Show, using isocost/isoquant analysis, how firms in Canada use relatively less labour and relatively more land than Japan for the production of similar goods, yet both production choices are economically efficient.
4. Demonstrate the difference between economic efficiency and technical efficiency, using the isocost/isoquant analysis.
5. Draw a hypothetical isocost curve and an isoquant curve tangent to the isocost curve. Label the combination of inputs that represents an efficient use of resources.
   a. How does a technological innovation affect your analysis?
   b. How does the increase in the price of the input on the *x*-axis affect your analysis?
6. Show graphically the analysis of the example in Figure A10-5 if the price of labour falls to $3. Demonstrate that the firm can increase production given the same budget.
7. Show graphically the analysis of the example in Figure A10-5 if the price of machines rises to $5. Demonstrate that the firm must reduce production if it keeps the same budget.

Perfect Competition

11

After reading this chapter, you should be able to:

- List the six conditions for a perfectly competitive market.
- Explain why producing an output at which marginal cost equals price maximizes total profit for a perfect competitor.
- Demonstrate why the marginal cost curve is the supply curve for a perfectly competitive firm.
- Determine the output and profit of a perfect competitor graphically and numerically.
- Construct a market supply curve by adding together the marginal cost curves of individual firms.
- Explain why perfectly competitive firms make zero economic profit in the long run.
- Explain the adjustment process from short-run equilibrium to long-run equilibrium.

There's no resting place for an enterprise in a competitive economy.

Alfred P. Sloan

The concept *competition* is used in two ways in economics. *Competition as a process* is a rivalry among firms and is prevalent throughout our economy. It involves one firm trying to take away market share from another firm. The other use of *competition* is as a *market structure*, which describes the characteristics of a particular market. It is this use that is the subject of this chapter. The *Perfectly Competitive* market structure is one which has highly restrictive assumptions, but which provides us with a reference point we can use to think about various market structures and competitive processes. Why is such a reference point important? Think of the following analogy.

The process of competition involves a rivalry among firms and is prevalent throughout our economy. The state of competition is the end result of the competitive process under certain conditions.

In physics when you study the laws of gravity, you initially study what would happen in a vacuum. Perfect vacuums don't exist, but talking about what would happen if you dropped an object in a perfect vacuum makes the analysis easier. So too with economics. Our equivalent of a perfect vacuum is perfect competition. In perfect competition the invisible hand of the market operates unimpeded. In this chapter we'll consider how perfectly competitive markets work and see how to apply the cost analysis developed in Chapters 9 and 10.

A PERFECTLY COMPETITIVE MARKET

A **perfectly competitive** market is *a market in which economic forces operate unimpeded.* For a market to be *perfectly competitive*, it must meet some stringent conditions:

A perfectly competitive market is one in which economic forces operate unimpeded.

1. The number of firms is large.
2. Firms' products are identical.
3. There is free entry and free exit; that is, there are no barriers to entry.
4. There is complete information.
5. Selling firms are profit-maximizing entrepreneurial firms.
6. Both buyers and sellers are price takers.

These conditions are needed to ensure that economic forces operate instantaneously and are unimpeded by political and social forces. For example, if there weren't a large number of firms, the few firms in the industry would have an incentive to get together and limit output so they could get a higher price. They would stop the invisible hand from working.

The Necessary Conditions for Perfect Competition

To give you a sense of these conditions, let's consider each a bit more carefully.

Six conditions for a market to be perfectly competitive are:
1. Large number of firms.
2. Identical product.
3. Free entry and free exit.
4. Complete information.
5. Profit-maximizing entrepreneurial firms.
6. Both buyers and sellers are price takers.

1. *The number of firms is large. Large* number of firms means that any one firm's output compared to the market output is so small that it has no influence on total market quantity or market price.
2. *Firms' products are identical.* This requirement means that each firm's output is indistinguishable from any other firm's output. They sell a homogeneous product. Corn bought by the bushel is relatively homogeneous. One kernel is indistinguishable from any other kernel. In contrast, you can buy 30 different brands of many goods—soft drinks, for instance: Pepsi, Coke, 7UP, and so on. They are all slightly different from one another and thus not homogeneous.
3. *There is free entry and free exit.* Firms are free to *enter a market or to expand within a market in response to market signals* such as price and profit. There are no barriers to entry. **Barriers to entry** are *social, political, or economic impediments that prevent other firms from entering a market.* They might be legal barriers, such as exist when firms acquire a patent to produce a certain product. Barriers might be technological, such as when the minimum efficient scale of production allows only one, or a few, firms to produce at the lowest average total cost. There must also be **free exit**. If a firm cannot *exit from an industry without incurring a*

Q-1 Why is the assumption of free entry/free exit necessary for the existence of perfect competition?

substantial loss on its investment (sunk costs), then it may hesitate to enter the industry in the first place; without free exit, there cannot be free entry.

4. *There is complete information*. In a perfectly competitive market, firms and consumers know all there is to know about the market—prices, products, and available technology, to name a few aspects. If any firm experiences a technological breakthrough, all firms know about it and are able to use the same technology instantaneously. No firm or consumer has a competitive edge over another.
5. *Selling firms are profit-maximizing entrepreneurial firms*. Firms can have many goals and be organized in a variety of ways. For perfect competition to exist, firms must seek maximum profit and only profit. Firms compete only on the basis of price; there is no non-price competition (quality, services, brand name, store hours, and other non-price matters).
6. *Both buyers and sellers are price takers*. A **price taker** is *a firm or individual who takes the price determined by market supply and demand as given*. Neither supplier nor buyer possesses market power, which means that no one individual buyer or seller has control over price. In a perfectly competitive market, market supply and demand determine the price; both firms and consumers take the market price as given.

The Definition of Supply and Perfect Competition

Homogeneous product is characteristic of perfect competition.

These are enormously strong conditions which are seldom met simultaneously. But they are necessary for a perfectly competitive market to exist. Combined, they create an environment in which each firm, following its own self-interest, will offer goods to the market in a predictable way. If these conditions hold, we can talk formally about the supply of a produced good and how it relates to costs. This follows from the definition of supply we gave in Chapter 4:

> Supply is a schedule of quantities of goods that will be offered to the market at various prices.

This definition requires the supplier to be a price taker. In almost all other market structures, firms are not price takers; they are price makers. They don't ask, "How much should I supply, given the market price?" Instead they ask, "Given a demand curve, how much should I produce and what price should I charge?" In other market structures, the supplier sets the quantity and price, based on costs, at whatever level is best for it.[1]

The first condition—that the number of firms is large—is necessary so that firms have no ability to *collude* (to act together with other firms to control price or market share). Conditions 2 through 4 make it impossible for any firm to forget about the hundreds of other firms out there just waiting to replace their supply. Condition 5 tells us a firm's goals. If we didn't know the goals, we wouldn't know how firms would react when faced with the given price.

What's nice about these conditions is that they allow us to formally relate supply to marginal cost. If the conditions for perfect competition hold, a firm's supply curve will be that portion of the firm's short-run marginal cost curve above the average variable cost curve, as we'll see shortly.

---

[1]A firm's ability to set price doesn't mean that it can choose just any price it pleases. Other market structures can be highly competitive, so the range of prices a firm can charge and still stay in business is often limited. Such highly competitive industries are not perfectly competitive—firms still set price rather than supply a certain quantity and accept whatever price they get.

APPLYING THE TOOLS

The Internet and the Perfectly Competitive Model

Recent technological developments are making the perfectly competitive model more directly relevant to our economy. Specifically, the Internet has eliminated the spatial dimension of competition (except for shipping), allowing individuals to compete globally rather than locally. When you see a bid on the Internet, you don't care where the supplier is (as long as you do not have to pay shipping fees). Because it allows access to so many buyers and sellers, the Internet reduces the number of seller-set posted price markets (such as found in retail stores), and replaces them with auction markets.

The Internet has had its biggest impact in firms' buying practices. When firms want to buy standardized products in the 2000s, they will often post their technical requirements for desired components on the net and allow suppliers from all over the world to bid to fill their orders. Firms have found that buying in this fashion over the Internet has on average lowered the prices they pay by over 10%.

Similar changes are occurring in consumer markets. With sites like Priceline.com, individuals can set the price they are willing to pay for goods and services (such as hotel rooms and airline tickets) and see if anyone wants to supply them. With sites such as eBay you can buy and sell almost anything.

In short, with the Internet, entry and exit are much easier than in traditional "brick and mortar" business, and that makes the market more like a perfectly competitive market. As Internet search engines become better designed for commerce, and as more people develop Internet skills, the economy will more and more closely resemble the perfectly competitive model.

If the conditions for perfect competition aren't met, we can still talk *informally* about the supply of produced goods and cost conditions. Even if the conditions for perfect competition don't fully exist, supply forces are still strong and many of the insights of the competitive model can be applied to firm behaviour in other market structures.

see page 254

Even if the conditions for a perfectly competitive market are not met, supply forces are still strong and many of the insights of the competitive model carry over.

Demand Curves for the Firm and the Industry

Now that we've introduced the supply side, let's turn our attention to the competitive demand curve for the firm. The demand curve for the *industry* is downward sloping as in Figure 11-1(a), but the demand curve for the *firm* is horizontal (perfectly elastic), as in Figure 11-1(b).

Q-2 How can the demand curve for the market be downward sloping but the demand curve for a competitive firm be perfectly elastic?

Why the difference? It's a difference in perspective. Each firm in a competitive industry is so small that even if it produces as much as it can, it does not need to lower the price in order to sell the additional output. Its actions will not affect the price it can get for its product; price is the same no matter how much it produces. Think of an individual firm's actions as adding one piece of sand to a beach. Does that raise the level of the beach? We can assume it doesn't. Similarly for a perfectly competitive firm. That is why we consider the demand curve facing the firm to be horizontal.

The price the firm can get is determined by the market supply and demand curves shown in Figure 11-1(a). Market price is $7, and the firm represented in Figure 11-1(b) will get $7 for each unit of its product whether it produces 10 units (point *A*), 20 units (point *B*), or 30 units (point *C*). Its demand curve is perfectly elastic even though the demand curve for the market is downward sloping.

This difference in perception is extremely important. It means that firms will increase their output in response to an increase in market demand even though that increase in output will cause price to fall and can make all firms collectively worse off.

A perfectly competitive firm's demand schedule is perfectly elastic even though the demand curve for the market is downward sloping.

Figure 11-1 (a and b) **MARKET DEMAND CURVE VERSUS INDIVIDUAL FIRM DEMAND CURVE**

Even though the demand curve for the market is downward sloping, the demand curve of an individual firm is perfectly elastic because each firm is so small relative to the market that its actions do not affect price.

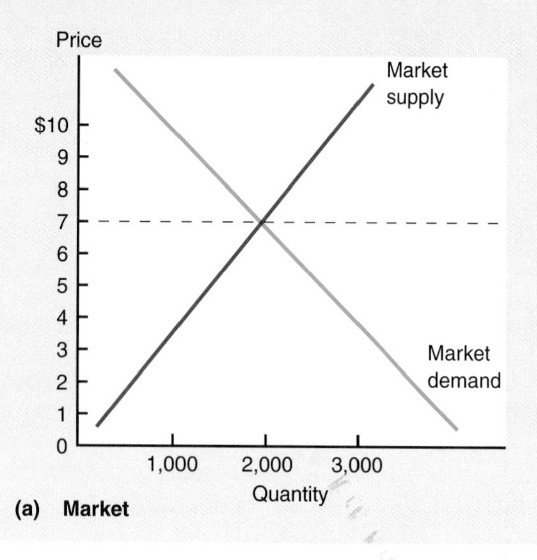

(a) Market

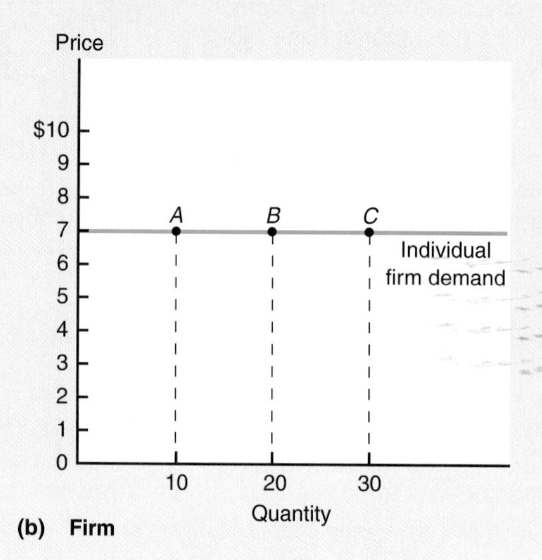

(b) Firm

THE PROFIT-MAXIMIZING LEVEL OF OUTPUT

The goal of a firm is to maximize profits. So when it decides what quantity to produce, it will continually ask the question "What will changes in quantity do to profit?" Since profit is the difference between total revenue and total cost, what happens to profit in response to a change in output is determined by **marginal revenue (MR),** *the change in total revenue associated with a change in quantity,* and **marginal cost (MC),** *the change in total cost associated with a change in quantity.* Marginal revenue and marginal cost are key concepts in determining the profit-maximizing or loss-minimizing level of output of any firm.

To determine the profit-maximizing output, all you need to know is *MC* and *MR*.

To emphasize the importance of *MR* and MC, those are the only cost and revenue figures shown in Figure 11-2. Notice that we don't illustrate profit at all. We'll calculate profit later. All we want to determine now is the profit-maximizing level of output. To do this you need only know MC and *MR*. Specifically, a firm maximizes profit when MC = *MR*. To see why, let's first look at MC and *MR* more closely.

Marginal Revenue

Let's first consider marginal revenue. Since a perfect competitor accepts the market price as given, marginal revenue is simply the market price. For example, if the firm increases output from 2 to 3, its revenue rises by $35 (from $70 to $105). So its marginal revenue is $35, the price of the good. Since at a price of $35 it can sell as much as it wants, for a perfectly competitive firm, *MR* = *P*.

For a perfectly competitive firm, *MR* = *P*.

Marginal revenue is given in column 1 of Figure 11-2(a). As you can see, *MR* equals $35 for all levels of output. But that's what we saw in Figure 11-1, which showed that the demand curve for a perfect competitor is perfectly elastic at the market price. For a perfect competitor, the marginal revenue curve and demand curve it faces are the same.

Figure 11-2 (a and b) **MARGINAL COST, MARGINAL REVENUE, AND PRICE**

The profit-maximizing output for a firm occurs where marginal cost equals marginal revenue. Since for a competitive firm $P = MR$, its profit-maximizing output is where $MC = P$. Any other output is less profitable.

Price = *MR*	Quantity produced	Total Cost	Marginal cost
$35.00	0	$ 40.00	
			$28.00
35.00	1	68.00	
			20.00
35.00	2	88.00	
			16.00
35.00	3	104.00	
			14.00
35.00	4	118.00	
			12.00
35.00	5	130.00	
			17.00
35.00	6	147.00	
			22.00
35.00	7	169.00	
			30.00
35.00	8	199.00	
			40.00
35.00	9	239.00	
			54.00
35.00	10	293.00	

(a) MC/MR (price) table

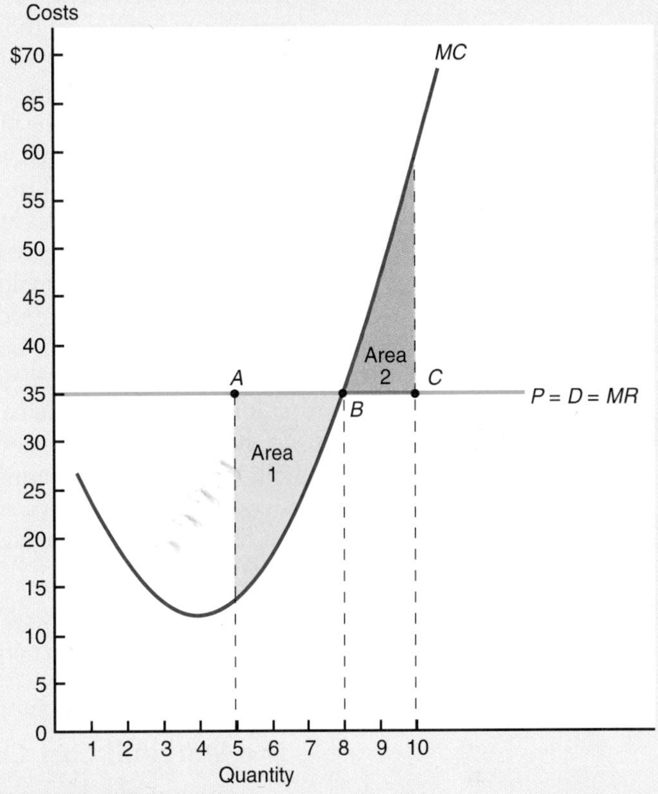

(b) MC/MR (price) graph

Marginal Cost

Marginal cost is the change in total cost that accompanies a one unit change in output. Figure 11-2(a) shows marginal cost in column 4. The marginal cost of increasing output from 1 to 2 is $20, and the marginal cost of increasing output from 2 to 3 is $16. The marginal cost right at 2 (which the marginal cost graph shows) would be between $20 and $16 at approximately $18. Notice that initially in this example, marginal cost is falling, but after the fifth unit of output, it's increasing. This is consistent with our discussion in earlier chapters.

Profit Maximization: MC = MR

To maximize profit, a firm should produce where marginal cost equals marginal revenue. Looking at Figure 11-2(b), a firm following that rule should produce at an output of 8 where $MC = MR = \$35$.

Let's say that initially the firm decides to produce 5, placing it at point A in Figure 11-2(b). At output A, the firm gets $35 for each widget but its marginal cost of

Q-3 What are the two things you must know to determine the profit-maximizing output?

Producers maximize profits by producing the quantity where MC = MR (= P).

Profit-maximizing condition for a competitive firm: $MC = MR$ (= P).

If marginal revenue does not equal marginal cost, a firm can increase profit by changing output. Therefore, profit is maximized when $MC = MR$ (= P).

increasing output is $17. We don't yet know the firm's total profit, but we do know how changing output will affect profit. For example, say the firm increases production from 5 to 6. Its revenue will rise by $35. (In other words, its marginal revenue is $35.) Its marginal cost of increasing output is $17. Since $MR > MC$, profit increases by $18 (the difference between MR, $35, and MC, $17), therefore, it makes sense (meaning the firm can increase its profit) to increase output from 5 to 6 units. It makes sense to increase output as long as the marginal revenue exceeds the marginal cost. The blue shaded area (1) represents the entire increase in profit the firm can get by increasing output.

Now let's say that the firm decides to produce 10 widgets, placing it at point C. Here the firm gets $35 for each widget. The marginal cost of producing that 10th unit is $54. So, $MC > MR$. If the firm decreases production by one unit, its cost decreases by $54 and its revenue decreases by $35. Profit increases by $19 ($54 − $35 = $19), so at point C, it makes sense to decrease output. This reasoning holds true as long as the marginal cost is above the marginal revenue. The red shaded area (2) represents the increase in profits the firm can get by decreasing output.

At point *B* (output = 8) the firm gets $35 for each widget, and its marginal cost is $35, as you can see in Figure 11-2(b). The marginal cost of increasing output by one unit is $40 and the marginal revenue of selling one more unit is $35, so its profit falls by $5. If the firm decreases output by one unit, its MC is $30 and its *MR* is $35, so its profit falls by $5. Either increasing or decreasing production will decrease profit, so at point *B*, an output of 8, the firm is maximizing profit.

Since *MR* is just market price, we can state the **profit-maximizing condition** of a competitive firm:

Produce the output where $MC = MR$ (= P).

If marginal revenue isn't equal to marginal cost, a firm can increase profit by changing output.

The Marginal Cost Curve Is the Supply Curve

Because the marginal cost curve tells us how much of a produced good a firm will supply at a given price, the marginal cost curve is the firm's supply curve.

Now let's consider again the definition of the supply curve as a schedule of quantities of goods that will be offered to the market at various prices. Notice that the marginal cost curve fits that definition. It tells how much the firm will supply at a given price. Figure 11-3, which repeats the MC curve from Figure 11-2(b), shows the various quantities the firm will supply at different market prices. If the price is $35, we showed that the firm would supply 8 (point *A*). If the price had been $19.50, the firm would have supplied 6 (point *B*); if the price had been $61, the firm would have supplied 10 (point C). Because the marginal cost curve tells us how much of a produced good a firm will supply at a given price, *the marginal cost curve is the firm's supply curve*. The MC curve tells the competitive firm how much it should produce at a given price. Specifically, the marginal cost curve is the firm's supply curve only if price exceeds average variable cost.

Firms Maximize Total Profit

Q-4 Why do firms maximize total profit rather than profit per unit?

Notice that when firms maximize profit, they maximize *total profit*, not profit per unit. Profit per unit would be maximized at a much lower output level than is total profit. Firms don't care about profit per unit; as long as an increase in output yields even a small amount of additional profit, it will increase total profits, and a profit-maximizing firm will increase output.

Figure 11-3 **THE MARGINAL COST CURVE IS A FIRM'S SUPPLY CURVE**

A perfectly competitive firm maximizes profit by producing the quantity at which marginal cost equals marginal revenue equals price. A perfectly competitive firm produces the quantity at which market price intersects the marginal cost curve. Since the marginal cost curve tells the firm how much to produce at each price, the marginal cost curve is the perfectly competitive firm's supply curve. This figure shows three points on a firm's supply curve; as you can see the quantity the firm chooses to supply depends on the price. For example, if market price is $19.50 the firm produces 6 units.

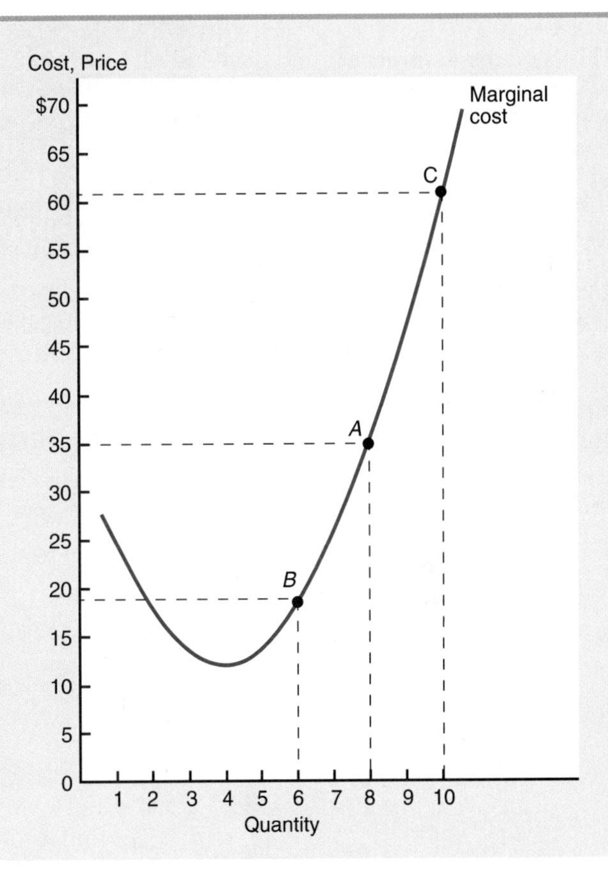

Profit Maximization Using Total Revenue and Total Cost

An alternative method of determining the profit-maximizing level of output is to look at the total revenue and total cost curves directly. Figure 11-4 (on page 240) shows total cost and total revenue for the firm we're considering so far. The table in Figure 11-4(a) shows total revenue in column 2, which is just the number of units sold times market price. Total cost is in column 3. Total cost is the cumulative sum of the marginal costs from Figure 11-2(a) plus a fixed cost of $40. Total profit (column 4) is the difference between total revenue and total cost. The firm is interested in maximizing total profit. Looking down column 4 of Figure 11-4(a), you can quickly see that the profit-maximizing level of output is 8, as it was using the $MR = MC$ rule, since total profit is highest at an output of 8.

Price is the signal for firms to increase supply.

In Figure 11-4(b) we plot the firm's total revenue and total cost curves from the table in Figure 11-4(a). The total revenue curve is a straight line; each additional unit sold increases revenue by the same amount, $35. The total cost curve is bowed upward at most quantities, reflecting the increasing marginal cost at different levels of output. The firm's profit is represented by the vertical distance between the total revenue curve and the total cost curve. For example, at an output of 5, the firm makes $45 in profit ($175 – $130).

Total profit is maximized where the vertical distance between total revenue and total cost is greatest. In this example total profit is maximized at an output of 8, just as in the alternative approach. At that output, marginal revenue (the slope of the total revenue curve) and marginal cost (the slope of the total cost curve) are equal.

APPLYING THE TOOLS

The Broader Importance of the $MR = MC$ Equilibrium Condition

This marginal revenue = marginal cost equilibrium condition is simple, but it's enormously powerful. As we'll see, it carries over to other market structures. If you replace revenue with benefits, it also forms the basis of economic reasoning. With whom should you go out? What's the marginal benefit? What's the marginal cost? Should you marry Pat? What's the marginal benefit? What's the marginal cost? As we discussed in Chapter 1, thinking like an economist requires thinking in these marginal terms and applying this marginal reasoning to a wide variety of activities. Understanding this condition is to economics what understanding gravity is to physics. It gives you a sense of if, how, and why prices and quantities will move.

Figure 11-4 (a and b) **DETERMINATION OF PROFITS BY TOTAL COST AND TOTAL REVENUE CURVES**

The profit-maximizing output level can also be seen by considering the total cost curve and the total revenue curve. Profit is maximized at the output where total revenue exceeds total cost by the largest amount. This occurs at an output of 8.

Quantity	Total revenue	Total cost	Total profit
0	$ 0	$ 40	$−40
1	35	68	−33
2	70	88	−18
3	105	104	1
4	140	118	22
5	175	130	45
6	210	147	63
7	245	169	76
8	280	199	81
9	315	239	76
10	350	293	57

(a) Total revenue and total cost table

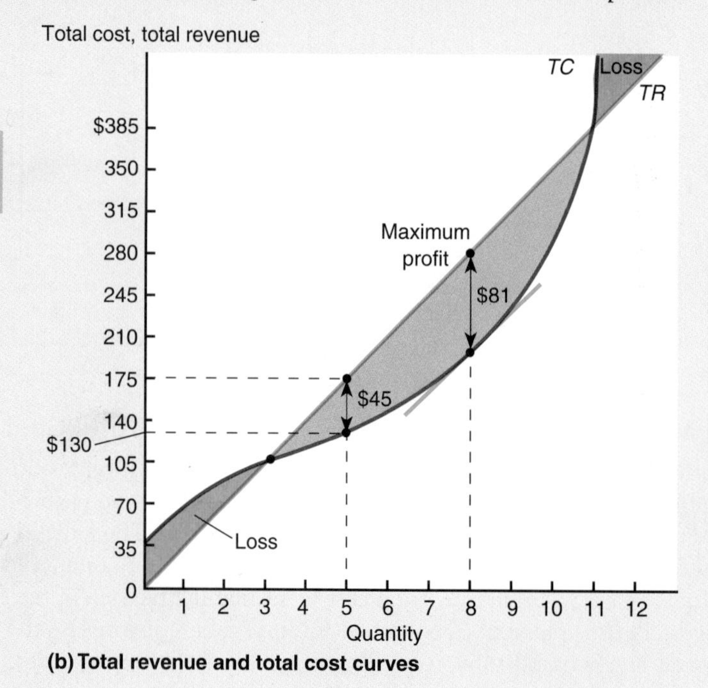

(b) Total revenue and total cost curves

TOTAL PROFIT AT THE PROFIT-MAXIMIZING LEVEL OF OUTPUT

Determining Profit from a Table of Costs and Revenue

l cost is all that is determine a perfectly firm's supply curve.

The initial discussion of the firm's choice of output, given price, presented only marginal cost and price. The $P = MR = MC$ condition tells us how much output a competitive firm should produce to maximize profit. It does not tell us the profit the firm

TABLE 11-1 Costs Relevant to a Firm

	1	2	3	4	5	6	7	8	9	10
Price= Marginal Revenue	Quantity Produced	Total Fixed Cost	Average Fixed Cost	Total Variable Cost	Average Variable Cost	Total Cost	Marginal Cost	Average Total Cost	Total Revenue	Total Profit
$35.00	0	$40.00	—	0	—	$ 40.00		—	0	$−40.00
35.00	1	40.00	$40.00	$ 28.00	$28.00	68.00	$28.00	$68.00	$ 35.00	−33.00
35.00	2	40.00	20.00	48.00	24.00	88.00	20.00	44.00	70.00	−18.00
35.00	3	40.00	13.33	64.00	21.33	104.00	16.00	34.67	105.00	1.00
35.00	4	40.00	10.00	78.00	19.50	118.00	14.00	29.50	140.00	22.00
35.00	5	40.00	8.00	90.00	18.00	130.00	12.00	26.00	175.00	45.00
35.00	6	40.00	6.67	107.00	17.83	147.00	17.00	24.50	210.00	63.00
35.00	7	40.00	5.71	129.00	18.43	169.00	22.00	24.14	245.00	76.00
35.00	8	40.00	5.00	159.00	19.88	199.00	30.00	24.88	280.00	81.00
35.00	9	40.00	4.44	199.00	22.11	239.00	40.00	26.56	315.00	76.00
35.00	10	40.00	4.00	253.00	25.30	293.00	54.00	29.30	350.00	57.00

makes. Profit is determined by total revenue minus total cost. Table 11-1 expands Figure 11-2(a) and presents a table of all the costs relevant to the firm.

The firm is interested in maximizing profit. Looking at Table 11-1 you can quickly see that the profit-maximizing position is 8, as it was before, since at an output of 8, total profit is highest.

Using the $MC = MR = P$ rule, you can also see that the profit-maximizing level of output is 8. Increasing output from 7 to 8 has a marginal cost of $30, which is less than $35, so it makes sense to increase output. Increasing output from 8 to 9 has a marginal cost of $40, which is more than $35, so it does not make sense to increase output. The output 8 is the profit-maximizing output. At that profit-maximizing level of output, the firm earns a profit of $81, which is calculated by subtracting total cost of $199 from total revenue of $280. Notice also that average total cost is lowest at an output of about 7, and the average variable cost is lowest at an output of about 6.[2] Thus, the profit-maximizing position (which is 8) is *not* necessarily a position that minimizes either average variable cost or average total cost. It is only the position that maximizes total profit.

Determining Profit from a Graph

These relationships can be seen in a graph. In Figure 11-5(a), on page 243, we add the average total cost and average variable cost curves to the graph of marginal cost and

[2]We say "about 6" and "about 7" because the table gives only whole numbers. The actual minimum point occurs at 5.55 for average variable cost and 6.55 for average total cost. The nearest whole numbers to these are 6 and 7.

The profit-maximizing output can be determined in a table (as in Table 11-1) or in a graph (as in Figure 11-5).

Q-5 If the firm described in Figure 11-5 is producing 4 units, what would you advise it to do, and why?

When the marginal revenue curve is above the ATC curve, the firm makes a profit. When the marginal revenue curve is below the ATC curve, the firm incurs a loss.

To determine maximum profit, you must first determine what output the firm will choose to produce by seeing where *MC* equals *MR*.

price first presented in Figure 11-2. Notice that the marginal cost curve goes through the lowest points of both average cost curves.

Find Output Where MC = MR To find profit graphically, first find the point where MC = *MR* (point *A*). That intersection determines the quantity the firm will produce if it wants to maximize profit. Why? Because the vertical distance between a point on the marginal cost curve and a point on the marginal revenue curve represents the additional profit the firm can make by changing output. For example, if it increases production from 6 to 7, its marginal cost is $22 and its marginal revenue is $35. By increasing output it can increase profit by $13 (from $63 to $76). The same reasoning holds true for any output less than 8. For outputs higher than 8, the opposite reasoning holds true. Marginal cost exceeds marginal revenue, so it pays to decrease output. So to maximize profit, the firm must see that there is no distance between the two curves—it must see where they intersect.

Find Profit Per Unit Where MC = MR After having determined the profit-maximizing quantity, determine what average total cost is at that output level, which is given by point *B* on the *ATC* curve ($25). The profit per unit is the difference between the price the firm receives per unit (price) and its cost per unit (average total cost). Since the firm earns a profit of P – ATC = 35 – 25, or distance *AB*, on each unit sold, profit per unit is distance *AB*. The firm sells a quantity of 8 (represented by distance *BC*), so multiplying profit per unit (distance *AB*) by number of units sold (distance *BC*) yields total profit (area ABCD). This corresponds to the calculation of profit:

$$\pi = (P - ATC) \times Q$$

Expanding,

$$\pi = P \times Q - ATC \times Q$$

Therefore,

$$\pi = TR - TC.$$

Notice that at the profit-maximizing position, the profit per unit isn't at its highest because average total cost is *not* at its minimum point. Profit per unit of output would be highest at point *E*. A common mistake is to draw a line up from point E to find profits. *To determine maximum profit you must first determine what output the firm will choose to produce by seeing where* MC *equals* MR *and then calculate profit.* By choosing quantity *E*, the firm forgoes the profits it would have earned on the eighth unit if they chose to produce at *B*. So E cannot maximize profit.

Zero Profit or Loss Where MC = MR Notice also that as the curves in Figure 11-5(a) are drawn, *ATC* at the profit-maximizing position is below the price, and the firm makes a profit per unit of a little over $10. The choice of short-run average total cost curves was arbitrary and doesn't affect the firm's profit-maximizing condition: MC = *MR*. It could have been assumed that fixed cost was higher, which would have shifted the *ATC* curve up. In Figure 11-5(b) it's assumed that fixed cost is $81 higher than in Figure 11-5(a). Instead of $40, it's $121. The appropriate average total cost curve for a fixed cost of $121 is drawn in Figure 11-5(b). Notice that in this case economic profit is zero and the marginal cost curve intersects the minimum point of the average total cost curve at an output of 8 and a price of $35. In this case, the firm is making **zero economic** profit. The cost curves reflect all relevant costs, both explicit and implicit, which means that cost of labour and materials, and electricity and insurance, are all included as costs and reflected in the cost curves: *ATC*, *AVC*, and MC. Also included are the

Figure 11-5 (a, b, and c) DETERMINING PROFITS GRAPHICALLY

The profit-maximizing output depends *only* on where the MC and MR curves intersect. The total amount of profit or loss that a firm makes depends on the price it receives and its average total cost of producing the profit-maximizing output. This figure shows the case of (a) a positive economic profit, (b) a zero economic profit, and (c) an economic loss.

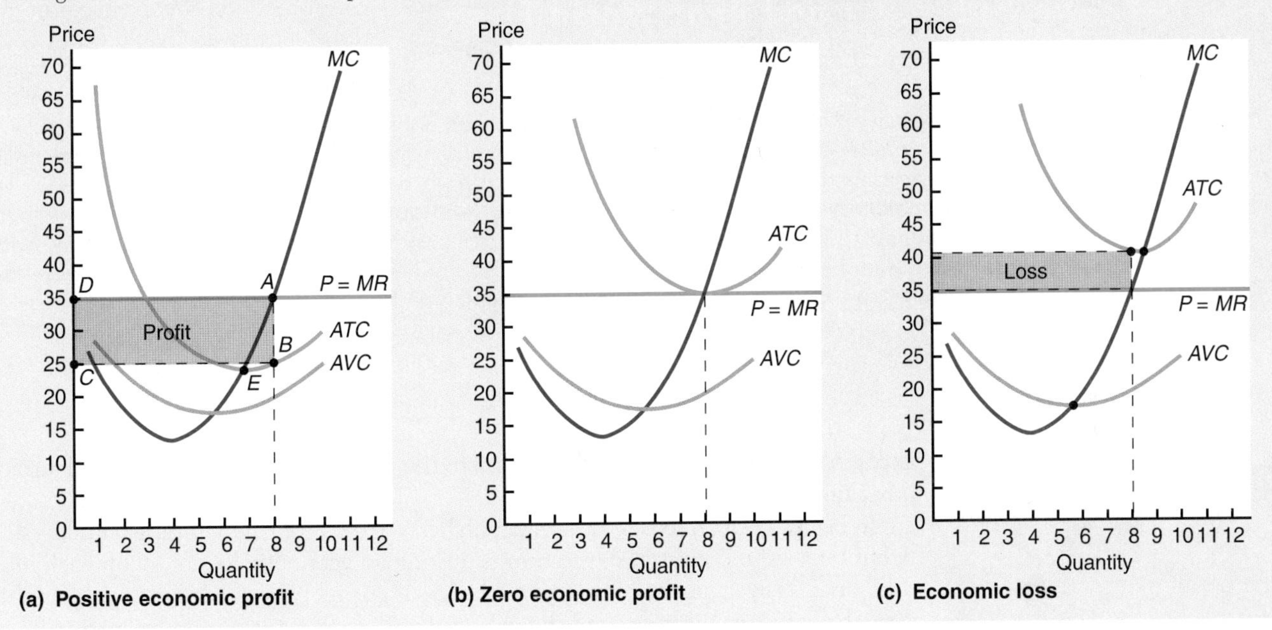

opportunity cost of the entrepreneur, and a rate of return to cover the risk inherent in running a firm. These must be considered costs because they must be covered by revenues. If these costs are not covered, the resources used in the firm have better opportunities elsewhere and should be reallocated to a better use. When all explicit and implicit costs are covered by revenues, that is, P = ATC, the firm is said to be earning **normal profit**. This means that *revenues are sufficient to cover all opportunity costs, and the firm is breaking even.* Profit *greater than normal profit* is **economic profit**, and is in

Q-6 What is wrong with the following diagram?

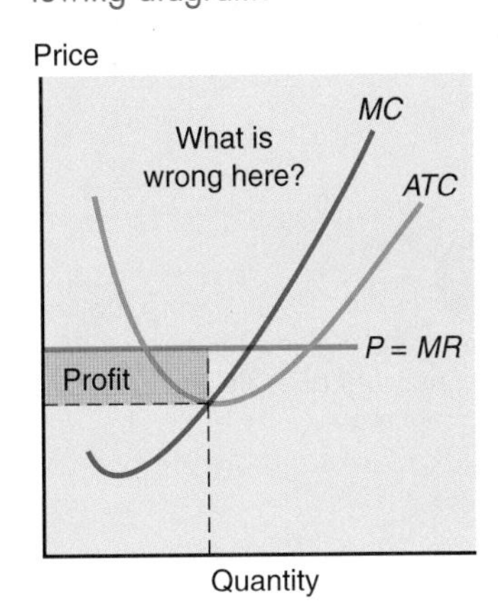

TABLE 11-2 The Role of Profits as Market Signals.

Profit Calculation	Type of Profit	Market Signal
$\pi > 0$	Positive economic profit, or Economic profit	Entry. Resources are drawn into the industry as existing firms expand output, and new firms enter the industry.
$\pi = 0$	Zero economic profit, Zero profit, or Normal profit	Static. The industry is in long run equilibrium, as resources are not flowing in or out of the industry.
$\pi < 0$	Economic loss	Exit. Resources leave the industry as existing firms reduce output, or exit the industry.

APPLYING THE TOOLS

Profit Maximization and Real-World Firms

Most real-world firms do not have profit as their only goal. The reason is that, in the real world, the decision-maker's income is part of the cost of production. For example, a paid manager has an incentive to reduce costs, but has little incentive to reduce his own income which, for the firm, is a cost. Alternatively, say that a firm is a worker-managed firm. If workers receive a share of the profits, they'll push for higher profits, but they'll also see to it that in the process of maximizing profits they don't hurt their own interest—maximizing their wages.

A manager-managed firm will push for high profits but will see to it that it doesn't achieve those profits by hurting the manager's interests. Managers' pay will be high. In short, real-world firms will hold down the costs of factors of production *except* the cost of the decision maker.

In real life, this problem of the lack of incentives to hold down costs is important. For example, firms' managerial expenses often grow larger even as firms are cutting "costs." Similarly, CEOs and other high-ranking officers of the firm often have enormously high salaries. How and why the lack of incentives to hold down costs affects the economy is best seen by first considering the nature of an economy with incentives to hold down all costs. That's why we use the profit-maximizing firm as our standard model.

This firm has reached its shutdown point. In the short run, firms require price greater than or equal to average variable cost, but in the long run, firms require price equal to or above average total cost.

excess of what is required to keep the firm in the industry. Economic profits will attract other firms to enter the industry.

In Figure 11-5(c), fixed cost is much higher—$169. Profit-maximizing output is still 8, but now at an output of 8, the firm is making an economic loss of $6 on each unit sold, since its average total cost is $41. The loss is given by the shaded rectangle. In this case, the profit-maximizing condition is actually a loss-minimizing condition. So MC = MR = P is both a *profit-maximizing condition* and a *loss-minimizing condition*.

These three cases, summarized in Table 11-2 on page 243, emphasize that the profit-maximizing output level doesn't depend on fixed cost or average total cost. It depends only on where marginal cost equals price.

The Shutdown Point

The supply curve of a competitive firm is its marginal cost curve above the average variable cost curve.

Let's consider Figure 11-6(a)—a reproduction of Figure 11-5(c)—and the firm's decision at various prices. At a price of $35, it's producing 8 units, and incurring a loss of $6 per unit. If it's making a loss, why doesn't it shut down? The answer lies in the fixed costs. There's no use crying over spilt milk. In the short run a firm knows it must pay these fixed costs regardless of whether or not it produces. The firm only considers the costs it can save by stopping production, and those costs are its variable costs. As long as a firm is covering its variable costs, it pays to keep on producing. By producing, its loss is $48; if it stopped producing, its loss would be all the fixed costs ($169). So it makes a smaller loss by producing (loss minimization).

Q-7 In the 1990s, many airlines were making losses, yet they continued to operate. Why?

When price equals average variable cost, the firm is earning enough to pay (only) the variable costs of production, but none of the fixed costs. If the firm produces 6 units, its loss is equal to the amount of fixed cost ($169); if it shuts down, its loss is equal to the amount of fixed cost, since the fixed cost must still be paid. It is always presumed that a profit-maximizing firm wishes to remain a going concern, so it will choose to continue producing, hoping that conditions will improve and the firm will return to normal profits.

The shutdown point is the point at which the firm will be better off if it shuts down than it will if it stays in business.

However, once the price falls below average variable costs (below $17.80), it will be cheaper to shut down (point *A* in Figure 11-6(a)). In that case the firm's loss from

Figure 11-6 **THE SHUTDOWN DECISION AND LONG-RUN EQUILIBRIUM**

A firm should continue to produce as long as price exceeds average variable cost. Once price falls below that, it will do better by temporarily shutting down and saving the variable costs. This occurs at point A in (**a**). In (**b**), the long-run equilibrium position for a perfectly competitive firm in an industry occurs after competitive entry has driven price down as low as possible, to P=ATC. In long-run equilibrium, only normal profits are made.

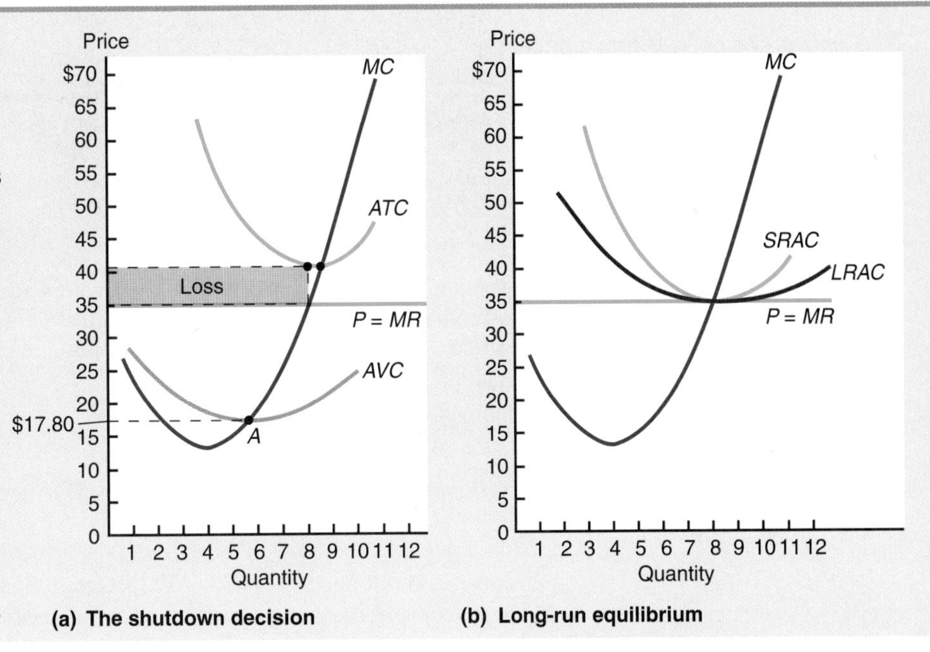

(a) The shutdown decision **(b) Long-run equilibrium**

producing would be more than $169, and it would do better to simply stop producing temporarily and avoid paying the variable cost. Thus, the point at which price equals AVC is the **shutdown point**—*that point at which the firm will be better off if it temporarily shuts down than it will if it stays in business*. When price falls below the shutdown point, the average variable costs the firm can avoid paying by shutting down exceed the price it would get for selling the good. When price is above average variable cost, in the short run a firm should keep on producing even though it's making a loss. As long as a firm's revenue is covering its variable cost, temporarily producing at a loss is the firm's best strategy because it's making a smaller loss than it would make if it were to shut down.

If $P \geq$ minimum of *AVC*, the firm will continue to produce in the short run. If $P <$ minimum of *AVC*, the firm will shut down.

SHORT-RUN MARKET SUPPLY AND DEMAND

Most of the preceding discussion has focussed on supply and demand analysis of a firm. Now let's consider supply and demand for the industry. Even though the demand curve faced by the firm is perfectly elastic, the industry demand curve is downward sloping.

How about the industry supply curve? The supply curve for a competitive firm is that firm's marginal cost curve above the average variable cost curve. In the short run when the number of firms in the market is fixed, the **market supply curve** is just the *horizontal sum of all the firms' marginal cost curves*. To move from individual firms' marginal cost curves or supply curves to the market supply curve we add the quantities all firms will supply at each possible price. Since all firms have identical marginal cost curves, a quick way of summing the quantities is to multiply the quantities from the marginal cost curve of a representative firm at each price by the number of firms in the market. As the short run evolves into the long run, the number of firms in the market can change in response to market signals, such as price and profit. For example, as more firms enter the market in response to economic profits being earned by existing firms (P> ATC), the market supply curve shifts to the right because more firms are supplying the quantity indicated by the representative marginal cost curve. Likewise, as the number of firms in the market

The market supply curve is the horizontal sum of all the firms' marginal cost curves, taking into account any changes in input prices that might occur.

To find a competitive firm's price, level of output, and profit, given a firm's marginal cost curve and average total cost curve, use the following three steps:

1. Determine the market price at which market supply and demand curves intersect. This is the price the "price-taking" perfectly competitive firm accepts for its products. Draw the horizontal marginal revenue (*MR*) curve at the market price.
2. Determine the profit-maximizing level of output by finding the level of output where the *MR* and *MC* curves intersect.
3. Determine profit by subtracting average total costs at the profit-maximizing level of output from the price and multiplying by the firm's output.

If you are demonstrating profit graphically, find the point at which *MC* = *MR*. Extend a line down to the *ATC* curve. Extend a line from this point to the vertical axis. To complete the box indicating profit, go up the vertical axis to the market price.

declines as economic losses (P < AVC) force some firms to exit, the market supply curve shifts to the left. Knowing how the number of firms in the market affects the market supply curve is important to understanding long-run equilibrium in perfectly competitive markets.

LONG-RUN COMPETITIVE EQUILIBRIUM

The analysis of the competitive firm consists of two parts: the short-run analysis just presented and the long-run analysis. In the short run the number of firms is fixed and the firm can either earn economic profit or incur economic loss. In the long run, firms enter and exit the market and neither economic profits nor economic losses are possible. In the long run, firms make zero economic profit. Thus, in the long run, only the zero profit equilibrium shown in Figure 11-6(b) is possible. As you can see, at that long-run equilibrium, the firm is at the minimum of both the short-run and the long-run average total cost curves.

Since profits create incentives for new firms to enter, output will increase, and the price will fall until zero profits are being made.

Why can't firms earn economic profit or make economic losses in the long run? Because of the entry and exit of firms: If there are economic profits, firms would enter the market, shifting the market supply curve to the right. As market supply increased, the market price would decline and reduce profits for each firm. Firms would continue to enter the market and the market price would continue to decline until the incentive of economic profits was eliminated. At that price, all firms are earning zero economic profit. Similarly, if the price was lower than the price necessary to earn a profit, firms incurring losses would leave the market and the market supply curve would shift to the left. As market supply shifts to the left, market price will rise. Firms will continue to exit the market and market price will continue to rise until all remaining firms are no longer incurring losses and are earning zero economic profit. Only at zero profit do entry and exit stop.

11.2 
see page 254

Zero profit does not mean that entrepreneurs don't get anything for their efforts. The entrepreneur is an input to production just like any other factor of production. In order to stay in the business the entrepreneur must receive their opportunity cost, or **normal profit** (*the amount the owners of business would have received in the next-best alternative*). That normal profit is built into the costs of the firm; **economic profits** are *profits above normal profits*.

APPLYING THE TOOLS

The Shutdown Decision and the Relevant Costs

Chapters 9 and 10 emphasized that it is vital to choose the relevant costs to the decision at hand. Discussing the shutdown decision gives us a chance to demonstrate the importance of those choices. Say the firm leases a large computer it needs to operate. The rental cost of that computer is a fixed cost for most decisions, if, as long as the firm keeps the computer, the rent must be paid whether or not the computer is used. However, if the firm can end the rental contract at any time, and thereby save the rental cost, the computer is not a fixed cost. But neither is it your normal variable cost. Since the firm can end the rental contract and save the cost only if it shuts down, that rental cost of the computer is an *indivisible setup cost.* For the shutdown decision, the computer cost is a variable cost. For other decisions about changing quantity, it's a fixed cost.

The moral: The relevant cost can change with the decision at hand, so when you apply the analysis to real-world situations, be sure to think carefully about what the *relevant cost* is.

What if one firm has superefficient workers or machinery? Won't the firm make a profit in the long run? The answer is, again, no. In a long-run competitive market, other firms will see the value of those workers and machines and will compete to get them for themselves. As firms compete for the superefficient factors of production, the prices of those specialized inputs will rise until all ecnomic profits are eliminated. However, those factors will receive what are called *rents* for their superior ability. For example, say the average worker receives $400 per week, but Sarah, because she's such a good worker, receives $600. So $200 of the $600 she receives is a rent to her specialized ability. Either her existing firm matches that $600 wage or she will change employment.

Q-8 If a competitive firm makes zero profit, why does it stay in business?

The zero profit condition is enormously powerful; it makes the analysis of competitive markets far more applicable to the real world than can a strict application of the assumption of perfect competition. If economic profit is being made, firms will enter and compete that profit away. Price will be pushed down to the average total cost of production as long as there is free entry (and exit). As we'll see in later chapters, in their analysis of whether markets are competitive, many economists focus primarily on whether barriers to entry exist.

The zero profit condition is enormously powerful; it makes the analysis of competitive markets far more applicable to the real world than would otherwise be the case.

Economic profit attracts resources from other industries.

ADJUSTMENT FROM THE SHORT RUN TO THE LONG RUN

Using the basics of the perfectly competitive supply and demand curves, we'll see how the adjustment to long-run equilibrium takes place in the firm and in the market.

An Increase in Demand

In Figure 11-7(a), the market is initially in long-run equilibrium at point A. Figure 11-7(b) shows a representative in long-run equilibrium at point A. Market equilibrium occurs at a price of $7 and market quantity supplied of 700 thousand units (point A in (a)), with each of 70 firms producing 10 thousand units (point *a* in (b)). Firms are in long-run equilibrium, making zero profit because $P = ATC$.

Q-9 If berets suddenly became the "in" thing to wear, what would you expect to happen to the price in the short run? In the long run?

Now suppose that consumer tastes change, and there is an increased demand for the product. Market demand shifts from D_0 to D_1; market equilibrium price (and quantity) rises. Firms see the market price increasing and increase their output until they're once again at a position where $MC = P$. This occurs at point *b* at a firm output of 12 in (b), which results in an overall market output of 840 thousand units at point *B* in (a). In the short run

Figure 11-7 (a and b) MARKET RESPONSE TO AN INCREASE IN DEMAND

Faced with an increase in demand which it sees as an increase in price and hence profits, a competitive firm will respond by increasing output (from *A* to *B*) in order to maximize profit. The market response is shown in (**a**); the firm's response is shown in (**b**). As all firms increase output and as new firms enter, price will fall until all profit is competed away. Thus the long-run market supply curve will be perfectly elastic, as is S_{LR} in (**a**). The final equilibrium will be the original price but a higher output. The original firms return to their original output (*A*), but since there are more firms in the market, the market output increases to *C*.

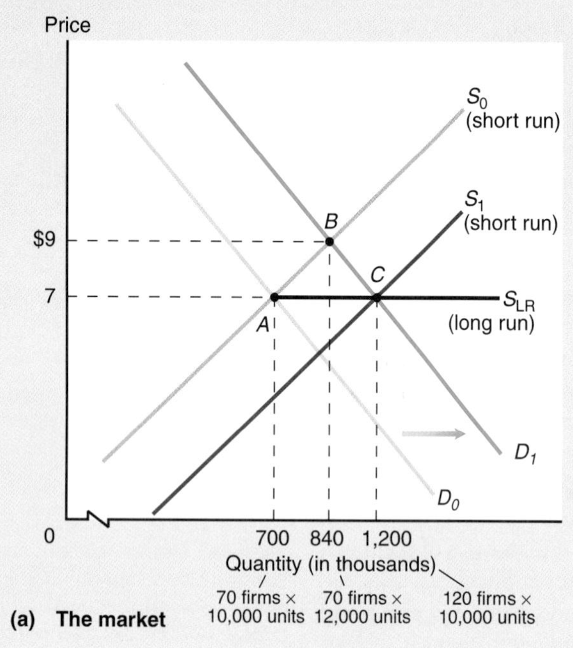

(a) The market

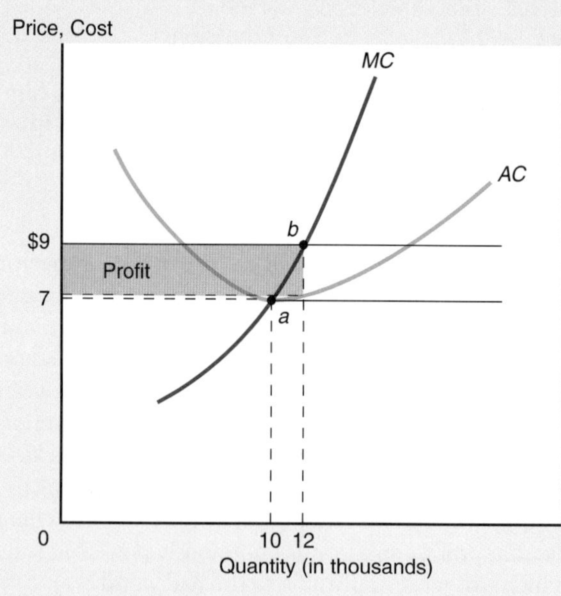

(b) One representative firm

In the long run firms earn zero profits.

the 70 existing firms each make an economic profit (the shaded area in Figure 11-7(b)). Price has risen to $9, but average cost is only $7.10, so with the price at $9 each firm is making a profit of $1.90 per unit. But price cannot remain at $9, since each firm will have an incentive to expand output and new firms will have an incentive to enter the market.

As existing firms expand and new firms enter, the short-run market supply curve shifts from S_0 to S_1 and the market price returns to $7. As market price falls, from $9 to $7, each individual firm reduces its output from 12,000 to 10,000. However, the entry of 50 new firms provides the additional output in this example, bringing market output to 1.2 million units sold for $7 apiece. The final equilibrium will be at a higher market output but the same price.

Long-Run Market Supply

The long-run market supply curve is a schedule of quantities supplied when firms are no longer entering or exiting the market. This occurs when firms are earning zero profit. In this case, the long-run supply curve is created by extending to the right the line connecting points A and C. Since equilibrium price remains at $7, the long-run supply curve is perfectly elastic. The long-run supply curve is horizontal because factor prices are constant. That is, factor prices do not increase as industry output increases. Economists call this a *constant-cost industry*. Two other possibilities exist: an *increasing-cost industry* (in which factor prices rise as more firms enter the market and existing firms expand production and compete for the available resources) and a *decreasing-cost industry* (in which factor prices fall as industry output expands).

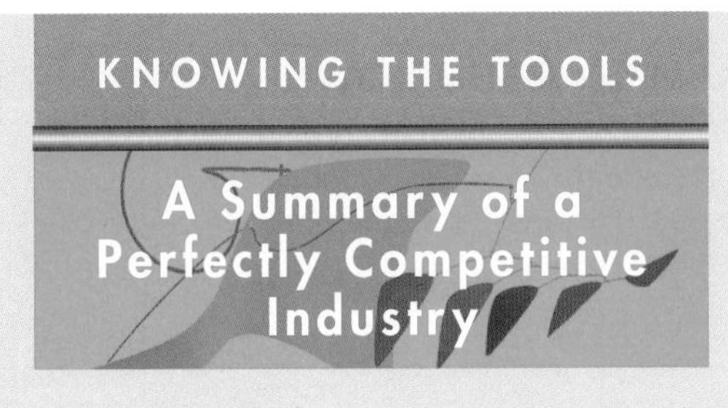

Four things to remember when considering a perfectly competitive industry are

1. The profit-maximizing condition for perfectly competitive firms is $MC = MR = P$.
2. To determine profit or loss at the profit-maximizing level of output, subtract the average total cost at that level of output from the price and multiply the result by the output level.
3. Firms will shut down production in the long run if price is equal to or falls below the minimum of their average variable costs.
4. A perfectly competitive firm is in long-run equilibrium only when it is earning zero economic profit, or when price equals the minimum of long-run average total costs.

Factor prices are likely to rise when industry output increases if the factors of production are specialized. An increase in the demand for the factors of production that accompanies an increase in output, in this case, will bid up factor prices. The effect on long-run supply is the following: The rise in factor prices raises costs for each individual firm and increases the price at which firms earn zero profit (break even). Firms will stop entering the market and expanding production at a higher equilibrium price since the price at which zero profit is made has risen, so market price cannot fall to the original price of $7. Therefore, in increasing cost industries, the long-run supply curve is upward sloping. Input costs would also rise if there are diseconomies of scale. The long-run equilibrium price would be higher and output would be lower than if input prices remained constant.[3]

The other possibility is a decreasing-cost industry. If factor prices decline when industry output expands, individual firms' cost curves shift down. As they do, the price at which the zero profit condition occurs falls and the price at which firms cease to enter the market also falls. In this case, the long-run market supply curve is downward sloping. Factor prices may decline as output rises when new entrants make it more cost-effective for other firms to provide services to all firms in the area. The supply of factors of production expands and reduces the price of inputs to production.

Notice that in the long-run equilibrium, once again zero profit is being made. Long-run equilibrium is defined by zero economic profit. Notice also that the long-run supply curve is more elastic than the short-run supply curve. That's because output changes are much less costly in the long run than in the short run. *In the short run, the price does more of the adjusting. In the long run, more of the adjustment is done by quantity.*

In the short run, the price does more of the adjusting. In the long run, more of the adjustment is done by quantity.

Examples in the Real World

Consider the Canadian retailing industry. During the 1990s the Canadian retail scene repeatedly illustrated how a dynamic competitive market adjusts to changing market conditions. In the last decade or so, shoppers have experienced a big change in the Canadian retail landscape, as well-known and lesser-known retailers were lost or absorbed by competitors: Eaton's, Bretton's, Pascal's, Robinson's, K-Mart, Canary Island, and many others.

[3]To check your understanding, ask yourself the following question: What if there had been economies of scale? If you answered, "There couldn't have been," you're really into economic thinking. (For those of you who aren't all that heavily into economic thinking, the reason is that if there had been economies of scale, the market structure would not have been perfectly competitive. One firm would have kept expanding and expanding and, as it did, its costs would have kept falling.

Figure 11-8 **A REAL-WORLD EXAMPLE: A SHUTDOWN DECISION**

As price falls below *ATC*, firms can produce in the short run as long as they are able to cover their variable costs. However, in the long run, all costs must be covered, so unless price covers minimum *ATC*, this firm will exit.

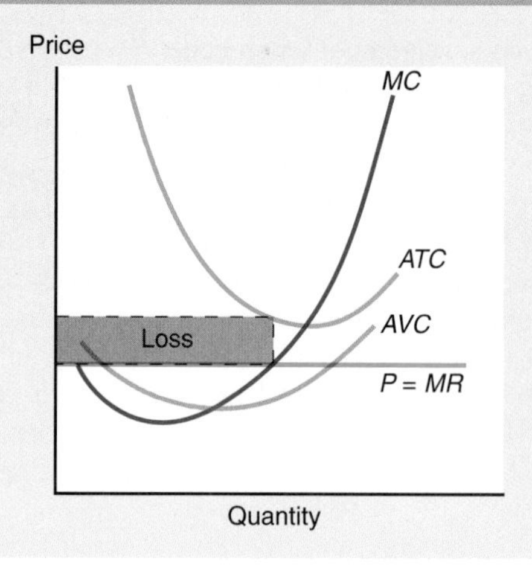

Figure 11-8 demonstrates what happened to many of these firms. Initially, these firms saw their losses as the temporary result of reduced demand in a slowing economy. As prices for products fell, P = MR fell below many of these firms' ATC. However, since P > AVC, many of the firms simply closed their less-profitable locations in an attempt to lower costs and continued to operate, even though they were making an economic loss.

Supply and demand curves can be used to describe most real-world events.

But the Canadian economy continued its weak performance, and the demand for the firms' products did not recover. Firms began to run out of options: low rent leases, extensions on credit lines, and favourable payment arrangements on loans could not be extended any further. As the firms moved from the short run into the long run, many were forced to decide whether to shut down. Without the prospect of quick recovery in the economy, these firms exited the Canadian retail industry.

Q-10 In the early 1990s, demand for hotel rooms in Hawaii decreased substantially because the value of the Japanese yen fell. In the short run, what would you expect to happen to the prices of these hotel rooms? How about in the long run?

The perfectly competitive model and the reasoning behind it are extremely powerful tools. Although the conditions of perfect competition are not fulfilled completely in the Canadian retail sector, the conclusions remain valid nonetheless. The persistence of economic losses in any particular industry will tend to force resources to move out of the industry, as we saw with the exit of many household names in retailing. Of course, as the economy recovered, we saw expansion by those retailers who had weathered the 1990s, as well as entry by new retailers. So, once again, we are seeing competitive market adjustments to changes in economic conditions.

CONCLUSION

The conditions for an industry to be perfectly competitive are so restrictive—large number of firms, homogeneous product, free entry and free exit, complete information, price competition only, and price-taking—that it is very likely that no industry satisfies all of the conditions. However, the perfectly competitive market structure is an excellent benchmark against which to evaluate industries. The key result in perfect competition is that price competition bids the equilibrium market price down to the lowest possible price, the price which just covers the opportunity cost of all inputs to the production process. In other market structures, such as monopoly and oligopoly, firms have some degree of market power which allows them to prevent the equilibrium market price from falling to the competitive level. In addition, all of the profit-maximizing results from Perfect Competition also apply to the other market structures.

When firms in an industry are experiencing positive economic profits, they are earning enough revenues to pay all the opportunity costs of their inputs (that is, all explicit and implicit costs), plus there is extra money left over. The excess profit attracts entry into the industry. New firms ("entry") or existing firms ("expansion") bring resources in from other industries, and the size of the industry grows. In industries which are experiencing losses, firms will continue to operate if their revenues cover variable costs. Firms can often negotiate lower payments for their fixed costs in the short run. Rent can be reduced, loan payments can be suspended and the amount owing can be rolled into new financing, and mortgage payments can be delayed. All of the firms that are owed a fixed cost payment have an incentive to see the troubled firm succeed. The bank earns its profits by lending money; it is not in the bank's best interest to foreclose, and the mall owner would rather have a store remain open and pay no rent at all than have an empty space papered over. In the short run, a firm can struggle to remain open while covering variable costs; however, in the long run, all firms must cover total costs. This is a valuable lesson. The inflow and outflow of resources to different industries is an important part of economic growth. It may be painful sometimes, but it is natural and desirable for some industries to contract while others expand. Because the growing industry is able to pay resources their opportunity cost, resources will move to the higher valued use, and this process creates value and growth in the economy overall. The result is a larger economic pie that we all can share.

Chapter Summary

- The necessary conditions for perfect competition are that the number of firms be large, firms' products be homogeneous (identical), there be free entry and free exit, there be complete information, sellers be profit-maximizing entrepreneurial firms, and buyers and sellers be price takers.
- A perfectly competitive firm maximizes profit by producing the output where marginal revenue equals marginal cost.
- The short run supply curve of a perfectly competitive firm is its marginal cost curve above AVC.
- To find the profit-maximizing level of output for a perfectly competitive firm, find the level of output where MC = *MR*. Profit is price less average total cost times output at the profit-maximizing level of output.
- Economic profit occurs when P > ATC. New firms enter the industry and existing firms expand production in response to economic profit.
- Economic loss occurs when P < ATC; some firms exit the industry.
- In the short run, competitive firms can make a profit or loss. In the long run, they make zero economic profits.
- The shutdown price for a perfectly competitive firm is a price below the minimum point of the average variable cost curve.
- The short-run market supply curve is the horizontal summation of the marginal cost curves for all firms in the market. An increase in the number of firms in the market shifts the market supply curve to the right, while a decrease shifts it to the left.
- Perfectly competitive firms make zero profit in the long run because if profit were being made, new firms would enter and the market price would decline, eliminating the profit. If losses were being made, firms would exit and the market price would rise, eliminating the loss.
- The long-run supply curve is a schedule of quantities supplied where firms are making zero profit. The slope of the long-run supply curve depends on what happens to factor prices when output increases. Constant-cost industries have horizontal long-run supply curves. Increasing-cost industries have upward-sloping long-run supply curves, and decreasing-cost industries have downward-sloping long-run supply curves.

Key Terms

barriers to entry *(233)*
economic profit *(246)*
free entry *(233)*
free exit *(233)*
marginal cost *(236)*
marginal revenue *(236)*
market supply curve *(245)*
normal profit *(246)*
perfectly competitive *(233)*
price taker *(234)*
profit-maximizing condition *(238)*
shutdown point *(245)*
zero economic profit *(242)*

Questions for Thought and Review

1. Why must buyers and sellers be price takers for a market to be perfectly competitive?
2. Draw marginal cost, marginal revenue, and average total cost curves for a typical perfectly competitive firm. Choose a market price above ATC, and indicate the profit-maximizing level of output and total profit for that firm. Is the firm in long-run equilibrium? Why or why not?
3. Draw marginal cost, marginal revenue, and average total cost curves for a typical perfectly competitive firm in long-run equilibrium and indicate the profit-maximizing level of output and total profit for that firm.
4. What portion of the marginal cost curve is the firm's supply curve? How is a firm's marginal cost curve related to the market supply curve?
5. Draw the *ATC*, *AVC*, and MC curves for a typical firm. Label the price at which the firm would shut down temporarily and the price at which the firm would exit the market in the long run.
6. Under what cost condition is the shutdown point the same as the point at which a firm exits the market?
7. Why is the long-run market supply curve upward sloping in an increasing-cost industry, downward sloping in a decreasing-cost industry, and horizontal in a constant-cost industry?
8. What will be the effect of a technological development that reduces variable costs in a competitive market on short-run price, quantity, and profit?
9. You're thinking of buying one of two firms. One has a profit margin of $8 per unit; the other has a profit margin of $4 per unit. Which should you buy? Why?
10. If marginal cost is four times the quantity produced and the price is $20, how much should the firm produce? Why?
11. Find three events in the newspaper that can be explained or interpreted with supply/demand analysis.
12. State what is wrong with each of the graphs.

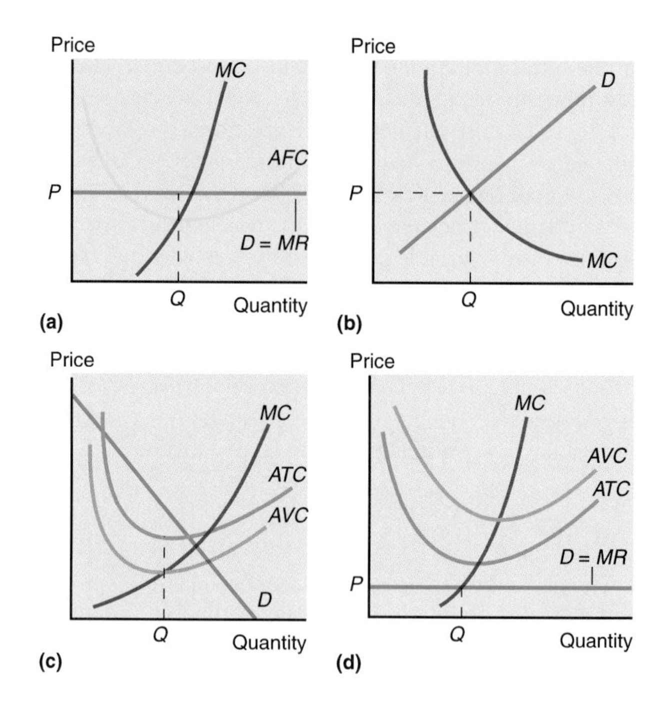

13. In the late 1990s, hundreds of music stores closed in the face of stagnant demand for CDs and new competitors—online music vendors and discount retailers. Explain how price competition from these new sources would cause a retail store to close. In the long run, what effect will new entrants have on the price of CDs?

Problems and Exercises

1. a. Based on the following table, what is the profit-maximizing output?

Output	Price	Total Costs
0	$10	$ 30
1	10	40
2	10	45
3	10	48
4	10	55
5	10	65
6	10	80
7	10	100
8	10	140
9	10	220
10	10	340

   b. How would your answer change if, in response to an increase in demand, the price of the good increased to $15?
2. A profit-maximizing firm has an average total cost of $4, but it gets a price of $3 for each good it sells.
   a. What would you advise the firm to do?
   b. What would you advise the firm to do if you knew average variable costs were $3.50?
3. Say that half of the cost of producing wheat is the rental cost of land (a fixed cost) and half is the cost of labour and machines (a variable cost). If the average total cost of producing wheat is $8 and the price of wheat is $6, what would you advise the farmer to do? ("Grow something else" is not allowed.)

4. Use the accompanying graph, which shows the marginal cost and average total cost curves for the shoe store, Zapateria, a perfectly competitive firm.

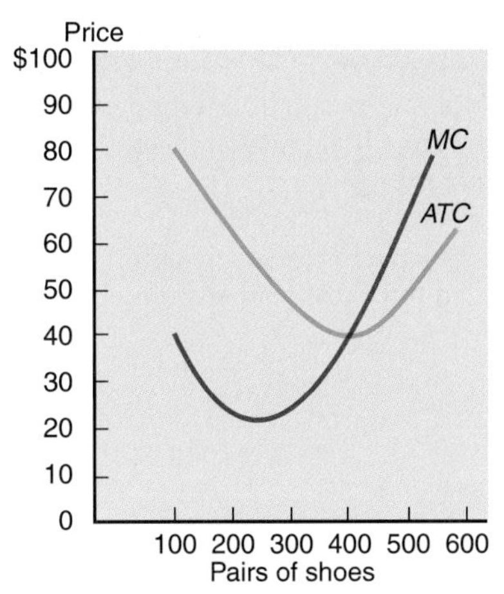

a. How many pairs of shoes will Zapateria produce if the market price of shoes is $70 a pair?
b. What is the total profit Zapateria will earn if the market price of shoes is $70 a pair?
c. Should Zapateria expect more shoe stores to enter this market? Why or why not?
d. What is the long-run equilibrium price in the shoe market assuming it is a constant-cost industry?

5. Each of 10 firms in a given industry has the costs given in the left-hand table. The market demand schedule is given in the right-hand table.

Quantity	Total Cost
0	12
1	24
2	27
3	31
4	39
5	53
6	73
7	99

Price	Quantity Demanded
2	110
4	100
6	90
8	80
10	70
12	60
14	50
16	40

a. What is the market equilibrium price and the price each firm gets for its product?
b. What is the equilibrium market quantity and the quantity each firm produces?
c. What profit is each firm making?
d. Below what price will firms begin to exit the market?

6. Suppose an increasing-cost industry is in both long-run and short-run equilibrium. Explain what will happen to the following in the long run if the demand for that product declines:
a. Price.
b. Quantity.
c. Number of firms in the market.
d. Profit.

7. Graphically demonstrate the quantity and price of a perfectly competitive firm.
a. Explain why a slightly larger quantity would not be preferred.
b. Explain why a slightly lower quantity would not be preferred.
c. Label the shutdown point in your diagram.
d. You have just discovered that shutting down means that you would lose your land zoning permit, which is required to start operating again. How does that change your answer to *c*?

8. A biotechnology firm has created a tomato that will not rot for weeks. It designed such a fruit by changing the genetic structure of the tomato. What effect will this technological change have:
a. On the price of tomatoes?
b. On farmers who grow tomatoes?
c. On the geographic areas where tomatoes are grown?
d. On where tomatoes are generally placed on salad bars in winter?

9. Currently central banks (banks of governments) hold 35,000 tons of gold—one-third of the world's supply. This is the equivalent of 17 years' production. In the 1990s there was discussion about the central banks selling off their gold, since it is no longer tied to money supplies. Assuming they did sell it:
a. Demonstrate, using supply and demand analysis, the effect on the price of gold in the long run and the short run.
b. If you were an economist advising the central banks and you believed that selling off the gold made sense, would you advise them to do it quickly or slowly? Why?

10. If the perfectly competitive model suggests that firms shut down when revenues do not cover variable costs ($P < AVC$),
a. Why don't stores close early (shut down) when there are only a few shoppers in the store?
b. Stores in shopping malls sometimes have no customers at all in their shop. Does this change your answer to *a*?

Web Questions

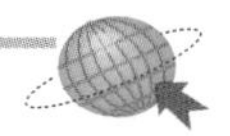

1. One new Internet market is eBay. Check out eBay at www.ebay.com and explain whether you believe that the eBay Internet auction market is perfectly competitive. Be sure to explain which of the six conditions are met and how.
2. Go to Expedia.ca and click on the "hotels" tab. Select a Canadian city and fill in the details for two adults requiring a hotel room for one week.

   a How many results did your search return? What is the range of hotel prices? What factors appear to influence the price?

   b. Which of the six conditions for perfect competition does the city's hotel market meet?

   c. Is your answer to question *b* consistent with your answer to *a*?

   d. Choose a large European city, then choose a small Canadian town. Do your answers change?

Answers to Margin Questions

1. Without the assumption of free entry, firms could make a profit by raising price; hence, their demand curve would not be perfectly elastic and, hence, perfect competition would not exist. *(233)*
2. The competitive firm is such a small portion of the total market that even if it produces and brings to market its entire capacity, it can have no effect on price. Consequently it takes the price as given, and hence its perceived demand curve is perfectly elastic. *(235)*
3. To determine the profit-maximizing output of a competitive firm, you must know price and marginal cost. *(238)*
4. Firms are interested in maximizing total profit. Maximizing profit per unit might yield very small total profits if the quantity sold was small. *(238)*
5. If the firm in Figure 11-5 were producing 4 units, you would explain to it that the marginal cost of increasing output is only $12 and the marginal revenue is $35, so they should significantly expand output until 8, where the marginal cost equals the marginal revenue, or price. *(242)*
6. The diagram is drawn with the wrong profit-maximizing output and hence the wrong profit. Output is determined where marginal cost equals price and profit is the difference between the average total cost and price at that output, not at the output where marginal cost equals average total cost. The correct diagram is shown above. *(243)*

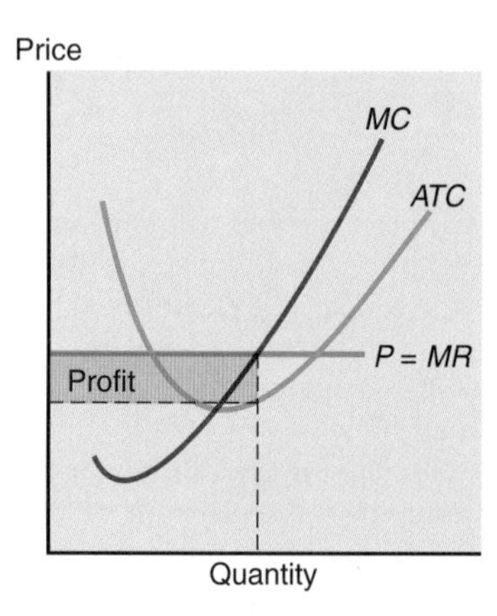

7. The marginal cost for airlines is significantly below average total cost. Since they're recovering their average variable cost, they continue to operate. In the long run, if this continues, some airlines will be forced out of business (forced to exit). *(244)*
8. The costs for a firm include the normal costs, which in turn include a return for all factors. Thus it is worthwhile for a competitive firm to stay in business, since it is doing better, or at least as well, as it could in any other activity. *(247)*
9. Suddenly becoming the "in" thing to wear would cause the demand for berets to shift rightward, pushing the price up in the short run. In the long run expansion by new and existing firms will increase the supply and price will fall back to the original level. (247)
10. A decline in demand should push the short-run price of these hotel rooms down. In the long run, however, once a number of hotels go out of business, the price of hotel rooms should eventually move back to approximately where it was before the decline, assuming constant returns to scale. *(250)*

Monopoly

12

After reading this chapter, you should be able to:

- Summarize how and why the decisions facing a monopolist differ from the collective decisions of competing firms.
- Explain why $MC = MR$ maximizes total profit for a monopolist.
- Determine a monopolist's price, output, and profit graphically and numerically.
- Explain why a price-discriminating monopolist will earn more profit than a normal monopolist.
- Show graphically the welfare loss from monopoly.
- Explain why there would be no monopoly without barriers to entry.
- List three normative arguments against monopoly.

Monopoly is business at the end of its journey.

Henry Demarest Lloyd

In Chapter 11 we considered the perfect competition market structure. We now move to the other end of the spectrum: monopoly. **Monopoly** is *a market structure in which one firm makes up the entire supply side of the market*. It is the polar opposite to competition.

Monopolies exist because of barriers to entry into a market that prevent entry by new firms. These can be legal barriers (as in the case where a firm has a patent that prevents other firms from entering) or natural barriers, such as where the size of the market can support only one firm.

THE KEY DIFFERENCE BETWEEN A MONOPOLIST AND A PERFECT COMPETITOR

For a competitive firm, marginal revenue equals price. For a monopolist, it does not. The monopolist must take into account the fact that its production decision affects price.

A key question we will answer in this chapter is: How does a monopolist's profit maximizing decision differ from the profit maximizing decision of perfectly competitive firm? Futhermore, how does the monopoly market solution differ from the competitive solution? Answering these questions brings out a key difference between perfect competition and monopoly. Since a competitive firm is too small to affect the price, it does not have to take into account the effect of its output decision on the price it receives. A competitive firm's marginal revenue (the additional revenue it receives from selling an additional unit of output) is the given market price. A monopolistic firm must take into account that its output decision will affect price; its marginal revenue is not its price. Since the monopolist is the only supplier, it faces the entire market demand curve. Because the market demand curve slopes downward, the monopolist can only sell more if it lowers the price. This was not the case for the perfectly competitive firm; it could sell its entire production capacity at the given market price. It was too small to affect price. As we will see, the production decision of the monopolist depends on the downward slope of the demand curve.

Q-1 Why must the monopolist lower price in order to sell more product?

In perfect competition, each individual competitive firm behaves in its own self-interest, which is not necessarily in the best interest of the firms collectively. In perfectly competitive markets, as one supplier is pitted against another, consumers benefit as the largest quantity of product is supplied at the lowest possible price. In monopolistic markets, the firm faces no competitors and this gives the monopoly firm a great deal of control over its choice of profit-maximizing price. Monopolists see to it that the monopolist, not the consumers, benefit; perfectly competitive firms cannot.

Monopolists prefer that the monopolist, not consumers, benefit.

A MODEL OF MONOPOLY

How much should the monopolistic firm choose to produce if it wants to maximize profit? To answer that we have to consider more carefully the effect that changing output has on the total profit of the monopolist. The perfectly competitive firm took market price as given (price taker), so its profit-maximizing decision involved choosing the correct quantity. The monopolist employs a two-step profit-maximizing process; it chooses quantity *and* price. First, we consider a numerical example; then we consider that same example graphically. The relevant information for our example is presented in Table 12-1.

Determining the Monopolist's Price and Output Numerically

Table 12-1 shows the price, total revenue, marginal revenue, total cost, marginal cost, average total cost, and profit at various levels of production. It's similar to the table in the last chapter where we determined a perfectly competitive firm's output. The big difference is that marginal revenue changes as output changes and is not equal to the price. Why?

First, let's remember the definition of marginal revenue: Marginal revenue is the change in total revenue associated with a change in quantity. In this example, if a monopolist increases output from 4 to 5, the price it can charge falls from $24 to $21 and its revenue increases from $96 to $105, so marginal revenue is $9. Marginal revenue of increasing output from 4 to 5 for the monopolist reflects two changes: a $21 gain in

TABLE 12-1 A Monopolist's Profit Maximization

1 Quantity	2 Price	3 Total Revenue	4 Marginal Revenue	5 Total Cost	6 Marginal Cost	7 Average Total Cost	8 Profit
0	$36	$ 0		$ 47			$−47
			$33		$ 1		
1	33	33		48		$48.00	−15
			27		2		
2	30	60		50		25.00	10
			21		4		
3	27	81		54		18.00	27
			15		8		
4	24	96		62		15.50	34
			9		16		
5	21	105		78		15.60	27
			3		24		
6	18	108		102		17.00	6
			−3		40		
7	15	105		142		20.29	−37
			−9		56		
8	12	96		198		24.75	−102
			−15		80		
9	9	81		278		30.89	−197

revenue from selling the 5th unit and a $12 decline in revenue because the monopolist must lower the price on the previous 4 units it produces by $3 a unit, from $24 to $21. This highlights the key characteristic of a monopolist—its output decision affects its price. Because an increase in output lowers the price on all previous units, a monopolist's marginal revenue is always below its price. Comparing columns 2 and 4, you can confirm that this is true.

Would the monopolist will increase production from 4 to 5 units? The marginal revenue of increasing output from 4 to 5 is $9, and the marginal cost of doing so is $13. Since marginal cost exceeds marginal revenue, increasing production from 4 to 5 will reduce total profit and the monopolist will not increase production. If it decreases output from 4 to 3, where $MC < MR$, revenue falls by $15 and costs rise by $8. It will not reduce output from 4 to 3. Since it cannot increase total profit by increasing output to 5 or decreasing output to 3, it is maximizing output at 4 units.

As you can tell from the table, profits are highest ($34) at 4 units of output and a price of $24. At 3 units of output and a price of $27, the firm has total revenue of $81 and total cost of $54, yielding a profit of $27. At 5 units of output and a price of $21, the firm has a total revenue of $105 and a total cost of $78, also for a profit of $27. The highest profit it can make is $34, which the firm earns when it produces 4 units. This is its profit-maximizing level.

Q-2 In Table 12-1, explain why 4 is the profit-maximizing output.

Determining the Monopolist's Price and Output Graphically

The monopolist's output decision can also be seen graphically. Figure 12-1 (on page 259) graphs the table's information into a demand curve, a marginal revenue curve, and a marginal cost curve. The marginal cost curve is a graph of the change in the firm's total cost

KNOWING THE TOOLS

A Trick in Graphing the Marginal Revenue Curve

Here's a trick to help you graph the marginal revenue curve. The *MR* line starts at the same point on the price axis as a linear demand curve, but it falls twice as fast as the demand curve, cutting the horizontal space in half, and intersecting the quantity axis at a point half the distance from where the demand curve intersects the quantity axis. If the demand curve is not linear, the marginal revenue curve still cuts the horizontal space in half, but it will follow the shape of the demand curve. If the demand curve does not intersect either axis, simply extend the demand curve to the two axes and measure halfway on the quantity axis (3 in the graph below). Then draw a line from where the demand curve intersects the price axis to that halfway mark. That line is the marginal revenue curve.

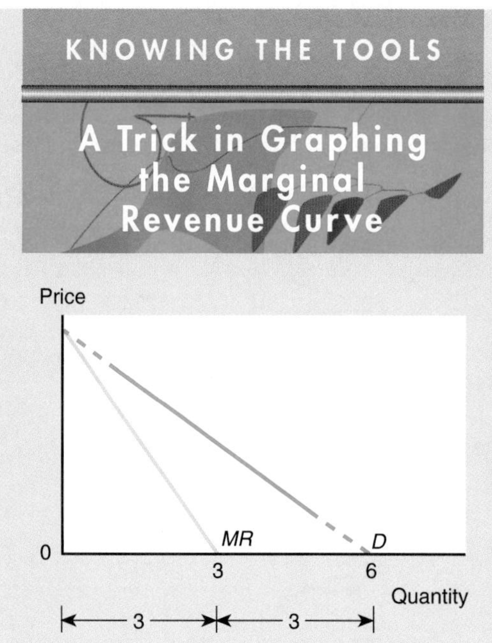

as it changes output. It's the same curve as we saw in our discussion of perfect competition. The marginal revenue curve tells us the change in total revenue when quantity changes. It is graphed by plotting and connecting the values from Table 12-1. Since marginal revenue occurs as output changes, it is plotted at the quantity midpoints.

The marginal revenue curve for a monopolist is new, so let's consider it a bit more carefully. It tells us the additional revenue the firm will get by expanding output. It is a downward-sloping curve that begins at the same point as the demand curve but has a steeper slope. In this example, marginal revenue is positive up until the firm produces 6 units, at which point marginal revenue equals zero (i.e. intersects the quantity axis at a price of zero.) Then marginal revenue is negative. This means that after 6 units the monopolist's total revenue decreases when it increases output.

Q-3 In the graph below, indicate the monopolist's profit-maximizing level of output and the price it would charge.

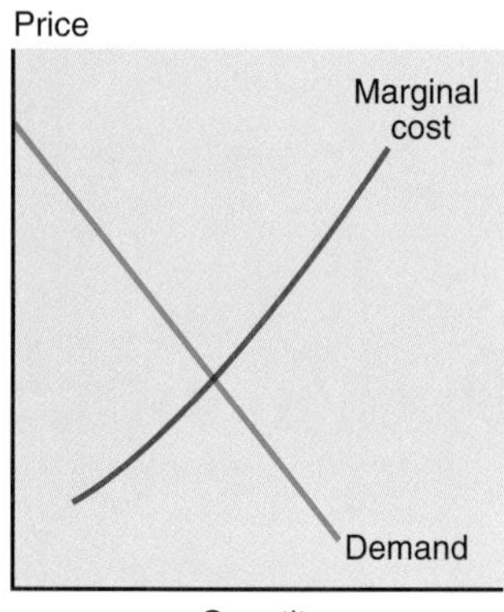

Notice specifically the relationship between the demand curve (which is the average revenue curve) and the marginal revenue curve. Since the demand curve is downward sloping, the marginal revenue curve is below the average revenue curve. (Remember, if the average curve is falling, the marginal curve must be below it.)

The Monopolist's Two-Step Profit-Maximizing Decision

The monopolist's profit-maximizing decision is a two-step process: First, it finds the profit-maximizing output, then it sets the price for that level of output. So, what output should the monopolist produce, and what price can it charge? To answer those questions, the key curves to look at are the marginal cost curve and the marginal revenue curve.

Step 1: MR = MC Determines the Profit-Maximizing Output The monopolist uses the general rule that any firm must follow to maximize profit: Produce the quantity at which MC = *MR*. If you think about it, it makes sense that the point where marginal

Figure 12-1 **DETERMINING THE MONOPOLIST'S PRICE AND OUTPUT GRAPHICALLY**

The profit-maximizing output is determined where the MC curve intersects the *MR* curve. To determine the price for this quantity, extend a line (vertically) up to the demand curve, which contains the information about how much consumers are willing to pay, in this case a price of $24. This price is higher than the competitive price, $20.50, and the quantity, 4, is lower than the competitor's quantity, 5.17.

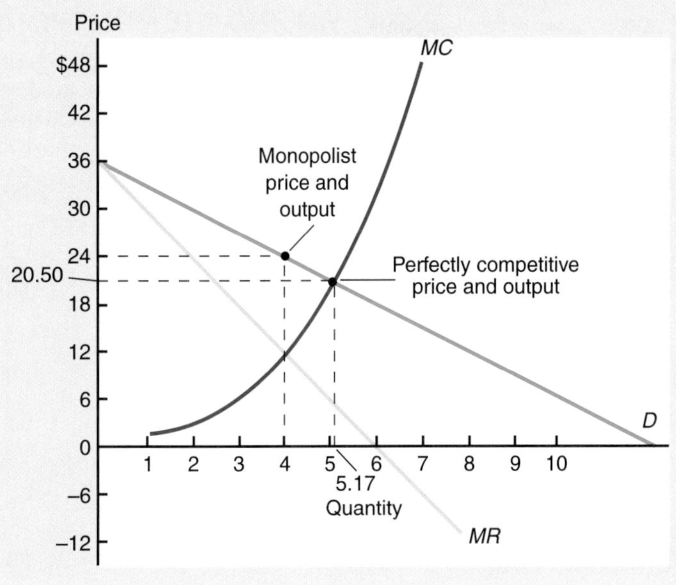

revenue equals marginal cost determines the profit-maximizing output. If the marginal revenue is below the marginal cost, it makes sense to reduce production. Reducing production decreases marginal cost and increases marginal revenue. When *MR* < *MC*, reducing output increases total profit. If marginal revenue is above marginal cost, the firm should increase production because total profit will rise. If the marginal revenue is equal to marginal cost, increasing production and decreasing production both lead to lower profits, so it does not make sense to increase or reduce production. So the monopolist will produce at the output level where MC = *MR*. As you can see, the output the monopolist chooses is 4 units, the same output that we determined numerically.[1] This leads to the following insights:

If *MR* > MC, the monopolist gains profit by increasing output.

If *MR* < MC, the monopolist gains profit by decreasing output.

If MC = *MR*, the monopolist is maximizing profit.

Thus, *MR* = MC is the profit-maximizing rule for a monopolist.

The general rule that any firm must follow to maximize profit is: Produce the output level at which *MC* = *MR*.

Step 2: Output Determines the Price a Monopolist Will Charge The *MR* = MC condition determines the quantity a monopolist produces; in turn, that quantity determines the price the firm will charge. A monopolist will charge the maximum price consumers are willing to pay for that quantity. Since the demand curve tells us what consumers will pay for a given quantity, to find the price a monopolist will charge, you must extend the quantity line up to the demand curve. We do so in Figure 12-1 and see that the profit-maximizing output level of 4 allows a monopolist to charge a price of $24.

[1]This could not be seen precisely in Table 12-1 since the table is for discrete jumps and does not tell us the marginal cost and marginal revenue exactly at 4; it only tells us the marginal cost and marginal revenue ($8 and $15, respectively) of moving from 3 to 4 and the marginal cost and marginal revenue ($16 and $9, respectively) of moving from 4 to 5. If small adjustments (1/100 of a unit or so) were possible, the marginal cost and marginal revenue precisely at 4 would be $12. Because drawing the curve implicitly assumes we can make very small changes, the graphs of the marginal revenue curve and marginal cost curve will intersect at an output of 4 and a marginal cost and marginal revenue of $12.

Figure 12-2 (a, b, c, and d) FINDING THE MONOPOLIST'S PRICE AND OUTPUT

Determining a monopolist's price and output is a two-step process: find quantity then find price. From the supply side (MC) and demand side (*D*) information in (a), create the monopolist's marginal revenue curve (b). In (c), the monopolist chooses its profit-maximizing output by finding the quantity where MC = *MR*. In (d), the price charged for Q_M is determined from the demand curve, which tells the monopolist how much consumers are willing to pay.

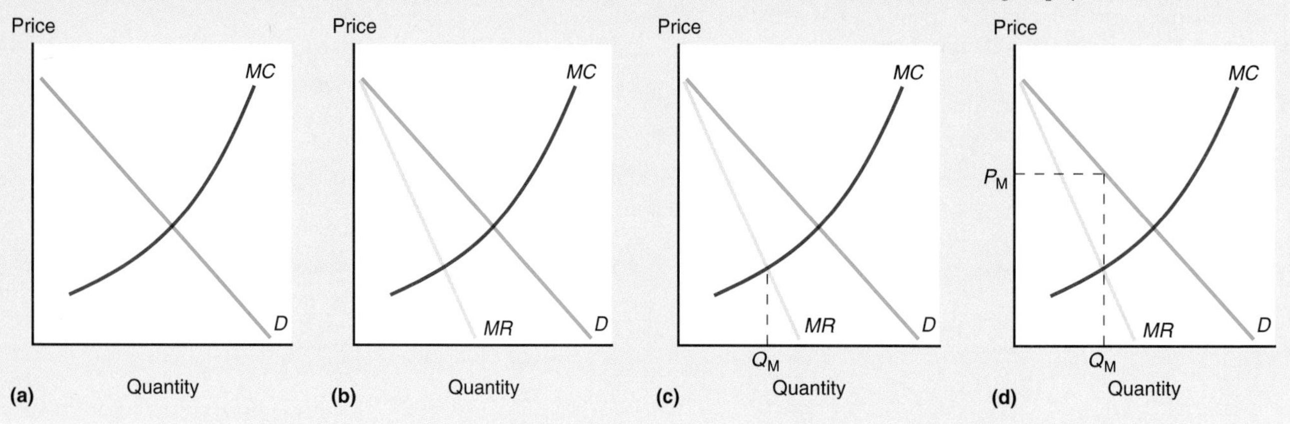

Comparing Monopoly and Perfect Competition

For a competitive industry, the horizontal summation of firms' marginal cost curves is the supply curve. Output for a perfectly competitive industry would be 5.17, and price would be $20.50, as Figure 12-1 shows. The monopolist's output was 4 and its price was $24. So, if a competitive market is made into a monopoly, you can see that output would be lower and price would be higher. The reason is that the monopolist is able to take into account the effect that restricting output has on price.

Q-4 Why does a monopolist produce less output than perfectly competitive firms would produce in the same industry?

Profit-maximizing output for the monopolist, like profit maximizing output for the competitor, is determined by the MC = *MR* condition, but because the monopolist's marginal revenue is below its price, its profit-maximizing output is different from a perfectly competitive market.

An Example of Finding Output and Price

A monopolist with marginal cost curve MC faces a demand curve *D* in Figure 12-2(a). Determine the price and output the monopolist would choose.

First, draw the marginal revenue curve, since we know that a monopolist's profit-maximizing output level is determined where MC = *MR*. This is done in Figure 12-2(b), by extending the demand curve back to the vertical and horizontal axes and then bisecting the horizontal axis.

Next, determine where MC = *MR*. Drop a line down to the quantity axis to determine the output the monopolist chooses, Q_M. This is done in Figure 12-2(c).

Finally, find where the quantity line intersects the demand curve. Then we extend a horizontal line from that point to the price axis, as in Figure 12-2(d), to determine the price the monopolist will charge, P_M.

PROFITS AND MONOPOLY

The monopolist's profit is determined by comparing average total cost to price. So before we can determine profit, we need to add another curve: the average total cost curve. The average total cost comes from the same cost information as marginal cost. (We can also calculate average variable and average fixed costs from the same cost information.)

To determine the monopolist's profit (loss), subtract average total cost from average revenue (P) at that level of output and multiply by the chosen output. For any firm, profit is calculated as $\pi = (P - ATC) \times Q$. If price exceeds average total cost at the output it chooses, the monopolist will make a positive economic profit. If price equals average total cost, the monopolist will make a normal profit. If price is less than average total cost, the monopolist will incur a loss.

A Monopolist Making a Profit

The monopolist's demand, marginal cost, and average total cost curves are presented in Figure 12-3(a). Draw the marginal revenue curve (Figure 12-3(b)). Then find the output level at which marginal cost equals marginal revenue. That intersection of $MR = MC$ tells us the monopolist's output, Q_M in Figure 12-3(b). To find what price the monopolist will charge at that output, go from the quantity Q_M up to the demand curve (point A) and then over to the price axis. Doing so gives price, P_M. We have now determined P_M and Q_M. What does it cost to produce Q_M? This is determined from the ATC curve for the quantity Q_M, so go up from Q_M to the average total cost curve (point B). The monopolist's average total cost of producing output Q_M is C_M. The monopolist earns $P_M - C_M$ on each Q_M unit sold, so profit is represented by the shaded rectangle in Figure 12-3(c), $(P_M - C_M)$ times Q_M. Mathematically, its profit is calculated the same way:

$$\pi = (P - ATC) \times Q$$

Substituting in the correct values, this monopolist earns

$$\pi = (P_M - C_M) \times Q_M.$$

Q-5 Indicate the profit that the monopolist shown in the graph below earns.

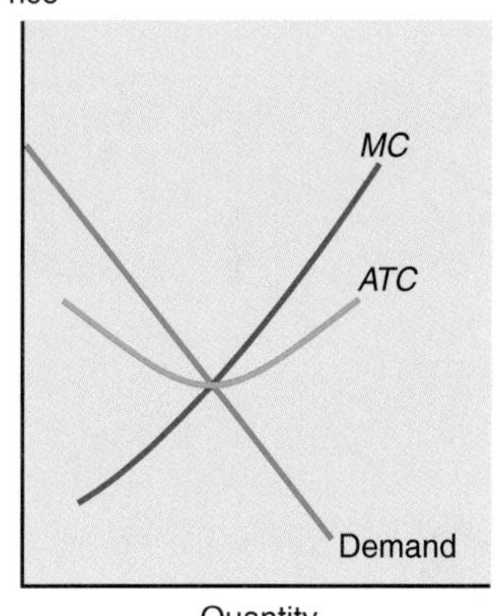

A Monopolist Breaking Even and Making a Loss

In Figure 12-4(a) and (b) we consider two other average total cost curves to show you that a monopolist may make a normal profit or a loss as well as an economic profit. In Figure 12-4(a) the monopolist is making zero economic profit; in Figure 12-4(b) it's making a loss. Whether a firm is making a profit, zero economic profit, or a loss depends on the level of its average total costs relative to price. So clearly in the short run a monopolist can be making either a profit or a loss, or it can be breaking even.

Figure 12-3 (a, b, and c) THE MONOPOLIST MAKES A PROFIT

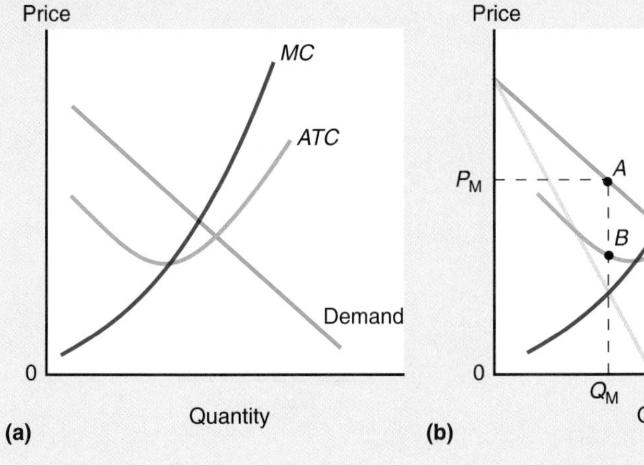

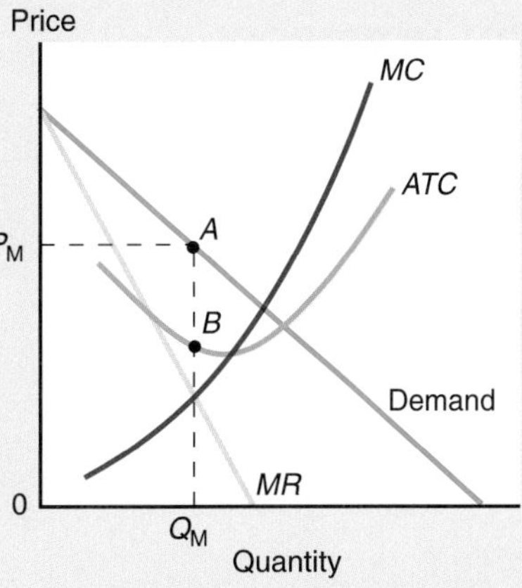

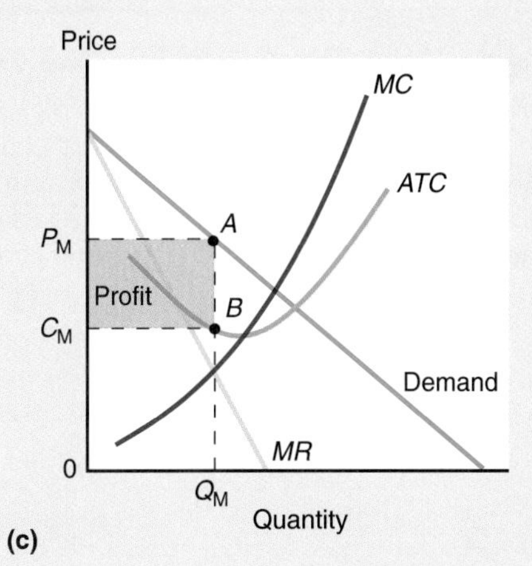

Figure 12-4 (a and b) **OTHER MONOPOLY CASES**

Depending on where the *ATC* curve falls, a monopolist can make a profit, break even (as in (**a**)), or make a loss (as in (**b**)) in the short run. In the long run, a monopolist that is making a loss will exit the industry.

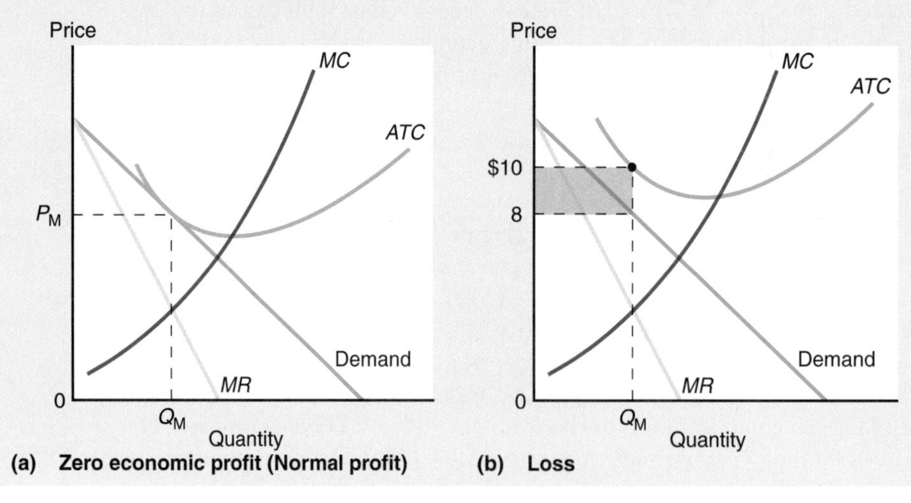

THE WELFARE LOSS FROM MONOPOLY

Q-6 Why is area *C* in Figure 12-5 not considered a loss from monopoly?

Let us compare graphically the **single price monopolist** equilibrium and perfectly competitive equilibrium in reference to producer and consumer surplus. This we do in Figure 12-5. In a competitive equilibrium, the total consumer and producer surplus is the area between the demand curve and the marginal cost curve up to market equilibrium quantity Q_C. The monopolist reduces output to Q_M and raises price to P_M. The benefit lost to society from reducing output from Q_C to Q_M is measured by the area under the demand curve between output levels Q_C and Q_M. That area is represented by the shaded areas labelled *A*, *B*, and *D*. Some of that loss is regained. Society gains the opportunity cost of the resources that are freed up from reducing production—the value of the resources in their next-best use, indicated by the shaded area A. So the net cost to society

12.1 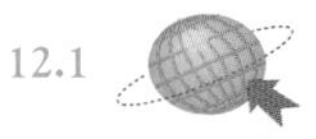
see page 274

Figure 12-5 **THE WELFARE LOSS FROM MONOPOLY**

The **welfare loss** from a monopoly is represented by the triangles *B* and *D*. The rectangle C is a transfer of consumer surplus to the monopolist. The area A represents the opportunity cost of resources diverted to other productive uses. This is not a loss to society, since the resources will be used in producing other goods.

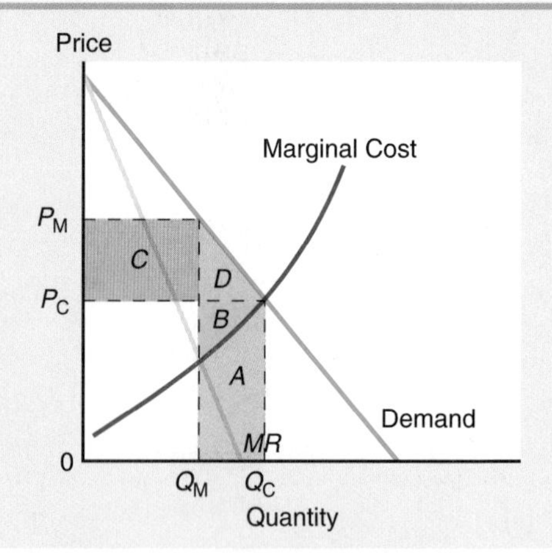

To find a monopolist's level of output, price, and profit, follow these four steps:

1. Draw the marginal revenue curve.
2. Determine the output the monopolist will produce: The profit-maximizing level of output is where *MR* and *MC* curves intersect.
3. Determine the price the monopolist will charge: Extend a line from the quantity up to the demandcurve. Where this line intersects the demand curve is the monopolist's price.
4. Determine the profit the monopolist will earn: Subtract the *ATC* from price at the profit-maximizing level of output to get profit per unit. Multiply profit per unit by quantity of output to get total profit.

of decreasing output from Q_C to Q_M is represented by areas *B* and *D*. (Area C is the monopolist's profit. It is neither a gain nor a loss to society. It represents a transfer of income from the consumer to the monopolist that occurs with the rise in price. Since both monopolist and consumer are members of society, the gain and loss offset each other. The net loss is zero.) The triangular areas *B* and *D* are the net loss to society from the existence of monopoly.

As discussed in Chapter 7, this area designated by *B* and *D* is often called the *deadweight loss* or *welfare loss triangle*. That welfare cost of monopoly is one of the reasons economists oppose monopoly. That cost can be summarized as follows: Because monopolies charge a price higher than marginal cost, people's decisions don't reflect the true cost to society. Because price exceeds marginal cost, people's choices are distorted; they choose to consume less of the monopolist's output and more of some other output than they would if markets were competitive. That distinction means that the marginal benefit of increasing output is higher than the marginal cost of increasing output, so output should be increased. But it isn't, so there's a welfare loss.

The welfare loss from monopoly is a triangle, as in the graph below.

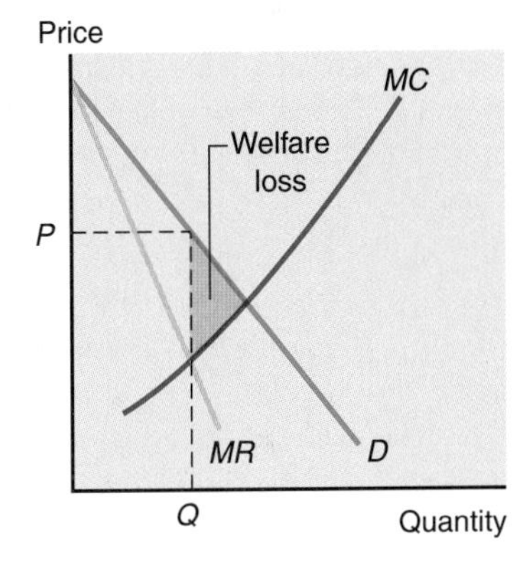

THE PRICE-DISCRIMINATING MONOPOLIST

So far we've considered monopolists that charge the same price to all consumers. Let's consider what would happen if our monopolist suddenly gained the ability to **price discriminate**—*to charge different prices to different individuals or groups of individuals* (for example, students as compared to businesspeople). If a monopolist can identify groups of customers who have different elasticities of demand, separate them in some way, and limit their ability to resell its product between groups, it can charge each group a different price. Specifically, it could charge consumers with less elastic demands a higher price and individuals with more elastic demands a lower price. By doing so, it will increase total profit. Suppose, for instance, Megamovie knew that at $12 it would sell 1,000 movie tickets and at $8 a ticket it would sell 1,500 tickets. Assuming Megamovie could show the film without cost, it would maximize profits by charging $12 to 1,000 moviegoers, earning a total profit of $12,000. If, however, it could somehow attract the additional 500 viewers at $8 a ticket without reducing the price to the first 1,000 moviegoers, it could raise its profit by $4,000, to $16,000. As you can see, a price-discriminating monopolist increases its profit.

Monopoly occurs when there is a single seller; there are also markets in which there is a single buyer. Such markets are called monopsonies. An example of a monopsony is a "company town" in which a single firm is the only employer. Whereas a monopolist takes into account the fact that if it sells more it will lower the market price, a monopsonist takes into account the fact that it will raise the market prices (wages) if it buys more (hires more people). Thus, it buys (hires) less and pays less than would a market which had an equivalent number of competitive buyers.

PERFECT PRICE DISCRIMINATION

Single-price monopolist: a firm that charges all customers the same price.

When a monopolist price discriminates, it charges individuals high up on the demand curve higher prices and those low on the demand curve lower prices.

Cars are seldom sold at list price. It is generally only after negotiation between the customer and the dealership's sales manager that the actual price is determined. In this case of price discrimination, the salesperson is able to determine the customer's elasticity, and is able to get the highest price from those people whose demands are least elastic. Buyers who are prepared to shop around and compare prices have relatively elastic demands, and generally pay less for the same vehicle than those who do not search out all the alternatives. In perfect price discrimination, the monopolist is able to identify every individual consumer's position on the demand curve. As a result, the monopolist can charge each person the highest price each would pay, thereby generating a larger profit than a single-price monopolist.

How does it work? In order for a monopolist to successfully price discriminate, it must be able to identify each person's willingness to pay, charge different people different amounts, and, very importantly, prohibit the resale of the product between people.

Consider Figure 12-6. The demand curve reflects consumers' willingness to pay. Consumer 1 is willing to pay $10, consumer 2 is willing to pay $9, consumer 3 is willing to pay $8, and so on. If the monopolist is charging a single price of $6, consumer 1 pays $6 and

Figure 12-6 PERFECT PRICE DISCRIMINATION

A single-price monopolist sells 5 customers one unit of its product, charging each person $6 for it. Consumer 1 is willing to pay $10; the consumer surplus for the first consumer is shown by the vertical arrow from $6 to $10. Consumer 2 is willing to pay $9; the arrow shows his consumer surplus as $3. Under perfect price discrimination, each consumer is charged the amount he or she is willing to pay (given by the height of the demand curve), so the monopolist takes all of the consumer surplus as profit. In addition, the perfect price discriminating monopolist sells to consumers 6 and 7, too, and the deadweight loss associated with single-price monopolies disappears, which is a net gain to society. The result of perfect price discrimination is larger output, no deadweight loss, no consumer surplus, and a larger profit for the monopolist.

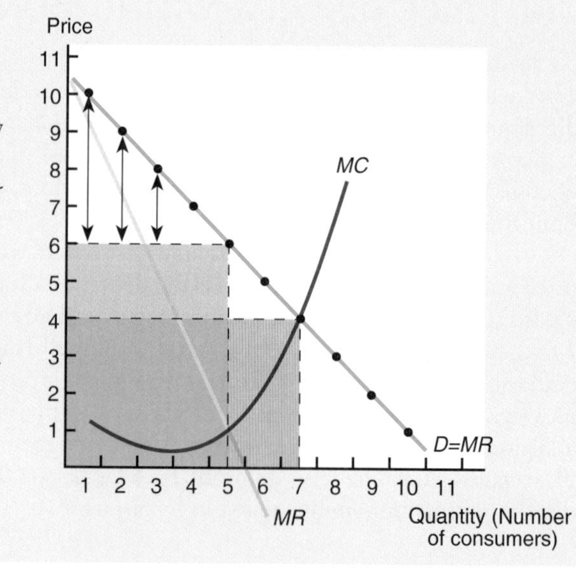

receives $4 of consumer surplus. Consumer 2 also pays $6, but receives only $3 of consumer surplus. The last person to buy the product at $6 is willing to pay $6, and receives no consumer surplus. The person willing to pay $5 does not buy the product if it is priced at $6. If the monopolist is able to identify each person according to willingness to pay, it can charge each person the full amount each is willing to pay. Consumer 1 is willing to pay $10, so a perfect price discriminating monopolist will charge him $10, and the $4 of consumer surplus that person received when the price was $6 simply becomes profit for the monopolist. Consumer 2 is charged the $9 he is willing to pay, and the monopolist captures his $3 of (previous) consumer surplus, and so on. Consumer 5 previously paid $6 and received no consumer surplus; the perfect price discriminating monopolist also charges him $6 now. Notice that, by charging in this way, the demand curve becomes the monopolist's marginal revenue curve under perfect price discrimination.

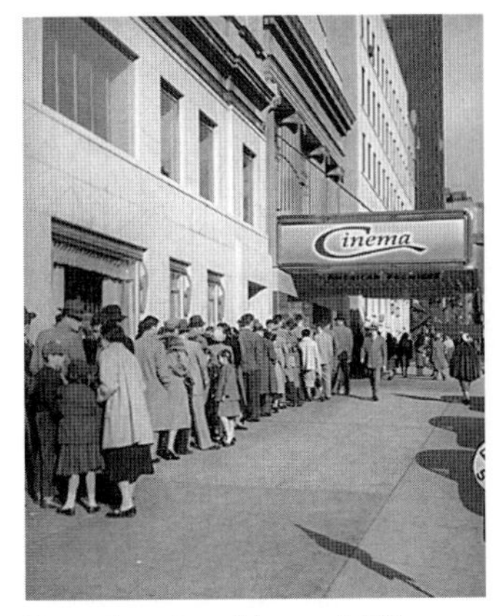

Price elasticity of demand differs among consumers. When consumers can be grouped according to their elasticities and resale of the good or service can be prevented, firms can improve their profits by price discriminating.

What happens to consumer 6, who was only willing to pay $5? If the monopolist follows its standard profit-maximizing rule to produce the output where MC=MR, then for consumer 6, the monopolist's marginal cost is about $2.50, while it can charge (marginal revenue) $5. Since *MR*>MC, the monopolist will sell to (the previously unserved) consumer 6 at a price consumer 6 is willing to pay. Similarly, consumer 7, who was only willing to pay $4 and therefore was not served earlier by the single-price monopolist, will buy from the price discriminating monopolist at a price of $4, since at the quantity of 7, *MR*=MC. The perfect price discriminating monopolist will stop expanding output when it reaches this point, which corresponds to the perfectly competitive output.

Therefore, as a result of perfect price discrimination, each buyer pays the maximum price he or she is willing to pay, all consumer surplus is transferred to the monopolist, the monopolist increases its profits, and output increases. The consumer surplus lost by consumers is redistributed to the monopolist as profit, and thus the surplus and the profit offset each other. There is no net loss or gain. The expansion of output to the perfectly competitive level, however, results in consumers being able to purchase items which they were not able to buy under a single-price monopoly. The deadweight loss associated with the single-price monopoly market structure is eliminated under perfect price discrimination. This is a net benefit.

Q.7 Why does a price-discriminating monopolist make a higher profit than a single-price monopolist?

In reality, price discrimination is usually unable to perfectly differentiate between consumers, but there are many markets in which the price charged depends on what user group a particular customer falls into. For example, it is easy for theatres to distinguish Monday or Tuesday night patrons from Friday night patrons, whose demand is relatively inelastic: resale of dated tickets is impossible. Haircuts for seniors and kids, whose demand is elastic, are easy to differentiate, and impossible to resell. Airlines try to prevent the resale of the cheaper advance-booking ("seat sale") tickets to those passengers whose travel needs are less elastic, such as business travelers. Generally, we tend to observe price discrimination in markets where consumers can be easily distinguished by their elasticities, and where products are difficult to resell.

BARRIERS TO ENTRY AND MONOPOLY

What prevents other firms from entering the monopolist's market in response to the positive economic profits the monopolist earns? Barriers to entry allow the monopolist to retain its dominance in a market. Barriers to entry may result from technical production conditions or may be the result of specific actions by the monopoly firm. In the absence of barriers to entry, the monopoly would face competition from other firms, which would erode its monopoly position.

If there were no barriers to entry, profit-maximizing firms would always compete away monopoly profits.

Economies of scale are relatively small; however, advertising forms a formidable barrier to entry.

In a natural monopoly a single firm can produce at a lower unit cost than can two or more firms.

Q-8 Why is the competitive price impossible for an industry that exhibits strong economies of scale?

Economies of Scale

The strongest barrier to entry arises as a result of economies of scale. When production is characterized by increasing returns to scale, the larger a firm becomes the lower its per unit costs become. The firm gains an absolute cost advantage. Smaller firms are unable to compete because, at a smaller output, these firms cannot capture enough economies to compete with the larger firm. In fact, if sufficiently large economies of scale exist in an industry, it would be inefficient to force the industry to have two producers. If each produced half of the output, neither could take advantage of economies of scale. Such industries are called natural monopolies. A **natural monopoly** is *an industry in which a single firm can produce at a lower cost than can two or more firms.* A natural monopoly will occur when the technology is such that the minimum efficient scale of production is so large that average total costs fall within the range of potential output, as illustrated in Figure 12-7 (a).

In Figure 12-7(a), if one firm produces Q_1, its cost per unit is C_1. If two firms each produce half that amount, $Q_{1/2}$, the cost per unit will be C_2, which is significantly higher than C_1. In cases of natural monopoly, as the number of firms in the industry increases, the average total cost of producing the market output increases. For example, if there were three equal-sized firms in the industry, each firm would produce a third of the output ($Q_{1/3}$), and have an average cost of C_3. Clearly, when production is characterized by significant economies of scale over the relevant market quantity, monopoly is the most economically efficient market structure. In other words, the market, given economies of scale, is only large enough to support one firm.

Until the 1990s local telephone service was a real-world example of such a natural monopoly. It makes little sense to have two sets of telephone lines going into people's houses. Technology changes and with wireless communications and cable connections, the technical conditions that made local telephone service a natural monopoly are changing. Such change is typical; natural monopolies are only natural given a technology.

A natural monopoly can also occur when a single industry standard owned by one firm is more efficient than multiple standards. An example is the operating system of computers. It is much more efficient (because the communication among users is easier) for there to be a single standard rather than multiple standards.

From a consumer welfare standpoint natural monopolies are different than other types of monopolies. In the case of a natural monopoly, even if a single firm makes some monopoly profit, the price it charges may still be lower than the price two firms making normal profit would charge because its average total costs will be lower. In the case of a natural monopoly not only is there no welfare loss from monopoly, there can actually be a welfare gain since a single firm producing is so much more efficient than many firms producing. Such natural monopolies are often organized as public utilities. For example, most towns have a single water department supplying water to residents.

Figure 12-7(b) shows the profit-maximizing level of output and price that a natural monopolist would choose. To show the profit-maximizing level of output, we've added a marginal cost curve that is below the average total cost curve and also falling. A natural monopolist uses the same $MC = MR$ rule that a monopolist uses to determine output. The monopolist will produce Q_M and charge a price P_M. Average total costs are C_M and the natural monopolist earns a profit shown by the red shaded box.

Where natural monopoly exists, the perfectly competitive solution is impossible, since average total costs are not covered where $MC = P$. A monopolist required by government to charge the competitive price P_C, where $P = MC$, will incur a loss shown by the blue shaded box because marginal cost is always below average total cost. Some output restriction is necessary in order for production to be feasible. In such

Figure 12-7 (a and b) **A NATURAL MONOPOLIST**

The graph in (**a**) shows the average cost curve for a natural monopoly. One firm serving the entire market, producing Q_1 would have average cost of C_1. If market demand is Q_1 and another firm enters the market, sharing quantity produced, each firm would produce $Q_{1/2}$ goods at average cost C_2. If three firms each produced $Q_{1/3}$, the average cost for each would be C_3. In a case of a natural monopoly, as the number of firms in the industry increases, the average cost of producing a fixed number of units increases.

The graph in (**b**) shows that a natural monopolist would produce Q_M and charge a price P_M. It will earn a profit shown by the red shaded box. If the monopolist were required to charge a price equal to marginal cost, P_C, it would incur a loss shown by the blue shaded box.

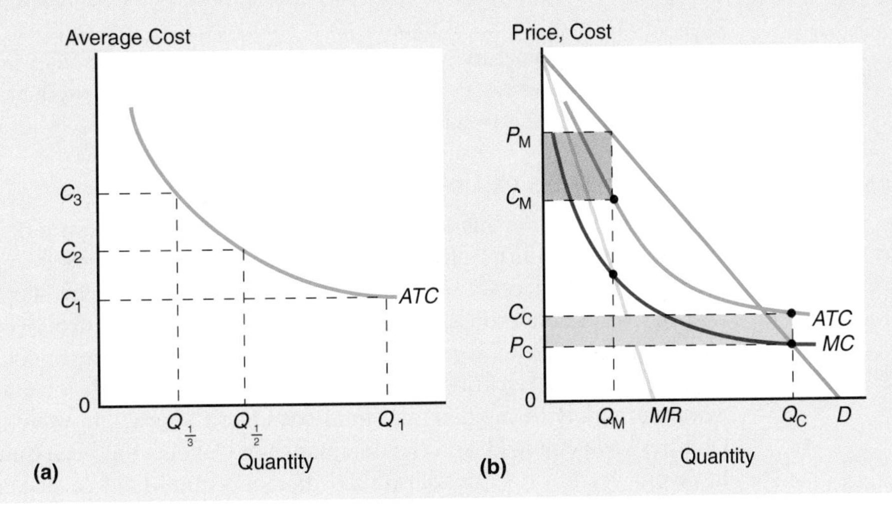

cases, monopolies are often preferred by the public as long as they are regulated by government. The issues involved in regulating natural monopolies are discueed in a later chapter.

Setup Costs

In many industries, high setup costs characterize production. The industry may be highly capital-intensive, requiring a large investment in very expensive, but very productive, specialized capital, such as an oil refinery, diamond mine, or automobile plant. Production technology characterized by these kinds of large investments can also be seen in terms of economies of scale, where only a large output (by a large firm) can justify the investment. Or, the industry may be one in which large amounts of money are spent on advertising, such as the perfume industry. Although economies of scale may not characterize the production technology of the perfume industry, heavy advertising creates a significant barrier to entry. Significant advertising expenses also characterize the automobile industry, and may prevent smaller producers from gaining much territory in the automobile industry.

Legislation

Monopolies can also exist as a result of government charter. Many Canadian utilities were originally incorporated as monopolies, because it was more efficient to have one firm supply the market than to have several competing firms. In the early days of the telephone industry, several competing firms served some of the larger cities: In order to phone someone, a person had to subscribe to the same company as the person they wished to speak to. People needed several telephone lines to their residence in order to be able to phone all the people they would wish to call. Clearly, this was inefficient, so in cases where competition was failing, government stepped in and allocated monopolies in telephone service. Provincial monopoly was common in Canada, although in many parts of the country, small telephone monopolies exist along with the provincial monopolies such as Bell and Telus to serve specific areas. Thunder Bay, Ontario, Edmonton, Alberta, and Prince Rupert, British Columbia, all have their own telephone systems.

Q.9 If a patent is a monopoly, why does the government give out patents?

Patents are another way in which government can grant a company a monopoly. A **patent** is *a legal protection of a technical innovation that gives the person holding the patent a monopoly on using that innovation*. To encourage research and development of new products, government gives out patents for a wide variety of innovations, such as genetic engineering, Xerox machines, and cans that can be opened without a can opener. Research and development may also help a firm develop a superior production technique that gives it an absolute cost advantage.

Other Barriers to Entry

Sometimes one company is able to gain ownership of some essential aspect of the production process; a firm may possess a unique input (as Microsoft may argue), or a firm may foreclose on access to crucial inputs through contracts with suppliers or distributors (as Microsoft's detractors may argue), or a firm may have control over a raw resource required for production. A good example is DeBeers. By owning or controlling the worldwide distribution network for diamonds, DeBeers enjoyed, for many years, a virtual world monopoly in the diamond industry. Over the years, however, other distribution facilities have emerged, and the dominance of DeBeers has been diminished, especially in recent years. The Canadian Ekati diamond mine markets the majority of its diamonds outside the DeBeers distribution system.

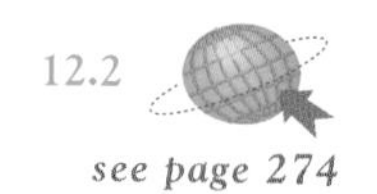
12.2 see page 274

Learning by doing is another way in which firms can gain a superior production technique, which confers an advantage to the existing firm, giving it a cost advantage over new entrants.

Companies may also engage in anti-competitive behaviour in order to keep out entrants and maintain their dominant position. The existence of positive economic profits is a strong attraction to new firms, which might be able to easily enter an industry that had comparatively weak barriers to entry. In such a case, companies already there may resort to illegal pricing strategies, such as predatory pricing, where the firm drops its price below cost in response to a threat of entry. When the new entrant is gone, the firm raises price to its previous level, and the firm continues to enjoy economic profits. While anticompetitive practices such as predatory pricing are violations of the Canadian *Competition Act*, proving that a company is engaging in these activities is very difficult.

NORMATIVE VIEWS OF MONOPOLY

Many laypeople's views of government-created monopoly reflect the same normative judgments that Classical economists made. Classical economists considered such monopolies unfair and inconsistent with liberty. Monopolies prevent people from being free to enter whatever business they want and are undesirable on normative grounds. In this view, government-created monopolies are simply wrong.

A second normative argument against monopoly involves the income distribution effects associated with monopoly. Although monopolists do not always earn an economic profit, they usually do, which means that there is a transfer of consumer surplus to the monopolist. Value that was originally consumer surplus is now profit for the monopoly. This way of seeing the distributional effect of monopoly is based on normative

APPLYING THE TOOLS

Can Price Controls Increase Output and Lower Market Price?

Chapter 7 explained how effective price ceilings increase market price, reduce output, and reduce the welfare of society. With any type of price control in a competitive market, some trades that individual buyers and sellers would like to have made are prevented. With competitive markets, price controls of any type generally have income distribution effects.

When there is monopoly the argument is not so simple. The monopoly price is higher than the marginal cost and society loses out; monopolies create their own deadweight loss. In the case of monopoly, price controls can actually lower price, increase output, and reduce deadweight loss.

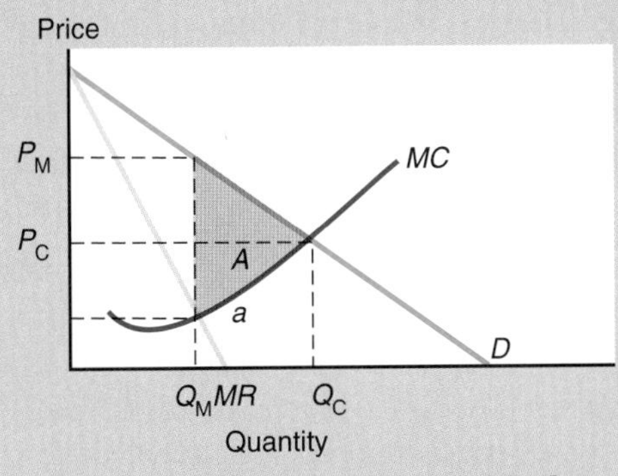

The monopoly sets its price at $MR = MC$. Output is Q_M and price is P_M; the welfare loss is the shaded triangle A. Now say that the government comes in and places a price ceiling on the monopolist at the competitive price, P_C. Since the monopolist is compelled by law to charge price P_C, it no longer has an incentive to restrict output. Put another way, the price ceiling—the dashed line P_C—becomes the monopolist's demand curve and marginal revenue curve. (Remember, when the demand curve is horizontal, the marginal revenue curve is identical to the demand curve.) Given the law, the monopolist's best option still is to produce where $MC = MR$, but that means $MC = MR = P_C$ so the monopolist maximizes profit by charging price P_C and increasing output to Q_C. Thus, the price ceiling causes output to rise and price to fall.

If, when there is monopoly, price controls can increase efficiency, why don't economists advocate price controls more than they do? Let's review four reasons why.

1. For a price control to increase output and lower price, it has to be set within the right price range—below the monopolist's price and above the price where the monopolist's marginal cost and marginal revenue curves intersect. It is unclear politically that such a price could be chosen. Even if regulators could pick the right price initially, markets may change. Demand may increase or decrease, putting the price control outside the desired range.
2. All markets are dynamic. The very existence of monopoly profits will encourage other firms in other industries to try to break into that market, keeping the existing monopolist on its toes. This is desirable. Because of this dynamic element, no market is ever a pure textbook monopoly.
3. Price controls create their own deadweight loss in the form of rent seeking. Price controls do not eliminate monopoly pressures. The monopolist has a big incentive to regain its ability to set its own price and will lobby hard to remove price controls. Economists see resources spent to regain their monopoly price as socially wasteful.
4. Governments have their own political agendas—there is no guarantee that governments will set the price at the competitive level.

The arguments are, of course, more complicated, and will be discussed in more detail, but this should provide a good preview of some of the policy issues to come in later chapters.

views of who deserves income; it transfers income from "deserving" consumers to "undeserving" monopolists.

A third normative reason people oppose government-created monopoly, that isn't captured by the standard model of monopoly, is that the possibility of government-created monopoly encourages people to spend a lot of resources in political pursuits trying to get the government to favour them with a monopoly, and less time doing "productive" things. It causes *rent-seeking* activities, in which people spend resources to gain monopolies for themselves.

Possible economic profits from monopoly lead potential monopolists to spend money to lobby government to give them a monopoly.

Each of these arguments probably plays a role in the public's dislike of monopoly. As you can see, these real-world arguments blend normative judgments with objective analysis, making it difficult to arrive at definite conclusions. Most real-world problems require this blending, making applied economic analysis difficult. The economist must interpret the normative judgments about what people want to achieve and explain how public policy can be designed to achieve those desired ends.

Let's now consider how economic theory might be used to analyze monopoly and to suggest how government might deal with monopoly.

GOVERNMENT POLICY AND MONOPOLY: AIDS DRUGS

Let's consider the problem of acquired immune deficiency syndrome (AIDS) and the combination of medicinal drugs, including azidothymidine (AZT), used to treat it. AZT, used in combination with other drugs in mixtures called cocktails, is believed to arrest AIDS completely. These drugs were developed by a small group of pharmaceutical companies, which have **patents** on them. Those patents give the drug companies a monopoly. Patents are given on medicine to encourage firms to spend enormous amounts of resources to find cures for various diseases. The monopoly the patent gives them lets them charge a high price so that the firms can expect to make a profit from their research. Whether such patents are in the public interest isn't an issue, since the patent has already been granted.

Q-10 The medicinal drug tetracycline sold for animals costs about 1/20 the cost of the same drug sold for human beings. What is the likely explanation?

What is an issue is what to do about these drugs. Currently demand for them is highly inelastic, so the price pharmaceutical companies can charge is high, even though their marginal cost of producing them is low. Whether they are making a profit depends on their cost of development. But since that cost is already spent, that's irrelevant to the current marginal cost; development cost affects their *ATC* curve, not their marginal cost curve. Thus, the pharmaceutical companies are charging an enormously high price for drugs that may help save people's lives, that cost them a very small amount of variable cost to produce.

What, if anything, should the government do? Some people have suggested that the government come in and regulate the price of the drugs, requiring the firms to charge only their marginal cost. This would make society better off. But most economists have a problem with that policy. They point out that doing so will significantly reduce the incentives for drug companies to research new drugs. One reason drug companies spend billions of dollars for drug research is their expectation that they'll be able to make large profits if they're successful. If drug companies expect the government to come in and take away their monopoly when they're successful, they won't search for cures. So forcing these pharmaceuticals to charge a low price for their drugs would help AIDS victims, but it would hurt people suffering from diseases that are currently being researched and that might be researched in the future. So there's a strong argument not to regulate.

But the thought of people dying when a cheap cure—or at least a partially effective treatment—is available is repulsive to many people. African countries, where 70 percent of all people infected with the virus that causes AIDS live, have threatened to license production of these drugs to local manufacturers and make them available at cost. U.S. pharmaceutical companies pressured the United States government to cut off foreign aid if they did so. Indeed, some companies have felt the pressure and have made drugs for AIDS available to AIDS patients in poor nations at a much lower price than they do to others (an example of price discrimination). In 1998 Glaxo Wellcome and several other pharmaceutical companies agreed to cut their prices 50 to 75 percent in developing countries. Still, the cost of the drug regimen is between $200 and $400 a month—far above monthly incomes in those countries.

An alternative policy suggested by economic theory is for the government to buy the patents and allow anyone to make the drugs so their price would approach their marginal cost. Admittedly this would be expensive. It would cause negative incentive effects, as the government would have to increase taxes to cover the buyout's costs. But this approach would avoid the problem of the regulatory approach and achieve the same ends. However, it would also introduce new problems, such as determining which patents the government should buy.

Whether such a buyout policy makes sense remains to be seen, but in debating such issues the power of the simple monopoly model becomes apparent.

CONCLUSION

The main element of monopoly, producing at $MR=MC$ and charging $P>MC$, means that there is some inefficiency associated with the monopoly market structure. Too little of the good is produced and the price charged is too high, relative to the perfectly competitive outcome.

Foreign competition may encourage a monopolist to use its profits for research and development.

Is monopoly undesirable then? The answer depends on the reason for the monopoly's existence. In many cases where a single large firm controls the supply side of the market (or where a few large firms dominate), the least cost production method may require a very large scale of production. If one firm producing at minimum efficient scale is the least cost means of production, the industry is a natural monopoly. Forcing competition would decrease efficiency and increase production costs. Economies of scale therefore provide the monopolist with a very effective barrier to entry.

If, however, the monopoly exists as a result of actions taken by the monopolist, such as extensive patenting, or foreclosure of markets through predatory pricing or through ownership or control over some critical aspect of production, there is less agreement that monopoly is the market structure that would naturally emerge in the market. Competition policy and other government legislation, such as patent legislation, can be used to mitigate the monopoly's abuse of its dominant position in the market.

The economic profit earned by monopoly in the long run is another source of concern; however, in today's global economy, no domestic markets are very well protected from foreign competition. In order to prevent foreign competitors from gaining access to the monopoly's domestic market, the monopolist may have to undertake large amounts of research and development in order to maintain its competitive edge in a rapidly advancing world technology scene. The monopolist can do so, since it earns a profit which is greater than normal profit. The monopolist has both the means (economic profits) and the incentive (foreign competition) to innovate in order to retain its market position. In fact, when we examine the Canadian economy, many industries are populated by one or a few very large firms; however, when we look at Canada's position in the world economy, even our biggest firms are small relative to their U.S. or European competitors.

What does this imply for Canadian industrial policy? The greatest difficulty Canadian policy makers face is how to keep Canadian industries relatively competitive domestically — in structure and in performance — while at the same time ensuring that our Canadian companies have the size and strength to survive and compete effectively at the global level.

Chapter Summary

- The price a monopolist charges is higher than that of a competitive market due to the restriction of output; a monopolist can make a profit in the long run.
- A monopolist's profit-maximizing output is where marginal revenue equals marginal cost.
- A monopolist can charge the maximum price consumers are willing to pay for the quantity the monopolist produces. This is given by the demand curve.
- To determine a monopolist's profit, first determine its output (where $MC = MR$). Then determine its price and average total cost at that output level. The difference between price and average total cost at the profit-maximizing level of output is profit per unit. Multiply this by output to find total profit.
- If a monopolist can (1) identify groups of customers who have different elasticities of demand, (2) separate them in some way, and (3) limit their ability to resell its product between groups, it can price discriminate.
- A price-discriminating monopolist earns more profit than a single-price monopolist because it can charge a higher price to those with less elastic demands and a lower price to those with more elastic demands.
- Because monopolists reduce output and charge a price that is higher than marginal cost, monopolies create a welfare loss to society.
- Important barriers to entry are increasing returns to scale, setup costs, government restrictions, and learning by doing.
- Natural monopolies exist in industries with strong economies of scale. Because their average total costs are falling over the relevant market quantity, it is more efficient for one firm to produce all the output.
- The competitive price is impossible in a natural monopoly because marginal cost is always below average total cost. No firm would enter an industry where not even normal (zero economic) profit can be made.
- Normative arguments against monopoly include the following: (1) monopolies are inconsistent with freedom, (2) the distributional effects of monopoly are unfair, and (3) monopolies encourage people to waste time and money trying to get monopolies.

Key Terms

barriers to entry *(265)*
monopoly *(255)*
natural monopoly *(266)*
patent *(270)*
price discriminate *(263)*
set-up cost *(267)*
single-price monopoly *(262)*
welfare loss *(262)*

Questions for Thought and Review

1. Demonstrate graphically the profit-maximizing positions for a perfectly competitive firm and for a monopolist. How do they differ?
2. Demonstrate graphically market equilibrium for a perfectly competitive market and for a perfectly price discriminating monopolist.
3. Monopolists differ from perfect competitors because monopolists make a profit. True or false? Why?
4. Why is marginal revenue below average revenue for a monopolist?
5. Say you place a lump sum tax (a tax that is treated as a fixed cost) on a monopolist. Illustrate how this will affect its output and pricing decisions. Illustrate what has happened to profits.
6. A monopolist is selling fish. But if the fish don't sell, they rot. What will be the likely elasticity at the point on the demand curve at which the monopolist sets the price?
7. Does price discrimination explain why tire companies have sales roughly half the time? Explain.
8. Airlines are always running sales. On closer look, however, existing fares can be cheaper than restricted flights. What accounts for the practice of advertising "bargain" fares that may not be the lowest fare available? What conditions in the airline market make this practice possible?
9. Demonstrate the welfare loss created by a monopoly.
10. Will the welfare loss from a monopolist with a perfectly elastic marginal cost curve be greater or less than the welfare loss from a monopolist with an upward-sloping marginal cost curve?
11. Why would airlines block new carriers at major airports? What effect does this have on fares and the number of flights at those airports? How much are airlines willing to spend to control the use of gates to block new carriers?

12. Copyrights provide authors with a monopoly. What effect would eliminating copyrights have on the price and output of books? Should copyrights be eliminated?

13. How is efficiency related to the number of firms in an industry characterized by strong economies of scale?

Problems and Exercises

1. A monopolist with a straight-line demand curve finds that it can sell two units at $12 each or 12 units at $2 each. Its fixed cost is $20 and its marginal cost is constant at $3 per unit.
   a. Draw the MC, ATC, MR, and demand curves for this monopolist.
   b. At what output level would the monopolist produce?
   c. At what output level would a perfectly competitive firm produce?
2. State what's wrong with the following graphs:

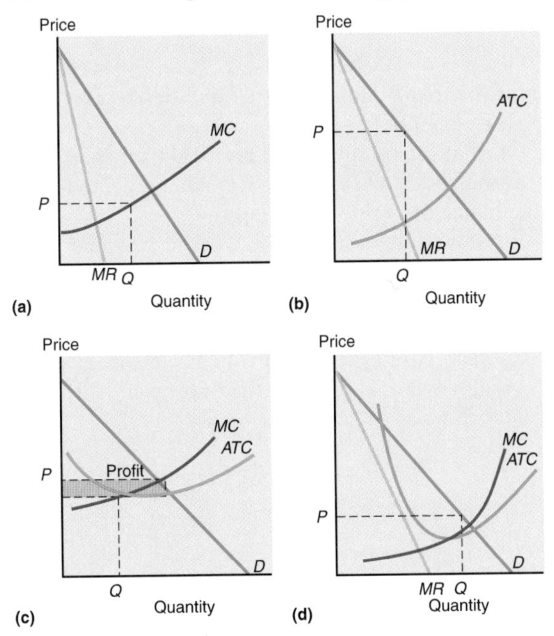

3. Wyeth-Ayerst Laboratories developed Norplant, a long-acting contraceptive, in the early 1990s. In the United States, the firm priced the contraceptive at $350, and in other countries, the firm priced it at $23.
   a. Why would the firm price it differently in different countries?
   b. Was the pricing fair?
   c. What do you think will happen to the price over time? Why?
4. Assume your city government has been contracting with a single garbage collection firm that has been granted an exclusive franchise, or sole right to pick up trash within the entire city limits. However, it has been proposed that the companies be allowed to compete for business with residents on an individual basis. The city government has estimated the price residents are willing to pay for various numbers of garbage collections per month and the total costs per resident as shown in the following table.

Pickup (*Q*)	Price per Pickup (Demand)	Total Revenue (*TR*)	Marginal Revenue (*MR*)	Total Cost (*TC*)	Marginal Cost (*MC*)	Average Total Cost (*ATC*)
0	$4.20	0	—	$ 3.20	—	—
1	$3.80	___	___	$ 4.20	___	___
2	$3.40	___	___	$ 5.60	___	___
3	$3.00	___	___	$ 7.80	___	___
4	$2.60	___	___	$10.40	___	___
5	$2.20	___	___	$13.40	___	___
6	$1.90	___	___	$16.80	___	___

   a. What are the fixed costs per month of garbage collection per resident?
   b. Considering that the current garbage collection firm the city has contracted with has a monopoly in garbage collection services, what is the current number of collections residents receive per month and the price charged residents for each collection? What is the economic profit received from each resident by the monopoly firm?
   c. If competitive bidding were allowed and therefore a competitive market for garbage collection services developed, what would be the number of collections per month and the price charged residents per collection? What is the economic profit received from each resident by the competitive firms?
   d. Based on the above analysis, should the city government allow competitive bidding? Why? Would you expect there to be any quality differences between the monopolistic and competitive trash collection firms?
5. Econocompany is under investigation by the Competition Bureau for violating competition laws. The government decides that Econocompany has a natural monopoly and that, if it is to keep its business, it must sell at a price equal to marginal cost. Econocompany says that it can't do that and hires you to explain to the government why it can't.
   a. You do so in reference to the following graph.
   b. What price would it charge if it were unregulated?

c. What price would you advise that it should be allowed to charge?

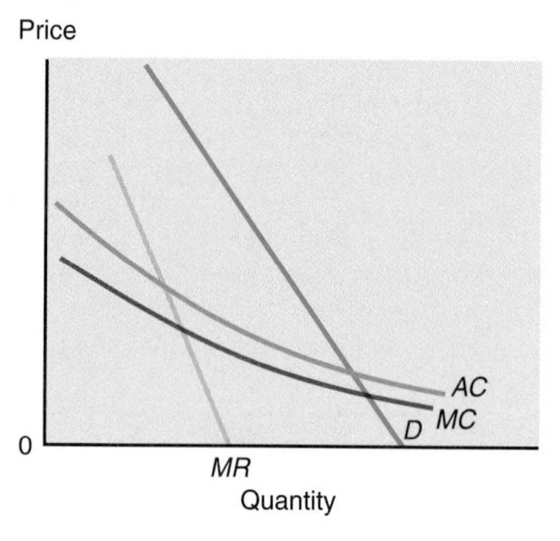

6. New York City has issued 11,787 taxi licenses, called *medallions*, and has not changed that number since 1937.
   a. What does that limitation likely do to the price of taxi medallions?
   b. In the early 1990s, the New York City Taxi Commission promulgated a rule that required single-cab medallion owners to drive their cabs full-time. What will that rule do to the price of the medallion?
   c. If New York City increased the number of medallions by 1,000, selling the additional 1,000 at the market rate, and gave half the proceeds to owners of existing medallions, what would happen to the price of medallions?
   d. What would happen to the wealth of existing medallion owners?

Web Questions

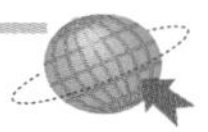

1. Research the Dairy Farmers of Canada Web site at www.dairyfarmers.org to answer the following questions:
   a. What is the mission of the Dairy Farmers of Canada? What is their logo?
   b. What kinds of activities does the Dairy Farmers of Canada organization perform on behalf of the dairy farmers it represents? How do these activities affect the price of milk?
   c. According to the standard model, what are the welfare effects of the dairy farmers' organization? What might the standard model be missing?
2. At the *Report on Business* Web site, www.robmagazine.com, find the article entitled "The Company that Built a Country," by Michael Bliss (Friday, August 31, 2001).
   a. Of what importance was a national Canadian railway in the 1800s? How did the Canadian Pacific Railway enhance the value of the railway?
   b. What does the article mean when it says that proven managerial excellence was transferable? How does it apply to Canadian Pacific's growth?
   c. Why were the Canadian National Railway and Trans-Canada Airlines (now Air Canada) created? What was the eventual result in the railway and airline markets? Why?

Answers to Margin Questions

1. The monopolist must lower price in order to sell more product because it faces the entire, downward-sloping market demand curve. By the Law of Demand, higher quantities are associated with lower prices, so if the monopolist wishes to sell a higher quantity, it must offer it at a lower price. *(256)*
2. At output 4, the marginal cost of $12 (between $8 and $16) equals the marginal revenue of $12 (between $15 and $9), making it the profit-maximizing output. It has the highest total profit, $34. *(257)*
3. To determine the profit-maximizing price and output, one must determine where the marginal revenue curve equals marginal cost. So one must first draw the marginal revenue curve and see where it intersects marginal cost. That intersection determines the quantity, as in the graph below. Carrying the line up to the demand curve determines the price. *(258)*

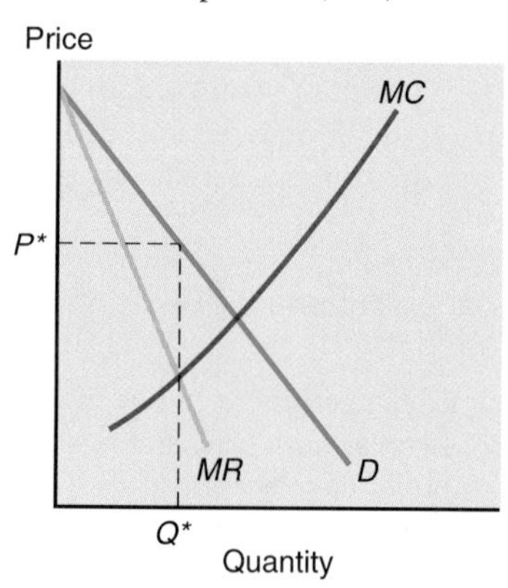

4. A monopolist produces less output than a perfectly competitive firm because it takes into account the fact that increasing output will lower the price of all previous units. *(260)*
5. To determine profit, follow the following four steps: (1) draw the marginal revenue curve, (2) find the level of output where MC = *MR* indicated in the graph below by Q^*, (3) find the price the monopolist would charge indicated by P^* and extend a horizontal line from the demand curve at that price to the price axis, (4) determine the average total cost at Q^* shown by C^* and extend a horizontal line from the *ATC* curve at that cost to the price axis. The box created is the monopolist's profit. The profit is the shaded box shown in the graph below. *(261)*

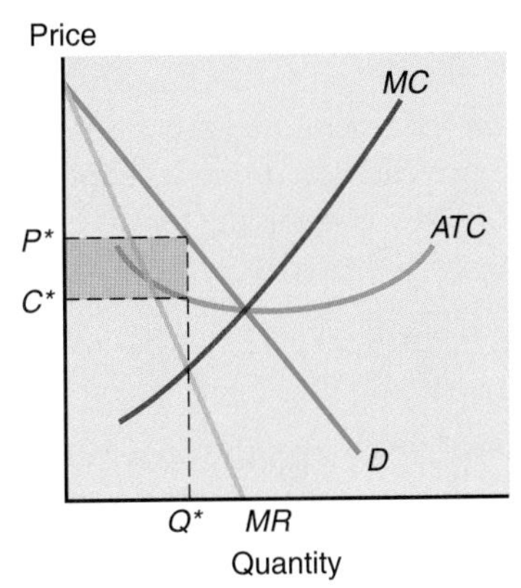

6. Area C represents the profit going to a monopolist. It is not considered a loss since, while consumers lose it, monopolists gain it. It is a redistribution of resources rather than an efficiency loss. *(262)*
7. A price-discriminating monopolist makes a greater profit than a single-price monopolist because a price-discriminating monopolist is able to charge a higher price to those consumers who have less elastic demands. *(265)*
8. The marginal cost curve for an industry that exhibits strong economies of scale is always below average total costs. Therefore, the competitive price, where *P* = MC, will always result in losses for firms. Firms would not enter such an industry and there would be no supply. *(266)*
9. The government gives out patents to encourage research and development of new products. This suggests that the public and government believe that certain monopolies have overriding social value. *(268)*
10. A likely explanation for medicinal drugs being sold at a much lower cost for animals than for human beings is differing elasticities of demand. The demand for drugs for human beings is highly inelastic, whereas the demand for medicinal drugs for animals is elastic. When there is a price-discriminating monopolist for these drugs, those with more inelastic demands are charged higher prices. *(270)*

The Algebra of Competitive and Monopolistic Firms

In the Appendix to Chapter 5, we presented the algebra relevant to supply and demand. To relate that algebra to competitive firms, all you must remember is that the market supply curve equals the marginal cost curve for the competitive industry. Let's review it briefly.

Say that marginal costs, and thus market supply, for the industry is given by

$$P = 2Q_S + 4$$

Let's also say that the market demand curve is

$$Q_D = 28 - \tfrac{1}{4}P$$

To determine equilibrium price and quantity in a competitive market, you must equate quantity supplied and quantity demanded and solve for price. First, rewrite the marginal cost equation with quantity supplied on the left:

$$Q_S = \tfrac{1}{2}MC - 2$$

Then set quantity demanded equal to quantity supplied and MC = *P*. Then solve for equilibrium price:

$$Q_S = Q_D \Rightarrow 28 - \tfrac{1}{4}P = \tfrac{1}{2}P - 2$$

$$112 - P = 2P - 8$$

$$3P = 120$$

$$P = 40$$

Thus, the equilibrium price is \$40. Competitive firms take this price as given and produce up until their marginal cost equals price. The industry as a whole produces 18 units.

Now let's consider the algebra relevant for a monopolistic firm. In the monopolistic case, supply and demand are not enough to determine where the monopolist will produce. The monopolist will produce where marginal revenue equals marginal cost. But, for the monopolist, the industry demand curve is the demand curve, which means that in order to determine where the monopolist will produce, we must determine the marginal revenue curve that goes along with the above demand curve. There are two ways to do that.

First, if you know calculus you can determine the marginal revenue curve in the following manner: Since marginal revenue tells us how much total revenue will change with each additional unit produced, you first specify the demand curve in terms of quantity produced.

$$P = 112 - 4Q$$

Since $TR = PQ$ we can multiply this by Q to get total revenue. Doing so gives us:

$$TR = PQ = 112Q - 4Q^2$$

To find marginal revenue, take the first derivative of total revenue with respect to Q.

$$P = 112 - 8Q$$

Second, if you don't know calculus, all you need to remember is the trick shown in a box in the chapter on how to graph the marginal revenue curve. Remember, the marginal revenue curve starts at the same price as the demand curve and falls twice as fast as the market demand curve. That is, its slope is twice the slope of the market demand curve.

Knowing that its slope is twice the market demand curve slope, you can write the marginal revenue curve with the same price axis intercept as the demand curve and a slope of two times the slope of the demand curve. (Warning: this only works with linear demand curves.) The price-axis intercept of the demand curve is the value of P where Q equals 0: 112. This is the same point for the marginal revenue curve. The quantity-axis intercept of the demand curve is the value of Q where P equals 0: 28. The marginal revenue curve has a quantity-axis intercept at half of this, at 14. Mathematically, such a curve is represented by

$$P = 112 - (112/14)P$$

or

$$P = 112 - 8Q$$

Now that we've determined the monopolist's marginal revenue curve, we can determine its equilibrium quantity by setting $MR = MC$ and solving for Q. Doing so gives us:

$$112 - 8Q = 2Q + 4$$
$$-10Q = -108$$
$$Q = 10.8$$

The monopolist then charges the price consumers are willing to pay for that quantity. Mathematically, substitute 10.8 into the demand equation and solve for price:

$$P = 112 - 4(10.8)$$
$$P = \$68.80$$

Comparing the price and quantity produced by a monopolist and those of a competitive industry shows that the monopolist charges a higher price and produces a lower output.

Questions for Thought and Review

1. The market demand curve is $Q_D = 50 - P$. The marginal cost curve is $MC = 4Q + 6$.
   a. Assuming the marginal cost curve is for a competitive industry as a whole, find the profit-maximizing level of output and price.
   b. Assuming the marginal cost curve is for only one firm which comprises the entire market, find the profit-maximizing level of output and price.
   c. Compare the two results.
2. The market demand curve is $Q_D = 160 - 4P$. A monopolist's total cost curve is $TC = 6Q^2 + 15Q + 50$.
   a. Find the profit-maximizing level of output and price for a monopolist.
   b. Find its average cost at that level of output.
   c. Find its profit at that level of output.
3. Suppose fixed costs for the monopolist in question 2 increases by 52.
   a. Find the profit-maximizing level of output and price for a monopolist.
   b. Find its average cost at that level of output.
   c. Find its profit at that level of output.
4. The market demand curve is $Q_D = 12 - \frac{1}{3}P$. Costs do not vary with output.
   a. Find the profit-maximizing level of output and price for a monopolist.
   b. Find the profit-maximizing level of output and price for a competitive industry.

Monopolistic Competition, Oligopoly, and Strategic Pricing

13

After reading this chapter, you should be able to:

- Describe two methods of determining market structure.
- List the four distinguishing characteristics of monopolistic competition.
- Demonstrate graphically the equilibrium of a monopolistic competitor.
- State the central element of oligopoly.
- Explain why decisions in the cartel model depend on market share and decisions in the contestable market model depend on barriers to entry.
- Illustrate a strategic decision facing a duopolist using the prisoner's dilemma.

Competition, you know, is a lot like chastity.
It is widely praised, but alas, too little practiced.

Carol Tucker

As soon as economists start talking about real-world competition, market structure becomes a focus of the discussion. **Market structure** refers to *the physical characteristics of the market within which firms interact*. It involves the number of firms in the market and the barriers to entry. Monopoly and competition are the two polar cases of market structure. Real-world markets generally fall in between, and it is essential to analyze the two market structures that fall between perfect competition and monopoly: monopolistic competition and oligopoly.

In monopolistic competition, firms' decisions are independent; in oligopoly, firms' decisions are interdependent.

Perfect competition has an almost infinite number of firms; monopoly has one firm. **Monopolistic competition** is *a market structure in which there are many firms selling differentiated products; there are few barriers to entry.* **Oligopoly** is *a market structure in which there are only a few firms; there are significant barriers to entry.*

THE PROBLEMS OF DETERMINING MARKET STRUCTURE

Any estimate of the distribution of market structures must be treated with care. Defining an industry is a complicated task—inevitably, numerous arbitrary decisions must be made. Similarly, defining the relevant market of a given industry is complicated. For example, there may be thousands of hardware stores in the country; however, within a particular geographic area, there may only be a few competitors. Or consider corner grocery stores: thousands of them exist in a city, yet, since there is usually only one located in any particular neighbourhood, each store has a (small) geographic monopoly. The same argument exists when we think of international competition. Many firms sell in international markets and, while a group of firms may compose an oligopoly in Canada, the international market might be more accurately characterized as monopolistic competition.

What industry does this firm belong to? Would its competition be local or global?

Another dimension of the definitional problem concerns deciding what is to be included in an industry. If you define the industry as "the transportation industry," there are many firms. If you define it as "the urban transit industry," there are fewer firms; if you define it as "the commuter rail industry," there are still fewer firms. Similarly with the geographic dimension of industry. There's more competition in the global market than in the local market. The narrower the definition, the fewer the firms.

Classifying Industries

One of the ways that economists classify markets is by cross-price elasticities (the responsiveness of the change in the demand for a good to change in the price of a related good). Industrial organization economist F. M. Sherer has suggested the following rule of thumb: When two goods have a cross-price elasticity greater than or equal to 3, they can be regarded as belonging to the same market.

Q.1 Which would have more output: the two-digit industry 21 or the four-digit industry 2111? Explain your reasoning.

The **North American Industry Classification System (NAICS)** is *an industry classification that categorizes firms by type of economic activity and groups firms with like production processes*. It was adopted by Canada, the United States, and Mexico in 1997 and replaces the Standardized Industrial Classification (SIC) codes developed in the 1930s. All firms are placed into 20 broadly defined two-digit sectors. These two-digit sectors are further subdivided into three-digit subsectors, four-digit industry groupings, five-digit industries, and six-digit national industry groupings. Each subgrouping becomes more and more narrowly defined. Table 13-1 lists the 20 sectors and shows the subgroupings for one sector, Information, to give you an idea of what's included in each.

When economists talk about industry structure, they generally talk about industries in the four- to six-digit subsector groupings. This is a convention. Economists are often called on to give expert testimony in court cases, and if an economist wants to argue that an industry is more competitive than its opponents say it is, he or she challenges this convention of using a four- to six-digit classification of industry, asserting that the classification is arbitrary (which it is) and that the relevant market should be the two- to three-digit classification.

13.1 see page 298

TABLE 13-1 Industry Groupings in the North American Industry Classification System

	Two-Digit Sectors	Three- to Six-Digit Subsectors
11	Agriculture, forestry, fishing, and hunting	
21	Mining	
22	Utilities	
23	Construction	
31–33	Manufacturing	
42	Wholesale trade	
44–45	Retail trade	
48–49	Transportation and warehousing	513 Broadcasting and telecommunications
51	Information	5133 Telecommunications
52	Finance and insurance	51332 Wireless telecommunications carriers, except satellite
53	Real estate and rental and leasing	513321 Paging
54	Professional, scientific, and technical services	
	Management of companies and enterprises	
56	Administrative and support, and waste management and remediation services	
	Education services	
	Health care and social assistance	
71	Arts, entertainment, and recreation	
	Accommodation and food services	
81	Other services (except public administration)	
	Public administration	

Measuring Industry Structure

To measure industry structure, economists use one of two methods: the concentration ratio or a Herfindahl index. These measure the market share held by a certain number of firms in an industry and provide an indication of industry competitiveness.

A **concentration ratio** is *the percentage of the total industry sales by the top few firms of the industry*. The most commonly used concentration ratio is the four-firm concentration ratio. For example, a four-firm concentration ratio, CR4, of 60 tells you that the top four firms in the industry produce 60 percent of the industry's output. The higher the ratio, the closer the industry is to an oligopolistic or monopolistic type of market structure. The concentration ratio only concerns the market shares of the top few companies of an industry. Other concentration ratios, such as the CR3 and CR5, are also reported. A CR3 of 60 means that the industry is more concentrated than an industry with a CR4 of 60.

A Herfindahl index is a method used by economists to classify how competitive an industry is.

The Herfindahl index takes into account the market shares of *all* of the firms in the industry. The **Herfindahl index** is *an index of market concentration calculated by adding the squared value of the individual market shares of all the firms in the industry*. For example, say that 10 firms in the industry each have 10 percent of the market:

$$\text{Herfindahl index} = 10^2+10^2+10^2+10^2+10^2+10^2+10^2+10^2+10^2+10^2 = 1{,}000$$

The Herfindahl index gives higher weights to the largest firms in the industry because it squares market shares. For five firms each with 20 percent of the market, the Herfindahl index is 2000 (5×20^2).

Because it squares market shares, the Herfindahl index gives more weight to firms with large market shares than to smaller firms.

The two measures can differ because of their construction, but generally if the concentration ratio is high, so is the Herfindahl index.

The Herfindahl index plays an important role in government policy; it is used as a rule of thumb in determining whether an industry is sufficiently competitive to allow a merger between two large firms in the industry. If the Herfindahl index is less than 1,000, the industry is considered to be sufficiently competitive, and the merger is not closely scrutinized.

Q-2 If the four-firm concentration ratio of an industry is 60 percent, what is the highest Herfindahl index that industry could have? What is the lowest?

Measuring the Size of Conglomerate Firms

Neither the four-firm concentration ratio nor the Herfindahl index gives us a picture of corporations' bigness. That's because many corporations are conglomerates—companies that span a variety of unrelated industries. For example, Alcan is generally thought of as an aluminum company, and the concentration ratio or the Herfindahl index for the aluminum industry would include Alcan's market share. However, the greatest proportion of Alcan's profits do not come from its aluminum division, but rather, from its world wide packaging division. Measuring the size of Alcan, and other conglomerate companies, when the companies' activities span two, or more, industries creates a completely different problem for economists and policy makers.

Introduction to Monopolistic Competition and Oligopoly

Generally speaking, less concentrated industries are more likely to behave like perfectly competitive markets. The number of firms in an industry plays an important role in determining whether firms explicitly take other firms' actions into account. In monopolistic competition, there are so many firms that individual firms do not explicitly take into account rival firms' likely responses to their decisions. Because monopolistically competitive firms tend to act independently, collusion is difficult. In oligopoly, however, there are only a few firms, and each firm is more likely to engage in **strategic decision making**—*taking explicit account of a rival's expected response to a decision it is making*. In oligopolies, all decisions, including pricing decisions, are strategic decisions. Because there are so few firms, and because of the firms' interdependence, collusion is much easier. Thus, we distinguish between monopolistic competition and oligopoly by whether or not firms must explicitly take into account competitors' reactions to their decisions.

Oligopolies take into account the reactions of other firms; monopolistic competitors do not.

Q-3 Your study partner, Jean, has just said that monopolistic competitors use strategic decision making. How would you respond?

CHARACTERISTICS OF MONOPOLISTIC COMPETITION

The four distinguishing characteristics of monopolistic competition are:

1. Many sellers.
2. Differentiated products.
3. Multiple dimensions of competition.
4. Easy entry of new firms in the long run.

Let's consider each in turn.

Many Sellers

There are many sellers in monopolistic competition; however, despite this, each firm is able to identify a particular small market segment as its own. Each firm can see its market share.

The fact that there are many sellers in monopolistic competition means that each firm competes, making decisions independently of its rivals. Collusion is difficult since,

APPLYING THE TOOLS

Foreign Competitive Oligopolies

Market structures change over time. Take, for instance, the automobile industry, which has always been used as the classic oligopoly model. Starting in the 1970s, however, foreign automakers have made large inroads into the North American market and have added new competition to it. Foreign companies such as Honda, Nissan, and Toyota have entered the North American market, as seen in the accompanying pie chart, which lists major automobile companies and their market shares.

The four-firm concentration ratio is over 75 percent, so the industry is still classified as an oligopoly. GM still considers what Ford's and Chrysler's reactions will be, but with the addition of foreign competition, there are getting to be too many firms for one firm to consider the reactions of all the other firms. The auto industry is becoming more monopolistically competitive.

Such change in industry structure is to be expected. Monopoly and oligopoly create the possibility that firms can make above-normal profits. Above-normal profits invite entry, and unless there are entry barriers, the result will likely be a breakdown in that monopoly or oligopoly.

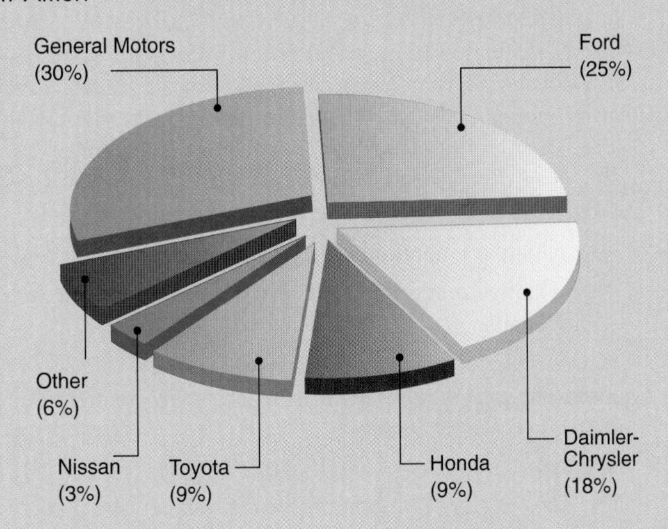

Market Shares in the Automotive Industry
Source: *Automotive News, 2000.*

when there are many firms, getting all of them to act as one is difficult. In economists' model of monopolistic competition, firms are assumed to act independently.

Product Differentiation

The "many sellers" characteristic gives monopolistic competition its competitive aspect. Product differentiation gives it its monopolistic aspect. In a monopolistically competitive market, competitors produce many close substitutes. The goods are not homogeneous, as in perfect competition; they are differentiated slightly. Irish Spring soap is slightly different from Ivory, which in turn is slightly different from Yardley's Old English.

So in one sense each firm has a small monopoly in the good it sells. But that monopoly is fleeting; it is based on advertising to let people know, and to convince them, that one firm's good is different from the goods of competitors. Because consumers have preferences for one or another firm's product, this confers a small degree of market power on the monopolistically competitive firms. The good may or may not really be different. Product differentiation can be based on real differences in product characteristics, or can be based on consumers' perceptions about product differences. Generally, monopolistically competitive industries are characterized by significant expenditures on advertising, which acts as an important barrier to entry in most monopolistically competitive industries. Any industry where brand proliferation exists is likely to be monopolistically competitive, for example, soap, jeans, cookies, and games.

Consumer preferences can be influenced. Differentiated products give firms some control over the price of the product.

Because a monopolistic competitor has some monopoly power, advertising to increase that monopoly power (and hence increase the firm's profits) makes sense as long as the marginal benefit of advertising exceeds the marginal cost. Despite the fact that their goods are similar but differentiated, monopolistically competitive firms make their decisions as if they had no effect on other firms.

Multiple Dimensions of Competition

In monopolistic competition, competition takes many forms.

In perfect competition, price is the only dimension on which firms compete; in monopolistic competition, competition takes many forms. Product differentiation reflects firms' attempts to compete on perceived attributes; advertising helps to differentiate firms' products. Other dimensions of competition include service and distribution outlets. These multiple dimensions of competition make it much harder to analyze a specific industry, but the alternative methods of competition follow the same two general decision rules as price competition:

- Compare marginal costs and marginal benefits; and
- Change that dimension of competition until marginal costs equal marginal benefits.

Advertising can be very expensive, and adds to cost, sometimes significantly raising costs. The goal of advertising, however, is to create or enlarge a firm's market share.

Ease of Entry of New Firms in the Long Run

The last condition a monopolistically competitive market must meet is that entry must be relatively easy; that is, there must be no significant entry barriers. Barriers to entry create the potential for long-run economic profit and prevent competitive pressures from pushing price down to average total cost. In monopolistic competition the existence of economic profits induces other firms to enter until no economic profit exists. In some monopolistically competitive manufacturing industries, firms are able to capture some small economies of scale, but generally, the largest barrier to entry is the expenditures on advertising to enhance product differentiation.

OUTPUT, PRICE, AND PROFIT OF A MONOPOLISTIC COMPETITOR

Let's consider the four characteristics of monopolistic competition and see what implication they have for analysis. First, we recognize that the firm has some monopoly power; therefore, a monopolistic competitor faces a downward-sloping demand curve which gives the firm control over price. However, although downward sloping, the monopolistically competitive firm's demand curve is highly elastic. This limits the firm's pricing power. Although consumers may prefer a particular brand of toothpaste, and may buy that brand even if it is more expensive than competing brands, if the company raises the price "too much," consumers will switch to a competitor's cheaper product. The downward-sloping demand curve means that in making decisions about output, the monopolistic competitor will, as will a monopolist, use a marginal revenue curve that is below price. At its profit-maximizing output, marginal cost will be less than price (not equal to price as it would be for a perfect competitor). We consider that case in Figure 13-1(a).

The monopolistic competitor faces the demand curve D, marginal revenue curve MR, and marginal cost curve MC. Since each firm can identify its own market share, this demand curve is *its* portion of the total market demand curve. Using the $MC = MR$

rule, the firm will choose output level Q_1 (because that's the level of output at which marginal revenue intersects marginal cost). Having determined output, the firm will charge what consumers are willing to pay for output Q_1, so the firm will set a price equal to P_1. This price exceeds marginal cost. This replicates the monopolist's decision.

Where does the competition come in? Competition implies zero economic profit in the long run. If price exceeds average total cost, the firm will earn positive economic profits. These profits attract entry, and because barriers to entry are low and entry is easy, the monopolistically competitive industry will see entry by new firms. Some customers of the existing firm switch to become customers of the new firms. Entry causes our firm's demand curve to shift left as it loses customers. To try to protect its profits, the firm would likely increase expenditures on product differentiation and advertising to offset that entry. (There would be an All New, Really New, Widget campaign.) These expenditures would shift its average total cost curve up. These two adjustments would continue until the economic profits disappeared and entry stopped. This occurs where the new demand curve is tangent to the new average total cost curve. A monopolistically competitive firm earns normal profit in the long run. Economic profits are determined by *ATC*, not by *MC*, so the competition part of monopolistic competition tells us where the average total cost curve must be at the long-run equilibrium output. It must be equal to price, and

Q-4 How do the equilibrium for a monopoly and for a monopolistic competitor differ?

Figure 13-1 (a, b, and c) MONOPOLISTIC COMPETITION

In (**a**) you can see that a monopolistically competitive firm prices in the same manner as a monopolist. It sets quantity where marginal revenue equals marginal cost and earns positive economic profit. In (**b**) you can see that the monopolistic competitor is not only a monopolist but also a competitor. The economic profit attracts entry, and some of the firm's customers defect to the new firm. As the demand for its product falls (D_1 shifts left to D_2), the firm counters by increasing its advertising, which raises its cost of production (ATC_1 shifts up to ATC_2). The firm experiences a profit squeeze: Notice that falling price (P_1 to P_2), rising cost (C_1 to C_2), and falling quantity (Q_1 to Q_2) all help reduce the firm's profit. (**c**) Entry will continue while economic profit exists. Competition implies zero economic profit in the long run. Economic profits are determined by the average total cost (*ATC*) curve. In long-run equilibrium, *ATC* must be equal to price. It will be equal to price only if the *ATC* curve is tangent to the demand curve at the output the firm chooses.

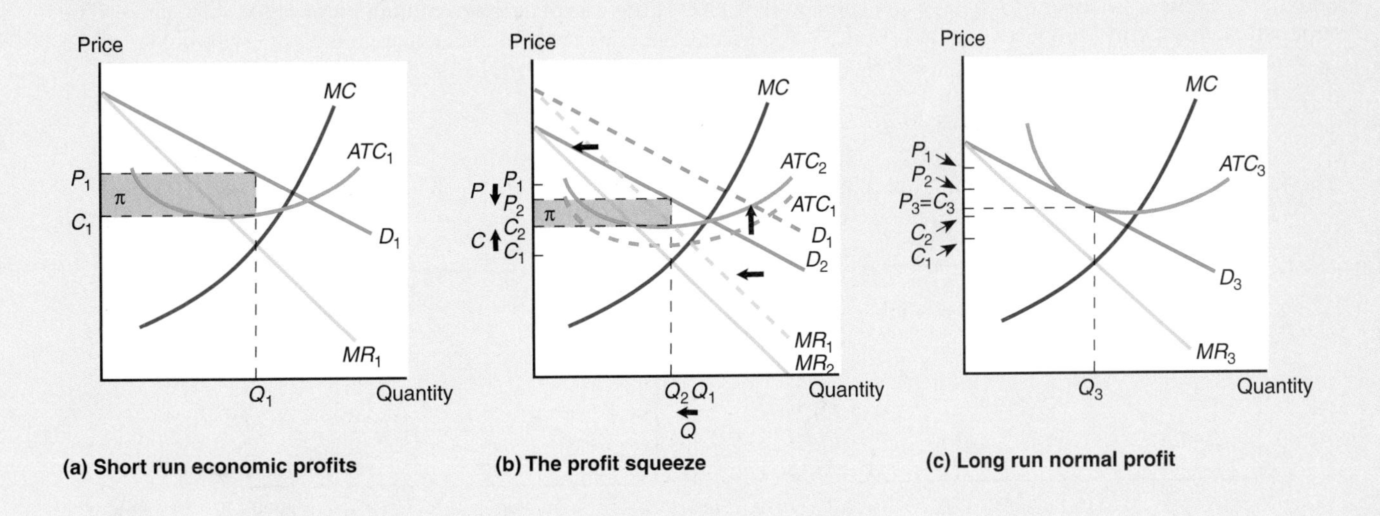

it will be equal to price only if the *ATC* curve is tangent to the demand curve at the output the firm chooses.

Comparing Monopolistic Competition with Perfect Competition

If both the monopolistic competitor and the perfect competitor make zero economic profit in the long run, it might seem that, in the long run at least, they're identical. They aren't, however. The perfect competitor perceives its demand curve as perfectly elastic, and the zero economic profit condition means that it produces at the minimum of the average total cost curve, where the marginal cost curve equals price. We demonstrate that case in Figure 13-2(a).

The monopolistic competitor faces a downward-sloping demand curve for its differentiated product. It produces where the marginal cost curve equals the marginal revenue curve, and not where MC equals price. In equilibrium, price exceeds marginal cost. The average total cost curve of a monopolistic competitor is tangent to the demand curve at that output level, which cannot be at the minimum point of the average total cost curve since the demand curve is sloping downward. The minimum point of the average total cost curve (where a perfect competitor produces) is at a higher output (Q_C) than that of the monopolistic competitor (Q_{MC}). We demonstrate the monopolistically competitive equilibrium in Figure 13-2(b) to allow you to compare monopolistic competition with perfect competition.

The difference between a monopolist and a monopolistic competitor is in the zero economic profit result.

The perfect competitor in long-run equilibrium produces at a point where MC = *P* = *ATC*. At that point, *ATC* is at its minimum. A monopolistic competitor produces at

Figure 13-2 (a and b) A COMPARISON OF PERFECT AND MONOPOLISTIC COMPETITION

The perfect competitor perceives its demand curve as perfectly elastic, and zero economic profit means that it produces at the minimum of the *ATC* curve, as represented in **(a)**. A monopolistic competitor, on the other hand, faces a downward-sloping demand curve and produces where marginal cost equals marginal revenue, as represented in **(b)**. In long-run equilibrium the *ATC* curve is tangent to the demand curve at that level, which is *not* at the minimum point of the *ATC* curve. The monopolistic competitor produces Q_{MC} at price P_{MC}. A perfect competitor with the same marginal cost curve would produce Q_C at price P_C.

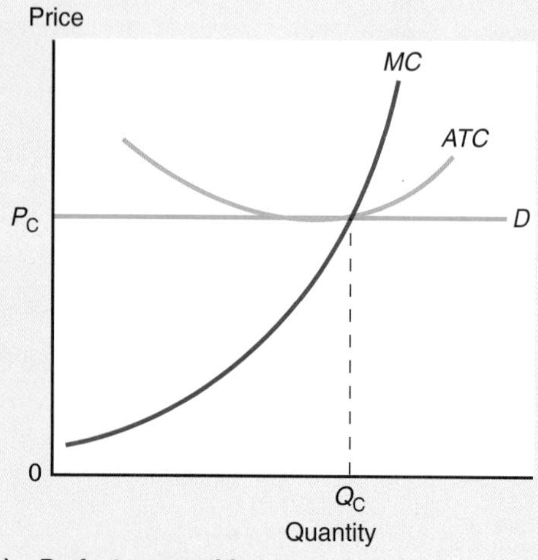

(a) Perfect competition

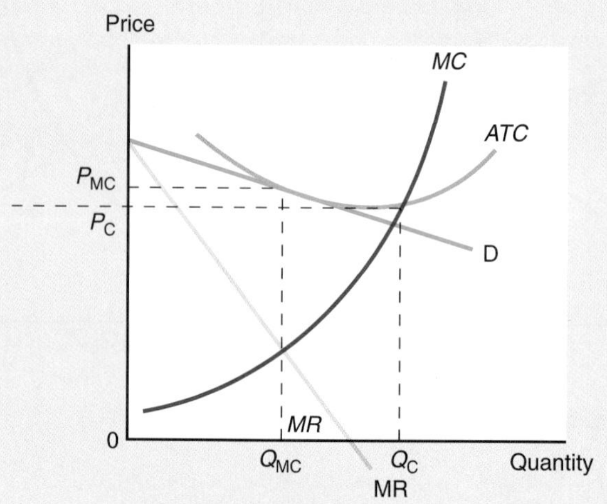

(b) Monopolistic competition

a point where $MC = MR$. Price is higher than marginal cost. For a monopolistic competitor in long-run equilibrium:

$$(P = ATC) \geq (MC = MR)$$

For a monopolistic competitor in long-run equilibrium, $(P = ATC) \geq (MC = MR)$.

At that point, *ATC* is *not* at its minimum, which means that there is some degree of productive and allocative inefficiency. The price paid for the product (*P*) exceeds the opportunity cost of resources used to produce it (MC); therefore, since $P > MC$, more should be produced. But it is not, since the firm is already maximizing profit at its $MR = MC$ output, and $P = ATC$, which implies normal profit.

What does this distinction between a monopolistically competitive industry and a perfectly competitive industry mean in practice? It means that for a monopolistic competitor, since increasing output lowers average cost, increasing market share is a relevant concern. If only the monopolistic competitor could expand its market, it could do better. For a perfect competitor, increasing output offers no benefit in the form of lower average cost. A perfect competitor would have no concern about market share (the firm's percentage of total sales in the market) since it can sell all it wishes to at the going market prices.

Comparing Monopolistic Competition with Monopoly

An important difference between a monopolist and a monopolistic competitor is in the position of the average total cost curve in long-run equilibrium. The monopolist generally makes a long-run economic profit since entry is prevented by significant barriers to entry. For the monopolistic competitor, barriers to entry are low, and entry means that no long-run economic profit is possible.

Advertising and Monopolistic Competition

While firms in a perfectly competitive market have no incentive to advertise (since they can sell all they want at the market price), monopolistic competitors have a strong incentive. That's because their products are differentiated from the others; advertising plays an important role in providing that differentiation.

Goals of Advertising The goals of advertising are to increase demand for the firm's product and to increase customer loyalty to the product. This will shift the firm's demand curve to the right and make it more inelastic. Advertising provides consumers with information about the firm's product and convinces them to prefer a specific brand. That allows the firm to sell more, and to charge a higher price. Advertising not only shifts the demand curve to the right, and it shifts the average total cost curve up.

Goals of advertising include shifting the firm's demand curve to the right and making it more inelastic.

Of course, when many firms are advertising, the advertising is done less to shift the demand curve out than to keep the demand curve where it is—to stop consumers from switching to a competitor's product.

Does Advertising Help or Hurt Society? Our perception of products (the degree of trust we put in them) is significantly influenced by advertising. Think of the following pairs of goods:

Rolex	Timex
Cheerios	Oat Circles
BMW	Chevrolet
Birkenstock	Keds

Each of these brand names conveys a sense of what it is and how much trust we put in the product, and that helps determine how much we're willing to pay for it. For example, most people would pay more for Cheerios than for Oat Circles. Each year firms spend billions on advertising. A 30-second commercial during the Super Bowl can cost as much as $3 million. Advertising increases firms' costs but also differentiates their products.

Are we as consumers better off or worse off with differentiated products? That's difficult to say. There's a certain waste in much of the differentiation that occurs. It shows up in the graph by the fact that monopolistic competitors don't produce at the minimum point of their average total cost curve. But there's also utility in having goods that are slightly different from one another. Consumers like choice. Henry Ford offered his Model T in any colour the customer wanted, as long as it was black.

Q-5 Why is it that monopolistically competitive firms advertise and perfect competitors do not?

Edward Chamberlin, who, together with Joan Robinson, was the originator of the description of monopolistic competition, believed that the difference between the cost of a perfect competitor and the cost of a monopolistic competitor was the cost of what he called "differentness."[1] If consumers are willing to pay that cost, then it's not a waste but rather a benefit to them.

Be careful about drawing implications from this analysis. Average total cost for a monopolistically competitive firm includes advertising and costs of differentiating a product. It's debatable whether we as consumers are better off with as much differentiation as we have, or whether we'd all be better off if all firms produced identical products at a lower cost.

CHARACTERISTICS OF OLIGOPOLY

The central element of oligopoly is that the industry is made up of a small number of very large firms, so that, in any decision it makes, each firm must take into account the expected reaction of other firms. Oligopolistic firms are *mutually interdependent* and can be collusive or noncollusive. Depending on the industry, the product may be homogeneous (oil) or may be differentiated (automobiles).

Oligopolistic firms are mutually interdependent.

One third of industries in Canada can be considered oligopolistic, for example, air transportation, rail transportation, oil refining, and gold mining. In particular, most industries can be considered geographic oligopolies. For example, most retail stores that you deal with are oligopolistic in your neighbourhood or town, although by national standards they may be quite competitive. For example, how many grocery stores do you shop at? Do you think they keep track of what their competitors are doing? You bet. They keep a close eye on their competitors' prices and set their own accordingly.

MODELS OF OLIGOPOLY BEHAVIOUR

Oligopolistic industries are difficult to characterize. Some oligopolies are characterized by 4 or 5 or so equal-sized firms (oil refining), some have one large company but a competitive fringe (airlines), some have 3 large firms and many small- to medium-sized competitors (mining and automobile industries). Because each oligopolistic industry is different, no single general model of oligopoly behaviour exists. Depending on the industry configuration, an oligopolist can decide on pricing and output strategy in many possible ways. Although there are five or six formal models, We'll focus on two informal

In many Canadian industries, the presence of economies of scale means that the entire market output will be supplied by a few very large firms.

[1]Joan Robinson, a Cambridge, England, economist, called this the theory of imperfect competition, rather than the theory of monopolistic competition.

models of oligopoly behaviour, the cartel model and the contestable market model, which should give you a sense of how real-world oligopolistic pricing takes place.

The nature of the interdependence of oligopolists differs in different industries so there is no simple formal model of oligopoly. Since there are few competitors, what one firm does specifically influences what other firms do, so an oligopolist's plan must always be a contingency or strategic plan: If my competitors act one way, I'll do X, but if they act another way, I'll do Y. In addition, strategic interactions have a variety of potential outcomes rather than a single outcome such as shown in formal models of other market structures. An oligopolist spends enormous amounts of time guessing what its competitors will do, and it develops a strategy of how it will act accordingly.

The Cartel Model

A **cartel** is *a combination of firms that acts as if it were a single firm.* If oligopolies can limit entry by other firms, they have a strong incentive to collude and cartelize the industry and to act as a monopolist would, restricting output to a level that maximizes profit to the combination of firms. Thus, the **cartel model of oligopoly** is *a model that assumes that oligopolies act as if they were monopolists that have assigned output quotas to individual member firms of the oligopoly so that total output is consistent with joint profit maximization.* All firms follow a uniform pricing policy that serves their collective interest.

If oligopolies can limit the entry of other firms and form a cartel, they increase the profits for the firms in the cartel.

Since a monopolist makes the most profit that can be squeezed from a market, cartelization is the best strategy for an oligopoly. It requires each oligopolist to hold its production below what would be in its own interest were it not to collude with the others. Such explicit formal collusion is against the law in Canada, but informal collusion is allowed and oligopolies have developed a variety of methods to collude implicitly. Thus, the cartel model has some relevance.

13.2
see page 298

Colluding and cartelization has some problems, however. For example, various firms' interests often differ significantly from one another, so it isn't clear what the collective interest of the firms in the industry is. In many cases a single firm, often the largest or dominant firm, takes the lead in pricing and output decisions, and the other firms (which are often called *fringe firms*) follow suit, even though they might have preferred to adopt a different strategy.

Q.6 Why is it difficult for firms in an industry to maintain a cartel?

To collude is to get together with other firms to set price or allocate market share.

This dominant-firm cartel model works only if the smaller firms face barriers to entry, or the dominant firm has significantly lower cost conditions. If that were not the case, the smaller firms would pick up an increasing share of the market, eroding the dominant firm's monopoly.

In other cases, the various firms meet—sometimes only by coincidence, at the golf course or at a trade association gathering—and arrive at a collective decision. In Canada, meetings for this purpose are illegal, but they do occur. In yet other cases the firms engage in **implicit collusion**—*multiple firms make the same pricing decisions even though they have not explicitly consulted with one another.* They "just happen" to come to a collective decision.

In some cases firms collude implicitly—they just happen to make the same pricing decisions. This is not illegal.

Implicit Price Collusion Implicit price collusion, in which firms just happen to charge the same price but didn't meet to discuss price strategy, isn't against the law. Oligopolies often operate as close to the fine edge of the law as they can. For example, many oligopolistic industries allow a price leader to set the price, and then the others follow suit. The airline and steel industries take that route. Firms just happen to charge the same price or very close to the same price. In some cases, the leadership rotates, with first one firm, and then the next, assuming the leadership role.

When one firm raises its price, what do the others do?

Cartels and Technological Change Even if all firms in the industry cooperate, other firms, unless they are prevented from doing so, can always enter the market with a technologically superior new product at the same price or with the same good at a lower price. It is important to remember that technological changes are constantly occurring, and that a successful cartel with high profits will provide significant incentives for technological change, which can eliminate demand for its monopolized product.

Kinked Demand: Why Are Prices Sticky? Informal collusion happens all the time in oligopolies. One characteristic of informal collusive behaviour is that prices tend to be sticky.

Firms have certain expectations of other firms' reactions. Specifically, if a firm increases its price, and it believes that other firms won't also raise price, its perceived demand curve for increasing price will be very elastic (D_1 in Figure 13-3). It will lose lots of business to the other firms that haven't raised their prices. The relevant portions of its demand curve and its marginal revenue curve are shown in blue in Figure 13-3.

If it decreased its price, however, the firm assumes that all other firms would immediately match that decrease, so it would gain very few, if any, additional sales. A large fall in price would result in only a small increase in sales, so its demand is very inelastic (D_2 in Figure 13-3). This less elastic portion of the demand curve and the corresponding marginal revenue curve are shown in red in Figure 13-3. Given the firm's expectations about whether rivals will match its price increases (no) or decreases (yes), the relevant portions of D_1 and D_2 form the firm's effective demand curve, which is "kinked" as a result of rival firms' expected behaviour.

Q-7 Is the demand curve as perceived by an oligopolist likely to be more or less elastic for a price increase or a price decrease?

Notice that when you put these two curves together you get a rather strange demand curve (it's kinked) and an even stranger marginal revenue curve (one with a gap). When the demand curve has a kink, the marginal revenue curve must have a gap.

When the demand curve has a kink, the marginal revenue curve must have a gap.

If firms do indeed perceive their demand curves to be kinked at the market price, we have another explanation of why prices tend to be sticky. In Figure 13-3, if we visually "complete" the marginal revenue curve by joining the vertical segment from *c* to *d*, firms following the profit-maximizing rule of producing where $MR = MC$ all produce the same quantity, *Q*, and all charge the same price, *P*, if their MC lies between points *c* and *d*. Shifts in marginal cost (such as MC_0 to MC_1) will not change the firm's profit maximization position. Usually lower cost producers will charge a lower price, thereby

Figure 13-3 THE KINKED DEMAND CURVE

One explanation of why prices are sticky is that firms face a kinked demand curve. When we draw the relevant marginal revenue curve for the kinked demand we see that the corresponding *MR* curve is discontinuous. It has a gap in it. Shifts in marginal costs between c and *d* will not change the price or the output that maximizes profits.

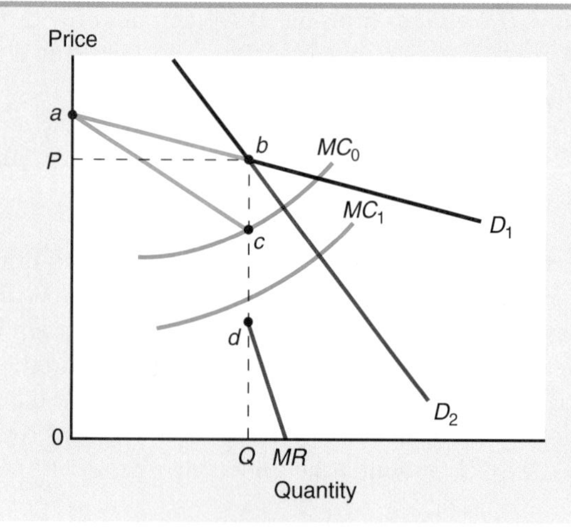

providing some discipline in a market. Here, this doesn't happen. A large shift in marginal cost is required before firms will change their price. Why should this be the case? The intuitive answer lies in the reason behind the kink. If the firm raises its price, other firms won't go along, so it will lose lots of market share. However, when the firm lowers price, other firms will go along and the firm won't gain market share. Thus, the firm has strong reasons not to change its price in either direction.

We should emphasize that the kinked demand curve is not a theory of oligopoly pricing. It does not say why the original price is what it is; the kinked demand curve is simply a theory of sticky prices.

The Contestable Market Model

A second model of oligopoly is the *contestable market model*. The **contestable market model** is *a model of oligopoly in which barriers to entry and barriers to exit, not the structure of the market, determine a firm's price and output decisions*. Thus, it places the emphasis on entry and exit conditions, and says that the price that an oligopoly will charge will exceed the cost of production only if new firms cannot exit and enter the market. The higher the barriers, the more the price exceeds cost. Without barriers to entry or exit, the price an oligopolist sets will be equivalent to the competitive price. Thus, an industry that structurally looks like an oligopoly because there are only 5 firms, may price competitively if barriers to entry are low and entry is likely.

In the contestable market model of oligopoly, pricing and entry decisions are based only on barriers to entry and exit, not on market structure. Thus, even if the industry contains only one firm, it could still be a competitive market if entry is open.

Comparison of the Contestable Market Model and the Cartel Model

Because each oligopolistic industry is different, strategies of oligopolies differ, and no one "oligopolistic model" will explain all oligopolies. Oligopolies with a stronger ability to collude are able to closely approximate the monopolist solution. Equilibrium of oligopolies with less ability to prevent new entry come close to the perfectly competitive solution.

An oligopoly model can take two extremes: (1) the cartel model, in which the members of the oligopoly collude and set a monopoly price; and (2) the contestable market model, in which an oligopoly with no barriers to entry sets a competitive price. Thus, we can say that an oligopoly's price will be somewhere between the competitive price and the monopolistic price. Other models of oligopolies give results in between these two.

Q-8 What are the two extremes an oligopoly model can take?

Much of what happens in oligopoly pricing is highly dependent on the specific legal structure within which firms interact. In Japan, where large firms are specifically allowed to collude, we see Japanese goods selling for a much higher price than those same Japanese goods sell for in North America. From the behaviour of Japanese firms, we get a sense of what pricing strategy Canadian oligopolists would follow in the absence of the restrictions placed on them by law.

Strategic Pricing and Oligopoly

Notice that both the cartel model and the contestable market model use **strategic pricing** decisions—*firms set their price based on the expected reactions of other firms*. Strategic pricing and interdependence are central characteristics of oligopoly.

Strategic pricing and interdependence are central characteristics of an oligopoly.

One can see the results of strategic decision making all the time. For example, consider a firm that announces that it will not be undersold—that it will match any competitor's lower price and will even go under it. Is that a pro-competitive strategy, leading to a low price? Or is it a strategy to increase collusive information and thereby prevent other firms from breaking implicit pricing agreements? Recent work in economics suggests that it is the latter.

Let's now see how a specific consideration of strategic pricing decisions shows that the cartel model and the contestable market model are related.

New Entry as a Limit on the Cartelization Strategy One of the things that limits oligopolies from acting as a cartel is the threat from outside competition. In many industries, the outside competition often comes from international firms. A cartel with few barriers to entry faces a long-run demand curve that's very elastic. This means that its price will be very close to its marginal cost and average cost. This is the same prediction that came from the contestable market theory.

Price Wars Whenever there's strategic decision making, there's the possibility of a price war. Price wars are the result of strategic pricing decisions breaking down. Thus, in any oligopoly it's possible that firms can enter into a price war where prices will fall below average total cost.

The reasons for such wars are varied. Since oligopolistic firms know their competitors, they can personally dislike them; sometimes a firm's goal can be simply to drive a disliked competitor out of business, even if that process hurts the firm itself. Interpersonal and interfirm relations are important in oligopoly.

Alternatively, a firm might follow a predatory pricing strategy—a strategy of pushing the price temporarily down below cost to drive the other firm out of business. If the predatory pricing strategy is successful, the firm can charge an even higher price and earn even greater long term profit because potential entrants know that the existing firm will drive them out if they try to enter. It's this continual possibility that strategies can change that makes oligopoly prices so hard to predict.

GAME THEORY, OLIGOPOLY, AND STRATEGIC DECISION MAKING

The inability to come to an explicit conclusion about what price and quantity an oligopoly will choose doesn't mean that economic reasoning and principles don't apply to oligopoly. They do. Most oligopolistic strategic decision making is carried out with the implicit or explicit use of **game theory** *(the application of economic principles to interdependent situations)*. Game theory is economic reasoning applied to decision making.[2]

An example of game theory is the **prisoner's dilemma,** *a well-known game that demonstrates the difficulty of cooperative behaviour in certain circumstances*. The standard prisoner's dilemma can be seen in the following example: Two suspects are caught and are interrogated separately. Each prisoner is offered the following options:

- If neither prisoner confesses, each will be given a 6-month sentence on a minor charge.
- If one prisoner confesses and the other does not, the one who confesses will go free and the other will be given a 10-year sentence.
- If they both confess, they'll each get a 5-year sentence.

In the prisoner's dilemma, where mutual trust gets each one out of the dilemma, confessing is the rational choice.

What strategy will each choose? Clearly, the best option is for both to keep quiet and receive the shortest sentence. Or is it? Given the uncertainty of the situation, this cooperative strategy does not dominate. The incentive to confess (go free) is strong, and

[2]If you've seen the movie *A Beautiful Mind*, this will sound familiar to you.

since neither prisoner can trust the other not to confess, the optimal strategy (the one that maximizes expected benefits) will be for each to confess, because each must assume the other will do the same. Confessing is the rational thing for each prisoner to do. That's why it's called the *prisoner's dilemma*. Trust gets one out of the prisoner's dilemma. If the prisoners can trust each other, the optimal strategy is not to confess, and they both get only a light sentence. But trust is a hard commodity to come by without an explicit enforcement mechanism. In trying to avoid the 10 year sentence, the prisoners settle for the 5-year sentence each.

Prisoner's Dilemma and a Duopoly Example

The prisoner's dilemma has its simplest application to oligopoly when the oligopoly consists of only two firms. So let us consider the strategic decisions facing a "foam peanut" (packing material) **duopoly**—*an oligopoly with only two firms*. Let us assume that the average total cost and marginal cost of producing foam peanuts are the same for both firms, and are such that only two firms can exist in the industry. These costs are shown in Figure 13-4(a).

Assume that a production facility with a minimum efficient scale of 4,000 tons is the smallest that can be built. In Figure 13-4(b), the marginal costs are summed and the industry demand curve is drawn. The competitive price is $500 per ton and the competitive output is 8,000 tons. The relevant industry marginal revenue curve is also drawn.

If the firms collude fully, they will act as a joint monopolist setting total output at 6,000 tons where $MR = MC$ (3,000 tons each). This gives them a price of $600 with a cost of $575 per ton, for a joint economic profit of $150,000, or $75,000 each. The firms prefer this equilibrium to the competitive equilibrium where they earn zero economic profit.

If they can ensure that they will both abide by the agreement, the monopolist output will be the joint profit-maximizing output. But what if one firm cheats? What if one

Figure 13-4 (a and b) FIRM AND INDUSTRY DUOPOLY COOPERATIVE EQUILIBRIUM

In (**a**) we show the marginal and average total cost curve for either identical firm in the duopoly. Thus, to get the average and marginal cost for the industry, double each. In (**b**) the industry marginal cost curve (the horizontal sum of the individual firms' marginal cost curves) is combined with the industry demand and marginal revenue curves. At the competitive solution for the industry, output is 8,000 and price is $500. As you can see in (**a**), at that price economic profits are zero. At the monopolistic solution, output is 6,000 and price is $600. As you can see in (**a**), *ATC* is $575 at an industry output of 6,000 (firm output of 3,000), so each firm's profit is $25 × 3,000 = $75,000 (the shaded area in (**a**)).

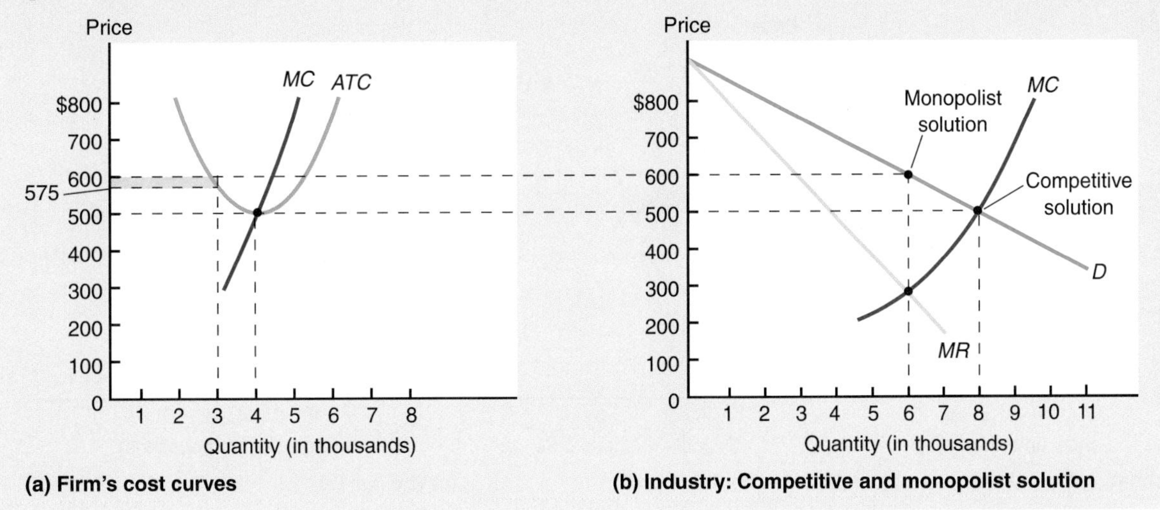

firm produces 4,000 tons (cheats by 1,000 tons)? The additional 1,000 tons in output will cause the price to fall to $550 per ton. The cheating firm's average total costs fall to $500 as its output rises to 4,000, so its profit rises to $200,000. The noncheating firm's profit moves in the opposite direction. Its average total costs remain $575, but the price it receives falls to $550, so it loses $75,000 instead of making $75,000. This gives it a large incentive to cheat also. The division of profits and output split is shown in Figure 13-5. If the noncheating firm decides to become a cheating firm, it eliminates its loss and the other firm's profit, and the duopoly moves to a zero profit position.

In Figure 13-5(a), you can see that the firm that abides by the agreement and produces 3,000 units makes a loss of $75,000; its average total costs are $575 and the price it receives is $550. In Figure 13-5(b) you can see that the cheating firm makes a profit of $200,000; its average costs are $500, so it is doing much better than when it did not cheat. The combined profit of the cheating and the noncheating firms is $200,000 − $75,000 = $125,000, which is lower than if they cooperated. By cheating, the firm has essentially transferred $125,000 of the other firm's profit to itself and has reduced their combined profit by $25,000. Figure 13-5(c) shows how output is split between the two firms. If both firms cheat, the equilibrium output moves to the competitive output, 8,000, and both of the firms make zero profit.

It is precisely to provide insight into this type of strategic situation that game theory was developed. It does so by analyzing the strategies of both firms under all circumstances

Figure 13-5 (a, b, and c) **FIRM AND INDUSTRY DUOPOLY EQUILIBRIUM WHEN ONE FIRM CHEATS**

This Figure demonstrates the three different outcomes. Figures **(a)** and **(b)** show the noncheating and the cheating firms' output and profit, respectively, while **(c)** shows the industry output and price.

If they both cheat on their collusive agreement to restrict output, the price is $500 and output is 8,000 (4,000 per firm) (point *A* in **(a, b,** and **c)**). Both firms make zero profit since their average total costs of $500 equal the price they receive.

If neither cheats, the industry output is 6,000, the price is $600, and their *ATC* is $575. This outcome gives them a profit of $75,000 each and would place them at point C in **(c)**. This outcome was considered in Figure 13-4.

If one firm cheats and the other does not, the output is 7,000 and the industry price is $550 (point *B* in **(c)**). The noncheating firm's loss is shown by the shaded area in **(a)**; its costs are $575, its output is 3,000, the price it receives is $550, and its loss is $75,000. The cheating firm's profit is shown by the shaded area in **(b)**. Its average total costs are $500, the price it receives is $550, and its output is 4,000, so its profit is $200,000. So if one firm is cheating, it pays to be that firm; it doesn't pay to be honest when the other firm cheats.

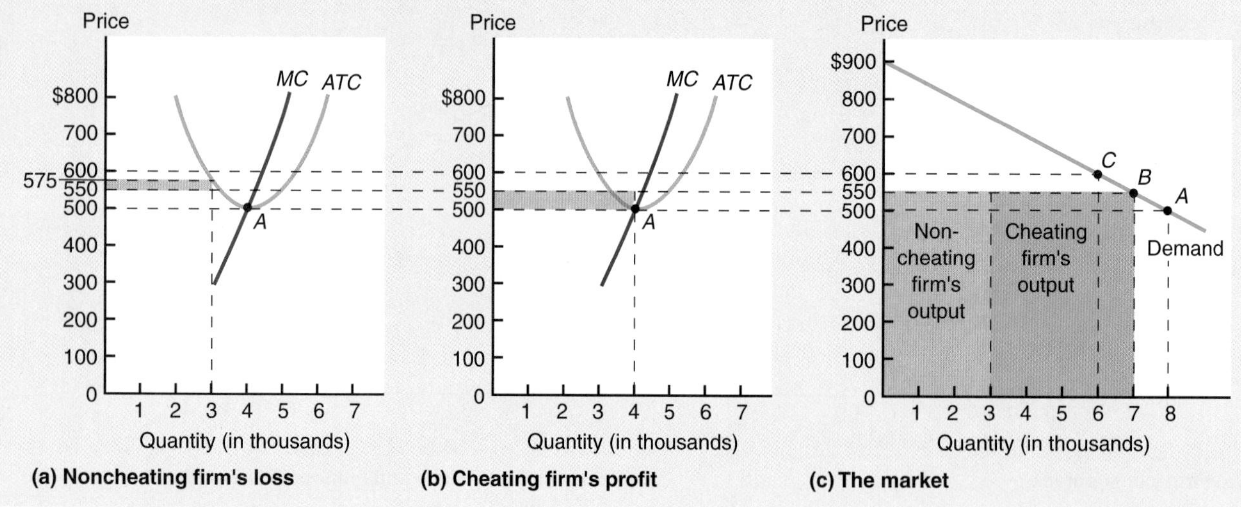

and placing the combination in a payoff matrix—a box that contains the outcomes of a strategic game under various circumstances.

Duopoly and a Payoff Matrix

The duopoly presented above is a variation of the prisoner's dilemma game. The results can be presented in a payoff matrix that captures the essence of the prisoner's dilemma. In Figure 13-6, each square shows the payoff from a pair of decisions listed in the columns and rows. The blue triangles show A's profit; the gold triangles show B's profit. For example, if neither cheats, the result for both is shown in the upper left-hand square, and if they both cheat, the result is shown in the lower right-hand square.

Notice the dilemma they are in if detecting cheating is impossible. If they can't detect whether the other one cheated and each believes the other is maximizing profit, each must expect the other one to cheat. But if firm A expects firm B to cheat, the relevant payoffs are in the second row. Given this expectation, if firm A doesn't cheat, it loses $75,000. So firm A's optimal strategy is to cheat. Similarly for firm B. If it expects firm A to cheat, its relevant payoffs are in the second column. Firm B's optimal strategy is to cheat. But if they both cheat they end up in the lower right-hand square with zero profit.

In reality, of course, cheating is partially detectable, and even though explicit collusion is illegal in Canada, implicit collusion is not. Moreover, in markets where similar conditions hold for some length of time, the cooperative solution is more likely, since each firm will acquire a reputation based on its past actions, and firms can retaliate against other firms that cheat. But the basic dilemma remains for firms and tends to push oligopolies toward a zero profit competitive solution.

Oligopoly Models, Structure, and Performance

Structure refers to the number, size, and interrelationship of firms in an industry. A monopoly (one firm) is considered the least competitive; perfectly competitive industries (an almost infinite number of firms) are considered the most competitive. Classification by structure is easy and accords nicely with intuition. The cartelization model fits best with this classification system because it assumes the structure of the market (the number of firms) is directly related to the price a firm charges. It predicts the behaviour that oligopolies charge higher prices than do monopolistic competitors.

Figure 13-6 THE PAYOFF MATRIX OF STRATEGIC PRICING DUOPOLY

The strategic dilemma facing each firm in a duopoly can be shown in a payoff matrix that captures the four possible outcomes. **A**'s strategies are listed horizontally; **B**'s strategies are listed vertically. The payoffs of the combined strategies for both firms are shown in the four boxes of the matrix, with **B**'s payoff shown in the gold shaded triangles and **A**'s payoff shown in the blue shaded triangles. For example, if **A** cheats but **B** doesn't, **A** makes a profit of $200,000 but **B** loses $75,000.

Their combined optimal strategy is to cartelize and achieve the monopoly payoff, with both firms receiving a profit of $75,000. However, each must expect that if it doesn't cheat and the other does cheat, it will lose $75,000. To avoid losing that $75,000, both firms will cheat, which leads them to the payoff in the lower right-hand corner—the competitive solution with zero profit for each firm.

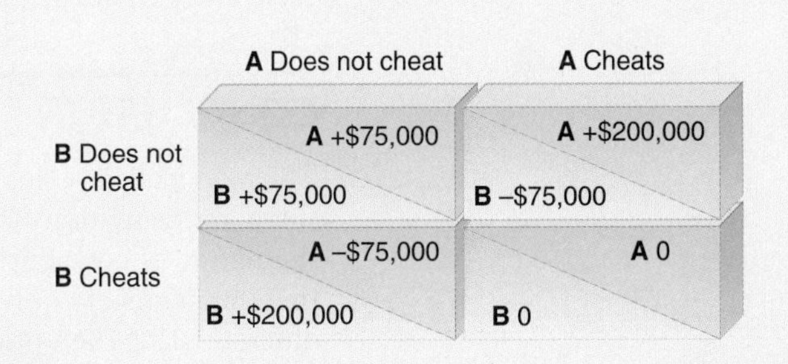

BEYOND THE TOOLS

Game Theory and Experimental Economics

Game theory has offered significant insight into the structure of economic problems but arrives at the conclusion that a number of alternative solutions are possible. A branch of economics—*experimental economics*—offers insight into which outcome will occur.

When game theorists have done experiments, generally they have found that people believe that the others in the game will work toward a cooperative solution. Thus, when the gains from cheating are not too great, often people do not choose the individual profit-maximizing position, but instead choose a more cooperative strategy, at least initially. Such cooperative solutions tend to break down, however, as the benefits of cheating become larger. Additionally, the larger the number of participants gets, the less likely it is that the cooperative solution will be chosen and the more likely it is that competitive solutions will be chosen.

Experimental economists have also found that the structure of the game plays an important role in deciding the solution. For example, posted-price markets, in which the prices are explicitly announced, are more likely to reach a collusive result than are non-posted or uncertain-price markets, where actual sale prices are not known.

Q-9 The Herfindahl index is 1,500. Using a contestable market approach, what would you conclude about this industry?

The contestable market model gives far less weight to market structure. According to it, markets that structurally look highly oligopolistic could actually be highly competitive—much more so than markets that structurally look more competitive. This contestable market model view judges markets by performance, not by structure. Performance includes such results as ratio of price to marginal cost, output, allocative and productive efficiency, product variety, innovation rate, and profits.

To see the implications of the contestable market approach, let's consider an oligopoly with a four-firm concentration ratio of 60 percent and a Herfindahl index of 1,500. Using the structural approach we would say that, because of the importance of interdependance in oligopoly, it's not quite certain what price firms in this industry would charge, but it seems reasonable to assume that there would be some implicit collusion and that the price would be closer to a monopolist price than to a competitive price. If that same market had a four-firm concentration ratio of 30 percent and a Herfindahl index of 700, the structural approach would suggest that the industry would be more likely to have a competitive price.

Q-10 The Herfindahl index is 1,500. Using a structural analysis of markets approach, what would you conclude about this industry?

A contestable market model advocate would disagree, arguing that barriers to entry and exit are what's important. If no significant barriers to entry exist in the first case but significant barriers to entry exist in the second case, the second case would be less competitive than the first. There's a similarity, however, in the two views, structure versus performance. Often barriers to entry are the reason there are only a few firms in an industry. And when there are many firms, it tends to be because that industry has only a few barriers to entry. In most situations, the two approaches come to the same conclusion.

CONCLUSION

The monopolistic competition market structure is similar to perfect competition in that there are many firms competing, and entry is easy. The difference, however, lies in the nature of competition. Perfectly competitive firms sell an identical product and therefore compete only on the basis of price. Monopolistic competitors compete by differentiating their products; this is the source of their (small) monopoly power. So, monopolistic competition is a blending of perfect competition and monopoly elements.

Oligopoly is much more like monopoly; however, rather than there being only one firm, there are a few very large firms. As one would imagine, dividing the market amongst four or five firms means that each firm wields considerable market power. Each

KNOWING THE TOOLS

A Comparison of Various Market Structures

Structure / Characteristics	Monopoly	Oligopoly	Monopolistic competition	Perfect competition
Number of firms	1	Few	Many	Almost infinite
Barriers to entry	Significant	Significant	Few	None
Pricing decisions	$MC = MR$	Strategic pricing, between monopoly and perfect competition	$MC = MR$	$MC = MR = P$
Output decisions	Most output restriction	Output somewhat restricted	Output restricted somewhat by product differentiation	No output restriction
Interdependence	Only firm in market, not concerned about competitors	Interdependent strategic pricing and output decision	Each firm acts independently	Each firm acts independently
Profit	Possibility of long-run economic profit	Some long-run economic profit possible	No long-run economic profit possible	No long-run economic profit possible
P* and *MC	$P > MC$	$P > MC$	$P > MC$	$P = MC$

firm certainly has less market power than a monopolist, but the firms together have substantial control over market price and quantity. Oligopolistic firms recognize the size and strength of their rivals, and this recognition affects an oligopolistic firm's decision making process.

The four market structures — perfect competition, monopolistic competition, oligopoly, and monopoly — are generally used to discuss the competitive behaviour of firms and the price and output performance of markets. Although all industries can be placed into one of the four market structures, some industries can be very difficult to categorize. In determining public policy, the performance of markets is often compared to the perfectly competitive outcome, although the conditions for perfect competition are not met by the market in question. In addition to price and output, other policy questions regarding product variety, innovation and technology adoption, and allocative efficiency are also important. However, the market structure framework provides a good starting point for analyzing real-world markets.

Chapter Summary

- Industries are classified by economic activity in the North American Industry Classification System (NAICS). Industry structures are measured by concentration ratios and Herfindahl indexes.
- A concentration ratio is the sum of the market shares of the firms with the largest market shares in a particular industry.
- A Herfindahl index is the sum of the squares of the individual market shares of all firms in an industry.
- Conglomerates operate in a variety of different industries. Industry concentration measures do not assess the bigness of these conglomerates.
- Monopolistic competition is characterized by (1) many sellers, (2) differentiated products, (3) multiple dimensions of competition, and (4) ease of entry for new firms.
- The central characteristic of oligopoly is that there are a small number of interdependent firms. The products can be homogeneous or differentiated.
- In monopolistic competition firms act independently; in an oligopoly they take into account each other's actions.
- Monopolistic competitors differ from perfect competitors in that monopolistically competitive firms face a highly elastic but downward-sloping demand curve.
- A monopolistic competitor differs from a monopolist in that a monopolistic competitor makes zero economic profit in long-run equilibrium.
- An oligopolist's price will be somewhere between the competitive price and the monopolistic price.
- Game theory and the prisoner's dilemma can explain strategic pricing decisions.
- A contestable market theory of oligopoly judges an industry's competitiveness more by performance and barriers to entry than by structure. Cartel models of oligopoly concentrate on market structure.

Key Terms

cartel *(287)*
cartel model of oligopoly *(287)*
concentration ratio *(279)*
contestable market model *(288)*
duopoly *(291)*
game theory *(290)*
Herfindahl index *(279)*
implicit collusion *(287)*
market structure *(277)*
monopolistic competition *(278)*
North American Industry Classification System (NAICS) *(278)*
oligopoly *(278)*
prisoner's dilemma *(290)*
strategic decision making *(280)*
strategic pricing *(289)*

Questions for Thought and Review

1. Which industry is more highly concentrated: one with a Herfindahl index of 1,200 or one with a four-firm concentration ratio of 55 percent?
2. What are the ways in which a firm can differentiate its product from that of its competitors? What is the overriding objective of product differentiation?
3. What are the "monopolistic" and what are the "competitive" elements of monopolistic competition?
4. Does the product differentiation in monopolistic competition make us better or worse off? Explain?
5. Both a perfect competitor and a monopolistic competitor choose output where $MC = MR$, and neither makes a profit in the long run. Why, then, does the monopolistic competitor produce less than a perfect competitor?
6. If a monopolistic competitor is able to restrict output, why doesn't it earn economic profits?
7. What are some of the barriers to entry in the restaurant industry? In the automobile industry?
8. Is an oligopolist more or less likely to engage in strategic pricing compared to a monopolistic competitor?
9. What is the difference between the contestable market model and the cartel model of oligopoly? How are they related?
10. Is a contestable model or cartel model more likely to judge an industry by performance? Explain your answer.
11. What did Adam Smith mean when he wrote, "Seldom do businessmen of the same trade get together but that it results in some detriment to the general public"?
12. Describe a situation you have faced in your lifetime that can be characterized as a prisoner's dilemma.
13. In the late 1990s, Kellogg's, which controlled 32 percent of the breakfast cereal market, cut the prices of some of its best-selling brands of cereal to regain market share lost to Post, which controlled 20 percent of the market. General Mills had 24 percent of the market. The price cuts were expected to trigger a price war. Based on this information, what market structure best characterizes the market for breakfast cereal?

Problems and Exercises

1. Suppose a monopolistic competitor has a constant marginal cost of $6 and faces the demand curve given in the following table:

Q	20	18	16	14	12	10	8	6
P	$ 2	4	6	8	10	12	14	16

   a. What output will the firm choose?
   b. What will be the monopolistic competitor's average fixed cost at the output it chooses?

2. A firm is convinced that if it lowers its price, no other firm in the industry will change price; however, it believes that if it raises its price, some other firms will match its increase, making its demand curve more inelastic. The current price is $8 and its marginal cost is constant at $4.
   a. Sketch the general shape of the firm's *MR*, *MC*, and demand curves.
   b. If the marginal cost falls to $3, what would you predict would happen to price?
   c. If the marginal cost rises to $5, what would you predict would happen to price?

3. You're the manager of a firm that has constant marginal cost of $6. Fixed cost is zero. The market structure is monopolistically competitive. You're faced with the following demand curve:

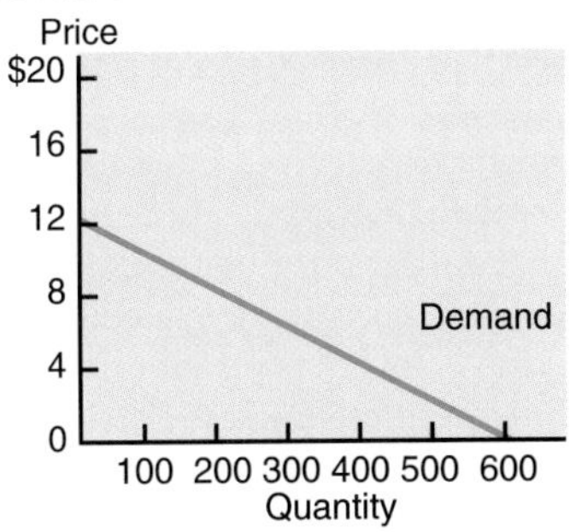

   a. Determine graphically the profit-maximizing price and output for your firm in the short run. Demonstrate what profit or loss you'll be making.
   b. Do the same for the long run.
   c. Thanks to a technological innovation, you have zero marginal cost. Demonstrate the new profit-maximizing price and output in the short run. Demonstrate graphically the short-run profit at that new profit-maximizing output.

4. The pizza market is divided as follows:

Pizza Hut	20.7%
Domino's	17.0
Little Caesars	6.7
Pizzazz	2.2
Triangles Pizza Takeout	2.0
All others	51.4

   a. How would you describe its market structure?
   b. What is the approximate Herfindahl index?
   c. What is the four-firm concentration ratio?

5. The retail grocery store market in Atlantic Canada is divided as follows:

Empire/Sobey's	34.5%
Loblaws	33.3%
Atlantic Coop	10.8%
Other grocery stores	2.4%
Other food distribution channels	18.9%

   a. How would you describe this market structure?
   b. Calculate the CR3 for the industry. Would the CR4 be much different? Explain.
   c. Loblaws focusses on its own private label products rather than carrying a traditional selection of national brands such as the other stores offer. What strategy is Loblaws following? What is the ultimate effect on competition in the market? Does your answer to (a) change as a result of knowing this?

6. Two firms, TwiddleDee and TwiddleDum, make up the entire market for wodgets. They have identical costs. They are currently colluding explicitly and are making $2 million each. TwiddleDee has a new CEO, Mr. Notsonice, who is considering cheating. He has been informed by his able assistant that if he cheats he can increase the firm's profit to $3 million, but that cheating will reduce TwiddleDum's profit to $1 million. You have been hired to advise Mr. Notsonice.
   a. Construct a payoff matrix for him that captures the essence of the decision.
   b. If the game is only played once, what strategy would you advise?
   c. How would your answer to *b* change if the game were to be played many times?
   d. What change in the profit made when colluding (currently $2 million) would be needed to change your advice in *b*?

7. In 1993, the infant/preschool toy market four-firm concentration ratio was 72 percent. With 8 percent of the market, Mattel was the fourth largest firm in that market. Mattel proposed to buy Fisher-Price, the market leader with 27 percent.
   a. Why would Mattel want to buy Fisher-Price?
   b. What arguments can you think of in favour of allowing this acquisition?
   c. What arguments can you think of against allowing this acquisition?
   d. How do you think the four-firm concentration ratio for the entire toy industry would compare to this infant/preschool toy market concentration ratio?

Web Questions

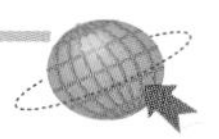

1. Canada, the United States, and Mexico have a system for categorizing production, the North American Industry Classification System (NAICS). Go to the Statistics Canada Web site at www.statcan.ca and use the search engine to research NAICS. Click on www.statcan.ca/english/concepts/SNA/naics.htm and www.statcan.ca/english/subjects/std/prefac.htm to find the answers to these questions:
   a. The classification system is supply-based. What does this mean, and why was a supply-based classification system chosen?
   b. Will the new classification system be more or less helpful than the SIC (Standard Industrial Classification) previously used in Canada for describing market structure?
   c. Describe the numbering system for NAICS. Which categories are new?
   d. In what ways does NAICS more suitably represent Canada's new economy? What does NAICS consider to be the major input in its Professional, Scientific, and Technical Services sector? Why?
2. The Organization of Petroleum Exporting Countries (OPEC) is an international cartel. Go to its home page at www.opec.org to answer the following questions:
   a. What are OPEC's objectives? How does it meet those objectives?
   b. What countries are members of OPEC? What percentage of world oil production comes from these nations? In what way is OPEC a cartel?
   c. What significant oil-exporting countries are not members? What has OPEC done to limit the effect of nonmember production on its pricing decisions?

Answers to Margin Questions

1. The smaller the number of digits, the more inclusive the classification. Therefore, the two-digit industry would have significantly more output. *(278)*
2. The highest Herfindahl index for this industry would occur if one firm had the entire 60 percent, and all other firms had an infinitesimal amount, making the Herfindahl index slightly over 3,600. The lowest Herfindahl index this industry could have would occur if each of the top four firms had 15 percent of the market, yielding a Herfindahl index of 900. *(280)*
3. We would respond that monopolistic competitors, by definition, do not take into account the expected reactions of competitors to their decisions; therefore, they cannot use strategic decision making. We would tell Jean she probably meant, "*Oligopolies* use strategic decision making." *(280)*
4. Both a monopoly and a monopolistic competitor produce where marginal cost equals marginal revenue. The difference is in the positioning of the average total cost curve. For a monopolistic competitor, that average total cost curve must be tangent to the demand curve because a monopolistic competitor makes no profits in the long run. A monopoly can make profits in the long run, so its average total cost can be below the price. *(283)*
5. Monopolistically competitive firms advertise because their products are differentiated from others. Advertising can convince people that a firm's product is better than that of other firms and increase the demand curve it faces. Perfect competitors, in contrast, have no incentive to advertise since their products are the same as every other firm's product and they can sell all they want at the market price. *(286)*
6. Maintaining a cartel requires firms to make decisions that are not in their individual best interests. Such decisions are hard to enforce unless there is an explicit enforcement mechanism, which is difficult in a cartel. *(287)*
7. The demand curve perceived by an oligopolist is more elastic above the current price because it believes that others will not follow price increases. If it increased price, it would see quantity demanded fall by a lot. The opposite is true below the current price. The demand curve below current price is less elastic. Price declines would be matched by competitors and the oligopolist would see little change in quantity demanded with a price decline. *(288)*
8. The two extremes an oligopoly model can take are: (1) a cartel model, which is the equivalent of a monopoly; and (2) a contestable market model, which, if there are no barriers to entry, is the equivalent of a competitive industry. *(289)*
9. The contestable markets approach looks at barriers to entry, not structure. Therefore, we can conclude nothing about the industry from the Herfindahl index. *(294)*
10. In a market with a Herfindahl index of 1,500, the largest firm would have, at most, slightly under 40 percent of the market. The least concentrated such an industry could be would be if seven firms each had between 14 and 15 percent of the market. In either of these two cases, the industry would probably be an oligopolistic industry and could border on monopoly. *(294)*

Globalization, Technology, and Real-World Competition

14

After reading this chapter, you should be able to:

- Define the monitoring problem and state its implications for economics.
- Explain how corporate takeovers can limit X-inefficiency.
- Discuss why competition should be seen as a process, not a state.
- List two actions firms take to break down monopoly and three they take to protect monopoly.
- Explain how globalization and technology have changed competition.
- Discuss why oligopoly is the best market structure for technological change.

It is ridiculous to call this an industry. This is rat eat rat; dog eat dog. I'll kill 'em, and I'm going to kill 'em before they kill me. You're talking about the American way of survival of the fittest.

Ray Kroc (founder of McDonald's)

Earlier chapters presented some nice, neat models, but as discussed in Chapter 13, often these models don't fit reality directly. Real-world markets aren't perfectly monopolistic; they aren't perfectly competitive either. They're somewhere between the two. The monopolistic competition and oligopoly models in Chapter 13 come closer to reality and provide some important insights into the "in-between" markets, but, like any abstraction, they, too, fail to capture important aspects of the actual nature of competition. This chapter discusses two issues that are very much in the news—technology and globalization—and relates these concepts to the models we developed earlier.

When reading this chapter, think about the two uses of competition discussed in Chapter 11: competition as a process, the end state of which is zero profits, and competition as a market structure. In this chapter the focus is on competition as a process—it is a rivalry between firms and between individuals. This competitive process is active in all market forms and is key to understanding real-world competition.

THE GOALS OF REAL-WORLD FIRMS AND THE MONITORING PROBLEM

The best place to start the analysis is with the basic assumption that firms are profit maximizers. There's a certain reasonableness to this assumption; firms definitely are concerned about profit, but are they trying to maximize profit? The answer is: It depends.

Short-Run versus Long-Run Profit

Q-1 What are two reasons why real-world firms are not pure profit maximizers?

The first insight is that if firms are profit maximizers, they aren't just concerned with short-run profit; most are concerned with long-run profit. Thus, even if they can, they may not take full advantage of a potential monopolistic situation now, in order to strengthen their long-run position. For example, many stores have liberal return policies, such as the Sears guarantee, "Satisfaction or your Money Refunded." Similarly, many firms spend millions of dollars improving their reputations. Most firms want to be known as good corporate citizens. Such expenditures on reputation and goodwill can increase long-run profit, even if they reduce short-run profit. (New Internet firms focus almost entirely on the long run, and many of the best known firms have yet to make a profit.)

14.1 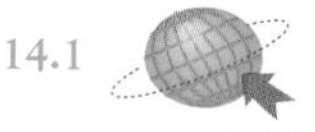
see page 318

The Problem with Profit Maximization

Most real-world production doesn't take place in owner-operated businesses; it takes place in large corporations.

A second insight into how real-world firms differ from the model is that the decision-makers' income is often a cost to the firm. Most real-world production doesn't take place in owner-operated businesses; it takes place in large corporations with eight or nine levels of management, thousands of shareholders whose shares are often held in trust for them, and a board of directors, chosen by management, overseeing the company by meeting two or three times a year. Signing a proxy statement is as close as most stockowners get to directing "their company" to maximize profit.

Managers' Incentives Why is the management structure of the firm important to the analysis? Because economic theory tells us that, unless someone is seeing to it that they do, self-interested decision makers have little incentive to hold down their pay. But their pay is a cost of the firm. And if their pay isn't held down, the firm's profit will be lower than otherwise. Most firms succeed in putting some pressure on managers to make at least a certain level of profit. (If you ask managers, they'll tell you that they face enormous pressure to do so.) So the profit motive certainly plays a role—but to say that profit plays a role is not to say that firms maximize profit. There are enormous wastes and inefficiencies in many businesses.

This structure presents a problem in applying the model to the real world. The general economic model assumes individuals are utility maximizers—that they're motivated

TABLE 14-1 Compensation of CEOs of Selected Companies

The chief executive officers took home an average pay increase of 42.9% for the year 2000. The figures represent direct compensation, which includes salary, bonuses, and all long-term incentives granted in the year. One of the driving forces behind higher Canadian CEO salaries is rising compensation levels in the U.S.

CEO	Company	Direct Compensation
John Roth	Nortel Networks	$70,754,000
Jacques Bougie	Alcan Aluminum	23,863,000
Dominic D'Alessandro	Manulife Financial Ltd.	17,833,000
Richard Harrington	Thomson Corp.	17,291,000
John Hunkin	CIBC	12,153,000
Jim Shaw	Shaw Communications	10,679,000
Charles Baillie	TD Bank	10,013,000
Ted Rogers	Rogers Communications	8,864,000
John Cleghorn	Royal Bank of Canada	8,746,000
Eugene Polistuk	Celestica	8,628,000

Source: Financial Post, May 22, 2001 (page C1).

by self-interest. Then, in the standard model of the firm, the assumption is made that firms, composed of self-interest-seeking individuals, are profit-seeking firms, without explaining how self-interest-seeking individuals who manage real-world corporations will find it in their interest to maximize profit for the firm. Economists recognize this problem. It's an example of the **monitoring problem**—*the need to oversee employees to ensure that their actions are in the best interest of the firm*.

The monitoring problem is that employees' incentives differ from the owner's (shareholders') incentives.

Need for Monitoring Monitoring is required because employees' incentives differ from the owner's incentives, and it's costly to make sure that the employee does the owner's bidding. This is the monitoring problem, a central problem studied by economists who specialize in industrial organization. They analyse internal structures of firms and look for a contract that managers can be given which will align the self-interest of the manager with the interests of the owners of the firm. A successful contract makes the manager's incentives compatible with the profit-maximizing goal of the firm's owners; an **incentive-compatible contract** is one in which *the incentives of each of the two parties to the contract are made to correspond as closely as possible*. Shareholders (owners) of the firm are interested strictly in profit maximizing; however, self-interested managers are interested in maximizing the firm's profit only if the structure of the firm requires them to do so.

Self-interested managers are interested in maximizing firm profit only if the structure of the firm requires them to do so.

When appropriate monitoring doesn't take place, high-level managers can pay themselves very well. The management of some large companies are very well paid and, as can be seen in Table 14-1, many receive multimillion-dollar salaries. But are these salaries too high? That's a difficult question. Most of the high salaries are not pure salaries, but include stock options and bonuses for performance.

One way to get an idea about an answer is to compare Canadian managers' salaries with those in Japan, where the control of firms is different. Banks in Japan have significant control over the operations of firms, and they closely monitor their performance. The result is that, in Japan, high-level managers on average earn about one-quarter of what their Canadian counterparts make, while wages of low-level workers are roughly comparable to those of low-level workers in Canada. Given Japanese companies' success

Q-2 Why would most economists be concerned about third-party payment systems in which the consumer and the payer are different?

in competing with Canadian companies, this suggests that high managerial pay in Canada reflects a monitoring problem inherent in the structure of corporations. There are, of course, other perspectives. Considering what some sports, film, and music stars receive places the high salaries of company managers in a different light.

What Do Real-World Firms Maximize?

If firms don't maximize profit, what do they maximize? What are their goals? The answer again is: it depends.

Real-world firms often have a set of complicated goals that reflect the organizational structure and incentives built into the system. Clearly, profit is one of their goals. Firms spend a lot of time designing incentives to get managers to focus on profit.

Although profit is one goal of a firm, often firms focus on other intermediate goals such as cost and sales.

But often intermediate goals become the focus of firms. For example, many real-world firms focus on growth in sales; at other times they institute a cost-reduction program to increase long-run profit. At still other times they may simply take it easy and not push hard at all, enjoying the position they find themselves in—being what British economist Joan Robinson called **lazy monopolists**—*firms that do not push for efficiency, but merely enjoy the position they are already in*. This term describes many, but not all, real-world corporations. When Robinson coined the term, firms faced mostly domestic competition. Today, with firms facing more and more global competition, firms are a bit less lazy than they were.

MONOPOLY AND INEFFICIENCY

Lazy monopolists are not profit maximizers; they see to it that they make enough profit so that the stockholders aren't squealing, but don't push as hard as they could to hold their costs down. They perform as efficiently as is necessary. The result is what economists call **X-inefficiency** (*firms operating far less efficiently than they technically could*). Such firms hold monopoly positions, but they don't earn large monopoly profits. Instead, their costs rise because of inefficiency; they may simply make a normal level of profit or, if X-inefficiency becomes bad enough, a loss.

Q-3 Why doesn't a manager have the same incentive to hold costs down as an owner does?

The standard model avoids dealing with the monitoring problem by assuming that the owner of the firm makes all the decisions. The owners of firms who receive profit, and only profit, would like to see that all the firm's costs are held down. Unfortunately, very few real-world firms operate that way. In reality owners seldom make the operating decisions. They hire or appoint managers to make those decisions. The managers they hire don't have that same incentive to hold costs down. Therefore, it isn't surprising to many economists that managers' pay is usually high and that high-level managers see to it that they have "perks" such as chauffeurs, jet planes, ritzy offices, and assistants to do as much of their work as possible.

Consider Figure 14-1. A profit-maximizing monopolist would produce at price P_M and quantity Q_M. Average total cost would be C_M, so the monopolist's profit would be the entire shaded rectangle (areas *A* and *B*). The lazy monopolist would allow costs to increase even up to the point where the firm only earned a normal profit. In Figure 14-1, costs rise to C_{LM}. The economic profit of the lazy monopolist is area *B*. The remainder of the potential profit (Area A) is eaten up in cost inefficiencies.

The competitive pressures a firm faces limit its laziness.

What places a limit on firms' laziness is the degree of competitive pressure they face. All economic institutions must have sufficient revenue coming in to cover costs, so all economic institutions have a limit on how lazy and inefficient they can become. They can translate all of the monopoly profit into X-inefficiency (perks for management and

Figure 14-1 TRUE COST EFFICIENCY AND THE LAZY MONOPOLIST

A profit-maximizing monopolist producing efficiently would have costs C_M and would produce at price P_M and quantity Q_M. A lazy monopolist, in contrast, would let costs rise until nearly the minimum level of profit is reached, at C_{LM}. Profit for the monopolist is represented by the entire shaded area, whereas profit for the lazy monopolist is reduced to area *B*.

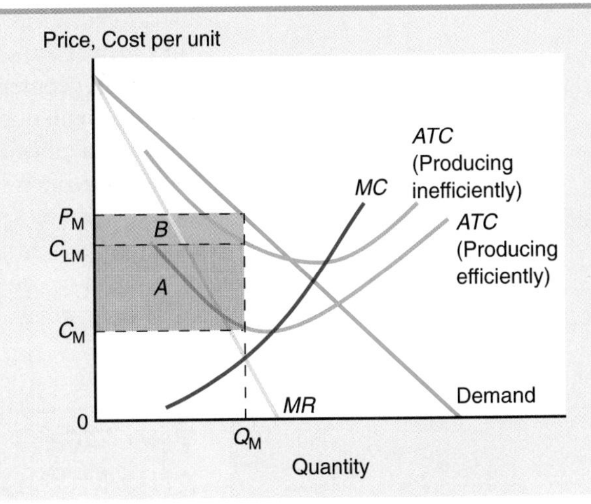

the employees), thereby benefiting the managers and workers in the firm, but once they've done so, they can't be more inefficient or they will go out of business.

How Competition Limits the Lazy Monopolist

If all firms in the industry are lazy, then laziness becomes the norm and competitive pressures don't work to reduce their profits. Laziness is relative, not absolute. But if a new firm enters or if an industry is opened up to international competition, the lazy monopolist can be squeezed and often must undergo massive restructuring to become competitive again. In the 1990's, many Canadian firms underwent such restructuring in order to make themselves internationally competitive.

A second way in which competitive pressure is placed on a lazy monopolist is by a **corporate takeover,** in which *another firm or a group of individuals issues a tender offer (that is, offers to buy up the stock of a company to gain control and to install its own managers)*. Usually such tender offers are financed by large amounts of debt, which means that if the takeover is successful, the firm will need to make large profits just to cover the interest payments on the debt.

A corporate takeover, or simply the threat of a takeover, can improve a firm's efficiency.

Managers don't like takeovers. A takeover may mean losing their jobs and the perks that go along with the jobs, so they have an incentive to restructure the company on their own to reduce the threat of takeover. Such restructuring frequently means incurring large amounts of debt to finance a large payment to stockholders. These payments put more pressure on management to operate efficiently. Thus, the threat of a corporate takeover places competitive pressure on firms to maximize profits.

Q-4 In what way does the threat of a corporate takeover place competitive pressures on a firm?

Were profit not a motive at all, one would expect the lazy monopolist syndrome to take precedence. Thus, it's not surprising that nonprofit organizations often display these lazy monopolist tendencies. For example, some colleges, schools, libraries, and nonprofit hospitals have a number of rules and ways of doing things that, upon reflection, benefit the employees of the institution rather than the customers. Studying these incentive-compatible problems is what management courses are all about.

Motivations for Efficiency Other Than the Profit Incentive

The drive for profit isn't the only drive that pushes for efficiency. Some individuals derive pleasure from efficiently run organizations. Such individuals don't need to be monitored. Thus, if administrators are well intentioned, they'll hold down costs even if they aren't monitored. In such cases, monitoring (creating an organization and structure that gives people incentives which are consistent with profits) can actually reduce efficiency! Some nonprofit organizations operate efficiently. Their success is built on their employees' pride in their jobs, not on their profit motive.

Individuals have complicated motives; some simply have a preference for efficiency.

Most economists don't deny that such inherently efficient individuals exist, and that everyone derives pleasure from doing a good job, but they believe that it's hard to maintain that push for efficiency year in, year out, when some of your colleagues are lazy monopolists enjoying the fruits of your efficiency. Holding down costs without the profit motive, however, takes stronger willpower than many people have.

THE FIGHT BETWEEN COMPETITIVE AND MONOPOLISTIC FORCES

Even if all the assumptions for perfect competition could hold true, it's unlikely that real-world markets would be perfectly competitive. The reason is that perfect competition assumes that individuals accept a competitive institutional structure, even though changing that structure could result in significant gains for sellers or buyers. The simple fact is that *self-interest-seeking individuals don't like competition for themselves* (although they do like it for others), and when competitive pressures get strong and the invisible hand's push turns to shove, individuals often shove back, using either social or political means. That's why you can understand real-world competition only if you understand how the invisible hand, social forces, and political pressures push against each other to create real-world economic institutions. Real-world competition should be seen as a process—a fight between the forces of monopolization and the forces of competition.

When competitive pressures get strong, individuals often fight back through social and political pressures. Competition is a process—a fight between the forces of monopolization and the forces of competition.

Economic Theories and Real-World Competition

Q-5 Name three goals, other than efficiency, that are important to Canadians.

The extreme rarity of perfectly competitive markets *should not* make you think that economics is irrelevant to the real world. Far from it. In fact, the movement away from perfectly competitive markets could have been predicted by economic theory.

Consider Figure 14-2. Competitive markets will exist only if suppliers or consumers don't collude. If the suppliers producing 0*G* can get together and restrict entry, preventing suppliers who would produce *GH* from entering the industry, the remaining suppliers can raise their price from P_H to P_G, giving them the shaded area *A* in additional income. If the cost of their colluding and preventing entry is less than that amount, economic theory predicts that these individuals will collude. The suppliers kept out of the market lose only area *C*, so they don't have much incentive to fight the restrictions on entry. Consumers lose the areas *A* plus *B*, so they have a strong incentive to fight. However, often the consumers' cost and difficulty (given their large numbers) of organizing a protest is higher than the suppliers' cost of collusion, so consumers are forced to accept the restrictions.

How Competitive Forces Affect Monopoly

Q-6 Why is it almost impossible for a perfect monopoly to exist?

Don't think that because perfect competition doesn't exist, competition doesn't exist. In the real world, competition is fierce; the invisible hand is no weakling. It holds its own against other forces in the economy.

Competition is so strong that it makes the other extreme (perfect monopolies) as rare as perfect competition. For a monopoly to last, other firms must be prevented from

Figure 14-2 MOVEMENT AWAY FROM COMPETITIVE MARKETS

In the case where suppliers of 0G can restrict suppliers of GH from entering the market, they can raise the price of the good from P_H to P_G, giving the suppliers of 0G area A in additional income. The suppliers kept out of the market lose area C in income. The consumers, however, lose both areas A and B, giving them strong incentive to fight collusion. Often the costs of organizing consumers are much higher than the costs for the suppliers, so consumers must accept the market restrictions.

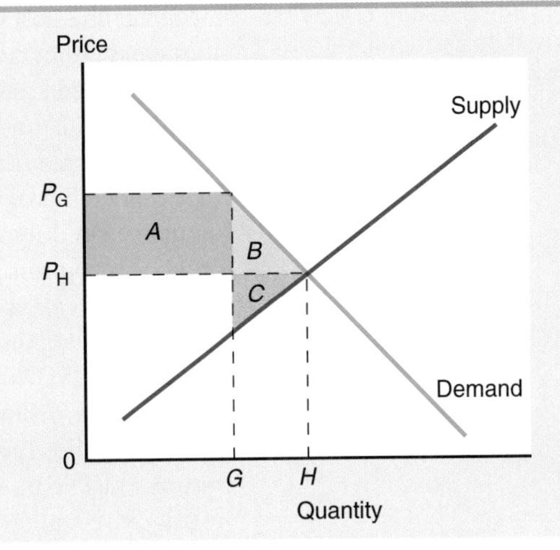

entering the market. In reality it's almost impossible to prevent entry totally, especially over a long period of time, and therefore it's almost impossible for perfect monopoly to exist. Monopoly profits (Area A) signal other firms to enter the industry to obtain some of that profit for themselves.

Breaking Down Monopoly Firms can break down a monopoly through political or economic means. If the monopoly is a legal monopoly, high profit will lead potential competitors to lobby to change the law that maintains that monopoly. If the law can't be changed—say, the monopolist has a **patent** *(a legal right to be the sole supplier of a good)*—potential competitors will generally get around the obstacle by developing a slightly different product or by working on a new technology which is also capable of serving the relevant need.

Establishing an initial presence in a market can be more effective than obtaining a patent when trying to extract monopoly profit.

Say, for example, that you've just discovered the proverbial better mousetrap. You patent it and prepare to enjoy the life of a monopolist. But to patent your mousetrap, you must submit the technical drawings of how your better mousetrap works to the patent office. That gives all potential competitors (some of whom have better financing and already existing distribution systems) a chance to study your idea and see if they can think of a slightly different way (a way sufficiently different to avoid being accused of infringing on your patent) to achieve the same end. They often succeed—so in addition to protecting their ideas with patents, many firms try to establish an initial presence in the market and rely on inertia to protect what little monopoly profit they can extract.

Reverse Engineering Going to the patent office isn't the only way competitors gather information about competing products. One of the other ways routinely used by firms is called **reverse engineering**—*the process in which a firm buys another firm's product, disassembles it, figures out what's special about it, and then copies it within the limits of the law.*

Competition and Natural Monopoly

The view one takes of the fight between competitive and monopolistic forces influences one's view of what government policy should be in relation to natural monopolies—industries whose average total cost is falling as output increases. Natural monopolies can make large profits and consequently there have been significant calls for government regulation of these monopolies to prevent their "exploitation" of the consumer.

New technologies can compete with and undermine natural monopolies.

Over the past decade, as the rate of technological change has accelerated, economists and policymakers have become less supportive of such regulation. They argue that even in these cases of natural monopoly, competition works in other ways. High monopoly profits fund research on alternative ways of supplying the product, such as sending TV signals through electrical lines or sending phone messages by satellite. New technologies provide competition to existing firms. Sometimes, however, when this competition doesn't work fast enough, people direct their efforts toward government, and political pressure is brought to bear either to control the monopoly through regulation or to break up the monopoly.

Regulating Natural Monopolies In the past, the pressure to regulate natural monopolies has been stronger than the competitive pressure that lowers prices. Regulated natural monopolies have been given the exclusive right to operate in an industry but, in return, they've had to agree to have the price they charge and the services they provide regulated. Regulatory boards control the price that natural monopolies charge so that it will be a "fair price," which they generally define as a price that includes all costs plus a normal return on capital investment (a normal profit, but no economic profit).

When firms are allowed to pass on all cost increases to earn a normal profit on those costs, they have little or no incentive to hold down costs.

When firms are allowed to pass on all cost increases to earn a normal profit on those costs, they have little or no incentive to hold down costs. In such cases, X-inefficiency develops, and such monopolies look for capital-intensive projects that will increase their rate bases. To fight such tendencies, regulatory boards must screen every cost and determine which costs are appropriate and which aren't—an almost impossible job. Once regulation gets so specific that it's scrutinizing every cost, the regulatory process becomes extremely bureaucratic, which itself increases the cost. Moreover, to regulate effectively, the regulators must have independent information and must have a sophisticated understanding of economics, cost accounting, and engineering. In most cases, the regulatory board relies on cost data provided by the regulated firm. As is often the case in economics, there's no easy answer to the problem. Because of the problems with regulation, more and more economists argue that even in the case of a natural monopoly, no regulation is desirable, and that society would be better off relying on direct competitive forces.

Q-7 What is the problem with regulations that set prices relative to costs?

Deregulating Natural Monopolies In the 1980s and 1990s such views led to the deregulation and competitive supply of both electric power and telephone services. Regulators are making these markets competitive by breaking down the layers of the industry into subindustries and deregulating those subindustries that can be competitive. For example, the telephone industry can be divided into the telephone line and switching equipment industry, the residential (and business) equipment industry, the pay phone industry, and the directory information industry. By dividing up the industry, regulators can carve out that part that has the characteristics of a natural monopoly and open the remaining parts to competition.

Let's take a closer look at the electric industry. It used to be that electricity was supplied by independent firms, each providing electricity for its own customers. Today, however, electricity is supplied through a large grid that connects to many regions of the country. (This grid was developed to provide backup power to different sections of the country.) Given this grid, electricity generated in one area can easily be sent all over the country. Now that many suppliers can compete for customers, a reasonable competition in power supply is feasible.

While the natural monopoly sectors remain under regulation, the competitive sectors of a natural monopoly can be deregulated.

The power line industry, however, cannot be competitive. It would be extremely costly for each company to run a separate power line into your house. That is, the power line industry exhibits *economies of scale*. Because of the economies of scale, the power line industry is the natural monopoly aspect of electrical power supply. The deregulation of electricity involves splitting off the production of electricity from the maintenance of the line—and choosing an appropriate charge for electric line maintenance. When newspapers report that the electrical power industry is being deregulated, that is not quite correct: only those portions of the market where competition can exist are being deregulated.

Economies of scale can create natural monopolies.

HOW FIRMS PROTECT THEIR MONOPOLIES

The image of competition being motivated by profits is a useful one. It shows how a market economy adjusts to ever-changing technology and demands in the real world. Competition is a dynamic, not a static, force.

Firms do not sit idly by and accept competition. They fight it. How do monopolies fight real-world competition? By spending money on maintaining their monopoly. By advertising. By lobbying. By producing products that are difficult to copy. By not taking full advantage of their monopoly position, which means charging a low enough price to discourage entry. Often firms could make higher short-run profits by charging a higher price, but they forgo the short-run profits in order to strengthen their long-run position in the industry.

Firms do not sit idly by and accept competition. They fight it.

Cost-Benefit Analysis of Creating and Maintaining Monopolies

Preventing real-world competition costs money. Monopolies are expensive to create and maintain. Economic theory predicts that if firms have to spend money on creating and protecting their monopoly, they're going to "buy" less monopoly power than if it were free. How much will they buy? They will buy monopoly power until the marginal cost of such power equals the marginal benefit. Thus, they'll reason:

Q-8 What decision rule does a firm use when deciding whether to create or maintain a monopoly?

- Does it makes sense for us to hire a lobbyist to fight against this law that will reduce our monopoly power? Here is the probability that a lobbyist will be effective, here is the marginal cost, and here is the marginal benefit.
- Does it make sense for us to buy this machine? If we do, we'll be the only one to have it and are likely to get this much business. Here is the marginal cost, and here is the marginal benefit.
- Does it make sense for us to advertise to further our market penetration? Here are the likely various marginal benefits; here are the likely marginal costs.

Examples of firms spending money to protect or create monopolies are in the news all the time. The farm lobby fights to keep quotas and farm support programs. Drug companies spend a lot of resources to discover new drugs they can patent. A vivid example of the length to which firms will go to create a monopoly position is Owens-Corning's fight to trademark its hue of pink fibreglass. Owens-Corning has spent more than $200 million to advertise and promote its colour "pink" and millions more in court to protect its right to sole use of that hue. Owens-Corning has weighed the costs and benefits and believes that its pink colour provides sufficient brand recognition to warrant spending millions to protect it.

Establishing Market Position

In "winner-take-all" markets, the initial competition is on establishing market position.

Some economists, such as Robert Frank, have argued that today's economy is becoming more and more like a monopoly economy. Modern competition, he argues, is a winner-take-all competition. In such a competition, the winner (established because of brand loyalty, patent protection, or simply consumer laziness) achieves a monopoly and can charge significantly higher prices than its costs without facing competition. The initial competition, however, focussing on establishing market position, is intense.

To see how important establishing a market position is in today's economy, consider the initial public offerings (IPOs) of the new firms that were so highly valued by Wall Street, and to a lesser extent, Bay Street, in the late 1990s. Many of these new firms had no profits and no likelihood of profits for a number of years, but they were selling at extraordinarily high stock prices. Why? The reasoning was that these companies were spending money to establish brand names. As their names became better known, they would establish a monopoly position, and eventually their monopoly positions would be so strong that they couldn't help but make a profit. (It was true for Microsoft.) The problem is that for most firms, it will not be true—in any competitive process there are winners and losers, and unfortunately, most people have no way of differentiating between the two.

DRIVING FORCES IN TODAY'S ECONOMY

Modern competition is different because of globalization and technology.

Many economists consider competition today to be different from competition 10 or 20 years ago. The differences may be simply a matter of degree, but the reasons are closely tied to globalization and technology.

Globalization

Over the past 20 years, markets have become more global in a number of ways. Through the General Agreement on Tariffs and Trade and the World Trade Organization, tariffs and other barriers to trade have been lowered significantly. Political tensions among world superpowers have eased, leading firms to be more confident that they can diversify their production and their sources of inputs. Communication costs have fallen dramatically, which has lowered the cost of searching for the lowest-cost source of inputs and location for production. The global economy gives firms a lot more places to search, and the Internet reduces the costs of that search significantly.

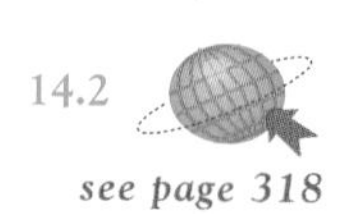

see page 318

Companies' strategic decision makers must take into account more than just domestic competitors. They face competition from foreign firms and most companies see their potential market as the world, not just the domestic market.

Globalization has two effects on firms. The first is positive; it increases the size of the gain to the winner. Because the world economy is so much larger than the domestic economy, the rewards for winning globally are much larger than the rewards for winning domestically. The second effect is negative; globalization makes it much harder to win, or even to stay in business. A company may be the low-cost producer in a particular country yet may face foreign competitors that can undersell it. The global economy increases the number of competitors for the firm. Consider the automobile industry. Three companies are headquartered in North America, but more than 20 automobile companies operate worldwide. North American automakers face stiff competition from foreign automakers; unless they meet that competition, they will not survive.

The global economy increases the number of competitors for the firm.

These two effects are, of course, related. When you compete in a larger market, you have to be better to survive, but if you do survive the rewards are greater.

Globalization increases competition by allowing greater specialization and division of labour, which, as Adam Smith first observed in *The Wealth of Nations*, increases growth and improves the standard of living for everyone. Thus, in many ways, globalization is simply another name for increased specialization. Globalization allows (indeed, forces) companies to move operations to countries with a comparative advantage. As they do so, they lower costs of production. Globalization can lead to companies specializing in smaller segments of the production process because the potential market is not just one country but the world. Such specialization can lead to increased productivity as firms learn by doing.

Globalization leads to specialization, which leads to increased productivity.

Surviving in a Global Economy In order to survive in the global economy, companies must be continually looking for ways to organize production that will improve efficiency and lower costs. One important way they are doing this is by breaking down the production process into its component parts and considering new ways of organizing production, as well as the firm.

Q-9 How does globalization reduce the cost of production?

Specialization in the modern economy involves dividing production (which includes manufacturing, distribution, sales, research, management, and advertising) into component parts, and searching for the cheapest method of producing each component. Firms may keep most of the production within the company but parcel out portions of it to divisions located in different parts of the world. Or they may simply outsource (buy from another firm) the part of production that can be done more cheaply by other domestic or foreign firms. Globalization lets companies take advantage of cost differentials across countries for different aspects of production. An example of an important cost differential is the cost of labour. Labour is a significant input to production and its costs differ widely among countries. Table 14-2 lists hourly compensation for workers in various countries.

For example, average hourly wages in manufacturing are $23.17 in Canada but only $8.35 in Taiwan. Labour-intensive parts of production might be relocated to Taiwan. Keep in mind that labour costs are only one part of the decision to relocate production. Firms also take into account differences in the quality of labour, the infrastructure, and regulations within countries. The fact remains, however, that barriers among countries have been lowered, allowing companies to consider relocating parts of their operations

TABLE 14-2 Hourly Compensation for Production Workers

Country	Cost in Cdn Dollars
Canada	$23.17
United States	28.53
Mexico	3.15
Japan	31.04
Korea	9.97
Taiwan	8.35
Germany	38.90
Hong Kong	8.08

Source: International Comparisons of Manufacturing Hourly Compensation Costs, 1999. Bureau of Labor Statistics, (http://stats.bls.gov) and author's calculations.

across the world. Firms are using comparative advantage not just in producing an entire product but in producing portions of the product. Firms are reevaluating the entire production process, asking where the production process for each component should occur.

Production within companies is often divided among many countries.

Consider automobiles again. GM, Ford, and Daimler-Chrysler all have production facilities throughout the world. When you buy a GM car, its components may have come from 50 different companies, and a Toyota car may have more North American-made components than a GM car. Components of the U.S. Air Force's F-16 fighter aircraft are produced in 11 countries; the aircraft is assembled in 4 countries.

This specialization and division of labour due to globalization means that competition takes place not just in the market for finished products but in numerous layers of the production process. Global competition helps hold down prices and wages firms pay to the factors of production. Those layers that do not face significant global competition may still be able to use lower-cost inputs to production, while maintaining their own high product prices. For example, legal work, as yet, faces very little international competition because (1) services are difficult to transport cheaply, (2) countries regulate the practice of law within their borders, and (3) laws tend to be country-specific. Hence the prices lawyers get for their services remain high. The manufacturing layer of production, however, is quite susceptible to foreign competition. Particularly susceptible are those industries which are labour-intensive, or in which, although significant capital investment may characterize production in Canada, the product can be manufactured using much cheaper labour-intensive production techniques elsewhere. Although it may appear to be a "step backwards" to return to a more labour-intensive production method, the cost advantage of doing so may dominate. That is, at least until competitive pressures raise wage rates in some of the developing countries where labour is abundant. This competition has resulted in strong downward pressure on worldwide manufacturing prices. Faced with increasing foreign competition, companies are outsourcing the manufacturing to foreign countries where costs are lower.

A finished product can be exported, or the production itself can be relocated.

Consider the letters for the popular game, Scrabble. Until recently the wooden tiles were made in a small Vermont town. In 1998, a Hong Kong company offered to produce the tiles at a lower price, and now Vermont lumber is shipped to Hong Kong, where the Scrabble tiles are made, and the tiles are shipped back to the United States. Hasbro, which makes Scrabble, is still a Canadian company; its marketing and distribution is run from the United States, but the production of its tiles is outsourced to a foreign firm.

Does Globalization Eliminate Jobs? It is sometimes argued that global competition eliminates jobs in Canada. To some degree this is absolutely true, but it is important to see global trade as a vehicle that increases total production and simultaneously creates jobs. Lowering costs of production can lead to lower product prices, which benefits consumers. Lower prices boost sales and, as sales rise, demand for the inputs to production rises, which can raise the prices of inputs and/or profits. This potentially benefits labourers and others who supply the various inputs to production. Consumers, producers, and suppliers of the factors of production all can potentially benefit.

Globalization eliminates jobs, but it also creates jobs.

Canadian firms, even small ones, more and more see themselves as global companies and are structuring themselves to compete in the global market. As they do so, their costs of production fall or the design of their products improves, putting them in a better position to sell their products. This increases the demand for those aspects of production—generally advertising, marketing, financial services, management, and distribution—which are still Canadian-based. This has helped keep the demand high for Canadian goods and, hence, increased the demand for Canadian labour.

Until recently, European firms have led a far more sheltered life. European governments' movement toward the European Union (EU) and the euro, the common European currency, are both attempts to broaden competition within European markets and to better prepare European firms to compete in the global economy. Trade restrictions between EU members have been eliminated, and with the euro the only currency among member nations (since 2002), comparison shopping will be much easier.

Technology

The second major driving force in the economy in recent years has been **technological development**—*the discovery of new or improved products or methods of production*. Technological advance lowers the costs of production and makes economies more efficient—producing more output with the same number of inputs.

Technological advance and globalization go together; technological advance requires large investments of time and money in very specialized areas. Globalization allows for that specialization and the possibility for large revenues to fund research. In addition, specialization allows producers to learn more about the particular aspects of production in which they specialize. As they learn more, they not only become more productive but are more likely to produce technological advances because they gain a deeper understanding of their specialty.

Technological advance and globalization go together; globalization leads to specialization, which leads to technological advance.

For example, instead of producing an entire line of clothing, companies might specialize in the production of certain types of carbon-based fibres and explore ways of making more useful material. The result of such specialization can be a technological advance, such as Gore-Tex—a material that insulates but also "breathes," and thus keeps individuals dry and cool on warm rainy days, and dry and warm on cold rainy or snowy days. Instead of spreading resources to the entire process of making a jacket, a company can concentrate on just one aspect—fibres.

This is what Canadian fashion designers did when the Free Trade Agreement between Canada and the United States was first introduced in 1989. Generally, it was felt that the Canadian (mostly Quebec) textile industry would be one of the industries that would not survive in a more competitive market and would have to be sacrificed in order for the rest of Canadian industries to benefit from the greater trade opportunities with the U.S. However, the Canadian textile industry focussed on what they were good at producing: lower volume specialty products. The industry has since won numerous awards for their work, and Canadians now have a worldwide reputation for quality of design.

TECHNOLOGY, EFFICIENCY, AND MARKET STRUCTURE

Given the significance of technology, an important question is: What causes technology to advance? Market incentives are an important part of the answer. Before markets existed, economies grew slowly. After markets came into existence in the 1700s, technology advanced more rapidly because individuals gained incentives, in the form of profits, to discover new and cheaper ways of doing things. Globalization of our economy provides an even greater incentive to develop new technologies, because the revenue that can be captured from a global market with 6 billion people (world population) is much greater than the revenue that can be generated from 30.5 million people (Canada's population).

Are some market structures more conducive to growth than others? The answer economists have come to is a tentative yes, and it is an answer that makes certain market structures look better than the way they were presented earlier. In the basic supply/demand framework, perfect competition is seen as the benchmark—it leads to efficient outcomes. All other market structures lead to some deadweight loss. But the supply/demand framework only considers questions of efficiency, it does not evaluate technological issues. It implicitly assumes that technology is fixed. If market structure does affect technological advance, another type of efficiency must be considered, dynamic efficiency. **Dynamic efficiency** refers to *a market's ability to promote cost-reducing or product-enhancing technological change*. Market structures that best promote technological change are dynamically efficient. Despite its problems, oligopoly provides the best market structure for technological advance. To see why, let's look at the four market structures: perfect competition, monopolistic competition, monopoly, and oligopoly.

In considering market structures, dynamic efficiency must be considered as well as static efficiency.

Perfect Competition and Technology

Perfectly competitive firms have no incentive to develop new technologies. Moreover, perfect competitors earn normal profits and consequently are generally unable to devote funds to the research and development that leads to technological change. Even if they did, they would gain little from it. A perfectly competitive market would quickly compete away the gains of the innovation, transferring many of the rewards of innovation to other firms, making it prohibitively difficult for the innovating firm to recoup the costs of developing the new technology.

Monopolistic Competition and Technology

Monopolistic competition is somewhat more conducive to technological change because firms have some market power. The promise of gaining additional market power—pricing or market share—provides the incentive to fund research in new technologies. But, as we learned earlier, competition among firms results in monopolistic competitors having no long-run profits. Easy entry limits their ability to recoup their investment in technological innovation. Eventually, their increased market share will deteriorate and they will return to earning normal profits. However, unlike perfect competition, there is at least a short-run incentive to innovate in monopolistic competition.

Through its support of patents, Canada does provide incentives to innovate. Patents allow the development of new products through the promise of monopoly profits for a specified period of time.

Monopoly and Technology

At the other end of the spectrum is pure monopoly. Monopolies may earn the profits needed to fund research and development, but they often do not have the incentive to innovate. Since a monopolist's market is protected from entry, the easiest path is the lazy monopolist path. Since almost all monopolies are created by government (the government gives a monopoly to a specific company), pure monopolists don't face the threat of new competitors. Until recently, European telephone companies and European domestic airlines were monopolies. These industries developed far fewer innovations than did the equivalent U.S. firms that faced more competition, and European industry prices were much higher than those in the United States. A European phone call, for instance, could cost five times as much as a U.S. phone call. The question is not so easily answered, however. In Canada, Canadian telephone companies were not competitive: they were mostly regulated provincial monopolies, such as Bell Canada and B.C. Tel, or crown corporations, such as Alberta Government Telephone. As provincial monopolies, Canadian telephone companies did not face any domestic competition, yet Canada has become a world leader in telecommunications, renowned for its technology. The telecommunications giant, Nortel, began as Bell Northern Research under the protection of the regulated monopoly Bell Canada.

Generally, in Europe, in Canada, and in many countries around the world, privatization and competition has been embraced. For example, European governments have moved in this direction. Both telecommunications and domestic airlines have been privatized, and their monopolies are slowly being removed. The result has been a decrease in prices of their products. Whether this privatization will lead to more technological progress remains to be seen, but many economists and policy makers generally believe that it will.

Oligopoly and Technology

Oligopoly is the market structure that is most conducive to technological change. Since the typical oligopolist realizes an ongoing economic profit, it has the necessarily large funds required to carry out significant research and development. Moreover, the belief that its competitors are innovating also forces it to do so. Oligopolists are constantly searching for ways to get an edge on competitors, so most technological advance takes place in oligopolistic industries.

Oligopoly tends to be most conducive to technological change.

Q-10 Why is oligopoly the best market structure for technological advance?

The computer industry is an example of an oligopolistic market that has demonstrated tremendous innovation. Technological progress has been rapid, following *Moore's law*—every 18 months the cost of computer speed is cut in half.

Other Views

Some economists, especially those who believe that the threat of competition is enough to keep a firm behaving competitively (a contestable market approach), argue that market structure does not matter for technological progress. It is the conditions of entry that matter. In addition, it can be argued that it is primarily developments in pure science that lead to technological advances. Businesses sometimes form partnerships with research facilities (such as universities) to work towards technological advances and then develop those that have market potential. It can also be argued that technological advances lead to the formation of oligopolies; oligopolies don't necessarily lead to technological advances. The cigarette industry and the aluminum industry are highly oligopolistic but have had little technological advance. In the steel industry, companies outside the group of existing producers started mini-mills that led to technological advance. The process did not originate with the oligopolistic steel companies.

Network Externalities, Standards, and Technological Lock-In

In support of the view that technology determines market structure, economists have focussed on those aspects of production that involve *network* externalities. An externality is an effect of a decision on a third party that is not taken into account by the decision maker. A positive **network externality** occurs *when greater use of a product increases the benefit of that product to everyone*. Telephones exhibit network externalities. If you were the only person in the world with a telephone, it would be useless. As the number of people with telephones increases, the telephone's value to the communication process becomes enormous. Another example of a product with network externalities is the Windows operating system. It is of much more use to you if many other

Network externalities lead to market standards and affect market structure.

people use it too, because you can then communicate with other Windows users and purchase software based on that platform.

Network externalities are important to market structure because they lead to the development of industry standards. Standards become important because network externalities involve interaction among many different individuals and processes. Many examples of the development of industry standards exist. Some are television broadcast standards (they differ in North America and Europe, which is why North American TVs cannot be used in Europe), building standards (there is a standard size of doors and windows), and electrical current standards (220 or 110; AC or DC).

Standards and Winner-Takes-All Industries Network externalities have two implications for the economic process. First, they increase the likelihood that an industry will become a winner-takes-all industry. Early in the development of new products, there may be two or three competing standards, any one of which could be a significant improvement over what existed before. As network externalities broaden the use of a product, the need for a single standard becomes more important and eventually one standard wins out. The firm that gets its standard accepted as the industry standard gains an enormous advantage over the other firms. This firm will dominate the market. Microsoft and its Windows operating system is an example. Once a standard develops, even if other firms try to enter with a better technological standard, they will have a hard time competing, because everyone is already committed to the existing industry standard. Deviating from that standard will reduce the benefits of the network externality.

The first-mover advantage partly helps explain the high stock prices of start-up technology companies.

First-Mover Advantage Firms in an industry developing a standard have a strong incentive to be the first to market with the product; they will be willing to incur large losses initially in their attempt to set the industry standard. The "first-mover advantage" helps explain why the stock of small technology companies sold for extremely high prices even though they were having large losses. (The other part of the explanation is investor greed.) The start-up firms incurred large losses because the firms were spending money to secure market share so that their products would become the industry standard. If the firm is successful in getting its product accepted as the standard, the demand for the product will rise and it will have enormous profits in the future.

QWERTY is a metaphor for technological lock-in.

Technological Lock-In The second implication of network externalities is that the market might not naturally adopt the most efficient standard. Standards can be inefficient and yet be maintained by the first-mover advantage. Some economists argue that the inefficiency can be quite large. Research by Paul David showed that the arrangement of the keys in the QWERTY computer keyboard originally was designed to slow down people's typing so that the keys would not stick on the early mechanical typewriters. As the technology of typewriters improved, the need to slow down typing soon ended, but because the QWERTY keyboard was introduced first, it had become the standard. Other, more efficient keyboards have been developed but not adopted. The QWERTY keyboard has dominated, even with its built-in inefficiencies. David suggested that QWERTY is a metaphor for **technological lock-in**—*when widespread use of a prior technology makes the adoption of subsequent technologies difficult*.

David's technological lock-in argument suggests that many of our adopted institutions and technologies may be inefficient. The QWERTY debate is a part of a larger debate about the competitive process and government involvement in that process. The issues are somewhat the same as they were in the earlier discussion of government regulation of natural monopolies. Many economists see government involvement as necessary to protect the economy and the consumer. They advocate what economist Brian Arthur calls "a nudging hand" approach, in which the government keeps the initial competition fair.

Modern debates about policy regarding competition take dynamic issues into account, but still leave open a debate about what the role of government should be.

Other economists see monopoly as part of the competitive process—something that will be eliminated as competitive forces act against it. Standards will develop, but they will be temporary. If the standards are sufficiently inefficient, they will be replaced, or an entirely new product will come along that makes the old standard irrelevant. For example, the QWERTY problem is solved by voice recognition software. For such economists, neither natural monopoly nor technological lock-in is a reason for government interference. Government interference, even the nudging hand, would stop the competitive process and make the society worse off.

Who is right? Whether the competitive process or a nudge here or there, is better, as Canadians, we generally view strict competition as incompatible with many of our most cherished values. In any event, even in cases where explicit regulation is not called for, the government plays a critical role in setting up and maintaining the appropriate rules and property rights in order to keep the competitive playing field reasonably level.

CONCLUSION

The stories of competition and monopoly have no end. Both are continuous processes. Monopolies create competition. Out of the competitive struggle, other monopolies emerge, only to be beaten down by competition. Globalization and technology are big parts of that struggle. Individuals and firms, motivated by self-interest, try to use the changes brought by globalization and technology to their benefit. By doing so they change both the global nature of the economy and the direction of technological change itself.

Chapter Summary

- The goals of real-world firms are many. Profit plays a role, but the actual goals depend on the incentive structure embodied in the structure of the firm.
- The monitoring problem arises because the incentives faced by managers are not always to maximize the profit of the firm. Economists have helped design incentive-compatible contracts to help alleviate the monitoring problem.
- Monopolists facing no competition can become lazy and not hold down costs as much as they can. X-inefficiency refers to firms operating less efficiently than they could technically.
- The competitive process involves a continual fight between monopolization and competition. Suppliers are willing to pay an amount equal to the additional profit gained from the restriction. Consumers are willing to pay an amount equal to the additional cost of products to avoid a restriction. Consumers, however, generally face prohibitively higher costs of organizing their efforts.
- Firms compete against patents that create monopolies by making slight modifications to existing patents and engaging in reverse engineering to copy other firms' products within the limits of the law.
- Natural monopolies are being deregulated by dividing the firms into various subindustries, protecting that part that exhibits the characteristics of a natural monopoly, and opening the remaining parts to competition.
- Firms protect their monopolies by such means as advertising, lobbying, and producing products that are difficult for other firms to copy.
- Firms will spend money on monopolization until the marginal cost equals the marginal benefit.

- X-inefficiency can be limited by the threat of competition or corporate takeovers. Takeovers usually result in a change in management.
- Globalization increases competition by providing more competition for domestic firms at all levels of production and by allowing firms to specialize. Globalization also increases the gain to the industry leader by reducing costs of production and by increasing the size of the market.
- Oligopoly provides the best market structure for technological advance because oligopolists have an incentive to innovate in the form of additional profits and because they have the long run economic profits to devote to investing in the research and development of new technologies.

Key Terms

corporate takeover *(303)*
dynamic efficiency *(312)*
incentive-compatible contract *(301)*
lazy monopolist *(302)*
monitoring problem *(301)*
network externality *(313)*
patent *(305)*
reverse engineering *(305)*
technological development *(311)*
technological lock-in *(314)*
X-inefficiency *(302)*

Questions for Thought and Review

1. Describe the monitoring problem. How does an incentive-compatible contract address the monitoring problem?
2. Are managers and high-level company officials paid high salaries because they're worth it to the firm, or because they're simply extracting profit from the company to give themselves? How would you tell whether you're correct?
3. Define *X-inefficiency*. Can a perfect competitor be X-inefficient? Explain why or why not.
4. Some analysts have argued that competition will eliminate X-inefficiency from firms. Will it? Why?
5. Nonprofit colleges must be operating relatively efficiently. Otherwise for-profit colleges would develop and force existing colleges out of business. True or false? Why?
6. If it were easier for consumers to collude than for suppliers to collude, there would often be shortages of goods. True or false? Why?
7. If it were easier for consumers to collude than for suppliers to collude, the price of goods would be lower than the competitive price. True or false? Why?
8. Monopolies are bad; patents give firms monopoly; therefore, patents are bad. True or false? Why?
9. Natural monopolies should be broken up to improve competition. True or false? Why?
10. Technically competent firms will succeed. True or false? Why?
11. Monsanto Corporation lost its U.S. patent protection for its highly successful herbicide Roundup in the year 2000. What do you suppose will be Monsanto's strategy for Roundup in the short run? In the long run?
12. Why would a company want to sacrifice short-run profits to establish market position?
13. What effect has globalization had on the ability of firms to specialize? How has this affected the competitive process?
14. Why would globalization lead to greater technological advance?
15. What two characteristics does a market structure need to have for firms in that industry to engage in technological advance?
16. Taking into consideration changing technologies, why might the basic supply/demand framework not lead to the most efficient outcome?
17. How do network externalities increase the winner-take-all nature of a market?

Problems and Exercises

1. The title of an article in *The Wall Street Journal* was "Pricing of Products Is Still an Art, Often Having Little Link to Costs." In the article, the following cases were cited:
   - Vodka pricing: All vodkas are essentially indistinguishable—colourless, tasteless, and odourless—and the cost of producing vodka is independent of brand name, yet prices differ substantially.
   - Perfume: A $100 bottle of perfume may contain $4 to $6 worth of ingredients.
   - Jean and shirts with company logos: The "plain pocket" jeans and shirts often cost 40 percent less than the brand-name items, yet they are essentially identical to the name-branded items.

   a. Discuss whether these differences undermine economists' analysis of pricing.
   b. What do each of these examples likely imply about fixed costs and variable costs?
   c. What do they likely imply about costs of production versus costs of selling?
   d. As what type of market would you characterize each of the above examples?

2. Demonstrate graphically the net gain to producers and the net loss to consumers if suppliers are able to restrict their output to Q_r in the following graph. Demonstrate the net deadweight loss to society.

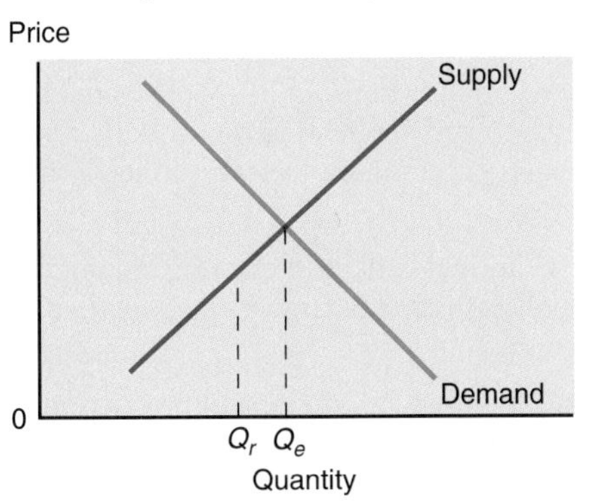

3. Airlines and hotels have many frequent flier and frequent visitor programs in which individuals who fly on the airline or stay at the hotel receive bonuses that are the equivalent of discounts.

   a. Give two reasons why these companies have such programs rather than simply offer lower prices.
   b. Can you give other examples of such programs?
   c. What is a likely reason why firms don't monitor these programs?
   d. Should the benefits of these programs be taxable?

4. Up to how much is the monopolist depicted in the accompanying graph willing to spend to protect its market position? Demonstrate your answer graphically.

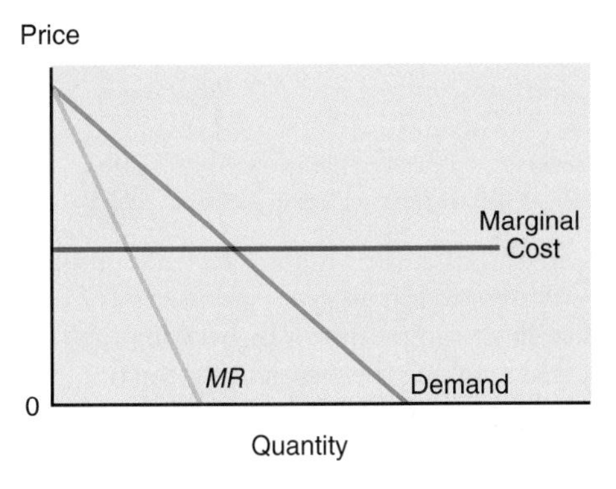

5. In 1999 the hourly cost to employers per German industrial worker was Cdn $38.90. The hourly cost to employers per Canadian industrial worker was $23.17, while the average cost per Taiwanese industrial worker was Cdn $8.35.

   a. Give three reasons why firms produce in Germany rather than in a lower-wage country.
   b. Germany has just entered into an agreement with other EU countries that allows people in any EU country, including Greece and Italy, which have lower wage rates, to travel and work in any EU country, including high-wage countries. Would you expect a significant movement of workers from Greece and Italy to Germany right away? Why or why not?
   c. Workers in Thailand are paid significantly less than workers in Taiwan. If you were a company CEO, what other information would you want before you decided where to establish a new production facility?

Web Questions

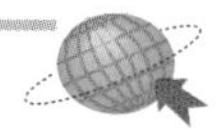

1. Nike coordinates the production and sales of its products worldwide. Go to Nike's homepage at www.nikebiz.com and answer the following questions:
   a. How much did the "Swoosh" design cost Nike? How much is it worth?
   b. Why did Nike launch a new logo? What products and services does the new logo appear on?
   c. How many factories does Nike have worldwide? How many countries supply Nike with product? Click on a country to view the profile of Nike business in that country.
   d. How have technology and globalization affected Nike since its inception as "Blue Ribbon Sports?"
2. Research Dell Corporation's history since its inception in 1984 by going to its home page at www.dell.com and clicking on "About Dell." Answer the following questions:
   a. What product does Dell sell? What was Dell's innovation in this market?
   b. In what countries does Dell sell its products? Which geographic area provides the largest source of revenue?
   c. In what countries does Dell produce its product?
   d. How does Dell plan to continue its growth in the future?

Answers to Margin Questions

1. Firms are not interested in just short-run profits. They are also interested in long-run profits. So, a firm might sacrifice short-run profits for higher long-run profits. Also, those making the decisions for the firm are not always those who own the firm. *(300)*
2. Most economists are concerned about third-party payment systems because of the problems of monitoring. It is the consumers who have the strongest incentive to make sure that they are getting value for their money. Any third-party payment system reduces the consumers' vigilance and therefore puts less pressure on holding costs down. *(301)*
3. A manager does not have the same incentive to hold costs down as an owner does because when an owner holds costs down, the owner's profits are increased, but when a manager holds costs down, the increased profits accrue to the owner, not the manager. Thus the manager has less direct motivation to hold costs down than an owner does. This is especially true if the costs being held down are the manager's perks and pay. *(302)*
4. The threat of a corporate takeover places competitive pressures on firms because it creates the possibility of the managers being replaced and losing all their perks and above-market-equilibrium pay. *(303)*
5. Many equity issues are of great importance to Canadians. In essence it is our belief in these types of issues that make us Canadian. A partial list would include: equitable income distribution (see Chapter 15); access to education and medical care; environmental issues, such as forest renewal, cleaning up the Great Lakes, air pollution abatement, and lowering greenhouse gas emissions to reduce global warming; sustainable economic growth and development, such as in fishing and mining; safe working conditions; fair wages; sense of community; and safe neighbourhoods. As Canadians, our list of social goals is long; we, as a nation, care. And Canadians have a worldwide reputation for caring. *(304)*
6. It is almost impossible for perfect monopoly to exist because completely preventing entry is nearly impossible. Monopoly rents are a signal to potential entrants to get the barriers of entry removed. *(304)*
7. The problem with cost-based regulation that sets prices relative to costs is that this removes the incentive for firms to hold down costs and can lead to X-inefficiency. While, in theory, regulators could scrutinize every cost, in practice that is impossible—there would have to be a regulatory board duplicating the work that a firm facing direct market pressure undertakes in its normal activities. *(306)*
8. If the additional benefit of creating or maintaining a monopoly exceeds the cost of doing so, do it. If it doesn't, don't. *(307)*
9. Globalization reduces the cost of production in two ways. First, it allows companies to specialize in smaller segments of the production process, which increases competition and lowers cost at all levels of the production process. Second, it allows companies to locate parts of the production process in those countries with comparative advantage in that segment of the production process. *(309)*
10. Oligopoly is the best market structure for technological advance because oligopolists have the profits to devote to research and development and have the incentive to innovate. Innovation may provide the oligopolist with a way to increase market share. *(313)*

Government Policy and Market Failures

15

After reading this chapter, you should be able to:

- Explain what an externality is and show how it affects the market outcome.
- Describe three methods of dealing with externalities.
- Define a public good and explain the problem with determining the value of a public good to society.
- Explain how informational problems can lead to market failure.
- List five reasons why government's solution to a market failure could worsen the market failure.

The business of government is to keep the government out of business—that is, unless business needs government aid.

Will Rogers

Now that we've been through the foundations of supply and demand and various market structures, we're ready to explore economic policy questions more deeply and develop a fuller understanding of some of the roles of government first presented in Chapter 5. There are two cases where the market does not yield an appropriate outcome. The first is **market failure**, where *the natural market forces of demand and supply do not function to create an efficient market equilibrium*. This is the topic of Chapters 15 and 16. The second is **market outcome failure**, where *the market forces produce an efficient equilibrium, but society does not prefer the equilibrium that naturally emerges*. This is the topic of Chapter 17.

KNOWING THE TOOLS

Pareto Optimality and the Perfectly Competitive Benchmark

Perfect competition serves as a benchmark for judging policies. A foundation for this benchmark is the work of Stanford economist Kenneth Arrow, who showed that the market translates self-interest into society's interest. (Arrow was given a Nobel Prize in 1972 for this work.) Arrow's ideas are based on many assumptions; however, we will discuss one, the interpretation of the term *society's welfare*. In the economic framework, society's welfare is interpreted as coming as close as one can to a *Pareto optimal position*—a position from which no person can be made better off without making another person worse off. (Pareto optimal policies will be discussed more in Chapter 17.)

Let's briefly consider what Arrow proved. He showed that if the market was perfectly competitive, and if there was a complete set of markets (a market for every possible good) now and in the future, the invisible hand would guide the economy to a Pareto optimal position. If these assumptions hold true, the supply curve (which represents the marginal cost to the suppliers) would represent the marginal cost to society. Similarly, the demand curve (which represents the marginal benefit to consumers) would represent the marginal benefit to society. In a supply/demand equilibrium, not only would an individual be as well off as he or she possibly could be, given where he or she started from, but so too would society. A perfectly competitive market equilibrium would be in a Pareto optimal position.

A number of criticisms exist to using perfect competition as a benchmark:

1. *The Nirvana criticism:* A perfectly competitive equilibrium is highly unstable. It's usually in some person's interest to restrict entry by others, and, when a market is close to a competitive equilibrium, it is in few people's interest to stop such restrictions. Thus, perfect competition will never exist in the real world. Comparing reality to a situation that cannot occur (i.e., to Nirvana) is not helpful because it leads to attempts to achieve the unachievable. A better benchmark would be a comparison with workable competition—a state of competition that one might reasonably hope could exist.
2. *The second-best criticism:* The conditions that allow the conclusion that perfect competition leads to a Pareto optimal position are so restrictive that they are never even approached in reality. If the economy deviates in hundreds of ways from perfect competition, how are we to know whether a movement toward a competitive equilibrium in one of those ways will be an improvement?
3. *The normative criticism:* Even if the previous two criticisms didn't exist, the perfect competition benchmark still isn't appropriate because there is nothing necessarily wonderful about Pareto optimality. A Pareto optimal position could be a horrendous position, depending on the starting position. For example, say the starting position is the following: One person has all the world's income and all the other people are starving. If that rich person would be made worse off by having some money taken from him and given to the starving poor, that starting position would be Pareto optimal. By most people's normative criteria, it would also be an unacceptable position.

Critics of the use of the perfect competition benchmark argue that society has a variety of goals. Pareto optimality may be one of them, but it's only one. They argue that economists should take into account all of society's goals—not just Pareto optimality—when determining a benchmark for judging policies.

The private market economic framework discussed so far in the book can be called the *invisible hand framework*. It says that if markets are perfectly competitive they will lead individuals to make voluntary choices which maximize their own benefit, which also happen to correspond to society's interest. The result of individuals acting in their own best interest is to create an outcome which is also in society's best interest is a very interesting consequence of the market economy. It is as if individuals are guided by an invisible hand to do what society wants them to do.

MARKET FAILURES

As a mechanism for allocating all kinds of goods, the private market economy functions well, with the invisible hand generally leading to a reasonably efficient and desirable equilibrium price and quantity. For the invisible hand to guide private actions toward the social good, however, a number of conditions must be met. When those conditions are not met, economists say that there is a **market failure**—*a situation in which the invisible hand pushes in such a way that individual decisions do not lead to socially desirable outcomes*. Two issues, the failure of the market and the failure of the market outcome, are the topics of the next three chapters. Chapter 15 begins the evaluation of market equilibrium with a discussion of three sources of market failures: externalities, public goods, and asymetric information. Chapter 16 continues with a discussion of market failure due to monopoly and other impediments to competition in the market.

Three important sources of market failure are externalities, public goods, and asymmetric information.

Any time a market failure exists there is a reason for possible government intervention to improve the outcome. But it is important to remember that even if these market failures exist, it is not clear that government action will improve the result, since the politics of implementing the solution often lead to further problems. These problems of government intervention are often called *government failures*, so after discussing the three sources of market failures, we will discuss government failures. The economic policy debate will then be framed as a matter of choosing which failure is likely to be the lesser of two evils.

Economic policy is often a choice between market failure and government failure.

Chapter 17 continues the discussion of government intervention by considering cases where the market yields an efficient equilibrium, but the equilibrium is not consistent with other goals of society, such as equitable income distribution. Sometimes, the equilibrium that emerges has undesirable characteristics which do not meet society's standards. Even when the market outcome is efficient, for example, issues of equity may be more important. Chapter 17 deals with the failure of the market *outcome*.

EXTERNALITIES

An important requirement for the invisible hand to guide markets in society's interest is that market transactions between two individuals have no side effects on anyone else. Such side effects are called **externalities**—*the effect of a decision on a third party that is not taken into account by the decision maker*. Externalities can be either positive or negative. Secondhand smoke and carbon monoxide emissions are examples of **negative externalities**, which occur *when a market transaction has a detrimental effect on others*. **Positive externalities** occur *when the market transaction has a beneficial effect on others*. An example is education. When you purchase a college education, not only you benefit but others as well benefit as a result of your education. Innovation is another example. The invention of the personal computer has had greater beneficial effects on society than were anticipated by the inventors. When there are externalities, the supply and/or demand curves no longer represent the marginal cost and marginal benefit curves to society.

An externality is an effect of a decision on a third party not taken into account by the decision maker.

A Negative Externality Example

Say that you and I agree that I'll produce steel for you. I'll build my steel plant on land I own, and start producing. We both believe our welfare will improve. But what about my neighbours? The resulting smoke will pollute the air they breathe. The people involved in the market transaction (you and I) are made better off, but people external to the trade are made worse off. Thus, there is a negative externality. My production of steel has a cost to society that neither you nor I take into account.

Q-1 Why does the existence of an externality prevent the market from working properly?

The effect of a negative externality is shown in Figure 15-1. The supply curve *S* represents the **private marginal cost** to society of producing steel. The demand curve *D* represents the social marginal benefit of consuming the steel. When there are no externalities, the private marginal costs and benefits represent the social marginal costs and benefits, so the supply/demand equilibrium (P_0, Q_0) represents the point where the

When there are externalities the social marginal cost differs from the private marginal cost.

Figure 15-1 THE EFFECT OF A NEGATIVE EXTERNALITY

When there is a negative externality, the social marginal cost will be above the private marginal cost and the competitive price will be too low to maximize social welfare.

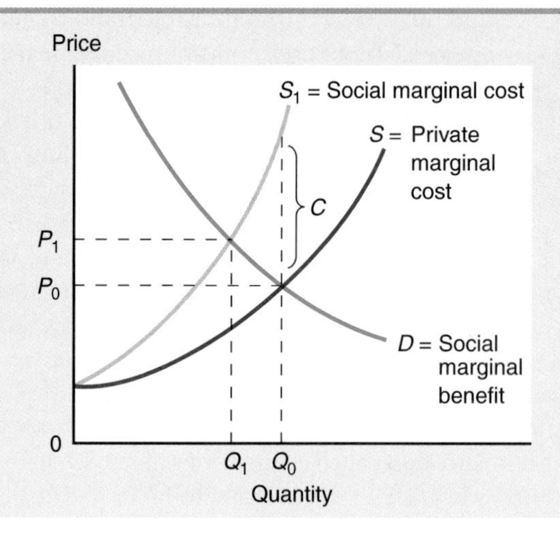

social marginal benefit equals the social marginal cost. At that point society is as well off as possible.

But now consider what happens when production results in negative externalities. In that case people not involved in production also incur costs. This means that the supply curve no longer represents the total cost (private and social marginal costs) of supplying the good. Social marginal cost is greater than the private marginal cost. This case can be represented by adding a curve in Figure 15-1 called the *social marginal cost curve*. The **social marginal cost** includes all the marginal costs that society bears—or *the private marginal costs of production plus the cost of the negative externalities associated with that production*.

Social marginal cost is the sum of private marginal cost plus the cost of negative externalities.

Since in this case the externality represents an additional cost to society, the social marginal cost curve lies above the private marginal cost curve. The distance between the two curves represents the additional marginal cost of the externality. For example, at quantity Q_0, the private marginal cost faced by the firm is P_0. The marginal cost from the externality at quantity Q_0 is shown by distance C. The externality cost is not taken into account, and the supply/demand equilibrium is at too high a quantity, Q_0, and at too low a price, P_0.

Notice that the market solution results in a level of steel production that exceeds the level that equates the social marginal costs with the social marginal benefits. If the market is to maximize welfare, some type of government intervention may be needed to account for the true costs of production, and reduce production from Q_0 to Q_1 and raise price from P_0 to P_1.

A Positive Externality Example

Q-2 If a positive externality exists, does that mean that the market works better than if no externality exists?

Private trades can also benefit third parties not involved in the trade. These are positive externalities. Again, an example is education. Consider a person who is working and takes a class at night. He or she will bring the knowledge from class back to co-workers through day-to-day interaction on projects. The co-workers will be learning the material from the class indirectly. They are outside the initial decision to take the class, and consequently do not pay any part of the tuition or other costs, but they benefit nonetheless.

In the case of positive externalities, the market will not provide enough of the good. Let's see how. In Figure 15-2, we start again with the standard demand and supply curves. The supply curve S represents the private marginal cost of the course. The demand curve

APPLYING THE TOOLS

Common Resources and the Tragedy of the Commons

Individuals tend to overuse commonly owned goods. Let's consider an example—say that grazing land is held in common. Individuals are free to bring their sheep to graze on the land. What is likely to happen? Each grazing sheep will reduce the amount of grass for other sheep. If individuals don't have to pay for grazing, when deciding how much to graze their sheep they will not take into account the cost to others of their sheep's grazing. The result may be overgrazing—killing the grass and destroying the grazing land. This is known as the *tragedy of the commons*, so named after an article[1] by biologist Garrett Hardin, who analyzed the common property problems associated with English and Early American grazing practices. A more contemporary example of the tragedy of the commons is fishing. The sea is a common resource; no one owns it, and whenever people catch fish, they reduce the number of fish that others can catch. The result will likely be overfishing.

The tragedy of the commons is an example of the problems posed by externalities. Catching fish imposes a negative externality. Because of the negative effect on others, the social cost of catching a fish is greater than the private cost. Overfishing has been a problem in Canada and throughout the world. Thus, the tragedy of the commons is caused by individuals not taking into account the negative externalities of their actions.

Why doesn't the market solve the externality problem? Some economists argue that in the tragedy of the commons examples it would, if given a chance. The problem is a lack of property rights (lack of ownership). If rights to all goods were defined, the tragedy of the commons would disappear. In the fishing example, if someone owned the sea, he or she would charge individuals to fish. By charging for fishing rights the owner would internalize the externality and thus avoid the tragedy of the commons.

[1] "The Tragedy of the Commons," *Science*, December, 1968, 1243-8.

D_0 is the **private marginal benefit** to those who take the course. Since others not taking the course also benefit, the social marginal benefit, shown by D_1, is above the private marginal benefit. The **social marginal benefit** equals *the private marginal benefit of consuming a good plus the benefits of the positive externalities resulting from consuming that good.* The vertical distance between D_0 and D_1 is the additional benefit that others who are not party to the transaction receive at each quantity. At quantity Q_0, the market equilibrium, the marginal benefit of the externality is shown by distance B. At this quantity, the social marginal benefit exceeds the social marginal cost. The market provides too

Positive externalities make the private marginal benefit below the social marginal benefit.

Figure 15-2 A POSITIVE EXTERNALITY

When there is a positive externality, the social marginal benefit will be above the private marginal benefit and the market price and quantity will be too low to maximize social welfare.

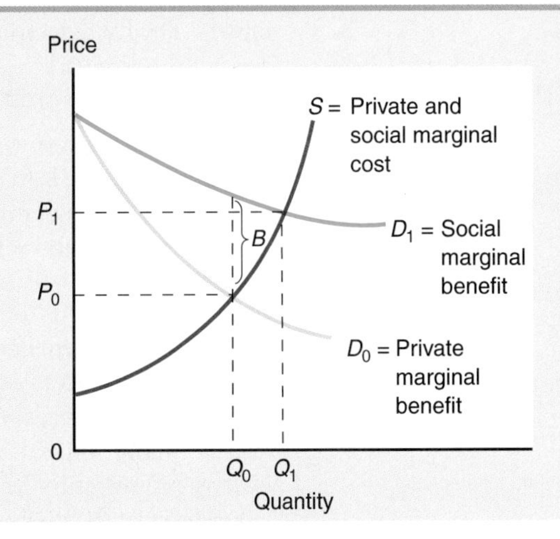

little of the good. The optimal price and quantity for society are P_1 and Q_1, respectively. Again, some type of intervention to increase quantity and price may be warranted.

ALTERNATIVE METHODS OF DEALING WITH EXTERNALITIES

Externalities can be dealt with by means of:

1. direct regulation,
2. incentive policies, and
3. voluntary solutions.

Ways to deal with externalities include (1) direct regulation, (2) incentive policies (tax incentive policies and market incentive policies), and (3) voluntary solutions.

Direct Regulation

In a program of **direct regulation,** *the amount of a good people are allowed to use is directly limited by the government*. Let's consider an example. Say we have two individuals, Ms. Thrifty, who uses 10 litres of gasoline a day, and Mr. Big, who uses 20 litres of gas a day. Say we have decided that we want to reduce total daily gas consumption by 10 percent, or 3 litres. The regulatory solution might require both individuals to reduce consumption by some specified amount. Likely direct regulatory reduction strategies would be to require an equal quantity reduction (each consumer reducing consumption by 1.5 litres) or an equal percentage reduction (each consumer reducing consumption by 10 percent).

Both of those strategies would work, but neither would be **efficient** *(achieving a goal at the lowest cost in total resources)*. This is because direct regulation does not take into account that the costs of reducing consumption may differ among individuals. Say, for example, that Ms. Thrifty could easily (i.e., almost costlessly) reduce consumption by 3 litres while Mr. Big would find it very difficult (costly) to reduce consumption by even 0.5 litre. In that case, either regulatory solution would be **inefficient** *(achieving a goal in a more costly manner than necessary)*. It would be less costly (more efficient) to have Ms. Thrifty undertake most of the reduction. A policy that would automatically make the person who has the lower cost of reduction *choose* (as opposed to being *required*) to undertake the most reduction would achieve the same level of reduction at a lower cost. In this case the efficient policy would get Ms. Thrifty to choose to undertake the majority of the reduction.

Q-3 It is sometimes said that there is a trade-off between fairness and efficiency. Explain one way in which that is true and one way in which that is false.

Note that it is not necessarily the case that the person consuming the smaller amount is the one who can most easily reduce consumption. Ms. Thrifty may run a delivery business which inelastically demands 10 litres a day, whereas Mr. Big may go for long leisurely drives in the countryside (elastic demand). The opportunity cost involved in Ms. Thrifty reducing gasoline consumption may be much higher than for Mr. Big.

Incentive Policies

Economists tend to like incentive policies to deal with externalities.

Two types of incentive policies would each get Ms. Thrifty to undertake the larger share of reduction. One is to create a tax incentive to achieve the desired reduction; the other is to create a type of property right embodied in a permit or certificate, and to allow individuals to buy and sell those property rights freely.

Tax Incentive Policies Let's say that the government imposes a tax on gasoline consumption of 50 cents per litre. This would be an example of a **tax incentive program** *(a program using a tax to create incentives for individuals to structure their activities in a way that is consistent with the desired ends)*. Since Ms. Thrifty can almost costlessly reduce her gasoline consumption, she will likely respond to the tax by reducing gasoline consumption, say, by 2.75 litres to 7.25 litres. She pays only \$3.63 (.50 × 7.25 litres) in tax but undertakes most of the conservation. Since Mr. Big finds it very costly to reduce his consump-

Figure 15-3 REGULATION THROUGH TAXATION

If the government sets a tax equal to a negative externality, individuals will respond by reducing the quantity of the pollution-causing activity supplied to a level that individuals would have supplied had they included the negative externality in their decision.

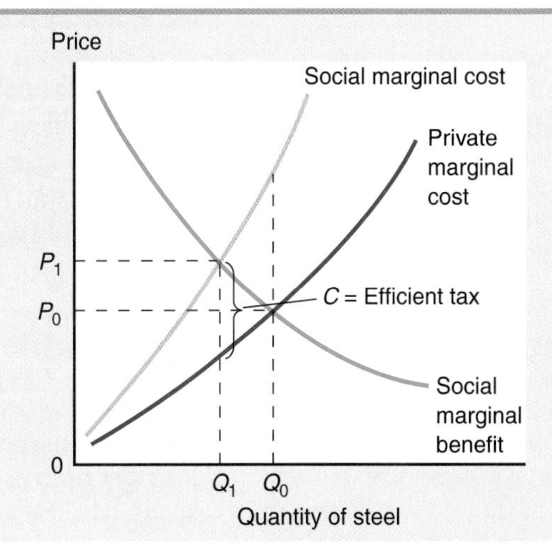

tion of gasoline, he will likely respond by reducing gasoline consumption by very little, say by 0.25 litre. He pays $9.88 (.50 × 19.75 litres) in tax but does little of the conservation.

In this example, the tax has achieved the desired end in a more efficient manner than would the regulatory solution—the person for whom the reduction is least costly cuts consumption the most. Why? Because the incentive to reduce is embodied in the price, and individuals choose how much to change their consumption. The tax has made them internalize the externality. The solution also has a significant element of fairness about it. The person who conserves the most pays the least tax. This solution, however, relies on a very important assumption, that each individual has an equal ability to pay.

Q-4 In what sense is the tax incentive approach fair?

Let's now consider how the tax incentive solution will solve the problem in our earlier example of steel production creating an externality. Figure 15-3 shows the situation. Say the government determines that the additional social marginal cost of producing steel equals C percent, which is equal to C dollars at output Q_1. If the government sets the pollution tax on steel production at C, the firm will reduce its output to Q_1 on its own. Such taxes on externalities are often called **effluent fees**—*charges imposed by government on the level of pollution created.* The efficient tax equals the additional cost imposed on society which is not taken into account by the decision maker. With such a tax, the cost the suppliers face is the social cost of supplying the good. With the tax, the invisible hand guides the traders to equate the social marginal cost to the social marginal benefit and the equilibrium is socially optimal.

Market Incentive Policies A second incentive policy that gets individuals to internalize an externality is a **market incentive plan** *(a plan requiring market participants to certify that they have reduced total consumption—not necessarily their own individual consumption—by a specified amount)*. Such a program would be close to the regulatory solution but involves a major difference. If individuals choose to reduce consumption by more than the required amount, they will be given a marketable certificate that they can sell to someone who has chosen to reduce consumption by less than the required amount. By buying that certificate, the person who has not personally reduced consumption by the requisite amount will have met the program's requirements. Let's see how the program would work with Mr. Big and Ms. Thrifty.

Marketable permits are being used more and more to deal with environmental issues. In many areas, air quality permits are required of firms before they may begin production. The only way a new firm may start production is if it buys permits from existing firms, which are allowed to sell permits only if they decrease air pollution by the amount of the permit.

Another example of the use of marketable permits is a program instituted in California to deal with drought. The only way a construction firm is allowed to build a new house is if it reduces existing water usage by a specific amount. To do this, the firm goes around to existing home owners, offering to put in water-saving devices for free. In some areas, more than 60 percent of the existing homes have introduced these water-saving devices, which were paid for by the construction firms and hence indirectly by the new home buyers.

On the national level, the United States introduced Project XL, which lets firms reduce air pollutants below existing requirements and, in doing so, free themselves from some specific regulations. This allows the firms to trade some pollution internally for others, achieving some of the flexibility that the marketable permits provide on an inter-firm level.

In our example, Mr. Big finds it very costly to reduce consumption while Ms. Thrifty finds it easy. So we can expect that Mr. Big won't reduce consumption much and will instead buy certificates from Ms. Thrifty, who will choose to undertake significant reduction in her consumption to generate the certificates, assuming she can sell them to Mr. Big for a high enough price to make that reduction worth her while. The price of the certificates will lie somewhere between Ms. Thrifty's valuation of the certificate (low) and the value Mr. Big places on the right to consume gasoline (high). So, as was the case in the tax incentive program, Ms. Thrifty undertakes most of the conservation and receives a financial benefit for it.

Incentive policies are more efficient than direct regulatory policies.

Obviously there are enormous questions about the administrative feasibility (cost) of these types of proposals, but what's important to understand here is not the specifics of the proposals but the way in which incentive policies lead to the *efficient* outcome. *Efficient* means *least costly* in terms of resources, with no consideration paid to who is bearing those costs. Incorporating the incentive into a price and then letting individuals choose how to respond to that incentive lets those who find it least costly undertake most of the adjustment.

More and more, governments are exploring market incentive policies for solving problems. Sin taxes (taxes on goods government believes to be harmful, such as alcohol, cigarettes, and gasoline) are an example of the tax incentive approach. (These will be discussed further in Chapter 17.) Marketable permits for pollution, discussed in the accompanying box, are an example of the marketable certificate approach.

Voluntary Reductions

A third possibility to address externalities is to make the reduction voluntary, leaving individuals free to choose whether to follow what is socially optimal or what is privately optimal. Let's consider how a voluntary program might work in our Mr. Big and Ms. Thrifty example. Let's say that Ms. Thrifty has a social conscience and undertakes most of the reduction while Mr. Big has no social conscience and does not reduce consumption significantly. It seems that this is a reasonably efficient solution. But what if the costs were reversed and Mr. Big had the low cost of reduction and Ms. Thrifty had the high cost? Then the voluntary solution would not be so efficient. Of course, it could be argued that when people choose to do something voluntarily it makes them better off. So one could argue that even in the case where Ms. Thrifty has a high cost of reduction, she may voluntarily undertake most of the reduction. This would imply however, that she receives a high benefit from reducing her consumption, high enough to compensate her for her high cost of reduction.

The largest problem with voluntary solutions is that a person's willingness to do things for the good of society generally depends on that person's belief that others will also be helping.

Q-5 What are two reasons to be wary of solutions based on voluntary action that is not in people's self-interest?

If a socially conscious person comes to believe that a large number of other people won't contribute, he or she will often lose that social conscience: Why should I do what's good for society if others won't? This is an example of the **free rider problem** (*individuals' unwillingness to share in the cost of a public good*), which economists believe will often limit, and eventually undermine, social actions based on voluntary contributions. Even a small number of free riders can undermine the social consciousness of many in the society and eventually the voluntary policy will fail.

Economists believe that a small number of free riders can undermine the social consciousness of many in the society and that eventually a voluntary policy will fail.

There are exceptions. During times of war and extreme crisis, voluntary programs are often successful. For example, during the Second World War the war effort was financed in part through successful voluntary programs. But for other long-term social problems that involve individuals accepting significant changes in their actions, generally the results of voluntary programs have not been positive.

The Optimal Policy

An **optimal policy** is *one in which the marginal cost of undertaking the policy equals the marginal benefit of that policy*. If a policy isn't optimal (that is, either the marginal cost exceeds the marginal benefit or the marginal benefit exceeds the marginal cost), resources are being misallocated because the savings from a reduction of expenditures on a program will be worth more than the gains that would be lost from reducing the program, or the benefit from spending more on a program will be worth more than the cost of expanding the program.

If a policy isn't optimal, resources are being misallocated because the savings from reduction of expenditures on a program will be worth more than the gains that will be lost from reducing the program.

Let's consider an example. Say the marginal benefit of a program significantly exceeds it marginal cost. That would seem good. But that would mean that we could expand the program by decreasing spending on some other program or activity whose marginal benefit is less than its marginal cost, with a net gain in benefits to society. To spend too little on a beneficial program is as inefficient as spending too much on a non-beneficial program. This accords with the previous discussion on marginal costs and marginal benefits. If the marginal benefit exceeds marginal cost, increase production. If marginal benefit is less than marginal cost, decrease production. Here, we account for *all* the costs and benefits to society, regardless of who pays the costs or receives the benefits.

This concept of optimality carries over to economists' view of most problems. For example, some environmentalists would like to completely rid the economy of pollution. Most economists believe that doing so is costly and that since it's costly, one would want to take into account those costs. That means that society should reduce pollution only to the point where the marginal cost of reducing pollution equals the marginal benefit. That point is called the *optimal level of pollution—the amount of pollution at which the marginal benefit of reducing pollution equals the marginal cost*. To reduce pollution below that level would make society as a whole worse off.

Some environmentalists want to rid the world completely of all pollution while most economists want to reduce pollution to the point where society agrees the marginal cost of reducing pollution equals the marginal benefit.

Property Rights

The examples involving Mr. Big and Ms. Thrifty started with the assumption that everyone had the *property right* to consume as much gasoline as he or she wished. Then, that right was taken away from them. **Property rights** are a *set of use and ownership rules in society which dictate who may use or enjoy a particular resource* (without adversely affecting others' use and enjoyment of their resources).

Property rights define the ownership and use of resources in society.

When the property right (to unlimited gasoline) was lost, in order to consume excess amounts of gasoline, Mr. Big and Ms. Thrifty had to pay for the privilege. The result could have been achieved without changing the allocation of property rights in the society. Mr. Big and Ms. Thrifty could have retained their right to consume unlimited amounts of gasoline. Instead, society could have paid them to reduce consumption. For example, we could have implemented a system of tax credits for voluntary consumption reduction. With this solution, Mr. Big and Ms. Thrifty retain the right to consume as much gasoline as they wish, and the taxpayers must pay them to reduce their consumption.

The result (reduced gasoline consumption) would be the same regardless of the assignment of property rights. It does not matter whether Mr. Big and Ms. Thrifty are not allowed to consume the excess gasoline and must pay to do so, or whether they are allowed to consume the excess gasoline and must be paid to reduce consumption, the result will still be the same reduction of gasoline consumption.

Coase Theorem states that the initial assignment of property rights does not affect the market outcome.

This identical result is known as Coase Theorem, named after the economist who first described this result in 1960, Ronald Coase.[1] Coase Theorem states that an optimal allocation of resources (efficient market solution) can always be achieved through market forces regardless of the initial assignment of property rights.

TYPES OF GOODS

Most goods are produced by private profit-maximizing corporations, and exchanged in freely functioning markets. As we saw, when externalities are present, the performance of the market economy to allocate goods efficiently can be impaired. There are other conditions as well that can reduce the effectiveness of the market. Table 15-1 presents four types of goods, along with the characteristics of the market for each type of good.

Rival goods have the characteristic that if one person consumes one unit of a good, another person cannot consume the same unit.

Excludable goods are goods where access can be controlled. These goods are traded in an ordinary market economy.

Ordinary Market Economy In all cases of ordinary market goods, property rights are well-defined, so that a company that produces a good is able to sell it at a price greater than or equal to marginal cost to someone who is able to consume it, such that marginal utility is greater than or equal to price. Goods which exchange freely in markets possess two important characteristics: they are rival and excludable. **Rival** goods are *goods in which one person's consumption diminishes another person's ability to consume the good;* for example, if I eat an ice cream cone, you cannot. You must choose another unit of the good. With rival goods, other people are automatically prevented from consuming the good. Most goods are rival goods. Most goods are also **excludable**, which means that *access to the good can be controlled.* This has a very important implication: nonpayers can be excluded from consuming the good. The ice cream cone is both rival and excludable, and therefore is well suited to being exchanged in an ordinary market.

Nonrival goods are goods that can be jointly consumed.

Congestion Other categories of goods do not trade as well in an ordinary market, for example, nonrival excludable goods. **Nonrival goods** are goods which *can be jointly consumed*; that is, two people can both consume the same unit of the good, for example, a piece of artwork or a concert. Nonrival excludable goods are those which can be jointly consumed and to which *access can be controlled.* Concerts can be enjoyed by all who have paid to see them. Roads can be enjoyed by all who "pay" to drive; however, since almost all roads are offered at zero marginal cost to each additional driver, these types of goods sometimes can suffer from congestion problems. The solution to congestion is generally to charge a positive price for access; for example, to create toll roads.

[1]Coase Theorem requires perfect information and costless transactions, conditions which are not normally met in real life.

TABLE 15-1 Four Types of Goods

	Rival	Nonrival (Jointly consumable)
Excludable	Ordinary market economy *examples: shirts, haircuts*	Congestion *examples: art exhibits, roads and bridges*
Nonexcludable	Common Property *examples: fishery, atmosphere*	Public Goods *examples: National Defense, National Parks*

Common Property **Nonexcludable goods** (*goods to which access cannot be controlled*) are difficult and sometimes impossible for the market to handle. Because nonpayers cannot be excluded, private producers have no incentive to produce or to manage these goods. Nonexcludable goods are common property; everyone can use them. Nonexcludable rival goods are goods to which access cannot be controlled; moreover, one person's use of the good precludes the use of it by another. The common property problem usually emerges with goods having these two characteristics.

Nonexcludable goods are goods to which access cannot be controlled.

The most important Canadian common property problem is the fishery. No one has property rights over (owns) the fish; no one can stop nonpayers from fishing (setting aside fishing licences for the moment); no one can charge for fishing; so no one has an incentive to manage the fish stocks. Even worse, the fish are rival goods: if one person takes a fish, no one else can have it. This creates a huge incentive to take the fish before someone else does. There is no incentive to wait; in fact, waiting imposes a penalty: You lose the fish.

Public Goods Nonexcludable, nonrival goods are also an interesting group of goods. Their use cannot be controlled, and their use is not diminished by use by others. These types of goods are known as public goods.

PUBLIC GOODS

A **public good** is *a good that is nonexcludable (no one can be excluded from its benefits) and nonrival (consumption by one does not preclude consumption by others)*. In reality there likely is no such thing as a pure public good, but many of the goods that government provides—education, defense, roads, and legal systems—have public good aspects to them. Probably the closest example we have of a pure public good is national defense. A single individual cannot protect himself or herself from a foreign invasion without protecting his or her neighbors as well. Protection for one person means that many others are also protected. Governments generally provide goods with significant public aspects to them because private businesses will not supply them, unless they transform the good into a mostly private good.

What is and is not considered a public good depends to a large extent on technology. Consider roads—at one point roads were often privately supplied, since with horses and buggies and few roads to travel (few substitutes), the road owners could charge tolls relatively easily. Then, with the increased speed of the automobile, collecting tolls on most roads became too time-consuming. At that point the nonexcludable public good aspect of roads became dominant—once a road was built, it was most efficiently supplied to everyone at a zero cost—and government became the provider of most roads.

Because public goods are nonexcludable and nonrival, they are often provided by the government.

Today, with modern computer technology, sensors that monitor road use can be placed on roads and in cars. Charging for roads has once again become more feasible. In the future we may again see more private provision of roads. Some economists have even called for privatization of existing roads, and private roads are being built in California and in Bangkok, Thailand.

One of the reasons that pure public goods are sufficiently interesting to warrant a separate discussion is that a modification of the supply/demand model can be used to neatly contrast the efficient supply of a private good with the efficient supply of a public good. The key to understanding the difference is to recognize that once a pure public good is supplied to one individual, it is simultaneously supplied to all, whereas it is possible to supply a private good only to the individual who purchased it. For example, if the price of an apple is 50 cents, the efficient purchase rule is for individuals to buy apples until the marginal benefit of the last apple consumed is equal to 50 cents. The analysis focusses on the individual. If the equilibrium price is 50 cents, the marginal benefit of the last apple sold in the market is equal to 50 cents. That benefit is paid for by one individual and is enjoyed by one individual. Since only one individual can consume the apple, his benefit equals society's benefit.

15.1 see page 339

Now consider a public good. Say that the marginal benefit of an additional missile for national defense is 50 cents to one individual and 25 cents to another. In this case the value of providing one missile provides 75 cents (25 + 50) of total social benefit. With a public good the focus is on the group. The societal benefit in the case of a public good is the *sum* of the individual benefits (since each individual gets the benefit of the good). With private goods, we count only the benefit to the person buying the good, since only one person gets it.

The above reasoning can be translated into supply and demand curves. The market demand curve represents the marginal benefit of a good to society. As we saw in Chapter 4 in the case of a private good, the market demand curve is the *horizontal sum* of the individual demand curves. This is because the total amount of a private good supplied is divided amongst many buyers. While the market demand curve for a private good is constructed by adding all the quantities demanded at every price, the market demand curve in the case of public goods is the *vertical sum* of the individual demand curves at every quantity. The full benefit of the total output is received by everyone, so we must add up the value of the good to everyone, which is given by the price each is willing to pay.

With private goods you sum demand curves horizontally; with public goods you sum them vertically.

Figure 15-4 gives an example of a public good. In it we assume that society consists of only two households—A and B, with demand curves D_A and D_B. To arrive at the market demand curve for the public good, we vertically add the price that each individual is willing to pay for each unit, since both receive a benefit when the good is supplied. Thus, at quantity 1 we add \$0.60 to \$0.50. We arrive at \$1.10, the marginal benefit of providing the first missile. By adding together the willingness to pay by individuals A and B for quantities 2 and 3, we generate the market demand curve for missiles. Extending this example from two individuals to the economy as a whole you can see that, even though the benefit of a public good is small to each person, the total benefit is large. With 31 million people in Canada, the benefit of that missile would be \$15.5 million, even if each person valued it at 50 cents.

If people think they will have to pay for a public good, they may not reveal their willingness to pay.

Adding demand curves vertically is easy to do in textbooks, but not in practice. With private-good demand curves individuals reveal their demand when they buy a good. If they don't buy it, it wasn't worth the price. Since individuals do not purchase public goods, their demand is not revealed by their actions. Government must estimate their value. If a public good is to be financed by a tax on the citizens who benefit from it, individuals have an incentive to conceal their willingness to pay for it. This is why in the supply of public goods we see the free rider problem. The self-interested citizen

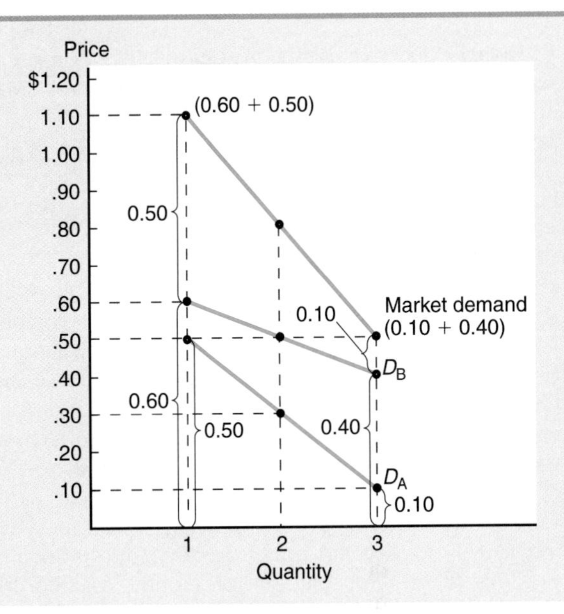

Figure 15-4 THE MARKET VALUE OF A PUBLIC GOOD

The market demand curve for a public good is constructed differently than for a private good. Since a public good is enjoyed by many people without diminishing its value to others, the market demand curve is constructed by adding the marginal benefit each individual receives from the public good at each quantity. For example, the value of the first unit to the market is $1.10, the sum of individual A's value ($0.50) and individual B's value ($0.60). In other words, vertically sum the individual demand curves to construct the market demand curve for a public good.

wants to benefit from the public good without bearing the cost of providing it. Similarly, if respondents think they will not be taxed, but will benefit from the public good, they have an incentive to exaggerate their willingness to pay.

Q-6 Why is it so difficult for government to decide the efficient quantity of a public good to provide?

Ottawa is home to the historic Rideau Canal (originally built to transport military goods) which flows through the middle of the city. In the summer, it is a recreational focal point for boaters and city residents alike. In the winter, the Rideau Canal is transformed into the world's longest skating rink. A number of years ago, the National Capital Commission proposed to charge skaters a fee to help defray the cost of maintaining the ice surface. There was a public outcry at this proposal: people who skated the canal on their commute to work, students who commuted between the two universities (conveniently located at either end of the canal), and people who skated for their daily exercise, as well as people involved in the tourist industry, all objected loudly. While some people felt that a small charge would be appropriate, there still remained a very big problem. The Rideau Canal is a nonrival nonexcludable good; everyone enjoys it together. It is a public good with a huge Free Rider problem. No one wished to see fencing erected along the entire length of the canal to control access, and the idea of patrolling the facility to check for payment was unpleasant and the cost of doing so prohibitively high. The solution? The National Capital Commission decided to rely on people's sense of community, so donation boxes were placed at various spots along the length of the Rideau Canal to accommodate people's social conscience.

Each year, Winterlude celebrations and the Rideau Canal Skateway attract over 600,000 visitors to the Ottawa/Hull area, many of whom enjoy a cool 7.8 kilometre skate on the world's longest skating rink (a nonrival, nonexcludable, public good).

INFORMATION PROBLEMS

The perfectly competitive model assumes that individuals have perfect information about what they are buying. So, if they voluntarily buy a good, it is a reasonable presumption that they expect that they are making themselves better off by doing so. But what happens if the buyer doesn't have perfect information? Say someone convinces you that he is selling you an expensive diamond, but it is actually glass. You are convinced, so you buy it. Or alternatively, say someone convinces you her used car is a cherry (in perfect condition). You buy it only to discover later that it is a lemon (faulty) and won't run no matter what you do to it.

APPLYING THE TOOLS

Public Goods and Free Riders

Stanley Park is an oasis of natural beauty in the heart of the noisy, crowded, busy city of Vancouver. A proposal to charge an access fee to Stanley Park was met with objections from many different people, ranging from those who used the park on a daily basis to those who had travelled specifically to visit the famous park. A survey of park users revealed tremendous differences in people's response to the proposed charge, and in their willingness to pay. When surveyed about paying a fee for using the park, those people who had travelled a great distance to visit the park—some from as far away as Europe—revealed the greatest willingness to pay, on average. Some even offered to pay, embarrassed that they had somehow walked right in without paying. Why would they be willing to pay to visit the park? Some visitors had already paid a substantial amount of money for the trip to Vancouver, and an extra $20 was insignificant compared to the expense they had already incurred. Local residents, however, who used the park every day, saw the issue in terms of a reallocation of property rights. Whereas they initially had the property rights granting them access to the park, the new fee would remove their right of access, making them pay for the privilege. Many local residents' willingness to pay was zero, and their average willingness to pay was very close to zero.

In economic terms, Vancouver's Stanley Park is a nonrival nonexcludable good, a public good characterized by the Free Rider problem. The park is already there, so the marginal cost of providing the park to another person is zero. In essence, Stanley Park fits into the lower right-hand box of Table 15-1. To collect a fee, large or small, would be uneconomic. This is why, in most cases of public goods, they are provided by the government, and collectively paid for by taxes.

Asymmetric information occurs when one person has information which is relevant to the exchange, but the other person does not. This information asymmetry (imbalance) can be a cause of market failure.

Real-world markets often involve inaccurate or concealed information, and outright cheating and deception. For example, car dealers usually know about defects in the cars they sell but do not always reveal those defects to consumers. Another example: Consumers who want life insurance sometimes do not reveal their health problems to the insurance company. In both cases, it is in the interest of the knowledgeable person to conceal information that the other person or firm would need to know to make an informed decision about the transaction. Hence, **asymmetric information** can be a cause of market failure. Transactions may take place when they otherwise would not if both parties had equal access to the available information (the insurance example), or transactions which would be socially desirable may not occur due to the asymmetry in information (the car example).

When *buyers and sellers don't have equal information*, markets for some goods may not work well. Consider the market for used cars. Owners of used cars generally know much more about their cars than buyers do. If sellers are profit maximizers, they will reveal as little as possible about the car's defects and emphasize its good qualities.

The characteristics of an experience good are often learned by using the good, sometimes over a period of time.

To be specific, let's say that only two types of cars exist—poor quality "lemons" that are worth $4,000 and good quality "cherries" that are worth $8,000. The market initially consists of equal quantities of lemons and cherries. Buyers cannot distinguish between lemons and cherries simply by looking at the vehicles. To find out whether a particular car is a "lemon" or a "cherry" requires a person to drive it for a while. It is an **experience good**—*a person learns the characteristics of a good through experience*. What will happen? Individuals, knowing that they have a 50 percent chance of buying a lemon, may offer $6,000 (the average of $4,000 and $8,000). Given that price, individuals with cherries will withdraw their cars from the market, not wishing to sell an $8000 car for $6000, and individuals with lemons, valued at $4000, will be happy to sell. Eventually, buyers will recognize that all the sellers of cherries have left the market. If the market unravels completely, in the end only lemons will be offered for sale, and buyers will only offer $4,000 with the correct expectation that all the cars offered would be lemons.

When the cherries—good used cars—have disappeared from the market, the result is a market failure.

Such a market failure is called an **adverse selection problem**—*a problem that occurs when buyers and sellers have different amounts of information about the good for sale*. In the case of adverse selection, only lemons are selected to remain in the market.

Adverse selection problems can occur when buyers and sellers have different amounts of information about the good for sale.

Insurance providers need to make a profit. To do so, they set rates that reflect their estimate of the costs of providing insurance. Consider auto insurance. The problem is that individuals have better information about their driving habits than does the insurance provider. Insurers want a diverse group of drivers to spread out the costs, but face a greater demand among those with the worst driving records. Seeing these drivers have more claims than average, insurance providers raise the rates. Those who have few claims find those charges to be too high and reduce the quantity of insurance they purchase. The providers, therefore, are left with a group with an even higher incidence of claims and higher costs than the general population of drivers. Less than the desired amount of low-cost insurance exists for good drivers. Public provision is often the answer. By government providing the insurance, all drivers go into the same insurance pool, so the low risk drivers balance the high risk ones. The market does not suffer adverse selection, and the result is a lower price and better coverage for all.

Q-7 Why do sellers sometimes offer free samples?

Workplace safety is another example of asymmetric information causing market failure. Although businesses have an incentive to provide a safe working environment, to limit costs, they may not choose a level of safety that would be preferred by employees. If the employer does not disclose unsafe working conditions, and those conditions cannot be easily identified by workers, there is an asymmetric information problem. The result is that employees may not be adequately compensated for the risks they face.

Q-8 How would you expect travel insurance rates to change if travel insurers could use information contained in DNA to predict the likelihood of major medical illnesses?

POLICIES TO DEAL WITH INFORMATION PROBLEMS

What should society do about informational problems that lead to market failures? One answer is to regulate the market and see that individuals provide the right information. Another is for the government to license individuals in the market, requiring those with licenses to reveal full information about the good being sold. Government has set up numerous regulatory commissions and passed laws that require full disclosure of information. The Competition Bureau, the Canadian Labour Relations Board, and the National Transportation Agency, as well as provincial licensing boards, are all examples of regulatory solutions designed to partially offset informational market failures.

But these regulatory solutions have problems of their own. The commissions and their regulations introduce restrictions on individuals that can potentially slow down the economic process, and can prevent trades that people want to make. Consider as an example the U.S. Food and Drug Administration (FDA). It restricts what drugs may be sold until sufficient information about the drugs' effects can be disclosed. The FDA testing and approval process can take 5 or 10 years, is extraordinarily costly, and raises the price of drugs. The delays have caused some people to break the law by taking the drugs before they are approved.

see page 339

A Market for Information

Economists who lean away from government regulation suggest that the market failures presented above are the result of the lack of a market in information. What does this mean? Information is valuable, and is an economic product in its own right. Left on

Some market failures may be due to the lack of a market in information.

"Lemon" or "cherry"? In what ways does the market help a buyer determine a car's characteristics?

their own, markets will develop to provide the information that people need, and are willing to pay for. For example, a large number of consumer magazines provide such information. In the car example, the buyer can hire a mechanic who can test the car with sophisticated diagnostic techniques and determine whether it is likely a cherry or a lemon. Firms can offer guarantees that will provide buyers with assurance that they can either return the car or have it fixed if the car is a lemon. There are many variations of such market solutions. If the government regulates information, these markets may not develop; people might rely on government instead of markets. This solution is appropriate for many markets in which asymmetric information leads to some degree of market failure, such as used car sales.

Licensing of Doctors

Let's consider another informational problem that contrasts the market approach with the regulatory approach: medical licensing.[2] Currently all doctors are required to be licensed in order to practice, but this was not always the case.

In the early 1800s, medical licenses were not required to practice medicine, so anyone who wanted to could set up shop as a physician. However, it is illegal to practice medicine in Canada today without a license.

Licensing of doctors is justified by information problems. Since individuals often don't have an accurate way of deciding whether a doctor is good, government intervention is necessary. The information problem is solved because licensing requires that all doctors have at least a minimum competency. People have the *information* that a doctor must be minimally competent because they see the license framed and hanging on the doctor's office wall.

Some economists argue that licensure laws were established to restrict supply, not to help the consumer.

A small number of economists, of whom Milton Friedman is the best known, have proposed that licensure laws be eliminated, leaving the medical field unlicensed. They argue that licensure was instituted as much, or more, to restrict supply as it was to help the consumer.

Even the strongest critics of licensure agree that, in the case of doctors, the informational argument for government intervention is strong. But the question is whether licensure is the right form of government intervention. Why doesn't the government simply provide the public with information about which treatments work and which don't? That would give the freest rein to *consumer sovereignty (the right of the individual to make choices about what is consumed and produced)*. If people have the necessary information but still choose to treat cancer with laetrile or treat influenza with massive doses of vitamin C, why should the government tell them they can't?

If the informational alternative is preferable to licensure, why didn't the government choose it? Friedman argues that government didn't follow that path because the licensing was done as much for the doctors as for the general public. Licensure has led to a monopoly position for doctors. They can restrict supply and increase price and thereby significantly increase their incomes.

Let's now take a closer look at the informational alternative.

The Information Alternative to Licensure

The informational alternative is to allow anyone to practice medicine, but to have the government certify doctors' backgrounds and qualifications. The government would require that doctors' backgrounds be made public knowledge. Each doctor would have to post the following information prominently in his or her office:

[2]The arguments presented here about licensing doctors also apply to dentists, lawyers, teachers, financial analysts, stockbrokers, realtors, and other professional groups.

1. Grades in college.
2. Grades in medical school.
3. Success rate for various procedures.
4. References.
5. Medical philosophy.

According to supporters of the informational alternative, these data would allow individuals to make informed decisions about their medical care. Like all informed decisions, they would be complicated. For instance, doctors who take only patients with minor problems can show high "success rates," while doctors who are actually more skilled but who take on problem patients may have to provide more extensive information so people can see why their success rates shouldn't be compared to those of the doctors who take just easy patients. But despite the problems, supporters of the informational alternative argue that it's better than the current situation.

Current licensure laws don't provide any of this information to the public. All a patient knows is that a doctor has managed to get through medical school and has passed the medical exams. The doctor may have done all this 30 years ago, but, once licensed, a doctor is a doctor for life.

The biggest problem in this case, and in many other cases involving asymmetric information, is that people do not possess the skill and training to accurately assess the information, here, the doctor's qualifications and the merits of alternate treatments. Information is expensive, requiring the expenditure of resources—time, money, and effort—to collect and analyze the information. In addition, you can't be an expert in everything; there are many decisions to make.

GOVERNMENT FAILURE AND MARKET FAILURES

The above three types of market failure—externalities, public goods, and asymmetric information problems—give you a good sense of how markets can fail to provide an efficient outcome, in which the social marginal benefit equals the social marginal cost. All real-world markets fail in some way. But the point of the above discussions was to provide you not only with a sense of the way in which markets fail but also with a sense that economists know that markets fail. A market failure may indicate a role for government intervention, but this must be carefully evaluated also. Why? The reason can be called **government failure**—when *the government intervention in the market to improve the market failure actually makes the situation worse*.

Q.9 Would an economist necessarily believe that we should simply let the market deal with a pollution problem?

Why are there government failures? Let's briefly list some important reasons:

1. *Government doesn't have an incentive to correct the problem*. Government reflects politics, which reflects individuals' interests in trying to gain more for themselves. Political pressures to benefit some group or another will often dominate doing the general good.
2. *Governments don't have enough information to deal with the problem*. Regulating is a difficult business. To intervene effectively, even if it wants to, government must have good information, but just as the market often lacks adequate information, so too does the government. Asymmetric information may also be a problem for the government.
3. *Intervention in markets is almost always more complicated than it initially seems*. Almost all actions have unintended consequences. Government attempts to

offset market failures can prevent the market from dealing with the problem more effectively. The difficulty is that generally the market's ways of dealing with problems work only in the long run. As government deals with the short-run problems, it eliminates the incentives that would have brought about a long-run market solution.

Q-10 If one accepts the three reasons for market failure, how might one still oppose government intervention?

4. *The bureaucratic nature of government intervention does not allow fine-tuning.* When market conditions change, the government solution often responds far more slowly.
5. *Government intervention leads to more government intervention.* Given the nature of the political process, allowing the government to intervene in one market often brings the government to intervene in other areas where intervention is harmful. Even in those cases where government action may seem likely to do some good, it might be best not to intervene, if that intervention will lead to additional government action in cases where it is not likely to do good.

CONCLUSION

Most Canadian goods and services are produced and exchanged in the framework of a market economy; however, in some cases, market forces are not sufficient to generate the desired results. Where there are externalities or where property rights are poorly defined, markets have difficulty allocating resources appropriately. In such cases, we can define a role for government. Where a market fails completely, government provision of the good is often prescribed. Where a market fails to achieve an efficient result, the government may step in and regulate the market. For example, when (negative) externalities are present, an industry will tend to overproduce, so the government may impose regulations which force the firms in the industry to take into account all costs of production, thereby suppressing the incentives to overproduce. Or, in the case of information asymmetries, government may assist the functioning of the market by imposing licensing requirements, which forces market participants to reveal their information (as in mutual fund sales).

Should the government intervene in the market? It depends.

Whether society supports a role for government in a particular market depends on the characteristics of the market, and on the goals of society. As we saw, rival, excludable goods have characteristics that allow for their efficient production and exchange under a market system. Generally, government intervention is not required in these markets, except, as we will see in chapter 17, where issues other than efficiency arise. So, should the government intervene in a market? The answer is "it depends."

Chapter Summary

- Three sources of market failure are externalities, public goods, and asymmetric information.
- An externality is the effect of a decision on a third party that is not taken into account by the decision maker. Positive externalities provide benefits to third parties. Negative externalities impose costs on third parties.
- The markets for goods with negative externalities produce too much of the good for too low of a price. The markets for goods with positive externalities produce too little of the good for too high a price.
- Economists generally prefer incentive-based programs over regulatory programs because incentive-based programs are more efficient. An example of an incentive-based program is to tax the producer of a good that results in a negative externality by the amount of the externality.
- Voluntary solutions are difficult to maintain for long periods of time because other people have an incentive to be free riders—to enjoy the benefits of others' volunteer efforts without putting forth effort themselves.

- An optimal policy is one in which the marginal cost of undertaking the policy equals its marginal benefit.
- Market goods are excludable. Their access can be controlled, so firms willingly produce and sell these products.
- Public goods are nonexcludable and nonrival. It is difficult to measure the benefits of public goods because people do not reveal their preferences by purchasing them in the marketplace.
- Theoretically, the value of a public good can be calculated by summing the value that each individual places on every quantity. This is done by vertically summing individual demand curves.
- Individuals have an incentive to withhold information that will move the price against their interest (lower price if one is a seller and a higher price if one is a consumer). Because of this incentive to withhold information, the markets for some goods unravel. Such market failures are a subset of asymmetric information problems known as adverse selection problems.
- Licensing and full disclosure are two solutions to the asymmetric information problem.
- Government intervention may worsen the problem created by the market failure. Government failure occurs because (1) governments don't have an incentive to correct the problem, (2) governments don't have enough information to deal with the problem, (3) intervention is more complicated than it initially seems, (4) the bureaucratic nature of government precludes fine-tuning, and (5) government intervention often leads to more government intervention.

Key Terms

adverse selection problem *(333)*
asymmetric information *(332)*
direct regulation *(324)*
efficient *(324)*
effluent fees *(325)*
experience good *(332)*
externality *(321)*
free rider problem *(327)*
government failure *(335)*
inefficient *(324)*
excludable *(328)*
market failure *(319)*
market incentive plan *(325)*
market outcome failure *(319)*
negative externality *(321)*
nonexcludable *(329)*
nonrival *(328)*
optimal policy *(327)*
positive externality *(321)*
private marginal benefit *(323)*
private marginal cost *(321)*
property rights *(327)*
public good *(329)*
rival *(328)*
social marginal benefit *(323)*
social marginal cost *(322)*
tax incentive program *(324)*

Questions for Thought and Review

1. State three reasons for a potentially beneficial role of government intervention.
2. Explain why a market incentive program is more efficient than a direct regulatory program.
3. How would an economist likely respond to the statement "There is no such thing as an acceptable level of pollution"?
4. Would a high tax on oil significantly reduce the amount of pollution coming from the use of oil? Why or why not?
5. Would a high tax on oil significantly reduce the total amount of pollution in the environment?
6. What is the allocation of property rights in a city where pedestrians have the "right-of-way" in crosswalks? What do the laws prohibiting "jay-walking" (crossing the street in the middle of a block instead of at the intersection) imply about these property rights?
7. If you are willing to pay $1,000 for a used stereo that is a "cherry" and $200 for a used stereo that is a "lemon," how much will you be willing to offer to purchase a stereo if there is a 50 percent chance that the stereo is a lemon? If owners of cherry stereos want $700 for their cherries, how will your estimate of the chance of getting a cherry change?
8. Define the adverse selection problem. Does your understanding of adverse selection change your view of commercial dating services? If so, how?
9. If neither buyers nor sellers could distinguish between "lemons" and "cherries" in the used-car market, what would you expect to be the mix of lemons and cherries for sale?
10. Automobile insurance companies charge lower rates to married individuals than they do to unmarried individuals. What economic reason is there for such a practice? Is it fair?
11. Should government eliminate Health Canada's role in restricting which drugs may be marketed? Why or why not?
12. List five ways you are affected on a daily basis by government intervention in the market. For what reason might government be involved? Is that reason justified?

Problems and Exercises

1. There's a gas shortage in Gasland. You're presented with two proposals that will achieve the same level of reduction in the use of gas. Proposal A would force everybody to reduce their gas consumption by 5 percent. Proposal B would impose a 50-cent tax on the consumption of a gallon of gas, which would also achieve a 5 percent reduction. Demand curves for two groups are shown below.
   a. Show the effect of both proposals on each group.
   b. Which group would support a regulatory policy? Which would support a tax policy?

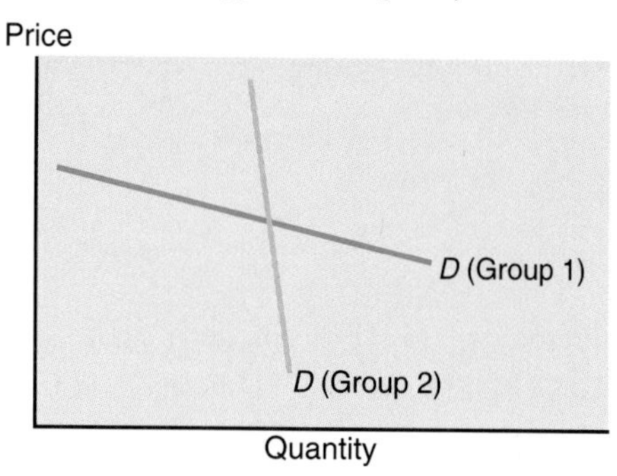

2. The marginal cost, social marginal cost, and demand for fish are represented by the curves in the following graph. Suppose that there are no restrictions on fishing.
   a. Assuming perfect competition, what is the catch going to be, and at what price will it be sold?
   b. What are the socially efficient price and output?
   c. Some sports fishers propose a ban on commercial fishing. As the community's economic adviser, you're asked to comment on it at a public forum. What do you say?

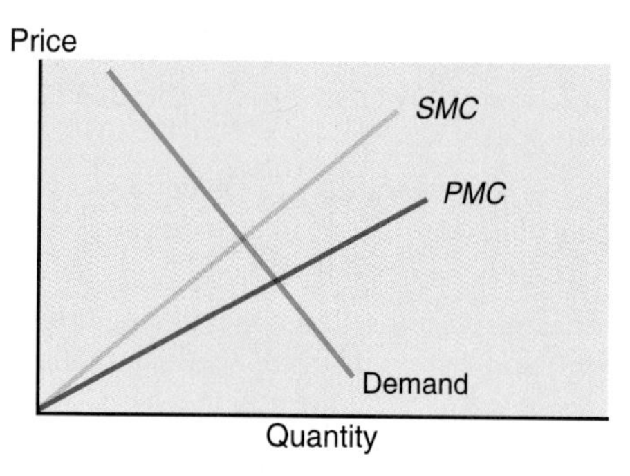

3. You are in Seattle, Washington, watching the Seattle Stomp—a dance home owners do in their trash cans.
   a. What can you say about trash fees in Seattle? Be as specific as possible.
   b. What change in fee structure might eliminate the Seattle Stomp?
   c. In 2001, the city of Burnaby, B.C., disallowed garbage cans larger than 77 litres and heavier than 20 kilograms. Do Burnaby residents do a "Burnaby Bounce" with their garbage cans? Explain.
   d. Do you give your garbage the Hometown Heel? Why or why not?
4. In *At the Hand of Man*, Raymond Bonner argues that Africa should promote hunting, charging large fees for permits to kill animals (for example, $7,500 for a permit to shoot an elephant).
   a. What are some arguments in favour of this proposal?
   b. What are some arguments against it?
5. California passed an air quality law that required 10 percent of all the cars sold in the state to emit zero pollution by 2003.
   a. What was the likely impact of this law?
   b. Can you think of any way in which this law might actually increase pollution rather than decrease it?
   c. How might an economist suggest modifying this law to better achieve economic efficiency?
6. Economics professors Thomas Hopkins and Arthur Gosnell of the Rochester Institute of Technology estimated that in the year 2000, regulations cost the United States $662 billion, or about $5,700 per family.
   a. Do their findings mean that the United States has too many regulations?
   b. How would an economist decide which regulations to keep and which to do away with?
7. In the 1990s a debate about dairy products concerned the labelling of milk produced from cows that have been injected with the hormone BST, which significantly increases their milk production. Since this synthetically produced copy of a milk hormone is indistinguishable from the hormone produced naturally by the cow, and milk from cows treated with BST is indistinguishable from milk from untreated cows, some people have argued that no labelling requirement is necessary. Others argue that the consumer has a right to know.
   a. Where do you think most dairy farmers stand on this labelling issue?
   b. If consumers have a right to know, should labels inform them of other drugs, such as antibiotics, normally given to cows?
   c. Do you think dairy farmers who support BST labelling also support the broader labelling law that would be needed if other drugs were included? Why?
8. Financial analysts are currently required to be licensed. Should they be licensed? Why or why not?
9. An advanced degree is required in order to teach at most postsecondary institutions. In what sense is this a form of restricting entry through licensing?
10. Who would benefit and who would lose if an informational alternative to licensing doctors were introduced?

Web Questions

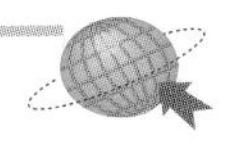

1. Under the "Communities and Schools" series at the Caledon Institute of Social Policy, at www.caledoninst.org, click on "Education and the Public Good."
   a. What is a public good? What concerns led to the creation of Canada's system of social security?
   b. In what sense is support for public goods an investment in Canada's economic future? In what sense is education an asset?
   c. How does John Ralston Saul see the relationship between education and democracy? Why is he worried?
2. Go to the Nobel Foundation Web site at www.nobel.se and select the Economics tab.
   a. According to Assar Lindbeck, what criteria have been used to award the prizes for Economics?
   b. In what year did A. Michael Spence (Stanford University), George A. Akerlof (University of California, Berkeley), and Joseph E. Stiglitz (Columbia University) jointly win the Bank of Sweden Prize in Economic Sciences in Memory of Alfred Nobel? Briefly describe the "contribution" each made to the field of economics. What is Spence's "Canadian connection"?
   c. In what year did Ronald Coase receive his Prize? Briefly describe his contribution to economics.
   d. The 2001 movie, "A Beautiful Mind," relates the remarkable story of mathematician/economist John F. Nash. In what year was Nash chosen to become a Nobel laureate? What was Nash's contribution to economics? What information assumptions does Nash equilibrium require?

Answers to Margin Questions

1. An externality is an effect of a decision not taken into account by the decision maker. When there are externalities, the private price no longer necessarily reflects the social price, and therefore the market may not work properly. *(321)*
2. No. The existence of a positive externality does not mean that the market works better than if no externality existed. It means that the market is not supplying a sufficient amount of the resource or activity, and insufficient supply can be as inefficient as an oversupply. *(322)*
3. An example of the trade-off between fairness and efficiency is whether to allow nontaxpayers to enjoy the benefits of a public park maintained through local taxes. It would cost too much to exclude them from enjoying the park, so the exclusion is inefficient, but not to exclude that person is unfair to the taxpayers who pay to maintain the park. An example of a policy that might be seen as both fair and efficient is a gas tax designed to deter pollution. Consumers choose to reduce their gas use based on the new price, so the solution is efficient. The solution has an element of fairness in it since those causing the pollution are those paying more. *(324)*
4. The tax incentive approach to pollution is fair in the following sense: individuals whose actions result in more pollution pay more. Individuals whose actions result in less pollution pay less. In some broader sense this may not be fair if one takes into account the initial positions of those polluting. For example, the poor may have older cars which get fewer miles per gallon and have to pay a higher cost of pollution resulting from gasoline use. *(325)*
5. Voluntary actions that are not in people's self-interest may not work in large groups because individuals will rely on others to volunteer. There is also a potential lack of efficiency in voluntary solutions since the person who voluntarily reduces consumption may not be the person who faces the least cost of doing so. *(327)*
6. It is difficult for government to decide the efficient quantity of a public good because public goods are not purchased by individuals in markets, so individuals do not reveal the value they place on public goods. Individuals also face incentives to overstate the value they place on public goods if they do not have to pay for them, and to understate the value if they do have to share the cost. *(331)*
7. Sellers do this because customers cannot get enough relevant information simply by looking at the product. This is very common with food items—they are experience goods. *(333)*
8. Since adverse selection is a problem in the travel insurance industry, with fuller information, we would expect that average rates would decline since the adverse selection problem would disappear. Insurers would be able to offer lower-cost insurance to people who are less likely to get sick and who perhaps choose not to be covered at higher rates. *(333)*
9. An economist would not necessarily believe that we should simply let the market deal with the pollution problem. Pollution clearly involves externalities. Where economists differ from many laypeople is how they would handle the problem. An economist is likely to look more carefully into the costs, try to build price incentives into whatever program is designed, and make the marginal private cost equal the marginal social cost. *(335)*
10. One can accept all three explanations for market failure and still oppose government intervention if one believes that government intervention will cause worse problems than the market failure. *(336)*

Competition Policy

16

After reading this chapter, you should be able to:

- Explain why a country needs competition policy, and describe the objectives of Canadian competition policy.
- Describe the main anticompetitive offences and how they are dealt with in the 1986 Canadian *Competition Act*.
- Explain the difference between the structure and performance methods of judging competition.
- Give a brief history of Canadian competition policy.
- State the issues involved in the Microsoft antitrust case.
- Differentiate among horizontal, vertical, and conglomerate mergers.
- List five reasons why unrelated firms would want to merge.
- Compare Canadian competition policy with competition policy of the U.S. and other countries.
- List three alternatives to competition policy that government can use to affect the competitive process.
- Describe the impact of globalization on competitiveness issues.

We have always known that
heedless self-interest was bad morals;
we now know that it is bad economics.

Franklin Delano Roosevelt

The relatively small size of the Canadian market suggests that in some industries, efficient production requires a high degree of concentration, since firms must be large enough to benefit from economies of scale and other cost advantages. The importance of a competitive position in world markets has been enhanced by the convergence of global competitive forces. The momentum of globalization has substantially affected Canadians' views of competition. Generally, it was recognized that in a small open economy such as Canada's, international competition can operate to discipline anti-competitive behaviour. The Canadian experience, however, suggests that the disciplinary effects of international competition have been relatively small and that the dominance of firms within a relatively small domestic market has had the greater influence. This has led to the need for a uniquely Canadian competition policy.

Q.1 What factors influence the level of competitiveness in Canadian industries?

COMPETITIVE MARKETS

We saw in Chapter 11 that perfectly competitive markets were characterized by a very large number of small firms, each producing a homogeneous (identical) product, facing a perfectly elastic demand curve, and competing on the basis of price alone. No firm held any degree of market power, and no differentiation of product between firms was possible. This led to the desirable twin outcomes of productive and allocative efficiency. Productive efficiency refers to the occurrence of production at lowest cost, while allocative efficiency refers to the equality of market price with marginal cost. We considered the perfectly competitive market structure to be a benchmark against which we would compare other market structures.

CONTESTABLE MARKETS

Q.2 What is meant by "costless" exit in a contestable market?

However, the conditions that characterize a perfectly competitive market are not generally found in industry. So the efficiencies achieved by perfectly competitive markets are also difficult to find in industry. What we do tend to find, however, are conditions that would make a market *contestable*. A **contestable market** is one where entry is free and exit is costless, so that the existence of profits will attract entry. **Sunk costs** are costs that cannot be transferred to a new use, and are therefore lost upon exit. Contestability implies that even a highly concentrated industry can operate in a relatively competitive manner. However, the higher the sunk costs, the less contestable — and the less competitive — a market will be. Production in many Canadian industries involves significant sunk investments, which, by raising the cost of exit, lower the probability of entry in response to economic profits.

COMPETITIVE MARKETS, CONTESTABLE MARKETS, AND THE ROLE OF COMPETITION POLICY

The perfectly competitive model, with its market characteristics of many firms, all selling an identical product, facing a perfectly elastic demand curve where even the slightest increase in price by a firm will send consumers fleeing to the competition, is clearly an abstract economic construct. In fact, although we refer to the efficiency outcomes of the perfectly competitive model, it cannot represent valid criteria for government policy in the real world. The assumptions of the perfectly competitive model are unrealistic and inappropriate as guidelines for government intervention in markets.

Contestable markets, on the other hand, are characterized by a credible threat of entry, which tends to reduce the extent to which existing firms exert their market dominance and earn economic profits. However, since production generally is characterized by at least some amount of sunk cost, and some Canadian industries have significant sunk costs, market discipline by means of contestability is less effective than Canadian consumers wish. Because of high sunk costs, in some industries, price must be significantly higher than marginal cost for entry to be a threat to existing firms.

Because many Canadian industries have large sunk costs, price can be much higher than cost without attracting entry.

Since perfectly competitive markets do not exist, and since contestability does not characterize many Canadian industries, Canadian firms are not necessarily constrained to behave competitively. Because profit incentives are strong and since non-competitive strategies can lead to larger profits, as we saw in Chapter 13, firms may not behave in ways that maximize consumer surplus. When this happens, we can define a role for government in generating competitive market outcomes by defining the conditions under which firms compete in the marketplace.

COMPETITION POLICY AND ECONOMIC GOALS

What we desire is an economy that performs in the most efficient way possible. We want to see the economy using its available resources in a way that contributes most to increasing the total value of output. We would like to see the free movement of resources from one use to another, better use as demands change and technology advances. In Canada, competition generally has been a less effective means of controlling firm behaviour and balancing market power than consumers would hope. So, when natural market forces in the economy fail to generate the desired competitive outcome, this usually indicates a role for government intervention to set the rules of competitive behaviour to level the playing field so that all firms may compete on an equal basis. Since many corporations wield considerable power and command vast resources, their economic and social responsibilities in the marketplace need to be defined.

In addition to economic efficiency, other important objectives include diffusing economic power and equalizing income distribution. Canadian competition policy concentrates on issues of economic efficiency, primarily price stability and rapid economic growth. The free operation of market forces allows the productive process to operate effectively to achieve efficiencies and thereby improve growth prospects for the industry and for the country. Improved operation of the market creates market signals which result in a pattern of output that is more closely related to consumer demand. Competition policy's emphasis on efficiency may also promote price stability by reducing the opportunity for firms to exercise market power, making it more difficult to pass rising costs on to consumers. Enhanced competition encourages firms to search for more efficient ways of operating, which reduces costs and, therefore, prices.

The goals of competition policy include economic efficiency—stable prices and rapid economic growth—as well as full employment, viable balance of payments, and the ability of Canadian firms to respond to global competitive forces.

Secondary benefits include full employment and a viable balance of international payments. Although fiscal policies (for example, tax policies) and monetary policies (for example, interest rate setting) are the main policy instruments for achieving the economic goal of full employment, competition policy indirectly helps the economy move towards full employment by moderating prices and encouraging economic growth in general. And a higher degree of competitiveness in Canadian industries enables the Canadian economy to better deal with competition at an international level; this aspect is currently gaining importance due to globalization.

RULES FOR CANADIAN COMPETITION POLICY

Judgment by Performance or Structure?

Competition policy is *the government's policy toward the competitive process*. It's the government's rulebook for carrying out its role as referee. In business a referee is needed for such questions as: When can two companies merge? What competitive practices are legal? When is a company too big? To what extent is it fair for two companies to coordinate their pricing policies? When is a market sufficiently competitive or too monopolistic?

Competition policy is government's policy toward the competitive process.

Canada has seen wide swings in economists' prescriptions concerning such questions, depending on which of the two views of competition has held sway. The two competing views are:

1. **Judgment by performance:** *We should judge the competitiveness of markets by the* performance *(behaviour) of firms in that market.*
2. **Judgment by structure:** *We should judge the competitiveness of markets by the* structure *of the industry.*

Judgment by performance is the view that competitiveness of a market should be judged by the behaviour of firms in that market; judgment by structure is the view that competitiveness of a market should be judged by the structure of that market.

To show how the Canadian government has applied these two views of competition in promoting workable and effective competition, this chapter considers government's application of competition laws to regulate business. It then considers how recent structural changes in the economy are altering the government's role in refereeing the market.

The Role for Rules

Large firms and monopolies are sometimes the result of mergers that reflect technological changes in production and expanding transportation systems which have made increased economies of scale more important. Despite their size, we would expect these companies to behave competitively.

If competition was strong, it would limit the profit large firms and monopolies made and force them to charge the competitive price. This would depend on how fast economic forces operated and how fragile competition was. This reflects the performance viewpoint—that competition should be relied on to break down monopolies. Bigness doesn't necessarily imply the absence of market competition, and the government's role should merely be to make sure that no significant barriers to entry are created.

Economists reflecting the performance viewpoint argued that competition was strong and would ultimately limit monopolies. Economists reflecting the structure viewpoint argued that monopolies should be broken up by government.

The structure viewpoint argues that competition is fragile and that it wouldn't operate unless there were a large number of small firms. Large firms and monopolies (even if they don't charge monopolistic prices) are bad, and should be broken up by government, and laws should not allow new monopolies to be formed.

Judging Markets by Structure and Performance: The Reality

Judgment by structure seems unfair on a gut level. After all, in economics the purpose of competition is to motivate firms to produce better goods than their competitors are producing, and to do so at lower cost. If a firm is competing so successfully that all the other firms leave the industry, the successful firm will be a monopolist, and on the basis of judgment by structure will be guilty of anticompetition violations. Under the judgment-by-structure criterion, a firm is breaking the law if it does what it's supposed to be doing: producing the best product it can at the lowest possible cost.

Supporters of the judgment-by-structure criterion recognize this problem but nonetheless favour the structure criterion. An important reason for this is practicality.

An important reason supporting the structure criterion is practicality.

APPLYING THE TOOLS

Wal-Mart and Competition

Wal-Mart expanded aggressively throughout the United States in the 1980s and early 1990s, reaching a total of about 3,000 stores in 1996. Then it moved into Canada. Not surprisingly, in its first seven years, Wal-Mart established itself as a major force in Canadian retailing, rising to number one merchandiser with $8 billion in sales from 178 stores. Although Canadian retailers may not wish it, Wal-Mart believes that Canadian consumers will support 300 Wal-Mart stores, which would represent a larger presence in Canadian markets than in its home market of the U.S. It does not deny that it, like many other stores, sells some goods at below cost. But it argues that when it sells below cost it does not do so to "destroy competition" or "injure competitors," but rather to maintain low prices for consumers. It claims that its pricing policies promote, not destroy, competition.

In principle, most economists agree with Wal-Mart; new competition, whether domestic or foreign, by its very nature, hurts existing businesses—that's the way the market competitive process works. Those who don't sell for the lowest price lose, and those who sell for the lowest price gain. But most economists also recognize that Wal-Mart's brand of competition can have externalities affecting the social fabric of small-town economies. A new Wal-Mart store can undermine the town centres and replace them with commercial sprawl on the outskirts of these towns. Whether these externalities are a reason to limit Wal-Mart's aggressive pricing policies is a debatable question.

Canadian competitors, such as The Bay, Sears, and Loblaw, are subject to the discipline of Wal-Mart's aggressive pricing strategies, and must exploit their own competitive strategies to attract the price-conscious shopper. Sears and The Bay in particular, have been diligently marketing customer loyalty programs, while Loblaw has been pricing aggressively and expanding its product selection in order to stave off further encroachment on its territory by Wal-Mart.

Contextual Judgments and the Capabilities of the Courts Judgment by performance requires that each action of a firm be analyzed on a case-by-case basis. Doing that is enormously time-consuming and expensive. In some interpretations, actions of a firm might be considered appropriate competitive behaviour; in other interpretations, the same actions might be considered inappropriate.

Another argument in favour of judging competitiveness by structure is that structure can be a predictor of future performance. Advocates of the structure criterion argue that a monopolist may be pricing low now, but it is, after all, a monopolist, and it won't price low in the future. The low price will eliminate competition now, and, once the competition is gone, the firm will not be able to resist the temptation to use its monopoly power. There is a fairly high correlation between market concentration (market structure) and firm performance.

Choosing the relevant market when evaluating competitiveness is difficult to do.

Determining the Relevant Market and Industry Supporters of the performance criterion admit that this standard has problems, but they point out that the structure criterion also has problems. As you saw in Chapter 13, it's difficult to determine the relevant market (local, national, or international) and the relevant industry (three-digit or five-digit NAICS code) necessary to identify the structural competitiveness of any industry.

Both structure and performance criteria have ambiguities, and in the real world there are no definitive criteria for judging whether a firm has violated the competition laws.

What should one make of this debate? The bottom line is that both structure and performance criteria are important in judging whether a firm has violated competition laws.

HISTORY OF COMPETITION POLICY IN CANADA

Competition policy is the regulatory framework that government uses to encourage or prohibit certain market practices. Canadian competition regulation began in 1889, with the first legislation directed at criminalizing conspiracy. In 1910, with the *Anti-combines Act*, merger and monopoly provisions were included in the legislation. A **combine** was defined as *any trust, merger, or monopoly which operated or was likely to operate to the detriment or against the interest of the public*. Under the legislation, six persons could apply to a judge for an order directing an investigation into an alleged combine. In 1923, the *Combines Investigation Act* was created, and this *Act* (plus amendments to it in 1960) was the law until 1986, when the new *Competition Act* was introduced.

The *Combines Investigation Act* of 1960

see page 369

Significant changes were made to the *Combines Investigation Act* in 1960. The term "combine" was dropped, but prohibitions against monopoly and merger offences were retained in Section 33. The criminal code section of the original 1889 conspiracy act was brought in as Section 32(1). Other *Combines Investigation Act* sections dealt with such anticompetitive behaviours as collusive agreements or conspiracy (Section 32), price discrimination (Section 34), misleading advertising (Sections 36 and 37), and resale price maintenance (Section 38), in which a supplier required a distributor to sell its product at a predetermined price. Remedies included fines, loss of tariffs, loss of patent protection, and publicity.

Conspiracy, price discrimination, misleading advertising, and resale price maintenance were offences under the *Combines Investigation Act*.

The *Combines Investigation Act* was moderately successful, with some sections of the *Act* achieving high success in prosecution, while other sections, notably the merger and monopoly section, saw poor results. From 1890 to 1983, 75% of the cases brought under Section 32(1) (Conspiracy) were won by the Crown, 25% of cases brought under Section 34 (Price Discrimination) were won, and 77% of cases brought under Section 38 (Resale Price Maintenance) were won. In cases prosecuted under the Conspiracy section, the Crown had to establish that the firms had the *intent* to lessen competition in order to win a conviction. In addition, Section 31, containing the merger and monopoly provisions, had to be prosecuted under criminal code standards, which meant that the burden of proof was much higher than under civil law. Between 1910 and 1986, only nine merger-related cases were brought to court by the Crown. Of those contested in court, the Crown won none; two pleaded guilty.

Competition Act (1986)

A major revision of Canadian competition policy was achieved in 1986 with the introduction of the *Competition Act*. Many of the deficiencies of the previous *Act* were corrected in the new *Competition Act*. The *Competition Act* embodies current Canadian competition law. The purpose of the *Act* is

Many of the deficiencies of the 1960 *Combines Investigation Act* were corrected in the *Competition Act* of 1986.

> "to maintain and encourage competition in Canada in order to promote the efficiency and adaptability of the Canadian economy, in order to expand opportunities for Canadian participation in world markets while at the same time recognizing the role of foreign competition in Canada, in order to ensure that small and medium-sized enterprises have an equitable opportunity to participate in the Canadian economy and in order to provide consumers with competitive prices and product choices." (*Competition Act*, R.S.C.1985, c.19 (2nd supplement), Part I section 1.1)

The *Competition Act* contains both a criminal and a civil section dealing with a wide variety of potentially anticompetitive behaviours. The Commissioner of Competition

Civil offences are dealt with through the Competition Tribunal, which is a court of public record designed specifically to deal with complex legal and economic issues.

is responsible for investigating alleged violations of the *Competition Act*, and has various investigative powers, including the power of search and seizure under court order. For criminal provisions, which include conspiracy, bid rigging, price maintenance, price discrimination, predatory pricing, and misleading advertising and other deceptive marketing practices, the Commissioner must refer the evidence to the Attorney General of Canada, who is responsible for prosecution. Upon conviction, the courts may impose fines, imprisonment, or prohibition orders against the offending parties.

For civil offences, which include mergers and abuse of dominant position, the Commissioner applies to the Competition Tribunal for remedy. The **Competition Tribunal** is *a public court of record which is designed to deal with complex legal, economic, and business issues, informally and expeditiously*. The Tribunal is made up of four Federal Court judges and other lay persons who generally have an economics or business background. The Tribunal has broad powers to issue orders to overcome anticompetitive effects in a market, including divestiture of assets. An order of **divestiture** requires a firm to sell off some portion of its assets so that its market power in some area is reduced. For example, in the Loblaw case, the Tribunal required the company to sell some of its stores so that it did not hold a dominant position. The Tribunal's judgments are based on both economics and Canadian law.

Divest: to sell off a portion of the assets of a company or its subsidiary.

Criminal Provisions of the *Competition Act*

Parts VI and VII contain the offences which are referred by the Commissioner of Competition to the Attorney General of Canada for review or prosecution. Here is a brief outline of some of the main sections of the *Competition Act*, and some of the economic arguments regarding the offences.

Misleading advertising distorts the consumer's ability to make rational choices and thus harms competitors engaging in honest promotions.

Sections 52 to 60: Misleading Advertising and Deceptive Marketing Common violations under these sections include misleading warranties, unsubstantiated performance and durability claims for products, and misrepresentation with respect to what is the "regular price" for an item. These offences are treated under criminal law because they can have serious economic consequences and can cause significant harm to competing firms which engage in honest promotions. The negative effects of these misleading and deceptive practices are especially serious when directed at large consumer markets and for long periods of time. Competitive conditions in a market require that consumers have access to relevant information about products, and not be subject to misleading representations, so they can make rational decisions. Both companies and individuals can be charged, and penalties include fines and imprisonment for up to five years. The highest fines imposed so far have been $1 million against a company and $500,000 against an individual, and the longest jail term has been one year.

Section 61: Price Maintenance Section 61 is designed to ensure that firms set their own prices under competitive conditions. This section prohibits anyone from influencing upward, or discouraging the reduction of the price of a product, by means of an agreement, threat, promise, or otherwise. The section also prevents a supplier from refusing to supply to a firm because of that firm's low pricing policy. Independent gasoline retailers in many markets complain that large integrated gas refiners "run out of product" before scheduled deliveries when the independent's retail prices are "too low."

Section 50: Predatory Pricing Section 50 of the *Competition Act* makes it an offence to engage in a policy of selling products at unreasonably low prices, having the effect of substantially lessening competition or eliminating a competitor. What is an

APPLYING THE TOOLS

Competition or Collusion?

Canadian competition laws concern far more than mergers and market structure; they also place legal restrictions on certain business practices, such as price-fixing. By law, firms are not allowed to *explicitly* collude in order to fix prices above the competitive level. A key aspect of the law is the explicit nature of the collusion that is disallowed. Airlines, gas stations, and firms in many other industries have prices that generally move in tandem—when one firm changes its price, others seem to follow. Such practices would suggest that these firms are implicitly colluding, but they are not violating the law unless there is explicit collusion.

To prove explicit collusion is difficult—there must be a smoking gun, and there is seldom sufficient evidence of explicit collusion to prosecute businesses. The Competition Bureau routinely receives complaints about gasoline prices appearing to be set by means other than market forces. In Saskatchewan, in October 1998, the Bureau was asked to investigate whether the major integrated refiner-marketing gasoline firms had violated criminal Sections 45 (conspiracy), 50 (price discrimination and predatory pricing), or 61 (resale price maintenance), or civil Section 79 (abuse of dominant position). The complaint followed the purchase in July 1998 of Western Canada's largest independent gasoline retailer, Mohawk Oil, by regional integrated Husky Oil. Were the smaller independent retailers being squeezed out?

In its decision, the Bureau considered such factors as national, regional, and local market shares, and the change in those market shares after the merger, prices in a cross-section of markets, product substitutability, price elasticity of demand, customer switching costs, and productivity and rationalization of operations.

The investigation found no evidence to support the allegations. The Bureau indicated that similar or identical price movements do not prove that an agreement exists. In retail gas sales, prices are visibly posted and consumers believe that the various companies' products are identical, so gasoline retailers cannot sell at prices higher than a competitor's without losing customers. Therefore, there would be a tendency for prices in a market to be similar or even identical. Furthermore, competing firms tried to differentiate their products by using coupons, rebates, and other loyalty programs, rather than competing on price. The Bureau found no evidence that any agreement or arrangement between gasoline retailers existed to prevent or lessen competition unduly. Predatory pricing was also dismissed since the low prices appeared to be the result of competitive market conditions, rather than a policy of unreasonably low prices by a firm. It was claimed that the integrated refiners "ran out of product" for an independent who was "too competitive." The refusal to supply allegation could not be substantiated; and the abuse of dominant position investigation failed to show that the firms engaged in anything beyond conscious parallelism, which is not illegal.

So, while it may appear that the gasoline retailers are engaging in anticompetitive activities, it is imperative to obtain the evidence required to prove the case. In many cases, the "smoking gun" either doesn't exist (which means the firms are not guilty), or simply cannot be found (you decide).

unreasonably low price? Under the guidelines of the *Act*, "unreasonably low" means prices which fall below a firm's average variable cost; also, prices which fall between a firm's average total and average variable cost *could* be considered unreasonably low. Predatory pricing has turned out to be a relatively rare event in Canada, since the predatory firm must be able to recover the losses it incurred during the predatory pricing episode. This means that the firm must have substantial market power and the industry must have significant barriers to entry to enable the firm to raise price above cost without entry succeeding. This section of the *Competition Act* is careful, however, to differentiate between legitimate, vigorous price competition, which would harm the profits of one or more rival firms, and predatory conduct, which undermines the entire competitive process.

The predatory pricing offence is careful to distinguish between legitimate vigorous competition, which harms rivals' profits, and predatory pricing, which harms the competitive process.

Section 45: Conspiracy This offence is the most serious of the criminal offences of the *Competition Act.* Upon conviction, the court may levy fines of up to $10 million or imprisonment for up to five years, or both. Prosecution under Section 45 requires the Crown to prove beyond a reasonable doubt that the *accused parties intended to enter into an agreement, and that they knew it would have the effect of lessening competition unduly*. As discussed in Chapter 13, when only a few large firms dominate an industry, their actions are interdependent. Because of this interdependence, and because they are few in number, it may be profitable for them to collude.

Q.3 Is a conspiracy that does not lessen competition or conscious parallelism considered to be an offence under the *Competition Act*?

Conscious parallelism: when (oligopolistic) firms set similar prices and behave similarly with respect to other business practices without an explicit agreement.

Illegal conspiracy agreements often are proven by documents, such as memos or letters, and by testimonial evidence of overt acts, such as meetings or telephone conversations. Circumstantial evidence may also be admitted, and a conviction obtained on the basis of it. However, similar prices and similar movements in prices are not necessarily evidence of collusion. When concentrated industries are characterized by a homogeneous (identical) product which is produced under similar and slowly changing cost conditions, it may be the case that firms charge identical prices because each knows that if it dropped its price, the others would all follow, resulting in lower profits for all. This consciously parallel behaviour can occur without explicit agreement, and without an explicit agreement, *conscious parallelism* is not an offence. In addition, conspiracies which have no anticompetitive effects are not offences under the *Competition Act*.

***Per se* offence:** A violation of the *Competition Act* in which proof of undue lessening or prevention of competition is not required; the injury to competition is *presumed*.

Section 49: *Per Se* Offences for Federal Financial Institutions A *per se* offence is a violation of the *Competition Act* in which proof of undue lessening or prevention of competition is not required; the injury to competition is *presumed* by the firm's action. It is interesting to note that, under the old *Combines Investigation Act*, a violation of the law occurred by virtue of the firm simply *being* a monopoly. The new *Competition Act* lists several *per se* offences pertaining to financial institutions. Under this section, the occurrence of the activity presumes that undue lessening of competition has occurred; therefore, engaging in the activity is an offence. This section prohibits every federal financial institution, including banks, trust companies, and insurance companies, from

> "making an agreement or arrangement with another federal financial institution with respect to
> (a) the rate of interest on a deposit,
> (b) the rate of interest or the charges on a loan,
> (c) the amount or kind of any charge for a service provided to a customer,
> (d) the amount or kind of a loan to a customer,
> (e) the amount or kind of service to be provided to a customer, or
> (f) (determining who gets loans)."

The penalty for this indictable offence is a fine of up to $10 million or imprisonment for up to five years, or both.

Civil Provisions of the *Competition Act*

The burden of proof is less for civil offences than for criminal offences.

The infractions under the civil portion of the *Competition Act*, Part VIII — Matters Reviewable by Tribunal — are dealt with by the Competition Tribunal, and are subject to civil remedies. The burden of proof is somewhat less than under a criminal proceeding.

Section 75: Refusal to Deal This section applies when a firm, or individual, is substantially affected in its ability to carry on its business as result of its inability to obtain supplies from a supplier. The individual meets the usual terms of trade with the supplier, the product is available to supply, and the individual has pursued all other means to obtain supplies. As a remedy, the Tribunal may order one or more suppliers to accept the firm or individual as a customer.

Section 77: Exclusive Dealing, Tied Selling, and Market Restriction These types of offences are vertical restraints. Vertical restraints link the upstream segment of a market to the downstream segment of the market. **Exclusive dealing** occurs when a *supplier requires the purchaser to deal with only the seller's particular product.* **Tied selling** occurs when the *supplier requires the buyer to purchase another of the supplier's products as a condition of obtaining the desired product.* **Market restriction** occurs when the *supplier requires the buyer to restrict his operations to a particular geographic market.* In these cases, the Bureau's task is two-fold. First, it must establish that the firm has engaged in the activity in question. Second, it must demonstrate that the activity will have anticompetitive effects in the market. To fulfill the "competition test," it must be shown that the practice is widespread or is done by a major supplier, and that the practice is likely to impede entry or expansion by a firm, or have other anticompetitive effects, and that as a result of the actions, competition is, or is likely to be, lessened substantially. If the Tribunal finds that this has happened, it may issue an order prohibiting the action.

Exclusive dealing, tied selling, and market restriction are offences if the practice is widespread or involves a major supplier and impedes entry or expansion by a competitor.

Section 78 and 79: Abuse of Dominant Position This section relies heavily upon economic reasoning for its interpretation. The Tribunal may issue an order if it finds that *one or more firms substantially or completely control a class or species of business, and have engaged in anticompetitive acts that have, or are like to have, the effect of preventing or lessening competition substantially in a market.* The types of anticompetitive acts are those which involve the use of market power by the dominant firm to prevent or impede entry or expansion by rivals, and includes squeezing out nonintegrated (independent) competitors, acquiring potential customers of another supplier, setting restrictive contracts which prevent customers from switching suppliers, buying up slow-selling products to prevent erosion of existing prices, adopting product specifications which are incompatible with a competitor, and selling products at prices lower than cost in order to discipline a competitor or eliminate him from the market.

Aggressive, pro-competitive behaviour which results from the superior competitive ability of the dominant firm, however, is not considered an offence. In its judgment, the Tribunal is not to consider the size or market share of the dominant firm in the market; rather, it is to consider the extent of the firm's market power. Market share is a necessary, but not sufficient, condition for market power. For investigative purposes, the Bureau uses a 35 / 65 guideline: firms whose market share is less than 35% of a particular product or geographic market likely do not possess market power; firms whose market shares exceed 65% likely possess market power. These percentages are rough guidelines only, and a number of other factors must also be considered: for example, barriers to entry, such as tariffs or government regulations, lack of substitute products, insufficient number of competitors, and low levels of innovation in the industry. If the Tribunal finds that abuse of dominant power has occurred, it may issue an order prohibiting the firm from engaging in the anticompetitive practice, or it may issue an order requesting a divestiture of the firm's assets or shares in order to overcome the anticompetitive effects.

Agressive, highly competitive behaviour which results from superior ability is not considered abuse of dominant position.

Sections 85 to 90: Specialization Agreements **Specialization Agreements** are *agreements made by two or more Canadian firms to divide up production between them* so that each can concentrate on, and become more efficient at, producing only a segment of the production. These Agreements are exempt from the conspiracy and exclusive dealing sections provided that the arrangement creates gains in efficiency and has been approved in advance by the Competition Tribunal. Sections 86(2) and 86(3) stipulate that the efficiency gains must result in

"(a) a significant increase in the real value of exports; or

(b) a significant substitution of domestic articles or services to imported articles or services"

and cannot simply represent a redistribution of income between individuals.

This section of the Canadian competition legislation is welcome news for Canadian firms operating within the small Canadian market, since it allows firms to concentrate on a smaller area of production and rationalize their operations, in some cases achieving minimum efficient scale.

In its decision, the Competition Tribunal looks at a number of issues surrounding a merger. Which elements of Canadian competition policy were applied to the banks' proposed mergers?

Sections 91 to 103: Mergers Formerly, mergers fell under the jurisdiction of criminal law, which is generally felt to be inappropriate for this type of offence. An individual convicted of a merger offence carried a criminal record, and the burden of proof was much higher under criminal law than under civil law. Under the *Competition Act*, merger is now a **Civil Reviewable Matter** adjudicated by the Competition Tribunal. This change recognizes that the possession of a monopoly position is not a criminal offence; however, the abuse of that position is an offence.

The Tribunal can prevent or dissolve a merger where the merger "prevents or lessens, or is likely to prevent or lessen, competition significantly." In its decision, the Tribunal must take into account a number of factors:

1. the existence of Canadian or foreign substitutes.
2. whether a party to the merger is about to fail.
3. barriers to entry before and after the merger.
4. whether the merger removes a vigorous competitor.
5. the nature and extent of innovation in the market.

The section also allows the merging firms to present an **efficiency defense**. *If gains in efficiency offset the competition concerns, then the merger can be allowed.* Specifically, Section 96(1) states that the Tribunal shall not disallow a merger if it

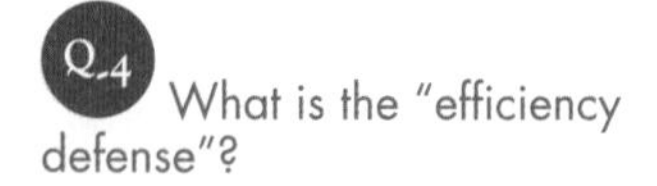

What is the "efficiency defense"?

"has brought about or is likely to bring about gains in efficiency that will be greater than, and will offset, the effects of any prevention or lessening of competition that will result or is likely to result from the merger or proposed merger and that the gains in efficiency would not likely be attained if the order were made."

No order against the firm will be made by the Tribunal if the lessening of competition is due to the firm's superior economic efficiency. Dominance in the context of Canadian industry is common, and should not be an offence *per se*. It is the *abuse* of this dominant position that Canadian competition policy seeks to prevent.

Mergers between large firms are subject to the prenotification requirement.

Notifiable Transactions Mergers over a certain asset or sales value are now subject to a **prenotification requirement** in Part IX of the *Competition Act*, which means that *they must make their intention to merge known to the Competition Bureau ahead of time*. Since it can be difficult, costly, and even impossible to reverse a merger, in cases involving assets or sales over $400 million, prenotification is required. During the prenotification process, the advantages and disadvantages of the proposed merger are assessed before the merging firms proceed with irreversible financial investments or other irreversible commitments required to complete the merger. In 1998, four Canadian banks initiated the prenotification process.

MERGERS, ACQUISITIONS, AND TAKEOVERS

Industrial structure changes as firms increasingly look for alternative ways of structuring themselves so that they can achieve the economies of scope and economies of scale that go along with specialization. Firms are simultaneously merging and breaking up. Firms are allowed to break up as much as they want; when they merge, however, they must ensure that any merger falls within the competition guidelines. Let's consider the various subcategories and types of mergers that are possible, in order to put recent merger activity into perspective.

Mergers, Acquisitions, and Takeovers

Merger is *a general term meaning the act of combining two firms*. In a merger, two firms combine to form one firm. The result is often an increase in market concentration.

Another kind of merger is an **acquisition**—*a merger in which a company buys another company and the purchaser has the right of direct control over the resulting operation (but does not always exercise that right)*. It is a merger, but it is not a merger of equals, and the acquiring firm does not necessarily take over direct control of the acquired firm's operations. In a merger of equals, neither firm takes over the other, and it's not clear who will be in charge after the merger.

Takeovers and acquisitions are said to be *friendly* or *hostile*. In a friendly takeover, one corporation is willing to be acquired by the other. A **hostile takeover** is *a merger in which the firm being taken over doesn't want to be taken over*. How can that happen?

Remember the discussion of corporations from Chapter 3. Corporations are owned by shareholders, but are managed by a different group of individuals. The two groups' interests do not necessarily coincide. When it is said that a corporation doesn't want to be taken over, that means that the corporation's managers don't want the company to be taken over. In a hostile takeover, the management of each corporation presents its side to the shareholders of both corporations. The shareholders of the corporation that is the takeover target ultimately decide whether or not to sell their shares. If enough shareholders sell, the takeover succeeds.

In a hostile takeover, the shareholders ultimately decide whether to sell their shares.

Horizontal Mergers Most Canadian competition policy concerns **horizontal mergers**—*the combining of two companies in the same industry*. Mergers of companies with substantial market shares in the same industry are most likely to be prohibited.

Horizontal mergers are companies in the same industry merging together.

Vertical Mergers A **vertical merger** is *a combination of two companies that are involved in different phases of producing a product*, one company being a buyer of products the other company supplies. For example, if a computer company merges with an electronic chip company, a vertical merger has taken place. The concern with vertical mergers is that the merged firm is able to limit access of other buyers or sellers to the market.

Vertical mergers are combinations of two companies, one of which supplies inputs to the other's production.

Conglomerate Mergers A third type of merger is a conglomerate merger. **Conglomerate mergers** involve *the merging of two relatively unrelated businesses*. Conglomerate mergers are generally approved by Canadian competition laws under the assumption that they do not significantly restrict competition.

Why would two unrelated firms want to merge? Or why would one want to be bought out by another? There are five general reasons:

1. *To achieve economies of scope*. Although the businesses are unrelated, some overlap is almost inevitable, so economies of scope are likely. For example, one

Q-5 If Dairyland, a maker of ice cream, bought a dairy farm, what type of merger would it be?

firm's technical or marketing expertise may be helpful to the other firm, or the conglomerate's increased size may give it better bargaining power with its suppliers.

2. *To get a good buy*. Firms are always on the lookout for good buys. If a firm believes that another firm's stock is significantly undervalued, it can buy that stock at its low price and then sell it at a profit later when the stock is no longer undervalued.
3. *To diversify*. Many industries are cyclical. In some parts of the business cycle they do poorly; in other parts of the business cycle they prosper. Buying an unrelated company allows a firm to diversify and thereby to even out the cyclical fluctuation in its profits.
4. *To ward off a takeover bid*. Firms are always susceptible to being bought out by someone else. Sometimes they prevent an unwanted buyout by merging with another firm in order to become so large that they're indigestible.
5. *To strengthen their political-economic influence*. The bigger you are, the more influence you have. Individuals who run companies like to have and use influence. Merging can increase their net influence considerably.

Conglomerate mergers are combinations of unrelated businesses. Five reasons why unrelated firms merge are:

1. To achieve economies of scope;
2. To get a good buy;
3. To diversify;
4. To ward off a takeover bid;
5. To strengthen their political-economic influence.

Mergers have been increasing over the last decade or so. The primary reasons for the increase in the number of mergers are globalization, deregulation, and technological change. Globalization leads to mergers because firms can gain instant foreign distribution networks and knowledge of local markets from mergers and can also lower costs by restructuring production to low cost areas. Deregulation of the telecommunications, electricity, and financial industries has encouraged mergers that take advantage of economies of scale and scope. Bank mergers and media company mergers are examples.

The acceleration of technological change in recent years is another contributor to merger activity. Firms are looking for ways to develop new technologies or take advantage of new technologies, and merging with another company is one way to acquire a new technology. The merger of Bell Canada and CTV is an example of a communications (technology) company combining with a "content" company.

At the same time that these mergers are taking place, firms are also engaging in **deacquisitions**—*one company's sale of either parts of another company it has bought, or parts of itself*. Sometimes regulators require such deacquisitions, known as divestiture, as a condition of a merger. Deacquisition also occurs as firms focus on those areas where they have comparative advantage and where growth is highest, and sell off aspects of their firms that are not part of their core business. For example, BCE Enterprises voluntarily split into two companies—Bell Canada, which focusses on telecommunications, and Nortel, which focusses on the computer aspects of their business. The motto of the 2000s is that firms have to continually reinvent themselves.

The Canadian market structure is a continually changing landscape.

Voluntarily breaking up companies is becoming much more prominent as firms try to find their niche in the global marketplace. Companies are continually spinning off portions of their business where they do not believe they have a comparative advantage, and buying businesses where they think they do have a comparative advantage. This process is likely to continue, making the Canadian market structure a continually changing landscape.

BEYOND THE TOOLS

White Knights and Golden Parachutes

The specific nature of a merger is of great interest to people working for the corporations involved. It determines whether employees will move up or down in the corporate pecking order, or even whether they'll keep their jobs at all. When a firm is threatened by a hostile takeover, top management is likely to hunt for another firm, a friendly one that will be a "white knight" coming to the rescue. The white knight firm, it is hoped, will keep all the management employees in their jobs. When their jobs are at stake, it is often questionable whether management searches out the firm that will pay shareholders the best price. Management's interest is likely to be in keeping their jobs, not in making money for the shareholders.

Strategies have been developed to protect shareholders from management's interest overriding shareholders' interest. One such strategy is the *golden parachute*. The board of directors of the corporation provides that if the corporation should be acquired by another corporation, people holding top management jobs are guaranteed large payments of money. This is a golden parachute. It means that management can feel comfortable when the corporation is approached by a potential buyer and will make decisions based on the shareholders' interest, not on whether management jobs will be left undisturbed.

Golden parachutes afford shareholders some assurance that if their firm receives an acquisition or takeover offer, management will consider it. Golden parachutes protect top managers because either they'll keep their jobs or, if they lose their jobs, their pain will be soothed by compensation in big bucks. Golden parachutes do nothing for low-level managers and ordinary workers, who may suffer demotions or layoffs after two firms have merged.

THE BANK MERGERS

On January 23, 1998, the Royal Bank of Canada (now RBC Financial) and the Bank of Montreal notified the Competition Bureau, under the prenotification requirement, of their intention to merge. Shortly after, the Canadian Imperial Bank of Commerce and the Toronto-Dominion Bank announced their intention to merge. The Bureau began a review of the proposed mergers immediately. Since the two mergers involved four of the five big Canadian banks, the competitive effects of the mergers were considered together. The review took approximately ten months to complete, due to the volume of documents to consider (roughly 400,000 pages of documents) and the complexity of the issues involved in the financial services industry. The question for the Commissioner to answer was whether the proposed mergers would likely lessen or prevent competition substantially, by raising prices, reducing choices, or reducing service. If so, the mergers would be prohibited unless efficiencies created by the mergers outweighed the negative impact on competition.

The Royal Bank is the largest and the Bank of Montreal is the third largest bank in Canada, and their proposed merger represented a combination of $315 billion in assets at the time of their merger proposal.[1] The Bureau's investigation covered a wide range of products and services and geographic areas served by the competing banks throughout Canada. The first step for the Bureau was to define the relevant product and geographic markets. A relevant market was determined to exist where a group of products were close substitutes. The banks' market shares were then calculated for each relevant market to determine whether the competitive threshold was exceeded, so in markets

[1]Figures as of December 31, 1997.

APPLYING THE TOOLS

Canada's Largest Supermarket Chain

In the fall of 1998, the grocery giant Loblaw Companies Ltd. announced that it had acquired the Atlantic Canada assets of The Oshawa Group Ltd., including three food warehouses in Nova Scotia and Newfoundland and Labrabor, and the Agora Food Merchants' supply contracts for 80 retailers operating under the names of IGA, Omni, and Traditions in 68 markets. Prior to the acquisition, Loblaw was the second largest food distributor in Atlantic Canada, supplying the Great Atlantic Superstore, Save-Easy, Supervalu, Red & White, and Foodmaster stores, plus Dominion stores in Newfoundland and Labrador only.

Atlantic Canada Pre-merger Grocery Market Shares

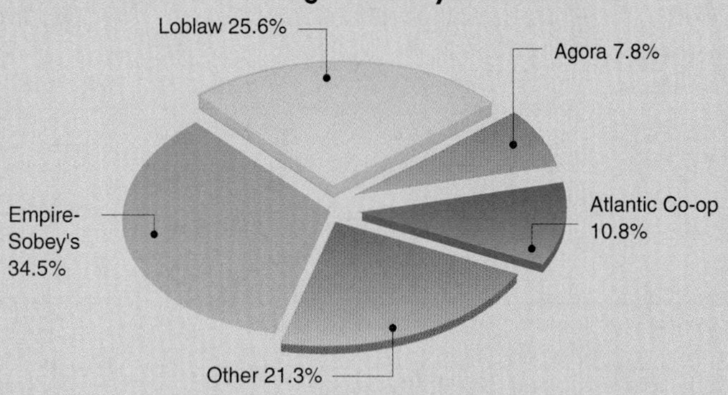

Atlantic Canada Post-merger Grocery Market Shares

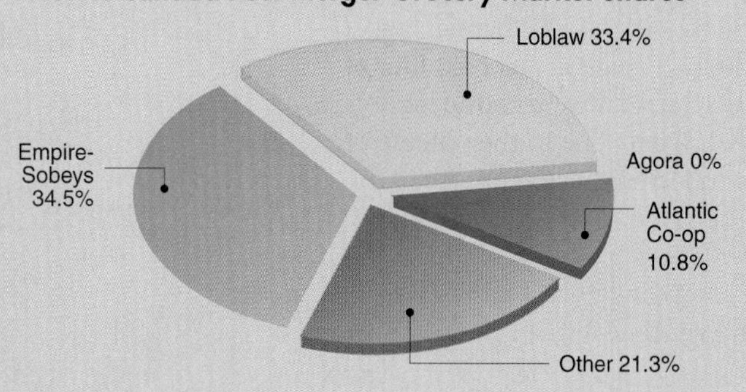

The merger raised some concerns. Analyzing the case on a regional basis, the Competition Bureau considered market share and concentration levels, the nature and extent of barriers to entry, the removal of the competitor (Agora) and the effectiveness of the remaining competition, and whether the increased concentration would likely lead to interdependent behaviour amongst remaining firms.

The merger was allowed to proceed. In its analysis, the Bureau found that Agora Foods had not been a vigorous competitor, generally pricing above the market leaders, Sobeys and Loblaw. The Bureau also noted that Loblaw was the new entrant in 54 of the 68 markets, and considered the impact on competition of mass merchandisers and food club stores. Competition concerns emerged in four markets: Halifax, Dartmouth, and New Minas in Nova Scotia, and St. John's in Newfoundland. The solution was divestiture of one store in each of these four markets. Loblaw also sold its interest in its Summerside, Prince Edward Island, store.

Interestingly, the Bureau recognized Loblaw's competitive strategy of emphasizing its own private label products, which resulted in a competitive advantage for it. Loblaw's competitors relied more on national brands, and the Bureau noted that the increased variety of product would benefit consumers as well as reduce the likelihood of interdependent behaviour among the firms.

Since 1998, Loblaw has increasingly taken on more competitors. In its recent efforts, Loblaw has moved towards a greater variety of products, expanding existing product lines, such as clothing and housewares, and moving into new market segments, such as cosmetics, bed and bath products, and electronics. It has plans to open a series of large-format general merchandise emporiums, similar to the Wal-Mart Supercenters in the U.S. The first emporium opened in Ontario in 2001, and resembles the Real Canadian Superstores operated by Loblaw in western Canada. Loblaw is well adapted to competing in many product lines, having experienced tremendous success in marketing perishable goods. It is likely that concern over threat of entry by a company such as Wal-Mart, which may bring its big box food-merchandise combination retailing to Canada, has driven the competitive actions of Loblaw stores.

where the share of the banks was 35% or greater, or if the combined share of the four largest competitors was 65% or greater,[2] the Bureau made a detailed assessment. A

[2]and the share of the merging banks was greater than 10%.

variety of competitive considerations were evaluated for each market, requiring detailed assessment, including barriers to entry and sunk costs, effectiveness of domestic and foreign competition, and technological advances in the financial services industry.

The Competition Tribunal selected three areas for review: branch banking, credit cards, and securities.

Three broad markets were selected for detailed review: branch banking, including both personal and business services,[3] credit cards,[4] and securities.[5] While the banks argued that the geographic markets relevant for the securities business and for branch banking were national, after examining bank records and interviewing various experts, the Bureau deemed these markets to be local in nature, citing that customers wanted investment advice provided by a conveniently located facility, close to where they lived or worked. Consumers also wished to be able to access their accounts at their local branch, while many businesses required frequent service, even daily services, such as night deposits, and cash and coin services. This was especially true for small and medium-sized businesses. The credit card market was determined to be national.

Using a two-year framework, the Bureau analyzed many different facets of competition. In considering the impact of barriers to entry in the banking industry, the Bureau concluded that barriers for new entrants were high. The extensive branch networks of the five major banks represented significant sunk investments, which had an estimated five- to seven-year breakeven period for a new entrant to recoup the costs of developing its own network. In addition, the incumbent banks already had branches at the best locations.

Significant barriers to entry exist in the banking industry, including the branch network, economies of scale and scope, and regulatory barriers.

Economies of scale and scope in the banking industry are large: computers and the wide range of products and services offered by the banks precludes entry on a small scale. An entrant would have to offer the same level of services in order to induce people to switch from their current financial institution. However, due to such conveniences as payroll deposit and automatic bill payment, switching imposes set-up costs on the customer and makes it unlikely that he or she will switch. In addition, the banks each have a strong brand name which has been developed over years of advertising and local branch presence. Customer loyalty is very strong.

Significant regulatory barriers also exist. Taxes on capital investment must be paid, even if the corporation is not earning a profit. Provincial regulation impedes the competitive discipline imposed by the smaller financial institutions, and Canadian ownership regulations—for example, the majority of the board of directors are required to be Canadian—limit the entry by foreign banks into the Canadian banking sector. Domestic and foreign competitive discipline is small.

Technology: Competitive element or barrier to entry?

Although the introduction of new technologies, such as automated banking machines, debit cards, and telephone and Internet banking, has transformed the financial services industry, the Bureau found that these developments did not eliminate the need for bank branches. Personal contact, including advice and problem solving, remains a key element in many financial products and services, even though channels such as banking machines and the Internet may become important too. The banks argued that technology had lowered barriers to entry; the Bureau believes that the opposite may have happened. Now a new entrant must be prepared to offer branch services as well as a host of advanced technology services.

[3]Personal transaction accounts, residential mortgages, personal loans, personal lines of credit, business transaction accounts, and operating loans.

[4]The credit card network, personal credit cards, Visa and MasterCard merchant acquiring, and primary merchant acquiring.

[5]Full-service brokerage and equity underwriting only.

The Bureau also considered the level of effective competition remaining in the relevant markets. The Bank of Nova Scotia, which did not choose a partner for the 1998 merger dance, was seen to represent relatively ineffective competition against the merged entities, as its size would be less than half the size of either of the combined banks. Canada Trust, which was an important regional competitor in Ontario, Alberta, and British Columbia at the time of the merger announcement, is now TD-Canada Trust, having merged with the Toronto-Dominion Bank in 2001. The Hongkong Bank of Canada, with its 117 branches, has meaningful branch presence only in British Columbia. Quebec presents only a small problem, since the National Bank of Canada and Le Mouvement des Caisses Desjardins are effective competitors in that province. Credit unions in Saskatchewan, Manitoba, and British Columbia provide alternatives, but the Bureau noted that *all* of the competitors to the merged banks are much smaller and, with their small size, face significant cost disadvantages. Furthermore, the Royal Bank and the Bank of Montreal competed against each other. The Bank of Montreal was a particularly vigorous competitor, leading other banks in lowering lending rates, aggressively pricing mortgages, and introducing a new electronic banking service, Mbanx. Allowing the merger of the Royal Bank and the Bank of Montreal would remove a vigorous competitor from the market.

The Bureau considered such factors as existence of competitors (Canadian or foreign), barriers to entry, innovation, and whether the merger removed a vigorous competitor.

In the credit card business, loss of the Bank of Montreal's MasterCard network would transfer significant market power to Visa, possibly leading to the elimination of MasterCard, while in the securities market, the bank merger would combine the largest and second-largest full-service brokerages, Nesbitt Burns and RBC Dominion Securities. Together, their investment advisors would number almost 3000, nearly equalling the combined total of Wood Gundy's 700 advisers, Scotia McLeod's 800 advisers, Toronto-Dominion's 425 advisers, and Merrill-Lynch's 1350 advisers. However, Levesque Beaubien Geoffrion, with 1300 advisers, was considered an effective competitor, particularly in Quebec.

According to Section 93(2) of the *Competition Act*, although market share is not a determining factor on its own, the proposed mergers would significantly increase concentration in an already concentrated industry, as the number of banks declined from five to three. In any merger case, the Tribunal must determine whether the firms' interdependent behaviour would likely lead to substantial lessening of competition. Generally, high barriers to entry, homogeneous products, demand predictability, stability of costs, and the existence of industry associations facilitate interdependent behaviour. The Bureau noted that *all* of these factors were present in the banking industry, and was concerned that the proposed mergers would tend to reduce the vigour of competition in the industry.

The four banks employed the efficiency defense.

The banks argued that the mergers should proceed on the grounds of efficiency gains. They estimated that they would save approximately $1 billion annually from eliminating duplication of the branch network and from the lowering of overhead and administrative costs. In addition, cost savings would be generated in technology spending. The banks argued that greater scale economies, needed to compete internationally, would occur as they consolidated payments services, deposit inquiries and accounting, and loan monitoring activities. Economies of scope could be reaped as the merged banks increased their product variety, and even more cost reductions could result from the adoption of each other's best practices.

In assessing the efficiency defense, the Bureau looked at recent bank mergers in the United States. Several features of these mergers were relevant for the Canadian situation. The U.S. data showed that mergers tended to be more successful if the merging partners had previous experience with large mergers. The Canadian banks only had experience with small acquisitions. The U.S. data also suggested that about half of the mergers reduced costs, while the other half realized no change in costs or incurred higher costs. It was unclear that the suggested efficiencies would be realized in the Canadian situation. If the merger were to proceed, in order to ensure that substantial lessening of competition in all markets did not occur, certain structural remedies would have to be imposed, including divestiture of branches and brand names, and the conversion of the Royal Bank's Visa portfolio to MasterCard.

Q-6 What factors did the Competition Bureau consider in its investigation of the 1998 proposed bank mergers?

The Competition Bureau declined to allow the mergers to take place. In its letter[6] to John E. Cleghorn and Matthew W. Barrett, chairmen and Chief Executive Officers of the Royal Bank and the Bank of Montreal, respectively, the Bureau stated that the merger was likely to lead to a substantial lessening or prevention of competition that would cause higher prices and lower levels of service and choice for several key banking services in Canada. The offerings of the merging parties were often the first and second choices of customers, and the Bureau found that the anticompetitive effects of the merger were not balanced by efficiency gains.

The Canadian financial services industry is characterized by a high level of concentration in small, segmented geographic markets; however, compared to international competitors, Canadian banks are relatively small. The decision on bank mergers reflects the difficulty that Canadian policy makers face in drafting legislation beneficial to Canadian economic interests. On the one hand, it is necessary that Canadian competition policy encourage domestic competition by rewarding efficiency and innovation, but it is also necessary that economic policies allow firms to take steps which will make them internationally competitive. This often means that firms need to get bigger, much bigger in some cases, in order to capture the relevant economies of scale and scope, and other size-related cost advantages that are necessary in order to compete. The *Competition Act* endeavours to discourage and prohibit business practices that prevent and lessen domestic competition, without unnecessarily hampering the ability of Canadian firms to compete internationally. The decision on the proposed bank mergers can be viewed as one that tries to balance a number of uniquely Canadian goals, efficiency, innovation, and international competitiveness.

Canadian competition policy must balance domestic competition issues with international competition issues.

INTERNATIONAL COMPETITION AND COMPETITION POLICY IN OTHER COUNTRIES

The nature of competition is changing in Canada. Ten or 20 years ago, when people talked about competition, they meant competition among Canadian firms. Now they often mean international competition.

Q-7 Does the U.S. have stricter or less strict antitrust laws than Canada has?

Because of this internationalization of competition, the political climate in Canada is changing. More and more, Canadian policymakers see the international market as the relevant market.

Other countries' approaches toward anticompetitive behaviour parallels the Canadian approach. Many countries also see the international market as the relevant market and have designed their competition laws accordingly. Their domestic markets are also too small to take advantage of economies of scale.

[6]With a copy to the then Minister of Finance, The Honourable Paul Martin.

APPLYING THE TOOLS

Competition Agencies in Other Countries

***Britain:* British Monopoly Commission**
While it has the power to recommend structural reorganization, the British Monopoly Commission generally has not done so. Instead, it has pushed for price reductions in certain industries. After the Second World War, a number of major industries were nationalized, putting the government in direct control of prices, but many of these industries have been privatized in the past few years.

***Japan:* Fair Trade Commission**
The Japanese Fair Trade Commission is weak and subordinated to the Ministry of International Trade and Industry and other government agencies. In retailing, small firms continue to dominate with the support of government. But the 1980s saw the beginnings of a retailing system with large stores, like the system in North America. The Fair Trade Commission may take a role in suppressing or slowing that development.

***Germany:* Federal Cartel Office**
The Federal Cartel Office is relatively small. Often it allows and even encourages cartels. It does have the authority to push for price reductions if it determines that the cartel has abused its power.

***France:* Commission on Competition**
The Commission on Competition has been very weak and has often advocated mergers. In the 1960s, France actively promoted large-scale mergers, and during that period the government nationalized large industries without hearing objections from the Commission on Competition.

***European Union:* The European Commission**
In 1990, the EU's rules regarding mergers officially took precedence over member nations' local rules in those cases in which the merger involved significant activities in more than one member state. There were many such mergers; as expected, the expansion of the market that was made possible by the integration of the EU economies in 1992 generated significant merger activity as firms consolidated and tried to take advantage of new trading possibilities.

It wasn't only EU member firms that were merging. The largest percentage of growth in EU merger activity has been between EU firms and non-EU, especially U.S., firms. At times this has resulted in conflict between the EU Commission and other countries' competition agencies. For instance, the Commission had serious concerns regarding the 1997 merger between Boeing and McDonnell Douglas, but backed down after the United States pressured them to allow the merger. This case was unusual; for the most part the EU Commission has been very lenient in allowing mergers and, from 1991 to 1998, the Commission disallowed only 10 mergers.

Countries oppose competition laws because of economies of scale, lack of strong ideology supporting competition, and strong cultural ties between government and business.

One important reason countries may oppose competition regulation is that they recognize the importance of economies of scale. In many countries that have only small markets, the minimum efficient production level often requires high levels of market concentration. A second reason is their history. Their ideology with respect to competition allows for a range of market models, including individualistic competition based on small producers. However, bigness is not necessarily bad. A third reason that other countries may not have strong competition laws is cultural. At one extreme, in the United States, government and business are often seen as enemies of each other. At the other extreme, in cultures such as Japan's or Germany's, government and business are seen as allies, working together to increase exports and compete internationally.

HISTORY OF U.S. ANTITRUST LAWS

U.S. ideology towards business and competition differs from Canadians' view. In the U.S., there is a strong bias towards *laissez-faire* and government noninvolvement in business.

However, there is simultaneously a populist (pro-people) sensibility that fears bigness and monopoly. These fears of bigness and monopoly burst forth in the late 1800s as many firms were merging or organizing together to form trusts, or cartels. As stated in Chapter 13, a *cartel*, or a *trust*, is a combination of firms in which the firms haven't actually merged but nonetheless act essentially as a single entity. A trust sets common prices and governs the output of individual member firms. A trust can, and often does, act like a monopolist.

In the 1870s and 1880s, trusts were forming in a number of industries, including railroads, steel, tobacco, and oil. Some of these trusts' actions are typified by John D. Rockefeller's Standard Oil. Standard Oil demanded that railroads pay it kickbacks on freight rates. These payments allowed Standard Oil to set lower prices for its products than other companies, which had to pay the railroads full price on freight. Standard Oil thus could sell at lower prices than its competitors.

If prices had remained low, this would have had a positive effect on consumers and a negative effect on Standard Oil's competitors. But that's not what happened. By 1882, Standard Oil had driven many of its competitors out of business. Standard Oil created a trust (cartel) and invited its few surviving competitors to join. Then Standard Oil Trust used the monopoly power it had gained to close down refineries, raise prices, and limit the production of oil. The price of oil rose from a competitive level to a monopolistic level, and the consumer, as well as Standard Oil's competitors, ended up suffering.

The Sherman Antitrust Act Public outrage against trusts like Standard Oil's was high. The organizers of the trusts were widely known as *robber barons* because of their exploitation of natural resources and their other unethical behaviour. In response, the U.S. Congress passed the **Sherman Antitrust Act** of 1890—*a law designed to regulate the competitive process*.

Public outrage at the formation and activities of trusts such as Standard Oil led to the passage of the *Sherman Act*, the *Clayton Act*, and the *Federal Trade Commission Act*.

The *Sherman Antitrust Act* contained two main sections:

> Section 1: Every contract, combination in the form of trust or otherwise, or conspiracy in restraint of trade or commerce among the several States, or with foreign nations, is hereby declared to be illegal.
>
> Section 2: Every person who shall monopolize, or attempt to monopolize, or combine or conspire with any other person or persons, to monopolize any part of the trade or commerce among the several States, or with foreign nations, shall be deemed guilty of a misdemeanor, and, on conviction thereof, shall be punished by a fine not exceeding five thousand dollars, or by imprisonment not exceeding one year, or by both said punishments, in the discretion of the court.

Q-8 What were the two provisions of the *Sherman Antitrust Act*?

The *Sherman Act* was broad and sweeping, but vague.

The *Sherman Act* was meant to be as sweeping and broad as its language sounds. After all, it was passed in response to a public outcry against trusts. But if you look at it carefully, in some respects it is vague and weak. For example, offenses under Section 2 were initially only misdemeanours, not felonies.[7] It's unclear what constitutes "restraint of trade." Moreover, although the act prohibits monopolization, it does not explicitly prohibit monopolies. In short, the *Sherman Act* let the courts decide U.S. antitrust policy.[8]

The following story summarizes the courts' role in antitrust policy. Three umpires are describing their job. The youngest of the three says, "I call them as I see them." The middle-aged umpire says, "No, that's not what an umpire does. An umpire calls them

[7]Under U.S. federal law, a misdemeanour is any misconduct punishable by only a fine or by a jail sentence of a year or less. A felony requires a sentence of more than a year.

[8]Subsequent amendments to the *Sherman Act* have strengthened it. For example, offenses under Section 2 are now felonies, not misdemeanours.

the way they are." The senior umpire says, "You're both wrong. They're nothing until I call them." And that's how it is with the courts and monopoly. Whether a firm is behaving monopolistically isn't known until the court makes its decision.

The Clayton Act and the Federal Trade Commission Act In an attempt to give more guidance to the courts and to provide for more vigorous enforcement of the antitrust provisions, in 1914 the U.S. passed the *Clayton Antitrust Act* and the *Federal Trade Commission Act*.

The **Clayton Antitrust Act** is *a law that made four specific monopolistic practices illegal* when their effect was to lessen competition:

The *Clayton Antitrust Act* made four specific monopolistic practices illegal:

1. Price discrimination.
2. Tie-in contracts.
3. Interlocking directorships.
4. Purchase of a competitor's stock.

1. Price discrimination, that is, selling identical goods to different customers at different prices.
2. Tie-in contracts, in which the buyer must agree to deal exclusively with one seller and not to purchase goods from competing sellers.
3. Interlocking directorships, in which memberships of boards of directors of two or more firms are almost identical.
4. Buying stock in a competitor's company when the purpose of buying that stock is to reduce competition.

As you can see, the U.S. laws are designed to address the same types of anticompetitive behaviour that the Canadian competition laws do. Both countries' competition laws give the government the power to regulate competition and police markets against unfair methods of competition and unfair or deceptive business practices.

THE MICROSOFT CASE

First Came IBM

In 1967 the U.S. Department of Justice sued IBM for violating antitrust laws. The department argued that the company had a 72 percent share of the general-purpose electronic digital computing industry, and that it had acquired that market share because of unfair business practices such as bundling of hardware, software, and maintenance services at a single price (that is, requiring customers to buy all three together). If you wanted IBM equipment (hardware), you also had to take IBM service and software, whether you wanted them or not. Other companies had little chance to compete. Moreover, the department argued that IBM constantly redesigned its computers, making it impossible for other companies to keep up and compete fairly on the sale of any IBM mainframe-compatible item.

In technology industries the market is continually changing.

IBM argued that the relevant market was broader, that it included all types of computers, such as military computers, programmable calculators, and other information-processing products. It further claimed that its so-called unfair practices were simply a reflection of efficient computer technology. Fast-moving technological developments required it to continually redesign its products merely to provide its customers with the latest, best equipment. And, it said, the only way to provide the best level of service to its customers was to include its maintenance services in the price of its products.

By 1982 the market had changed. Many of the government's objections had become moot, and the U.S. government withdrew its case. Mainframe computers were being replaced with personal computers, and the globalization of the computer industry significantly reduced IBM's dominance.

About the same time as the case was at its height, IBM was negotiating with a young upstart company about an operating system for a small part of its market—the personal computer (PC) market, which was just developing. Bill Gates, the president of the young company, offered to sell its disk operating system (DOS) to IBM for $75,000. IBM refused to buy it,[9] and by the early 1990s, the cost of that decision was clear. The mainframe market was dying, Bill Gates had become a multibillionaire, and his company—Microsoft—had become a controlling force in the PC market.

The IBM case was dropped by the United States, but the prosecution likely led to IBM's problems in the 1990s. It won but it also lost.

Q.9 What was the resolution of the IBM case?

Then Came Microsoft

In the courtroom everyone waited for the Microsoft star witness, an economist who argued that Microsoft did not have a monopoly; much as IBM had argued previously. "The market is too dynamic, too much in flux for a monopoly to exist."

"But how about the fact that Microsoft intentionally tied its Web browser to its Windows operating system in order to harm Netscape's ability to compete?" the government lawyer asked.

"Those were technical decisions," the economist answered, "necessary to make the Web browser operate efficiently."

"Yeah, right," replied the government lawyer.

One of the U.S.'s most important antitrust cases brought in the 1990s was the Microsoft case. This is an extremely interesting case to consider both because of its similarities to the IBM case and because of the issues it raises about competition, the competitive process, and government's role in that competitive process.

Microsoft makes computer software. From its small start some 20 years ago, it now has about 50 percent of the world's total software market. It has an even larger share of the world's operating system software market. Between 80 and 90 percent of all computers use its Windows or DOS operating systems.

Since all software must be compatible with an operating system, the widespread use of Microsoft operating systems gave Microsoft enormous power—power that competitors claim it has used to gain competitive advantage for its other divisions. Competitors' calls for action, and reports of monopolistically abusive acts by Microsoft, led the U.S. Department of Justice in 1998 to charge Microsoft with violating antitrust laws.

The government suit against Microsoft charged the company with being a monopoly and using that monopoly power in a predatory way. Specifically, the charges were as follows:

1. Possessing monopoly power in the market for personal computer operating systems.
2. Tying other Microsoft software products to its Windows operating system.
3. Entering into agreements that keep computer manufacturers that install Windows from offering competing software.

Microsoft's 80–90 percent market share in PC operating systems has remained unchanged for about a decade. The U.S. Department of Justice argued that this longstanding monopoly position was the result of unfair business practices. Microsoft argued that Windows sold so well because it was a superior product. Microsoft further argued that, because it faced competition from technological change, it was not a monopolist.

[9]To have bought DOS would have given IBM greater control over the PC market, and would have made a court-ordered breakup of IBM more likely.

You will notice that these arguments also correspond to the provisions of Canadian Competition Policy.

Is Microsoft a Monopolist? The computer software industry is a market with barriers to entry that originate from two sources—network externalities and economies of scale. Network externalities exist because as the number of applications supported by a single platform increase, the value of the platform also increases. Economies of scale exist because the cost of developing a new platform and new software is significant, while the cost of producing it is minimal. It is a potential candidate for monopoly.

Whether one sees Microsoft as a monopolist depends on whether one views it in a static or dynamic framework.

Is Microsoft a monopoly in the market for operating systems? Looking only at the market within a static framework, Microsoft, given its stable 90 percent share, almost definitely has a monopoly. Looking at the market from a dynamic perspective, the issue is much more complicated. Competing operating systems exist—Macintosh (developed by Apple), OS/2 (developed by IBM), and Linux are all competitors to Windows.

Another potential competitive force is the merging of software and hardware. PCs—machines which handle a variety of tasks—could become obsolete, and instead the market may consist of inexpensive machines that perform specific tasks more efficiently than a multipurpose machine like the PC can. These changes could eliminate Microsoft's monopoly advantage. In this dynamic view of the market, Microsoft's monopoly is temporary, and will survive only if it out-competes the other technologies.

Is Microsoft a Predatory Monopolist? The U.S. Department of Justice argued that Microsoft used its monopoly in the operating systems market to gain a larger share of the software market and engaged in unfair practices against its competitors to maintain the barriers to entry in the operating system market.

Competing software companies alleged that companies like Corel—which had the leading word-processing software, WordPerfect—were put at a significant disadvantage because Microsoft combined its own software with the Windows operating system. Not surprisingly, Microsoft's word-processing system, Word, has become the dominant one. By directing the development of new software to favour Windows, Microsoft strengthened the barrier to entry created by network externalities. Microsoft also penalized computer manufacturers that installed Windows if they installed competing software. IBM, for example, was denied Windows 95 when it decided to pre-install its PCs with Lotus 1-2-3, a direct competitor to Microsoft's Excel.

Microsoft used its power in one market to give it advantages in other markets.

Microsoft was also alleged to have engaged in unfair practices in addressing the threat of competition in the operating system market. For example, Netscape designed and marketed a very popular Web browser called Netscape Navigator. Navigator posed a threat to Microsoft not only because it could serve as a platform for other software applications and circumvent the need for Windows but also because Navigator could work on many operating systems, increasing the ability of software to work on systems other than Windows. In response to that threat, Microsoft attempted to get Netscape to agree to stop developing Navigator as a competing platform in exchange for a "special relationship" with Microsoft. Netscape wouldn't agree. In response, Microsoft withheld for three months the source code Netscape needed to provide its browser on the Windows platform. This gave Microsoft an advantage to offer its own Web browser, Internet Explorer, with Windows 95. Microsoft then bundled its browser to Windows (essentially offering it at no cost) and made it virtually impossible for consumers to remove the Internet Explorer icon from the PC screen. Installing Windows 95 actually disabled competing Internet browsers. Microsoft also prohibited computer manufacturers such as

Compaq from offering Netscape as an alternative browser when they sold PCs with Windows.

The government argued that this competition was unfair and predatory. Microsoft argued that Internet Explorer was an improvement to Windows 95; it was part of Windows. Netscape Navigator, which had been more popular than Internet Explorer and had seen its sales climbing rapidly, was in a matter of a few years nearly replaced by the Microsoft browser.

Resolution of the Microsoft Case As all this was going on, the competitive nature of the market changed enormously. America Online (AOL), which was the world's largest Internet service provider, bought Netscape. AOL then merged with Time Warner, in a move that combined a bricks and mortar content company that was a major force in traditional media with a major Internet company. Microsoft argued that, even if it had been a monopoly before (which it claims it was not), these changes reduced that monopoly, and eliminated the need for any drastic policy settlement, such as breaking up Microsoft.

So, is Microsoft a monopoly? And has it been involved in anticompetitive practices? The answer Judge Thomas Penfield Jackson gave in 2001 was yes. The judge concluded that Microsoft violated Section 2 of the *Sherman Act* by attempting to maintain its monopoly power by anti-competitive means. He also ruled that Microsoft violated Section 1 of the *Sherman Act* by unlawfully tying its Web browser to its operating system. In a strongly worded decision he stated that "Microsoft mounted a deliberate assault upon entrepreneurial efforts that, left to rise or fall on their own merits, could well have enabled the introduction of competition into the market . . . ." As a remedy the government proposed breaking up Microsoft into two companies. However, on June 28, 2001, the U.S. Court of Appeals overturned the lower court's antitrust decision against Microsoft. While the Appeals Court maintained Judge Penfield Jackson's finding that Microsoft indeed did abuse its monopoly position in computer operating systems in order to suppress competition, the Court did not support the structural remedy of divestiture which would have broken the software giant into two companies, one for the operating system and one for the software applications. Microsoft has appealed the ruling that it is an illegal monopoly to the U.S. Supreme Court.

CHANGING VIEWS OF COMPETITION

In recent years, the approach to competition law by Canada and by other countries has changed. In the 1950s and 1960s the prevailing ideology saw big business as "bad," but by the 1990s the view had changed to one that was more complex—big business was seen as a combination of good and bad.

As Canada became more integrated into the global economy, big business faced significant international competition and hence competition created by the Canadian market structure became less important. As technologies became more complicated, the issues in competition enforcement also became more complicated.

COMPETITIVE EFFECTS OF GLOBALIZATION

Trade liberalization lowers trade barriers, opens up foreign markets to Canadian firms, and leads to the integration of world markets.

Canada is a small, open economy, which is particularly susceptible to the effects of globalization and free trade. Trade liberalization lowers trade barriers, enhances Canadian firms' access to foreign markets, enhances foreign firms' access to Canadian markets, and leads to the integration of the global economy. The impact on Canadian industry is significant, and there are implications for competition policy.

With increased **globalization**, *the integration of world markets*, the relevant market expands beyond the domestic economy. Cross-border competition and the threat of foreign entry, through acquisition, merger, or investment, can exert a significant disciplinary force on Canadian firms' competitive behaviour, making even highly concentrated markets contestable. By allowing freer flows of factors of production, trade liberalization reduces the ability of Canadian firms to maintain cartels and to raise domestic prices. The economic profits quickly attract entry from abroad, and enforce a competitive solution in the market.

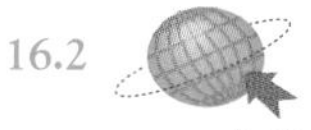
16.2 *see page 369*

With globalization, all firms need to lower costs to be internationally competitive. Being the lowest cost Canadian producer is no longer sufficient; companies need to be able to match prices in global markets. In order to avoid being at a competitive disadvantage globally, Canadian companies must aggressively address cost issues, such as economies of scale and scope, technology adoption, and training and education of their labour force. Integration into the global economy also means that Canadian firms can take advantage of specialization, and concentrate on a particular aspect of their business, rather than the whole spectrum. They are able to do this because they now can sell into a much larger market, where the demand for the specialized product is great enough to support a high degree of specialization by the firm. The Canadian market is simply not large enough to enable the firm to specialize profitably.

Globalization encourages Canadian firms to specialize and innovate. In many Canadian industries, the domestic market is small, so firms must produce for world markets in order to achieve minimum efficient scale.

The need for strong Canadian competition policy in light of stronger global competitive pressure is diminished; Canadian firms are responding well to international competition, and since the 1989 *Free Trade Agreement* with the United States, have been making investments to meet the rising forces of international competition. Canadian competition policy has helped, too. Export agreements, research and development joint ventures, and specialization agreements all assist Canadian firms to meet the rigours of global competition. Foreign competition and static and dynamic efficiencies are important considerations in competition policy, while structural factors, such as market share and industry concentration, are becoming less relevant. Both globalization and Canadian competition law encourage firms to take steps towards achieving superior competitive performance, to take risks in innovation, and to succeed in a dynamic world market.

However, this global competitive environment produces domestic competitive discipline only to a certain degree. At some point, the compatibility between competition law and globalization disappears. Consider the case of the 1998 proposed bank mergers. Global competitive forces have urged the Canadian banks to become superior performers in anticipation of their eventual entry into the global financial marketplace. The Canadian banks have already successfully expanded their various operations, for example commercial banking and brokerages, into the U.S. markets, and are now ambitiously preparing to enter the broader international financial markets, where the major firms are many times the size of a Canadian bank. Unfortunately, the degree of concentration in the Canadian banking industry is already high, even though the "big" five banks are small by international standards. Were the Canadian banks to achieve (for example, by merging) the size necessary to achieve economies of scale and to take advantage of cost-saving technologies and other cost advantages, this would have significant repercussions on competition in the domestic market.

The competitive discipline imposed by global market forces on Canadian industries can lead to unanticipated, and sometimes undesirable, changes in the domestic economy. In such a case, when the forces of globalization and Canadian competition policy lead to different outcomes, Canadians must decide which result they prefer.

Globalization can lead to unanticipated and undesirable changes in the Canadian economy.

Globalization is not an issue that will go away. Innovation and technological change, increased opportunities in international markets, the information technology revolution, tariff and trade policies worldwide, regulatory reform, privatization, and Canadian preferences towards business issues, must be frankly and ethically evaluated as Canadians make important decisions regarding the future of Canadian industrial policy. Canadians need to ensure that Canada's competition law meets the needs of Canadian businesses and Canadian consumers, as well as meeting the challenges of a fiercely competitive world market.

REGULATION, GOVERNMENT OWNERSHIP, AND INDUSTRIAL POLICIES

Competition policy is not the only way governments affect the competitive process. Other ways include (1) regulating the activities of firms, (2) government ownership—taking charge of the firms and operating them directly, and (3) industrial policy—influencing firms with laws and taxes. This section discusses some of the central elements of these three ways of affecting the competitive process.

The government can also affect the competitive process by (1) regulation, (2) government ownership, and (3) industrial policy.

Regulation

Regulation involves the setting of the rules that firms must follow if they are to conduct business. There are two types of regulation—price regulation and social regulation. *Price regulation* is regulation directed at industries that have natural monopoly elements. In order to allow them to take advantage of the economies of scale, firms are given an exclusive right to conduct business, but are subject to pricing controls; for example, the Canadian Radio-television and Telecommunications Commission (CRTC), which regulates cable television, telephones, television, and other communications areas. *Social regulation* is concerned with the conditions under which goods and services are produced, the safety of those goods, and the side effects of production on society; for example, Health Canada.

Two types of regulation are price regulation and social regulation.

Price Regulation Price regulation is usually imposed in those industries where there seems to be a natural monopoly. In such cases, a single producer is most efficient because two firms dividing the market could not take advantage of the economies of scale. But if the firm is to have the monopoly, then it will be able to charge high prices to consumers and transfer the consumer surplus into its profit. Thus, when such monopolies are granted, government must also regulate the price they charge. Usually this has taken the form of requiring the firm to charge its average total cost plus a profit margin.

A natural monopoly is an industry in which significant economies of scale make the existence of more than one firm inefficient.

That sounds reasonable in theory, but the practice has problems. The first is that the regulated firm does not have an incentive to hold down costs. Cost increases lead directly to price increases. X-inefficiency will exist. Regulatory boards have tried to counteract that tendency of rising costs by permitting firms to pass on only "legitimate" costs, but generally the people on the regulatory boards rely on the cost information provided

Price regulated firms often do not have an incentive to hold down costs.

by the regulated firm to determine whether costs are legitimate or not, and the companies have a strong incentive to make all costs look legitimate.

Another problem critics have seen in regulation is that, once established, regulation may tend to extend far beyond natural monopolies and be introduced into industries where competition could work. Still another problem is that regulation tends to continue even after technological change has created competitive market conditions.

Social Regulation[10] Social regulation differs from pricing regulation in that:

1. Social regulation applies to most firms, and is not designed specifically for a natural monopoly; for example, the Workers' Compensation Board regulations regarding safety in the workplace.
2. Social regulation affects large aspects of business: working conditions, the quality of the products, and the production processes firms are allowed to use.

Whereas pricing regulation has declined in the last 20 years, social regulation has increased. Critics of social regulation point out that regulation has high administrative and compliance costs and that those costs hurt consumers more than the regulation benefits them. They believe that this occurs because the social regulation laws are too often poorly written and ambiguous, and put into law without information on what is reasonable and feasible. The result is higher prices, less technical progress than there otherwise would have been, and fewer new entrants into a field as the regulatory burdens become unbearable for small firms.

Problems with price regulation have led to deregulation in recent years.

Advocates of social regulation argue that the benefits of social regulation are worth the costs, and that the objections are simply a call for better regulation. They argue that social regulation has made manufacturing much safer in Canada, has improved the quality of life and the environment enormously, introduced far more justice into the economy, and reduced discrimination against minorities.

Judging between these two views is difficult because measurement of both the costs and benefits is difficult or impossible. In such cases, the economic cost/benefit framework cannot provide a definitive answer.

Government Ownership

Instead of regulation, an alternative way of dealing with the problems of natural monopolies is for the government to own the firms itself. European countries have used this approach more often than has Canada. Instead of regulating the telephone (or other natural monopoly) industry, governments took it over and ran it. Since the 1980s most countries have been selling off government-owned businesses to private owners. Why? Most governments have found that government-owned firms did not have an incentive to hold costs down or to introduce new technology. Workers in government-owned firms, who were guaranteed jobs, used political threats to hold their wages high. Since the government firms faced little competition, they could raise prices and pass on the higher costs to consumers. The result was that European prices for telephone service, airline travel, and electricity were much higher than in North America. Economic integration in Europe has been accompanied by privatization of many of the formerly government-owned industries; the hope is that the larger markets will provide the needed competition.

Government-owned firms tend not to have an incentive to hold costs down.

Q-10 Why have European countries recently privatized many government-owned firms in the telecommunications, electricity, and airline industries?

[10] Social regulation is discussed more fully in Chapter 17.

Industrial Policies

An **industrial policy** is *a formal policy that government takes toward business*. In thinking about government's relation with business, it is important to remember that, in actual fact, a country's industrial policy is embodied in its tax code, its laws, and its regulatory structure, and in the positions the government takes in international negotiations about tariffs and trade. An example is the government's strong support of international copyrights and patents, which prevent foreign firms from making copies of products without paying a royalty to the firm which owns the idea. The policy is, however, an implicit policy of working with business, not an explicit policy of directing business.

An industrial policy is a formal policy that government takes toward business.

CONCLUSION

The aim of Canadian competition policy is to promote behaviour that leads to efficient outcomes in the marketplace. Market-based economies rely on market incentives to ensure competitive firm behaviour; however, these incentives do not always produce the desired result. Perfectly competitive markets are an ideal; contestable markets generally don't provide sufficient discipline for firms to behave competitively; and countries usually have social and economic goals besides efficiency. Even international competition does not adequately discipline domestic firms. Therefore, Canada, along with most other industrialized countries, relies on competition policy to help encourage competitive firm behaviour.

What conclusion should we come to regarding market structure and government policy toward the competitive process? That's a tough question, because the problem has so many dimensions. What we can say is that market structure is important, and generally more competition is preferred to less competition. We can also say that, based on experience, government-created and protected monopolies have not been the optimal solution, especially when industries are experiencing technological change. But how government should deal with monopolies that develop as part of the competitive process is less clear. Competition has both dynamic elements and market structure elements, and often monopolies that develop as part of the competitive process are only temporary and may disappear as changes in technology push changes in production methods and, subsequently, in industry structure. Thus the debate about government entering into the market to protect competition has no single answer. Once again, the answer is, "it depends."

Chapter 15 introduced specific problems in markets, such as externalities, and Chapter 16 considered broader issues of market functioning, and whether there was a role for government to improve market outcomes. The Canadian economy is a mixed economy; private market incentives are combined with government intervention to create an economy in which a variety of social and economic objectives are met. Chapter 17 continues the evaluation of the role of government in the Canadian economy by exploring how the many other goals of society, such as income distribution and social responsibility, are handled by our economy.

Chapter Summary

- Judgment by performance means judging the competitiveness of markets by the behaviour of firms in that market. Judgment by structure means judging the competitiveness of markets by how many firms operate in the industry and their market shares.
- There is a debate on whether markets should be judged on the basis of structure or on the basis of performance.
- Competition policy is the government's policy toward the competitive process.
- When contestable markets do not function as competitively as desired, the Canadian *Competition Act* steps in to regulate firm behaviour.
- The *Competition Act* focusses on achieving efficiency in the marketplace, rather than goals such as full employment or equitable income distribution.
- The *Competition Act* divides anticompetitive behaviour into two categories: criminal offences, which include misleading advertising, predatory pricing, and conspiracy; and civil offences, which include refusal to deal, abuse of dominant position, and merger.
- The U.S. antitrust suit against IBM filed in 1967 was withdrawn in 1982 because the computer market had changed, making the charges against IBM moot.
- In 1998 the U.S. government filed an antitrust suit against Microsoft, alleging that Microsoft abused its market power to maintain its monopoly position. In 1999 the courts found that Microsoft has a monopoly that is protected by barriers to entry and that Microsoft engaged in practices to maintain that monopoly power.
- Three types of mergers are horizontal, vertical, and conglomerate.
- A horizontal merger is the combination of two companies in the same industry, a vertical merger is the combination of two companies in different industries, and a conglomerate merger is the combination of two companies in relatively unrelated industries.
- Five reasons that two unrelated firms would want to merge are economies of scope, a good buy, diversification, warding off a takeover bid, and strengthening political-economic influence.
- The increasing internationalization of the Canadian market has changed Canadian competition policy from looking at just domestic competition to considering international competition.
- Three alternatives to competition policy are regulation, government ownership, and industrial policy.

Key Terms

acquisition *(351)*
Civil Reviewable Matter *(350)*
Clayton Antitrust Act *(360)*
Combines Investigation Act *(345)*
Competition Act *(345)*
competition policy *(343)*
Competition Tribunal *(346)*
conglomerate merger *(352)*
conscious parallelism *(348)*
conspiracy *(348)*
deacquisition *(352)*
divestiture *(346)*
efficiency defense *(350)*
exclusive dealing *(349)*
globalization *(364)*
horizontal merger *(352)*
hostile takeover *(351)*
industrial policy *(367)*
judgment by performance *(343)*
judgment by structure *(343)*
market restriction *(349)*
merger *(351)*
per se offence *(348)*
prenotification *(351)*
Sherman Antitrust Act *(359)*
takeover *(351)*
tied selling *(349)*
vertical merger *(352)*

Questions for Thought and Review

1. What is the difference between judgment by performance and judgment by structure?
2. Distinguish the basis of judgment for the bank mergers.
3. How would the U.S. economy likely differ today if Standard Oil had not been broken up?
4. How did the 1986 *Competition Act* clarify the *Combines Investigation Act*?
5. Universities require that students take certain courses at that university in order to get a degree. Is that an example of a tie-in contract that limits consumers' choices? If so, should it be against the law?

6. Universities give financial aid to certain students. Is this price discrimination? If so, should it be against the law?
7. Should interlocking directorships be against the law? Why or why not?
8. If you were an economist for a firm that wanted to merge, would you argue that the three-digit or five-digit NAICS industry is the relevant market? Why?
9. If you were an economist for Mattel, manufacturer of the Barbie doll, which was making an unsolicited bid to take over Hasbro, manufacturer of G.I. Joe, would you argue that the relevant market is dolls, preschool toys, or all toys including video games? Why? Would your answer change if you were working for Hasbro?
10. Has telephone service improved since telephone deregulation? What does this imply about competition laws?
11. How did the antitrust suit against IBM affect IBM's future business?
12. In what market did Microsoft have a monopoly in the late 1990s? What technological advances threatened that monopoly?
13. Under the Competition Bureau's guidelines, would a merger be allowed between a firm that had 15% market share and one that had 12% market share? The post-merger four-firm concentration ratio (CR4) would be 60%.
14. Should Canada have a policy against conglomerate mergers? Why or why not?
15. How has the globalization of the Canadian economy changed Canadian competition policy?
16. What two methods does government have for dealing with natural monopolies? What problems are associated with each?
17. How would you design an industrial policy to avoid the problems inherent in industrial policies?

Problems and Exercises

1. You're working at the Bureau of Competition Policy. Ms. Ecofame has just brought in a new index, the Ecofame index, which she argues is preferable to the Herfindahl index. The Ecofame index is calculated by cubing the market share of the top 10 firms in the industry.
   a. Calculate an Ecofame guideline that would correspond to the Competition Bureau guidelines.
   b. State the advantages and disadvantages of the Ecofame index as compared to the Herfindahl index.
2. Using a monopolistic competition model, a cartel model of oligopoly, and a contestable market of oligopoly, discuss and demonstrate graphically, where possible, the effect of competition policy.
3. In 1993 Mattel proposed acquiring Fisher-Price for $1.2 billion. In the toy industry, Mattel is a major player with 11 percent of the market. Fisher-Price has 4 percent. The other two large firms are Tyco, with a 5 percent share, and Hasbro, with a 15 percent share. In the infant/preschool toy market, Mattel has an 8 percent share and Fisher-Price has a 27 percent share, the largest. The other two large firms are Hasbro, with a 25 percent share, and Rubbermaid, with a 12 percent share.
   a. What are the approximate Herfindahl and four-firm concentration ratios for these firms in each industry?
   b. If you were Mattel's economist, which industry definition would you suggest using in court if you were challenged by the government?
   c. Give an argument why the merger might decrease competition.
   d. Give an argument why the merger might increase competition.
4. In 2001, Aardvark Airlines offered a 50-percent-off sale on its fares. In 2002, Flyhigh Airlines sued Aardvark Airlines over this action.
   a. What was the likely basis of the suit?
   b. How does the knowledge that Flyhigh was in serious financial trouble play a role in the suit?
5. Demonstrate graphically how regulating the price of a monopolist can both increase quantity and decrease price.
   a. Why did the regulation have the effect it did?
   b. How relevant to the real world do you believe this result is in the "contestable markets" view of the competitive process?
   c. How relevant to the real world do you believe this result is in the "cartel" view of the competitive process?

Web Questions

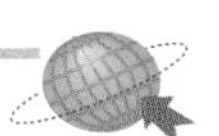

1. Go to strategis.ic.gc.ca and research the following questions.
   a. Which four Acts fall under jurisdiction of the Competition Bureau? How does a person file a deceptive business practice complaint under one of these Acts? Is your complaint confidential? How is a civil complaint handled? How is a criminal complaint handled?
   b. What consumer information topics are listed? What items are listed under "Recent News and Warnings?" Choose one, and briefly summarizes the issue.

c. What is "Green Marketing?" In its "Principles and Guidelines for Environmental Labelling and Advertising," what actions are required to ensure responsible labelling and advertising? Describe Industry Canada's guiding principles.

2. Go to www.internationalcompetitionnetwork.org and answer the following questions:
   a. What is the International Competition Network and why was it formed?
   b. Who are the member countries of the International Competition Network? How many continents do they represent? What is the International Competition Network's mission? Why is competition advocacy an important function of a competition agency?
   c. Who represents Canada at the International Competition Network? What is his "job" in Canada?
   d. What is the mandate of the Merger Review Working Group? Briefly describe the issues involved.
   e. What is the mandate of the Advocacy Working Group? Briefly describe their work.

Answers to Margin Questions 

1. Degree of concentration and firm interdependence as well as international competition are major factors influencing competitiveness in Canadian industries. *(341)*
2. Exit is not entirely free of cost, but there are no artificial barriers to exit. The potential entrant has no sunk costs; all fixed costs are recoverable if the firm exits, even in industries characterized by economies of scale or large initial capital investments. *(341)*
3. No, neither one is illegal. *(348)*
4. This defense is captured in section 96(1) of the *Competition Act*, and states that a merger shall be allowed if gains in efficiency resulting from the merger are great enough to offset the anticompetitive effects of the merger. *(350)*
5. If Dairyland bought a dairy farm, it would be a vertical merger because Dairyland would be buying one of its suppliers. *(352)*
6. In its investigation, the Bureau considered market concentration, barriers to entry, sunk costs, economies of scale, number of remaining competitors, range of products and services, regulations governing entry, interdependent behaviour, and the likelihood of collusion, as well as the efficiency arguments presented by the banks. *(357)*
7. The United States has antitrust laws that are more strict than those of Canada. *(357)*
8. The *Sherman Antitrust Act* contained two main sections. The first stated that every contract, combination, or conspiracy in restraint of trade was illegal. The second stated that every person who shall monopolize or attempt to monopolize shall be deemed guilty of a misdemeanour. These provisions, while sounding strong, were so broad that they were almost unenforceable, and the interpretation was left to the courts. *(359)*
9. In the late 1960s, the Department of Justice filed suit against IBM for violating the antitrust laws. It alleged that IBM had a monopoly of the general-purpose electronic digital computing industry and that it had acquired its market share because of unfair business practices. The case dragged on for 13 years but never went to court. In 1982, the government withdrew its lawsuit. The antitrust case, however, had significant effects on IBM. It is likely that the experience caused IBM to shy away from the then-small personal computer market. This decision by IBM very likely was the beginning of the serious problems that IBM faced in the 1990s. *(361)*
10. Many countries, including Canada and many European countries have been privatizing firms in telecommunications, electricity, and airline industries because the government-owned firms were inefficient, resulting in higher costs for consumers and less innovation. *(366)*

Microeconomic Policy, Economic Reasoning, and Beyond

17

After reading this chapter, you should be able to:

- List three reasons why economists sometimes differ in their views on social policy.
- Explain why liberal and conservative economists often agree in their views on social policy.
- Explain the cost/benefit approach the typical economist takes to analyze regulations.
- Describe three types of failure of market outcomes.
- Explain why economists are doubtful government can correct failure of market outcomes.
- Explain how politics sometimes makes a positive and sometimes a negative contribution to economic policy.

If an economist becomes certain of the solution of any problem, he can be equally certain that his solution is wrong.

H. A. Innis

One important job of economists is to give advice to politicians and other policy-makers on a variety of questions relating to social policy: How should unemployment be dealt with? How can society distribute income fairly? Should the government redistribute income? Would a program of equal pay for jobs of comparable worth (a pay equity program) make economic sense? Should the minimum wage be increased? These are tough questions.

Chapters 15 and 16 introduced the concept of market failure, and considered various situations of market failure and their possible solutions. Chapter 15 discussed the role of government in three situations where private markets fail to generate the efficient market equilibrium: externalities, public goods, and asymmetric information. Direct government participation, as in the case of publicly provided national parks, or government regulation, as in the case of licensing, were discussed as possible solutions to the market failures.

Chapter 16 considered the case where monopoly and other impediments to competition lead to varying degrees of market failure. The role of government in encouraging competition is to lay out the rules of play in the market economy; the Canadian *Competition Act* is the set of rules by which firms can participate in the market. Violations of the rules of competition bring forth direct government intervention, through the Bureau of Competition Policy.

Chapter 17 considers yet another set of conditions where government involvement in the market economy may be desirable. Although the market economy may generate an equilibrium result that is efficient, there may be other characteristics about the outcome that are undesirable. Society's goals include efficiency, but society also values other goals. Equity of income distribution, fair access to available benefits of a civilized society (schooling, libraries, swimming pools), and social responsibility are other important goals. This chapter presents an evaluation of the competitive market outcome in terms of criteria other than efficiency. We consider various social goals, and how these goals may or may not be met by a freely functioning market economy, and how government may sometimes be able to improve on the market outcome.

The supply/demand model is a tool for analysis, not a rule.

Economic reasoning, and the supply/demand model are tools for analysis. They are not rules. To draw policy implications from the supply/demand model requires it to be placed in context. Used in the proper context, the supply/demand model is enormously strong. Used out of context, it can lead to conclusions that don't seem right, and that maybe are not right. The economic model focusses on efficiency, and efficiency is only one of society's goals.

The chapter is divided into two parts. The first part of the chapter extends the supply/demand model to a broader cost/benefit framework, tying together the discussion we had about economic reasoning in the introductory chapters with the chapters that developed the foundations of the supply/demand model. The second part of the chapter discusses how markets that are working perfectly may still lead to outcomes that are undesirable.

ECONOMISTS' DIFFERING VIEWS ON SOCIAL POLICY

Economists' views on social policy differ widely because (1) their objective economic analyses are influenced by their subjective value judgments; (2) their interpretations of economic evidence and of how political and social institutions work vary widely; and (3) their proposals are often based on various models that focus on different aspects of problems.

Economists have many different views on social policy because:

1. Economists' suggestions for social policy are determined by their subjective value judgments (normative views) as well as by their objective economic analyses.
2. Policy proposals must be based on imprecise empirical evidence, so there's considerable room for differences of interpretation not only about economic issues but also about how political and social institutions work. Economic policy is an art, not a science.
3. Policy proposals are based on various models that focus on different aspects of a problem.

All three reasons directly concern policy making in economics. Any policy proposal must combine both economic analysis and value judgments because the goals of policy reflect value judgments. When an economist makes a policy proposal, it's of this type: "If A, B, and C are your goals, then you should undertake policies D, E, and F to achieve those goals most efficiently." In making these policy suggestions, the economist's role is much the same as an engineer's: he or she is simply telling someone else how to achieve desired ends most efficiently. Ideally the economist provides as objective a solution as possible to achieve the desired goals.

How Economists' Value Judgments Creep into Policy Proposals

Even though economists attempt to be as objective as possible, value judgments enter into their analyses in three ways: interpretation of policymakers' goals, interpretation of empirical evidence, and choice of economic models.

Interpretation of the Policymaker's Values In practice, social goals are seldom clear; they're vaguely understood and vaguely expressed. For instance, "We want to make the poor better off" doesn't define what *poor* or *better off* mean. Nor is it clear how judgments should be made when a policy will benefit some individuals at the expense of others, as real-world policies inevitably do.

In practice, social goals are seldom clear; they're generally vaguely understood and vaguely expressed.

Faced with this problem, some academic economists have argued that economists should recommend only **Pareto optimal policies**—*policies that benefit some people and hurt no one*. The policies are named in honor of the famous Italian economist Wilfredo Pareto, who first suggested that kind of criterion for judging social change.[1] It's hard to object to the notion of Pareto optimal policies because, by definition, they improve life for some people while hurting no one.

Pareto optimal policies are policies that benefit some people and hurt no one.

Every policy inevitably has some side effect of hurting, or at least seeming to hurt, somebody. In the real world, Pareto optimal policies don't exist. Any economist who has advised governments on real-world problems knows that all real-world policies make some people better off and, at the same time, some people worse off.

Economists' policy proposals try to spell out the effect of a policy on aspects other than efficiency: for example, the distribution of income and wealth, whether a policy will help the majority of people, who those people are, and whether the policy is consistent with the policymaker's value judgments. Economic models are able to distinguish quite clearly the efficiency effects associated with a particular policy; however, the equity issues often prove difficult to evaluate. Because the policymaker's desires are often expressed in vague statements about wishing to do good, they must be interpreted by the economist. And that interpretation can be difficult to confirm.

Interpretation of Empirical Evidence Empirical evidence is almost always imprecise. For example, an economist assessing the elasticity of a product's demand in the relevant price range can't run an experiment to isolate prices and quantities demanded; instead she must look at events in which hundreds of other things changed, and do her best to identify what caused what. Economists are trained to be as objective as they can be, but pure objectivity is impossible.

Q-1 If someone suggests that economists should focus only on Pareto optimal policies, how would you respond?

Consider an example. In the 1980s, some politicians proposed that a large tax be imposed on sales of disposable diapers, citing studies that suggested disposable diapers made up between 15 and 30 percent of the garbage in a landfill. Other studies showed

[1]Pareto, in his famous book *Mind and Society*, suggested this criterion as an analytic approach for theory, not as a criterion for real-world policy. He recognized the importance of the art of economics and that real-world policy has to be judged by much broader criteria than efficiency alone.

disposable diapers made up only 1 or 2 percent of the refuse going into landfills. Such differences in empirical estimates are the norm, not the exception. Inevitably, if precise estimates are wanted, more studies are necessary. (In this case, the further studies showed that the lower estimates were correct.) But policy debates don't wait for further studies. Economists' value judgments influence which incomplete study they choose to believe is more accurate.

Choice of Economic Models A model, because it can only focus on certain aspects of economic reality, necessarily reflects certain value judgments, so an economist's choice of model must also reflect certain values.

This book presents the mainstream economic model. That model concentrates on market outcomes and market applications, and directs us to certain conclusions regarding efficiency. Two other general models that some economists follow are a **Marxian model** which is *a model that focusses on equitable distribution of power, rights, and income among social classes* and a **public choice model** which is *a model that focusses on economic incentives as applied within a political framework*. These two models, by emphasizing different aspects of economic interrelationships, especially those regarding equity issues and politics, sometimes direct us to other conclusions. These models provide solutions which integrate efficiency and socio-political equity issues. Equity issues account for the majority of economic debate, especially those presented in the various media.

Q-2 How does a Marxist analysis of labour markets differ from a mainstream analysis?

Let's consider an example. Mainstream economic analysis directs us to look at how the invisible hand achieves an efficient equilibrium through the market. Thus, when mainstream economists look at labour markets, they generally see supply and demand forces leading to equilibrium. If the wage rate is "low," it is the result of market forces operating on demand and supply, and is strictly an efficient outcome. Notice that there is a strong reliance on value judgment here too—the efficient market outcome is "best," a judgement which is questioned by other models. When Marxist economists look at labour markets and see the same "low" wage, their model focusses on the dominance of producers relative to employees, and they generally see exploitation of workers by capitalists. When public choice economists look at labour markets, they see individuals using government to protect their monopolies. Their model focusses on political restrictions that provide rents to various groups. Each model captures different aspects of reality. That's why it's important to be familiar with as many different models as possible, and to recognize the contribution each one makes to our understanding of our economy.

Each model captures different aspects of reality. That's why it's important to be familiar with as many different models as possible.

The Need for a Worldview

17.1

see page 387

John Maynard Keynes, an economist who gained fame in the 1930s, once said that economists should be seen in the same light as dentists—as competent technicians. In dealing with real-world economic policy, however, Keynes was no mere technician. He had a definite worldview, which he shared with many of the policymakers he advised. An economist who is to play a role in policy formation must be willing to combine value judgments and technical knowledge. That worldview determines how and when a particular economic model will be applied.

AGREEMENT AMONG ECONOMISTS ON SOCIAL POLICY

Despite their widely varying values, both liberal and conservative economists tend to agree on policy prescriptions. Economic models focus on certain issues—specifically on incentives and individual choice. Economists believe economic incentives are important, and most economists tend to give significant weight to individuals' ability to choose reasonably. This leads economists, both liberal and conservative, to look at problems differently than other people do.

Liberal and conservative economists agree on many policy prescriptions because they use the same models, which focus on incentives and individual choice.

Many people think economists look at the world coldheartedly. The mainstream economic model, which relies on the demand and supply framework, measures the efficiency of outcomes of various policies; additional analysis is required to evaluate the equity outcomes of the policies. Demand and supply simply generate the equilibrium price and quantity from a particular policy; other than the outcome (in a free market) being efficient, the desirability of other aspects of economic life (health, income distribution, education) is not mentioned, so we must broaden our analysis if we are to evaluate our economy more completely. Efficiency, as seen in chapter 15, is only one of many important elements of people's (economic) lives.

Q-3 In terms of economic policy, when can "being mean" actually be "being nice"?

ECONOMISTS' COST/BENEFIT APPROACH TO GOVERNMENT REGULATION

Economists differ in their views on government regulation, but they generally adopt a **cost/benefit approach** to problems—*assigning costs and benefits, and making decisions on the basis of the relevant costs and benefits*—which requires them to determine a quantitative cost and benefit for everything, including life. What's the value of a human life? All of us would like to answer, "Infinite. Each human life is beyond price." But if that's true, then in a cost/benefit framework, everything of value should be spent on preventing death. People should take no chances. They should drive at no more than 50 kilometres per hour with airbags, triple-cushioned bumpers, and double roll bars.

Many regulations are formulated for political reasons and do not reflect cost/benefit considerations.

Cost/benefit analysis is analysis in which one assigns a cost and benefit to alternatives, and draws a conclusion on the basis of those costs and benefits.

It might be possible for manufacturers to make a car in which no one would die as the result of an accident. But people don't buy the auto safety accessories that are already available, and many drivers ignore the present speed limit. Instead, many people want cars with style and speed.

The Value of Life

Far from regarding human life as priceless, people make decisions every day that reflect the valuations they place on their own lives. Table 17-1 presents one economist's estimates of some of these quantitative decisions. These values are calculated by looking at people's revealed preferences (the choices they make when they must pay the costs). To find them, economists calculate how much people will pay to reduce the possibility of their death by a certain percentage. If that's what people will pay to avoid death, the cost to them of death can be calculated by multiplying the inverse of the reduction in the probability of death by the amount they pay. (What is relevant for these calculations is not the actual probabilities but the decision maker's estimate of the probabilities.)

For example, say someone will buy a car whose airbags add \$500 to the vehicle's cost. Also say that the buyer believes that an airbag will reduce the chance of dying in an automobile accident by 1/720. That means that to increase the likelihood of surviving an auto accident by 1/720, the buyer will pay \$500. That also means that the buyer is implicitly valuing his or her life at roughly \$360,000 (720 × \$500 = \$360,000).

Q-4 If Table 17-1 correctly describes the valuation individuals place on life with regard to air bag purchases and seat belt usage, how would you advise them to alter their behaviour in order to maximize utility?

Another way of determining the value that society places on life is to look at awards juries give for the loss of life. One study looking at such awards found that juries on average value life at about \$3.5 million.

No one can say whether people know what they're doing in making these valuations, although the inconsistencies in the valuations people place on their lives suggest

TABLE 17-1 Value of Life

Such figures are increasingly being used in court to support claims for loss of enjoyment of life.

Basis for Calculation	Value of Life (in 2000 U.S. dollars)
Automotive air bag purchases	$ 516,000
Smoke detector purchases	542,000
Sulphur scrubbers	724,000
Wage premiums for dangerous police work	1,231,000
Auto safety features	3,620,000
Regulations of radium content in water	3,620,000
Wage premiums for dangerous factory jobs	4,633,000
Seat belt usage	4,640,000
Rules for workplace safety	5,212,000
Premium tire usage	5,200,000

Source: Stan V. Smith, Ph.D., adjunct professor, DePaul College of Law, and president, Corporate Financial Group, Chicago. Used by permission. (Adjustments made by author.)

Economists argue that individuals' revealed choices are the best estimate that society can have of the value of life.

that to some degree they don't, or that other considerations are entering into their decisions. But even given the inconsistencies, it's clear that people are placing a finite value on life.

Assessing the Costs and Benefits

The costs and benefits are not always clear, but economists measure and estimate as many of the values as they can. Identifying which items are costs, and to whom, and which items are benefits, and, again, to whom, is only the beginning. Once the costs and benefits have been identified, an estimate of the values involved must be made. Consider an example. A manufacturing plant moves into an industrial park, and begins production. The benefits are many: increased tax base for the community (which increases benefits to its citizens as a direct result), jobs, (at the new plant and at related firms, such as suppliers and distributors) greater product availability, and a higher profile for the industrial park (perhaps resulting in a greater demand for its services, and corresponding secondary benefits to the greater community—taxes, jobs, etc.). Some of the benefits are easily measured, some are very difficult to measure, and some can only be estimated. On the other hand, the arrival of the manufacturing plant brings costs: increased traffic congestion, pollution, and noise. In addition, the bakery and coffee shop located beside the manufacturing plant are affected (probably both negatively and positively) by the operations of the plant; this also must be included in the analysis.

Comparing Costs and Benefits of Different Dimensions

After the marginal cost and marginal benefit data have been gathered and processed, one is ready to make an informed decision. Will the cost of a new regulation outweigh the benefit, or vice versa? Here again, economists find themselves in a difficult position in evaluating a regulation, for example, about airplane safety. Many of the costs of regulation are small but occur in large numbers. But when those costs are compared to the benefits of avoiding a major accident, the dimensions of comparison are often wrong.

Cost/benefit analysis sometimes leads to uncomfortable results.

For example, say it is discovered that a loose bolt was the probable cause of a plane crash. A regulation requiring airline mechanics to check whether that bolt is tightened and, to ensure that they do so, requiring them to fill out a form each time the check is made, might cost $1. How can we compare $1 to the $600 million cost of the crash?

Such a regulation obviously makes sense from the perspective of gaining a $600 million benefit from $1 of cost.

But wait. Each plane might have 4,000 similar bolts, each of which is equally likely to cause an accident if it isn't tightened. If it makes sense to check that one bolt, it makes sense to check all 4,000. And the bolts must be checked on each of the 4,000 flights per day. All of this increases the cost of tightening bolts to $16 million per day. But the comparison shouldn't be between $16 million and $600 million. The comparison should be between the marginal cost ($16 million) and the marginal benefit, which depends on how much tightening bolts will contribute to preventing an accident.

Let's say that having the bolts checked daily reduces the probability of having an accident by 0.001. This means that the check will prevent one out of a thousand accidents that otherwise would have happened. The marginal benefit of checking a particular bolt isn't $600 million (which it would be if you knew a bolt was going to be loose), but is:

$$0.001 \times \$600 \text{ million} = \$600{,}000$$

That $600,000 is the marginal benefit that must be compared to the marginal cost of $16 million.

Cautions with Cost-Benefit Analysis

Although cost-benefit analysis can assist our decision-making process by providing not only an analytic framework, but calculations of the magnitudes of the costs and benefits involved in the decision, it has weaknesses.

Revealed Preference

One problem in using a cost-benefit approach to analyze a decision lies with the use of revealed preferences to indicate consumer surplus. Remember that consumers reveal their preferences for particular goods and services through (the height of) their demand curve. There are two big difficulties with relying on revealed preferences. The first is that consumers' decisions are constrained by their incomes. A person wishing to build a fence around his yard to keep his kids safe faces a tradeoff: The fence boards each cost $3 at the local lumber yard. A jug of milk at the grocery store costs $3. Therefore, each fence board costs one jug of milk. The income constraint plays a very important role in consumers' decisions; poor people don't value the safety of their kids any less than a rich person does, but the income constraint forces them to make choices they otherwise would not make. Poor people don't drive unsafe cars because they don't value safety; on their incomes, they can't afford safety. The point is that it is income-constrained preferences which are revealed in the marketplace.

The market outcome relies on revealed preferences, which are determined by real-world considerations such as consumer incomes, available product substitutes, and consumer assessment of relevant risks.

The second problem with using revealed preferences in cost-benefit analyses is that people's understanding of risk is poor. Even without considering the requirement of full information for rational decision-making, people's knowledge of the risks involved in many of their decisions is weak. Utility maximization under full information is complicated enough that people sometimes rely on alternative decision mechanisms, such as habit, but decision-making under conditions of risk or uncertainty is even more complex. How important are seat belts in saving lives? What about speeding? Our behaviour often indicates that our understanding of the relative risks of various activities is erratic and unreliable.

Bias Towards Quantifiable Costs and Benefits

The numbers in our plane crash example are hypothetical. The numbers used in real-world decision making are not hypothetical, but they are often difficult to estimate in

Q-5 Why should you be very careful about any cost/benefit analysis?

real life. Measuring costs, benefits, and probabilities is difficult, and economists often disagree on specific costs and benefits.

Cost/benefit analysis is often biased toward quantifiable costs.

Costs have many dimensions, some more quantifiable than others. Cost/benefit analysis is often biased toward quantifiable costs and away from nonquantifiable costs, or it involves enormous ambiguity as nonquantifiable costs are quantified. This is particularly true for situations involving personal and environmental impacts. For example, what is the impact of airport noise on nearby homeowners and wildlife?

The subjectivity and ambiguity of costs are one reason why economists differ in their views of regulation. But their reasoning process—comparing marginal costs and marginal benefits—is the same; they differ only on the estimates they calculate.

The Problem of Other Things Changing

One problem economists have concerns the "other things equal" assumption discussed in Chapters 4 and 5. Supply and demand analysis assumes that all other things remain equal. But in a large number of issues it is obvious that other things do not remain equal. However, it is complicated to sort out how they change, and the sorting-out process is subject to much debate. The larger the issue, the more other things change, and hence the more debate there is.

Q-6 When using marginal cost/marginal benefit analysis, do "other things remain constant"? Explain.

Let's consider the minimum wage example we discussed in earlier chapters. Suppose you can estimate the supply and demand elasticities for labour. Is that enough to enable you to estimate the number of people who will be made unemployed by a minimum wage? To answer that, ask yourself: Are other things likely to remain constant? The answer is: No; a chain of possible things can change. Say the firm decides to replace these workers with machines. So it will buy some machines. But machines are made by other workers, and so the demand for workers in the machine-making industry will rise. So the decrease in employment in the first industry may be offset by an increase in employment elsewhere.

But there are issues on the other side too. For example, workers who get the higher wage may not receive a net benefit if other things change. Say you had a firm that was paying a wage lower than the minimum wage but was providing lots of training, which was preparing people for better jobs in the future. Now the minimum wage goes into effect. The firm keeps hiring workers, but it eliminates the training. Its workers are actually worse off.

How important are such issues? That's a matter of empirical research, which is why empirical research is central to economics. Unfortunately, the data aren't very good, which is one of the reasons why there is so much debate about policy issues in economics.

The Cost/Benefit Approach in Context

Economics teaches people to be "reasonable."

Economics teaches people to be reasonable. The cost/benefit approach to problems (which pictures a world of individuals whose self-interested actions are limited only by competition) makes economists look for the self-interest behind individuals' actions, and for how competition can direct that self-interest towards the public interest.

In an economist's framework,

- Well-intentioned policies often are prevented by individuals' self-interest-seeking activities.
- Politicians have more incentive to act fast—to look as if they're doing something—than to do something that makes sense from a longer term view.

The marginal cost/marginal benefit story is embodied in the supply/demand framework.

The marginal cost/marginal benefit approach tells a story. That story is embodied in the supply/demand framework. Supply represents the marginal costs of a trade, and demand represents the marginal benefits of a trade. Equilibrium is where quantity supplied

KNOWING THE TOOLS

Economic Efficiency and the Goals of Society

Economic efficiency means achieving a goal at the lowest possible cost. For the definition to be meaningful, the goal must be specified. Efficiency in the pursuit of efficiency is meaningless. Thus, when we talk about economic efficiency, we must have some goal in mind. In the supply/demand framework, we *assume* the goal is to maximize total output. Each of the three failures of market outcomes that we discuss in this section represents a situation in which the goals of society cannot be captured by a single measure—where society's goal is more complicated than to maximize total output—and thus the assumed goal of efficiency—maximizing total output—is not the only goal of society.

equals quantity demanded—where marginal cost equals marginal benefit. That equilibrium maximizes the combination of consumer and producer surplus and leads to an efficient, or Pareto optimal, outcome. The argument for competitive markets within that supply/demand framework is that markets allow the society to achieve **economic efficiency**—*achieving a goal, in this case producing a specified amount of output, at the lowest possible cost*. Alternatively expressed, the story is that, given a set of resources, markets produce the greatest possible output. When the economy is efficient, it is on its production possibility curve, producing total output at its lowest opportunity cost.

Q-7 True or false? The goal of society is efficiency.

The supply/demand framework is logical, satisfying, and (given its definitions and assumptions) extraordinarily useful. It provides a framework for analyzing many policy problems. It tells us that every policy has a cost, every policy has a benefit, and if the assumptions are met, competition sees to it that the benefits to society are achieved at the lowest possible cost. Applied to policy issues, the supply and demand framework presents trade-offs in a simple, logical, useful way. It is what "thinking like an economist" is all about.

FAILURE OF MARKET OUTCOMES

The second part of this chapter discusses some implicit assumptions that the supply/demand framework suppresses. Sometimes a market can function efficiently, but the outcome has undesirable characteristics. Some important social goal has not been met. While there is no market failure, there is a failure of the market outcome. A **failure of market outcome** occurs *when, even though the market is functioning properly, it is not achieving society's goals*.

Failure of market outcomes occurs when, even though the market is functioning properly, it is not achieving society's goals.

Three separate types of failures of market outcomes will be considered:

1. *Failures due to distributional issues*: Whose surplus is the market maximizing?
2. *Failures due to rationality problems of individuals*: What if individuals don't know what is best for them?
3. *Failures due to violations of inalienable rights of individuals*: Are there certain rights that should not be for sale?

Distribution

Labour markets are generally thought of as reasonably competitive and relatively freely operating markets. However, in some cases, the result of market forces is that some people don't earn enough income to be able to survive—the demand for their labour intersects the supply for their labour at a wage of 25 cents an hour. Also assume there are no market failures, as described in Chapter 15. (Information is perfect, trades have no negative externalities, and all goods are private goods.)

The market forces yield an equilibrium wage that is so low the worker can't survive. Allowing the market solution coincides with a Darwinian "survival of the fittest" approach to social policy. Having a society where people work for an inadequate wage undermines our sense of civilization. Even though the market is doing precisely what it is supposed to be doing—equating quantity supplied and quantity demanded—most people would not find the outcome acceptable.

Implicit within the supply/demand framework is a "survival of the fittest" approach to social policy.

While the demand and supply framework is an important analytic tool in economics, economists also believe that the market solution does not provide satisfactory results in all situations. For example, issues of human dignity do not support the creation of efficient markets in humans or human organs.

Distribution of Total Surplus Now let's relate this distributional issue to the supply/demand framework by considering distribution of consumer and producer surplus. For most discussions of economic policy, an implicit assumption is that the goal of policy is to create as much total surplus as possible. In a world of only one good and one person, that goal would be clear. But with many goods and many people, distribution of the total surplus becomes an issue. In the wage example above, the reason most people do not like the market outcome is that they care about not only the size of the total surplus but also how total surplus is distributed. The supply/demand framework does not distinguish between those who get producer and consumer surplus, and while total surplus may be maximized, the distribution issue is not addressed.

Examples of Distributional Issues Let's consider two real-world examples where distributional issues are likely to play a significant role in value judgments about the market outcome. The problem is fundamental. Consumers maximize utility subject to an income constraint, and the income constraint matters when deciding what goods will be demanded and in what amounts. Our economy may produce $200-an-ounce olive oil, but may not provide a minimum level of health care (dental coverage, for example) for all. This happens because although the consumer is sovereign, the income distribution is highly unequal. The high income of the wealthy means that they are willing (and able) to pay, so there is demand for $200-an-ounce bottles of olive oil, and in a private market setting, businesses will establish production facilities to produce it (or any one of a million other luxury items), and it is sold in the market. Producer plus consumer surplus is maximized (marginal benefit equals marginal cost), and the outcome is efficient. But society's goal to maximize total consumer and producer surplus is only one of many goals. Given the distribution of income, the market finds it inefficient to produce health care for the poor. The poor just don't have sufficient income to demand it. The income constraint matters. Since they have little income, the poor are given little weight in the measure of consumer surplus. Canadians, in their just society, do not prefer the market solution to be applied in certain situations, for example, healthcare, education, and children's recreation.

A second example of where distribution of income likely makes a big difference in our normative judgments, and where we would likely not apply the consumer and producer surplus reasoning, concerns the demand for the AIDS drug cocktail. The cocktail can stop AIDS from killing people; thus, the desire for the AIDS cocktail among individuals with AIDS is high. The desire for the drug among those without AIDS is minimal.

In some African countries, 30 percent of the population has AIDS. Since consumer surplus reflects desire, one might think that in Africa the consumer surplus from the desire for the AIDS drug cocktail would be enormous. But it isn't. Most people in Africa have relatively little income; in fact, most have so little income that they cannot afford the cocktail at all. Since the price of the cocktail is above their total income, they get no consumer surplus from the cocktail at all in the supply/demand framework—it would

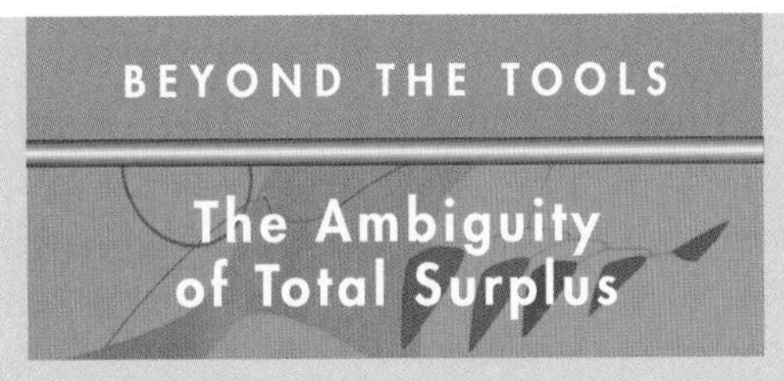

To talk about consumer and producer surplus we must figure out a way of taking into account people's valuation of surplus. The market provides one way—it allows each person to vote with their income in the buying of goods. This means that whoever has the income plays a role in determining how we measure total surplus. This presents a circular reasoning problem in moving from total surplus to social welfare, and in talking about efficiency as if it involves maximizing total surplus. If the distribution of income is one of society's goals, then efficiency cannot unambiguously be defined without also specifying distributional goals.

Let's consider an extreme example to show the problems this can present. Say we start with two individuals, Jules and Jim, and two goods, apples and oranges. Jules likes only oranges and Jim likes only apples. If Jim has all the income, only apples count for determining total output because Jim likes only apples. Since no oranges are traded, oranges are given no weight because Jules has no income. Now, say that Jules has all the income. In that case only oranges contribute to total output. More generally, when two individuals have different tastes, the way in which income is distributed can change the contribution of goods to total output.

Economists get around part of the problem theoretically either by assuming individuals have the same tastes or by assuming that income can be costlessly redistributed. This separates the issue of equity from the issue of efficiency. In practice, economists recognize that these conditions do not hold. They know that in the real world it is extraordinarily difficult to redistribute income. Massive redistributions of income generally require massive political upheaval. For this reason economists are careful to apply the producer and consumer surplus analysis only to those cases where the conditions are "reasonable" approximations of reality—where distributional and taste issues do not play a big role in a policy recommendation. Of course, economists may disagree on what are "reasonable" approximations of reality. That is why economic policy is an art, not a science.

be "inefficient" to supply it to them. In the supply/demand framework you can only have a demand for a good if you have the desire *and* the income to pay for it.

The point of these examples is not to convince you that the demand and supply framework is useless. Far from it. For the majority of goods, it is a useful analytical tool that demonstrates the power of competitive markets. The point of the examples is to show you the type of case where overriding the supply/demand framework in policy considerations may be socially desirable if society's goals include a particular distribution of consumer surplus. The sole purpose of society is not to maximize consumer and producer surplus. Society has other goals. Once these other goals are taken into account, the competitive result may not be the one that is desired.

For many goods maximizing total surplus in the demand and supply framework is a useful shorthand.

Societies integrate other goals into market economics by establishing social safety nets (programs such as welfare, employment insurance, and medical coverage). When individuals are below a certain income, what they receive does not depend solely on what they earn in the market. How high to set a given social safety net is a matter of debate, but favouring the market outcome in most situations should not be viewed as inconsistent with favouring a social safety net in others.

Consumer Sovereignty and Rationality Problems

Even when consumers have full information, they sometimes do not do what is in their own best interest. For example, people smoke or drink or drive without fastening their seatbelts. Although people may freely engage in these activities, it is not generally agreed in society that these actions by individuals are in their best interests. (In addition, there are also issues of negative externalities associated with each of these actions.) So, the market solution—let people enter freely into whatever trades they want to—is not necessarily the best solution. Again, the market is working, but the outcome may be a failure according to a broader set of social goals.

APPLYING THE TOOLS

Elasticity and Taxation to Change Behaviour

A good way to see how economists view the difference between the effect of a sin tax and the effect of a tax to raise revenue is to ask: Would a policymaker rather have an elastic or an inelastic demand curve for the good being taxed? If the purpose is to raise revenue while creating only a minimal amount of deadweight loss, an inelastic demand is preferable. If the purpose is to change behaviour, as it is in the example of an alcohol-dependent individual, a more elastic demand curve is better because a relatively small tax can cause a relatively large reduction in purchases.

Consider an example of taxation to reduce consumption. If government believes that smoking is bad for people, it can decrease the amount people smoke by placing a tax on cigarettes. If the demand for cigarettes is inelastic, then the tax will not significantly decrease smoking; but if the demand is elastic, then it will. If demand is inelastic, government may choose alternative methods of affecting behaviour, such as advertising campaigns. If the purpose of taxation is to raise taxes, an inelastic demand would be better; that's why most provinces rely on general sales taxes for revenue—such taxes allow them to raise revenue with relatively little effect on the efficiency of the market.

The following table provides a quick review of when a tax will be most effective, given a particular goal of government.

Goal of Government	Most Effective When
Raise revenue, limit efficiency loss	Demand or supply is inelastic
Reduce consumption	Demand or supply is elastic

Q-8 A cocaine addict purchases an ounce of cocaine from a drug dealer. Since this was a trade both individuals freely entered into, is society better off?

This problem is sometimes called *rationality failure of individuals*. The supply/demand framework starts with the proposition that individuals are completely **rational**—that *what individuals do is in their own best interest*. However, this is not always the case. Most of us are irrational at times; we sometimes can "want" something that we really "don't want," such as smoking, chocolate, or any other of our many vices.

Even if we don't have serious addictions, we may have minor ones. In addition, we are influenced by what people tell us we want. Businesses spend large sums of money on advertising to convince us that we want certain things. Individuals can be convinced they want something that, if they thought further about it, they would not want. The fact that individuals can be easily influenced can be a second reason for government intervention.

Let's look at an example: The government has taken the position that if people could be induced to stop smoking, they would be better off. **Sin taxes**—*taxes that discourage activities society believes are harmful (sinful)*—are meant to do just this. Based on the consumer surplus argument, a tax on smoking would create deadweight loss; it would reduce the combination of consumer and producer surplus. But in this case society believes consumer surplus does not reflect individuals' welfare.

Notice the difference between the argument for taxes to change behaviour (sin taxes) and the argument for taxes to raise revenue discussed in Chapter 7. When government wants to raise revenue it takes into account how much deadweight loss is created by the tax. With sin taxes, government is trying to discourage the use of the good that is being taxed. When society takes the position that individuals' demands in the marketplace do not reflect their true welfare, it is not at all clear that the market result is efficient. (See the box above.)

Inalienable Rights

Nice Guy wants to save his ailing son, who needs an operation that costs $300,000. He doesn't have that kind of money, but he knows that Slave Incorporated, a newly created company, has been offering $300,000 to the first person who agrees to become a slave for life. He enters into the contract, gets his money, and saves his son. The market is working just as it is supposed to. There's no information problem—Nice Guy knows what he's doing and Slave Inc. knows what it's doing. Both participants in the trade believe that the exchange is making them better off.

Many people's view of the trade will likely be different; they would regard such a market outcome—an outcome that allows slavery—as a market outcome failure. That is why governments have developed laws that make such trades illegal.

17.2

see page 387

As Amartya Sen pointed out (and won a Nobel Prize for doing so), most societies regard certain rights as inalienable. By definition, inalienable rights cannot be sold or given away. There can be no weighing of costs and benefits. For example, freedom is an inalienable right, so slavery is wrong, and any trade creating slavery should not be allowed, regardless of any issues of consumer and producer surplus.

The Need to Prioritize Rights To understand why market outcomes might be undesirable, we have to go back and consider markets in a broader perspective. Markets develop over time as individuals trade to make themselves better off. But markets don't just come into existence—they require the development of property rights for both suppliers and consumers. Each side must know what is being traded. So markets can exist only if there are property rights.

Q-9 True or false? If someone chooses to sell himself into slavery, the individual, and thus society, is better off.

Property rights, in turn, are included in a broader set of rights that are part of society's constitution—the right to vote, the right to free speech, the right to life. If one allocation of property rights conflicts with other rights, society must make a judgment about which right has priority. Thus, within the written or unwritten constitution of a society, rights must be prioritized.

Examples of Inalienable Rights Moral prohibitions that are related to inalienable rights include those against prostitution, selling body parts, and selling babies. The moral judgment our society has made about these rights may or may not be correct. Should such trades be subjected to the market? Society must make these judgments. Societies sometimes struggle for many decades trying to answer these fundamental questions about rights and freedoms. Such issues are moral questions which do not have to stand up to the consumer and producer surplus arguments. If something is wrong, it is wrong; whether it is efficient is irrelevant. So in every society, moral judgments must be made about where markets should exist.

Moral judgments underlie all policy prescriptions.

GOVERNMENT FAILURE

Distributional issues, issues of individuals' rationality, and the existence of inalienable rights are representative of the types of problems that can arise in the market. For most economists these issues play a role in interpreting the policy results that follow from the economic model presented, even when there is no market failure. But it is important to remember that even these failures of market outcomes do not necessarily imply a need for government action. The reason for that is government failure.

As discussed in Chapter 15, if the failure is to be corrected someone must formulate and enact the policy, and if we believe that government's attempt to correct it will do more harm than good, then we can still support the market as the lesser of two evils. For the government to correct the problem, it must

1. Recognize the problem.
2. Have *the will* to work towards solving the problem.

APPLYING THE TOOLS

The Conventional Policy Wisdom among Economists

Where do economists stand on whether government can correct a failure of market outcomes? The easy answer is that they conclude that to make a policy decision we must weigh the costs of market failure against the costs of government failure. But those costs are often poorly specified and difficult to estimate. Thus, policy considerations require subjective judgments. How do economists fit these broader considerations into their analysis?

Most economists downplay the distribution issues for the majority of goods, and consider distribution only in the most extreme situations. They believe that it is better to be open about the distributional goals and to give money directly to individuals, rather than to hide the redistribution by changing the pricing structure through subsidizing goods. Let's take an example: The European Union's agricultural policy currently provides large amounts of price supports for European agricultural production. To keep farmers in business, the prices of agricultural goods are kept high. If the social decision were to keep farmers in business, most economists, however, would prefer to see the EU provide direct subsidies to farmers. Then the policy of redistribution is clear to everyone, and is far less costly in terms of both efficiency and implementation.

The "rights argument" plays a role in all economists' policy arguments. Almost all economists oppose selling citizenship. All oppose slavery. All see economic policy as being conducted within a constitutional setting, and that means that inalienable rights come before market efficiency.

There are, of course, areas of ambiguity—allowing the regulated sale of body parts from individuals who have died is one such area. There is currently a world wide shortage of organs for transplants. When someone dies, from a medical perspective his or her organs usually can be "harvested" and used by someone else—but only if the deceased had signed a donor card. If the family of the deceased donor were given $5,000 for burial expenses, some economists argue, the shortage of transplant organs would disappear and everyone would be better off—the family could give the deceased a much nicer funeral and people needing the organs could live. Society is moving cautiously in that direction. The United States is much more market-oriented than Canada or the European countries. One U.S. State, Pennsylvania, has recently announced that it is giving $300 in funeral expenses to the survivors of those who donate organs. Economists are more open to such market solutions than the general public, but there is nothing in economics that requires such solutions.

The argument about problems arising from rationality issues is also accepted by most economists, but they downplay it for most nonaddictive goods. The reason is that while it is true that individuals may not know what they want, it is far less likely that the government will know better. Based on that view, on average, the acceptance of consumer sovereignty, and the market result, is probably warranted. Exceptions, however, include children and some elderly.

For the government to correct a problem it must

1. recognize the problem
2. have the will to deal with it and
3. have the ability to deal with it.

3. Have *the ability* to solve the problem.

Government seldom can do all three of these well. Often the result is that government action is directed at the wrong problem at the wrong time.

Probably the most vocal group of economists on the subject of government failure are *public choice economists*. This group, started by James Buchanan and Gordon Tullock, has pointed out that politicians are subject to the laws of supply and demand, like everyone else. Often the result of politics is that the redistribution of income that takes place does not redistribute income from rich to poor, but from one group of middle class to another group of middle class. Public choice economists argue that when the government intervenes in a market, the incentives are not to achieve its goal in the least-cost manner; the incentives are to provide a policy that its voting constituency likes. The result is larger and larger government involvement because the government wishes to be seen to be doing something about the problem; however, this activity may generate little benefit for society. Public choice economists advocate as little government intervention as possible regardless of whether there are market failures or failures of market outcomes.

POLITICS AND ECONOMIC POLICY

Economic policy must be applied within a political context. This means that political elements must be taken into account. Politics enters into the determination of economic policy in two ways, one positive and one negative. Its positive contribution is that politicians take market failures and failures of market outcomes into account when formulating policy. Ultimately the political system decides what externalities should be adjusted for, what is a desirable distribution, what rights rank above the market, and when people's demand does not reflect their true demand. To the extent that the government's political decisions reflect the will of society, government is making a positive contribution.

The negative contribution is that political decisions do not always reflect the will of society. The political reality is that, in the short run, government's first goal is to get re-elected. If they don't get re-elected, all of their longer term goals become irrelevant. Politicians and other policymakers know that; the laws and regulations they propose reflect such calculations. Politicians don't get elected and re-elected by constantly saying that all choices have costs and benefits. So while policymakers listen to the academic economists from whom they seek advice, and with whom in private they frequently agree, in practice they often choose to ignore that advice. Remember: sometimes what is good economics is not good politics. Conversely, what is good politics is not necessarily good economics.

Q-10 In what way does government positively contribute to economic policy? In what way does it negatively contribute to economic policy?

Because government both adjusts for failures of market outcomes and is subject to short-run political pressures, the way in which economic reasoning influences policy can be subtle. In short, economic policy made in the real world reflects a balancing of cost/benefit analysis and special interest group desires, within a political framework. Political economy is a more accurate description of real world economics.

Good economics is not necessarily good politics, and good politics is not necessarily good economics.

CONCLUSION

Adam Smith, the creator of modern economics, was a philosopher; his economics was part of his philosophy. Before he wrote the *Wealth of Nations*, in which he set out his argument for markets, he wrote a book called *The Theory of Moral Sentiments*, in which he laid out his broader philosophy. That foundation, in turn, was part of the Scottish Enlightenment, which spelled out what was meant by a good society, and how they believed the rights of individuals and society should be considered. Any economic policy issue must be interpreted within a broad philosophical framework. Clearly, an introductory course in economics cannot cover all of the broader philosophical and political issues. But it can highlight their importance, and emphasize that economic policy arguments must fit within that broader context.

This chapter was written to give you a sense of that broader context for economic policy. Cost/benefit analysis and the supply/demand framework are powerful tools for analyzing issues and determining policy conclusions. But to apply them successfully, they must be applied in context.

Economics provides the tools, not the rules, for policy.

Economics involves the thoughtful use of economic insights and empirical evidence. If this chapter gave you a sense of the nature of that thoughtful application along with the core of economic reasoning, then it succeeded in its purpose.

Economics involves the thoughtful use of economic insights and empirical evidence.

Chapter Summary

- Economists differ because of different underlying value judgments, because empirical evidence is subject to different interpretations, and because their underlying models differ.
- Value judgments inevitably work their way into policy advice, but good economists try to be objective.
- Economists tend to agree on certain issues because their academic training is similar. Economists use models that focus on economic incentives and rationality.
- The economic approach to analyzing issues is a cost/benefit approach.
- Economists generally have reservations about regulations.
- Collecting and using empirical evidence is an important part of economics.
- People make income-constrained choices every day that at least partially reveal the value that they place on their lives. The value of life is calculated by multiplying the inverse of the reduction in the probability of death by the amount individuals pay for that reduction.
- Collecting and interpreting empirical evidence is difficult, which contributes to disagreements among economists.
- Economics involves the thoughtful use of economic theory and empirical evidence.
- The cost/benefit approach and the supply/demand framework de-emphasize the possibility that market outcomes may be undesirable to society.
- Three failures of market outcomes are failures due to distributional issues, failures due to rationality problems of individuals, and failures due to violations of inalienable rights.
- Although an implicit assumption in most policy discussions is that the goal of policy is to maximize consumer and producer surplus, society does care about how that total surplus is distributed.
- The supply/demand framework assumes that individuals are rational. Individuals are not always rational in practice. Their actions are swayed by addictions, advertising, and other pressures.
- Some rights, called inalienable rights, cannot be bought and sold. What rights are inalienable are moral judgments decided by the members of society. These rights do not have to stand up to the same cost/benefit framework.
- Economics provides the tools, not the rules, for policy.

Key Terms

cost/benefit approach *(375)*
economic efficiency *(379)*
failure of market outcomes *(379)*
Marxian model *(374)*
Pareto optimal policies *(373)*
public choice model *(374)*
rational *(382)*
sin taxes *(382)*

Questions for Thought and Review

1. Could anyone object to a Pareto optimal policy? Why?
2. Would it be wrong for economists to propose only Pareto optimal policies?
3. Would all economists oppose price controls? Why or why not?
4. Should the sale of body organs be allowed? Why or why not?
5. In cost/benefit terms, explain your decision to take an economics course.
6. How much do you value your life in dollar terms? Are your decisions consistent in that valuation?
7. If someone offered you $1 million for one of your kidneys, would you sell it? Why or why not?
8. Why might an economist propose a policy that has little chance of adoption?
9. In the 1970s legislators had difficulty getting laws passed requiring people to wear seat belts. Now not only do most people wear seat belts, many cars have air bags too. Do people value their lives more now than in the 1970s?
10. What are economists' views of politicians?
11. If the hourly wage equivalent of welfare for a single mother with two children is $12.45 an hour, what problem emerges and what policy recommendations would you make?
12. Economist Steven D. Levitt estimated that, on average, for each additional criminal locked up in the United States, 15 crimes are eliminated. In addition, although it costs about $30,000 a year to keep a prisoner incarcerated, the average prisoner would have caused $53,900 worth of damage to society per year if free. If this estimation is correct, does it make economic sense to build more prisons?

Problems and Exercises

1. Say that the cost of a car crash is $8,000. Assume further that installing a safety device in a car at a cost of $12 will reduce the probability of an accident by 0.05 percent. The plant makes 1,000 cars each day.
   a. If the preceding are the only relevant costs, would you favour or oppose the installation of the safety device?
   b. What other costs might be relevant?
2. In a U.S. study of hospital births, the single most important prediction factor of the percentage of vaginal births as opposed to cesarean (C-section) births was mother's income.
   a. What implications about the health care debate can you draw from the above results?
   b. What issues would change the results?
3. The technology is now developing so that road use can be priced by computer. A computer in the surface of the road picks up a signal from your car and automatically charges you for the use of the road.
   a. How could this technological change contribute to ending bottlenecks and rush-hour congestion? Demonstrate graphically.
   b. How will people likely try to get around the system?
   c. If people know when the prices will change, what will likely happen immediately before? How might this be avoided?
4. In the early 1990s, the 14- to 17-year-old population fell because of low birth rates in the mid-1970s. Simultaneously, aging baby boomers decided to have kids and this increased the number of parents needing baby-sitters. What effect will these two events likely have on:
   a. The number of times parents go out without their children?
   b. The price of baby-sitters?
   c. The average age of baby-sitters?
5. As organ transplants become more successful, scientists are working on ways to transplant animal organs to humans. Pigs are the favourites as "donors" since their organs are about the same size as human organs.
   a. What would the development of such organ farms likely do to the price of pigs?
   b. If you were an economic adviser to the government, would you say that such a development would be Pareto optimal (for humans)?
   c. Currently, there is a black market in human organs. What would this development likely do to that market?
6. If one uses a willingness-to-pay measure in which life is valued at what people are willing to pay to avoid risks that might lead to death, the value of a U.S. citizen's life is $2.6 million, a Swede's life is worth $1.2 million, and a Portuguese's life is worth $20,000 (according to an article in *The Journal of Transport Economics and Policy*).
   a. What policy implications does this value schedule have?
   b. Say you operate an airline. Should you spend more on safety precautions in the U.S. than you do in Portugal?
   c. Should safety standards be lower in Portugal and Sweden than in the U.S.?
7. According to U.S. government statistics, the cost of averting a premature death differs among various regulations. Car seat belt standards cost $100,000 per premature death avoided, while hazardous waste landfill disposal bans cost $4.2 trillion per premature death avoided.
   a. If you were choosing between these two regulations, which would you choose? Why?
   b. If these figures are correct, should neither, one, the other, or both of these regulations be implemented?
8. A politician once proposed that the government sell the organs of dead welfare recipients to help pay off the welfare recipients' welfare costs and burial expenses.
   a. What was the likely effect of that proposal?
   b. Why would that be the effect?
9. Technology exists such that individuals can choose the sex of their offspring. Assume that 70 percent of the individuals choose male offspring.
   a. What effect will that have on social institutions such as families?
   b. What effect will it have on dowries—payments made by the bride's family to the groom—which are still used in a number of developing countries?
   c. Why might an economist suggest that if 70 percent male is the expectation, families would be wise to have daughters rather than sons?

Web Questions

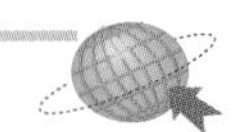

1. Go to www.caledoninst.org and study the article entitled, "Reclaiming our Humanity."
   a. Before the Great Depression, what role did social institutions play in supporting individuals in need? What is menat by "never again?"
   b. According to the *British North American Act* (BNA Act) and the *Constitution Act* of 1982 (The Constitution), over what areas of public spending does the federal government have jurisdiction? Over what areas and programs do the provinces preside? What prob-

lems have arisen as a result of the federal-provincial division of powers?

c. What are the major demographic, social, and economic forces which are putting pressure on Canadian social policy? Describe their influences.

d. What is meant by life-long learning? What recommendations does the article make regarding education and learning?

2. Go to the Nobel Foundation Web site at www.nobel.se, and enter the keyword "Amartya" into the search engine. Read the 1998 Press Release and Amartya Sen's autobiography.
   a. What is Amartya Sen's academic training? What economic issues interest Sen?
   b. How did Sen's education and experiences influence the type of academic questions he pursued over his career?

Answers to Margin Questions

1. In the real world, Pareto optimal policies don't exist, and all real-world policies make someone better off and someone worse off. In making real-world policy judgments, one cannot avoid the difficult distributional and broader questions. It is those more difficult questions, which are value-laden, that make economic policy an art rather than a science. *(373)*

2. Marxist analysis of the labour market differs from the mainstream analysis in that it emphasizes the imbalance of power among social classes. Thus, a Marxist analysis will likely see exploitation built into the institutional structure. Mainstream analysis is much more likely to take the institutional structure as given and not question it. *(374)*

3. Oftentimes being "mean" in the short run can actually involve being "nice" in the long run. The reason is that often policy effects that are beneficial in the long run have short-run costs, and people focussing on those short-run costs see the policy as "mean." *(375)*

4. To maximize life, one would expect that the marginal value per dollar spent should be equal in all activities. Thus, if Table 17-1 is correct, it would suggest that you should be far less concerned about seat belt usage and far more concerned about whether your automobile has air bags or not. *(375)*

5. Costs and benefits are ambiguous; economists often disagree enormously on how to quantify specific costs and benefits, or the costs and benefits are difficult or impossible to quantify. Thus, you should be extremely careful about using a cost/benefit analysis as anything more than an aid to your analysis of the situation. *(378)*

6. Other things do not always remain constant. The more macro the issue, the more things are likely to change. These changes must be brought back into the analysis, which complicates things enormously. *(378)*

7. False. Efficiency is achieving a goal as cheaply as possible. Stating efficiency as a goal does not make sense. *(379)*

8. No. The cocaine addict may be responding to the cravings created from the addiction, and not acting from any rational desire for more cocaine. Society may not be better off. *(382)*

9. False. Society may find that personal freedom is an inalienable right. Selling such a right would make society worse off. *(383)*

10. Government makes a positive contribution by adjusting for market failures and failures of market outcomes. Government may make a negative contribution because government decisions are influenced by short-run political pressures. *(385)*

Politics and Economics: The Case of Agricultural Markets

18

After reading this chapter, you should be able to:

- Describe the competitive nature of agricultural markets.
- Explain the short-run and long-run problems in agriculture.
- State the general rule of political economy in a democracy.
- Explain the unique problems associated with farming.
- Explain how a price support system works.
- Explain, using supply and demand curves, the distributional consequences of five alternative methods of price support.
- Discuss real-world influences which affect the design of agricultural policy.

A farmer is always going to be rich next year.

Philemon (300 B.C.)

The Canadian farming industry faces unique challenges moving into the new millenium. The solutions are complex, because the difficulties stem from multiple sources. In our analyses of the Canadian farming industry, we must take into account a number of considerations—some economic, some political, and some social.

Agricultural markets provide good examples of the interaction between the invisible hand, and social and political forces. Considering the economics of agricultural markets shows us how powerful a tool supply and demand analysis is in helping us understand not only the workings of perfectly competitive markets but also the effects of government intervention in a market.

APPLYING THE TOOLS

The Cost of a Box of Wheaties

When people think of agricultural products, they often think of the products they buy, like Wheaties. Doing so gives them the wrong impression of the cost of agricultural products. To see why, let's consider a 620 gram box of Wheaties that costs you, say, $3.35.

If you look at the ingredients, you'll see that you're buying wheat, sugar, salt, malt syrup, and corn syrup. So you're buying agricultural products, right? Well, a little bit. Actually, the total cost of those agricultural ingredients is probably somewhere around 25 cents, less than 10 percent of the cost of the box of Wheaties. What are you spending the other 90 percent on? Well, there's packaging, advertising, transporting the boxes, processing the ingredients, stocking the grocery store shelves, and profits. These are important components of Wheaties, but they aren't agricultural components.

The point of this example is simple: Much of our food expenditure isn't for agricultural goods; it's for the services that transform agricultural goods into processed foods, convince us we want to eat those foods, and get those foods to us.

While the chapter is about agricultural markets, please bear in mind that the lessons of the analysis are applicable to a wide variety of markets in which the invisible hand and political forces interact. As you read the chapter, applying the analysis to other markets will be a useful exercise.

THE NATURE OF AGRICULTURAL MARKETS

In many ways, agricultural markets fit the classic picture of perfect competition. First, there are many independent sellers who are generally *price takers*. Second, there are many buyers. Third, the products are interchangeable: Farm A's wheat can be readily substituted for farm B's wheat. And fourth, prices can, and do, vary considerably. On the basis of these inherent characteristics, it is reasonable to talk about agricultural markets as competitive markets.

Agricultural markets involve a constant stuggle between political and economic forces.

In other ways, however, agricultural markets are far from perfectly competitive. The competitiveness of many agricultural markets is influenced by government programs. In fact, neither Canada nor any other country allows the market, unhindered, to determine agricultural prices and output. In Canada, there are over a hundred agencies regulating Canadian agricultural markets.

The competitive market in agriculture is not a story of the invisible hand alone. It's the story of a constant struggle between political, social, and economic forces. Whenever the invisible hand pushes prices down, various political forces generally work to push them back up.

PROBLEMS IN AGRICULTURE

There are two main problems in agriculture: farmers' incomes are low, and their incomes fluctuate. The reasons for this are complex, because agriculture is a complicated business which faces some variables that other Canadian businesses do not.

SHORT-RUN FLUCTUATIONS IN SUPPLY

The risk-reward trade-off in farming is greater than in other, "riskier," industries.

There is a paradox in farming: when crops are good, prices for the farm output fall, and farmers' incomes drop. And when crops are poor, low supply tends to push up farm product prices, and farmers' incomes rise. So, good harvests often mean bad times and a fall in income, while poor harvests often mean a rise in income.

A fact of life that farmers must deal with is that agricultural production tends to be highly unstable because it depends on weather and luck. Crops can be affected by too little rain, too much rain, insects, frost, heat, wind, hail—none of which can be controlled. Say you're an apple grower and the spring weather is beautiful—until the week that your trees are blossoming, when it rains continually. Bees don't fly when it rains, so they don't pollinate your trees. No pollination, no apple crop. There goes your apple crop for this year, and there goes your income.

The paradox follows from the fact that the short-run demand for most agricultural goods is highly inelastic. Because short-run demand is so inelastic, short-run changes in supply can have a significant effect on price. The result is that good harvests for farmers in general can lower prices significantly, while poor harvests can raise prices significantly. Because the price effect overwhelms the quantity effect (as it does when demand is inelastic), farmers face a short-run paradox. With inelastic demand, a decrease in supply raises price, and also raises total revenue, and an increase in supply lowers price, and also lowers total revenue. So a plentiful crop usually does not translate into higher incomes for the farmer.

Consider Figure 18-1. In a "normal" year, supply is S_0 and equilibrium price in a competitive market is P_0. Selling amount Q_0 generates total revenue of P_0 times Q_0, which is represented as Area ABCD. A bountiful crop shifts the supply curve to S1 and drives the competitive market price down to P_1. Because demand is price inelastic, total revenue falls to P_1 times Q_1, or Area DCE. The farmer gains Area E, but loses Area AB. Clearly, a "good" year has a devastating effect on farmers' incomes. A poor crop, however, results in a lower supply, S_2, and although a smaller quantity, Q_2, is sold, total revenue is higher, P_2 times Q_2, or Area FAD. Here, the farmer gains because Area F is larger than Area BC.

Figure 18-1 **SHORT-RUN SUPPLY FLUCTUATIONS**

Yearly fluctuations in supply can lead to large changes in farmers' incomes. In a normal year supply is S_0, and revenues are P_0 times Q_0 (Area ABCD). If the supply is S_1, revenues fall to P_1 times Q_1 (Area DCE), and if supply is S_2, revenues rise to P_2 times Q_2 (Area FAD).

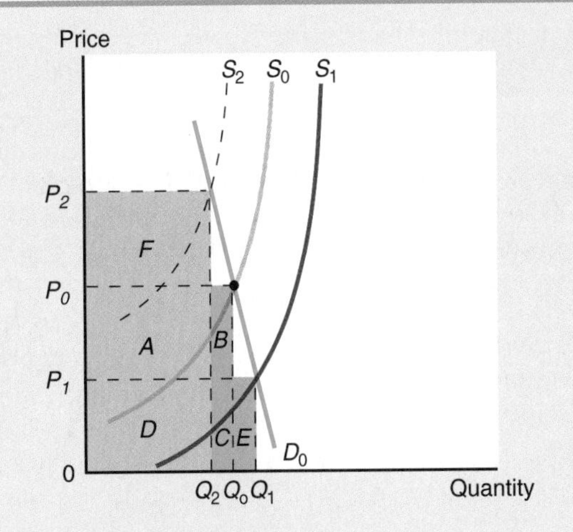

LONG-RUN EXPANSION OF SUPPLY

Technological Change and the Decline in Farming

Q-1 What technological advances have the agricultural industries experienced?

Rapid Productivity Gains Most countries, Canada included, began as predominantly agricultural societies. When Canada began a little more than 130 years ago, 97 percent of the labour force was engaged in farming. Today less than 3 percent of the Canadian labour force works in agriculture.

The decline in the number of farmers isn't the result of a failure of Canadian agriculture. Rather, it's the result of its tremendous success—the enormous increase in its productivity. It used to take the majority of the population to provide food for the country. Today it takes only a small proportion to produce more food than the Canadian population can consume.

Productivity gains result from the use of more effective fertilizers and pesticides, and from advances in farming techniques, such as crop rotation and greater use of irrigation. Technological gains relating to genetic development have already had a big impact on grain farming, and are now transforming other agri-food production. More advanced weather forecasting techniques allow farmers to take steps to protect themselves from such devastating events as an early frost; for example, the harvesting can be started earlier to avoid an incoming cold snap. Computers have made many aspects of farming more efficient, including record keeping and income tax filing. The Internet keeps isolated individuals in touch with recent developments, and, in some cases, feeding and other routine tasks are controlled by computers.

Q-2 What is the paradox in farming?

Slow Demand Growth Growth in world demand for many agricultural products expands at a rate not much greater than population growth. While incomes are growing worldwide, the income elasticity of demand for agricultural products is low, which means that as incomes grow, the demand for agricultural products grows less quickly. In addition, productivity advances in some countries have turned them from net buyers of agricultural products to net suppliers of these same products. China and Russia were once major consumers of Canadian wheat; however, both now sell wheat in the international markets. The impact of this is twofold: there is less demand and more supply, putting downward pressure on wheat prices.

Figure 18-2 shows how success can lead to problems. In the long run the demand for wheat is inelastic (i.e., the percentage change in quantity demanded is small relative to

Figure 18-2 A PARADOX

At price P_0 the quantity of wheat produced is Q_0. Total income is P_0Q_0. The supply of wheat increases from S_0 to S_1 due to increased productivity, and the demand for wheat grows slowly from D_0 to D_1, and the price of wheat will fall from P_0 to P_1 and quantity demanded will increase from Q_0 to Q_1. The increase in farmers' income (area C) is small. The decrease in farmers' income (area A) is large. Overall, the combination of increased productivity and slow demand growth has led to a decrease in farmers' incomes.

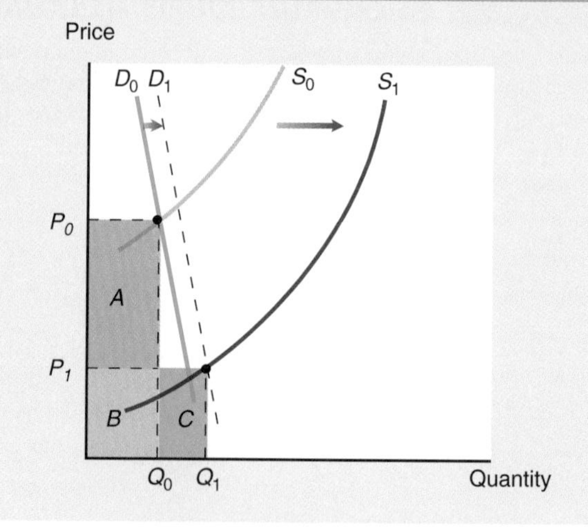

the percentage change in price), as it is for most agricultural products, so the figure shows an inelastic demand curve.

In this example, initially farmers are selling quantity Q_0 for price P_0. Their total income is P_0Q_0, shown by areas A and B. Now say that increases in productivity shift the supply curve out from S_0 to S_1. Output increases from Q_0 to Q_1, and price falls to P_1. Income falls to P_1Q_1, shown by areas B and C. Farmers have gained area C but lost area A. The net effect is the difference in size between the two areas. So in this example, the net effect is negative.

Due to competition among farmers, most benefits of productivity increases in agriculture have gone to consumers in the form of lower prices.

Productivity has increased but total revenue has fallen, and many farmers have stopped farming altogether. Whenever the supply curve shifts outward in the inelastic range of the demand curve, the result will be a greater decline in price than increase in quantity, which lowers total revenue. Due to competition among farmers, most benefits of productivity increases in agriculture have gone to consumers in the form of lower prices.

Q-3 How can it be in the interest of the agricultural industry to have a "bad year"?

Private and Public Solutions

The Difficulty of Coordinating Farm Production Competitive farmers take the market price as given. That's the definition of a competitive industry. While it is in the industry's interest to reduce total supply, it is in each individual farmer's interest to increase output even if the combination of *all* farmers increasing output would cause prices to fall.

It is, however, in farmers' joint interest to figure out ways to avoid supply fluctuations. Specifically, it's in their interest to figure out ways to limit the production of all farmers.

In a competitive industry, limiting production is easier said than done. It is difficult for farmers to limit production privately, among themselves, because there are too many of them to coordinate easily.

Q-4 What are two ways around the paradox?

The difficulty of organizing privately to limit supply can be avoided by organizing through government. Canadian political structure provides an alternative way for farmers (and other suppliers) to coordinate their actions and limit supply. Suppliers can organize and get government to establish programs to limit production or hold price high, thereby avoiding the problem of low and fluctuating incomes. Government programs have been a combination of **price stabilization programs**—*programs designed to eliminate short-run fluctuations in prices, while allowing prices to follow their long-run trend line* and **price support programs**—*programs designed to maintain prices at levels higher than the market prices.*

The General Rule of Political Economy If farmers are helped by farm programs, who is hurt? The answer is taxpayers and consumers. One would expect that these broad groups would strongly oppose farm programs because farm programs cost them in two ways: (1) higher taxes that government requires in order to buy up surplus farm output, and (2) higher prices for food.

Economists who specialize in the relationship between economics and politics (known as *public choice economists*) have suggested that the reasons for farm groups' success involve the nature of the benefits and costs. The groups that are hurt by agricultural subsidies are large, but the negative effect on each individual in that group is relatively small. Large groups that experience small costs per individual don't provide a strong political opposition to a small group that experiences large gains. This seems to reflect a **general rule of political economy** in a democracy: *When small groups are helped by a government action and large groups are hurt by that same action, the small group tends to*

The general rule of political economy states that small groups that are significantly affected by a government policy will lobby more effectively than large groups that are affected by that same policy.

lobby far more effectively than the large group; thus, policies tend to reflect the small group's interest, not the interest of the large group.

This bias in favour of farm programs is strengthened by the historical importance of agriculture. Benefits are concentrated while the costs are spread over a large group. Consumers and taxpayers in general, who would be hurt by price supports, generally lack the political organization necessary to make their will known and counter the pressure for price controls.

UNIQUE PROBLEMS IN AGRICULTURE

As mentioned earlier, farming is subject to some unique problems, which suggests that intervention in the agriculture markets may have merit.

Foreign farm subsidies contribute to low incomes for Canadian farmers.

First, farm incomes are low. This problem, in itself, is not necessarily a reason for government intervention in agricultural markets; however, there are other factors which combine with low incomes to make it an important issue. Farm production is subject to random negative impacts, such as weather and insect infestations, which have a major impact on farm revenues. Farm subsidies in other jurisdictions, particularly in Europe and the United States, have the effect of driving prices down below the social cost of production, which again has negative consequences on Canadian farm incomes. Both of these negative impacts on farm revenues are beyond the individual farmer's control. With low incomes, issues of poverty arise, and are especially important when children are involved. According to Canadian Customs and Revenue Agency figures for the 1997 tax year, 314,680 individuals reported net farming income as their major source of income. While the average income for individuals reporting employment income as their major source of income in the 1997 tax year was $28,089, individuals reporting farming income as their major source of income reported average net incomes of $ 4,362. Net business and net professional average incomes were $7,497 and $44,145, respectively. Net fishing average income came in at $9,965.

18.1 see page 406

So, with such low incomes, why would anyone want to be a farmer? It would be reasonable to expect people to leave farming and search out other employment. However, this often is not the result. Farms are generally located in rural districts, often with only a small community nearby, with few opportunities. Farms are difficult to sell because an alternative use for the land doesn't exist. Selling to another farmer is the only option, so there is no "exit." In addition, farmers often are very skilled people, performing a variety of tasks daily to keep the farm operational. They are mechanics, electricians, welders, horticulturalists, carpenters and more, yet because they usually do not have the formal qualifications for any of these occupations, they cannot find employment in these areas. So it is difficult to leave the farm.

Farmers' skills may not be transferable to other occupations because they may lack formal qualifications.

Second, farming is a way of life, passed down from one generation to the next. It is a lifestyle that the child inherits from the family, and the family farm is an important Canadian institution that Canadians do not wish to see decimated. These are serious issues. Forcing individuals to uproot their lives, relocate, and abandon their homes and villages might be efficient, but it doesn't fit very well with the Canadian sense of what is just.

Third, we must also be aware of particular economic factors which characterize farming. Farm production and, consequently, farm incomes are susceptible to random negative events that many other businesses do not have to contend with. Bad weather may negatively impact a delivery business by making it more difficult or expensive to deliver goods to customers; however, the same bad weather can completely destroy a

KNOWING THE TOOLS

Farm Incomes

Farmers in Canada work hard to produce a variety of products for consumption — wheat, barley, canola, apples, pears, grapes, cranberries, corn, vegetables, hogs, and cattle, to mention just a few. Although production techniques vary considerably across farm types, incomes mostly do not. For those individuals reporting farming income as their main source of income, average net farming income was $4,362. In comparison, net fishing income was $9,965, net business income was $7,479, and net professional income was $44,145. A breakdown of incomes into categories reveals that, while there are a few "rich farmers" out there, the vast majority have very modest incomes indeed. Here are the figures for individuals reporting net farming income as a source of income. Notice that, on average, farmers supplemented their farm incomes by $1,958 of other non-farm income.

20.5% (or 64,630 individuals) of the total number of individuals reporting some farming income reported a total income greater than $50,000, and a net farming income of $10,707. 38.8% reported a total income between $25,000 and $50,000, and a net farming income of $6,806, while the largest group, 40.1%, reported a total income below $25,000, and net farming income of $3,638, which amounts to the princely sum of $303 per month.

Total Income (Taxable Returns)	Farming Income Reported	Average Net Farming Income
> $250,000	1.0%	$28,897
150,000 – 250,000	1.2%	$25,848
100,000 – 150,000	2.1%	$15,234
90,000 – 100,000	1.1%	$12,834
80,000 – 90,000	1.5%	$10,691
70,000 – 80,000	2.4%	$8,416
60,000 – 70,000	4.2%	$7,148
50,000 – 60,000	6.9%	$6,798
40,000 – 50,000	9.9%	$5,925
30,000 – 40,000	16.9%	$7,367
20,000 – 30,000	22.4%	$5,515
10,000 – 20,000	24.1%	$3,278
1 – 10,000	6.1%	$4,244
All Incomes		$6,320

Figures from Canada Customs and Revenue Agency for the tax year 1997, and author's calculations.

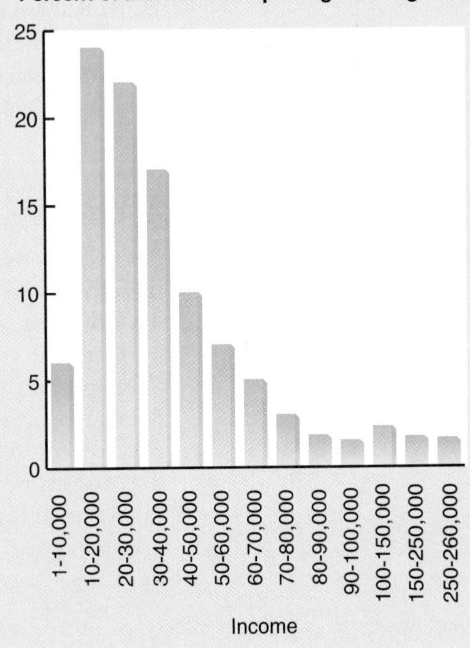

farmer's entire annual income. A hailstorm just before harvest time can pound a bumper crop of ripe grain until the stocks lie flat upon the ground, and cannot be harvested. The Manitoba floods in the spring of 1996 left the ground so wet that farmers could not get their equipment onto the land in time to plant. The machinery simply sank into the mud, and for many farmers the entire year's production effort was thwarted. Other risks borne by farmers which are generally not borne by other businesses include infestations of insects and disease. It is only possible to insure against some of these liabilities. The farmer bears the rest of the risk. Furthermore, there is usually a positive relationship between risk and reward in business. In the "high tech" companies, for instance, there is tremendous risk of failure; however, there is also the possibility of tremendous reward. In farming, unfortunately, the high risk is not balanced by a high reward.

Q.5 What unique elements characterize agricultural production?

Fourth, farming is characterized by high fixed costs, the most important of which is specialized machinery, which is very expensive to buy. There are also high variable costs. Not only is the machinery expensive to buy, it also costs money to operate and maintain. Other costs include the land, and the seed, fertilizer, and pesticides, as well as the time commitment (opportunity cost) of the farmer to the land. Farm revenues in some years do not cover the total costs of operating the farm, so many farmers subsidize their operations by working off the farm during the winter months or by having a second income.

Farming approximates perfect competition, but farmers must interact with oligopolistic or monopolistic markets.

Fifth, agricultural products are produced under conditions approximating perfect competition; however, when buying inputs and selling their output, farmers must deal with markets that are generally far from perfectly competitive. Many of the markets where farmers deal with for their inputs are oligopolistic, which means that the firms have tremendous market power against the individual farmer. Machinery is available from only a few manufacturers, and the local dealer is likely the only seller in the community, which means that the perfectly competitive farmer must deal with a monopoly. Fertilizer and pesticide producers also wield tremendous market power in contrast to the individual farmer.

Agricultural policies, while politically controversial, find their economic roots in these issues. If we agree that farming in Canada presents unique problems, how do we address the concerns of farmers, and other stakeholders, such as consumers and taxpayers?

FIVE PRICE SUPPORT OPTIONS

In a price support system, the government maintains a higher-than-equilibrium price.

Let's now consider the theory underlying some alternative farm price support options. In doing so, we'll try to understand which options, given the political realities, would have the best chance of being implemented, and why.

Consider Figure 18-3. Given that fluctuating supply leads to the problem of fluctuating incomes, and that the market equilibrium price, P_E, does not provide an adequate income to farmers, government intervention is warranted. In a price support system, the government maintains a higher-than-equilibrium price, such as P_1. At support price P_1, the quantity people want to supply is Q_S, but the quantity demanded at that price is Q_D.

At price P_1, there's excess supply, which exerts a downward pressure on price. To maintain price at P_1, some other force must be exerted; otherwise the invisible hand will force the price down.

The government has various options to offset the downward pressure on price:

Five price support options are:

a. Regulatory price floor.
b. Economic incentives to reduce supply.
c. Regulatory measures to reduce supply.
d. Subsidizing the sale of goods to consumers.
e. Buying up and storing, giving away, or destroying the good.

a. Using legal and regulatory force to prevent anyone from selling or buying at a lower price.
b. Providing economic incentives to reduce the supply enough to eliminate the downward pressure on price.
c. Imposing regulatory measures to reduce the supply and raise the price.
d. Subsidizing the sale of the good to consumers so that while suppliers get a high price, consumers have to pay only a low price.
e. Buying up and storing, giving away, or destroying enough of the good so that the total demand (including government's demand) increases enough to eliminate downward pressure on price.

These methods distribute the costs and benefits in slightly different ways. Let's consider each in detail.

Figure 18-3 A PRICE SUPPORT SYSTEM

In a price support system, the government maintains a higher-than-equilibrium price. At support price P_1, the quantity of product demanded is only Q_D, while the quantity supplied is Q_S. This causes downward pressures on the price, P_1, which must be offset by various government measures.

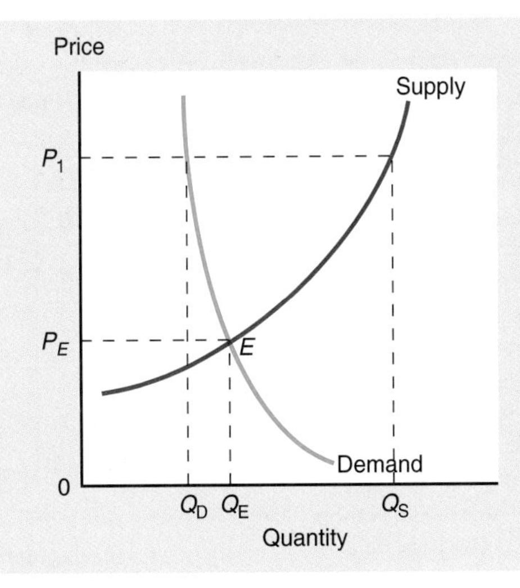

a. Supporting the Price by Regulatory Measures

Suppose the government simply passes a law saying that, from now on, the price of wheat will be at least $5 per bushel. No one may sell wheat at a lower price. If the competitive equilibrium price rises above $5, the law has no effect. When the competitive equilibrium is below the price floor (say the competitive equilibrium is $3.50 per bushel), the law limits suppliers from selling their wheat at that lower price.

The price floor helps some suppliers and hurts others. Those suppliers who are lucky enough to sell their wheat benefit. Those suppliers who aren't lucky and can't find buyers for their wheat are hurt. How many suppliers will be helped and how many will be hurt depends on the elasticities of supply and demand. When supply and demand are inelastic, a large change in price brings about a small change in quantity supplied, so the hurt group is relatively small. When the supply and demand are elastic, the hurt group is larger.

In Figure 18-4(a), at $5 suppliers would like to sell quantity Q_2 but they can sell only Q_1. They end up with a surplus of wheat, $Q_2 - Q_1$. Consumers, who must pay the higher price, $5, and receive only Q_1 rather than Q_e, are also hurt.

The Need for Rationing The law must establish some noneconomic method of rationing the limited demand among the suppliers. If individual farmers have a surplus, they may try to dispose of that surplus by selling it on the black market at a price below the legal price. To maintain the support price, the government will have to arrest farmers who sell below the legal price. If the number of producers is large, such a regulatory approach is likely to break down quickly, since individual incentives to sell illegally are great and the costs of enforcing the law are high.

In understanding who benefits and who is hurt by price floors, it's important to distinguish between two groups of farmers: the farmers who were producing before the law went into effect, and the farmers who entered the market afterward. In Figure 18-4(a), the first group supplies Q_e; the second group, which would want to enter the market when the price went up, would supply $Q_2 - Q_e$. Why must the second group be clearly identified? Because one relatively easily enforceable way to limit the quantity supplied is to forbid any new farmers to enter the market. Only people who were producing at the beginning of the support program will be allowed to produce. Restricting production to the existing suppliers will restrict the quantity supplied to Q_e, leaving only the reduction, $Q_e - Q_1$, to be rationed among suppliers.

Figure 18-4 (a, b, c, d, and e)

ALTERNATIVE METHODS OF GOVERNMENT PRICE SUPPORTS

Alternative methods have different distributional consequences. The consequences of direct regulatory measures are shown in (**a**); the consequences of providing economic incentives to reduce supply in (**b**); the consequences of reducing supply with a quota in (**c**); the consequences of subsidizing the sale in (**d**); and the consequences of buying up and storing the good in (**e**).

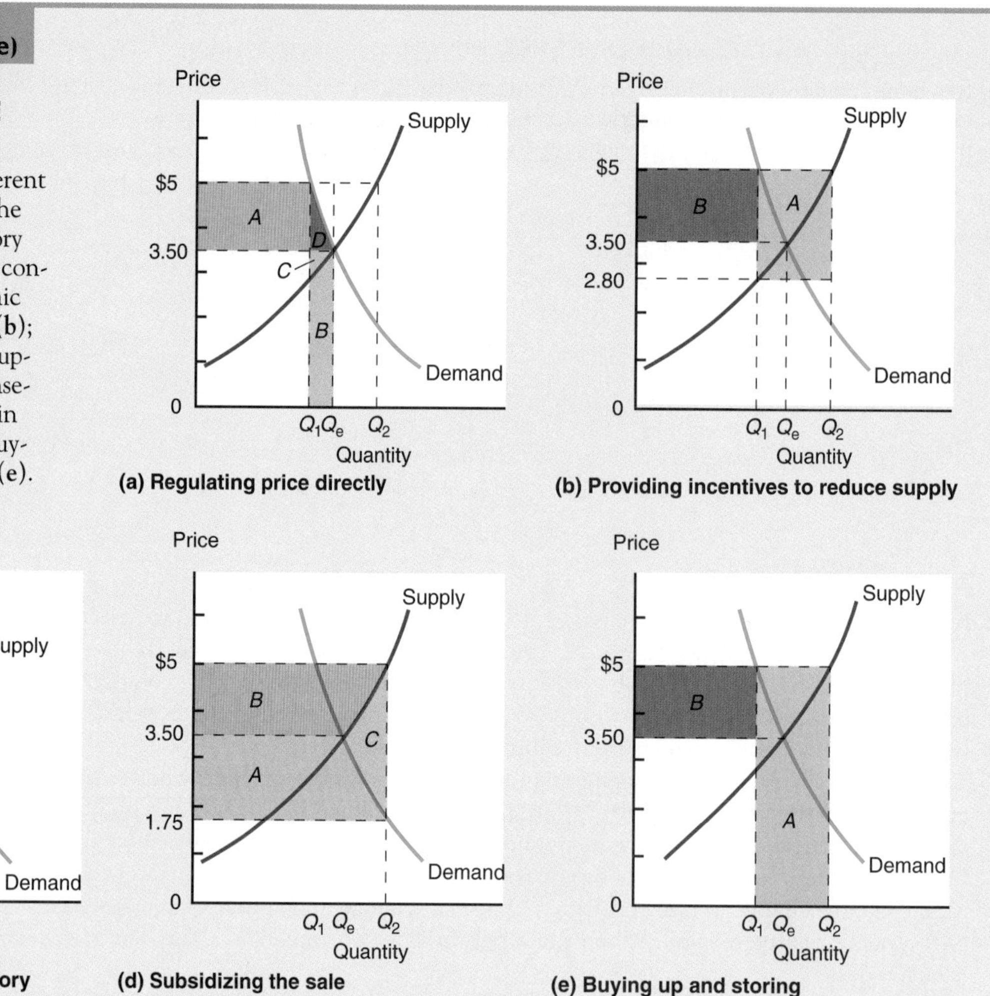

With a price floor, some method of nonprice rationing must determine how the limited demand will be distributed among suppliers.

To use this method of restriction is to **grandfather**—*to pass a law affecting a specific group but providing that those in the group before the law was passed are not subject to the law.* To grandfather in existing suppliers is one of the easiest provisions to enact into law and one of the easiest to enforce; thus, it is one of the most widely used. For example, when supply limitations are imposed, existing growers are allowed to continue planting land they were currently using for production. They could not, however, plant any new land. Generally, they are allowed to sell their acreage allocations, so that if old land is taken out of production, new land can be added.

When it comes to grandfathering groups out of production, foreign producers are perhaps the politically easiest targets. To keep the domestic price of a good up, foreign imports must be limited as well as domestic production. Canadian taxpayers might put up with subsidizing Canadian farmers, but they're likely to balk at subsidizing foreign farmers. So most farm subsidy programs are supplemented with tariffs and quotas on foreign imports of the same commodity. (See Chapter 19 for definitions and further discussion of tariffs and quotas.)

Distributional Consequences Notice that with the equilibrium in the inelastic portion of the demand curve, even though the average farmer is constrained as to how much can be sold, he or she is made better off by that constraint because the total revenue going to all farmers is higher than it would be if supply weren't constrained. The farmer's total revenue from this market increases by rectangle A in Figure 18-4(a) and decreases by the rectangle composed of the combined areas B and C. Of course, making the farmer better off is not cost-free. Consumers are made worse off because they must pay more for a smaller supply of wheat. There's no direct cost to taxpayers other than the cost of enforcing and administering the regulations.

Notice in the diagram the little triangle made up of areas C and D, which shows an amount of income that society loses but farmers don't get. It simply disappears. As discussed in Chapter 7, that little triangle is the welfare loss of producer and consumer surplus from the restriction.

b. Providing Economic Incentives to Reduce Supply

A second way in which government can keep a price high is to provide farmers with economic incentives to reduce supply.

Looking at Figure 18-4(b), you see that at the support price, $5 per bushel, the quantity of wheat supplied is Q_2 and quantity demanded is Q_1. To avoid a surplus, the government must somehow find a way to shift the quantity supplied back from Q_2 to Q_1. For example, it could pay farmers not to grow wheat. How much would such an economic incentive cost? Given the way the curves are drawn, to reduce the quantity supplied to Q_1, the government would have to pay farmers $2.20 ($5.00 − $2.80) for each bushel of wheat they didn't grow. This payment of $2.20 would induce suppliers producing $Q_2 - Q_1$ not to produce, reducing the quantity supplied to Q_1. The payment is shown by the A rectangle.

The Need for Rationing There is, however, a problem in identifying those individuals who would truly supply wheat at $5 a bushel. Knowing that the government is paying people not to grow wheat, people who otherwise had no interest in growing wheat will pretend that at $5 they would supply the wheat, simply to get the subsidy. To avoid this problem, often this incentive approach is combined with our first option, regulatory restrictions. Farmers who are already producing wheat at Q_e are grandfathered in; only these farmers are given economic incentives not to produce. All others are forbidden to produce.

Q-6 Which of the five methods of price support would farmers favour least? Why?

Distributional Consequences When economic incentives are supplied, the existing farmers do very well for themselves. Their income goes up for two reasons. They get rectangle A from the government in the form of payments not to grow wheat, and they get rectangle B from consumers in the form of higher prices for the wheat they do grow. Farmers are also free to use their land for other purposes, so their income rises by the amount they can earn from using the land taken out of wheat production for something other than growing wheat. Consumers are still being hurt as before: They are paying a higher price and getting less. In addition, they're being hurt in their role as taxpayers because the lightly shaded area (rectangle A) represents the taxes they must pay to finance the government's economic incentive program. Thus, this option is much more costly to taxpayers than the regulatory option.

c. Reducing Supply by Regulatory Measures

A third way in which government can keep price high is to impose regulatory measures to reduce supply. In a **quota** system, *the individual farmer is permitted to produce and sell any*

Q-7 Which of the five methods of price support would taxpayers favour least? Why?

quantity up to a maximum amount of product. This restricts the total supply of the produce, raising the price.

In Figure 18-4 (c), a quota of Q_1 rationed amongst farmers according to some criteria, means that the maximum quantity supplied to the market is Q_1. (You can imagine the supply curve becoming perfectly inelastic (vertical) at the price of $2.80, at Q_1. Higher prices do not increase the quantity supplied.) The intersection of the demand curve and the quota, Q_1 results in the quantity Q_1 sold at a market price of $5.00. The Canadian dairy industry uses a quota system to regulate the production of milk, and thus the industry is able to raise and maintain the market price of fluid milk above the free-market equilibrium price.

Q-8 Which of the five methods of price support would consumers favour least? Why?

Distributional Consequences The distributional consequences are somewhat different from the case where economic incentives are used to reduce supply. Farmers still gain area A, as price rises from $3.50 to $5.00; however, they lose areas *B* and C as output falls from Q_e to Q_1. Although taxpayers no longer must finance an incentive program, consumers still lose because they pay a higher price for a smaller quantity. Area *D* represents a deadweight loss to society since this quantity ($Q_e - Q_1$) is no longer produced and consumed.

d. Subsidizing the Sale of the Good

A fourth option is for the government to subsidize the sale of the good in order to hold down the price consumers pay but keep the amount suppliers receive high. Figure 18-4(d) shows how this works. Suppliers supply quantity Q_2 and are paid $5 per bushel. The government then turns around and sells that quantity at whatever price it can get—in this case, $1.75. No direct transfer takes place from the consumer to the supplier. Both are made better off. Consumers get more goods at a lower price. They are benefited by area *A*. Suppliers get a higher price and can supply all they want. They are benefited by area *B*. What's the catch? The catch, of course, is that taxpayers foot the entire bill, paying the difference between the $5 and the $1.75 ($3.25) for each bushel sold. The cost to taxpayers is represented by areas *A*, *B*, and *C*. This option costs taxpayers the most of any of the four options, but it also gives them the most product (which has value).

e. Buying Up and Storing, Giving Away, or Destroying the Good

The final option is for the government to buy up all the quantity supplied that consumers don't buy at the support price. This option is shown in Figure 18-4(e). Consumers buy Q_1 at price $5; the government buys $Q_2 - Q_1$ at price $5, paying the A rectangle.

Distributional Consequences In this case consumers transfer rectangle B to suppliers when they pay $5 rather than $3.50, the competitive equilibrium price. The government (i.e., the taxpayers) pays farmers rectangle A. The situation is very similar to our second option, in which the government provides suppliers with economic incentives not to produce. However, it's more expensive since the government must pay $5 per bushel rather than paying farmers $2.20 per bushel. In return for this higher payment, the government is getting something in return: $Q_2 - Q_1$ of wheat.

The Need to Dispose of Surplus Of course, if the government buys the surplus wheat, it takes on the problem of what to do with this surplus. Say the government decides to give it to the poor. Since the poor were already buying food, in response to a

APPLYING THE TOOLS

Subsidies Vary: Across Countries and Across Products

One of the difficulties Canadian farmers face is the high level of subsidies offered by foreign governments to their farmers. Government subsidies account for nearly 40% of OECD farm incomes, and exceed 60% in the highly subsidized South Korea, Switzerland, and Norway. Average farm subsidies in 2000 varied from 2% in New Zealand and 6% in Australia, to 24% in the United States, 49% in the European Union, and up to a whopping 65% in Japan.

Looking at individual products, subsidies vary significantly across regions. All figures represent the percentage value of the subsidy. As you can see, Canada's subsidy rates vary considerably by product:

While the trend in the 1990s reflected fewer subsidies and trade distortions, and greater market orientation, developments since 1998 suggest that agricultural reform and trade liberalization efforts have been derailed. In 2000, agricultural subsidies in OECD countries was estimated at US$300 billion, resulting in distorted international trade and depressed agricultural product prices.

	Percent Subsidy for							
Region	Wheat	Oilseeds	Barley	Beef	Poultry	Milk	Eggs	Total Agriculture
Canada	9	6	9	6	4	58	24	16
United States	38	11	38	4	2	61	3	22
European Union	56	49	68	62	15	57	6	45
Australia	4	2	4	4	3	31	6	7
OECD	45	22	55	34	8	58	12	37

Source: www.agsummit.gov.ab.ca.

free food program they will replace some of their purchases with the free food. This replacement brings about a drop in demand—which means that the government must buy even more surplus. Instead of giving it away, though, the government can burn the surplus or store it indefinitely in warehouses and grain elevators. Burning up the surplus or storing it, at least, doesn't increase the amount government must buy.

see page 406

Why, doesn't the government give the surplus to foreign countries as a type of humanitarian aid? The reason is that just as giving the surplus to our own poor creates problems in Canada, giving the surplus to the foreign poor creates problems in the countries involved. The foreign poor are likely to spend most of their income on food. Free food would replace some of their demand, thus lowering the price for those who previously sold food to them. Giving anything away destroys somebody's market.

Which Group Prefers Which Option?

The five price support options can be used in various combinations. It's a useful exercise at this point to think through which of the options farmers, taxpayers, and consumers would likely favour.

The first option, regulation, costs the government the least, but it benefits farmers the least. Since existing farmers are likely to be the group directly pushing for price supports, government is least likely to be asked to choose this approach. If it is chosen, most of the required reduction in quantity supplied will probably come from people who might enter farming at some time in the future, not from existing farmers.

The second option, economic incentives, costs the government more than the first and third options but less than the fourth and fifth options. Farmers are benefited by economic incentive programs in two ways. They get paid not to grow a certain crop, and they can sometimes get additional income from using the land for other purposes. When farmers aren't allowed to use their land for other purposes, they usually oppose this option, preferring the fourth or fifth option.

The third option, imposing regulations to reduce supply, harms consumers (through higher price and reduced quantity) and benefits farmers (since demand is inelastic).

The fourth option, subsidies on the sales to keep prices down, benefits both consumers (who get low prices) and farmers (who get high prices). Taxpayers are harmed the most by this option. They must finance the subsidy payments for all subsidized farm products.

The last option, buying up and storing or destroying the goods, costs taxpayers more than the first three options but less than the fourth, since consumers pay part of the cost. However, it leaves the government with a surplus to deal with. If there's a group who can take that surplus without significantly reducing their current demand, then that group is likely to support this option.

Interest Groups

The actual political debate is, of course, much more complicated than presented here. For example, other pressure groups are involved. Recently, farm groups and environmental groups have combined forces and have become more effective in shaping and supporting farm policy. Thus, recent new restrictions on supply in farming often operate in ways that environmentalists would favour, such as regulating the types of fertilizer and chemicals farmers can use.

Q.9 Are taxpayers, farmers, and consumers separate groups that are independent of each other?

Moreover, the three interest groups discussed here—farmers, taxpayers, and consumers—aren't entirely distinct one from another. Their memberships overlap. All taxpayers are also consumers, farmers are both taxpayers and consumers, and so on. Thus, much of the political debate is simply about from whose pocket the government is going to get money to help farmers. Shall it be the consumer's pocket (through higher prices)? Or the taxpayer's (through higher taxes)? That said, the political reality is that consumer and taxpayer interests and the lobbying groups that represent them generally examine only part of the picture—the part that directly affects them. Accordingly, politicians often act as if these groups had separate memberships. Politicians weigh the options by attempting to balance their view of the general good with the power and preferences of special interest groups.

International Issues

The final real-world complication that must be taken into account is the international dimension. If you think government is significantly involved in Canadian agriculture, you should see its role in other countries such as the members of the European Union (EU) and Japan. For example, more than half the EU's budget is devoted to farm subsidies, and most of its farms stay in business only because of protection. Our agricultural policy is, in part, determined by trade negotiations with these other countries. For example, a reduction in EU subsidies could bring about a reduction in our subsidies.

If you think government is significantly involved in Canadian agriculture, you should see its role in other countries, such as the members of the European Union and Japan.

The Canadian government has provided more than $11 billion to Saskatchewan grain and oilseed farmers since 1985, to help the farmers compete in world markets against highly subsidized European and U.S. producers. Market-distorting export and production subsidies have depressed world prices for agricultural products, making it uneconomic for Canadian farmers to produce without massive subsidies. In 1981, wheat sold for $6 per bushel; by 1993, the world price of wheat had fallen to $2 per bushel.

In response to the difficulties imposed by foreign subsidies, prairie farmers have made an effort to diversify their operations and move into higher value-added products. The three prairie provinces have diversified their agricultural base, moving away from reliance on the traditional grains and oilseeds, wheat and canola (rapeseed). While Alberta leads the country in agricultural diversification, Manitoba is now the second largest producer of potatoes (after Prince Edward Island). With four french fry plants, it is the main supplier of french fries for McDonald's restaurants in Chicago and the U.S. Midwest. As well, farmers are aggressively adopting large-scale hog, poultry, and cattle farming. A typical "big barn" hog farm in southern Ontario may house as many as 2000 hogs. This particular industrial trend is raising questions about the handling of the large quantity of manure produced by these types of intensive agricultural operations.

The early 2000s also saw major debates about genetically modified seeds. The EU was threatening to prevent import of any crop that was genetically modified. Such an import ban would represent a significant barrier to trade.

CONCLUSION

Agricultural markets are generally considered to closely approximate a perfectly competitive market — large number of suppliers, each selling a homogeneous product without the ability to distinguish one producer's product from another, which leads to price-taking behaviour by all suppliers, and zero long-run profits. However, as we saw, agricultural markets face some difficulties.

Oligopolistic farm equipment manufacturers possess tremendous market power.

The perfectly competitive agriculture markets must interface with other types of market structures where firms have a high degree of market power — for example, the markets where machinery and fertilizer are sold are oligopolies or local monopolies. Agricultural markets are also characterized by certain risks, for example, weather, insect infestation, and disease, that most other industries do not need to worry about. These factors are beyond the control of the farmer, yet can have a devastating effect on his operation. Agricultural markets also face inelastic demand, which means that year to year fluctuations in supply wreak havoc on farm incomes. In addition, world demand for agricultural products is income inelastic, and therefore grows slowly, roughly at a rate just above population growth. Many countries also subsidize agricultural production, which distorts international trade patterns and drives down the prices for agricultural products, creating pressure on the Canadian government to also provide support for Canadian farmers.

Q-10 Economic theory tells us that a Buy and Store strategy is preferable to direct regulation. True or false? Why?

Economics doesn't tell you whether government intervention or any particular policy is good or bad. That, you must decide for yourself. But what economics can do is pose the policy question in terms of gains and losses for particular groups. Posing the question in that framework often cuts through to the real reasons behind various groups' support for this or that policy. Often people support programs that transfer money from other taxpayers and consumers to themselves. They are, however, unlikely to say that is their motive.

The economic framework directs you to look beyond the reasons people say they support policies; it directs you to look for the self-interest. The supply/demand framework provides a neat graphical way to picture the relative gains and losses resulting from various policies.

But as usual there's an *on the other hand*. Just because some groups may support policies for self-serving reasons, it is not necessarily the case that the policies are bad or shouldn't be adopted. Reality is complicated, with many more grey areas than black-and-white choices.

Chapter Summary

- Agricultural markets have many qualities of perfectly competitive markets: sellers are price takers, there are many buyers, products are interchangeable, and prices vary considerably. The competitiveness of agricultural markets is affected by significant government intervention.
- The paradox in farming is the result of the inelastic demand in most agricultural markets. Increases in productivity increase supply; but, because demand is inelastic, the percentage decline in price is greater than the percentage increase in equilibrium quantity. Total revenue declines.
- Farming is characterized by a number of unique challenges that other types of businesses generally do not have to face. For example, farm output is subject to the ravages of weather, insects, and diseases, which are not under farmers' control.
- The family farm is a Canadian institution. The isolated location of much of Canada's farmland (which means there are few, if any, alternative land uses), a high fixed cost investment, and the difficulty of finding other types of work (because most trades require certification) keeps the farmer on the land, and makes "exit" from the farming industry unlikely.
- A general rule of political economy in a democracy is that policies tend to reflect small groups' interests, not the interests of large groups.
- A price support program works by government maintaining higher-than-equilibrium prices through regulations, economic incentives, quotas, subsidies, and buying up and storing or destroying.
- Regulatory price supports cost government the least, but benefit the farmers the least.
- Economic incentive price supports cost the government and taxpayers more than regulatory price supports, but less than subsidy price supports or buying up and storing the good.
- Subsidy price supports benefit consumers, who pay lower prices, and farmers, who receive higher prices. Subsidy price supports cost taxpayers the most.
- Buying up and storing the good gives government a surplus to deal with.
- Agricultural policy is affected by interest groups (consumers, taxpayers, and farmers) and international issues (farm policies of our trading partners).

Key Terms

general rule of political economy *(394)*
paradox *(392)*
grandfather *(398)*
price stabilization programs *(393)*
price support programs *(393)*
quota *(399)*

Questions for Thought and Review

1. If the demand for farm products were elastic rather than inelastic, would the paradox in farming still exist? Why or why not?
2. Demonstrate, using supply and demand curves, the distributional consequences of a price support system achieved through acreage restriction.
3. Which would a taxpayers' group prefer: price support achieved through buying up the surplus or through providing economic incentives for not producing? Why?
4. What is the most costly method of price support to the taxpayer? Demonstrate graphically.
5. What is the least costly method of price support to the taxpayer? Demonstrate graphically.
6. Why do tariffs and quotas generally accompany price support systems?
7. How does the elasticity of supply affect the cost of price supports in each of the four options?
8. Why is grandfathering an attractive option for governments when they institute price supports?
9. All government intervention in markets makes society worse off. True or false? Evaluate.
10. Why do governments around the world intervene in agricultural markets?

Problems and Exercises

1. Show graphically how the effects of an increase in supply will differ according to the elasticities of supply and demand.
   a. Specifically, demonstrate the following combinations:
      (1) An inelastic supply and an inelastic demand.
      (2) An elastic supply and an inelastic demand.
      (3) An elastic supply and an elastic demand.
      (4) An inelastic supply and an elastic demand.
   b. Demonstrate the effect of a government guarantee of the price in each of the four cases.
   c. If you were a farmer, which of the four combinations would you prefer?
2. Congratulations. You've been appointed finance minister of Farmingland. The Prime Minister wants to protect her political popularity by increasing farmers' incomes. She's considering two alternatives: (a) raising agricultural prices by adding governmental demand to private demand; and (b) giving farmers financial incentives to restrict supply and thereby increase price. She wants to use the measure that's least costly to the government. The conditions of supply and demand are illustrated in the accompanying diagram. (S_1 is what the restricted supply curve would look like. P_s is the price that the Prime Minister wants to establish.) Which measure would you advise?

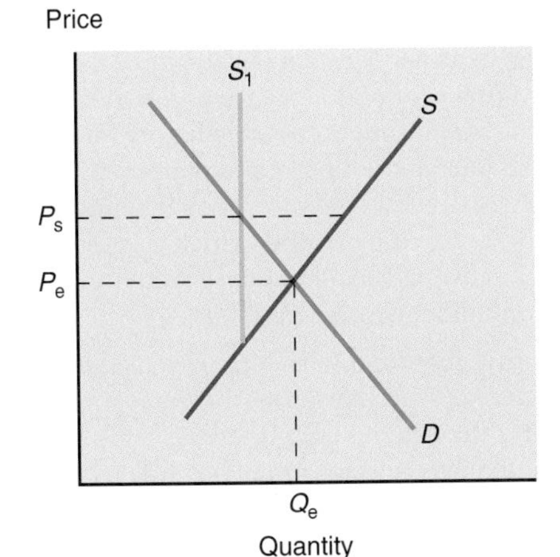

3. The U.S. *Pure Food and Drug Act* of 1906 is known as "Dr. Wiley's Law." It is generally regarded by non-economic historians as representing the triumph of consumer interests over producer interests.
   a. Why might an economist likely be somewhat wary of this interpretation?
   b. What evidence would a skeptical economist likely look for to determine the motives behind the passage of this law?
   c. What would be the significance of the fact that the *Pure Food and Drug Act* was passed in 1906, right when urbanization and technological change were fostering new products that competed significantly with existing producers' interests?
4. The U.S. government makes it against the law to grow peanuts unless the grower has been granted a government quota. It also essentially forbids peanut imports and sets a minimum price of peanuts at about 50 percent higher than the price of peanuts on the world market. This program costs the government $4 million a year in administrative costs.
   a. Are there likely any other costs associated with the program?
   b. Demonstrate graphically how to come up with about $250 million of additional costs.
   c. When "peanut land"—land with peanut quotas—is sold, what is the likely price of that land compared to equivalent land without a peanut quota?
   d. Say that, under the World Trade Organization, the U.S. agrees to allow open imports of peanuts into the U.S. and guarantees that all sellers will receive the existing price. What will happen to the governmental costs of the program?
   e. Say the government limits the guaranteed high price to U.S. producers. What will it have to do to make that guarantee succeed?
5. Say that a law, if passed, will reduce Mr. A's wealth by $100,000 and increase Mr. B's wealth by $100,000.
   a. How much would Mr. A be willing to spend to stop passage of the law?
   b. How much would Mr. B be willing to spend to ensure passage of the law?
   c. What implications for social policy do your answers to *a* and *b* have?
6. Suppose that a law, if passed, would reduce wealth for a group of one hundred people living in A-town, but would increase wealth for Mr. B, who lived just outside of A-town (on a farm!)
   a. How much would each person of A-town be willing to spend to stop the passage of the law?
   b. How much would Mr. B be willing to spend to ensure passage of the law?
   c. What would be the result? Would the law be passed?

Web Questions

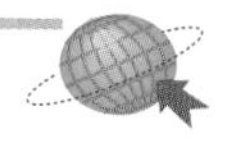

1. Go to the Web site of the Canadian Federation of Agriculture, at www.cfa-fca.ca and research the following questions.
   a. What are the five major agriculture sectors in Canada? Which is the largest and in what regions of the country is it found? Name the top three crops of this sector.
   b. How large is the dairy sector? Where are Canada's main dairy-producing areas? Explain why dairy output does not tend to fluctuate.
   c. According to the 1996 Census, how many farms were there in Canada? Over the last twenty years, what has happened to the number of farms, farm size, and total area cultivated?
2. At the Agriculture and Agri-Food Canada Web site at www.agr.gc.ca, click on Agriculture in Canada, then click on Agriculture: Food and Much More, and select Canadian Wheat Board.
   a. What does the Canadian Wheat Board do?
   b. According to 2000 OECD statistics, how much government support is offered per tonne of wheat produced in the U.S., European Union, Australia, and Canada?
   c. What has happened to the producer's share of production from the mid-1970s to the late 1990s? Which product has seen the biggest change? Explain.

Answers to Margin Questions

1. Agriculture has seen advances in farming techniques, equipment, fertilizers, and pesticides, genetic developments, and better weather forecasting made possible through the use of satellites. *(392)*
2. The paradox in farming is the phenomenon of doing poorly because you're doing well. It exists when demand for your product is inelastic. Specifically, as it applies to agriculture, it means that when most farmers produce a lot, prices are low and their net income drops. *(392)*
3. Because demand is inelastic, it is in the interest of the agricultural industry for the supply of agricultural goods to decline from bad weather or other supply disruptions. The percentage rise in price will be greater than the percentage decline in quantity demanded, and total revenue for the industry will rise. *(393)*
4. There are two ways around the paradox. One is for suppliers to coordinate their activity and limit supply. The second way is for suppliers to lobby and get government to establish programs to limit production, stabilizing the price and holding it high. Because of the difficulty of coordinating the large number of farmers, it is this second track that farmers have followed. *(393)*
5. Unique characteristics of agricultural production include exposure to risks such as weather, insect infestation, and disease, demand which is highly price and income inelastic, and annual fluctuations in supply which can cause large swings in prices and incomes of producers. *(395)*
6. Farmers are least likely to support the regulatory method of price support, in which regulatory force is used to prevent anyone from selling or buying at a lower price. Although such a policy benefits farmers, it benefits them far less than other price support policies. *(399)*
7. Taxpayers will likely least favour the price support method of subsidizing the sale of goods to consumers, because this method costs taxpayers the most. The low price paid by consumers and the high price received by farmers would necessitate large subsidies. *(400)*
8. Consumers would least favour the price support method of providing economic incentives to reduce supply and the price support method of regulatory force. Both these methods reduce the supply and push up the price. Some consumers would benefit from the buying up, giving away, or destroying method, which suggests that consumers on average would prefer this to the regulatory or the economic incentive method. *(400)*
9. While this chapter discusses taxpayers, farmers, and consumers as separate groups independent of each other, in reality they are not. Each individual is, generally, both a taxpayer and a consumer, while farmers are generally members of all three groups. It is nonetheless useful to treat them as separate groups because specific interests predominate: for example, farmers' interests as farmers significantly outweigh their interests as taxpayers or as consumers. *(402)*
10. False. Economic theory tells us nothing about what is preferable. Choices about what is preferable can only be made by specifying one's value judgments. Such choices belong in normative economics and represent the art of economics, where distributional effects, broader sociological issues, and value judgments are included in the analysis. *(403)*

International Trade Policy

19

After reading this chapter, you should be able to:

- Explain the principle of comparative advantage.
- Explain the principle of absolute advantage.
- List three determinants of the terms of trade.
- Discuss why countries impose trade restrictions.
- Summarize why economists generally oppose trade restrictions.
- Explain three policies countries use to restrict trade.
- Explain how free trade associations both help and hinder international trade.

One of the purest fallacies is that trade follows the flag. Trade follows the lowest price current. If a dealer in any colony wished to buy Union Jacks, he would order them from Britain's worst foe if he could save a sixpence.

Andrew Carnegie

If economists had a mantra, it would be "Trade is good." Trade allows specialization and division of labour and thereby promotes technological growth. Consistent with that mantra, most economists oppose trade restrictions. Not everyone agrees with economists; almost every day we hear calls from some sector of the economy to restrict foreign imports in order to save jobs and protect workers from unfair competition. In this chapter we consider why economists generally favour free trade, and why, despite what economists tell them, countries impose trade restrictions.

APPLYING THE TOOLS

The Gains to International Traders

Trade does not take place on its own—markets and trade require entrepreneurs to bring it about. The market is not about abstract forces; it is about real people operating to improve their position. Many of the gains from trade do not go to the countries involved but rather to the trader. And the gains that traders get can be enormous.

Consider, for example, the beautifully knit Peruvian sweaters often sold at art fairs and college campuses for $75 apiece. The Peruvian women knitting those sweaters are paid only a small fraction of that $75—say $6 apiece—and the trader makes the difference. So much of the benefits of trade do not go to the producer or the consumer; they go to the trader. Another example is the high-priced runners ($105) that many students wear. Those runners are likely made in China, and the cost of making a pair is about $8. The trader has other costs, of course; there are, for example, costs of transportation and advertising—someone has to convince you that you need those runners. (Just do it, right?) But that advertising is not done in China, and a portion of the benefits of the trade are accruing to advertising firms, which can pay more to creative people who think up those crazy ads.

THE PRINCIPLE OF COMPARATIVE ADVANTAGE

The principle of comparative advantage states that as long as the relative opportunity costs of producing goods differ among countries, then there are potential gains from trade.

The reason two countries trade is that trade can make both countries better off. This arises from the principle of comparative advantage presented in Chapter 2. The basic idea of the **principle of comparative advantage** is that *as long as the relative opportunity costs of producing goods (how much of one good must be given up in order to get more of another good) differ among countries, then there are potential gains from trade*. Let's review this principle by considering the story of I.T., an imaginary international trader, who convinces two countries to enter into trades by giving both countries some of the advantages of trade; he keeps the rest for himself.

The Gains from Trade

Here's the situation. On his trips to Canada and Saudi Arabia, I.T. noticed that no trade was taking place between them. He also noticed that the opportunity cost of producing a ton of food in Saudi Arabia was 10 barrels of oil and that the opportunity cost in Canada of producing a ton of food was 1/10 of a barrel of oil. At the time, Canada's production was 60 barrels of oil and 400 tons of food, while Saudi Arabia's production was 400 barrels of oil and 60 tons of food.

I.T. made Canada and Saudi Arabia the following offer: If Canada would specialize in food, devoting the resources then being used to produce 60 barrels of oil to producing food, it could increase its food production from 400 tons to 1,000 tons. I.T. would then give Canada 120 barrels of oil in exchange for 500 of those tons. That would leave 500 tons of food for Canada (100 tons more than it had before the deal) and give it double the amount of oil—120 barrels (compared to its pre-deal 60 barrels). By accepting this deal, Canada wound up with more of both commodities without increasing the resources it expended.

Q-1 In terms of oil and food, exactly how much richer is I.T. after the trade?

He told Saudi Arabia that if it would specialize in oil, devoting the resources used to produce 60 tons of food to producing oil and thus increasing its oil production from 400 barrels to 1,000 barrels, he would give it 120 tons of food—double the amount of food it had before the deal—in exchange for 500 barrels of that oil. Like Canada, Saudi Arabia wound up with more of both commodities without increasing the resources it expended. Thus, both countries ended up with more than they initially had of both goods.

BEYOND THE TOOLS

International Issues in Perspective

Over the past 50 years, international issues have become increasingly important. That statement would be correct even if the reference period went back as far as the 1600s.

In the 1600s, 1700s, and most of the 1800s, international trade was vital to the North American economy. Canada and the United States grew out of colonial possessions of England, France, and Spain. These "new world" colonial possessions were valued for their gold, agricultural produce, and natural resources. From a European standpoint, international trade was the colonies' reason for being.

International trade has social and cultural dimensions. While much of this chapter deals specifically with economic issues, we must also remember the cultural and social implications of trade.

Consider an example from history. In the Middle Ages, Greek ideas and philosophy were lost to Europe when hordes of barbarians swept over the continent. Europeans only rediscovered the ancient ideas and philosophy in the Renaissance as a by-product of trade between the Italian merchant cities and the Middle East. (The Greek ideas that had spread to the Middle East were protected there from European upheavals.) *Renaissance* means rebirth: a rebirth in Europe of Greek learning. Many of our traditions and sensibilities today are based on those of the Renaissance, and that Renaissance was caused, or at least significantly influenced, by international trade. Had there been no trade, our entire philosophy of life might be different than it is.

Another example is the major change in socialist countries in the 1990s. Through the 1960s China, the Soviet Union, and the Eastern European countries were relatively closed societies—behind the Iron Curtain. That changed in the 1970s and 1980s as these socialist countries opened up trade with the West as a way to speed up their own economic development. That trade, and the resulting increased contact with the West, gave the people of those countries a better sense of the material goods available in the West. That trade also spread Western ideas of the proper organization of government and the economy to these societies. A strong argument can be made that along with trade came the seeds of discontent that changed those societies and their economies forever.

Dividing Up the Gains from Trade

As the above story suggests, when countries avail themselves of comparative advantage there are high gains from trade to be made. Who gets these gains is unclear. The principle of comparative advantage doesn't determine how those gains from trade will be divided up among the countries involved and among traders who make the trade possible. While there are no definitive laws determining how real-world gains from trade will be apportioned, economists have developed some insights into how those gains are likely to be divided up.

The Trader's Gains

The first insight concerns how much the trader gets. The general rule is:

> The more competition that exists among traders, the less likely it is that the trader gets big gains from trade; more of the gains from trade will go to the citizens in the two countries, and less will go to the traders.

What this insight means is that where entry into trade is unimpeded, most of the gains of trade will pass from the trader to the countries. Thus, the trader's big gains from trade occur in markets that are newly opened.

Three determinants of the terms of trade are:

1. The more competition, the less the trader gets.
2. Smaller countries get a larger proportion of the gain than larger countries.
3. Countries producing goods with economies of scale get a larger gain from trade.

Trading companies understand this. Numerous import/export companies exist whose business is discovering possibilities for international trade in newly opened markets. Individuals representing trading companies promote projects or goods in countries. For example, in 1999 at the end of the NATO bombing campaign in Kosovo, what the business world calls the *import/export contingent* flew to Kosovo with offers of goods and services to sell. Many of these same individuals had been in Iraq and Iran in the early

Canada is a "small open economy." This means that trade is important to us. As you can see, Canada's main trading partner is the United States, taking 85% of Canadian exports, and sending us more than 75% of our imports. Our close geographic proximity and our social and cultural similarities are the main reasons for the high volume of trade between our two countries. The figures below show the relative importance of the United States in Canada's trade pattern.

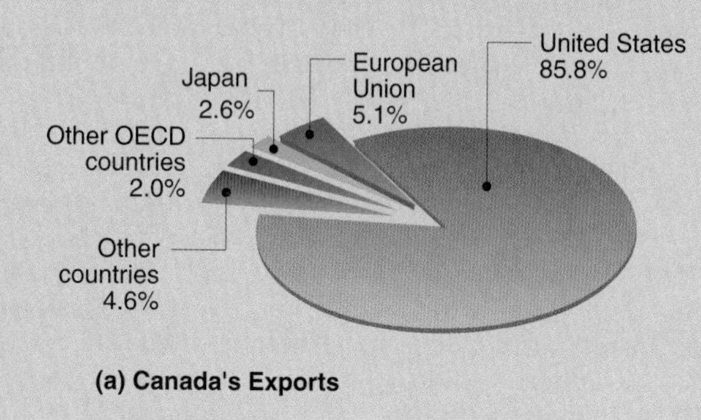

(a) Canada's Exports

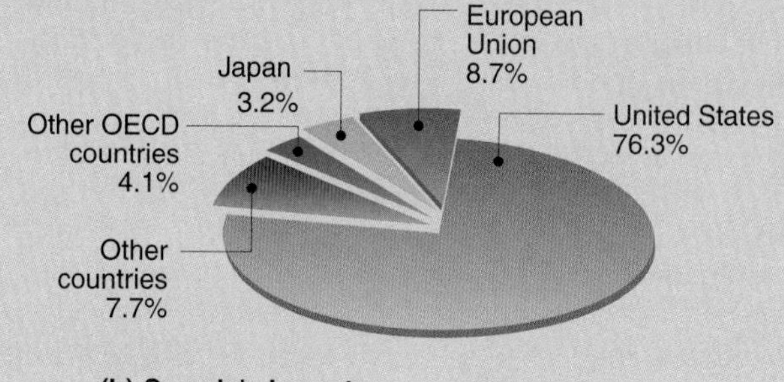

(b) Canada's Imports

Source: *www.wto.org*

1990s, in Saudi Arabia when oil prices rose in the 1970s, and in the Far East when China opened its doors to international trade in the 1980s.

Small Countries Gain

A second insight is:

> Once competition prevails, smaller countries tend to get a larger percentage of the gains from trade than do larger countries.

Q-2 In what circumstances would a small country not get the larger percentage of the gains from trade?

The reason, briefly, is that more opportunities are opened up for smaller countries by trade than for larger countries. The more opportunities, the larger the relative gains. Say, for instance, that Canada begins trade with Mali, a small country in Africa. Enormous new consumption possibilities are opened up for Mali—prices of all types of goods will fall. Assuming Mali has a comparative advantage in fish, before international trade began cars were probably extraordinarily expensive in Mali, while fish was cheap. With international trade, the price of cars in Mali falls substantially, so Mali gets the gains. Because the Canadian economy is so large compared to Mali's, the price of fish in Canada doesn't change noticeably. Mali's fish are just a drop in the bucket. The price ratio of cars to fish doesn't change much for Canada, so it doesn't get much of the gains from trade. Mali gets almost all the gains from trade. This is Canada's experience when it (a relatively small country) negotiated the original Free Trade Agreement with the United States (a country with an economy twelve times as large).

Trade benefits: What would the price be for a Canadian-grown coconut?

There's an important catch to this gains-from-trade argument. The argument holds only if competition among traders prevails. That means that Mali residents are sold cars at the same price (plus shipping costs) as Canadian residents. International traders in small countries often have little competition from other traders and keep large shares of the gains from trade for themselves. In the preceding example, Canada and Saudi Arabia didn't get a large share of the benefits. It was I.T. who got most of the benefits. Since the traders often come from the larger country, the smaller country doesn't get the benefits of the gains from trade; the larger country's international traders do.

Economies of Scale

A third insight is:

> Gains from trade go to the countries producing goods that exhibit economies of scale.

Trade allows an increase in production. If there are economies of scale, that increase can lower the average cost of production of a good. Hence, an increase in production can lower the price of the good in the producing country. The country producing the good with the larger economies of scale has its costs reduced by more, and hence gains more from trade than does its trading partner.

ABSOLUTE ADVANTAGE AND COMPARATIVE ADVANTAGE: GRAPHICAL ANALYSIS

Why Do Nations Trade?

International trade exists for the same reason that all trade exists: Party A has something that party B wants and party B has something that party A wants. Both parties can be made better off by trade. Economists generally use graphs to illustrate the gains from trade.

The Principle of Absolute Advantage

Trade between countries in different types of goods is relatively easy to explain. For example, trade in raw materials and agricultural goods for manufactured goods can be easily explained by the **principle of absolute advantage:**

A country that can produce a good at a lower resource cost than another country has an absolute advantage in producing that good.

> A country that can produce a good at a lower cost than another country is said to have an absolute advantage in the production of that good. When two countries have absolute advantages in different goods, there are gains of trade to be had.

The principle of absolute advantage explains trade of, say, oil from Saudi Arabia for food from Canada. Saudi Arabia has millions of barrels of easily available oil, but growing food in its desert climate and sandy soil is expensive. Canada can grow food cheaply in its temperate climate and fertile soil, but its oil isn't as easily available or as cheap to extract. Because it can produce a certain amount of oil with fewer resources, Saudi Arabia has an absolute advantage over Canada in producing oil. Because it can produce a certain amount of food with fewer resources, Canada has an absolute advantage over Saudi Arabia in producing food. When each country specializes in the good it has an absolute advantage in, both countries can gain from trade.

In Figure 19-1, consider a hypothetical numerical example that demonstrates how the principle of absolute advantage can lead to gains from trade. For simplicity, assume constant opportunity costs.

Figure 19-1(b) and (d) show the choices for Canada; Figure 19-1(a) and (c) show the choices for Saudi Arabia. In Figure 19-1(a) and (b) you see that Canada and Saudi Arabia can produce various combinations of food and oil by devoting differing percentages of their resources to producing each. Comparing the two tables and assuming the resources in the two countries are comparable, we see that Saudi Arabia has an absolute advantage in the production of oil and Canada has an absolute advantage in the production of food.

Figure 19-1 (a, b, c, and d) ABSOLUTE ADVANTAGE: CANADA AND SAUDI ARABIA

Looking at tables (a) and (b), you can see that if Saudi Arabia devotes all its resources to oil, it can produce 1,000 barrels of oil, but if it devotes all of its resources to food, it can produce only 100 tons of food. For Canada, the story is the opposite: devoting all of its resources to oil, Canada can only produce 100 barrels of oil—1/10 as much as Saudi Arabia—but if it devotes all of its resources to food, it can produce 1,000 tons of food—10 times as much as Saudi Arabia. Assuming resources are comparable, Saudi Arabia has an absolute advantage in the production of oil, and Canada has an absolute advantage in the production of food. The information in the tables is presented graphically in (c) and (d). These are the countries' production possibility curves without trade. Each point on each country's curve corresponds to a row on that country's table.

Percentage of resources devoted to oil	Oil produced (barrels)	Food produced (tons)	Row
100%	1,000	0	*A*
80	800	20	*B*
60	600	40	*C*
40	400	60	*D*
20	200	80	*E*
0	0	100	*F*

(a) Saudi Arabia's production possibility table

Percentage of resources devoted to oil	Oil produced (barrels)	Food produced (tons)	Row
100%	100	0	*A*
80	80	200	*B*
60	60	400	*C*
40	40	600	*D*
20	20	800	*E*
0	0	1,000	*F*

(b) Canada's production possibility table

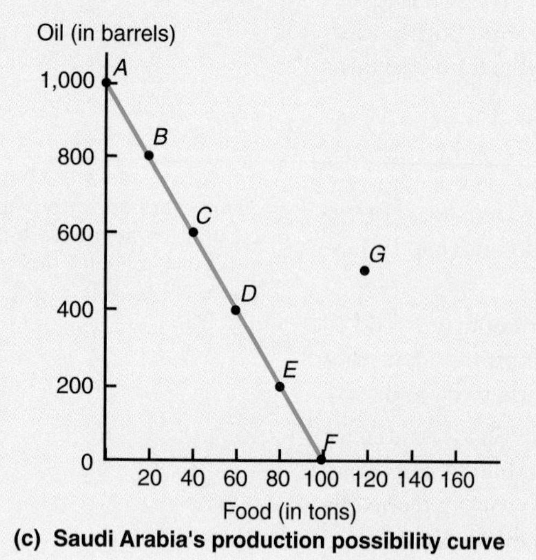

(c) Saudi Arabia's production possibility curve

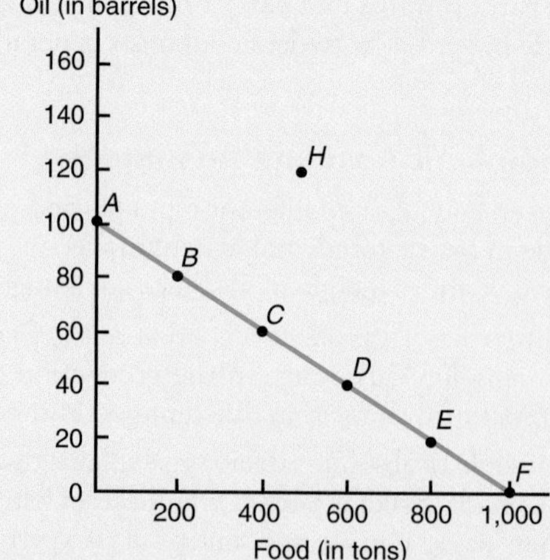

(d) Canada's production possibility curve

For example, when Canada and Saudi Arabia devote equal amounts of resources to oil production, Saudi Arabia can produce 10 times as much as Canada. Alternatively, when Canada devotes 60 percent of a given amount of resources to oil production, it gets 60 barrels of oil. But when Saudi Arabia devotes 60 percent of that same amount of resources to oil production, it gets 600 barrels of oil. The information in the tables is presented graphically in Figure 19-1(c) and (d). These graphs represent the two countries' production possibility curves without trade. Each combination of numbers in the table corresponds to a point on the curve. For example, point *B* in each graph corresponds to the entries in row *B*, columns 2 and 3, in the relevant table.

Let's assume that, without any international trade, Canada has chosen point C (production of 60 barrels of oil and 400 tons of food) and Saudi Arabia has chosen point *D* (production of 400 barrels of oil and 60 tons of food).

Now I.T. (International Trader), who understands the principle of absolute advantage, comes along and offers the following deal to Saudi Arabia:

> If you produce 1,000 barrels of oil and no food (point A) and give me 500 barrels of oil while keeping 500 barrels for yourself, I guarantee you 120 tons of food, double the amount of food you're now getting. I'll put you on point G, which is above your current production possibility curve. You'll get more oil and more food. It's an offer you can't refuse.

I.T. then flies off to Canada, where he makes the following offer:

> If you produce 1,000 tons of food and no oil (point *F*) and give me 500 tons of food while keeping 500 tons for yourself, I'll guarantee you 120 barrels of oil, double the amount you're now getting. I'll put you on point *H*, which is above your current production possibility curve. You'll get more oil and have more food. It's an offer you can't refuse.

Both countries accept; they'd be foolish not to. So the two countries' final consumption positions are as follows:

	Oil (barrels)	Food (tons)
Total production	1,000	1,000
Canadian consumption	120	500
Saudi consumption	500	120
I.T.'s profit	380	380

For arranging the trade, I.T. makes a handsome profit of 380 tons of food and 380 barrels of oil.

I.T. has become rich because he understands the principle of absolute advantage. Unfortunately for I.T., the principle of absolute advantage is easy to understand, which means that he will quickly face competition. Other international traders come in and offer the countries even better deals than I.T. offered, squeezing his profit share. With free entry and competition in international trade, eventually I.T.'s share is squeezed down to his costs plus a normal return for his efforts.

Now obviously this hypothetical example significantly overemphasizes the gains a trader makes. Generally the person arranging the trade must compete with other traders and offer both countries a better deal than the one presented here. But the person who first recognizes a trading opportunity often makes a sizable fortune. The second and third persons who recognize the opportunity make smaller fortunes. Once the insight is generally recognized, the possibility of making a fortune is gone. Traders still make their normal returns, and in the long run, benefits of trade go to the producers and consumers in the trading countries, not the traders.

The Principle of Comparative Advantage

Now consider the possibility of trade between Germany and Algeria in automobiles and food. No other traders have considered trade between these two countries because Germany is so much more productive than Algeria in all goods. No trade is currently taking place because Germany has an absolute advantage in production of both autos and food. Assuming the resources in the two countries are comparable, this case can be seen in Table 19-1.

Germany's opportunity cost of producing an auto is 2/1. That means Germany must give up 2 tons of food to get 1 additional auto. For example, if Germany is initially producing 60 autos and 80 tons of food, if it cuts production of autos by 20, it will increase

TABLE 19-1 Germany's Comparative Advantage over Algeria in the Production of Autos and Food

(a) Germany				(b) Algeria			
% of Resources Devoted to Autos	Autos Produced	Food Produced	Row	% of Resources Devoted to Autos	Autos Produced	Food Produced	Row
100%	100	0	*A*	100%	20	0	*A*
80	80	40	*B*	80	16	1	*B*
60	60	80	*C*	60	12	2	*C*
40	40	120	*D*	40	8	3	*D*
20	20	160	*E*	20	4	4	*E*
0	0	200	*F*	0	0	5	*F*

its food output by 40. For each car lost, Germany gains 2 tons of food. When Algeria cuts its production of autos by 4, it gains 1 ton of food. Algeria's opportunity cost of producing another auto is 1/4. It must give up 1 ton of food to get an additional 4 autos.

A country that can produce a good at a lower relative opportunity cost (i.e., give up the least amount of production of other goods) than another country has a comparative advantage in producing that good.

If Algeria needs to give up only 1/4 ton of food to get an auto while Germany needs to give up 2 tons of food to produce 1 auto, there are potential gains to be made, which can be split up among the countries and the International Trader. I.T. can make the countries offers they can't refuse. "*Absolute advantage* is not necessary for trade; *comparative advantage* is." This is known as the **principle of comparative advantage:** *As long as the relative opportunity costs of producing goods (how much of one good must be given up in order to get more of another good) differ among countries, there are potential gains from trade, even if one country has an absolute advantage in everything.*

It is comparative advantage, not absolute advantage, that forms the basis of trade. If one country has a comparative advantage in one good, the other country must, by definition, have a comparative advantage in the other good.

I.T. sees that, unexpectedly, Germany has a comparative advantage in producing food and Algeria has a comparative advantage in producing cars. I.T. makes the Germans the following offer:

> You're currently producing and consuming 60 autos and 80 tons of food (row C of Table 19-1(a)). If you'll produce only 48 autos but 104 tons of food and give me 22 tons of food, I'll guarantee you 13 autos for those 22 tons of food. You'll have more autos (61) and more food (82). It's an offer you can't refuse.

I.T. then goes to Algeria and presents the Algerians with the following offer:

> You're currently producing 4 tons of food and 4 automobiles (row *E* of Table 19-1(b)). If you'll produce only automobiles (row A) and turn out 20 autos, keeping 5 for yourself and giving 15 of them to me, I'll guarantee you 5 tons of food for the 15 autos. You'll have more autos (5) and more food (5). It's an offer you can't refuse.

Neither Germany nor Algeria can refuse. They both agree to I.T.'s offer. The final position appears as follows:

	Autos	Food
Total production	68	104
German consumption	61	82
Algerian consumption	5	5
I.T.'s profit	2	17

Competitiveness, Exchange Rates, and Comparative Advantage

In microeconomics, most of our analysis is conducted in real terms without reference to a numeraire (price level) or exchange rate. These financial issues are traditionally covered in macroeconomics, where the study of money and financial issues is a central focus. Money and financial markets are necessary to make trade and payment imbalances possible. However, it would be inappropriate to not mention those issues, since exchange rates play a central role in determining a country's absolute advantage. In fact, without implicit exchange rates, absolute advantage cannot be determined.

In turn, absolute advantage plays a big role in whether a country can have a temporary trade surplus or trade deficit.

The exchange rate is defined as the price of foreign currency. It is the amount of domestic currency required to buy one unit of a foreign currency. For example, if the Canada-U.S. exchange rate is $1.50, it means that it costs $1.50 Canadian to buy one U.S. dollar. Generally a high exchange rate (high price of foreign currency means low-valued domestic currency) encourages exports and discourages imports; a low exchange rate discourages exports and encourages imports. An example of the importance of exchange rates can be seen by considering Canada and Japan: in the mid-1980s a dollar bought 200 yen, but in the mid-2000s the dollar bought less than 100 yen. That change significantly decreased the absolute advantage of Japan and discouraged our consumption of Japanese products. The resurgence of the North American auto industry was in large part due to that change in the exchange rate.

The exchange rate is the price paid to obtain one unit of foreign currency.

REASONS FOR TRADE RESTRICTIONS

Let's now turn to a different question: If trade is beneficial, as it is in our examples of I.T., why do countries restrict trade? There are many reasons why:

1. Unequal internal distribution of the gains from trade.
2. Haggling by companies over the gains from trade.
3. Haggling by countries over trade restrictions.
4. Specialized production: learning by doing and economies of scale.
5. Macroeconomic aspects of trade.
6. National security.
7. International politics.
8. Increased revenue brought in by tariffs.

Reasons for restricting trade include:

1. Unequal internal distribution of the gains from trade.
2. Haggling by companies over the gains from trade.
3. Haggling by countries over trade restrictions.
4. Specialized production: learning by doing and economies of scale.
5. Macroeconomic aspects of trade.
6. National security.
7. International politics.
8. Increased revenue brought in by tariffs.

Unequal Internal Distribution of the Gains from Trade

In one example, I.T. persuaded Saudi Arabia to specialize in the production of oil rather than food, and persuaded Canada to produce more food than oil. That means, of course, that some Canadian oil workers will have to become farmers, and in Saudi Arabia some farmers will have to become oil producers.

Often people don't want to make radical changes in the kind of work they do—they want to keep on producing what they're already producing. So when these people see the same kinds of goods that they produce coming into their country from abroad, they lobby to prevent the foreign competition. Sometimes they're successful. A good example is the "voluntary" quotas—numerical limits—placed on Japanese cars exported to Canada in the 1980s. These quotas saved Canadian jobs, but, by reducing competition,

TABLE 19-2 Cost of Saving Jobs in Selected Industries

Industry	Cost of Production (per job saved)
Specialty steel	$1,000,000
Color TVs	420,000
Ceramic tiles	135,000
Clothing	36,000–82,000
Agriculture	20,000 (per farmer)
Dairy	1,800 (per cow)

Source: GATT, 1993 (http://www.wto.org).

forced consumers to pay higher prices for cars. Economists have estimated that it costs consumers, in higher car prices, about $100,000 for each job saved. Table 19-2 lists economists' estimates of the cost to consumers of saving a job in some other industries.

Had I.T. been open about the difficulties of trading, he would have warned the countries that change is hard. It has very real costs. But these costs of change are relatively small compared to the gains from trade. Moreover, they're short-run, temporary costs, whereas gains from trade are permanent, long-run gains. Once the adjustment has been made, the costs will be gone but the benefits will still be there.

For most goods, the benefits for the large majority of the population so outweigh the small costs to some individuals that, decided on a strict cost/benefit basis, international trade is still a deal you can't refuse. With benefits so outweighing costs, it would seem that transition costs could be forgotten. But they can't.

Benefits of trade are generally widely scattered among the entire population. In contrast, costs of free trade often fall on specific small groups.

Benefits of trade are generally widely scattered among the entire population. In contrast, costs of free trade often fall on small groups of people who loudly oppose the particular free trade that hurts them. Though the benefits of free trade to the country overall exceed the costs of free trade to the small group of individuals, the political push from the few (who are hurt) for trade restrictions often exceeds the political push from the many (who are helped) for free trade. The result is trade restrictions on a variety of products.

Q-3 Why does the EU place high barriers against agricultural products?

It isn't only in Canada that the push for trade restrictions focusses on the small costs and not on the large benefits. For example, the European Union (EU) places large restrictions on food imports from nonmember nations. If the EU removed those barriers, food prices in EU countries would decline significantly. For example, it is estimated that meat prices would fall by about 65 percent. Consumers would benefit, but farmers would be hurt. The farmers, however, have the political clout to see that the costs are considered and the benefits aren't. The result: The EU places high duties on foreign agricultural products.

Because eliminating trade restrictions often imposes high costs on a small group in society, government has instituted trade adjustment assistance programs to compensate for the losses.

The cost to society of relaxing trade restrictions has led to a number of programs to assist those who are hurt. Such programs are called **trade adjustment assistance programs**—*programs designed to compensate losers for reductions in trade restrictions*.

The argument for these programs is the following:

> Trade will make most members of society better off, but will make a particular subgroup in society worse off. Because of the country's political dynamics, this subgroup can prevent free trade. By structuring programs so that they transfer some of society's gains to individuals who are made worse off by trade, opposition to trade can be eliminated, and society will be better off.

Although trade adjustment programs compensate those who are hurt by trade, it does not eliminate the opposition to free trade.

Governments have tried to use trade adjustment assistance to facilitate free trade, but they've found that it's enormously difficult to limit the adjustment assistance to those who are actually hurt by international trade. As soon as people find that there's assistance for people injured by trade, they're likely to try to show that they too have been hurt and deserve assistance. Losses from free trade are exaggerated. Instead of only a small portion of the gains from trade being needed for trade adjustment assistance, much more is demanded and telling people who claim to be hurt that they aren't really being hurt isn't good politics.

Telling people who claim to be hurt that they aren't really being hurt isn't good politics.

Haggling by Companies over the Gains from Trade

Many naturally advantageous trades do not happen because each side wants a larger share of the gains from trade than the other side thinks should be allotted. This is another example of the prisoner's dilemma.

To see how companies haggling over the gains of trade can restrict trade, let's reconsider the original deal that I.T. proposed. I.T. got 380 tons of food and 380 barrels of oil. Canada got an additional 100 tons of food and 60 barrels of oil. Saudi Arabia got an additional 100 barrels of oil and 60 tons of food.

Suppose the Saudis had said, "Why should we be getting only 100 barrels of oil and 60 tons of food when I.T. is getting 380 barrels of oil and 380 tons of food? We want an additional 300 tons of food and another 300 barrels of oil, and we won't deal unless we get them." Similarly Canada might have said, "We want an additional 300 tons of food and an additional 300 barrels of oil, and we won't go through with the deal unless we get them." If either the Canadian or the Saudi Arabian company that was involved in the trade for its country (or both) takes this position, I.T. might just cancel the deal. Tough bargaining positions can make it almost impossible to achieve gains from trade.

The side that drives the hardest bargain usually gets the most gains from the deal, but it also risks making the deal fall through. Such strategic bargaining goes on all the time. **Strategic bargaining** means *demanding a larger share of the gains from trade than you can reasonably expect*. If you're successful, you get the lion's share; if you're not successful, the deal falls apart and everyone is worse off.

Strategic bargaining can lead to higher gains from trade for the side that drives the hardest bargain, but it can also make the deal fall through.

Q-4 In strategic trade bargaining, it is reasonable to be unreasonable. True or false? Explain.

Haggling by Countries over Trade Restrictions

Another type of trade bargaining that often limits trade is bargaining between countries. Trade restrictions and the threat of trade restrictions play an important role in that kind of haggling. Sometimes countries must go through with trade restrictions that they really don't want to impose, just to make their threats credible.

Once one country has imposed trade restrictions, other countries attempt to get those restrictions reduced by threatening to increase their own restrictions. Again, to make the threat credible, sometimes countries must impose or increase trade restrictions simply to show they're willing to do so.

Ultimately, strategic bargaining power depends on negotiators' skills and the underlying gains from trade that a country would receive. A country that would receive only a small portion of the gains from trade is in a much stronger bargaining position than a country that would receive significant gains because it's easier for them to walk away from trade.

The problem with strategic trade policies is that they can backfire. One rule of strategic bargaining is that the other side must believe that you'll go through with your threat. Thus, strategic trade policy can lead a country that actually supports free trade to impose trade restrictions, just to show how strongly it believes in free trade.

Even though most economists support free trade, they admit that in bargaining it may be necessary to adopt a strategic position. A country may threaten to impose trade restrictions if the other country does so. When such strategic trade policies are successful, they end up eliminating or reducing trade restrictions.

Specialized Production

Comparative advantage suggests that one country is inherently more productive than another country in producing certain goods. But when one looks at trading patterns, it's often not at all clear why particular countries have a comparative advantage in certain goods. There's no inherent reason for Switzerland to specialize in the production of watches or for South Korea to specialize in the production of cars. International trade cannot be explained by countries' resource endowments.

If they don't have inherent advantages, why are countries often so good at producing what they specialize in? Two important explanations are that they *learn by doing* and that *economies of scale* exist.

Learning by doing means becoming better at a task the more you perform it.

Learning by Doing **Learning by doing** means *becoming better at a task by performing it repeatedly*. Take watches in Switzerland. Initially production of watches in Switzerland may have been a coincidence; the person who started the watch business happened to live there. But then people in the area became skilled in producing watches. Their skill made it attractive for other watch companies to start up. As additional companies moved in, more and more members of the labour force became skilled at watchmaking, and word went out that Swiss watches were the best in the world. That reputation attracted even more producers, so Switzerland became the watchmaking capital of the world. Had the initial watch production occurred in Austria, not Switzerland, Austria might have become the watch capital of the world.

When there's learning by doing, it's much harder to attribute inherent comparative advantage to a country. One must always ask: Does country A have an inherent comparative advantage, or does it simply have more experience? Once country B gets the experience, will country A's comparative advantage disappear? If it will, then country B has a strong reason to initially limit trade with country A in a particular product in order to give its own workers time to learn by doing, and develop a comparative advantage.

In economies of scale, costs per unit of output go down as output increases.

Economies of Scale In determining whether an inherent comparative advantage exists, a second complication is **economies of scale**—*the situation in which costs per unit of output fall as output increases*. Many manufacturing industries (such as steel and automobiles) exhibit economies of scale. The existence of significant economies of scale means that costs are lower when one country specializes in one good and another country specializes in another good. But economies of scale do not indicate which country should specialize in what product. Producers in a country argue that if the government would establish barriers, they would be able to exploit economies of scale, lower their costs per unit, and eventually sell at a lower price than foreign producers.

Q.5 Is it efficient for a country to maintain a trade barrier in an industry that exhibits economies of scale?

A number of countries follow trade strategies to allow them to take advantage of economies of scale. For example, in the 1970s and 1980s Japan's government consciously directed investment into automobiles and high-tech consumer products, and significantly promoted exports in these goods to take advantage of economies of scale.

Most countries recognize the importance of learning by doing and economies of scale. A variety of trade restrictions are based on these two phenomena. The most common expression of the learning-by-doing and economies-of-scale insights is the **infant industry argument,** which says that *with initial protection, an industry will be able to become competitive*. Countries use this argument to justify many trade restrictions. They argue, "You may now have a comparative advantage, but that's simply because you've

The infant industry argument says that with initial protection, an industry will be able to become competitive.

been at it longer, or are experiencing significant economies of scale. We need trade restrictions on our _______ industry to give it a chance to catch up. Once an infant industry grows up, then we can talk about eliminating the restrictions."

Macroeconomic Aspects of Trade

The comparative advantage argument for free trade assumes that a country's resources are fully utilized. When countries don't have full employment, imports can decrease demand for domestic products and increase unemployment. Exports can stimulate domestic aggregate demand and decrease unemployment. Thus, when an economy is in a recession, there is a strong macroeconomic reason to limit imports and encourage exports. These macroeconomic effects of free trade play an important role in the public's view of imports and exports. When a country is in a recession, pressure to impose trade restrictions increases substantially.

National Security

A country may impose trade restrictions for reasons of national security.

Countries sometimes justify trade restrictions on grounds of national security. These restrictions take two forms:

1. Export restrictions on strategic materials and defense-related goods.
2. Import restrictions on defense-related goods. For example, in a war we don't want to be dependent on oil from abroad.

For a number of goods, national security considerations make sense. For example, most countries restrict the sale of certain military items to countries that are likely to be hostile someday. The problem is where to draw the line about goods having a national security consideration. Should countries protect domestic agriculture? All high technology items, since they might be useful in weapons? All chemicals? Steel? The national security argument has been extended at times to a wide variety of goods that have only an indirect connection to national security. When a country makes a national security argument for trade, we must be careful to consider whether a domestic political reason may be lurking behind that argument.

International Politics

International politics frequently provides another reason for trade restrictions. The argument is: Trade helps you, so we'll hurt you by stopping trade until you do what we want. So what if it hurts us too? It'll hurt you more than us.

Canada has imposed various trade and other economic sanctions in conjunction with directives from the United Nations. If the U.N. Security Council determines that a certain country poses a threat to peace or has undertaken an act of aggression, as a member of the United Nations, Canada has pledged[1] to support the effort to restore international peace and security. The measures generally involve economic and trade sanctions, rather than military force. In recent years, under the *United Nations Act*, Canada has levied a variety of trade sanctions against various countries. Canada prohibits trade in arms, petroleum, aircraft, diamonds, and mining equipment with Angola. Canada has imposed stringent economic sanctions against Iraq, including a ban on exports[2] and imports. Libya also faces various sanctions, including export restrictions on goods relating to the oil and gas industry; transactions with Libya currently require Ministerial Certificates.

[1]Through the *United Nations Act*.

[2]Goods authorized by the U.N. Sanctions Committee under the Food and Oil Programme require an exemption certificate from the Department of Foreign Affairs and International Trade.

Canada also uses economic sanctions, including trade restrictions, to avert serious international crises. Canada can impose trade sanctions through the *Special Economic Measures Act* (SEMA), which can restrict or prohibit Canadians from exporting, selling, or shipping goods to a country that has breached international peace and security conditions. SEMA can also restrict the transfer of technical data, provision of financial services, and travel to and from certain identified countries.[3]

Canada can also apply the *Export and Import Permits Act*. Canada currently controls the export of military goods and technology to countries that pose a threat to Canada, are involved in hostilities, or have a persistent record of serious human rights violations. Canada's *Export Control List* is a list of products which require export permits from the Minister of Foreign Affairs. Canada also maintains an *Import Control List*, which specifies products for which import is restricted. The *Area Control List* contains the names of countries that require permits for the export of *any* good. Two countries are currently[4] on Canada's *Area Control List*: Angola and Myanmar.

Increased Revenue Brought In by Tariffs

Revenues from tariffs are an important source of tax revenue for developing countries.

A final argument made for one particular type of trade restriction—a tariff—is that tariffs bring in revenues. They are less important as a source of revenue today for developed countries because those countries have instituted other forms of taxes. However, because many developing countries lack the tax infrastructure, tariffs remain a primary source of revenue for these countries. Tariffs are relatively easy to collect and are paid by people rich enough to afford imported goods. These countries justify many of their tariffs with the argument that they need the revenues.

WHY ECONOMISTS GENERALLY OPPOSE TRADE RESTRICTIONS

Economists generally oppose trade restrictions because:

1. From a global perspective, free trade increases total output.
2. Trade restrictions lead to retaliation.
3. International trade provides competition for domestic companies.

Each of the preceding arguments for trade restrictions has some validity, but most economists discount them and support free trade. The reason is that, in their considered judgment, the harm done by trade restrictions outweighs the benefits.

Larger Total Output

Economists' first argument for free trade is that, viewed from a global perspective, free trade increases total output. From a national perspective, economists agree that particular instances of trade restriction may actually help one nation, even as most other nations are hurt. But they argue that the country imposing trade restrictions can benefit *only if the other country doesn't retaliate* with trade restrictions of its own. Retaliation is the rule, not the exception, however, and when there is retaliation, the increased trade restrictions cause both countries to lose.

Q-6 What was the long-run effect on the steel industry of the trade restrictions on the import of steel during the 1950s and 1960s?

Increased Competition

A second reason most economists support free trade is that trade restrictions reduce international competition. International competition is desirable because it forces domestic companies to reduce costs and innovate to stay competitive. If trade restrictions on imports are imposed, domestic companies don't work as hard, and they become less efficient.

[3]Currently none.

[4]As of September, 2001.

Infant Industries Remain Infants

Economists dispose of the infant industry argument by reference to the historical record. In theory the argument makes sense. But very few of the infant industries protected by trade restrictions have ever grown up. What tends to happen instead is that infant industries become dependent on the trade restrictions and use political pressure to keep that protection. As a result, they often remain immature and internationally uncompetitive. Most economists would support the infant industry argument only if the trade restrictions included definite conditions under which the restrictions would end.

Very few of the infant industries protected by trade restrictions have ever grown up.

National Security Issues

Most economists agree with the national security argument for export restrictions on goods that are directly war related. Selling bombs to Iraq doesn't make much sense. Economists point out that the argument is often carried far beyond goods directly related to national security. In addition, trade restrictions on military sales can often be evaded. Countries simply have another country buy the goods for them. Such third-party sales—called *trans-shipments*—are common in international trade and limit the effectiveness of any absolute trade restrictions for national security purposes.

Economists also argue that by fostering international cooperation, international trade makes war less likely—a significant contribution to national security.

Addiction to Trade Restrictions

Economists' final argument against trade restrictions is: Yes, some trade restrictions might benefit a country, but almost no country can limit its restrictions to the beneficial ones. Trade restrictions are addictive—the more you have, the more you want. Thus, a majority of economists take the position that the best response to such addictive policies is "Just say no."

Q.7 How can free trade improve national security?

Yes, some restrictions might benefit a country, but almost no country can limit its restrictions to the beneficial ones.

VARIETIES OF TRADE RESTRICTIONS

Let's now turn to the policies countries can use to restrict trade. These include tariffs, quotas, voluntary restraint agreements, embargoes, regulatory trade restrictions, and nationalistic appeals.

Tariffs

Tariffs are *taxes governments place on internationally traded goods*—generally imports. (Tariffs are also called *customs duties*.) Tariffs are the most-used and most-familiar type of trade restriction. Tariffs operate in the same way a tax does: They make imported goods relatively more expensive than they otherwise would have been, and thereby encourage the consumption of domestically produced goods. On average, Canadian tariffs raise the price of imported goods by about 4 percent. Table 19-3 presents average tariff rates for industrial goods for a number of countries and the European Union.

Three policies used to restrict trade are:

1. Tariffs (taxes on internationally traded goods).
2. Quotas (quantity limits placed on imports).
3. Regulatory trade restrictions (government-imposed procedures that limit imports).

Our main trading partner, the United States, is generally considered to be a pro-competition economy; however, for a period in its history, the U.S. was extremely protectionist. Probably the most infamous tariff was the U.S.'s Smoot-Hawley Tariff of 1930, which raised tariffs on imported goods an average of 60 percent. It was passed at the height of the Great Depression in the United States in the hope of protecting American jobs. It didn't work. Canada and other countries responded with similar tariffs. As a result of the trade wars, international trade plummeted, unemployment worsened, and the international depression deepened. These massive negative effects of the

see page 431

TABLE 19-3 Trade-Weighted Tariff Averages by Country after the Uruguay Round

These rates will be continually changing as the changes negotiated by the WTO come into effect.

Country	%	Country	%
Argentina	30.9	Mexico	33.7
Australia	12.2	New Zealand	11.3
Austria	7.1	Norway	2.0
Brazil	27.0	Peru	29.4
Canada	4.8	Phillipines	22.2
Chile	24.9	Poland	9.9
Colombia	35.1	Romania	33.9
Costa Rica	44.1	Senegal	13.8
Czech Republic	3.8	Singapore	5.1
El Salvador	30.6	Slovak Republic	3.8
European Union	3.6	South Africa	17.2
Finland	3.8	Sri Lanka	28.1
Hong Kong	0.0	Sweden	3.1
Hungary	6.9	Switzerland	1.5
Iceland	11.5	Thailand	28.0
India	32.4	Tunisia	34.1
Indonesia	36.9	Turkey	22.3
Jamaica	50.0	United States	3.5
Japan	1.7	Uruguay	30.9
Korea	8.3	Venezuala	30.9
Macau	0.0	Zimbabwe	4.6
Malaysia	9.1		

Source: General Agreement on Tariffs and Trade (GATT), The Results of the Uruguay Round of Multilateral Trade Negotiations, November, 1994. (http://www.wto.org)

Great Depression tariffs provided convincing evidence that free trade is preferable to trade restrictions.

The far-reaching effects of the Smoot-Hawley Tariff was the main reason the **General Agreement on Tariffs and Trade (GATT),** *a regular international conference to reduce trade barriers*, was established in 1947 immediately following the Second World War. In 1995 GATT was replaced by the **World Trade Organization (WTO),** *an organization whose functions are generally the same as GATT's—to promote free and fair trade among countries*. Unlike GATT, the WTO is a permanent organization with an (albeit weak) enforcement system. Since then, multiple rounds of negotiations have resulted in a decline in worldwide tariffs.

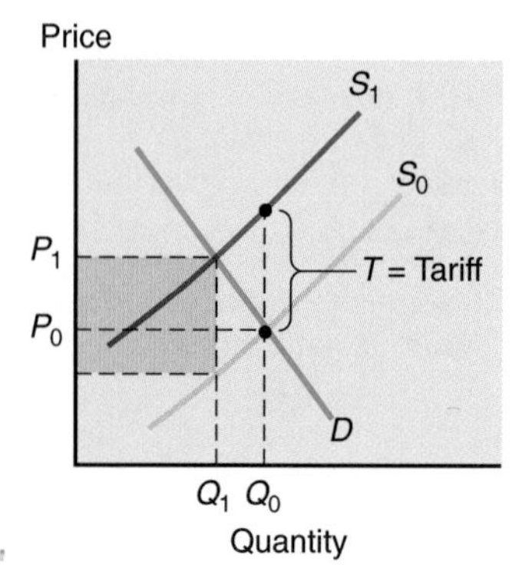

A quota limiting foreign quantity supplied to Q_1 is the equivalent of a tariff of T.

Quotas

Quotas are *quantity limits placed on imports*. Their effect in limiting trade is similar to the effect of a tariff. Chapter 5 showed the geometry of quotas and tariffs. The results of tariffs and quotas in terms of price and quantity are the same, as you can see in the graph in the margin.

The issues involved with tariffs and quotas are slightly different than in Chapter 5 because our country is small relative to the world economy, and imports generally compete with domestic production. The small-country assumption means that the supply from the world to this country is perfectly elastic at the world price, $2, as in Figure 19-2(a).

The world price of the good is unaffected by this country's demand. This assumption allows us to distinguish the world supply from domestic supply. In the absence of any trade restrictions, the world price of $2 would be the domestic price. Domestic low-cost

Figure 19-2 (a and b) FOREIGN SUPPLY OF A GOOD WHEN THE DOMESTIC COUNTRY IS SMALL

This exhibit shows the effects of a tariff in (a) and of a quota in (b) when the domestic country is small. The small-country assumption means that the world supply is perfectly elastic, in this case at $2.00 a unit. With a tariff of 50 cents, world supply shifts up by 50 cents. Domestic quantity demanded falls to 175 and domestic quantity supplied rises to 125. Foreign suppliers are left supplying the difference, 50 units. The domestic government collects revenue shown in the shaded area. The figure in (b) shows how the same result can be achieved with a quota of 50. Equilibrium price rises to $2.50. Domestic firms produce 125 units and consumers demand 175 units. The difference between the tariff and the quota is that, with a tariff, the domestic government collects the revenue from the higher price. With a quota, the benefits of the higher price accrue to the foreign and domestic producers.

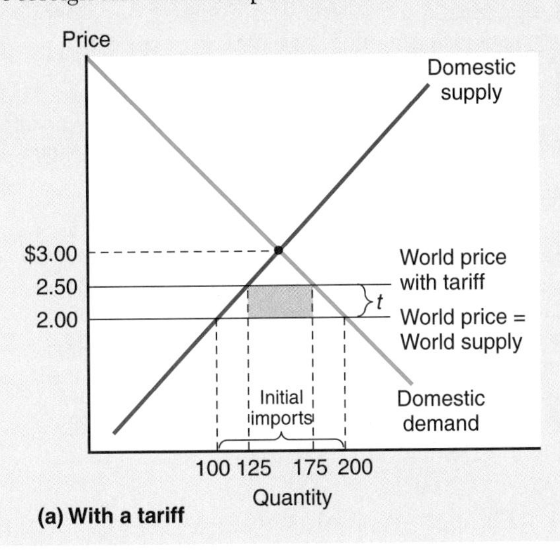

(a) With a tariff

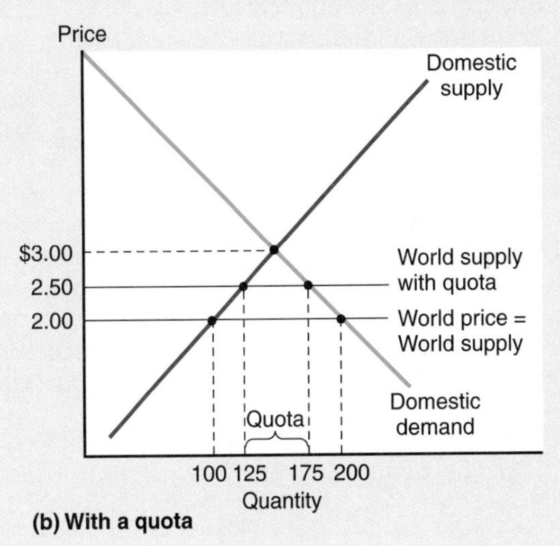

(b) With a quota

suppliers would supply 100 units of the good at $2. The remaining 100 units demanded are being imported.

Figure 19-2(a) shows the effect of a tariff of 50 cents placed on all imports. Since the world supply curve is perfectly elastic, all of this tax, shown by the shaded region, is borne by domestic consumers. Price rises to $2.50 and quantity demanded falls to 175. With a tariff, the rise in price will increase domestic quantity supplied from 100 to 125, and will reduce imports to 50. Now let's compare this situation with a quota of 50, shown in Figure 19-2(b). Under a quota of 50, the final price would be the same, but higher revenue would accrue not to the government but instead to foreign and domestic producers. One final difference: Any increase in demand under a quota would result in higher prices because it would have to be filled by domestic producers. Under a tariff, any increase in demand would not affect price.

Q-8 What is the difference between a tariff and a quota?

Voluntary Restraint Agreements

Voluntary restraint agreements are often not all that voluntary.

Imposing new tariffs and quotas is specifically ruled out by the WTO, but foreign countries know that WTO rules are voluntary and that, if a domestic industry brought sufficient political pressure on its government, the WTO would be forgotten. To avoid the imposition of new tariffs on their goods, countries often voluntarily restrict their exports. That's why Japan agreed informally to limit the number of cars it exported to Canada.

The effect of such voluntary restraint agreements is similar to the effect of quotas: They directly limit the quantity of imports, increasing the price of the good and helping domestic producers. For example, when Canada encouraged Japan to impose "voluntary"

APPLYING THE TOOLS

Canada's Trade Partners: Trading Unfairly

Despite our close ties—social, cultural, and economic—with the United States, every now and again, allegations fly. In the early 1990's, softwood lumber was an issue. U.S. softwood lumber producers accused Canada of unfairly subsidizing lumber production through low provincial stumpage rates. Stumpage fees are paid by Canadian lumber companies for the right to cut timber on Crown land, which is owned by the government. The U.S. producers argued that because they cut timber mostly on privately owned land, and must pay much higher market-based fees to cut timber, they were unable to compete against the Canadian producers. In 1996, the Softwood Lumber Agreement was signed, which limited the quantity of Canadian lumber exported to the United States. In early 2001, the Agreement expired, and once again U.S. lumber producers called for their government to impose countervailing duties on the low cost Canadian lumber. Countervailing duties are duties which are imposed to offset the effects of foreign subsidies.

The complaints, however, go both ways across the border. In 1997, the Canadian International Trade Tribunal, the body which handles disputes for the North American Free Trade Agreement (NAFTA), cited the U.S. for illegally dumping certain prepared baby foods, allegedly causing material injury to the Canadian baby food industry. Dumping occurs when a product is sold for a lower price in a foreign market than in its domestic market. The H.J. Heinz company, the sole Canadian producer of the baby food in question—preparations in jars for infants aged 4 to 18 months—claimed that the U.S. producer, Gerber, was unfairly pricing its product in Canada, resulting in lost sales and revenues for Heinz.

The result of the inquiry was the imposition of a 69% duty against Gerber baby food imports, which was later repealed due to Canadian competition concerns. Remember, Heinz was a monopoly.

Complaints against other countries abound. In February 2001, the Canada Customs and Revenue Agency (CCRA) received a complaint from Dofasco regarding dumping of steel products[1] by Brazil, Chinese Taipei, Macedonia, Italy, Luxembourg, Malaysia, China, Korea, and South Africa. Dofasco claimed harm in the form of lost market share, lowered prices, lost sales and revenues, and increased inventories.

During the period January–December 2000, CCRA found that 88% of goods were dumped, with margins ranging from 6.5% to 22.8%. CCRA also noted that market share of Canadian firms, Dofasco, Stelco, Algoma, and Ispat Sidbec, declined from 94% to 77% since 1996, steel prices fell, and inventories increased. As you can see in the table, CCRA estimated that 100% of the steel imported from Korea, Luxembourg, Malaysia, and South Africa was dumped, representing a total of approximately 27% of total steel imports.

Country	Estimated Percentage of Goods Dumped (by volume)	Estimated Percentage of Total Imports
Brazil	75	12.87
China	85.5	11.41
Chinese Taipei	78.1	9.41
Italy	99.8	1.74
Korea	100	21.85
Luxembourg	100	1.04
Macedonia	99.9	62
Malaysia	100	.84
South Africa	100	2.96

A wide variety of imported goods have been subject to antidumping measures or countervailing duties, or both:

Country	Products
United States	carpets, iceberg lettuce, potatoes, refrigerators, dishwashers, dryers, sugar
China	bicycles and frames, fresh garlic, footwear, portable file cases
Czech Republic	shot shells
France	cigarette tubes, steel
Mexico	steel
Spain	steel
Turkey	concrete reinforcing bar

What can we conclude? International trade confers benefits (gains from trade), but not without conflicts.

[1]Cold-rolled steel sheet products, used in the production of household appliances, drums and pails, and office furniture.

quotas on exports of its cars, Toyota benefited from the quotas because it could price its limited supply of cars higher than it could if it sent in a large number of cars, so profit per car would be high. Since Canadian car companies faced less competition, they also benefited. They could increase their prices because Toyota had done so.

Embargoes

An **embargo** is *a total restriction on import or export of a good*. Embargoes are usually established for international political reasons rather than for primarily economic reasons. For example, Canada has imposed arms embargoes against several countries, including Afghanistan, Angola, Eritrea, Iraq, Liberia, Libya, Rwanda, and Sierra Leone. The arms embargo against Libya has been in place since 1994, but Canadian sanctions also include prohibitions against aircraft, aeronautical products, and flights to and from Libya, prohibitions against technical assistance to any Libyan pilot, flight engineer, or maintenance personnel, and prohibitions against financing, extending credit, or advancing funds directly or indirectly to Libya. Penalties for Canadians participating in these activities range from fines of $200 to $5000, or prison terms from three months to five years.

An embargo is a total restriction on import or export of a good.

Prohibitions against Afghanistan's former Taliban government were made long before the attacks on New York City's World Trade Center and the Pentagon in Washington, D.C., on September 11, 2001. In November of 1999, under the *United Nations Act*, "United Nations Afghanistan Regulations,"[5] flights to Canada of Taliban-owned or -operated aircraft were forbidden,[6] and all financial and property transactions involving the Taliban were prohibited. Concerns regarding the Taliban's support of international terrorism were raised in December 2000, when the U.N. Security Council Resolution 1333 was adopted. The Resolution demanded that the Taliban cease providing sanctuary and training for terrorists, banned terrorist camps in Taliban territory, and demanded the surrender of indicted terrorist Osama bin Laden. The sanctions included an arms embargo, the closing of all offices of Ariana Afghan Airlines, and a ban on all flights to and from territory under Taliban control. At the time, however, although supporting the harsh economic measures, Canada raised concerns over the violations of human rights, especially those of women and girls, and over the humanitarian effect that such sanctions would have on the situation in Afghanistan.

Trade restrictions and other economic sanctions generally do not work quickly against developing critical events in international relations. The main effect is punishment, and, to some extent, deterrence, and the effects of the sanctions are realized over time. Economic sanctions can be considered a form of economic warfare. Their motivation and role is largely political, with the goal of providing a means of managing disturbances in world affairs without resorting to the use of armed force. Economic sanctions are most effective in achieving their objective when they are implemented by a group of countries acting collectively.

Are safety regulations designed to protect a country's citizens from harm, or to protect domestic producers from competition?

Regulatory Trade Restrictions

Tariffs, quotas, and embargoes are the primary *direct* methods to restrict international trade. There are also indirect methods that restrict trade. These are called **regulatory trade restrictions** (*government-imposed procedures that limit imports*). One type of regulatory trade restriction has to do with protecting the health and safety of a country's residents. For example, a country might restrict imports of all vegetables grown where

[5] P.C. 1999-2015 10 November, 1999.

[6] Unless for humanitarian reasons, approved in advance.

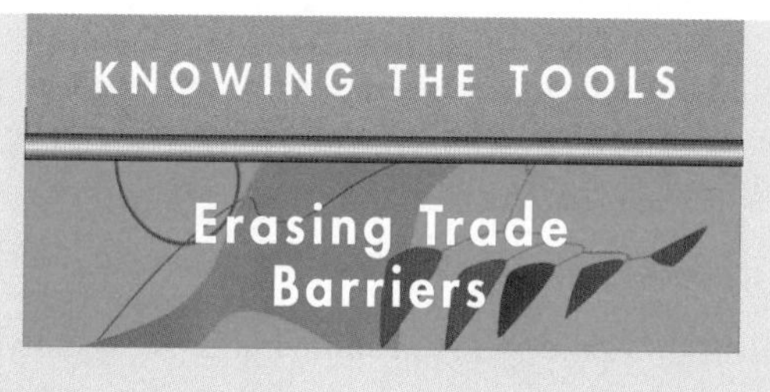

International borders between countries means that any exchange of goods or services results in "international trade." Free trade means that tariffs or other regulations do not restrict the flow of goods and services. The next step beyond free trade in goods and services is the free flow of the resources used to produce those goods. One region that has accomplished free trade in goods and services and in resources (including capital and labour) is the European Union. The European Union is a group of fifteen European countries[1] that eliminated their borders in 1995.

While the border *around* the European Union remains intact, *within* the fifteen countries people can travel freely; they can live in one country and work in another, and investment can flow freely to its most highly valued use. The economic benefits of the European Union are clear: the creation of a single market for goods and services enabled firms to capture cost savings through economies of scale, learning by doing, and specialization, and allowed a more efficient economy to develop.

Furthermore, the gains are not strictly economic. The freedom to live and work anywhere within the European Union is a benefit people value, particularly the highly mobile younger citizens, who study and work throughout the region. The region also reaps social and political benefits. Europe has lived through two World Wars, and with the formation of the European Union, an enhanced sense of belonging has developed. People are now citizens of the European Union—removing the physical barriers has also tended to remove the barriers in people's minds. Although people remain citizens of their own country, they are now also citizens of the remarkable European Union.

[1]Austria, Belgium, Denmark, Finland, France, Germany, Greece, Ireland, Italy, Luxembourg, The Netherlands, Portugal, Spain, Sweden, and the United Kingdom.

Q.9 How might a country benefit from having an inefficient customs agency?

Some regulatory restrictions are imposed for legitimate reasons; others are designed simply to make importing more difficult.

certain pesticides are used, knowing that most countries use those pesticides. The effect of such a regulation would be to halt the import of vegetables.

A second type of regulatory restriction involves making import and customs procedures so intricate and time-consuming that importers simply give up. Some regulatory restrictions are imposed for legitimate reasons; others are designed simply to make importing more difficult and hence protect domestic producers from international competition. It's often hard to tell the difference. A good example of this difficulty occurred in 1988, when the EU disallowed all imports of meat from animals that had been fed growth-inducing hormones.

Nationalistic Appeals

Finally, nationalistic appeals can help to restrict international trade. The "Buy Canadian" campaign is an example. Foreign companies often go to great lengths to get a MADE IN CANADA classification on goods they sell in Canada. For example, components for many autos are made in Japan but shipped to North America and assembled in Ontario or Tennessee so that the finished car can be called a North American product.

FREE TRADE ASSOCIATIONS

A free trade association is a group of countries that allows free trade among its members and puts up common barriers against all other countries' goods.

Most politicians feel a strong push from constituents to impose trade barriers. Even politicians who accept free trade in principle support restrictions in practice. They have, however, found it possible to support a limited type of free trade in which a few countries agree to have free trade with each other. A **free trade association** (or *customs union*) is a *group of countries that allows free trade among its members and, as a group, puts up common barriers against all other countries' goods*.

The European Union (EU) is a successful free trade association. All barriers to trade among the EU's member countries were removed in 1992. In the coming decade more

APPLYING THE TOOLS

Dumping

The WTO allows countries to impose trade restrictions on imports if they can show that the goods are being dumped. *Dumping* is selling a good in a foreign country at a lower price than it sells for in the country where it's produced. On the face of it, who could complain about someone who wants to sell you a good cheaply? Why not just take advantage of the bargain price? However, dumping by another country can force domestic producers out of business. Having eliminated the competition, the foreign producer then has the field to itself and can raise the price. Dumping can be a form of predatory pricing.

Dumping also has short-term macroeconomic and political effects in the importing country. Even if one believes that the dumping is not predatory pricing, it can displace workers in the importing country, causing political pressure on that government to institute trade restrictions. If that country's economy is in a recession, the resulting unemployment will have substantial macroeconomic repercussions, so pressure for trade restrictions will be amplified.

European countries can be expected to join the EU. In 1993, Canada, the United States and Mexico agreed to enter into a similar free trade union, and created the North American Free Trade Association (NAFTA). Under NAFTA, tariffs and other trade barriers among these countries are being gradually reduced. Some other trading associations include Mercosur (among South American countries) and Asean (among Southeast Asian countries).

see page 431

Economists have mixed reactions to free trade associations. They see free trade as beneficial, but are concerned about the possibility that these regional free trade associations will impose significant trade restrictions on nonmember countries. They also believe that bilateral negotiations between member nations will replace multilateral efforts among members and nonmembers. Whether the net effect of these bilateral negotiations will be positive or negative remains to be seen.

Q-10 What is economists' view of limited free trade associations such as the EU or NAFTA?

One way countries strengthen trading relationships among groups of countries is through a most-favoured-nation status. The term **most-favoured nation** refers to *a country that will be charged the lowest tariff on its exports*. Thus, if Canada lowers tariffs on goods imported from Japan, which has most-favoured-nation status with Canada, it must lower tariffs on those same types of goods imported from any other country with most-favoured-nation status.

A most-favoured nation is a country that will pay as low a tariff on its exports as will any other country.

CONCLUSION

International trade is very important to Canada. With international transportation and communication becoming easier and faster, and with other countries' economies growing, the Canadian economy will inevitably become more interdependent with the other economies of the world. As international trade becomes more important, the push for trade restrictions will likely increase. Various countries' strategic trade policies will likely conflict, and the world will find itself on the verge of an international trade war that would benefit no one.

Concern about that possibility leads most economists to favour free trade. But as often happens, good economic policies often are not good politics. Whether politicians follow economists' advice or whether they follow the politically popular policy will play a key role in determining the course of world trade in the 2000s.

Chapter Summary

- According to the principle of comparative advantage, as long as the relative opportunity costs of producing goods (what must be given up in one good in order to get another good) differ among countries, there are potential gains from trade, even if one country has an absolute advantage in everything.
- Three insights into the terms of trade include:
  1. The more competition exists in international trade, the less the trader gets and the more the trading countries get.
  2. Smaller countries tend to get a larger percentage of the gains from trade than do larger countries.
  3. Gains from trade go to countries that produce goods that exhibit economies of scale.
- Reasons that countries impose trade restrictions include unequal internal distribution of the gains from trade, haggling by companies over the gains from trade, haggling by countries over trade restrictions, learning by doing and economies of scale, macroeconomic aspects of trade, national security, international political reasons, and increased revenue brought in by tariffs.
- Economists generally oppose trade restrictions because of the history of trade restrictions and their understanding of the advantages of free trade.
- Trade restrictions include: tariffs, quotas, embargoes, voluntary restraint agreements, regulatory trade restrictions, and nationalistic appeals.
- Free trade associations help trade by reducing barriers to trade among member nations. Free trade associations may hinder trade by building up barriers to trade with nations outside the association; negotiations among members could replace multilateral efforts to reduce trade restrictions among members and nonmembers.

Key Terms

economies of scale *(418)*
embargo *(425)*
free trade association *(426)*
General Agreement on Tariffs and Trade (GATT) *(422)*
infant industry argument *(418)*
learning by doing *(418)*
most-favoured nation *(427)*
principle of absolute advantage *(411)*
principle of comparative advantage *(414)*
quotas *(422)*
regulatory trade restrictions *(425)*
strategic bargaining *(417)*
tariffs *(421)*
trade adjustment assistance programs *(416)*
World Trade Organization (WTO) *(422)*

Questions for Thought and Review

1. Widgetland has 60 workers. Each worker can produce 4 widgets or 4 wadgets. Each resident in Widgetland currently consumes 2 widgets and 2 wadgets. Wadgetland also has 60 workers. Each can produce 3 widgets or 12 wadgets. Wadgetland's residents consume 1 widget and 9 wadgets. Is there a basis for trade? If so, offer the countries a deal they can't refuse.
2. Suppose that two countries, Machineland and Farmland, have the following production possibility curves.

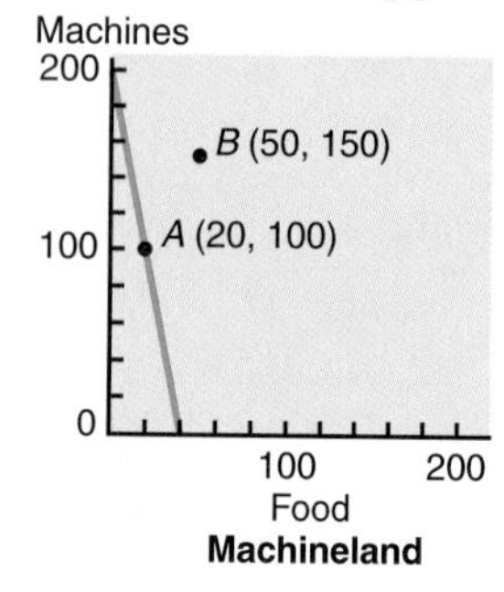

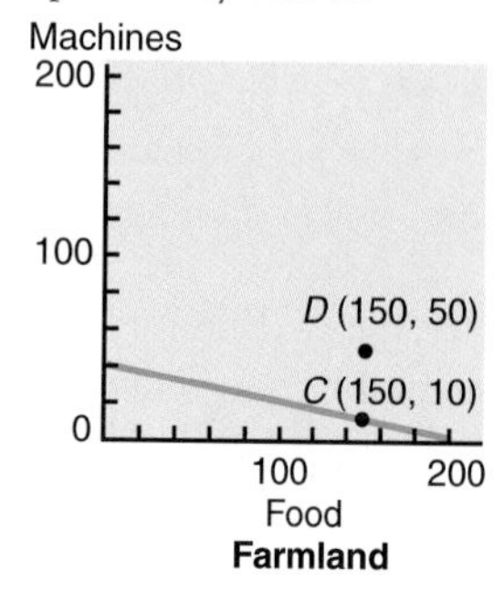

   a. Explain how these two countries can move from points A and C, where they currently are, to points B and D.
   b. If possible, state by how much total production for the two countries has risen.
   c. If you were a trader, how much of the gains from trade would you deserve for discovering this trade?
   d. If there were economies of scale in the production of both goods, how would your analysis change?
3. Suppose Greece and France each produced only Kalamata olives and Roquefort cheese. Their production possibility curves are given in the following tables:

Greece

% of Resources Devoted to Olives	Olives Produced (millions of olives)	Roquefort Produced (thousands of pounds)
100%	500	0
80	400	100
60	300	200
40	200	300
20	100	400
0	0	500

France

% of Resources Devoted to Olives	Olives Produced (millions of olives)	Roquefort Produced (thousands of pounds)
100%	200	0
80	160	10
60	120	20
40	80	30
20	40	40
0	0	50

Currently, contrary to expectations, Greece produces 300 million olives and 200,000 pounds of Roquefort, while France produces 40 million olives and 40,000 pounds of Roquefort.

a. What is the opportunity cost of producing olives in each country? What is the opportunity cost of producing Roquefort in each country?
b. The European Commission rules that traditional foods have trademark protection. Only Roquefort made in France can be Roquefort and only olives produced in Greece can be Kalamata olives. Now Greece and France, following the new ruling, must each produce their respective trademarked goods and trade. Are they made better off? Why?
c. Show your answer graphically by selecting points now attainable that were previously not attainable.

4. Suppose there are two countries, Busytown and Lazylake, with the following production possibility tables:

Busytown

% of Resources Devoted to Cars	Cars Produced (thousands)	Gourmet Meals Produced (thousands)
100%	60	0
80	48	10
60	36	20
40	24	30
20	12	40
0	0	50

Lazylake

% of Resources Devoted to Cars	Cars Produced (thousands)	Gourmet Meals Produced (thousands)
100%	50	0
80	40	10
60	30	20
40	20	30
20	10	40
0	0	50

a. Draw the production possibility curves for each country.
b. Which country has the absolute advantage in producing cars? In producing gourmet meals?
c. Which country has the comparative advantage in producing cars? In producing gourmet meals?
d. Suppose each country specializes in the production of one good. Explain how Busytown can end up with 36,000 cars and 22,000 meals and Lazylake can end up with 28,000 meals and 24,000 cars.

5. Why does competition among traders affect how much of the gains to trade are given to the countries involved in the trade?
6. Why do smaller countries usually get most of the gains from trade? What are some reasons why a small country might not get the gains of trade?
7. Which country will get the larger gain from trade—a country with economies of scale or diseconomies of scale? Explain your answer.
8. Suggest an equitable method of funding trade adjustment assistance programs. Why is it equitable? What problems might a politician have in implementing such a method?
9. If you were economic adviser to a country that was following your advice about trade restrictions and that country fell into a recession, would you change your advice? Why, or why not?
10. A country's trade balance improves when the economy goes into recession. Is this consistent with economic predictions? Why or why not?
11. What are two reasons economists support free trade?
12. Demonstrate graphically how the effects of a tariff differ from the effects of a quota.
13. How do the effects of voluntary restraint agreements differ from the effects of a tariff? How are they the same?
14. Mexico exports vegetables to Canada. These vegetables are grown using chemicals that are not allowed in Canadian vegetable agriculture. Should Canada restrict imports of Mexican vegetables? Why or why not?

15. Placing a price floor on imported fruit will provide strong relief for domestic fruit growers, and help preserve jobs in the industry. What costs arise from the price floor?

16. A study by the World Bank on the effects of Mercosur, a regional trade pact among four Latin American countries, concluded that free trade agreements "might confer significant benefits, but there are also significant dangers." What are those benefits and dangers?

Problems and Exercises

1. Suppose there are two countries that do not trade—Corny Island and Graneland. Each country produces the same two goods—corn and wheat. For Corny Island the opportunity cost of producing 1 bushel of wheat is 3 bushels of corn. For Graneland the opportunity cost of producing 1 bushel of corn is 3 bushels of wheat. At present, Corny Island produces 20 million bushels of wheat and 120 million bushels of corn, while Graneland produces 20 million bushels of corn and 120 million bushels of wheat.
   a. Explain how, with trade, Graneland can end up with 40 million bushels of wheat and 120 million bushels of corn while Corny Island can end up with 40 million bushels of corn and 120 million bushels of wheat.
   b. If the states ended up with the numbers given in *a*, how much would the trader get?
2. Country A can produce, at most, 40 olives or 20 pickles, or some combination of olives and pickles such as the 20 olives and 10 pickles it is currently producing. Country B can produce, at most, 120 olives or 60 pickles, or some combination of olives and pickles such as the 100 olives and 50 pickles it is currently producing.
   a. Is there a basis for trade? If so, offer the two countries a deal they can't refuse.
   b. How would your answer to *a* change if you knew that there were economies of scale in the production of pickles and olives rather than the production possibilities described in the question? Why? If your answer is yes, which country would you have produce which good?
3. The world price of textiles is P_w, as in the accompanying figure of the domestic supply and demand for textiles.

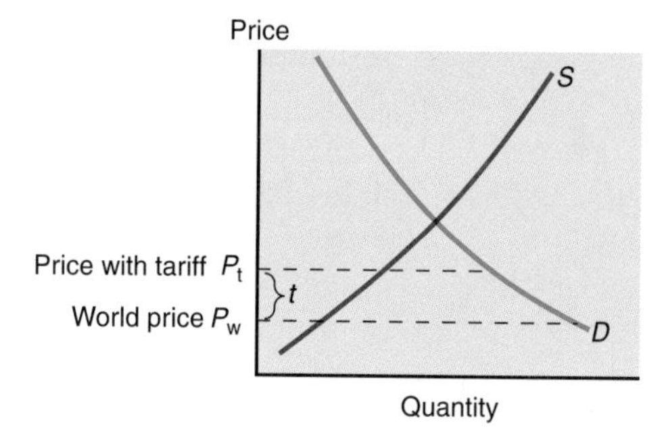

The government imposes a tariff t, to protect the domestic producers. For this tariff:
   a. Shade in the gains to domestic producers.
   b. Shade in the revenue to government.
   c. Shade in the costs to domestic producers.
   d. Are the gains greater than the costs? Why?
4. In 1999 the hourly cost to employers per German industrial worker was Cdn $38.90. The hourly cost to employers per Canadian industrial worker was $23.17, while the average cost per Taiwanese industrial worker was $8.35.
   a. Give three reasons why firms produce in Germany rather than in a lower-wage country.
   b. Germany has just entered into an agreement with other EU countries that allows people in any EU country, including Greece and Italy, which have lower wage rates, to travel and work in any EU country, including high-wage countries. Would you expect a significant movement of workers from Greece and Italy to Germany right away? Why or why not?
   c. Workers in Thailand are paid significantly less than workers in Taiwan. If you were a company CEO, what other information would you want before you decided where to establish a new production facility?
5. The director-general of GATT published a pamphlet on the costs of trade protection. He subtitled the pamphlet "The Sting: How Governments Buy Votes on Trade with the Consumer's Money."
   a. What likely is meant by this subtitle?
   b. If a government is out to increase votes with its trade policy, would it more likely institute tariffs or quotas? Why?
6. One of the basic economic laws is "the law of one price." It says that given certain assumptions one would expect that if free trade is allowed, the price of goods in countries should converge to a single price.
   a. Can you list what three of those assumptions likely are?
   b. Should the law of one price hold for labour also? Why or why not?
   c. Should it hold more so or less so for capital than for labour? Why or why not?

Web Questions

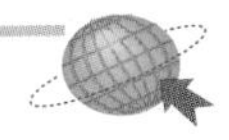

1. Go to the WTO's home page at www.wto.org to find out how trade disputes are settled.
   a. What is the procedure for settling disputes?
   b. What is the timetable for the settlement procedure?
   c. What happens if one of the countries does not abide by the settlement?
2. Two of the free trade associations mentioned in the chapter were Asean (www.aseansec.org) and Mercosur (www.americasnet.com/mauritz/mercosur/english/). Go to their home pages and answer the following questions for each:
   a. What countries belong to the trade association?
   b. When was the association established?
   c. What is the association's stated objective?
   d. What is the combined gross domestic product of all members?

Answers to Margin Questions

1. I.T. ended up with 380 tons of food and 380 barrels of oil after the trade. *(408)*
2. The percentage of gains from trade that goes to a country depends upon the change in the price of the goods being traded. If trade led to no change in prices in a small country, then that small country would get no gains from trade. Another case in which a small country gets a small percentage of the gains from trade would occur when its larger trading partner was producing a good with economies of scale and the small country was not. A third case is when the traders who extracted most of the surplus or gains from trade come from the larger country; then the small country would end up with few of the gains from trade. *(410)*
3. The EU places high barriers against agricultural products to protect its farmers. As is the case with many of the international trade barriers, this is primarily for political, not economic, purposes. *(416)*
4. True. In strategic trade bargaining it is reasonable to be unreasonable. The belief of the other bargainer that you will be unreasonable leads you to be able to extract larger gains from trade. *(417)*
5. Generally not. The larger the output, the lower the per unit cost of production. Trade allows a country to produce a large quantity which may not be supported by domestic demand, and to benefit from lower cost and gains from trade. *(418)*
6. The steel industry became internationally uncompetitive and produced overpriced steel. The restrictions produced an uncompetitive industry. *(420)*
7. Free trade fosters international cooperation, making war less likely. *(421)*
8. A tariff is a type of tax that the government places on internationally traded goods. A quota is a quantity restriction that a country places on internationally traded goods. *(424)*
9. An inefficient customs agency can operate with the same effect as a trade restriction, and if trade restrictions would help the country then it is possible that an inefficient customs agency could also help the country. *(426)*
10. Views are mixed on free trade associations such as NAFTA or the EU. While free trade is beneficial, there is concern over the possibility that these limited trade associations will impose trade restrictions on nonmember countries. Whether the net effect of these will be positive or negative is a complicated issue. *(427)*

Who Gets What? The Distribution of Income

20

After reading this chapter, you should be able to:

- State what a Lorenz curve is.
- Explain how the poverty definition is both an absolute and a relative measure.
- Discuss Canadian income inequality in a global context.
- Summarize the statistical findings on income and wealth distribution.
- State two alternative ways to describe income distribution.
- Explain three problems in determining whether an equal income distribution is fair.
- List three side effects of redistributing income.
- Summarize the Canadian tax and expenditure programs to redistribute income.

"God must love the poor," said Lincoln, "or he wouldn't have made so many of them." He must love the rich, or he wouldn't divide so much mazuma among so few of them.

H. L. Mencken

In 2000, John Roth, former Chief Executive Officer of telecommunications leader Nortel Networks, earned \$6.7 million in salary and bonuses, plus an additional \$135 million from exercising stock options. This amounts to \$2.725 million per week.

Assuming he worked 70 hours per week, that's $38,929 an hour, more than the yearly income for the average Canadian. While we generally believe that a person's pay reflects his or her contribution to production, there appears to be little correlation between executive pay and company performance, whether measured as increases in revenues, earnings, or shareholder returns.

Today, the average doctor earns $160,000 per year; that's $3,077 per week. Assuming she works 70 hours per week (she's conscientious, makes house calls, and spends time with her hospitalized patients), that's $44 per hour.

Chuck Dickens, a cashier in a fast-food restaurant, earns $8.00 per hour. The average Canadian works roughly 2000 hours a year, so Chuck's annual income is $16,000, or $308 a week.

Nguyen, a peasant in Vietnam, earns $250 a year; that's $4.80 per week. Assuming he works 70 hours per week (you have to work hard at that rate of pay just to keep from starving), that's 7 cents per hour.

Are such major differences typical of how income is distributed among people in general? Are such differences fair? And if they're unfair, what can be done about them? This chapter addresses such issues.

WAYS OF CONSIDERING THE DISTRIBUTION OF INCOME

Income represents the amount of money a person earns in a year; it is a "flow" concept. The total amount of money a person has accumulated is his or her **wealth**. Wealth is a "stock," and results mostly from working and earning income, but also results, for example, from investing in stocks or bonds or real estate, or from inheritance.

There are several different ways to look at income distribution. In the 1800s, economists were concerned with how income was divided among the owners of businesses (for whom profits were the source of income), the owners of land (who received rent), and workers (who earned wages). That concern reflected the relatively sharp distinctions among social classes that existed in capitalist societies at that time. Landowners, workers, and owners of businesses were separate groups, and few individuals moved from one group to another.

Time has changed that. Today workers, through their pension plans and investments in financial institutions, own shares issued on the stock exchanges. Landowners as a group represent only a small portion of total income. Companies are run, not by capitalists, but by managers who are, in a sense, workers. In short, the social lines have blurred.

This blurring of the lines between social classes doesn't mean that we can forget the question "Who gets what?" It simply means that our interest in who gets what has a different focus. We no longer focus on classification of income by source. Instead we look at the relative distribution of total income. How much income do the top 5 percent get? How much do the top 15 percent get? How much do the bottom 10 percent get? **Share distribution of income** is the name given to *the relative division of total income among income groups*.

The share distribution of income is the relative division or allocation of total income among income groups.

A second distributional issue economists are concerned with is the **socioeconomic distribution of income** (*the allocation of income among relevant socioeconomic groupings*). How much do the old get compared to the young? How much do women get compared to men?

The socioeconomic distribution of income is the relative division or allocation of total income among relevant socioeconomic groups.

THE SHARE DISTRIBUTION OF INCOME

The Canadian share distribution of income measures aggregate family income, from the poorest segment of society to the richest. It ranks people by their income and tells how much the richest 20 percent (a quintile) and the poorest 20 percent receive. For example, the poorest 20 percent might get 5 percent of the income and the richest 20 percent might get 40 percent.

The Lorenz Curve

A Lorenz curve is a geometric representation of the share distribution of income among families in a given country at a given time.

Q-1 When drawing a Lorenz curve, what do you put on the two axes?

Figure 20-1(a) presents the share distribution of income for Canada in 1996. In it you can see that the 20 percent of Canadians receiving the lowest level of income got 6.1 percent of the total income. The top 20 percent of Canadians received 40.6 percent of the total income. The ratio of the income of the top 20 percent compared to the income of the bottom 20 percent was about 7:1.

A **Lorenz curve** is *a geometric representation of the share distribution of income among families in a given country at a given time*. It measures the cumulative percentage of *families* on the horizontal axis, arranged from poorest to richest, and the cumulative percentage of *family income* on the vertical axis. Since the figure presents cumulative percentages (all of the families with income up to a certain level), both axes start at zero and end at 100 percent.

A perfectly equal distribution of income would be represented by a diagonal line like the one in Figure 20-1(b). That is, the poorest 20 percent of the families would have 20 percent of the total income (point *A*); the poorest 40 percent of the families would have 40 percent of the income (point *B*); and 100 percent of the families would have 100

Figure 20-1 (a and b) A LORENZ CURVE OF CANADIAN INCOME, 1996

If income were perfectly equally distributed, the Lorenz curve would be a diagonal line. In (**b**) we see the Canadian Lorenz curve based on the numbers in (**a**) compared to a Lorenz curve reflecting a perfectly equal distribution of income.

Income quintile	Percentage of total family income	Cumulative percentage of total family income
Lowest fifth	6.1 %	6.1%
Second fifth	11.9	18.0
Third fifth	17.4	35.4
Fourth fifth	24.0	59.4
Highest fifth	40.6	100.0

(a) Income shares

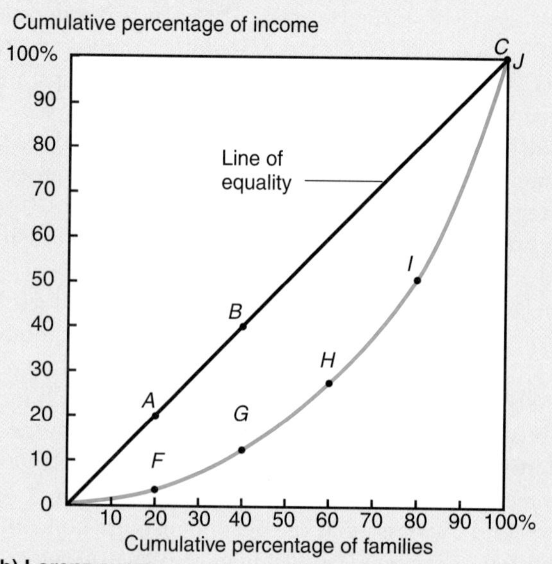

(b) Lorenz curve.

Source: Canadian Council on Social Development (www.ccsd.ca).

percent of the income (point C). An unequal distribution of income is represented by a Lorenz curve that lies below the diagonal line. All real-world Lorenz curves lie below the diagonal because income is distributed unequally everywhere in the real world.

The coloured line in Figure 20-1(b) represents a Lorenz curve of the Canadian income distribution (before taxes and transfers) presented in Figure 20-1(a)'s table. From Figure 20-1(a) you know that, in 1996, the bottom 20 percent of the families received 6.1 percent of the income. Point *F* in Figure 20-1(b) represents that combination of percentages (20 percent and 6.1 percent). To find what the bottom 40 percent received, we must add the income percentage of the bottom 20 percent and the income percentage of the next 20 percent. Doing so gives us 18.0 percent (6.1 plus 11.9 percent from column 2 of Figure 20-1(a)). Point G in Figure 20-1(b) represents the combination of percentages (40 percent and 18.0 percent). Continuing this process for points *H*, *I*, and *J*, you get a Lorenz curve that shows the share distribution of income in Canada in 1996.

Canadian Income Distribution over Time

Since the 1970s, Canada's highest quintile has been gaining income share at the expense of the lowest three quintiles. Through the turbulent decade of the 1970s, the opulent 1980s, the recession of 1991-92, and the slow-growing 1990s, Canada's highest income families gained a 2.3% share of income, while the middle quintile of Canadian families lost 1%, the second lowest quintile lost 1.1%, and the lowest quintile lost .2% of their income share. Figure 20-2 illustrates the transfer of income share from lower income Canadians to Canada's income elite, and the Lorenz Curves which demonstrate that this transfer of income has made the Canadian income distribution less equal. (The curve moves away from the line of equality.)

Figure 20-2 (a and b) CANADA'S CHANGING INCOME DISTRIBUTION

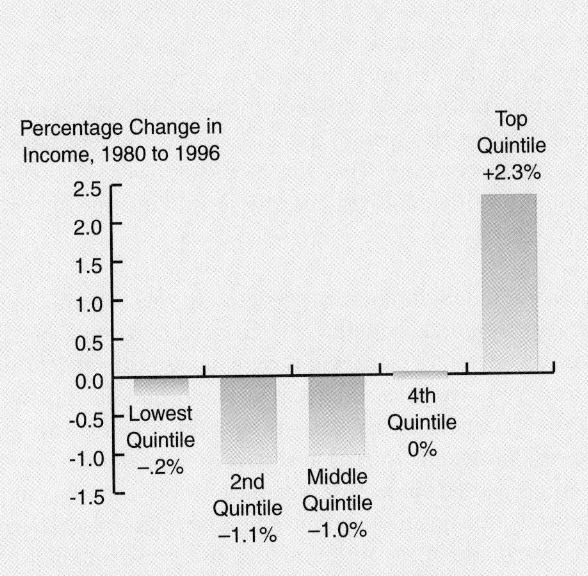

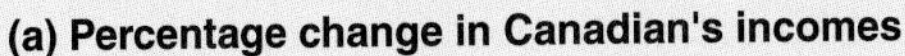

(a) Percentage change in Canadian's incomes

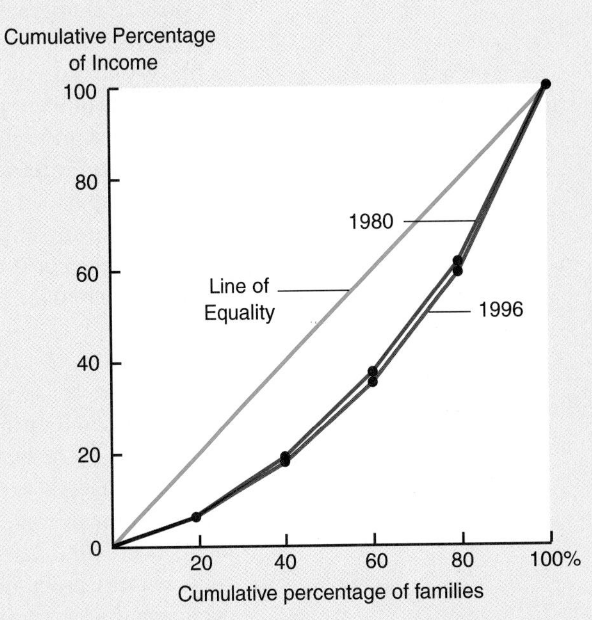

(b) Lorenz Curve

Despite Canada's economic prosperity, the distribution of income within Canada is unequal, particularly relative to other industrialized countries. Why are Canada's economic accomplishments not being shared evenly by all of its citizens? There are a number of reasons.

Economic Restructuring During the 1980s and 1990s, tremendous change occurred in the Canadian economy. Industry faced massive technological change and formidable new international competition, and consequently focussed on developing what was termed the "flexible work force," which, for many workers, meant nothing more than a move to part-time status, and a corresponding loss of job security. Companies downsized to become more competitive; they needed to be "leaner and meaner" in order to survive. In the 1990s, the information technology revolution began infiltrating Canadian industries in significant ways, again imposing change on the economic landscape. The service sector grew rapidly, drawing many workers into jobs characterized by low pay and low job security, which tended to be part-time rather than full-time. The nature of the firm evolved to embrace the changes in technology—outsourcing emerged, which enabled firms to access the best production methods in each field; multinational production strategies developed to capture the cheapest and best production techniques worldwide; and knowledge and information became valuable commodities.

Economic restructuring of the Canadian economy in the 1980s and 1990s reduced Canadian income distribution equality.

But some workers and some industries got left behind. Unable to participate in the technology revolution, they were not able to reap the benefits which sprang from the economic growth. The labour market became polarized, as jobs became identified as "good jobs" or "bad jobs." The 1970s, 1980s, and the 1990s all saw an increase in nonstandard employment—part-time, overtime, subcontracting, self-employment, and work-at-home—much of which involved nonpermanent positions. These decades were also characterized by relatively high levels of unemployment. Since most people's income results from wage earnings from a job, many of these people fell behind.

Household Restructuring Partly in response to changes in the economy, and partly in response to changes in social views and customs, households also underwent a restructuring process during the 1970s, 1980s, and 1990s. Dual income families became the norm. With relatively high inflation, two incomes were required in order to provide the "necessities" of life for the family, which have always included such items as soccer and piano lessons and schooling, but which also began to include such new amenities as computers and Internet access. Family formation and childbearing were delayed, as men and women pursued careers. These decades also saw an increase in singe-parent families. A single-income household cannot compete with the earning power of a two-income family. For these families, and for single individuals, the second income option is just not available.

Restructuring of Government Benefits Unemployment negatively affects family incomes, and reductions in employment insurance benefits and in social assistance (welfare) means that a period of prolonged unemployment will have a devastating effect on an individual's or family's long-term opportunities. Many lower and middle income households faced two injuries to their economic survival—their paid labour earnings shrank (through wage cuts or periods of unemployment) and government transfer payments decreased. The poorest families gained slightly as a result of more generous income security programs. For this lowest income group, government transfers offset their declining share of wage income. However, families in the middle and lower-middle income groups (second and third quintiles) faced increases in personal income taxes, and government transfers, while a critical component of these families' total income, did

Changes in taxes and government support programs have had a net negative impact on lower-middle and middle income (2nd and 3rd quintiles) Canadians.

Figure 20-3 SOURCES OF HOUSEHOLD INCOME BY QUINTILE

For households in the top four quintiles, market income represented more than 60% of total income, however, for the lowest quintile, market income represented roughly 10% of total income. Note that households in all quintiles received government transfers.

Source: Urban Poverty in Canada (www.ccsd.ca)

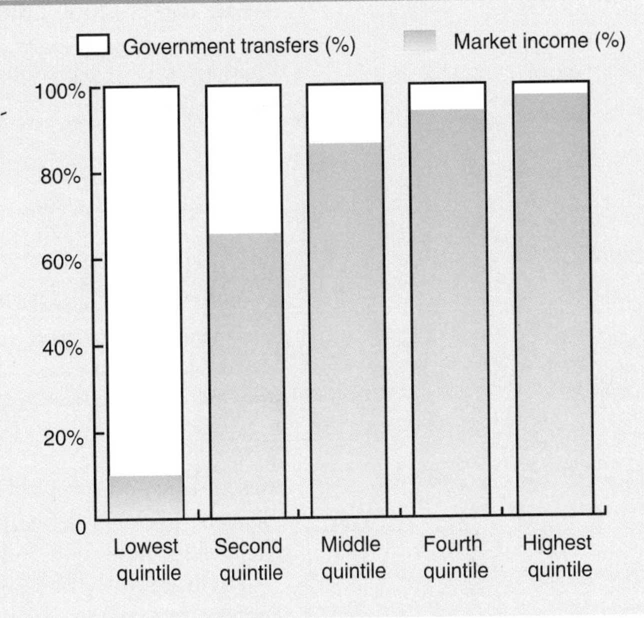

not rise sufficiently to compensate for the income lost as a result of higher taxes. They also saw less full-time and more part-time work, and less permanent and more temporary work. Figure 20-3 outlines Canadians' sources of income according to quintile.

DEFINING POVERTY

Much of the government's concern with income distribution has centered on the poorest group—those in poverty. Defining poverty is not easy. Do we want to define it as an absolute amount of real income? If poverty were defined as an *absolute* amount of real income, few in Canada would be in poverty today; most of today's poor have higher real incomes than the middle class had 50 or 60 years ago. Or do we want to define it as a *relative* concept that rises as the average income in the society rises? For example, anyone with an income of less than one-fifth of the average income could be defined as being in poverty. If that relative concept of poverty were chosen, then the proportion of people classified as poor would always be the same.

Defining Poverty

Canada does not have an official poverty line; however, the Low Income Cut-Off line (LICO), developed by Statistics Canada, is used to identify individuals and families living in poverty in Canada. The LICO is set as the income at which families spend 20 percent more of their income on food, shelter, and clothing than does the average Canadian family. To calculate income, Statistics Canada includes wages and salaries before deductions, net income from self-employment, investment income, pensions, alimony, and transfer payments such as Employment Insurance payments and refundable tax credits. Excluded from the income calculation are items such as capital gains, gambling income, inheritances, and free meals or produce from a home garden. Each year, Statistics Canada calculates the percentage of gross income spent on essentials—roughly 35 percent today—and adds 20 percent, so the LICO is the income where a

Q-2 Is the Canadian definition of poverty an absolute or a relative definition?

Although Canada has no official definition of poverty, the LICO is used as Canada's working definition of poverty.

The LICO is calculated as the percentage of the average family's income spent on food, shelter, and clothing, plus 20 percent.

TABLE 20-1 2000 Low Income Cut Offs

Number of Persons in Household	Community Population				
	>500,000	100,000 – 499,999	30,000 – 99,999	<30,000	Rural
1	$18,189	$15,600	$15,491	$14,414	$12,569
2	22,734	19,500	19,364	18,017	15,711
3	28,275	24,252	24,082	22,408	19,540
4	34,226	29,356	29,152	27,127	23,653
5	38,258	32,815	32,588	30,323	26,440
6	42,291	36,275	36,022	33,517	29,228
7 or more	46,324	39,735	39,457	36,713	32,015

Source: *The Canadian Fact Book on Poverty,* 2000 (www.ccsd.ca).

household spends roughly 55 percent of its income on food, shelter, and clothing. In 2000, the LICO for a family of three[1] living in a major metropolitan area was $28,275. The average household income in Canada was $61,014, which is 2.16 times the LICO.

There are thirty-five Low Income Cut Offs, varying by household size and size of the community in which the household resides. There are seven different household sizes, ranging from one person to seven or more persons, and five different community sizes, ranging from rural areas to cities with a population greater than 500,000. Table 20-1 contains all the LICOs for 2000.

Poverty can be measured in absolute or relative terms.

There are two ways to describe poverty. Poverty can be defined in absolute terms; for example, the **poverty line** represents *the minimum income required to purchase the basic necessities*. Anyone earning less than this amount of money is considered to be living in poverty. The main problem with this measure is how to define "necessities." The LICO defines poverty in relative terms, where the poverty line is set where income is "low" compared to average living standard in the community. Over the years, as Canada's standard of living has improved, adjustments to the LICO have reflected this, as a progressively smaller and smaller proportion of income is considered to be devoted to necessities.

The LICO measures poverty relative to the average Canadian household. Although Canadian living standards are much higher than in many countries around the world, we must look at "necessities" in relative terms. What would be considered necessities in an average Canadian household may be luxuries owned by only the richest families in some parts of the world.

Interestingly, the United States is one of the few industrialized countries which has an official poverty definition. It was developed in 1964, and is based on the cost of purchasing the food required for a minimum adequate diet. It was determined that a family would require an income three times greater than the cost of food in order to buy other necessities. Since 1964, the amounts have been adjusted for cost of living, but not for the standard of living, which, particularly in the U.S., has improved considerably since 1964.

The Costs of Poverty

Why is income inequality an issue? One answer is that society suffers when some of its people are in poverty, just as the entire family suffers when one member doesn't have enough to eat. Most people derive pleasure from knowing that others are not in poverty.

[1]The average Canadian family has 3.06 members.

Poverty brings significant costs to society. Poverty increases incentives for crime. In contrast, as people's incomes increase, they have more to lose by committing crimes, and therefore fewer crimes are committed. As the economy has boomed in the 1990s the crime rate has decreased.

Those who favour equality of income argue that the increased poverty in the late 1970s and 1980s represents a failure of the economic policies of that period. Others respond that the widening gap between rich and poor is not the result of government tax and spending policies. It has more to do with demographic changes. For example, the number of single-parent families increased dramatically during this period, while rapid growth of the labour force depressed wages for young unskilled workers.

Advocates of reducing poverty respond that this argument is unconvincing. They argue that the tax cuts of the 1990s favoured the rich, while decreased funding for government programs hurt the poor. To compensate, they argue, free day care should be provided for children so that single parents can work full-time, and government should supplement the low wages of the working poor. They argue that demographic changes are not a valid excuse for ducking a question of morality.

Poverty Issues

Expectations We all expect to live comfortably and to be able to afford to raise our children under decent circumstances. The home we live in should be large enough to accommodate a growing family, the neighbourhood should be safe, and our children should enjoy some of the amenities of modern life—toys, clothes, computer games, and a good education—beyond "food on the table." We want our kids to grow up healthy and well-adjusted, so they can face their challenges squarely. By giving kids a proper start in life, society can avoid higher social costs later, in areas such as health care and prison. People living below the poverty line, however, are less able to provide their children with a good start in life. The children struggle while they are still young, and they grow up with diminished expectations and facing fewer options to succeed. Education is often beyond family finances, so children of poor families many times do not proceed beyond high school. Obtaining a post-secondary education, while not a guarantee against poverty, is an important factor in avoiding the lowest quintile.

Lowered expectations are one of the costs of poverty.

Social Distress Poverty makes its presence felt in many different ways. Family problems, anxiety, alcoholism, crime, and work and school absenteeism can all result from the financial pressures of trying to survive on an inadequate income. Children living in poverty are disadvantaged academically, suffer poorer health than their wealthier friends and are a greater draw on the health care system, and see themselves facing the same limited opportunities as their parents.

In addition, poverty is often geographically concentrated, so the symptoms of poverty are compounded. Recreation centres and schools located in poor neighbourhoods tend to be less well equipped than those in richer neighbourhoods. The economic inequality often leads to social inequality. Neighbourhoods go into decline, and the effects of poverty deepen and spread. Table 20-2 (on the next page) illustrates that there is a strong geographic component to poverty.

Economic inequallity often leads to social inequality.

Differential Impacts Certain groups of people are more likely to find themselves below the poverty line. Aboriginals, recent immigrants, visible minorities, persons with disabilities, elderly women, youth, and single-parent families have higher incidences of poverty.

TABLE 20-2 A Sample of Canada's Neighbourhoods

City	Poor Neighbourhoods	Total Neighbourhoods	Proportion
St. John's	6	41	14.6%
Montreal	228	755	30.2%
Ottawa/Hull	28	213	13.1%
Kitchener	0	80	0%
Toronto	101	777	13.0%
Saskatoon	6	49	12.2%
Calgary	8	153	5.2%
Vancouver	20	297	6.7%

Source: *The Canadian Fact Book on Poverty*, 2000 (www.ccsd.ca).

Education and occupational skills are inversely related to poverty.

Education Reduced access to education remains a significant factor in poverty, and lack of education forms a sizable barrier against earning a good income. Education leads directly to enhanced job prospects and opportunities, but parents may be unable to financially help support their children's attempt to gain an education, and to break the cycle of poverty. There is a strong negative correlation between both education level and occupational skill level (human capital) and poverty.

20.1 
see page 455

Structural Component Although not a guarantee—there are no guarantees in life—good education, good job skills, and full-time employment tend to be critical factors in avoiding poverty. Poverty isn't just "had a bad year." Poverty tends to have a strong structural component, especially when geographically concentrated. Sporadic employment patterns become the norm, individuals' attachments to the labour force become less permanent, education is undervalued, future prospects are dismal, and there is prolonged reliance on welfare. People may be drawn into violence and crime as they try to dissipate financial stresses and discouragement arising from the lack of opportunity. Structural processes work against success. Joblessness results in social isolation, deterioration of skills, and lost access to the job information network. Poverty also contains a strong continuity factor—the first generation's success is strongly correlated with the second generation's success. To break the cycle of poverty, we need to look at why certain families find themselves living below the poverty line, and what reasonably can be done to help them improve their situations, for themselves and for their children.

INTERNATIONAL DIMENSIONS OF INCOME INEQUALITY

When considering income distribution, we usually are looking at conditions within a single country. There are other ways to look at income. We might judge income inequality in Canada relative to income inequality in other countries. Is the Canadian distribution of income more or less equal than another country's? We could also look at how income is distributed among countries. Even if income is relatively equally distributed within countries, it may be unequally distributed among countries.

Canada has less income inequality than most developing countries but more income inequality than many developed countries.

Comparing the Canadian Income Distribution with That in Other Countries

Figure 20-4 gives us a sense of how the distribution of income in Canada compares to that in other countries. We see that Canada has significantly more income inequality

APPLYING THE TOOLS

The Gini Coefficient

A second measure economists use to talk about the degree of income inequality is the Gini coefficient of inequality. The Gini coefficient is derived from the Lorenz curve by comparing the area between the Lorenz curve and the diagonal (area *A*) and the total area of the triangle below the diagonal (areas *A* and *B*). That is:

$$\text{Gini coefficient} = \frac{\text{Area } A}{(\text{Areas } A + B)}$$

A Gini coefficient of zero would be perfect equality, since area *A* is zero if income is perfectly equally distributed. The highest the Gini coefficient can go is 1. So all Gini coefficients must be between zero and one. The lower the Gini coefficient, the closer income distribution is to being equal. The Gini coefficient for Canada was 0.315 in 1998.

The following table gives Gini coefficients for a number of other countries. The Gini coefficients for transitional economies such as the Slovak Republic are most likely higher today because they are now market economies and their incomes are less equally distributed.

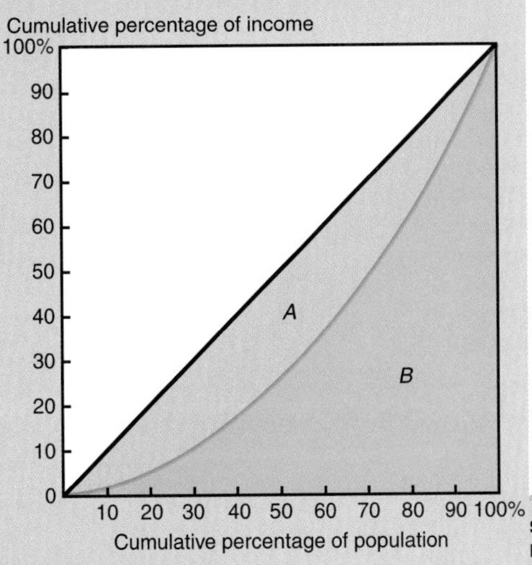

Gini Coefficients for Selected Countries

Country	Gini coefficient
Algeria	.353
Bangladesh	.283
Brazil	.601
Canada	.315
Czech Republic	.266
Denmark	.247
Germany	.281
Guatemala	.596
Hungary	.279
Indonesia	.342
Japan	.350
Latvia	.270
Netherlands	.315
Panama	.568
Philippines	.429
Romania	.255
Slovak Rep.	.195
South Africa	.584
Thailand	.462
United Kingdom	.326
United States	.460

Source: *World Development Report,* 1998/99, The World Bank (http://www. worldbank.org).

Figure 20-4 **CANADIAN INCOME DISTRIBUTION COMPARED TO THAT OF OTHER COUNTRIES**

Among countries of the world, Canada has neither the most equal nor the most unequal distribution of income.

Source: *World Development Report,* The World Bank, 1998/1999 (http://www.worldbank.org).

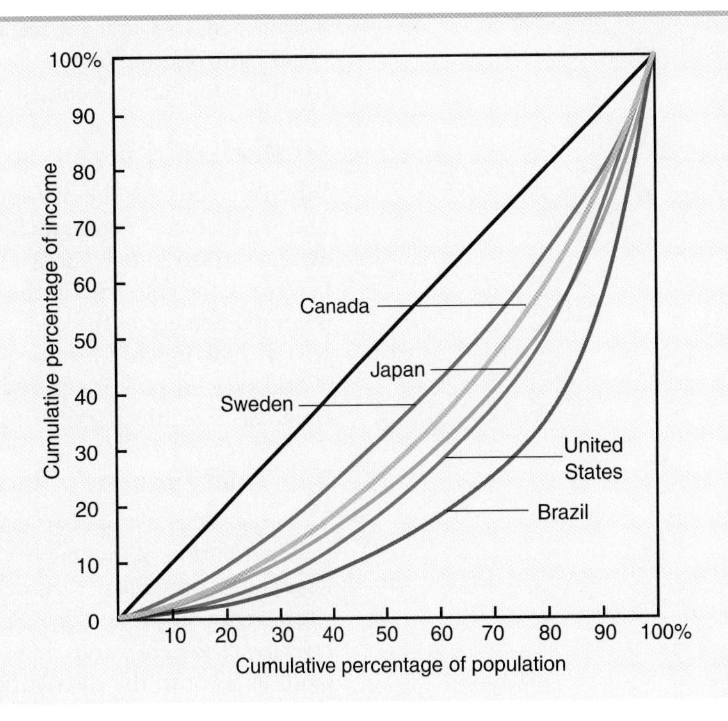

TABLE 20-3 Income Distribution for Selected Countries

Income inequality is reflected in the share of income received by the highest and lowest income segments of the population. As you can see, there are tremendous differences in distribution amongst countries.

Country	Percentage Share of Income			
	Lowest 10%	Lowest 20%	Highest 20%	Highest 10%
Brazil	1.0	2.6	63.0	46.7
Egypt	4.4	9.8	39.0	25.0
Guatemala	1.6	3.8	60.6	46.0
Hungary	4.1	10.0	34.4	20.5
India	3.5	8.1	46.1	33.5
Ireland	2.5	6.7	42.9	27.4
Malaysia	1.7	4.4	54.3	38.4
Mexico	1.6	4.0	56.7	41.1
Morocco	2.6	6.5	46.6	30.9
Nigeria	1.6	4.4	53.3	35.4
Norway	4.1	9.7	35.8	21.8
Russia	1.7	4.4	53.7	38.7

Source: *2001 World Development Indicators* (www.worldbank.org).

than Sweden, but significantly less than Brazil (or most other developing and newly industrialized countries).

Q-3 How does the income distribution in Canada compare with that in other countries?

An important reason why Canada has more income inequality than Sweden is that Sweden's tax system is more progressive. Until recently (when Sweden's socialist party lost power), the top marginal tax rate on the highest incomes in Sweden was 80 percent, compared to about 45 percent in Canada. Given this difference, it isn't surprising that Sweden has less income inequality. In a newly industrialized country like Brazil, where a few individuals, who own most of the businesses and the land, earn most of the income and, to a large degree, control the government, the government is not likely to begin redistributing income to achieve equality. Table 20-3 highlights the income distributions for various countries.

Income Distribution among Countries

When we consider the distribution of world income, the picture becomes even more unequal than the picture we see within countries. The reason is clear: Income is highly unequally distributed among countries. The average per capita income of the richest 5 percent of the countries in the world is more than 100 times the average income of the poorest 5 percent of the countries of the world. Thus, a Lorenz curve of world income would show much more inequality than the Lorenz curve for a particular country. Worldwide, income inequality is enormous. A minimum level of income in Canada would be a wealthy person's income in a poor country like Bangladesh.

The Total Amount of Income in Various Countries

To gain a better picture of income distribution problems, you need to consider not only the distribution of income but also the total amounts of income in various countries. Figure 20-5 presents per capita income (gross domestic product) for various countries, while Figure 20-6 (on page 444) shows the overall distribution of income worldwide. Looking at the enormous differences of income among countries, we must ask which is more important: the distribution of income or the absolute level of income? Which would you rather be: one of four members in a family that has an income of $3,000 a

Figure 20-5 PER CAPITA INCOME (GROSS DOMESTIC PRODUCT) IN VARIOUS COUNTRIES, 1999

Income is unequally distributed among the countries of the world. These relative comparisons change considerably over time as exchange rates fluctuate.

Source: CIA *World Factbook*, Central Intelligence Agency, 1999 (http://www.ocdi.gov) and various country home pages.

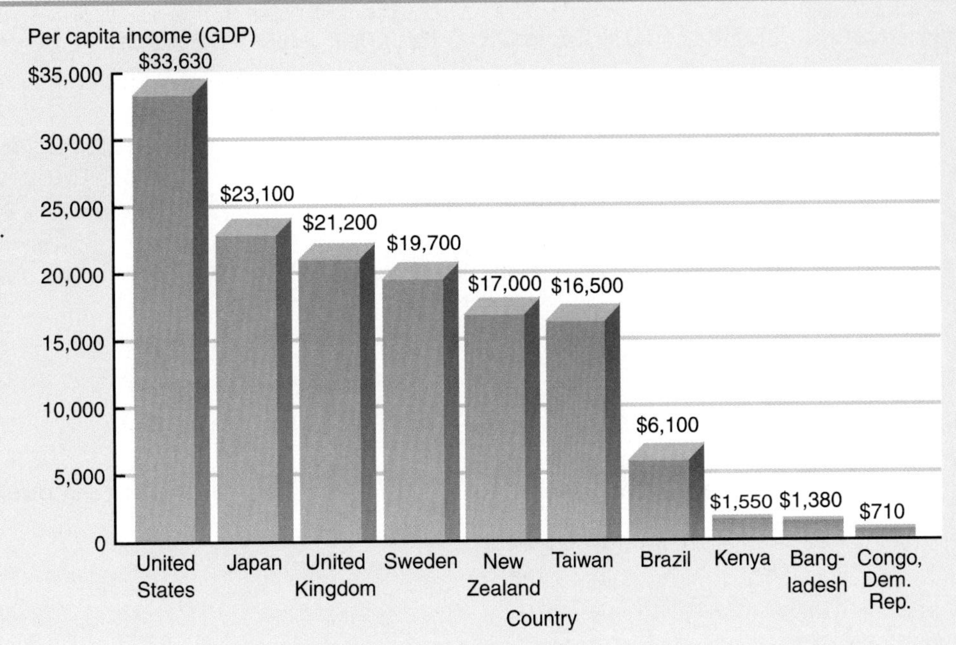

year, which places you in the top 10 percent of Bangladesh's income distribution, or one of four members of a family with an income of $12,000 (four times as much), which places you in the bottom 10 percent of the income earners in Canada?

THE DISTRIBUTION OF WEALTH

In considering equality, two measures are often used: *equality of wealth* and *equality of income*. **Wealth** is *the value of the things individuals own less the value of what they owe*. It is a *stock* concept representing the value of assets such as houses, buildings, and machines. For example, a farmer who owns a farm with a net worth of $1 million is wealthy compared to an investment banker with a net worth of $225,000.

Income is a flow; wealth is a stock.

Income is *payments received plus or minus changes in value in a person's assets in a specified time period*. In contrast to wealth, income is a *flow* concept. It's a stream through time. That farmer might have an income of $20,000 a year while the investment banker might have an income of $80,000 a year. The farmer, with $1 million worth of assets, is wealthier than the investment banker, but the investment banker has a higher income than the farmer.

Wealth is significantly more unequally distributed than income.

In Canada wealth is much more unequally distributed than income. One reason why income distribution is unequal in Canada is because wealth distribution is unequal. A wealthy individual often derives income—in the form of rent, interest, dividends, and profits—from his or her wealth, so unequal wealth distribution will tend to encourage unequal income distribution. Most of us have little chance of becoming one of the top 5 percent of the wealthholders in Canada. Once there was a time when people's ultimate financial goal was to be a millionaire. In the 2000s, the ultimate financial goal for the wealthiest people is to be a billionaire. The millionaire's club is no longer an exclusive club.

Of course, people in the club don't always stay there; the club is constantly changing. For example, a number of families who were in the club earlier are no longer in it.

Figure 20-6 **DISTRIBUTION OF WORLD INCOME AND POPULATION**

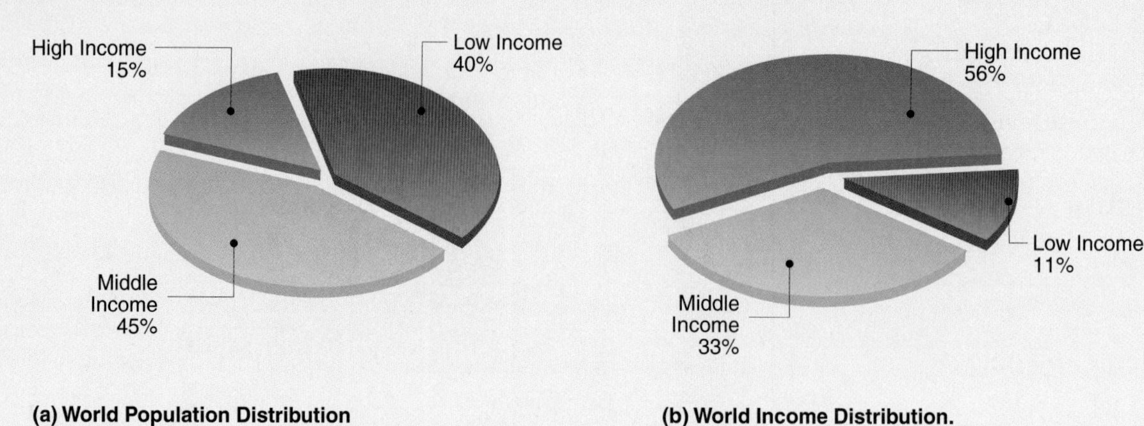

Source: *2001 World Development Indicators.* (www.worldbank.org)

Billionaires often lose a million here, gain a million there; sometimes they even become multibillionaires. Seldom do they become poor.

Many of the Japanese billionaires lost billions in the fall of the Japanese stock market, and fell off the top 10 in the world list of wealthiest people. Today, some of these people and families might only be multimillionaires. (It's tough to stay a billionaire.)

SOCIOECONOMIC DIMENSIONS OF INCOME INEQUALITY

The size distribution of inequality is only one of the dimensions that inequality of income and wealth can take. As mentioned before, the distribution of income according to source of income (wages, rents, and profits) was once considered important. Today's focus is on the distribution of income based on geographic region and other socioeconomic factors such as gender, education, and type of job.

Income Distribution According to Socioeconomic Characteristics

Table 20-4 gives an idea of the distribution of income according to socioeconomic characteristics. You can see that low levels of education and training, and certain individual characteristics are associated with very low incomes.

Income Distribution According to Class

Early economists focussed on the distribution of income by wages, profits, and rent because that division corresponded to their class analysis of society. Landowners received rent, capitalists received profit, and workers received wages. Tensions among these classes played an important part in economists' analyses of the economy and policy.

Canada has socioeconomic classes with some mobility among classes. This is not to say such classes should exist; it is only to say that they do exist.

Even though class divisions by income source have become blurred, other types of socioeconomic classes have taken their place. Class divisions are no longer determined by income source. For example, upper-class people do not necessarily receive their income from rent and profits. Today we have "upper-class" people who derive their

TABLE 20-4 Poverty Rate by Socioeconomic Variable.

Variable	Poverty Rate
EDUCATION	
Post-Secondary	15.9
High School	23.0
Less than High School	29.6
EMPLOYMENT	
Full-year full-time	7.5
Full-year part-time	18.7
Part-year part-time	25.4
No employment	46.2
OCCUPATIONAL SKILL	
High Skill (e.g., managers, professionals, technicians, supervisors, tradespeople)	11.0
Moderate Skill (e.g., clerical, sales and service, semiskilled manual workers)	18.2
Low Skill (e.g., sales and service, other manual workers)	25.9
INDIVIDUAL CHARACTERISTICS	
Aboriginals	55.6
Persons with Disabilities	36.1
Visible Minorities	37.6
Single-parent families	59.2
Unattached men	42.6
Unattached women	47.5
AGE	
Youth (age15-24)	30.7
Seniors (age 65 and over)	25.0
IMMIGRANTS	
Pre-1986	19.7
1986-1990	35.1
1991-1996	52.1

Source: *Urban Poverty in Canada.* (www.ccsd.ca)

income from wages and "lower-class" people who derive their income from profits (usually in the form of pensions, which depend on profits from the investment of pension funds in stocks and bonds).

The Importance of the Middle Class What has made the most difference in today's class structure in Canada compared to its class structure in earlier periods and to the structure in today's developing countries is the tremendous growth in the relative size of the middle class. The class structure used to be a pyramid. From a base composed of a large lower class, the pyramid tapered upward through a medium-sized middle class to a peak occupied by the upper class (Figure 20-7(a)). The class structure is still pyramidal in developing countries. In Canada and other developed countries, the pyramid has bulged out into a diamond, as shown in Figure 20-7(b). The middle class has become the largest class, while the upper and lower classes are smaller in relative terms.

In Canada the middle class is the largest class.

This enormous increase in the relative size of the middle class in developed countries has significantly blurred the distinction between capitalists and workers. In early capitalist society, the distributional dispute (the fight over relative income shares) was

Q-4 How have distributional disputes about income changed over time?

Figure 20-7 (a and b) THE CLASS SYSTEM AS A PYRAMID AND AS A DIAMOND

The class system in developing countries is a pyramid; in Canada the class system is more diamond-shaped.

Upper class
Middle class
Lower class

(a) Developing country's class system

Upper class
Middle class
Lower class

(b) Canadian class system

largely between workers and capitalists. In modern capitalist society, the distributional concerns are among various types of individuals. Union workers versus nonunion workers; salaried workers versus workers paid by the hour; old versus young; women versus men.

Distributional Questions and Tensions in Society While economists tend to focus on the size distribution of income, others tend to emphasize class and group structures in their analysis. They emphasize the control that the upper class has over the decision process and the political process. Conservative economists emphasize the role of special interests of all types in shaping government policy. Both socialist and conservative analyses bring out the tensions among classes in society much better than does the mainstream economic analysis.

Both socialist and conservative analyses bring out the tensions among classes in society much better than does mainstream economic analysis.

When people feel they belong to a particular class or group, they will often work to further the interests of that class or group. They also generally have stronger feelings about inequalities among classes or groups than when they lack that sense of class or group identity. Using a classless analysis means overlooking the implications of class and group solidarity in affecting the tensions in society.

20.2

see page 455

Those tensions show up every day in political disputes over the tax system and in the quiet fuming of individuals as they see someone else earning more for doing the same job. Such tensions exist in all countries. In some transitional and developing countries they break out into the open as political demonstrations or armed insurrections.

Those tensions have been kept to a minimum in Canadian society. A majority of Canadians believe that the income distribution is sufficiently fair for them to accept their share more or less contentedly. To remedy the unfairness that does exist, they don't demand that the entire system be replaced. Instead they work for change within the present system. They look to affirmative action laws, pay equity laws, minimum wage laws, and social welfare programs for any improvement they perceive to be necessary or desirable. There's much debate about whether these government actions have achieved the desired ends, but the process itself reduces tensions and has worked toward the maintenance of the entire system.

People's acceptance of the Canadian economic system is based not only on what the distribution of income is but also on what people think it should be, what they consider fair. It is to that question that we now turn.

INCOME DISTRIBUTION AND FAIRNESS

Judgments about whether the distribution of income is fair or should be changed are normative ones, based on the values the analyst applies to the situation. Value judgments necessarily underlie all policy prescriptions.

Philosophical Debates about Equality and Fairness

Depending on one's values, any income distribution can be justified. For example, Friedrich Nietzsche, the 19th-century Germany philosopher, argued that society's goal should be to support its supermen—its best and brightest. Lesser individuals' duty should be to work for the well-being of these supermen. Bertram de Juvenal, a 20th-century philosopher, has argued that a high level of income inequality is necessary to sustain the arts, beauty, education, and civilization. Wealth is required so that some people can own beautiful paintings and attend expensive opera productions; these valuable products cannot be supported by only average incomes. These philosophers argue that our lives are improved because some people do own such works of art and because opera performances exist. Inequality is necessary for these exquisite things to exist, and the fact they exist enriches the lives of everyone.

Q-5 Is it self-evident that greater equality of income would make the society a better place to live? Why?

Other philosophers disagree strongly. They argue that equality itself is the overriding goal. Canadians strongly believe that everyone should have a fair chance to develop their talents and to benefit from the advantages of the Canadian economy. This belief is inherent in Canada's many social programs.

Believing that equality is an overriding goal does not necessarily imply that income should be equally distributed. For example, John Rawls (a Harvard University professor who believes that equality is highly desirable and that society's goal should be to maximize the welfare of the least well-off) agrees that to meet that goal some inequality is necessary. Rawls argues that if, in pursuing equality, you actually make the least well-off worse off than they otherwise would have been, then you should not pursue equality any further. For example, say under one policy there would be perfect equality and everyone would receive $10,000 per year. Under another policy, the least well-off person receives $12,000 per year and all others receive $40,000. Rawls argues that the second policy is preferable to the first even though it involves more inequality.

While economists in their objective role limit themselves to explaining the effects of various policies on the distribution of income, individuals, in order to judge economic policies, must make value judgments and concern themselves with the distributional effects of economic policy.

Economists explain the effects that various policies will have on the distribution of income and they let the policymakers, who represent society's wishes, judge whether those effects are desirable.

Fairness and Equality

Canadians have a strong tendency to favour equality—equality is generally seen as fair. However, in some instances equality of income is not directly related to people's view of fairness. For example, consider this distribution of income between John and Fred:

John gets $50,000 a year.

Fred gets $12,000 a year.

Think a minute. Is that fair?

Q-6 You are dividing a pie among five individuals. What would be a fair distribution of that pie?

Here's some more information. Say that John gets that $50,000 for holding down three jobs at a time, while Fred gets his $12,000 for sitting around doing nothing. At this point, many of us would argue that it's possible John should be getting even more than $50,000 and Fred should be getting less than $12,000.

Fairness has many dimensions and it is often difficult to say what is fair and what isn't.

Three problems in determining whether an equal income distribution is fair are:

1. People don't start from equivalent positions.
2. People's needs differ.
3. People's efforts differ.

But wait! What if we discover that Fred is an invalid and unless his income increases to $15,000 a year he will die? Most of us would change our minds again and argue that Fred deserves more, regardless of how much John works.

But wait! How about if, after further digging, we discover that Fred is an invalid because he squandered his health on alcohol, drugs, and fried foods? In that case some people would likely change their minds again as to whether Fred deserves more.

Looking only at a person's income masks many dimensions that most people consider important in making value judgments about fairness.

Fairness as Equality of Opportunity

The concept of fairness is crucial and complicated, and it deserves deeper consideration than just a gut reaction.

When most people talk about believing in equality in income, they mean they believe in equality of opportunity for comparably endowed individuals to earn income. If equal opportunity of equals leads to inequality of income, that inequality in income is fair. Unfortunately, there's enormous latitude for debate on what constitutes equal opportunity of equals.

In the real world, needs differ, desires differ, and abilities differ. Should these differences be considered relevant differences in equality? You must answer that question before you can judge any economic policy, because to make a judgment on whether an economic policy should or should not be adopted, you must make a judgment about whether a policy's effect on income is fair. The concept of fairness is crucial and complicated, and it deserves deep consideration.

THE PROBLEMS OF REDISTRIBUTING INCOME

Let's now say that we have considered all the issues discussed so far in this chapter and have concluded that some redistribution of income from the rich to the poor is necessary if society is to meet our ideal of fairness. How do we go about redistributing income?

First, we must consider what programs exist and what their negative side effects might be. The side effects can be substantial and can subvert the intention of the program so that far less money is available overall for redistribution and inequality is reduced less than we might think.

Three Important Side Effects of Redistributive Programs

Three side effects of redistribution of income include:

1. The labour/leisure incentive effect.
2. The avoidance and evasion incentive effect.
3. The incentive effect to look more needy than you are.

Three important side effects that economists have found in programs to redistribute income are:

1. A tax may result in a switch from labour to leisure.
2. People may attempt to avoid or evade taxes, leading to a decrease in measured income.
3. Redistributing money may cause people to make themselves look as if they're more needy than they really are.

Q-7 When determining the effects of programs that redistribute income, can one reasonably assume that other things will remain equal?

All economists believe that people will change their behaviour in response to changes in taxation and income redistribution programs. These responses, called *incentive effects of taxation*, are important and must be taken into account in policymaking. Some economists believe that incentive effects are so significant that little taxation for redistribution purposes should take place. They argue that when the rich do well, the total pie is increased so much that the spillover benefits to the poor are greater than the proceeds they would get from redistribution. For example, supporters of this view argue that the growth in capitalist economies was made possible by entrepreneurs. Because those entrepreneurs invested in new technology, income in society grew. Moreover, those entrepreneurs paid taxes. The benefits resulting from entrepreneurial action spilled over to the poor, making the poor far better off than any redistribution would. The fact that some of those entrepreneurs became rich is irrelevant because all society was better off due to their actions.

History has shown that during periods of rapidly changing technology and rapid economic growth—for example, during the industrialization of North America or during the information technology revolution of the 1990s—income distribution shows that only some individuals are able to participate in the income growth. The economy of the early 1900s was dominated by large corporations and a few wealthy families; however, it is not clear that this type of concentrated economic power was necessary for the tremendous growth in production and incomes that occurred at that time. This is especially true during the more recent economic revolution of the 1990s.

Other economists believe that there should be significant taxation for redistribution. While they agree that sometimes the incentive effects are substantial, they see the benefits arising from equality overriding these effects. (Incentive effects are discussed in more detail in Chapter 21.)

Politics, Income Redistribution, and Fairness

We began this discussion of income distribution and fairness by making the assumption that our value judgments determine the taxes we pay—that if our values led us to the conclusion that the poor deserved more income, we could institute policies that would get more to the poor. Reality doesn't necessarily work that way. Often politics, not value judgments, play a central role in determining what taxes individuals will pay. The group that can deliver the most votes will elect lawmakers who will enact tax policies that benefit that group at the expense of groups with fewer votes.

Often politics, not value judgments, play a central role in determining what taxes an individual will pay.

On the surface, the democratic system of one person/one vote would seem to suggest that the politics of redistribution would favour the poor, but it doesn't. One would expect that the poor would use their votes to make sure income was redistributed to them from the rich. Why don't they? The answer is complicated.

One reason is that many people don't vote because they assume that one vote won't make much difference. Psychological factors may affect people's decisions. People like to associate with winners, so by voting for the winning political party, they may feel they belong to the ruling elite by virtue of having voted for the party, even though they may not benefit or may even be harmed by the policies enacted by the party. Furthermore, elections require financing. Much of that financing comes from the rich. The money is used for advertising and publicity aimed at convincing the poor that it's actually in their best interests to vote for a person who supports the rich. People can be misled by that kind of biased publicity. Unfortunately, reasonable-sounding arguments can be made to support just about any position.

INCOME REDISTRIBUTION POLICIES

The preceding discussion should have provided you with a general sense of the difficulty of redistributing income. Let's now consider briefly how income redistribution policies and programs have worked in the real world. In considering this, it is helpful to keep in mind that there are two direct methods and one indirect method through which government redistributes income. The direct methods are (1) *taxation* (policies that tax the

Direct methods of redistribution are taxation and expenditure programs.

rich more than the poor) and (2) *expenditures* (programs that help the poor more than the rich). The indirect method involves the establishment and protection of property rights. Let's first consider direct methods.

Taxation to Redistribute Income

The Canadian government gets its revenue from a variety of taxes. The largest source of revenue is the personal income tax, followed by the corporate income tax and various sales and other taxes.

Taxes can be progressive, proportional, or regressive.

Q-8 A progressive tax is preferable to a proportional tax. True or false? Why?

As stated in Chapter 7, tax systems can be progressive, proportional (sometimes called *flat rate*), or regressive. A **progressive tax** is one in which *the average tax rate increases with income*. It redistributes income from the rich to the poor. A **proportional tax** is one in which *the average rate of tax is constant regardless of income level*. It is neutral in regard to income distribution. A **regressive tax** is one in which *the average tax rate decreases as income increases*. It redistributes income from the poor to the rich.

Sales taxes are considered regressive because everyone pays the same rate, regardless of income. In addition, poor people spend a larger proportion of their total income and generally are not able to spend part of it in tax-free jurisdictions.

Governments get most of their income from the following sources:

1. Income taxes, which are generally somewhat progressive.
2. Sales taxes, which tend to be proportional (all people pay the same tax rate on what they spend) or slightly regressive. (Since poor people often spend a higher percentage of their incomes than rich people, poor people pay a higher average tax rate as a percentage of their incomes than rich people.)
3. Property taxes, which are taxes paid on the value of people's property (real estate). Since the value of people's property tends to be related to income, the property tax is considered to be roughly proportional.

When all the taxes paid by individuals to all levels of governments are combined, the conclusion that most researchers come to is that little income redistribution takes place on the tax side. The progressive taxes are offset by the regressive taxes, so the overall tax system is roughly proportional. That is, on average the tax rates individuals

KNOWING THE TOOLS

Federal Government's $100-billion 5-year Tax Reduction Plan.

On January 1, 2001, the federal government made changes to the Federal Income Tax system, resulting in an average income tax cut of 21%. A family of four with an income of $40,000 saves about $1100 in federal personal income tax, while a two-earner family of four earning $60,000 saves about $1000 in federal personal income tax.

Individuals earning an income of $30,000 enjoy approximately a 6% drop in their federal marginal income tax rate. Individuals earning an income of $60,000 see the 6% drop on their first $30,000, and enjoy an approximately 8% decrease in federal tax on their next $30,000. Finally, individuals earning an income of $90,000 benefit from the 6% decrease in federal tax on their first $30,000 and the 8% decrease on their next $31,000, plus see a decrease of approximately 10% in the tax rate applicable on the last $29,000 earned. In addition, the deficit reduction surtax levied on high income earners is eliminated. Other changes to the tax system include inflation indexation of the tax system, increases in both the Child Tax Benefit Credit and the Education Tax Credit for full- and part-time students, and a reduction in the capital gains inclusion rate, so that only half of capital gains are now taxable.

Income	Previous tax rate	New tax rate
First $30,000	17%	16%
$30,000 - $61,000	24%	22%
$61,000 - $100,000	29%	26%
> $100,000	29%	29%

pay are roughly equal. Recent changes in the tax laws have lowered the rate that most people pay.

Expenditure Programs to Redistribute Income

Taxation has not proved to be an effective means of redistributing income. However, the government expenditure system has been quite effective. Some of the federal government's expenditures that contribute to redistribution include the following:

Expenditure programs have been more successful than taxation for redistributing income.

Welfare. Welfare is a provincially administered program of income assistance to support individuals and families which have no other way of supporting themselves. If a person's living expenses exceed his or her financial resources, they are generally eligible to receive a monthly allowance which covers the cost of basic requirements, such as food, shelter, and clothing, as well as health care and employment-related expenses. Although the provinces manage their welfare programs, major funding comes from the federal government. Since April of 1996, the federal government contributes to provincial and territorial spending on health care, post-secondary education, and social services through the Canada Health and Social Transfer (CHST).

Old Age Security. Old Age Security (OAS) is Canada's largest public pension plan, providing a modest pension income to Canadians who have reached age 65. To qualify, a person must have been a resident of Canada for at least 10 years after having reached age 18. The maximum benefit—$440 per month in 2001—is payable to those who lived in Canada for at least 40 years after age 18; those pensioners who have less than 40 years' residency receive an amount equal to one fortieth of the full pension for every year they lived in Canada. The pension amount is adjusted for inflation four times a year. The pension is also taxable and this reduces the pension amount for people with incomes above approximately $55,000. As a result, 5 percent of seniors receive less than $440 per month, and 2 percent of seniors—those with net incomes in excess of approximately $90,000—lose the entire amount.

Guaranteed Income Supplement. The Guaranteed Income Supplement (GIS) provides an additional monthly benefit for low-income pensioners. This amount is not taxable, and can only be collected while residing in Canada. "Snowbirds"—Canadian seniors who spend their winters in warmer climates, such as Florida—lose the GIS if they remain out of Canada for longer than six months.

Canada Pension Plan. The Canada Pension Plan (CPP) is a retirement income plan and a death and disability insurance plan. Canadians who have paid into the plan receive a full pension starting at age 65; however, individuals may draw a reduced pension starting as early as age 60. In 2001, the maximum monthly pension was $775. CPP amounts are adjusted for inflation every January, and are taxable. Together, the OAS and the CPP form a significant portion of seniors' income. The strength of Canada's retirement system is that the risk is shared amongst individuals, employers, and the governments. In 1998, contribution rates to the CPP were increased to ensure sustainability of the pension plan. Each employee makes a tax deductible contribution on earnings between $3,500 and $38,300. The appropriate amount is calculated and deducted from paycheques; the employer makes an equal contribution. The CPP revenues are managed separately from other government revenues, and the CPP funds are used to pay benefits and administration costs, and to make investments on behalf of the program. Quebec runs a similar plan, the Quebec Pension Plan.

Q.9 The Canada Pension Plan is only a retirement system. True or false?

Employment Insurance **Employment insurance** is *short-term financial assistance, regardless of need, to eligible individuals who are temporarily out of work*. It is limited financial assistance to people who are out of work through no fault of their own and have worked in a covered occupation for a substantial number of weeks in the period just before they became unemployed. A person can't just quit a job and live on unemployment benefits. While receiving unemployment benefits, people are expected to actively search for work. The amount of the benefit is always considerably less than the amount the person earned when working; however, lower-income workers receive unemployment payments that are more nearly equal to their working wage than do higher-income workers.

How Successful Have Income Redistribution Programs Been?

Most government redistribution works through its expenditure programs, not through taxes.

Figure 20-8 shows Lorenz curves before and after taxes. As you can see, the after-tax income is more equally distributed. But because of the incentive effects of collecting and distributing the money, that redistribution has come at the cost of a reduction in the total amount of income earned by the society. The debate about whether the gain in equality of income is worth the cost in reduction of total income is likely to continue indefinitely.

While the direct methods of redistributing income get the most press and discussion, perhaps the most important redistribution decisions that the government makes involve an indirect method, the establishment and protection of property rights. Let's take an example: intellectual property rights. Intellectual property consists of things like a book you've written, a song you've composed, or a picture you've drawn. How these property rights are structured plays a fundamental role in determining the distribution of income.

In Peru, the key to economic prosperity is recognized as the development of enforceable property rights. Property rights give market value to the productive effort, which is the key to economic development. Property rights provide incentives for individuals to take risks to create and to produce, and to be rewarded for the risk taken.

Figure 20-8 DISTRIBUTION OF INCOME BEFORE AND AFTER TAXES

Some redistribution takes place through the tax system, making the after-tax distribution of income more equal than the before-tax distribution of income.

Source: *www.ccsd.ca*

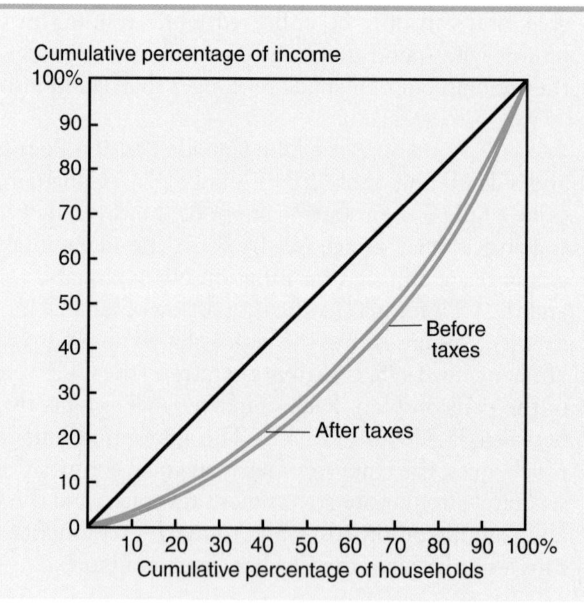

If strict private property rights are given for, say, a design for a computer screen (e.g., a neat little trash can in the corners and windows of various files) any user other than the designer herself will have to pay for the right to use it. The designer (or the person who gets the legal right to the design) becomes very rich. If no property rights are given for the design, then no payment is made and income is much more equally distributed. Of course, without a promise of high returns to designing a computer screen device, fewer resources will be invested in finding the ideal design. While most people agree that some incentive is appropriate, there is no consensus on whether the incentives embodied in our current property rights structure are too large. It is highly probable that the trash can (recycling bin) design, while ingenious, would have been arrived at with a much smaller incentive.

Q-10 Why are property rights important in the determination of whether any particular income distribution is fair?

The point of the above example is not that property rights in such ideas should not be given out. The point is that decisions on property rights issues have enormous distributional consequences that are often little discussed, even by economists. Ultimately, we can answer the question of whether income redistribution is fair only after we have answered the question of whether the initial property rights distribution is fair.

CONCLUSION

Income redistribution is an important but difficult question. Specifically, income distribution questions are integrally related to questions about the entire economic system. Supply and demand play a central role in the determination of the distribution of income, but they do so in an institutional and historical context. Thus, the analysis of income distribution must include that context as well as the analyst's ethical judgments about what is fair.

In Western economies, risk and entrepreneurial effort has been extremely well-rewarded. Enormous rewards have been won for braving relatively small risks. Bill Gates, born one generation earlier, probably would have been involved in the earliest stages of the technological revolution, where there was tremendous risk, along with great effort, brilliant insight, and modest monetary reward; born one generation later, he probably would have been designing Web pages!

Chapter Summary

- The Lorenz curve is a measure of the distribution of income among families in a country. The further the Lorenz curve is from the diagonal, the more unequally income is distributed.
- The LICO, Canada's unofficial definition of poverty, is a relative measure because it is set in relation to the average Canadian's standard of living.
- Income is less equally distributed in Canada than in some countries such as Sweden, but more equally distributed than in other countries such as Brazil. There is more income inequality among countries than income inequality within a country.
- Wealth is distributed less equally than income.
- Income differs substantially by class and by other socioeconomic characteristics such as age and gender.
- Fairness is a philosophical question. People must judge a program's fairness for themselves.
- Income is difficult to redistribute because of incentive effects of taxes, avoidance and evasion effects of taxes, and incentive effects of redistribution programs.
- On the whole, the Canadian tax system is roughly proportional, so it is not very effective as a means of redistributing income.
- Government spending programs are more effective than tax policy in reducing income inequality.

Key Terms

Canada Pension Plan *(451)*
Employment Insurance *(452)*
Guaranteed Income Supplement *(451)*
income *(434, 444)*
Lorenz curve *(434)*
Old Age Security *(451)*
poverty line *(438)*
progressive tax *(450)*
proportional tax *(450)*
regressive tax *(450)*
share distribution of income *(433)*
socioeconomic distribution of income *(433)*
wealth *(433)*

Questions for Thought and Review

1. Why are we concerned with the distribution of income between men and women, but not between redheads and blonds?
2. The Lorenz curve for Belarus looks like this:

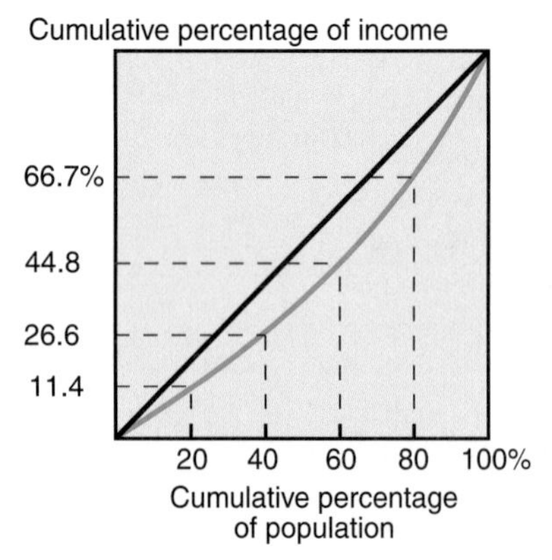

   How much income do the top 20 percent of individuals in Belarus receive?
3. If we were to draw a Lorenz curve for lawyers, what would it represent?
4. Should poverty be defined absolutely or relatively? Why?
5. Some economists argue that a class distinction should be made between managerial decision makers and other workers. Do you agree? Why or why not?
6. If a garbage collector earns more than an English teacher, does that mean something is wrong with the economy? Why or why not?
7. Is it ever appropriate for society to:
   Let someone starve?
   Let someone be homeless?
   Forbid someone to eat chocolate?
8. If you receive a paycheque, what percentage of it is withheld for taxes? What incentive effect does that have on your decision to work?
9. Which have been more successful in redistributing income—tax or expenditure programs?
10. There are many more poor people in Canada than there are rich people. If the poor wanted to, they could exercise their power to redistribute as much money as they please to themselves. They don't do that, so they must see the income distribution system as fair. Discuss.

Problems and Exercises

1. The accompanying table shows income distribution data for three countries.

	Percentage of Total Income		
Income quintile	India	Czech Republic	Mexico
Lowest 20%	8.5%	10.5%	4.1%
Second quintile	12.1	13.9	7.8
Third quintile	15.8	16.9	12.5
Fourth quintile	21.1	21.3	20.2
Highest 20%	42.5	37.4	55.4

   a. Using this information, draw a Lorenz curve for each country.
   b. Which country has the most equal distribution of income?
   c. Which country has the least equal?
   d. By looking at the three Lorenz curves, can you tell which country has the most progressive tax system? Why or why not?
2. "There are lies, damned lies, and statistics. Then, there are annual poverty figures." Both liberal and conservative economists believe Canadian poverty statistics are suspect. Here are some reasons:
   (1) They do not take into account such benefits as medical and dental coverage.
   (2) They do not consider regional cost-of-living differences.

(3) They do not take into account unreported income.
   a. What would the effect of correcting each of these be on measured poverty?
   b. Would making these changes be fair?
3. The dissident Russian writer A. Amalrik has written:

   > . . . The Russian people . . . have . . . one idea that appears positive: the idea of *justice* . . . In practice, "justice" involves the desire that "nobody should live better than I do" . . . The idea of justice is motivated by hatred of everything that is outstanding, which we make no effort to imitate but, on the contrary, try to bring down to our level, by hatred of any sense of initiative, of any higher or more dynamic way of life than the life we live ourselves.

   What implications would such a worldview have for the economy?
4. List four conditions you believe should hold before you would argue that two individuals should get the same amount of income.
   a. How would you apply the conditions to your views on welfare?
   b. How would you apply the conditions to your views on how progressive the income tax should be?
   c. If the income tax were made progressive in wage rates (tax rates increase as wage rates increase) rather than progressive in income, would your conditions be better met? Why?
5. In Taxland, a tax exemption is granted for the first $10,000 earned per year. Between $10,000.01 and $30,000, the tax rate is 25 percent. Between $30,000.01 and $50,000, it's 30 percent. Above $50,000, it's 35 percent. You're earning $75,000 a year.
   a. How much in taxes will you have to pay?
   b. What is your average tax rate? Your marginal tax rate?
   c. Taxland has just changed to a tax credit system in which, in lieu of any exemption, eligible individuals are given a cheque for $4,000. The two systems are designed to bring in the same amount of revenue. Would you favour or oppose the change? Why?
6. Some economists have proposed making the tax rate progressivity depend on the wage rate rather than the income level. Thus, an individual who works twice as long as another but who receives a lower wage would face a lower marginal tax rate.
   a. What effect would this change have on incentives to work?
   b. Would this system be fairer than our current system? Why or why not?
   c. If, simultaneously, the tax system were made regressive in hours worked so that individuals who work longer hours faced lower marginal tax rates, what effect would this change have on hours worked?
   d. What would be some of the administrative difficulties of instituting the above changes to our income tax code?

Web Questions

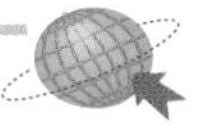

1. Go to the Web site: www.ccra-adcr.ca/tax/individuals.
   a. At what rate are capital gains income taxed?
   b. What is the alternative minimum tax?
   c. What are education credits?
2. The World Bank tracks the economic, social, and environmental state of the world. At their Web site, www.worldbank.org/wdr, find the *2001 World Development Report: Attacking Poverty*.
   a. What issues are highlighted in the 2001 World Development Report? What statistical information does the World Bank consider important according to its list of World Development Indicators?
   b. What economic policies drive economic growth?
   c. Peruse Tables 1 through 21 and 1a, which present the World Bank's indicators (variables). Choose two indicators that interest you, and write a 2-page report on the differences between some of the wealthier nations and some of the poorer nations with respect to the variables you chose.

Answers to Margin Questions

1. When drawing a Lorenz curve, you put the cumulative percentage of income on the vertical axis and the cumulative percentage of families (or population) on the horizontal axis. *(434)*
2. The Canadian definition of poverty is a relative measure. *(437)*
3. Canada has significantly more income inequality than Sweden, but significantly less than Brazil. *(442)*
4. In early capitalist society, the distributional fight was between workers and capitalists. In modern capitalist society, the distributional fight is more varied. For example, in Canada, we are concerned about such variables as age, gender, and education. *(445)*
5. No, it is not self-evident that greater equality of income would make society a better place to live. Unequal income distribution has its benefits. Still, most people

would prefer a somewhat more equal distribution of income than currently exists. *(447)*

6. What is fair is a very difficult concept. It depends on people's needs, people's wants, to what degree people are deserving, and other factors. Still, in the absence of any more information than is given in the question, we would divide the pie equally. *(447)*
7. No, one cannot reasonably assume other things remain constant. Redistributive programs have important side effects that can change the behaviour of individuals and subvert the intent of the program. Three important side effects include substituting leisure for labour, a decrease in measured income, and attempts to appear more needy. *(448)*
8. As a general statement, "A progressive tax is preferable to a proportional tax" is false. A progressive tax may well be preferable, but that is a normative judgment (just as its opposite would be). Moreover, taxes have incentive effects that must be considered. *(450)*
9. False. The Canada Pension Plan includes many other aspects, such as disability benefits and survivors' benefits. *(451)*
10. The initial distribution of property rights underlies the initial distribution of income. Those with the property rights will reap the returns from those rights. Ultimately, we can answer the question whether income distribution is fair only after we have answered whether the initial property rights distribution is fair. *(453)*

Work and the Labour Market

21

After reading this chapter, you should be able to:

- Use the theory of rational choice to explain why an increase in the marginal tax rate is likely to reduce the quantity of labour supplied.
- List four factors that influence the elasticity of market labour supply.
- Explain how the demand for labour is a derived demand.
- List four factors that influence the elasticity of market labour demand.
- Define *monopsonist* and *bilateral monopoly.*
- Discuss real-world characteristics of labour markets in terms of market, political, and social forces.
- List three types of discrimination.

Work banishes those three great evils: boredom, vice, and poverty.

Voltaire

Most of us earn our living by working. We supply labour (get a job) and get paid for doing things that other people tell us they want done. Even before we get a job, work is very much a part of our lives. We spend a large portion of our school years preparing for work. The courses you take are an investment in human capital (skills embodied in workers through experience, education, and on-the-job training).

People tend to define themselves by their work.

Work will occupy at least a third of your waking hours. To a great extent, your job will define you. When someone asks, "What do you do?" you won't answer, "I clip coupons, go out on dates, visit my children . . ." Instead you'll answer, "I work for the Blank Company" or "I'm an economist" or "I'm a teacher." Defining ourselves by our work means that work is more than the way we get income. It's a part of our social and cultural makeup. If we lose our jobs, we lose part of our identity.

There's no way this one chapter can discuss all the social, political, cultural, and economic dimensions of work and labour, but it's important to begin by at least pointing them out in order to put the discussion of labour markets into perspective. Labour is a factor of production, and the **labour market** is *a factor market in which individuals supply labour services for wages to other individuals and to firms that need (demand) labour services*. Because social and political pressures are particularly strong in labour markets, we can understand the nature of such markets only by considering how social and political forces interact with economic forces to determine our economic situation.

If the invisible hand were the only force operating, wages would be determined entirely by supply and demand. There's more to it than that, but, for the most part, the labour market is organized around the concepts of supply and demand.

THE SUPPLY OF LABOUR

The labour supply choice facing an individual (that is, the decisions of whether, how, and how much to work) can be seen as a choice between nonmarket activities and market activities. Nonmarket activities include sleeping, dating, studying, playing, cooking, cleaning, and gardening. Market activities include taking some type of paid job or working for oneself, directly supplying products or services to consumers.

Economists focus on the incentive effect when considering an individual's choice of whether and how much to work.

Many considerations are involved in individuals' choices of whether and how much to work and what kind of job to work at. Social background and conditioning are especially important, but the factor economists focus on is the **incentive effect** (*how much a person will change his or her hours worked in response to a change in the wage rate*). The incentive effect is determined by the value of supplying one's time to market activities relative to the value of supplying one's time to nonmarket activities. The normal relationship is:

Applying rational choice theory to the supply of labour tells us that the higher the wage, the higher the quantity of labour supplied.

The higher the wage, the higher the quantity of labour supplied.

This relationship between the wage rate and the quantity of labour supplied is shown in Figure 21-1. The wage rate is measured on the vertical axis; the quantity of labour supplied is measured on the horizontal axis. As you can see, the supply curve's upward slope indicates that as the wage rate increases, the quantity of labour supplied increases. Why is that the normal relationship? Because work involves opportunity cost. By working one hour more, you have one hour less to devote to other activities. Alternatively, if you devote the hour to other activities, you lose one hour's worth of income from working.

Q.1 Under the usual conditions of supply, what would you expect would happen to the amount of time you study if the wage of your part-time job rises?

Say, for example, that by working you would have made \$10 per hour. If you decide to work two hours less, you'll have \$20 less to spend, but two hours more available for other activities (including spending the smaller amount of money). When the wage rises, say to \$12 per hour, an hour of leisure has a higher opportunity cost. As the cost of leisure goes up, you buy less of it, meaning that you work more.

The incentive effects represented by the market supply curve result from individuals' decisions to enter or leave the labour market, and from individuals' decisions to work more, or fewer, hours. Given the institutional constraints in the labour market, which require many people to work a fixed set of hours if they work at all, much of the incentive effect of higher wages influences the either/or decisions of individuals. This affects the labour force participation rate (the number of people employed or looking for work as a percentage of people able to work) rather than adjusting the number of hours worked.

Figure 21-1 THE SUPPLY OF LABOUR

The supply of labour is generally considered to be upward sloping because the opportunity cost of not working increases as wages get higher.

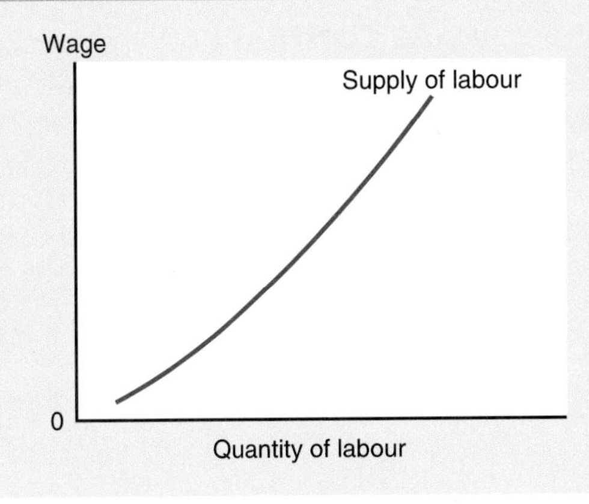

TABLE 21-1 Minimum Wage Rates, 2001

Minimum wage rates are set in response to the economic conditions of each province or territory. Assuming a person works approximately 2000 hours a year, the annual incomes of persons earning the minimum wage range from a low of $11,000 in Newfoundland to a high of $16,000 in British Columbia.

Location	Minimum Wage Rate	Location	Minimum Wage Rate
British Columbia	$8.00 / $6.00*	Alberta	$5.90
Saskatchewan	$6.00	Manitoba	$6.25
Ontario	$6.85	Quebec	$7.00
Newfoundland	$5.50	New Brunswick	$5.90
Nova Scotia	$5.80	Prince Edward Island	$5.80
Yukon	$7.20	Northwest Territories	$6.50
Nunavut	$6.50		

* Two-tiered minimum wage: $6.00 for the first 500 hours, or six months, whichever is less.

Source: www.ccsd.ca.

For example, when wages rise, retired workers may find it worthwhile to go back to work, and many teenagers may choose to find part-time jobs.

Real Wages and the Opportunity Cost of Work

The nominal wage is the actual number of dollars you are paid per hour. The real wage is your nominal wage adjusted for inflation. If nominal wages rise faster than the inflation rate, real wages rise.

What is the relationship between the real wage and work? The upward-sloping supply curve of labour tells you that, other things equal, as wages go up, the quantity of labour supplied goes up. But if you look at the historical record, you will see that over the last century real wages in Canada increased substantially but the average number of hours worked per person fell. See Table 21-1 for minimum wage rates across Canada. This difference is partly explained by the income effect. (See the box on page 461.) Higher incomes make people richer, and richer people choose more leisure.

Given that people are far richer today than they were 50 or 100 years ago, it isn't surprising that they work less. What's surprising is that they work as much as they do—eight hours a day rather than the four or so hours a day that would be enough to give

Figure 21-2 WAGE CONTRACT SETTLEMENTS, 1980-2001

When wages do not rise as quickly as inflation, real wages decrease and workers' purchasing power falls.

Source: www.statcan.ca/english/indepth/75-001/online/00901/kl-ic_a.html. Perspectives on Labour, September, 2001.

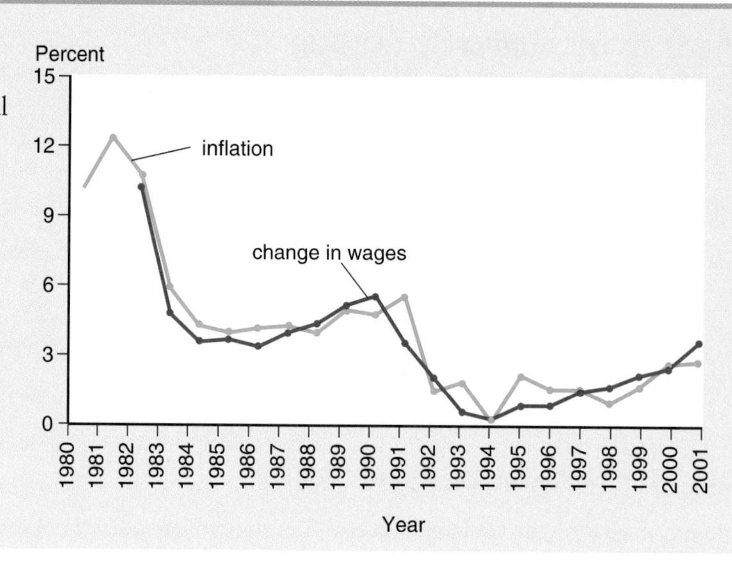

people the same income they had a century ago. Note that real wages have not risen across the board in the past two decades: see Figure 21-2.

The explanation of why people haven't reduced their hours of work more substantially can be found in how leisure has changed. A century ago, conversation was an art. People could use their time for long, leisurely conversations. Letter writing was a skill all educated people had, and cooking dinner was a three-hour event. Today if people were satisfied with spending their leisure time having long conversations, whittling, and spending quality time with their families rather than skiing, golfing, or travelling, they could get by with working perhaps only four or five hours per day instead of eight hours. But that isn't the case.

Modern gadgets increase the efficiency of leisure but cost money, which means people must work more to enjoy their leisure.

Today leisurely dinners, conversations about good books, and witty letters have been replaced by "efficient" leisure: a fast-food supper, a home movie, and the instant analysis of current events. Microwave ovens, frozen dinners, cellular telephones, the Internet—the list of products designed to save time is endless. All these products that increase the "efficiency" of leisure (increase the marginal utility per hour of leisure spent) cost money, which means people today must work more to enjoy their leisure! One reason people work hard is so that they can play hard (and expensively).

Economists do not try to answer the normative question of whether people are better off today, working hard to play hard, or simply are more harried.

The fast pace of modern society has led a number of people to question whether we, as a society, are better off working hard to play hard. Are we better off or simply more harried? Most economists don't try to answer this normative question; but they do point out that people are choosing their harried lifestyle, so to argue that people are worse off, one must argue that people are choosing something they don't really want. That may be true, but it's a tough argument to prove.

The Supply of Labour and Nonmarket Activities

In addition to leisure, labour supply issues and market incentives play an important role in other market activities. For example, there's a whole set of illegal activities, such as selling illegal drugs, that are alternatives to taking a legal job.

Let's say that an 18-year-old street kid figures he has only two options: Either he can work at a minimum wage job or he can deal drugs illegally. Dealing drugs involves

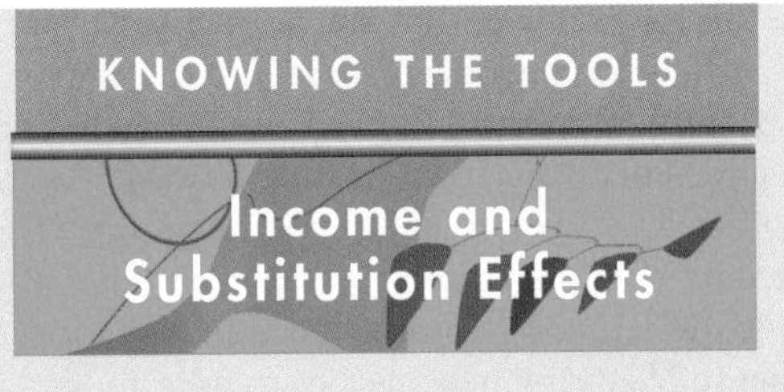

Because labour income is such an important component of most people's total income, when wages change other things often do not stay equal, and at times the effect can seem strange. For example, say that you earn $10 an hour and you decide to work eight hours per day. Suddenly demand for your services goes up and you find that you can receive $40 an hour. Will you decide to work more hours? According to the rational choice rule, you will, but you might also decide that at $40 an hour you'll work only six hours a day—$240 a day is enough; the rest of the day you want leisure time to spend your money. In such a case a higher wage means working less.

Does this violate the rational choice rule? The answer is no, because other things—specifically, your income—do not remain equal. The higher wage makes you decide to work more—as the rational choice rule says; but the effect of the higher wage is overwhelmed by the effect of the higher income that allows you to decide to work less.

To distinguish between these two effects, economists have given them names. The decision, based on the rational choice rule, to work more hours when your pay goes up is called the *substitution effect.* You substitute work for leisure because the price of leisure has risen. The decision to work fewer hours when your pay goes up, based on the fact that you're richer and therefore can live a better life, is called the *income effect.*

It's possible that the income effect can exceed the substitution effect, and a wage increase can cause a person to work less, but that possibility does not violate the rational choice rule, which refers to the substitution effect only. If you review the chapter on individual choice, a good exercise is to show the income and substitution effects with indifference curves and to demonstrate how it might be possible for an increase in the wage to lead to a decline in hours of work.

enormous risks of getting arrested or shot, but it also means earning $50 or $75 an hour. Given that choice, many risk-takers opt to sell drugs.

For middle-class individuals who have prospects for good jobs, the cost of being arrested can be high—an arrest can destroy their future prospects. For poor street kids with little chance of getting a good job, an arrest makes little difference to their future. For them the choice is easy. This is especially true for the entrepreneurial types—the risk-takers—who might have become the business leaders of the future. Prohibiting certain drugs leads to potentially high income from selling those drugs and has significant labour market effects. The incentive effects that prohibition has on the choices of jobs facing poor teenagers is one reason why some economists support the legalization of currently illegal drugs.

Income Taxation, Work, and Leisure

It is after-tax income, not before-tax income, that determines how much you work. Why? Because after-tax income is what you give up by not working. The government, not you, forgoes what you would have paid in taxes if you had worked. This means that when the government raises your marginal rate (the tax you pay on an additional dollar), it reduces your incentive to work. When the marginal tax rate gets high—say 50 percent—it can significantly reduce individuals' incentive to work and earn income.

One main reason why the government reduced marginal income tax rates in the 2000s was to reduce the negative incentive effects of high taxes. Reducing the marginal tax rate won't completely eliminate the problem of negative incentive effects on individuals' work effort. The reason is that the amounts people receive from many other government programs are tied to earned income. When your earned income goes up, your benefits from these other programs go down.

An increase in the marginal tax rate is likely to reduce the quantity of labour supplied because it reduces the net wage of individuals and hence, through individuals' incentive effect, causes them to work less.

Q-2 Why do income taxes reduce your incentive to work?

European countries, which have relatively high marginal tax rates, are struggling with the problem of providing incentives for people to work.

Say, for example, that you're getting welfare and you're deciding whether to take a $6-an-hour job. Income taxes reduce the amount you take home from the job by 20 percent, to $4.80 an hour. But you also know that the Welfare Department will reduce your welfare benefits by 50 cents for every dollar you take home. This means that you lose another $2.40 per hour, so the marginal tax rate on your $6-an-hour job isn't 20 percent; it's 60 percent. By working an hour, you've increased your net income by only $2.40. When you consider the transportation cost of getting to and from work, the expense of getting new clothes to wear to work, the cost of child care, and other job-associated expenses, the net gain in income is often minimal. Your effective marginal tax rate is almost 100 percent! At such rates, there's an enormous incentive either not to work or to work off the books (get paid in cash so you have no recorded income that the tax agent can trace).

Q-3 What is the irony of any need-based program?

The negative incentive effect can sometimes be even more indirect. For example, college bursaries are generally given on the basis of need. A person that earns more gets less financial aid; the amount by which the bursary is reduced as a person's income increases acts as a marginal tax on income. Why work hard to provide for yourself if a program will take care of you if you don't work hard? Hence, the irony in any need-based assistance program is that it reduces the people's incentive to prevent themselves from being needy. These negative incentive effects on labour supply that accompany any need-based program present a public policy dilemma for which there is no easy answer.

Elasticity of market supply depends on:

1. Individuals' opportunity cost of working.
2. The type of market being discussed.
3. The elasticity of individuals' supply curves.
4. Individuals entering and leaving the labour market.

The Elasticity of the Supply of Labour

Exactly how these various incentives affect the amount of labour an individual supplies is determined by the elasticity of the individual's supply curve of labour.

The elasticity of the market supply curve is determined by the elasticity of individuals' supply curves and by individuals entering and leaving the labour force. Both of these, in turn, are determined by individuals' opportunity cost of working. If a large number of people are willing to enter the labour market when wages rise, then the market labour supply will be highly elastic even if individuals' supply curves are inelastic.

The elasticity of supply also depends on the type of market being discussed. For example, the elasticity of the labour supply facing one firm in a small town will likely be far greater than the elasticity of the labour supply of all firms in that town. If only one firm raises its wage, it will attract workers away from other firms; if all the firms in town raise their wages, any increase in labour must come from increases in labour force participation, increases in hours worked per person, or in-migration (the movement of new workers into the town's labour market).

What is the opportunity cost of this worker? What is her elasticity of supply? What is the firm's elasticity of demand for her?

Existing workers prefer an inelastic labour supply because that means an increase in demand for labour will raise their wage more. Employers prefer an elastic supply because that means an increase in demand for labour doesn't result in large wage increases.

Because of the importance of the elasticity of labour supply, economists have spent a great deal of time and effort estimating it. Their best estimates of labour supply elasticities to market activities are about 0.1 for the primary income earner of the household and 1.1 for secondary income earners. These elasticity figures mean that a wage increase of 10 percent will increase the quantity of labour supplied by 1 percent for the primary income earner (an inelastic supply) and 11 percent for the secondary income earner in households (an elastic supply). Why the difference? Institutional factors. Hours of work are only slightly flexible. Since most heads of households are employed, they cannot significantly change their hours worked. The primary income earner must work to support the family; a 10 percent decrease in wages will only result in a 1 percent decrease in the quantity of labour supplied. For the secondary income earner in the household, employment is less essential, therefore, the elasticity of supply is greater.

Immigration and the International Supply of Labour

International limitations on the flow of people, and hence on the flow of labour, play an important role in elasticities of labour supply. In many industries, wages in developing countries are 1/10 or 1/20 the rate of wages in Canada. In the early 1990s, the European Union introduced open borders among member countries. That institutional change has brought about a more open flow of individuals into higher-wage EU countries from lower-wage EU countries, although other institutionalized restrictions on flows of people, such as language and culture barriers, prevented the EU from being a unified labour market through the early 2000s.

THE DERIVED DEMAND FOR LABOUR

The Firm's Decision to Hire

What determines a firm's decision to hire someone? The answer is simple. A profit-maximizing firm hires someone if it thinks there's money to be made by doing so. Unless there is, the firm won't hire the person. So for a firm to decide whether to hire someone, it must compare the worker's **marginal revenue product (MRP)** (*the marginal revenue it expects to earn from selling the additional worker's output*) with the wage that it expects to pay the additional worker. For a competitive firm (for which $P = MR$) marginal revenue product equals the worker's **value of marginal product (VMP)**—the worker's **marginal physical product (MPP)** (*the additional units of output that hiring an additional worker will bring about*) times the price (P) at which the firm can sell the additional product.

Marginal revenue product = $MPP \times P$

Say, for example, that by hiring another worker a firm can produce an additional 6 widgets an hour, which it can sell at $2 each. That means the firm can pay up to $12 per hour and still expect to make a profit. Notice that a key question for the firm is: How much additional product will we get from hiring another worker? A competitive firm can increase its profit by hiring another worker as long as the value of the worker's marginal product (which also equals her marginal revenue product) ($MPP \times P$) is greater than or equal to her wage.

To see whether you understand the principle, consider the example in Figure 21-3(a). Column 1 shows the number of workers, all of whom are assumed to be identical. Column 2 shows the total output of those workers. Column 3 shows the marginal physical product of an additional worker. This number is determined by looking at the change in the total product due to this person's work. For example, if the firm is currently employing 30 workers and it hires one more, the firm's total product or output will rise from 294 to 300, so the marginal product of moving from 30 to 31 workers is 6.

Notice that workers' marginal product decreases as more workers are hired. Why is this? Remember the assumption of fixed capital: More and more workers are working with the same amount of capital so there is diminishing marginal productivity.

Column 4 shows **labour productivity**—*the average output per worker*, which is a statistic commonly referred to in economic reports. It's determined by dividing the total output by the number of workers. Column 5 shows the additional worker's marginal revenue product, which, since the firm is assumed to be competitive, is determined by multiplying the price the firm receives for the product it sells ($2) by the worker's marginal physical product.

Column 5, the marginal revenue product, is of central importance to the firm. It tells the firm how much additional money it will make from hiring an additional

In the late 1800s, many workers worked in sweatshops; they often had quotas that required them to work 60 or more hours a week. Fines were imposed for such indiscretions as talking or smiling.

Figure 21-3 (a and b) **DETERMINING HOW MANY WORKERS TO HIRE AND THE FIRM'S DERIVED DEMAND FOR LABOUR**

The marginal revenue product is any firm's demand curve for labour. Since for a competitive firm $P = MR$, a competitive firm's derived demand curve is its value of the marginal product curve ($P \times MPP$). This curve tells us the additional revenue the firm gets from having an additional worker. From the chart in (**a**) we can see that when the firm increases from 27 to 28 workers, the marginal product per hour for each worker is 9. If the product sells for $2, then marginal revenue product is $18, which is one point on the demand curve for labour (point *A* in (**b**)). When the firm increases from 34 to 35 workers, the value of the marginal product is $4. This is another point on the firm's derived demand curve (point *B* in (**b**)). By connecting the two points, as shown in (**b**), you can see that the firm's derived demand curve for labour is downward sloping.

1 Number of workers	2 Total product per hour	3 Marginal physical product per hour	4 Average product per hour	5 Marginal revenue product (*MRP*)
27	270		10.00	
		9		$18
28	279		9.96	
		8		16
29	287		9.90	
		7		14
30	294		9.80	
		6		12
31	300		9.68	
		5		10
32	305		9.53	
		4		8
33	309		9.36	
		3		6
34	312		9.18	
		2		4
35	314		8.97	

(a) Calulating MPP and MRP

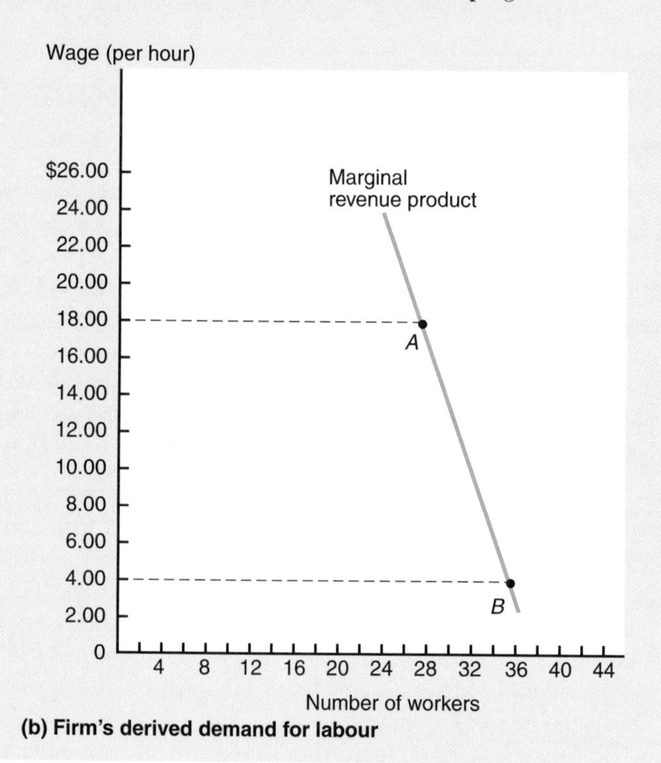

(b) Firm's derived demand for labour

worker. That marginal revenue product represents a competitive firm's demand for labour.

Figure 21-3(b) graphs the firm's derived demand for labour, based on the data in column 5 of Figure 21-3(a). The resulting curve is the firm's **derived demand curve for labour,** which *shows the maximum amount of labour, measured in labour hours, that a firm will hire*. To see this, let's assume that the wage is $9 and that the firm is hiring 30 workers. If it hires another worker so it has 31 workers, workers' marginal revenue product of $12 exceeds their wage of $9, so the firm can increase profits by doing so. It increases output and profits since the additional revenue the firm gets from increasing workers from 30 to 31 is $12 and the additional cost the firm incurs is the wage of $9.

Now say the firm has hired 4 additional workers so it has 34 workers. As the firm hires more workers, the marginal product of workers declines. As you can see from the graph in Figure 21-3(b), the marginal revenue product of the last worker hired (increasing from 33 to 34 workers) is $6. Since the workers' marginal revenue product of $6 is less than their wage of $9, now the firm can increase profits by laying off some

APPLYING THE TOOLS

Difficulties in Determining Marginal Productivities

The economic model of labour markets assumes that marginal productivities can be determined relatively easily. In reality they can't. They require guesses and estimates that are often influenced by a worker's interaction with the person doing the guessing and estimating. Thus, social interaction plays a role in determining wages. If you get along with your manager, his or her estimate of your marginal productivity is likely to be higher than if you don't. And for some reason, managers' estimates of their own marginal productivity tend to be high. In part because of difficulties in estimating marginal productivities, actual pay can often differ substantially from marginal productivities.

workers. Doing so decreases output but increases profit, because it significantly increases the average product of the remaining workers.

When a worker's wage of $9 equals the marginal revenue product, the firm has no incentive to change the number of employees. In this example, the wage ($9) equals workers' marginal revenue product at 32 workers. When the firm has 32 workers, either hiring a worker or laying off a worker will decrease profits. Decreasing from 32 to 31 workers saves $9.00 in wages, but loses $10 in revenue, and increasing from 32 to 33 workers gains $8 in revenue, but costs $9 in wages. Since the marginal revenue product curve tells the firm, given a wage, how many workers it should hire, *the marginal revenue product curve is the firm's demand curve for labour.*

The marginal revenue product curve is the firm's demand curve for labour.

The fact that the demand curve for labour is downward sloping means that as more workers are hired, workers' marginal product falls. This might tempt you to think that the last worker hired is inherently less productive than the first worker hired. But that simply can't be because, by assumption, the workers are identical. Thus, the marginal product of any worker must be identical to the marginal product of any other worker, given that a specified number of workers are working. What the falling marginal product means is that *when 30 rather than 25 workers are working,* the marginal product of any one of those 30 workers is less than the marginal product of any one of 25 of those workers when only 25 are working. When the other inputs are constant, hiring an additional worker lowers the marginal product not only of the last worker but also of any of the other workers.

To understand what's going on here you must remember that when marginal product is calculated, all other inputs are held constant—so if a firm hires another worker, that worker will have to share machines or tools with other workers. When you share tools, you start running into significant bottlenecks, which cause production to fall. That's why the marginal product of workers goes down when a new worker is hired. This assumption that all other factors of production are held constant is an important one. If all other factors of production are increased, workers' productivity will not necessarily fall as output increases.

Why does a firm hire another worker if doing so will lead to a decrease in other workers' productivity and, possibly, a fall in the average productivity of all workers? Because the firm is interested in total profit, not productivity. As long as hiring an extra worker increases revenue by more than the worker costs, the firm's total profit increases. A profit-maximizing firm would be crazy not to hire another worker, even if by doing so it lowers the marginal product of the workers.

FACTORS AFFECTING THE DEMAND FOR LABOUR

There are many technical issues that determine how the demand for products is translated through firms into a demand for labour (and other factors of production). Consider three general principles:

1. Changes in the demand for a firm's product will be reflected in changes in its demand for labour.
2. The structure of a firm plays an important role in determining its demand for labour.
3. A change in the other factors of production that a firm uses will change its demand for labour.

Let's consider each of these principles in turn.

Changes in the Firm's Demand

The first principle is almost self-evident. An increase in the demand for a product leads to an increase in demand for the labourers who produce that product. The increase in demand pushes the price up, raising the marginal revenue product of labour (which, you'll remember, for a competitive firm is the price of the firm's product times the marginal physical product of labour).

The implications of this first principle, however, are not so self-evident. Often people think of firms' interests and workers' interests as being counter to one another, but this principle tells us that in many ways they are not. What benefits the firm also benefits its workers. Their interests are in conflict only when it comes to deciding how to divide up the total revenues among the owners of the firm, the workers, and the other inputs. Thus, it's not uncommon to see a firm and its workers fighting each other at the bargaining table, but also working together to prevent imports that might compete with the firm's product or to support laws that may benefit the firm.

Therefore, the cost of labour to a firm should be modelled as if it is determined at the same time that its price and profitability are determined, not separately.

The Structure of the Firm and Its Demand for Labour

The way in which the demand for products is translated into a demand for labour is determined by the structure of the firm. For example, let's consider the difference between a monopolistic industry and a competitive industry. For both, the decision about whether to hire is based on whether the wage is below or above the marginal revenue product. But the firms that make up the two industries calculate their marginal revenue products differently.

The price of a competitive firm's output remains constant regardless of how many units it sells. Thus, its marginal revenue product equals the value of the marginal product. To calculate its marginal revenue product we simply multiply the price of the firm's product by the worker's marginal physical product. For a competitive firm:

Marginal revenue product of a worker =

Value of the worker's marginal product =

$MPP \times$ Price of product

The price of a monopolist's product decreases as more units are sold, since the monopolist faces a downward-sloping demand curve. The monopolist takes that into account. That's why it focusses on marginal revenue rather than price. As it hires more labour and produces more output, the price it charges for its product will fall. Thus, for a monopolist:

TABLE 21-2 The Effect of Monopoly and Firm Structure on the Demand for Labour

1	2	3	4	5	6	7
					Marginal Revenue Product	
Number of Workers	Wage	Price *P*	Marginal Revenue (Monopolist) *MR*	Marginal Physical Product *MPP*	Competitive (*MPP* × *P*)	Monopolist (*MPP* × *MR*)
5	$2.85	$1.00	$.75	5	$5.00	$3.75
6	2.85	.95	.65	3	2.85	1.95
7	2.85	.90	.55	1	.90	.55

Marginal revenue product of a worker =
MPP × Marginal revenue

Since a monopolist's marginal revenue is always less than price, a monopolist industry will always hire fewer workers than a comparable competitive industry, which is consistent with the result we discussed in Chapter 12: a monopolistic industry will always produce less than a competitive industry, other things equal.

To ensure that you understand the principle, let's consider the example in Table 21-2, a table of prices, wages, marginal revenues, marginal physical products, and marginal revenue products for a firm in a competitive industry and a monopolistic industry.

A firm in a competitive industry will hire up to the point where the wage equals *MPP* × *P* (columns 5 × 3). This occurs at 6 workers. Hiring either fewer or more workers would mean less profit for the firm.

Now let's compare the competitive industry with an equivalent monopolistic industry. Whereas the firm in the competitive industry did not take into account the effect an increase in output would have on prices, the monopolist must do so. It takes into account the fact that in order to sell the additional output of an additional worker, it must lower the price of all units of the good. The relevant marginal revenue product for the monopolist appears in column 7. At 6 workers, the worker's wage rate of $2.85 exceeds the worker's marginal revenue product of $1.95, which means that the monopolist would hire fewer than 6 workers—5 full-time workers and 1 part-time worker.

As a second example of how the nature of firms affects the translation of demand for products into demand for labour, consider what would happen if workers rather than independent profit-maximizing owners controlled the firms. You saw before that whenever another worker is hired, other inputs constant, the marginal physical product of all workers falls. That can contribute to a reduction in existing workers' wages. The profit-maximizing firm doesn't take into account that effect on existing workers' wages. In fact, it wants to hold its costs down. If existing workers are making the decisions about hiring, they'll take that wage decline into account. If they believe that hiring more workers will lower their own wage, they have an incentive to see that new workers aren't hired. Thus, like the monopolist, a worker-controlled firm will hire fewer workers than a competitive profit-maximizing firm.

There aren't many worker-controlled firms in Canada but a number of firms include existing workers' welfare in their decision processes. Moreover, with the growth of the team concept discussed in Chapter 9, existing workers' input into managerial decision making is increasing. In many firms, workers have some say in whether additional workers will be hired and at what wage they will be hired. Other firms have an implicit understanding or a written contract with existing workers that restricts hiring and firing decisions. Some firms, such as IBM, had never laid off a worker; if they had to reduce their workforce, they created early retirement incentives. Ultimately, however, if their business gets bad enough, the invisible hand wins out over the social forces, and they lay off workers. That happened for IBM, and many other large technology firms, in the 1990s.

Why do firms consider workers' welfare? They do so to be seen as a "good employer," which makes it easier for them to hire good people. Given the strong social and legal limitations on firms' hiring and firing decisions, one cannot simply apply marginal productivity theory to the real world. One must first understand the institutional and legal structures of the labour market. However, the existence of these other forces doesn't mean that the economic forces represented by marginal productivity don't exist. Rather, it means that firms struggle to find a wage policy that accommodates both economic and social forces in their wage-setting process. For example, in the 1980s and 1990s, a number of firms (such as airline and automobile firms) negotiated two-tier wage contracts. They continued to pay their existing workers a higher wage, but paid new workers a lower wage, even though old and new workers were doing identical jobs. These two-tier wage contracts were the result of the interactions of the social and market forces.

Changes in Other Factors of Production

A third principle determining the derived demand for labour is the amount of other factors of production that the firm has. Given a technology, an increase in other factors of production will increase the marginal physical product of existing workers. For example, let's say that a firm buys more machines so that each worker has more machines with which to work. The workers' marginal physical product increases, and the cost per unit of output for the firm decreases. The net effect on the demand for labour is unclear; it depends on how much the firm increases output, how much the firm's price is affected, and how easily one type of input can be substituted for another—or whether it must be used in conjunction with others.

A firm is producing at minimum cost if it gets the same marginal product per dollar spent on each input.

While we can't say what the final effect on demand will be, we can determine the firm's **cost minimization condition**—where *the ratio of marginal product to the price of an input is equal for all inputs*.[1] When a firm is using resources as efficiently as possible, and hence is minimizing costs, the marginal product of each factor of production divided by the price of that factor must equal that of all the other factors. Specifically, the *cost minimization condition* is:

$$\frac{MP_l}{w} = \frac{MP_m}{P_m} = \frac{MP_x}{P_x}$$

where

w = Wage rate

l = Labour

m = Machines

x = Any other input

If this cost minimization condition is not met, the firm could hire more of the input with the higher marginal product relative to price, and less of other inputs, and produce the same amount of output at a lower cost.

[1]This cost minimization condition was explicitly discussed in the appendix to Chapter 9.

Let's consider a numerical example. Say the marginal product of labour is 20 and the wage is $4, while the marginal product of machines is 30 and the rental price of machines is $4. You're called in to advise the firm. You say, "Fire one worker, which will decrease output by 20 and save $4; spend that $4 on machines, which will increase output by 30." Output has increased by 10 while costs have remained constant. As long as the marginal products divided by the prices of the various inputs are unequal, you can make such recommendations to lower cost.

If $\frac{MP_m}{P_m} > \frac{MP_l}{W}$ then substituting machines for labour will reduce costs.

Changes in these factors make demand for labour shift around a lot. This shifting introduces uncertainty into people's lives and into the economic system. Often people attempt to build up institutional barriers to reduce uncertainty—either through social or political forces. Thus, labour markets function under an enormous volume of rules and regulations and rules. We need to remember that while economic factors often lurk behind the scenes to determine pay and hiring decisions, these are often only part of the picture.

THE DEMAND FOR LABOUR

The demand for labour follows the basic law of demand:

The higher the wage, the lower the quantity of labour demanded.

This negative relationship between the wage rate and the quantity of labour demanded is shown by the blue line in Figure 21-4. Its downward slope shows that as the wage rate falls the quantity of labour demanded rises. Equilibrium is at wage W_e and quantity of labour exchanged is Q_e.

When individuals are self-employed (work for themselves), the demand for their labour is the demand for the product or service they supply—be it cutting hair, shampooing rugs, or filling teeth. You have an ability to do something, you offer to do it at a certain price, and you see who calls. You determine how many hours you work, what price you charge, and what jobs you take. The income you receive depends on the demand for the good or service you supply and your decision about how much labour you want to supply. In analyzing self-employed individuals, we can move directly from demand for the product or service to demand for labour.

Figure 21-4 EQUILIBRIUM IN THE LABOUR MARKET

When the supply and demand curves for labour are placed on the same graph, the equilibrium wage, W_e, is where the quantity supplied equals quantity demanded. At this wage, Q_e workers are hired.

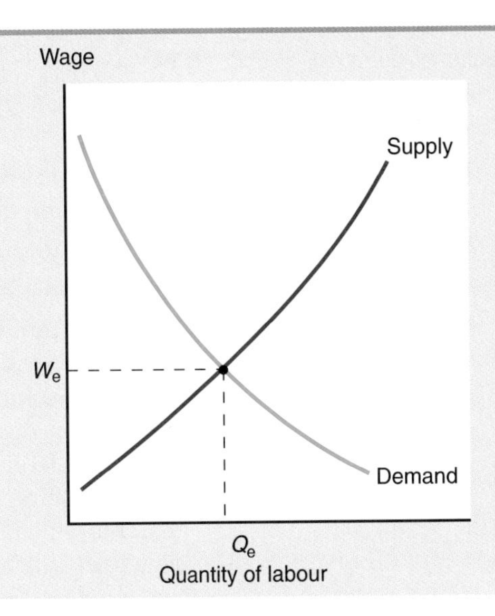

When a person is not self-employed, determining the demand for labour is a two-step process: Consumers demand products from firms; firms, in turn, demand labour and other factors of production. The demand for labour by firms is a **derived demand**—*the demand for factors of production by firms, depends on (is derived from) consumers' demand for the firm's products*. In other words, the demand for labour is derived from consumers' demand for the goods that the firm sells. Thus, you can't think of demand for a factor of production such as labour separately from demand for goods. Firms translate consumers' demands into a demand for factors of production.

Derived demand is the demand for factors of production by firms, which depends on consumers' demands for the firm's products.

Factors Influencing the Elasticity of Demand for Labour

Q.4 Name at least two factors that influence the elasticity of a firm's derived demand for labour.

The elasticity of the derived demand for labour, or for any other input, depends on a number of factors. One of the most important is (1) *the elasticity of demand for the firm's good*. The more elastic the demand for the good, the more elastic the derived demand for labour. Other factors influencing the elasticity of derived demand for labour include (2) *the relative importance of labour in the production process* (the more important, the less elastic is the derived demand); (3) *the possibility of, and cost of, input substitution in production* (the easier substitution is, the more elastic is the derived demand); and (4) *the degree to which the marginal productivity falls with an increase in labour* (the faster productivity falls, the less elastic is the derived demand).

Each of these relationships follows from the definition of *elasticity* (the percentage change in quantity divided by the percentage change in price) and a knowledge of production. To be sure you understand, ask yourself the following question: If all I knew about two firms was that one was a perfect competitor and the other was a monopolist, which firm would I say is likely to have the more elastic derived demand for labour? If your answer wasn't automatically "the competitive firm" (because its demand curve is perfectly elastic and hence more elastic than a monopolist's), you should review the discussion of factors influencing demand elasticity in Chapter 6 and relate that to this discussion. The two discussions are similar and serve as good reviews for each other.

Four factors that influence the elasticity of demand for labour are:

1. The elasticity of demand for the firm's good.
2. The relative importance of the factor in the production process.
3. The possibility of, and cost of, input substitution in production.
4. The degree to which marginal productivity falls with an increase in the factor.

Labour as a Factor of Production

The traditional factors of production are land, labour, capital, and entrepreneurship. When economists talk of the labour market, they're talking about two of these factors: labour and entrepreneurship. **Entrepreneurship** is *labour services that involve high degrees of organizational skills, responsibility, and creativity*. It is a type of creative labour.

Entrepreneurship is labour services that involve high degrees of organizational skills, responsibility, and creativity.

The reason for distinguishing between labour and entrepreneurship is that an hour of work is not simply an hour of work. If a high degree of organizational skill, responsibility, and creativity is exerted (which is what economists mean by *entrepreneurship*), one hour of such work can be equivalent to weeks of simple labour. That's one reason that pay often differs between jobs.

Shift Factors of Demand

Q.5 What would happen to the demand for labour if a firm's product became more popular?

Factors that shift the demand curve for labour will put pressure on the equilibrium price to change. Let's consider some examples. Say the cost of a competing factor of supply, such as a machine that also could do the job, rises. That would increase the demand for labour (shift the demand curve rightward), and put pressure on the wage to rise.

Alternatively, say a new technology develops that requires skills different from those currently being used—for instance, requiring knowing how to use a computer rather than knowing how to use a slide rule. The demand for individuals knowing how to use slide rules will decrease, and their wage will tend to fall.

Another example: Say an industry becomes more monopolistic. What will that do to the demand for labour in that industry? Since monopolies produce less output, the answer is that it would decrease the demand for workers, since the industry would hire fewer of them. The demand for workers would shift leftward and wages would tend to fall.

Finally, say the demand for the firm's good increases. Then the firm's demand for labour will also increase.

Technology and the Demand for Labour What effect will a change in technology have on the demand for labour? The simple reasoning often used by laypeople to argue that the development of new technology will decrease the demand for labour is wrong. Their simple reasoning is: "Technology makes it possible to replace workers with machines, so it will decrease the demand for labour." This is sometimes called *Luddite reasoning* because it's what drove the Luddites to go around smashing machines in early-19th-century England.

Luddites oppose technological advances.

What's wrong with Luddite reasoning? First, look at history. Technology has increased enormously, yet the demand for labour has not decreased; instead it has increased as output has increased. Luddite reasoning doesn't take into account the fact that total output has risen. A second problem with Luddite reasoning is that labour is necessary for building and maintaining the machines, so increased demand for machines should increase the demand for labour.

Luddite reasoning isn't *all* wrong. Technology can sometimes decrease the demand for certain types of skills. The computer has decreased demand for calligraphers; the automobile reduced demand for carriage makers. New technology changes the types of labour demanded. If you have the type of labour that will be made technologically obsolete, you can be hurt by technological change. However, technological change hasn't reduced the overall demand for labour; it has instead led to an increase in total output and a need for even more labourers to produce that output.

In the 21st century we're likely to see a continued increase in the use of robots to do many repetitive tasks that blue-collar workers formerly did. Thus, demand for manufacturing labour will likely continue to decline, but it will be accompanied by an increase in demand for service industry labour—designing and repairing robots and designing activities that will fill up people's free time.

International Competitiveness and a Country's Demand for Labour Many of the issues in the demand for labour concern one firm's or industry's demand for labour relative to another firm's or industry's demand. When we're talking about the demand for labour by the country as a whole—an issue fundamentally important to many of the policy issues being discussed today—we have to consider the country's overall international competitiveness. A central determinant of a country's competitiveness is the relative wage of labour in that country compared to the relative wage of labour in other countries.

Q-6 Name two factors besides relative wages that determine the demand for labour in one country compared to another.

Wages vary considerably among countries. For example, in 1999 workers in the manufacturing industry earned an average $23.17 an hour in Canada, C$38.90 an hour in Germany, and C$2.62 an hour in Mexico. Multinational corporations are continually making decisions about where to place production facilities, and labour costs—wage rates—play an important role in these decisions. The country's exchange rate also plays an important role in determining the demand for labour in a country. For example, in the early 1990s many Japanese automobile companies switched their production of cars to be sold in Canada from production facilities in Japan to facilities in Canada. Why? Because the rise in value of the yen, and fall in value of the dollar, meant that the hourly rate of labour in Canada was about $23 and the hourly rate in Japan was about C$28.

Wages aren't the only consideration when a firm decides where to locate production. Other factors include worker productivity, transportation costs, trade restrictions, and social institutions.

But why produce in Canada when the hourly rate in Taiwan, for example, was only 1/4 that in Canada? Or when the hourly rate in Mexico was only 1/10 that in Canada? The reasons are complicated, but include (1) differences in workers' productivity; (2) transportation costs—producing in the country to which you're selling keeps transportation costs down; (3) potential trade restrictions—Japan was under enormous pressure to reduce its trade surplus, and producing in Canada helped it avoid trade restrictions; (4) compatibility of production techniques with social institutions—production techniques must fit with a society's social institutions; and (5) the *focal point phenomenon*—a situation where a company chooses to move, or expand, production to another country because other companies have already moved or expanded there. A company can't consider all places, and it costs a lot of money to explore a country's potential as a possible host country. Japanese businesses know what to expect when they open a plant in Canada or the United States; they don't know in many other countries. So, Canada, the United States, and other countries that Japanese businesses have knowledge about, become focal points. They are considered as potential sites for business, while other, possibly equally good, countries are not. Combined, these reasons lead to a "follow-the-leader" system in which countries fall in and out of global companies' production plans. The focal-point countries expand and develop; the others don't. During the 1990s, especially for telecommunications and computer technologies, the U.S. became the focal point.

THE ROLE OF OTHER FORCES IN WAGE DETERMINATION

Supply and demand forces strongly influence wages, but they do not fully determine wages.

Supply and demand forces strongly influence wages, but they do not fully determine wages. Real-world labour markets are filled with examples of suppliers or demanders who resist these supply and demand pressures through organizations such as labour unions, professional associations, and agreements among demanders. But, supply and demand analysis is a useful framework for considering such resistance.

For example, say that you're advising a firm's workers on how to raise their wages. You point out that if workers want to increase their wages, they must figure out some way either to increase the demand for their services or to limit the amount of labour supplied to the firm. One way to limit the number of workers the firm will hire (and thus keep existing workers' wages high) is to have the firm agree to pay an above-equilibrium wage, as in Figure 21-5(a). Say that in their contract negotiations the workers get the firm to agree to pay a wage of W_1. At wage W_1, the quantity of labour supplied is Q_S and the quantity of labour demanded is Q_D. The difference, $Q_S - Q_D$, represents the number of people who want jobs at wage W_1 compared to the number who have them. In such a case, jobs must be rationed.

As a second example, consider what would happen if immigration laws were liberalized. If you say the supply curve of labour would shift rightward and the wage level would drop, you're right, as shown in Figure 21-5(b). In it, the supply of labour increases from S_0 to S_1. In response, the wage falls from W_0 to W_1 and the quantity of labour demanded increases from Q_0 to Q_1.

Q-7 How could an increase in the supply of labour lead to an increase in the demand for labour?

In analyzing the effect of such a major change in the labour supply, however, remember that the supply and demand framework is relevant only if the change in the supply of labour doesn't also affect the demand for labour. In reality, a change in immigration laws

Figure 21-5 THE LABOUR MARKET IN ACTION

In (**a**) you can see the effect of an above-equilibrium wage: If workers convince the firm to pay them a wage of W_1, more workers will be supplied (Q_S) than demanded (Q_D). With an excess supply of labour, jobs must be rationed. In (**b**) you can see the effect of an increase in the supply of labour—for example, because of liberalization in the immigration laws. The supply of labour curve shifts outward from S_0 to S_1. Assuming the demand for labour remains the same, the increase in the supply of labour will cause the wage level to drop from W_0 to W_1.

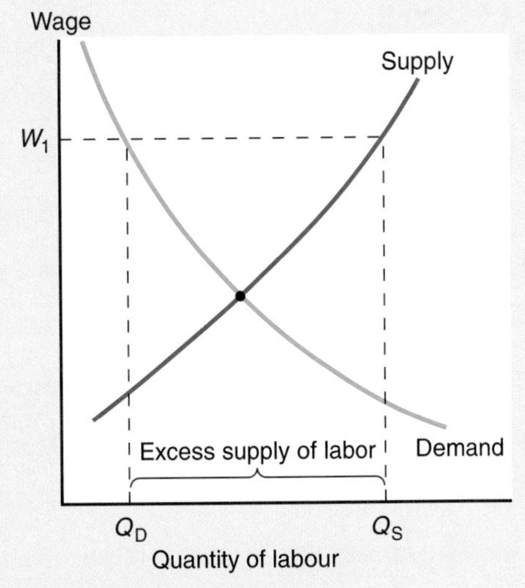

(a) Maintaining excess supply

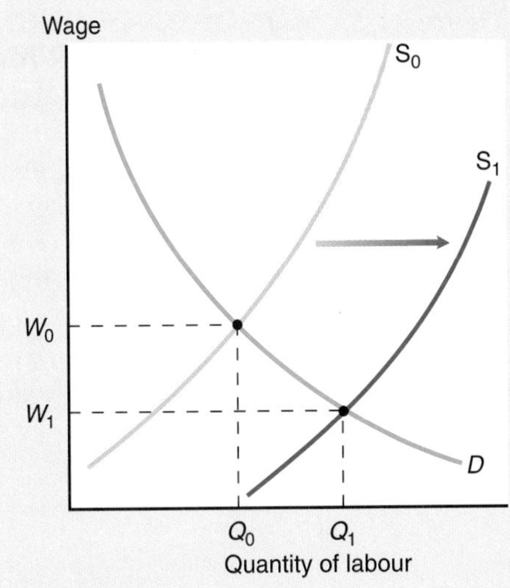

(b) An increase in supply

might increase the demand for products overall, thereby increasing the demand for labour and raising wages.

IMPERFECT COMPETITION AND THE LABOUR MARKET

Just as product markets can be imperfectly competitive, so too can labour markets. For example, there might be a **monopsony** (*a market in which a single firm is the only buyer*). Alternatively, workers might have organized a union that allows the workers to operate as if there were only a single seller of labour. In effect, the union could operate as a monopoly. Alternatively again, there might be a **bilateral monopoly** (*a market with only a single seller and a single buyer*). Let's briefly consider these three types of market imperfections.

A monopsony is a market in which a single firm is the only buyer. A bilateral monopoly is a market in which a single seller faces a single buyer.

Monopsony

When there's only one buyer of labour services, it makes sense for that buyer to take into account the fact that if it hires another worker, the equilibrium wage will rise and it will have to pay the higher wage to all of its workers. The choice facing a monopsonist can be seen in Figure 21-6, in which the supply curve of labour slopes upward so that the **marginal factor cost** (*the additional cost to a firm of hiring another worker*) is above the supply curve since the monopsonist takes into account the fact that hiring another worker will increase the wage rate it must pay to all workers.

A monopsonist takes into account the fact that hiring another worker will increase the wage rate it must pay all workers.

Instead of hiring Q_c workers at a wage of W_c, as would happen in a competitive labour market, the monopsonist hires Q_m workers and pays them a wage of W_m. (A good exercise to see that you understand the argument is to show that where there's a monopsonist, a minimum wage simultaneously can increase employment and raise the wage.)

Figure 21-6 MONOPSONY, UNION POWER, AND THE LABOUR MARKET

A monopsonist hires fewer workers and pays them less than would a set of competitive firms. The monopsonist determines the quantity of labour, Q_m, to hire at the point where the marginal factor cost curve intersects the demand curve. The monopsonist pays a wage of W_m. A union has a tendency to push for a higher wage, W_u, and a lower quantity of workers, Q_u.

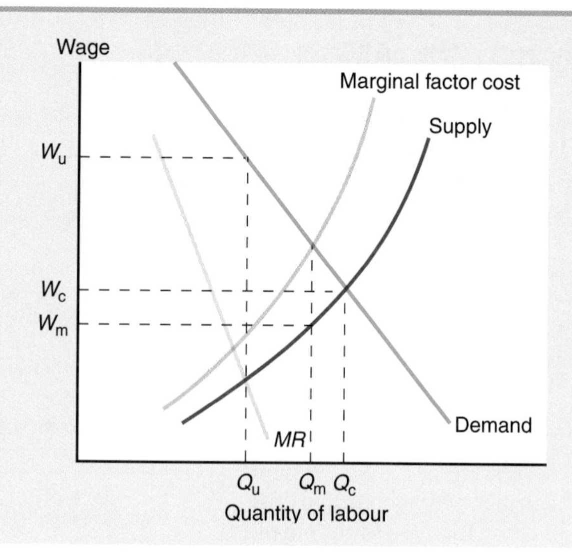

Union Monopoly Power

When a union exists, it will have an incentive to act as a monopolist, restricting supply in order to increase its members' wages. To do so it must have the power to restrict both supply and union membership. There would be a strong tendency for the union to act like a monopolist and reduce quantity similar to the monopsonist case, except for one important difference. The wage would be above the competitive wage at W_u, as in Figure 21-6. Faced with a wage of W_u, competitive firms will hire Q_u workers. Thus, with union monopoly power, the benefits of restricting supply accrue to the union members, not to the firm as in the monopsonist case.

Bilateral Monopoly

Finally, consider a bilateral monopoly in which a monopsonist faces a union with monopoly power. In this case, we can say that the equilibrium wage will be somewhere between the monopsonist wage W_m and the union monopoly wage W_u. The equilibrium quantity will be somewhere between Q_u and Q_m in Figure 21-6. Where in that range the wage and equilibrium quantity will be depends on the two sides' negotiating skills and other noneconomic forces.

Downsizing, Upsizing, and X-Inefficiency

During the 1980s and 1990s at the same time as employment grew, there was much concern in the nation about **downsizing**—*a reduction in the workforce*—of major corporations, especially at the level of middle management. Considering these two issues—increased total employment and the downsizing at particular firms—gives us a good sense of how supply and demand analysis carries over into the real world.

Q-8 How is it that total employment in the late 1990s grew and companies downsized at the same time?

First let's consider the increase in employment during this time period. What were the reasons total employment grew? First, the economy grew and the growth increased the demand for goods and services, and hence for labour. Second, the value of the dollar remained low relative to the value of the currencies of other major industrial countries. Demand for Canadian exports increased, and hence the demand for labour increased. Third, union power in Canada declined during this time period, lowering real wages (adjusted for inflation), thereby increasing the quantity of labour demanded. Combined, these three forces accounted for the rise in employment.

The downsizing was, in part, a way by which union power was reduced. Firms producing with a high-wage workforce had an incentive to shift their production to firms that had low-wage workforces—particularly to low-wage firms abroad. In addition, competitive forces led firms to specialize in what they could do well, and have other firms do what they could not. The term developed for this is **outsourcing**—*a firm shifting production from its own plant to other firms, often in the United States or abroad, where costs are lower.*

Some reasons employment grew in the 1990s include:
1. Economic growth.
2. Low value of the dollar.
3. Declining union power.

Why didn't high-wage firms simply cut wages? To some degree, they did, but cutting wages leads to dissatisfaction among remaining workers and lower productivity.

Such outsourcing was only part of the cause of downsizing. A second part was a reduction in X-inefficiency. As discussed in Chapter 14, for a variety of reasons, such as the monitoring problem, managers have a tendency to become lazy and produce inefficiently. When faced with significant international and domestic competition, as firms have faced in recent years, they must reduce that inefficiency or go out of business. Thus, downsizing is a normal part of the way the competitive process works. It results from a reduction in demand for labour from inefficient firms and a transfer of demand to more efficient firms, who, it might be said, are upsizing.

Downsizing is a normal part of the way the competitive process works.

POLITICAL AND SOCIAL FORCES AND THE LABOUR MARKET

Let's now consider some real-world characteristics of labour markets. For example:

1. English teachers are paid close to what economics teachers are paid even though the quantity of English teachers supplied significantly exceeds the quantity of English teachers demanded, while the quantity of economics teachers supplied is approximately equal to the quantity demanded.
2. On average, women earn about 65 cents for every $1 earned by men (see Figure 21-7).
3. Firms often pay higher than "market" wages.
4. Firms often don't lay off workers even when there is a decrease in the demand for their product.
5. It often seems that there are two categories of jobs: dead-end jobs and jobs with potential for career advancement. Once in a dead-end job, a person finds it almost impossible to switch to a job with potential.

Supply and demand analysis alone doesn't explain these phenomena. Each of them can, however, be explained as the result of market, political, and social forces. Thus, to understand real-world labour markets, it is necessary to broaden the analysis of labour markets to include other forces that limit the use of the market. These include legal and social limitations on the self-interest-seeking activities of firms and individuals. Let's consider a couple of the central issues of interaction among these forces and see how they affect the labour market.

To understand real-world labour markets, one must broaden the analysis.

Figure 21-7 WOMEN'S VERSUS MEN'S EARNINGS

Average earnings are lower for women than for men. Many factors may be affecting this, including type of occupation, wage rates, hours of work, and gender biases.

Source: www.statcan.ca/english/Pgdb/People/Labour/labor01a.htm , Average Earnings by Sex and Work Pattern, October, 2001.

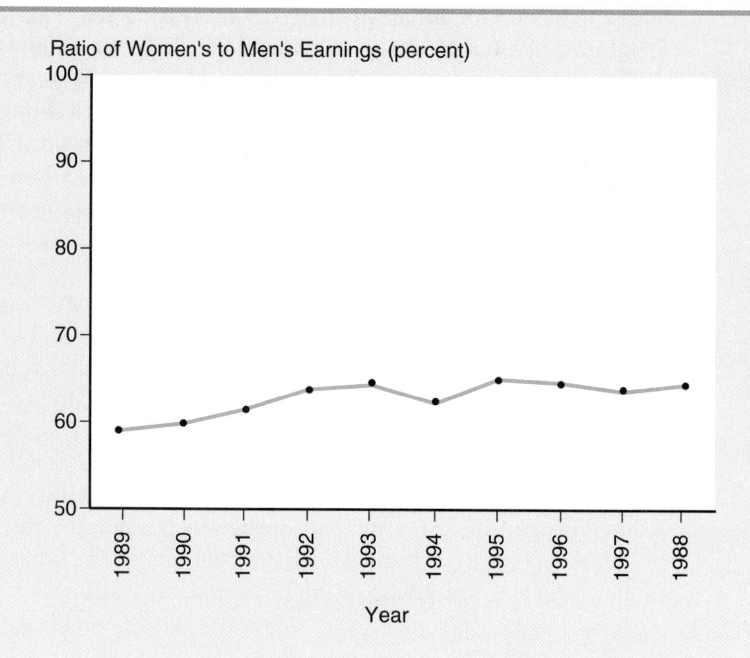

Fairness and the Labour Market

21.1 see page 485

People generally have an underlying view of what's fair. If they feel good about a job, then they will work hard; if they feel they're being taken advantage of, they can be highly disruptive.

On some assembly-line jobs, effort is relatively easy to monitor, so individuals can be—and in the past often were—treated like machines. Their feelings and emotions were ignored. Productivity was determined by the speed of the assembly line; if workers couldn't or wouldn't keep up the pace, they were fired.

Firms sometimes pay what's called efficiency wages to keep workers happy and productive.

Efficiency Wages Most modern jobs, however, require workers to make decisions and to determine how best to do a task. Today's managers are aware that workers' emotional state is important to whether they make sound decisions and do a good job. So most firms recognise that worker satisfaction will help improve the firm's profits, so they will try to keep their workers happy. It's in their own interest to do so. That might mean paying workers more than the going market wage, not laying them off even if layoffs would make sense economically, providing day care so the workers aren't worried about their children, or keeping wage differentials among workers small to limit internal rivalry. Such actions can often make long-run economic sense, even though they might cost the firm more in the short run. They are common enough that they have acquired a name—**efficiency wages**—*wages paid above the going market wage in order to keep workers happy and productive*.

Views of fairness also enter into wage determination through political channels. Social views of fairness influence government, which passes laws to implement those views. Minimum wage laws, pay equity laws, and antidiscrimination laws are examples.

Pay Equity Laws **Pay equity laws** are *laws mandating equal pay for equal work*—that is, mandatory "fairness." The problem is in defining what is equal work. Do you define equal work by the education it requires, by the effort the worker puts into it, or by other characteristics? Similarly with pay: Compensation has many dimensions and it is not at all clear which are the relevant ones, or whether the political system will focus on the relevant ones.

Economists who favour pay equity laws point out that social and intrafirm political issues are often the determining factors in setting pay. In fact, firms often have their own implicit or explicit comparable worth systems built into their structure. For example, seniority, not productivity, often determines pay. Bias against women and minorities and in favour of high-level management is sometimes built into firms' pay-setting institutions. In short, within firms, pay structure is influenced by, but is not determined by, supply and demand forces. Pay equity laws are designed to affect those institutional biases, and thus are not necessarily any more incompatible with supply and demand forces than are current pay-setting institutions.

Job Discrimination and the Labour Market

Discrimination exists in all walks of life: Women are paid less than men, and minorities are often directed into lower-paying jobs. Economists have done a lot of research in order to understand the facts about discrimination and what can be done about it. The first problem is to measure the amount of discrimination and get an idea of how much discrimination is caused by what. Let's consider discrimination against women.

Q-9 Economic theory argues that discrimination should be eliminated. True or false? Why?

On average, women receive somewhere around 65 percent of the pay that men receive. That has increased from about 60 percent in the 1980s. This pay gap suggests that discrimination is occurring. The economist's job is to figure out how much of this is caused by discrimination.

Analyzing the data, economists have found that somewhat more than half of the pay difference can be explained by causes other than discrimination, such as length of time on the job. But that still leaves a relatively large difference that can be attributed to discrimination.

Three Types of Direct Demand-Side Discrimination In analyzing discrimination, it's important to distinguish various types. The first is discrimination based on relevant individual characteristics. Firms commonly make decisions about employees based on individual characteristics that will affect job performance. For example, restaurants might avoid hiring (discriminate against) applicants with unpleasant personalities.

TABLE 21-3 Average Earnings, 2000

Average wages and average earnings are higher for men than for women, and are higher for union workers than for nonunion workers.

	Union Member	Nonunion
Average Hourly Wage	$19.46	$15.31
Men	$20.45	$17.32
Women	$18.35	$13.19
Average Weekly Earnings	$712.79	$568.23
Men	$794.59	$684.80
Women	$620.27	$445.87

Source: www.statcan.ca/english/indepth/75-001/online/00901/kl-ic_a.html. Perspectives on Labour, September, 2001.

BEYOND THE TOOLS

Democracy in the Workplace

Slavery is illegal. You cannot sell yourself to someone else, even if you want to. It's an unenforceable contract. But work, which might be considered a form of partial slavery, is legal. You can sell your labour services for a specific, limited period of time.

Is there any inherent reason that such partial slavery should be seen as acceptable? The answer to that question is complicated. It deals with the rights of workers, and is based on value judgments. You must answer it for yourself. It's also a good introduction to Karl Marx's analysis of the labour market (which deals with alienation) and to some recent arguments about democracy in the workplace.

Marx saw selling labour as immoral, just as slavery was immoral. He believed that capitalists exploited workers by alienating them from their labour. The underlying philosophical issues of Marx's concern are outside the scope of economics. But it's a useful exercise to think about why we treat the labour market as somehow different from other markets.

Some of Marx's philosophical tenets are shared by the modern democracy-in-the-workplace movement. In this view, a business isn't owned by a certain group, but rather is an association of individuals who have come together to produce a certain product. For one group—the shareholders—to have all the say as to how the business is run, and for another group—the workers—to have no say, is immoral in the same way that not having a democratic government is immoral. Often, workers' commitment to the company is much stronger and for a longer period of time than the average shareholders' commitment to the company. According to this view, work is as large a part of people's lives as is national or local politics, and a country can call itself a democracy only if it has democracy in the workplace.

As with most grandiose ideas, this one is complicated. Companies that close plants generally offer a severance package to departing workers which is based on years of service, so the longer the employee's commitment to the firm, the larger the severance package. This widespread practice reflects the view that workers have certain inalienable rights. If you know you must let a worker participate in decisions, you're going to be much more careful about whom you hire.

None of these considerations mean that democracy in the workplace can't work. There are examples of somewhat democratic "firms." Universities are run as partial democracies, with the faculty deciding what policies should be set. But as soon as you add worker democracy to production, more questions come up: What about consumers? Shouldn't they, too, have a voice in decisions? What about the community within which the firm is located?

Economics can't answer such questions. Economics can, however, be used to predict and analyze some of the difficulties such changes might bring about.

Three types of discrimination are:
1. Discrimination based on individual characteristics that will affect job performance.
2. Discrimination based on correctly perceived statistical characteristics of the group.
3. Discrimination based on individual characteristics that don't affect job performance or are incorrectly perceived.

Another example might be a firm hiring more young salespeople because its clients like to buy from younger rather than older employees. If that characteristic can be an identifying factor for a group of individuals, that discrimination becomes more visible.

A second type of discrimination is discrimination based on group characteristics. This occurs when firms make employment decisions about individuals because they are members of a group with statistical job characteristics that affect job performance. A firm may correctly perceive that young people in general have a lower probability of staying on a job than do older people and therefore may discriminate against younger people.

A third type of discrimination is discrimination based on irrelevant individual characteristics. This discrimination is based on individual characteristics that do not affect job performance or is based on incorrectly perceived statistical characteristics of groups. A firm might not hire people over 50 because the supervisor doesn't like working with older people, even though older people may be just as productive as, or even more productive than, younger people.

Of the three types, the third will be easiest to eliminate; it doesn't have an economic motivation. In fact, discrimination based on individual characteristics that don't affect job performance is costly to a firm, and market forces will work toward eliminating it.

An example of the success of a firm's policy to reduce discrimination is the decision by McDonald's to create a special program to hire workers with learning disabilities. Individuals who are learning disabled often make good employees. They tend to have lower turnover rates and follow procedures better than do many of the more transient employees McDonald's hires. Moreover, through their advertising, McDonald's helped change some negative stereotypes about people with disabilities. So in this case market forces and political forces are working together.

If the discrimination is of either of the first two types and is based on characteristics that do affect job performance, either directly or statistically, the discrimination will be harder to eliminate. Not discriminating can be costly to the firm, so political forces to eliminate discrimination will be working against market forces supporting discrimination. Whenever discrimination saves the firm money, the firm will have an economic incentive to discriminate. The firm may even appear to be complying with the law, even when it isn't. For example, when asked to explain why it isn't hiring an older person the firm will state a reason other than age.

Discrimination based on characteristics that affect job performance is hard to eliminate.

Institutional Discrimination Institutional discrimination is discrimination in which the structure of the job makes it difficult or impossible for certain groups of individuals to succeed. Consider the policies of colleges and universities. To succeed as a professor, administrator, or other professional in the academic market, one must devote an enormous amount of effort during one's 20s and 30s. But these years are precisely the years when, given genetics and culture, many women have major family responsibilities. This makes it difficult for women to succeed in the academic market. Were academic institutions different—say, a number of positions at universities were designed for high-level, part-time work during this period—it would be easier for women to advance.

Institutions can have built-in discrimination.

Requiring peak time commitment when women are also facing peak family responsibilities is the norm for many companies, too. Thus, women face significant institutional discrimination.

Whether this discrimination is embodied in the firm's structure or in the family is an open question. For example, sociologists have found that in personal relationships women tend to move to be with their partners more than men move to be with their partners. In addition, women in two-parent relationships do much more work around the house and take a greater responsibility for child rearing than men do even when both the man and the woman are employed.

How important are these sociological observations? Do you expect your personal relationship with your partner to be fully equal? The usual response is the following: 80 percent of women expect a fully equal relationship; 20 percent expect their partner's career to come first. Eighty percent of the men expect their own careers to come first; 20 percent expect an equal relationship. Somebody's expectations aren't going to be fulfilled. Put simply, most observers believe that institutional discrimination that occurs in interpersonal relationships is significant.

Economists have made adjustments for these sociological factors, and have found that institutional factors explain a portion of the lower pay that women receive but that other forms of workplace discrimination also explain a portion.

Whether prejudice should be allowed to affect the hiring decision is a normative question for society to settle. In answering these normative questions, our society has passed laws making it illegal for employers to discriminate on the basis of age, religion, disability, or national origin. The reason society has made it illegal lies in its ethical belief in equal opportunity for all.

Unions serve both employees and employers. In addition to their influence on wages, unions also provide valuable training and a safe work environment for young people to learn a trade in. The employer has the advantage of being able to tap into a trained work force, sometimes on short notice.

THE EVOLUTION OF LABOUR MARKETS

Labour markets as we now know them developed in the 1700s and 1800s. Given the political and social rules that operated at that time, the invisible hand was free to push wage rates right down to the subsistence level. Workweeks were long and working conditions were poor. Workers began to turn to other ways—besides the market—of influencing their wage. One way was to use political power to place legal restrictions on employers in their relationship with workers. A second way was to organize together—to unionize.

Evolving Labour Laws

Over the years, government has responded to workers' political pressure with a large number of laws that limit what can and what cannot be done in the various labour markets. For example, in many areas of production, laws limit the number of normal hours a person can work in a day to eight. The laws also prescribe the amount of extra pay an employee must receive when working more than the normal number of hours. (Generally it's time-and-a-half.) Similarly, the number and length of workers' coffee breaks are defined by law (one half hour break every six hours).

Child labour laws mandate that a person must be at least 16 years old in order to be hired. The safety and health conditions under which a person can work are regulated by laws. (For example, on a construction site all workers are required to wear hard hats.) Workers can be fired only for cause, and employers must show that they had cause to fire a worker. (For example, a 55-year-old employee cannot be fired simply because he or she is getting old. However, mandatory retirement at age 65 is common in many occupations. Why?) Employers must not allow sexual harassment in the workplace. (Bosses can't make sexual advances to employees and firms must make a good-faith attempt to see that employees don't sexually harass their co-workers.)

Combined, these laws play an enormously important role in the functioning of the labour market.

Unions and Collective Bargaining

Some of the most important labour laws concern workers' right to organize together in order to bargain collectively with employers. These laws also specify the tactics workers can use to achieve their ends. In the latter part of the 1800s, workers had few rights to organize themselves; now, however, almost a third of Canadian workers now are members of a union (see Figure 21-8). In the first half of 2001, an average of 12.6 million people had paid employment, up 373,000 from the previous year. Over that time, union membership grew from 3.7 million to 3.8 million workers. In the private sector, 18.1 percent of employees belong to a union, while in the public sector, which includes government departments, Crown corporations, schools, and hospitals, union membership is high — 71 percent of employees belong to a union. A slightly higher proportion of men (30.7 percent) than women (29.2 percent) are union members, and more full-time employees (31.5 percent) than part-time employees (23.2 percent) are union members.

Although average earnings are higher for union members, as shown in Table 21-3 on page 477 and Figure 21-9, other factors do influence wage rates. Union membership is stronger amongst men, older workers, and the more highly educated. Union

Figure 21-8 **UNION MEMBERSHIP BY PROVINCE, 2001**

Union membership rates vary from a low in Alberta (22.5 percent) to a high in Newfoundland (39.4 percent).

Source: www.statcan.ca/english/indepth/75-001/online/00901/kl-ic_a.html. Perspectives on Labour, September, 2001.

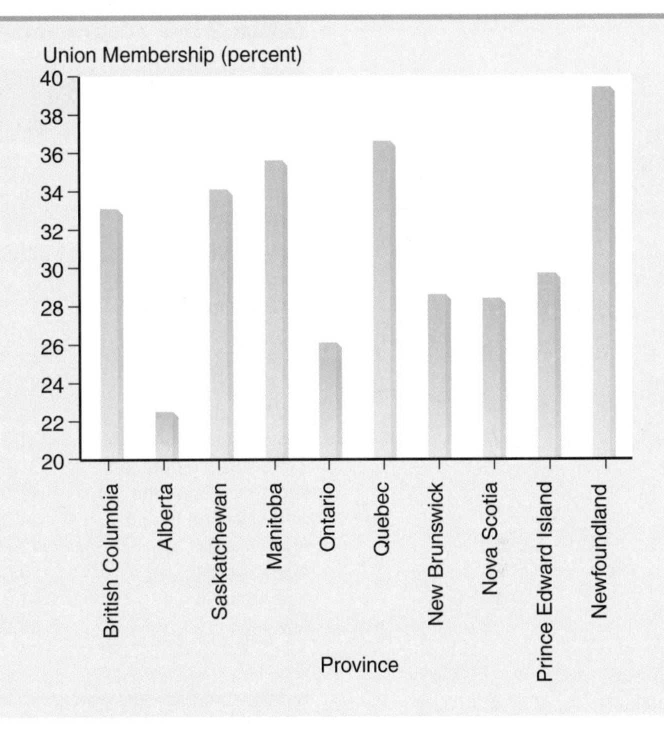

Figure 21-9 **EARNINGS BY PROVINCE**

Average weekly earnings for full-time employees vary across provinces.

Source: www.statcan.ca/english/indepth/75-001/online/00901/kl-ic_a.html. Perspectives on Labour, September, 2001.

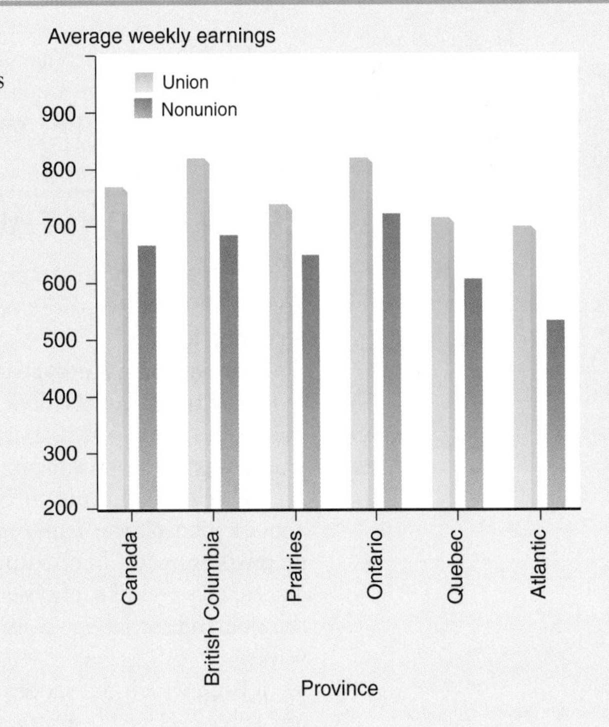

TABLE 21-4 Characteristics of Union Membership, 2001

Characteristic	Union Membership (percentage)	Characteristic	Union Membership (percentage)
Men	30.7	Finance, insurance, and real estate	8.2
Women	29.2	Accommodation & food services	6.9
Private sector	18.1	Clerical	26.8
Public sector	71.0	Management	9.6
Corporation Size		Public administration	66.6
< 20 employees	12.7	Nursing	81.7
20 to 99 employees	30.1	Elementary and secondary school teachers	86.5
100 to 500 employees	42.5	Retail	12.5
> 500 employees	54.0	Construction trades	38.8
1 to 12 months on the job	14.5	Permanent job status	30.7
1 to 5 years on the job	21.7	Nonpermanent job status	24.8
5 to 9 years on the job	31.6	Full-time employment	31.5
9 to 14 years on the job	42.0	Part-time employment	23.2
> 14 years on the job	54.5	Education	
Age 15 to 24	12.8	< Grade 9	30.4
Age 25 to 44	29.6	High school graduation	27.3
Age 45 to 54	41.8	Post secondary certificate or diploma	33.8
Age > 55	36.3	University degree	33.8

Source: www.statcan.ca/english/indepth/75-001/online/00901/kl-ic_a.html.
Perspectives on Labour, September, 2001.

membership is also stronger amongst those workers who have spent a longer time on the job and those who are employed by larger firms. All of these factors contribute to higher earnings. Table 21-4 lists other characteristics of union membership.

CONCLUSION: THE LABOUR MARKET AND YOU

Q-10 A person should take a job because it pays well. True or false? Why?

This chapter is meant to give you a sense of how the labour market works. But what does it all mean for those of you who'll soon be getting a job or are in the process of changing jobs?

Jobs requiring a university degree pay significantly more, on average, than do jobs requiring only a high school diploma. In recent years the income gap between the two groups has noticeably increased. So the answer to the question of whether it's worthwhile to pursue your education is probably yes.

21.2 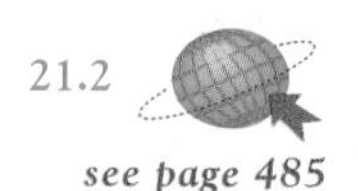

see page 485

There's much more to consider in a job than the salary. What's most important about a job isn't the wage, but whether you like what you're doing and the life that job is consistent with. In choosing a job, first and foremost pick a job that you enjoy. Then, among jobs you like, choose a job in a field in which the supply of labour is limited, or the demand for labour is increasing. Either of those trends is likely to lead to higher wages.

Jobs in which the supply will likely be limited are those in which social or political forces have placed restrictions on entry or those requiring special abilities. If you have some special ability, try to find a job you enjoy in which you can use that ability.

Your choice of jobs is one of the most important choices you'll be making in your life. Good luck.

Chapter Summary

- Incentive effects are important in labour supply decisions. The higher the wage, the higher the quantity supplied.
- Elasticity of market supply of labour depends on (1) individuals' opportunity cost of working, (2) the type of market being discussed, (3) the elasticity of individuals' supply curves, and (4) individuals entering and leaving the labour market.
- The demand for labour by firms is derived from the demand by consumers for goods and services. The demand for labour follows the basic law of demand—the higher the wage, the lower the quantity demanded.
- Elasticity of market demand for labour depends on (1) the elasticity of demand for the firm's good, (2) the relative importance of labour in production, (3) the possibility and cost of substitution in production, and (4) the degree to which marginal productivity falls with an increase in labour.
- Technological advances and changes in international competitiveness shift the demand for labour. Both have reduced demand for some types of labour and increased demand for other types. The net effect has been an increase in the demand for labour.
- A monopsony is a market in which a single firm is the only buyer. A monopsonist hires fewer workers at a lower wage compared to a competitive firm.
- A bilateral monopoly is a market in which there is a single seller and a single buyer. The wage and number of workers hired in a bilateral monopoly depends on the relative strength of the union and the monopsonist.
- Firms are aware of workers' well-being and will sometimes pay efficiency wages to keep workers happy and productive.
- Views of fairness in the labour market have led to laws that mandate comparable pay for comparable work.
- Discrimination may be based on (1) relevant individual characteristics, (2) group characteristics, and (3) irrelevant individual or group characteristics. The easiest to eliminate is discrimination based on irrelevant individual characteristics. The other two are motivated by market incentives.
- Labour markets have evolved and will continue to evolve. Since the 1980s, labour unions have been declining in importance.

Key Terms

bilateral monopoly *(473)*
cost minimization condition *(468)*
derived demand *(470)*
derived demand curve for labour *(464)*
downsizing *(474)*
efficiency wages *(476)*
entrepreneurship *(470)*
incentive effect *(458)*
labour market *(458)*
labour productivity *(463)*
marginal factor cost *(473)*
marginal physical product (*MPP*) *(463)*
marginal revenue product (*MRP*) *(463)*
monopsony *(473)*
outsourcing *(475)*
pay equity laws *(477)*
value of marginal product (*VMP*) *(463)*

Questions for Thought and Review

1. Why are social and political forces more active in the labour market than in most other markets?
2. Welfare laws are bad, not for society, but for the people they are meant to help. Discuss.
3. Which would you choose: selling illegal drugs at $75 an hour (20 percent chance per year of being arrested) or a $6-an-hour factory job? Why?
4. If the wage goes up 20 percent and the quantity of labour supplied increases by 5 percent, what's the elasticity of labour supply?
5. My brother was choosing between being a carpenter and being a plumber. I advised him to take up plumbing. Why?
6. Why might it be inappropriate to discuss the effect of immigration policy using supply and demand analysis?
7. Demonstrate graphically the effect of a minimum wage law. Does economic theory tell us such a law would be a bad idea?
8. Show graphically how a minimum wage can simultaneously increase employment and raise the wage rate.
9. Pay equity laws require employers to pay the same wage scale to workers who do comparable work or have comparable training. What likely effect would these laws have on the labour market?
10. Why is unemployment so high among youth? What should be done about the situation?
11. Give four reasons why women earn less than men. Which reasons do you believe are most responsible for the wage gap?

Problems and Exercises

1. Using the information in Figure 21-3, answer the following questions:
   a. If the market wage were $7 an hour, how many workers would the firm hire?
   b. If the price of the firm's product fell to $1, how would your answer to *a* change?
2. If firms were controlled by workers, would they likely hire more or fewer workers? Why?
3. In the 1980s and the 1990s farmers switched from small square hay bales, which they hired students on summer break to stack for them, to large round bales, which can be handled almost entirely by machines. What is the likely reason for the switch?
4. Should teachers be worried about the introduction of computer- and video-based teaching systems? Why or why not?
5. A competitive firm gets $3 per widget. A worker's average product is 4 and marginal product is 3. What is the maximum the firm should pay the worker?
6. How would your answer to question 5 change if the firm were a monopolist?
7. Fill in the following table for a competitive firm that has a $2 price for its goods.

Number of Workers	*TP*	*MPP*	*AP*	*MRP*
1	10		____	
		____		____
2	19			
		8		____
3	____		____	
				____
4	____		8.5	
		____		$12
5	____		____	

8. Your manager comes in with three sets of proposals for a new production process. Each process uses three inputs: land, labour, and capital. Under proposal A, the firm would be producing an output where the *MPP* of land is 30, labour is 42, and capital is 36. Under proposal B, at the output produced the *MPP* would be 20 for land, 35 for labour, and 96 for capital. Under proposal C, the *MPP* would be 40 for land, 56 for labour, and 36 for capital. Inputs' cost per hour is $5 for land, $7 for labour, and $6 for capital.
   a. Which proposal would you adopt?
   b. If the price of labour rises to $14, how will your answer change?
9. A study done by economists Daniel Hamermesh and Jeff Biddle found that people who are perceived as good-looking earn an average of 10 percent more than those who are perceived as homely and 5 percent more than people who are perceived as average-looking. The pay differential was found to be greater for men than for women.
   a. What conclusions can you draw from these findings?
   b. Do the findings necessarily mean that there is a "looks" discrimination?
   c. What might explain the larger pay penalty for males for looks?
10. If a teen training wage law was passed by which employers were allowed to pay teenagers less than the minimum wage, what effect would you predict this law would have, based on standard economic theory?
11. Economists Mark Blaug and Ruth Towse did a study of the market for economists in Britain. They found that the quantity demanded was about 150–200 a year, and that the quantity supplied was about 300 a year.
   a. What did they predict would happen to economists' salaries?
   b. What likely happens to the excess economists?
   c. Why doesn't the price change immediately to bring the quantity supplied and the quantity demanded into equilibrium?
12. Need-based bursaries work as an implicit tax on earnings and hence discourage saving for college and university.
   a. If the marginal tax rate parents face is 20 percent, and 5 percent of parents' assets will be deducted from a student's financial aid each year for the four years a child is in school, what is the implicit marginal tax on that portion of income that is saved? (For simplicity assume the interest rate is zero and that the parents' contribution is paid at the time the child enters college.)
   b. How would your answer differ if parents had two children, with the second entering college right after the first one graduated? (How about three?) (Remember that the assets will likely decrease with each child graduating.)
   c. When parents are divorced, how should the contribution of each parent be determined? If your school has need-based bursaries, how does it determine the expected contributions of divorced parents?
   d. Given the above, would you suggest moving to an (ability-based) scholarship program? Why or why not?
13. Explain each of the following phenomena using the invisible hand or social or political forces:
   a. Firms often pay higher than market wages.
   b. Wages don't fluctuate much as unemployment rises.

c. Pay among faculty in various disciplines at colleges and universities does not vary much although market conditions among disciplines vary significantly.

14. A recent study by the International Labour Organization estimates that 250 million children in developing countries between the ages of 5 and 14 are working either full- or part-time. The estimates of the percentage of children working is as high as 42 percent in Kenya. Among the reasons cited for the rise in child labour are population increases and poverty.
    a. Why do firms hire children as workers?
    b. Why do children work?
    c. What considerations should be taken into account by countries when deciding whether to implement an international ban on trade for products made with child labour?

15. Interview three married female and three married male professors at your school, asking them what percentage of work in the professor's household each adult household member does.
    a. Assuming your results can be extended to the population at large, what can you say about the existence of institutional discrimination?
    b. If gender-related salary data for individuals at your school are available, determine whether women or men of equal rank and experience receive higher average pay.
    c. Relate your findings in *a* and *b*.
    d. Does the existence of institutional discrimination suggest that no discrimination by employers exists? Why or why not?

Web Questions

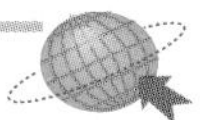

1. The Canadian Human Rights Commission provides information on pay equity. Go to its home page at www.chrc-ccdp.ca to answer the following questions.
    a. What is pay equity?
    b. What are the guiding principals in evaluating jobs for pay equity?
    c. Does the Commission believe that most of the pay gap is from women earning less in the same jobs or women working in lower-paying occupations? How does this affect the enforcement of antidiscrimination laws?

2. Go to Statistics Canada's Web site at www. statcan.ca and read the latest release of the Labour Force Survey.
    a. What kinds of information does the Labour Force Survey report?
    b. Describe the current labour market. What areas are experiencing job growth (or job losses)?
    c. Which industries would you find interesting to work in? How does the employment picture look for these industries?

Answers to Margin Questions

1. Under usual conditions of supply, one would expect that if the wage of your part-time job rises, the quantity of labour you supply in that part-time job also rises. Institutional constraints such as tax considerations or company rules might mean that the quantity of labour you supply doesn't change. However, under the usual conditions of supply, you will study less if the wage of your part-time job rises. *(458)*
2. Taxes reduce the opportunity cost, or relative price, of nonwork activities. So you will substitute leisure for labour as marginal tax rates increase. *(461)*
3. The irony of any need-based program is that such a program reduces people's incentive to prevent themselves from becoming needy. *(462)*
4. Some factors that influence the elasticity of a firm's derived demand curve include (1) the elasticity of demand for the firm's good; (2) the relative importance of the factor in the production process; (3) the possibility, and cost, of substitution in production; and (4) the degree to which the marginal productivity falls with an increase in the factor. *(470)*
5. The demand for labourers at that firm would shift rightward. *(470)*
6. Differences among countries in productivity, transportation costs, trade restrictions, and social institutions all determine the relative demand for labour in one country compared to another country. *(471)*
7. If the increase in labour supply leads to an increase in the demand for products in general, the increase in labour supply will also lead to an increase in labour demand. *(472)*
8. General growth in the economy, the fall in the value of the dollar, and the decline in the power of unions all led to more employment in the late 1990s. At the same time, increasing competitive pressures caused many inefficient firms to downsize. The net effect, however, was a rise in total employment. *(474)*

9. Economic theory does not argue that discrimination should be eliminated. Economic theory tries to stay positive. Discrimination is a normative issue. If one's normative views say that discrimination should be eliminated, economic theory might be useful to help do that most efficiently. *(477)*
10. There is more to life than income, so it does not necessarily follow that one should take the job that pays the highest wage. Each person must decide for him- or herself how to weigh the various dimensions of a job. *(481)*

Nonwage Income and Property Rights

The four traditional categories of income are wages, rent, profits, and interest. Wages, discussed in the chapter, are determined by economic factors (the forces of supply and demand), with strong influences by political and social forces, which often restrict entry or hold wages at non-market-clearing levels. Supply and demand determine price and income, given an institutional structure that includes property rights and a contractual legal system.

The same holds true for nonwage income: rent, profits, and interest. The forces of supply and demand also determine these forms of income. But, as we have emphasized throughout the book, supply and demand are not necessarily the end of the story. Supply and demand determine price and income, given an institutional structure that includes property rights—the rights given to people to use their property as they see fit—and the contractual legal system—the set of laws that govern economic behaviour of the society. If you change property rights, you change the distribution of income. Thus, in a larger sense, supply and demand don't determine the distribution of income; the distribution of property rights does.

The system of property rights and the contractual legal system that underlie the Canadian economy evolved over many years. Many people believe that property rights were unfairly distributed to begin with; if you believe that, you'll also believe that the distribution of income and the returns to those property rights are unfair. In other words, you can favour markets but object to the underlying property rights. Many political fights about income distribution concern disputes over property rights, not disagreements over the use of markets.

Such distributional fights have been going on for a long time. In feudal times much of the land was held communally; it belonged to everyone, or at least everyone used it. It was common land—a communally held resource. As the economy evolved into a market economy, that land was appropriated by individuals, and these individuals became landholders who could determine the use of the land and could receive rent for allowing other individuals to use that land. Supply and demand can explain how much rent will accrue to a landholder; it cannot explain the initial set of property rights.

The type of issues raised by looking at the underlying property rights are in large part academic for western societies. The property rights that exist, and the contractual legal system under which markets operate, are already determined. You're not going to see a new alternative set of property rights in which the ownership of property is transferred to someone else. The government may impose shifts at the margin; for example, new zoning laws—laws that set limits on the use of one's property—will modify property rights and create debates about whether society has the right to impose such laws. But there will be no wholesale change in property rights. Western economic thinking takes property rights as given.

But taking property rights as given isn't a reasonable assumption for the developing countries or the formerly socialist countries now in the process of establishing markets. They must decide what structure of property rights they want. Who should be given what was previously government land and property? Who should own the factories? Do those societies want land to be given to individuals in perpetuity, or do they want it given to individuals for, say, 100 years? As these questions have been raised, economists have redirected their analysis to look more closely at the underlying legal and philosophical basis of supply and demand. As they do so they are extending and modifying the economic theory of income distribution.

GLOSSARY

A

Ability-to-pay principle. Individuals who are most able to bear the burden of the tax should pay the tax.

Acquisition. A merger in which a company buys another company and the purchaser has the right of direct control over the resulting operation.

Adverse selection problem. Problem that occurs when buyers and sellers have different amounts of information about the good for sale.

Annuity rule. The present value of any annuity is the annual income it yields divided by the interest rate.

Art of economics. The application of the knowledge learned in the achievement of the goals one has determined in normative economics.

Average fixed cost. Fixed cost divided by quantity produced.

Average product. Output per worker.

Average total cost. Total cost divided by the quantity produced.

Average variable cost. Variable cost divided by quantity produced.

B

Balance of trade. The difference between the value of the goods and services a country imports and the value of the goods and services it exports.

Bar graph. Graph where the area under each point is filled in to look like a bar.

Barrier to entry. A social, political, or economic impediment that prevents firms from entering the market.

Benefit principle. Individuals who receive the benefit of a good or service should pay the tax necessary to supply that good.

Bilateral monopoly. Market with only a single seller and a single buyer.

Budget line. A curve that shows the various combinations of goods an individual can buy with a given income (budget). Also known as the income constraint.

Business cycle. The upward or downward movement of economic activity, or real GDP, that occurs around the growth trend.

Business. Private producing unit in our society.

C

Capitalism. An economic system based on private property and the market in which, in principle, individuals decide how, what, and for whom to produce.

Capitalists. Businesspeople who have acquired large amounts of

money and use it to invest in business.
Cartel. A combination of firms that acts as if it were a single firm.
Cartel model of oligopoly. A model that assumes that oligopolies act as if they were monopolists that have assigned output quotas to individual member firms of the oligopoly so that total output is consistent with joint profit maximization.
Clayton Antitrust Act. A U.S. law that outlawed four specific monopolistic practices: price discrimination, tie-in contracts, interlocking directorships, and buying stock in a competitor's company in order to reduce competition.
Closed shop. Firm where unions control the hiring.
Comparable worth laws. Laws mandating comparable pay for comparable work.
Comparative advantage. The ability to be better suited to the production of one good than to the production of another good.
Competition policy. The government policy toward the competitive process.
Complements. Goods that are used in conjunction with other goods.
Concentration ratio. The percentage of the total industry that the top firms of the industry have.
Conglomerate merger. The merging of two relatively unrelated businesses.
Constant returns to scale. Situation when long-run average total costs do not change with an increase in output. Also: Output will rise by the same proportionate increase as all inputs.
Consumer sovereignty. Principle that the consumer's wishes rule what's produced.
Consumer surplus. The value the consumer gets from buying a product less its price. Also: The difference between what consumers would have been willing to pay and what they actually pay.
Contestable market model. A model of oligopoly in which barriers to entry and barriers to exit, not the structure of the market, determine a firm's price and output decisions.
Coordinate system. Two-dimensional space in which one point represents two numbers.
Corporate takeover. A firm or a group of individuals issues a tender offer for another company (that is, offers to buy up the stock of a company) to gain control and to install its own managers.
Corporation. Business that is treated as a person, legally owned by its stockholders. Its stockholders are not liable for the actions of the corporate "person."
Cost minimization condition. Situation where the ratio of marginal product to the price of an input is equal for all inputs.
Cost/benefit approach. Assigning costs and benefits, and making decisions on the basis of the relevant costs and benefits.
Cross-price elasticity of demand. The percentage change in demand divided by the percentage change in the price of another good.

D

Deadweight loss. The loss of consumer and producer surplus from a tax.
Deacquisition. One company's sale of parts of either another company it has bought, or parts of itself.
Decision tree. A visual description of sequential choices.
Decreasing returns to scale. Output rises by a smaller proportionate increase than all inputs.
Demand. Schedule of quantities of a good that will be bought per unit of time at various prices, other things constant.
Demand curve. Graphic representation of the relationship between price and quantity demanded.
Demerit goods or activities. Goods or activities the government deems bad for people even though they choose to use the goods or engage in the activities.
Derived demand. The demand for factors of production by firms, which depends on consumers' demands.
Derived demand curve for labour. Curve that shows the maximum amount of labour, measured in labour hours, that a firm will hire.
Diminishing marginal productivity. Increasing one input, keeping all others constant, will lead to smaller and smaller gains in output.
Direct regulation. The amount of a good people are allowed to use is directly limited by the government.
Direct relationship. Relationship in which when one variable goes up, the other goes up too.
Diseconomies of scale. Situation when the long-run average total cost increases as output increases.
Divest *see* **Deacquisition.**
Division of labour. The splitting up of a task to allow for specialization of production.
Downsizing. A reduction in the workforce.
Duopoly. An oligopoly with only two firms.
Dynamic efficiency. A market's ability to promote cost-reducing or product-enhancing technological change.

E

Economic decision rule. If the marginal benefits of doing something exceed the marginal costs, do it. If the marginal costs of doing something exceed the marginal benefits, don't do it.

Economic efficiency. Achieving a goal at the lowest possible cost.
Economic forces. The necessary reactions to scarcity.
Economic model. Framework that places the generalized insights of the theory in a more specific contextual setting.
Economic policy. An action (or inaction) taken by government, to influence economic events.
Economic principle. Commonly held economic insight stated as a law or general assumption.
Economic profit. Explicit and implicit revenue minus explicit and implicit cost.
Economically efficient. Method that produces a given level of output at the lowest possible cost.
Economics. The study of how human beings coordinate their wants and desires, given the decision-making mechanisms, social customs, and political realities of the society.
Economies of scale. Situation when long-run average total costs decrease as output increases. Also: Situation in which costs per unit of output fall as output increases.
Economies of scope. Situation when the costs of producing products are interdependent so that it's less costly for a firm to produce one good when it's already producing another.
Efficiency. Achieving a goal as cheaply as possible (using as few inputs as possible).
Efficiency wages. Wages paid above-the-going-market wage in order to keep workers happy and productive.
Efficient. Achieving a goal at the lowest cost in total resources without consideration as to who pays those costs.
Effluent fees. Charges imposed by government on the level of pollution created.
Elastic. The percentage change in quantity is greater than the percentage change in price ($E > 1$).
Embargo. A total restriction on the import or export of a good.
Employment rate. Number of people who are working as a percentage of the labour force.
Engel curve. A curve which shows the relationship between income and the consumption of a particular good.
Entrepreneur. An individual who sees an opportunity to sell an item at a price higher than the average cost of producing it.
Entrepreneurship. The ability to organize and get something done. Also: Labour services that involve high degrees of organizational skills, concern, oversight responsibility, and creativity.
Equilibrium. A concept in which opposing dynamic forces cancel each other out.
Equilibrium price. The price toward which the invisible hand drives the market.
Equilibrium quantity. The amount bought and sold at the equilibrium price.
European Union (EU). An economic and political union of European countries that is both an economic free trade area and a loose political organization.
Excess demand. Quantity demanded is greater than quantity supplied.
Excess supply. Quantity supplied is greater than quantity demanded.
Exchange rate. The rate at which one country's currency can be traded for another country's currency.
Excise tax. A tax that is levied on a specific good.
Explicit costs. A firm's costs which are quantifiable and usually represent an actual payment made by the firm.
Exports. The value of goods sold abroad.
External economies. Cost savings which are due to factors outside the control of the individual firm.
Externality. An effect of a decision on a third party not taken into account by the decision maker.

F

Fallacy of composition. The false assumption that what is true for a part will also be true for the whole.
Feudalism. Economic system in which traditions rule.
Firm. An economic institution that transforms factors of production into consumer goods.
Fixed costs. Costs that are spent and cannot be changed in the period of time under consideration.
Foreign exchange market. Market in which one currency can be exchanged for another.
Free rider problem. Individuals' unwillingness to share in the cost of a public good.
Free rider. Person who participates in something for free because others have paid for it.
Free trade association. Group of countries that have reduced or eliminated trade barriers among themselves, and, as a group, puts up common barriers against all other countries' goods.

G

Game theory. The application of economic principles to interdependent situations.
General Agreement on Tariffs and Trade (GATT). Until recently, a regular international conference to reduce trade barriers. It has been replaced by the World Trade Organization (WTO).
General rule of political economy. When small groups are helped by a government action and large groups

are hurt by that same action, the small group tends to lobby far more effectively than the large group. Thus, policies tend to reflect the small group's interest, not the interest of the large group.

Giffen good. A good where the income effect dominates the substitution effect.

Global corporations. Corporations with substantial operations on both the production and sales sides in more than one country.

Good/bad paradox. Phenomenon of doing poorly because you're doing well.

Government failures. Situations where the government intervenes and makes things worse.

Grandfather in. To pass a law affecting a specific group but providing that those in the group before the law was passed are not subject to the law.

Graph. Picture of points in a coordinate system in which points denote relationships between numbers.

Gross domestic product (GDP). The total market value of all final goods and services produced in an economy in a one-year period.

Group of Five. Group that meets to promote negotiations and coordinate economic relations among countries. The Five are Japan, Germany, Britain, France, and the United States.

Group of Seven. (G7) Group that meets to promote negotiations and coordinate economic relations among countries. The Seven are Japan, Germany, Britain, France, Canada, Italy, and the United States.

Group of Eight. (G8) The Group of Seven, plus Russia.

H

Herfindahl index. An index of market concentration calculated by adding the squared value of the individual market shares of all firms in the industry.

Horizontal merger. The combining of two companies in the same industry.

Hostile takeover. A merger in which the firm being taken over doesn't want to be taken over.

Households. Groups of individuals living together and making joint decisions.

Human capital. The skills that are embodied in workers through experience, education, and on-the-job training, or, more simply, people's knowledge.

I

Implicit collusion. Multiple firms make the same pricing decisions even though they have not explicitly consulted with one another.

Implicit costs. A firm's expenses which are difficult to quantify because there is no explicit payment made for the use of the resource, measured as opportunity cost.

Imports. The value of goods purchased abroad.

Incentive effect. How much a person will change his or her hours worked in response to a change in the wage rate.

Incentive-compatible contract. The incentives of each of the two parties to the contract are made to correspond as closely as possible.

Income. Payments received plus or minus changes in value in one's assets in a specified time period.

Income constraint. A curve that shows us the various combinations of goods an individual can buy with a given income.

Income effect. As the price of a good falls, consumers' purchasing power rises so they buy more of it.

Income elasticity of demand. The percentage change in demand divided by the percentage change in income.

Income expansion path. A curve which shows how a consumer adjusts consumption of two goods, X and Y, as income changes. It traces out the tangencies between a consumer's indifference curves and budget lines, as income increases.

Increasing returns to scale. Output rises by a greater proportionate increase than all inputs.

Indifference curve. A curve that shows combinations of goods among which an individual is indifferent.

Indivisible setup cost. The cost of an indivisible input for which a certain minimum amount of production must be undertaken before the input becomes economically feasible to use.

Industrial policy. Formal policy that government takes toward business.

Industrial Revolution. A time when technology and machines rapidly modernized industrial production and mass-produced goods replaced handmade goods.

Inefficiency. Getting less output from inputs which, if devoted to some other activity, would produce more output.

Inefficient. Achieving a goal in a more costly manner than necessary.

Inelastic. The percentage change in quantity is less than the percentage change in price ($E < 1$).

Infant industry argument. With initial protection, an industry will be able to become competitive.

Inferior goods. Goods whose consumption decreases when income increases.

Input. What you put into a production process to achieve an output.

International Monetary Fund (IMF). A multinational, international financial institution concerned primarily with monetary issues.

Interpolation assumption. Assumption that the relationship between variables is the same between points as it is at the points.

Inverse relationship. A relationship between two variables in which when one goes up the other goes down.

Invisible hand. The price mechanism, the rise and fall of prices that guides our actions in a market.

Invisible hand theory. A market economy, through the price mechanism, will allocate resources efficiently.

Isocost line. A line that represents alternative combinations of factors of production that have the same costs.

Isoquant curve. A curve that represents combinations of factors of production that result in equal amounts of output.

Isoquant map. A set of isoquant curves that show technically efficient combinations of inputs that can produce different levels of output.

J

Judgment by performance. To judge the competitiveness of markets by the behaviour (performance) of firms in that market.

Judgment by structure. To judge the competitiveness of markets by the structure of the industry.

L

L. The broadest measure of money.

Labour market. Factor market in which individuals supply labour services for wages to other individuals and to firms that need (demand) labour services.

Labour productivity. The average output per worker.

Laissez-faire. Economic policy of leaving individuals' wants to be controlled by the market.

Law of demand. Quantity demanded rises as price falls, other things constant. Also can be stated as: Quantity demanded falls and price rises, other things constant.

Law of diminishing marginal productivity. As more and more of a variable input is added to an existing fixed input, eventually the additional output one gets from that additional input is going to fall.

Law of diminishing marginal rate of substitution. As you get more and more of a good, if some of that good is taken away, then the marginal addition of another good you need to keep you on your indifference curve gets less and less.

Law of supply. Quantity supplied rises as price rises. Also can be stated as: Quantity supplied falls as price falls.

Lazy monopolist. Firm that does not push for efficiency, but merely enjoys the position it is already in.

Learning by doing. As we do something, we learn what works and what doesn't, and over time we become more proficient at it. Also: To improve the methods of production through experience.

Learning curve. The process in which workers become better at doing a task by doing it over and over.

Limited liability. The liability of a stockholder (owner) in a corporation; it is limited to the amount the stockholder has invested in the company.

Line graph. Graph where the data are connected by a continuous line.

Linear curve. A curve that is drawn as a straight line.

Long-run decision. Decision in which a firm chooses among all possible production techniques.

Lorenz curve. A geometric representation of the share distribution of income among families in a given country at a given time.

Luxuries. Goods that have an income elasticity greater than 1.

M

Macroeconomic externality. Externality that affects the levels of unemployment, inflation, or growth in the economy as a whole.

Macroeconomics. The study of the economy as a whole, which includes inflation, unemployment, business cycles, and growth.

Marginal benefit. Additional benefit above what you've already derived.

Marginal cost. Additional cost to you over and above the costs you have already incurred. Also: Increase (decrease) in total cost from increasing (or decreasing) the level of output by one unit. Also: The change in total cost associated with a change in quantity.

Marginal factor cost. The additional cost to a firm of hiring another worker.

Marginal physical product (MPP). The additional units of output that hiring an additional worker will bring about.

Marginal product. The additional output that will be forthcoming from an additional worker, other inputs constant.

Marginal rate of substitution. The rate at which one good must be added when the other is taken away in order to keep the individual indifferent between the two combinations.

Marginal revenue. The change in total revenue associated with a change in quantity.

Marginal revenue product (MRP). The marginal revenue a firm expects to earn from selling an additional worker's output.

Marginal utility. The satisfaction one gets from consuming one additional unit of a product above and beyond what one has consumed up to that point.

Market demand curve. The horizontal sum of all individual demand curves.

Market failures. Situations where the market does not lead to a desired result. Also: Situation in which the invisible hand pushes in such a way that individual decisions do not lead to socially desirable outcomes.

Market force. Economic force that is given relatively free rein by society to work through the market.

Market incentive plan. A plan requiring market participants to certify that they have reduced total consumption—not necessarily their own individual consumption—by a specified amount.

Market structure. The physical characteristics of the market within which firms interact.

Market supply curve. Horizontal sum of all individual supply curves. Also: Horizontal sum of all the firms' marginal cost curves, taking account of any changes in input prices that might occur.

Marxian model. A model that focuses on equitable distribution of power, rights, and income among social classes.

Mercantilism. Economic system in which government determines the what, how, and for whom decisions by doling out the rights to undertake certain economic activities.

Merger. The act of combining two firms.

Merit goods or activities. Goods and activities that government believes are good for you, even though you may not choose to consume the goods or engage in the activities.

Microeconomics. The study of individual choice, and how that choice is influenced by economic forces.

Minimum efficient level of production. The amount of production that spreads setup costs out sufficiently for a firm to undertake production profitably.

Minimum wage law. Law specifying the lowest wage a firm can legally pay an employee.

Monitoring costs. Costs incurred by the organizer of production in seeing to it that the employees do what they're supposed to do.

Monitoring problem. The need to oversee employees to ensure that their actions are in the best interest of the firm.

Monopolistic competition. A market structure in which there are many firms selling differentiated products; there are few barriers to entry.

Monopoly. A market structure in which one firm makes up the entire market.

Monopoly power. Ability of individuals or firms currently in business to prevent other individuals or firms from entering the same kind of business.

Monopsony. Market in which a single firm is the only buyer.

Most-favoured nation. A country that will be charged as low a tariff on its exports as any other country.

Movement along a demand curve. The graphic representation of the effect of a change in price on the quantity demanded.

Movement along a supply curve. The graphic representation of the effect of a change in price on the quantity supplied.

N

Natural monopoly. An industry in which a single firm can produce at a lower cost than can two or more firms.

Necessity. A good that has an income elasticity less than 1.

Negative externality. The effect of a decision on a third party that is not taken into account by the decision maker is detrimental to others.

Network externality. Phenomenon that greater use of a product increases the benefit of that product to everyone.

NIMBY. Not In My Back Yard. A mindset of approving a project but not wanting it to be nearby.

Nonlinear curve. A curve that is drawn as a curved line.

Nontariff barriers. Indirect regulatory restrictions on exports and imports.

Normal goods. Goods whose consumption increases with an increase in income.

Normal profit. The amount the owners of business would have received in the next-best alternative.

Normative economics. The study of what the goals of the economy should be.

North American Free Trade Agreement (NAFTA). A U.S.-Canada-Mexico free trade zone that is phasing in reductions in tariffs.

North American Industry Classification System (NAICS). An industry classification that categorizes firms by type of economic activity and groups firms with like production processes.

O

Oligopoly. A market structure in which there are only a few firms; there are often significant barriers to entry.

Opportunity cost. The benefit forgone by undertaking a particular activity.

Optimal policy. Policy in which the marginal cost of undertaking the policy equals the marginal benefit of that policy.

Output. A result of a productive activity.

Outsourcing. A firm shifting production from its own plant to other firms, either in the United States or abroad, where wages are lower.

P

Pareto optimal policy. Policy that benefits some people and hurts no one.

Partnership. Business with two or more owners.

Patent. A legal right to be the sole supplier of a good. (Note: A patent is good for only a limited time.)

Perfectly competitive. Describes a market in which economic forces operate unimpeded.

Perfectly elastic. Quantity responds enormously to changes in price (E^D = infinity).

Perfectly inelastic. Quantity does not respond at all to changes in price ($E^D = 0$).

Pie chart. A circle divided into "pie pieces," where the individual pie represents the total amount and the pie pieces reflect the percentage of the whole pie that the various components make up.

Positive economics. The study of what is, and how the economy works.

Positive externality. The effect of a decision not taken into account by the decision maker is beneficial to others.

Poverty threshold. The income below which a family is considered to live in poverty.

Present value. A method of translating a flow of future income or saving into its current worth.

Price ceiling. A government-imposed limit on how high a price can be charged.

Price discriminate. To charge different prices to different individuals or groups of individuals.

Price elasticity of demand. The percentage change in quantity demanded divided by the percentage change in price.

Price elasticity of supply. The percentage change in quantity divided by the percentage change in price.

Price-expansion path. A curve which shows how a consumer adjusts consumption of two goods, X and Y, as the price of one of the goods changes.

Price floor. A government-imposed limit on how low a price can be charged.

Price stabilization program. Program designed to eliminate short-run fluctuations in prices, while allowing prices to follow their long-run trend line.

Price support program. Program designed to maintain prices at higher levels than the market prices.

Price taker. Firm or individual who takes the market price determined by market supply and demand as given.

Principle of absolute advantage. A country that can produce a good at a lower cost than another country is said to have an absolute advantage in the production of that good. When two countries have absolute advantages in different goods, there are gains of trade to be had.

Principle of comparative advantage. As long as the relative opportunity costs of producing goods (what must be given up in one good in order to get another good) differ among countries, then there are potential gains from trade, even if one country has an absolute advantage in everything.

Principle of diminishing marginal utility. As you consume more of a good, after some point the marginal utility received from each additional unit of a good decreases with each additional unit consumed.

Principle of increasing marginal opportunity cost. In order to get more of something, one must give up ever-increasing quantities of something else.

Principle of rational choice. Spend your money on those goods that give you the most marginal utility (MU) for your money.

Prisoner's dilemma. Well-known game that demonstrates the difficulty of cooperative behavior in certain circumstances.

Private good. A good that, when consumed by one individual, cannot be consumed by other individuals.

Private property rights. Control a private individual or a firm has over an asset or a right.

Producer surplus. Price the producer sells a product for less the cost of producing it.

Production. The transformation of the factors of production—land, labour, and capital—into goods.

Production function. The relationship between the inputs (factors of production) and outputs.

Production possibility curve. A curve measuring the maximum combination of outputs that can be obtained from a given number of inputs.

Production possibility table. Table that lists a choice's opportunity costs by summarizing what alternative outputs you can achieve with your inputs.

Production table. A table showing the output resulting from various combinations of factors of production or inputs.

Productive efficiency. Achieving as much output as possible from a given amount of inputs or resources.

Productivity. Output per unit of input.

Profit. A return on entrepreneurial activity and risk taking. Alternatively, what's left over from total revenues after all the appropriate costs have been subtracted. Also: Total revenue minus total cost.

Profit-maximizing condition. $MR = MC = P$.

Progressive tax. Tax whose rates increase as a person's income increases.

Proletariat. The working class.

Proportional tax. Tax whose rates are constant at all income levels, no matter what a taxpayer's total annual income is.

Public choice model. A model that focusses on economic incentives as applied to politicians.

Public choice economists. Economists who integrate an economic analysis of politics with their analysis of the economy.

Public good. A good that if supplied to one person must be supplied to all and whose consumption by one individual does not prevent its consumption by another individual.

Q

Quantity demanded. A specific amount that will be demanded per unit of time at a specific price, other things constant.

Quantity supplied. A specific amount that will be supplied at a specific price.

Quota. Limitation on how much of a good can be shipped into a country.

R

Rational. What individuals do is in their own best interest.

Regressive tax. Tax whose rates decrease as income rises.

Regulatory trade restrictions. Government-imposed procedural rules that limit imports.

Rent control. A price ceiling on rents, set by government.

Rent-seeking activities. Activities designed to transfer surplus from one group to another.

Reverse engineering. The process of a firm buying other firms' products, disassembling them, figuring out what's special about them, and then copying them within the limits of the law.

Rule of 72. The number of years it takes for a certain amount to double in value is equal to 72 divided by its annual rate of increase.

S

Scarcity. The goods available are too few to satisfy individuals' desires.

Share distribution of income. The relative division of total income among income groups.

Sherman Antitrust Act. A U.S. law designed to regulate the competitive process.

Shift in demand. The effect of anything other than price on demand.

Shift in supply. The graphic representation of the effect of a change in other than price on supply.

Short-run decision. Decision in which the firm is constrained in regard to what production decisions it can make.

Shutdown point. Point at which the firm will be better off if it temporarily shuts down than it will if it stays in business.

Sin tax. A tax that discourages activities society believes are harmful (sinful).

Slope. The change in the value on the vertical axis divided by the change in the value on the horizontal axis.

Social capital. The habitual way of doing things that guides people in how they approach production.

Social marginal benefit. The marginal private benefit of consuming a good plus the benefits of the positive externalities resulting from consuming that good.

Social marginal cost. The marginal private costs of production plus the cost of the negative externalities associated with that production.

Socialism. Economic system based on individuals' goodwill toward others, not on their own self-interest, and in which, in principle, society decides what, how, and for whom to produce.

Socioeconomic distribution of income. The allocation of income among relevant socioeconomic groupings.

Sole proprietorship. Business that has only one owner.

Soviet-style socialism. Economic system that uses administrative control or central planning to solve the coordination problems: what, how, and for whom.

Specialization. The concentration of individuals in certain aspects of production.

State socialism. Economic system in which government sees to it that people work for the common good until they can be relied upon to do that on their own.

Stock. Certificates of ownership in a company.

Strategic bargaining. Demanding a larger share of the gains from trade than you can reasonably expect.

Strategic decision making. Taking explicit account of a rival's expected response to a decision you are making.

Strategic pricing. Firms set their price based on the expected reactions of other firms.

Substitute. A good that can be used in place of another.

Substitution effect. As the price of a good falls, it becomes relatively cheaper so consumers buy more of it.

Sunk costs. Costs that have already been incurred and cannot be recovered.
Supply. A schedule of quantities a seller is willing to sell per unit of time at various prices, other things constant. Put another way, a schedule of quantities of goods that will be offered to the market at various prices, other things constant.
Supply curve. Graphical representation of the relationship between price and quantity supplied.

T

Takeover. The purchase of one firm by a shell firm that then takes direct control of all the purchased firm's operations.
Tariff. Tax on imports.
Tax incentive program. A program of using a tax to create incentives for individuals to structure their activities in a way that is consistent with the desired ends.
Tax incidence. Identifies who bears the burden of an excise tax.
Team spirit. The feelings of friendship and being part of a team that bring out people's best efforts.
Technical efficiency. As few inputs as possible are used to produce a given output.
Technological change. An increase in the range of production techniques that provides new ways of producing goods.
Technological development. The discovery of new or improved products or methods of production.
Technological lock-in. The use of a technology makes the adoption of subsequent technology difficult.
Technology. The way we make goods.
Total cost. Explicit payments to the factors of production plus the opportunity cost of the factors provided by the owners of the firm.
Total revenue. The amount a firm receives for selling its product or service plus any increase in the value of the assets owned by the firm.
Total utility. The total satisfaction one gets from consuming a product.
Trade adjustment assistance programs. Programs designed to compensate losers for reductions in trade restrictions.
Trade deficit. An excess of imports over exports.
Trade surplus. An excess of exports over imports.

U

Unemployment compensation. Short-term financial assistance, regardless of need, to eligible individuals who are temporarily out of work.
Union shop. Firm in which all workers must join the union.
Unit elastic. The percentage change in quantity is equal to the percentage change in price ($E = 1$).
Util. One unit of "satisfaction."
Utility. The pleasure or satisfaction that one expects to get from consuming a good or service.
Utility-maximizing rule. Utility is maximized when the ratios of the marginal utility to price of two goods are equal.

V

Value of marginal product (VMP). An additional worker's marginal physical product multiplied by the price at which the firm could sell that additional product.
Variable costs. Costs that change as output changes.
Vertical merger. A combination of two companies that are involved in different phases of producing a product.

W

Wage outsourcing. A firm shifting production from its own plant to other firms, either in Canada or abroad, where wages are lower.
Wealth. The value of the things individuals own less the value of what they owe.
Welfare capitalism. An economic system in which the market is allowed to operate but in which government plays dual roles in determining distribution and making the what, how, and for whom decisions.
Welfare loss triangle. A geometric representation of the welfare cost in terms of misallocated resources that are caused by a deviation from a supply/demand equilibrium.
World Bank. A multinational, international financial institution that works with developing countries to secure low-interest loans.
World Trade Organization (WTO). An organization whose functions are generally the same as GATT's were—to promote free and fair trade among countries. Also: Organization committed to getting countries to agree not to impose new tariffs or other trade restrictions except under certain limited conditions.

X

X-inefficiency. Firms operating far less efficiently than they could technically.

COLLOQUIAL GLOSSARY

A

Ads (noun). Short for "advertisements."

All the rage (descriptive phrase). Extremely popular, but the popularity is likely to be transitory.

Andy Warhol (proper name). American artist who flourished in the period 1960-1980. He was immensely popular and successful with art critics and the intelligentsia but, above all, he gained worldwide recognition in the same way and of the same quality as movie stars and sports athletes do. His renown has continued even after his death.

Armada (proper noun). Historic term for the Spanish navy. Now obsolete.

Automatic pilot (noun). To be on automatic pilot is to be acting without thinking.

B

Baby boom (noun). Any period when more than the statistically predicted number of babies is born. Originally referred to a specific group: those born in the years 1945-1964.

Baby boomers (descriptive phrase). Americans born in the years 1945 through 1964. An enormous and influential group of people whose large number is attributed to the "boom" in babies that occurred when military personnel, many of whom had been away from home for four or five years, were discharged from military service after the end of World War II.

Backfire (verb). To injure a person or entity who intended to inflict injury.

Balloon (verb). To expand enormously and suddenly.

Beluga caviar (noun). Best, most expensive, caviar.

Benchmark (noun). A point of reference from which measurement of any sort may be made.

Better mousetrap (noun). Comes from the proverb, "Invent a better mousetrap and the world will beat a path to your door."

Bidding (or bid) (verb sometimes used as a noun). Has two different meanings. (1) Making an offer, or a series of offers, to compete with others who are making offers. Also the offer itself. (2) Ordering or asking a person to take a specified action.

Big bucks (noun). Really, really large sum of money.

Big Mac (proper noun). Brand name of a kind of hamburger sold at McDonald's restaurants.

Blow it (verb; past tense: blew it). To do a poor job, to miss an opportunity, to perform unsatisfactorily.

Boost (verb and noun). To give a sudden impetus, or boost, to something or someone.

Botched up (adjective). Operated badly; spoiled.

Bring home (verb). To emphasize or convince.

Buffalo (adjective, as used in this book). "Buffalo chicken wings" are a variety of tempting food developed in, and hence associated with, the city of Buffalo. (Not all chicken wings are Buffalo chicken wings.)

Bus person (noun). Has no relation to transportation. It's a term for the person who clears the tables in a restaurant.

C

CEO (noun). Abbreviation of "chief executive officer."

Call (verb). In sports refereeing, one meaning of "to call" is for the referee to announce his or her decision on a specific point.

Carriage maker (noun). Person or firm that makes carriages, a type of horse-drawn conveyance almost never seen any more except in films. Members of the British royal family ride in carriages on important ceremonial occasions, such as weddings.

Cellophane (noun). A transparent wrapping material. It differs from plastic wrap in that it is made of cellulose, not plastic.

Central Park West (proper noun). A fashionable and expensive street in New York City.

Charade (noun). A pretense, usually designed to convince someone that you are doing something that you are definitely not doing.

Chit (noun). Type of IOU (which see) or coupon with a designated value that can be turned in toward the purchase or acquisition of some item.

Clear-cut (adjective). Precisely defined.

Clip coupons (verb). To cut coupons out of newspapers and magazines. The coupons give you a discount on the price of the item when you present the item and the coupon at the cashier's counter in a store. Sometimes you are directed to buy the item and then send the coupon and an identifying code from the item's package to the manufacturer, who will mail you the discount.

Clout (noun). Influence or power.

Coined (verb). "Invented" or "originated."

Coldhearted (adjective). Without any sympathy; aloof; inhuman.

Co-opted (adjective). Overwhelmed.

Cornrows (noun). Hair style in which hair is braided in shallow, narrow rows over the entire head.

Corvette (noun). A type of expensive sports car.

Couch (verb). To construct and present an argument.

Cry over spilt milk (verb). To indulge in useless complaint or regret. Note that there is a departure from standard English spelling in this phrase, which uses the spelling "spilt" instead of "spilled." Either is correct, but "spilt" is seldom used. (Another such variation is the rare "spelt" for usual "spelled.")

D

Deadbeat (noun). Lazy person who has no ambition, no money, and no prospects.

Deadweight (noun). Literally, the unrelieved weight of any inert mass (think of carrying a sack of bricks); hence, any oppressive burden.

Drop in the bucket (noun). Insignificant quantity compared to the total amount available.

Dyed-in-the-wool (adjective). Irretrievably convinced of the value of a particular course of action or of the truth of an opinion. Literally, wool that is dyed after it is shorn from the sheep but before it is spun into thread.

E

Establishment (noun and adjective). As a noun, the prevailing theory or practice. As an adjective, something that is used by people whose views prevail over other people's views.

F

Fake (verb). To fake is to pretend or deceive; to try to make people believe that you know what you're doing or talking about when you don't know or aren't sure.

Fire (verb). To discharge an employee permanently. It's different from "laying off" an employee, an action taken when a temporary situation makes the employee superfluous but the employer expects to take the employee back when the temporary situation is over.

Fix (verb). To prepare, as in "fixing a meal." This is only one of the multiplicity of meanings of this verb.

Fleeting (adverb). This word's usage is elegant and correct, but rare. It means transitory or short-lived.

Flop (noun). A dismal failure.

Follow suit (verb). To do the same thing you see others do. Comes from card games where if a card of a certain suit is played, the other players must play a card of that suit, if they have one.

Follow the flag (verb). To be committed to doing business only with firms that produce in your own country or in your "colonies"—that is, territories that belong to your country.

Follow the leader (noun). Name of a children's game. Metaphorically, it means to do what others are doing, usually without giving it much thought.

G

G.I. Joe (noun). A toy in the form of a boy (as "Barbie" is a girl). Original meaning was "government issue"—i.e., an item, such as a uniform, issued by the U.S. government to a member of the U.S. armed forces, and, by extension, the person to whom the item was issued.

GM (noun). The General Motors automobile company.

Gadget (noun). Generic term for any small, often novel, mechanical or electronic device or contrivance, usually designed for a specific purpose. For instance, the small wheel with serrated rim and an attached handle used to divide a pizza pie into slices is a gadget.

Get across (verb). To convince.

Get you down (descriptive phrase). Make you depressed about something or make you dismiss something altogether. (Do not confuse with "get it down," which means to understand fully).

Go-cart (noun). A small engine-powered vehicle that is used for racing and recreation.

Gold mine (noun). Metaphorically, any activity that results in making you a lot of money.

Good offices (descriptive phrase). An expression common in 18th century England, meaning "services."

Got it made (descriptive phrase). To succeed.

Grind (noun). Slang for necessary intense effort that may be painful but will likely benefit your understanding.

Groucho Marx (proper name). A famous U.S. comedian (1885-1977).

Gung-ho (adjective). Full of energy and eager to take action.

Guns and butter. (descriptive phrase). Metaphor describing the dilemma whether to devote resources to war or to peace.

H

Haggling (noun). Bargaining, usually in a petty and confrontational manner.

Handy (adjective). Convenient.

Hard liquor (noun). Alcoholic beverages with a high content of pure alcohol. Beer and wine are not "hard liquor" but most other alcoholic drinks are.

Hassle (noun). Unreasonable obstacle. As a verb, "to hassle" means to place unreasonable obstacles or arguments in the way of someone.

Hawking (adjective). Selling aggressively and widely.

Hefty (adjective). Large; substantial.

Hog bellies (noun). Commercial term for the part of a pig that becomes bacon and pork chops. (Also called "pork bellies.")

Holds its own (descriptive phrase). Refuses to give up, even in the face of adversity or opposition.

Home free (descriptive phrase). Safe and successful.

Hot dog (noun). A type of sausage.

How come (expression). Why? That is, "How has it come about that . . . ?"

I

"In" (preposition sometimes used as an adjective). Placed within quotation marks to show it is used with a special meaning. Here it is used as an adjective, to indicate: "fashionable or popular, usually just for a short period."

Incidentals (noun). Blanket term covering the world of small items a person uses on a daily basis as the need happens to arise—that is, needed per incident occurring. Examples are aspirin, combs, and picture postcards.

Iron Curtain (noun). Imaginary but daunting line between Western Europe and adjacent communist countries. After the political abandonment of Communism in these countries, the Curtain no longer exists.

It'll (contraction). "It will."

J

Jolt (noun). A sudden blow.

Junk food (noun). Food that tastes good but has little nutritional value and lots of calories. It is sometimes cheap, sometimes expensive, and it's quick and easy to buy and eat.

Just say no (admonition). Flatly refuse. This phrase became common in the 1970s after Nancy Reagan, the wife of the then-president of the United States, popularized it in a campaign against the use of addictive drugs.

K

Ketchup (noun). Spicy, thick tomato sauce used on, among other foods, hot dogs.

Kick in (verb). To activate; to start or begin. (Can also mean "to contribute to.")

Kickback (noun). A firm's giving part of the price it has received for its product or service back to the firm or individual who authorized the purchase of that product or service. In effect it is a type of bribe or blackmail demanded or expected by a purchaser's agent.

Klutz (noun). Awkward, incompetent person.

Knockoff (noun). A cheap imitation.

L

Late Victorian (adjective or noun). Embodying some concept typical of the late period of Queen Victoria. Also, a person from that period or who acts like someone from that period. (Queen Victoria was queen of England from 1837 to 1901.)

Lay off (verb). To discharge a worker temporarily.

Levi's (noun). Popular brand of jeans.

Lion's share (noun). By far the best part of a bargain.

Lobby (verb and noun). To lobby is to attempt by organized effort to influence legislation. As a noun, a lobby is an organized group formed to influence legislation. A lobbyist is a member of a lobby.

Lousy (adjective). Incompetent or distasteful.

M

MBA (noun). An academic degree: Master of Business Administration.

Mazuma (noun). U.S. slang term for money. It was used in the first half of the 20th century but is now rare, to say the least.

Messed up (adjective). Damaged or badly managed.

Mind your own business (admonition). Don't meddle in other people's affairs; don't ask intrusive questions.

Mind your Ps and Qs (expression). Pay close attention to distinctions. It comes from the similarity of the small printed letters "p" and "q" where the only visual distinction is the location of the downstroke. Also, the letters are right next to each other in our alphabet.

Moot (adjective). Irrelevant because the issue in question has already been decided.

Mousetrap (noun). Producing a better mousetrap is part of the saying, "Make a better mousetrap and the world will beat a path to your door." Metaphorically, producing a better mousetrap stands for doing anything better than it has previously been done.

N

NA (abbreviation). "Not available."

NASDAQ (also sometimes spelled "Nasdaq") (noun). Stock market operated by the National Association of Securities Dealers. The "AQ" stands for "Automated Quotations."

NATO (noun). North American Treaty Organization. Western alliance for joint economic and military cooperation. It includes the United States, Canada, and several European nations.

Nature of the beast (descriptive phrase). Character of whatever you are describing (need not have anything to do with a "beast").

Nirvana (noun). This word is adopted from Buddhism. Its religious meaning is complicated, but it is used colloquially to mean salvation, paradise, harmony, perfection.

No way (exclamation). Emphatic expression denoting refusal, denial, or extreme disapproval.

Nudge (noun and verb). A little push (noun); to give a little push (verb).

O

Off-the-cuff (adjective). A quick, unthinking answer for which the speaker has no valid authority (comes from the alleged practice of writing an abbreviated answer on the cuff of your shirt, to be glanced at during an examination).

Oliver Wendell Holmes (proper name). A justice of the U.S. Supreme Court, famous for his wit, his wisdom, his literary ability, his advocacy of civil rights, and his long life (1841-1935).

On their toes (descriptive phrase). Alert; ready for any eventuality.

P-Q

Ps and Qs. See under "Mind."

Park Avenue (noun). Expensive and fashionable street in New York City.

Pass the buck (descriptive phrase). Evade responsibility by forcing someone else to make the relevant decision.

Peanuts (noun). Slang for a small amount, usually money but sometimes anything with a small value.

Pecking order. Hierarchy.

Peer pressure (descriptive phrase). Push to do what everyone else in your particular group is doing.

Penny-pincher (noun). Person who is unusually careful with money, sometimes to the point of being stingy.

Perks (noun). Short for "perquisites."

Philharmonic (adjective). A philharmonic orchestra is an orchestra that specializes in classical music. Sometimes used as a noun, as in "I heard the Philharmonic."

Pie (noun). Metaphor for the total amount of a specific item that exists.

Populist (noun and adjective). As a noun, this means a member of a political party that purports to represent the rank and file of the people. As an adjective, it means a political party, a group, or an individual that purports to represent rank and file opinion.

Pound (noun). Unit of British currency.

Practice makes perfect (expression). The grammar of this phrase is illogical but the meaning is clear.

Premium tires all round (descriptive phrase). Premium tires are tires of superior quality. When all the tires on your vehicle are premium tires, you have them "all round."

Proxy (noun). A stockholder can give a "proxy" to the firm. It is an authorization that permits the firm's officials to vote for the proposition that the stockholder directs them to vote for. By extension, proxy means a substitute.

Pub (noun). Short for "public house," a commercial establishment where alcoholic drinks are served, usually with refreshments and occasionally with light meals.

R

Raise your eyebrows (verb). To express surprise, usually by a facial expression rather than vocally.

Red flag (noun). A red flag warns you to be very alert to a danger or perceived danger. (Ships in port that are loading fuel or ammunition raise a red flag to signal danger.)

Red-handed (adjective). Indisputably guilty. Comes from being found at a murder or injury scene with the blood of the victim on one's hands.

Right on! (exclamation). Expression of vigorous, often revolutionary, approval and encouragement.

Ritzy (adjective). Very expensive, fashionable, and ostentatious. Comes from the entrepreneur Caesar Ritz, a Swiss developer of expensive hotels, active in the first quarter of the 20th century.

Rule of thumb (descriptive phrase). An estimate that is quick and easy to make and is reliable enough for rough calculations. Comes from using the space from the tip of your thumb to the thumb's first joint to represent an inch.

S

Saks (proper name). A mid-size department store that sells expensive, fashionable items. There are very few stores in the Saks chain, and Saks stores are considered exclusive.

Savvy (adjective). Slang term meaning very knowledgeable. Adaptation of the French verb, "savoir," meaning "to know."

Scab (noun). Person who takes a job, or continues in a job, even though workers at that firm are on strike.

Scraps (noun). Little pieces of leftover food. Also little pieces of anything that is left over: for example, steel that is salvaged from a wrecked car.

Sears Catalog (proper noun). Sears is a large chain of stores that sells a wide variety of goods. Before shopping malls, interstate highways, and the Internet, Sears used to have a huge mailing list to which it sent enormous catalogs. A person receiving such a catalog would have information about, and access to, thousands of items, many of which the person might not have known existed before the catalog provided the prospect.

Set up shop (verb). To go into business.

Shorthand (noun). Any of several systems of abbreviated writing or writing that substitutes symbols for words and phrases. Shorthand was widely used in business until the introduction of mechanical and electronic devices for transmitting the human voice gradually made shorthand obsolete. Today it means to summarize very briefly or to substitute a short word or phrase for a long description.

Shy away (verb). To decisively refrain from something. (Comes from the world of horses, who are said to "shy at" things that startle them.)

Sixpence (noun). A British coin that is no longer in use. It represented six British pennies and its U.S. equivalent in the 2000s would be about a nickel.

Skin of one's teeth (descriptive phrase). To succeed, but just barely. A micromeasure less and one would not have succeeded.

Smoke screen (noun). Metaphorically, anything used intentionally to hide one's true intentions.

Smoking gun (noun). This term has come to stand for any indisputable evidence of guilt or misdeeds.

Soft drink (noun). Nonalcoholic beverage.

Sourpuss (noun). Dour; sulky; humourless. Derives from "sour," which is self-explanatory, and "puss," a slang word for "face."

Squash (verb). To crush or ruin.

Steady (noun). A person to whom you are romantically committed and with whom you spend a lot of time, especially in social activities.

Sticky (adjective). Resistant to change, as if glued on.

Strongarm (adjective). Repressive and violent.

Super Bowl (noun). Important football game played annually that attracts million of viewers (most of them see the game on TV).

Sucker (noun). A gullible person.

T

Tables were turned (descriptive phrase). The advantage of one side over the other reverses so that now the winner is the loser and the loser is the winner.

Tacky (adjective). In very poor taste.

Take the heat (verb). To accept all criticism of one's action or inaction, whether or not one is actually the person that should be blamed.

Time-and-a-half (noun). In labour law, 150 percent of the normal hourly wage.

Tombstone ad (noun). Newspaper advertisement announcing the completion of a stock or bond offering.

Ton (noun). A ton weighs 2,000 pounds and an English ton (often spelled "tonne") weighs 2,240 pounds. In this book the term is used most frequently to mean simply "a large quantity."

Tough (adjective). Very difficult.

Trendy (adjective). A phenomenon that is slightly ahead of traditional ways and indicates a trend. Something trendy may turn into something traditional, or it may fade away without ever becoming mainstream.

Truck (verb). To exchange one thing for another. This was Adam Smith's definition in 1776 and it is still one of the meanings of the verb.

Turf (noun). Territory, especially the figurative territory of a firm.

Turn of the century (expression). The few years at the end of an expiring century and the beginning of a new century. For example: 1998-2002.

Turn up one's nose (verb). To reject.

Twinkies (noun). Brand name of an inexpensive small cake.

U

Union Jack (noun). Nickname for the British flag.

Up in arms (adjective). Furious and loudly protesting. Comes from the use of "arms" to stand for "firearms."

V

Vanity license plate (descriptive phrase). One-of-a-kind motor vehicle license plate issued to your individual specification. It might have your name, your profession, or any individual set of letters and numbers you choose that will fit on the plate.

W

Wadget (noun). Term used by economists to stand for any manufactured good except goods designated as widgets, which see.

Wal-Mart (proper name). A very large store that sells thousands of inexpensive items. There are hundreds of Wal-Marts in the United States and the company is beginning to expand into foreign markets, such as Canada.

White knight (noun). A company that comes to the rescue of another company. The term comes from the game of chess—some chess sets have white pieces and black pieces—and from the children's book, *Alice Through the Looking Glass*, where the story is structured as a game of chess and a chess piece, the white knight, tries to rescue Alice.

Whiz (noun). An expert.

Whopper (proper noun). Brand name of a kind of hamburger sold at Burger King restaurants.

Widget (noun). The opposite of a wadget, which see.

Wild about (descriptive phrase). Extremely enthusiastic about undertaking a particular action or admiring a particular object or person.

Wind up (descriptive phrase). To discover that you have reached a particular conclusion or destination.

With-it (adjective). Current in one's knowledge.

Wodget (noun). A made-up term for a produced good. Variation of "widget," which see.

Wound up (past tense of verb "wind up"). To have found oneself in a particular situation after having taken particular actions.

Writing on the wall (descriptive phrase). To see the writing on the wall is to realize that a situation is inevitably going to end badly. It comes from the Biblical story that Nebuchadnezzar, king of Babylon, saw a fatal prediction written on a wall.

PHOTO CREDITS

Page 8, Bleichroeder Print Collection, Baker Library, Harvard Business School
Page 31, Copyright © Corbis
Page 31, Copyright © PhotoDisc
Page 37, Copyright © PhotoDisc
Page 56, Copyright © PhotoDisc
Page 90, Copyright © PhotoDisc
Page 95, Copyright © PhotoDisc
Page 106, Copyright © Eyewire
Page 121, Copyright © Eyewire
Page 142, Copyright © PhotoDisc
Page 143, Copyright © PhotoDisc
Page 146, Copyright © PhotoDisc
Page 149, Copyright © PhotoDisc
Page 153, Copyright © PhotoDisc
Page 160, Copyright © Corel
Page 161, Copyright © PhotoDisc
Page 167, Copyright © PhotoDisc
Page 167, Copyright © Corel
Page 172, Copyright © Corel
Page 178, Copyright © Eyewire
Page 179, Copyright © PhotoDisc
Page 182, Copyright © PhotoDisc
Page 186, Copyright © Eyewire
Page 199, Copyright © Canapress/Phill Snel
Page 201, Copyright © Eyewire
Page 205, Copyright © PhotoDisc
Page 206, Copyright © PhotoDisc
Page 206, Copyright © PhotoDisc
Page 215, Copyright © PhotoDisc
Page 216, Copyright © PhotoDisc
Page 218, Copyright © Eyewire
Page 221, Copyright © Canapress/Tannis Toohey
Page 222, Copyright © Bettmann/CORBIS
Page 223, Copyright © Canapress/Paul Warner
Page 225, Copyright © PhotoDisc
Page 234, Copyright © Eyewire
Page 238, Copyright © PhotoDisc
Page 239, Copyright © PhotoDisc
Page 244, Copyright © Canapress/Toronto Star
Page 247, Copyright © Canapress/Chuck Stoody
Page 265, Copyright © PhotoDisc
Page 266, Copyright © PhotoDisc
Page 271, Copyright © Canapress
Page 278, Copyright © PhotoDisc
Page 281, Copyright © PhotoDisc
Page 286, Copyright © PhotoDisc
Page 288, Copyright © PhotoDisc
Page 306, Copyright © PhotoDisc
Page 310, Copyright © PhotoDisc
Page 314, Copyright © PhotoDisc
Page 331, Copyright © Canapress/Jonathan Hayward
Page 334, Copyright © PhotoDisc
Page 350, Copyright © PhotoDisc
Page 364, Copyright © PhotoDisc
Page 377, Copyright © PhotoDisc
Page 380, Copyright © PhotoDisc
Page 391, Copyright © PhotoDisc
Page 403, Copyright © Corel
Page 410, Copyright © Corel
Page 425, Copyright © Corel
Page 450, Copyright © PhotoDisc
Page 462, Copyright © PhotoDisc
Page 463, Copyright © Bettmann/CORBIS
Page 480, Copyright © Canapress/Jonathan Hayward

INDEX

E